Springer Proceedings in Mathematics & Statistics

Volume 309

Springer Proceedings in Mathematics & Statistics

This book series features volumes composed of selected contributions from workshops and conferences in all areas of current research in mathematics and statistics, including operation research and optimization. In addition to an overall evaluation of the interest, scientific quality, and timeliness of each proposal at the hands of the publisher, individual contributions are all refereed to the high quality standards of leading journals in the field. Thus, this series provides the research community with well-edited, authoritative reports on developments in the most exciting areas of mathematical and statistical research today.

More information about this series at http://www.springer.com/series/10533

Takeo Ohsawa · Norihiko Minami
Editors

Bousfield Classes and Ohkawa's Theorem

Nagoya, Japan, August 28–30, 2015

Springer

Editors
Takeo Ohsawa
Graduate School of Mathematics
Nagoya University
Nagoya, Aichi, Japan

Norihiko Minami
Department of Computer Science
Nagoya Institute of Technology
Nagoya, Japan

ISSN 2194-1009 ISSN 2194-1017 (electronic)
Springer Proceedings in Mathematics & Statistics
ISBN 978-981-15-1587-3 ISBN 978-981-15-1588-0 (eBook)
https://doi.org/10.1007/978-981-15-1588-0

Mathematics Subject Classification (2010): 14Fxx, 14Lxx, 16Exx, 18Dxx, 18Fxx, 19Exx, 32Gxx, 55Nxx, 55Pxx

This Springer imprint is published by the registered company Springer Nature Singapore Pte Ltd.
The registered company address is: 152 Beach Road, #21-01/04 Gateway East, Singapore 189721, Singapore

Tetsusuke Ohkawa, 1977 (owned by Takeo Ohsawa)

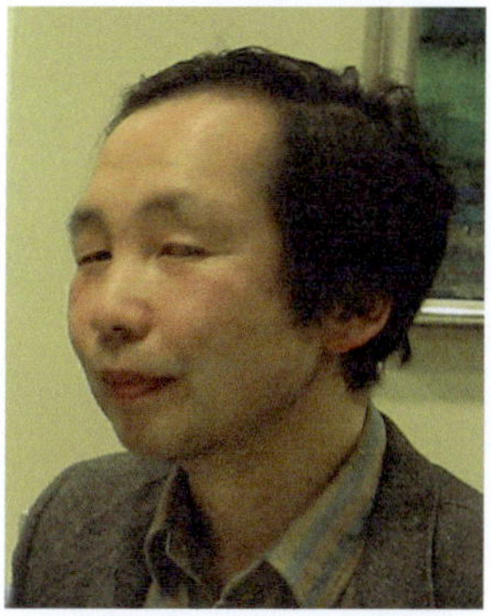

Tetsusuke Ohkawa, 2006 (owned by Masayuki Yamasaki)

Foreword

Since the extensive calculations of stable homotopy groups of spheres in the 1960s, algebraic topologists have recognized the extreme complexity of stable homotopy theory. This has prompted efforts to step back from the intricate details and seek a more global understanding. In the 1970s, we learned how to view stable homotopy through the "eyes" of a homology theory E_* using an E_*-localization that turned the E_*-equivalences of spectra into isomorphisms. For rational homology theory and much more richly for K-homology theory, this did indeed lead to an algebraic overview of stable homotopy theory. More relevantly, it led to a very different global approach in which we sought of classify spectra E according to the E_*-equivalences that they give. Thus, for a spectrum E, we considered the class $\langle E \rangle$ of all spectra F such that the F_*-equivalences are the same as the E_*-equivalences in the stable homotopy category. These classes not only inherit the wedge and smash product operations for spectra but also have a partial ordering with nice lattice properties. After our initial study of these classes in 1979, many interesting sorts of spectra were classified in this way. However, the most surprising and important general result on these classes came in 1989 when Tetsusuke Ohkawa demonstrated that they just form a set and thus an actual lattice. Although this lattice is still poorly understood, it does seem to provide a very fundamental overview of stable homotopy theory.

In the ensuing years, this classification system and Ohkawa's theorem have been generalized far beyond the original setting to other sorts of "stable" categories, beginning with the derived categories of commutative rings. We believe that Ohkawa's theorem will have many more ramifications, and the surveys in this volume should help to stimulate much more work in this area.

Chicago, USA
2016

A. K. Bousfield

Contents

Memories on Ohkawa's Mathematical Life in Hiroshima

Takao Matumoto

Abstract Some of Ohkawa's mathematical life in Hiroshima are suggested.

Keywords Biographies · Obituaries · Personalia · Bibliographies

1 Master Thesis

In the graduate school of Tokyo University he studied with Prof. Kazuhiko Aomoto at first and then with Prof. Mitsuyoshi Kato.

His Master thesis 'Group π with finite dimensional $K(\pi, 1)$' written in 1976 has three parts:

(I) Construction of $K(\pi, 1)$ complex,

(II) Analogy of Cartan's theorem for 2-dimensional simplicial complex,

(III) Grushko's theorem for an amalgamated product.

2 MathSciNet

In MathSciNet we can find seven papers written in English:

(1) The Matsumoto tripling for compact simply connected 4-manifolds. Tohoku Math. J. 31(1979), 525–535 (with M. Kato, S. Kojima and M. Yamasaki).

(2) Homological separation of 2-spheres in a 4-manifold. Topology 21(1982), 297–313.

(3) The pure braid groups and the Milnor $\bar{\mu}$-invariants of links. Hiroshima Math. J. 12(1982), 485–489.

T. Matumoto (✉)
Department of Mathematics, Graduate School of Science, Hiroshima University, 1-3-1 Kagamiyama, Higashi-Hiroshima 739-8526, Japan
e-mail: matumot1@amber.plala.or.jp

T. Ohsawa and N. Minami (eds.), *Bousfield Classes and Ohkawa's Theorem*, Springer Proceedings in Mathematics & Statistics 309,
https://doi.org/10.1007/978-981-15-1588-0_1

(4) The injective hull of homotopy types with respect to generalized homology functors. Hiroshima Math. J. 19(1989), 631–639 (Theme of the conference).

(5) A vanishing theorem of Araki-Yosimura-Bousfield-Kan spectral sequences. Hiroshima Math. J. 23(1993), 114 (Ph. D. thesis).

(6) A remark on homology localization. Hiroshima Math. J. 28(1998), 1–5.

(7) On epimorphisms and monomorphisms in the homotopy category of CW complexes. Japan. J. Math. (N.S.) 26(2000), 153–156 (with T. Matumoto).

3 RIMS Kokyuroku

In Research Institute for Mathematical Sciences Kokyuroku there are eight papers written in Japanese:

(J1) Analogy of Cartan's theorem for 2-dimensional simplicial complex. 268(1976), 69–74.

(J2) On 2-dimensional $K(\pi, 1)$. 283(1976), 52–57.

(J3) Cable knots of fibered knots are fibered. 309(1977), 80–85.

(J4) Higher separating of links. 346(1979), 80–87.

(J5) Separating problem of elements of $\pi_2(M^4)$. 369(1979), 122–127.

(J6) Pure braid groups and Milnor $\bar{\mu}$-invariants. 417(1981), 100–105.

(J7) On h_*-injective spectrum as analogy of injective module. 781(1992), 129–131.

(J8) On half localization. 838(1993), 50–54.

4 Some Comments

He was offered a job as an assistant at Hiroshima University from Prof. Masahiro Sugawara in 1978, when I moved to Hiroshima. All the papers except J1, J2, J3 and 1 were written in Hiroshima. He had many other talks, home pages and results.

For example, an important Remark 4.3 of 'On the set of free homotopy classes and Brown's construction. Hiroshima Math. J. 14(1984), 359–369 by T. Matumoto, N. Minami and M. Sugawara' is due to him. But he did not agree to be a coauthor.

He became associate professor at Hiroshima Institute of Technology in 1996.

Depth and Simplicity of Ohkawa's Argument

Carles Casacuberta

Abstract This is an expository article about Ohkawa's theorem stating that acyclic classes of representable homology theories form a set. We provide background in stable homotopy theory and an overview of subsequent advances in the study of Bousfield lattices. As a new result, we prove that there is a proper class of acyclic classes of nonrepresentable homology theories.

Keywords Spectra · Homology theories · Acyclicity · Bousfield classes

1 Introduction

The main purpose of this article is to present the statement and proof of Ohkawa's theorem [25, Theorem 2] without assuming expertise on the reader's part in homotopy theory. Thus in Sects. 2 and 3 we collect basic facts about homology theories, spectra, Spanier–Whitehead duality, and Adams representability.

Most of Ohkawa's article [25] was devoted to a discussion of injective hulls of spaces and spectra with respect to homology theories. After the publication of that article, it remained generally unnoticed that the proof of the fact that Bousfield classes of spectra form a set instead of a proper class did not depend on injective hulls—although it had likely been inspired by the study of those.

In fact, Ohkawa's theorem did not become widespread until Dwyer and Palmieri published in [10] another proof of the same result, motivated by earlier thoughts of Strickland [32], who studied jointly with Hovey and Palmieri [15, 17] the complete lattice resulting from the fact that Bousfield classes of spectra form a set. Their work triggered further progress in the understanding of chromatic homotopy theory [4, 37] and, more generally, tensor triangulated categories [11, 18, 36], including derived categories of commutative rings.

C. Casacuberta (✉)
Facultat de Matemàtiques i Informàtica, Universitat de Barcelona (UB), Gran Via de les Corts Catalanes 585, 08007 Barcelona, Spain
e-mail: carles.casacuberta@ub.edu

T. Ohsawa and N. Minami (eds.), *Bousfield Classes and Ohkawa's Theorem*, Springer Proceedings in Mathematics & Statistics 309,
https://doi.org/10.1007/978-981-15-1588-0_2

We present Ohkawa's proof of [25, Theorem 2] without changing anything substantial from the original argument, in order to illustrate both its simplicity and the far-reaching depth of the idea behind it. Recent generalizations of Ohkawa's theorem in the context of triangulated categories by Iyengar–Krause [18] and Stevenson [31] used different methods, but the general form of the same result described in [8] for non necessarily stable combinatorial model categories was proved using precisely a version of Ohkawa's argument.

All the variants of Ohkawa's theorem published so far include representability as a crucial ingredient. In its original formulation, it was indeed a statement about representable homology theories, whose featuring property is that they preserve coproducts and filtered colimits. This property is essential in the proof of Ohkawa's theorem given by Dwyer and Palmieri in [10], which is based on the fact that every CW-spectrum is a filtered union of its finite subspectra. Additivity and exactness are also fundamental hypotheses in [18, Theorem 2.3] for the validity of Ohkawa's theorem in well generated tensor triangulated categories.

The proof of the version of Ohkawa's theorem presented in [8] no longer requires additivity nor exactness—not even homotopy invariance—but it is a result about endofunctors in combinatorial model categories preserving λ-filtered colimits for some regular cardinal λ; see [9] in this volume for details.

One could ask if this assumption can be weakened further. In Sect. 7 we show that if one considers non necessarily representable homology theories without any extra assumption, then there is a proper class of distinct Bousfield classes of those.

2 Homology Theories

Generalized homology theories were studied by G. W. Whitehead in [34] after the discovery of K-theory and other functorial constructions on spaces that satisfied the Eilenberg–Steenrod axioms [12] except the dimension axiom. In order to state these axioms in a simple way, we will only consider reduced homology theories and restrict their scope to CW-complexes, that is, topological spaces constructed by successively attaching cells of increasing dimensions [35, Sect. 5].

For $n \geq 0$, the *n-skeleton* $X^{(n)}$ of a CW-complex X is the union of its cells of dimension lower than or equal to n. A *pointed* CW-complex is a pair consisting of a CW-complex X and a distinguished 0-cell x_0. Pointed CW-complexes form a category whose morphisms are continuous maps $f: X \to Y$ with $f(X^{(n)}) \subseteq f(Y^{(n)})$ for all n and $f(x_0) = y_0$.

A *reduced homology theory* is a collection of functors $\{h_n\}_{n\in\mathbb{Z}}$ from pointed CW-complexes to abelian groups with the following properties:

- *Homotopy invariance:* If two maps $f, g\colon X \to Y$ are homotopic, then the induced homomorphisms $h_n(f)$ and $h_n(g)$ coincide for all n.
- *Exactness:* Every inclusion $i\colon A \hookrightarrow X$ of a subcomplex induces, for all n, an exact sequence of abelian groups

$$h_n(A) \longrightarrow h_n(X) \longrightarrow h_n(X/A).$$

- *Suspension isomorphism:* There is a natural isomorphism $h_n(X) \cong h_{n+1}(\Sigma X)$ for all n and all X, where $\Sigma X = S^1 \wedge X$.

Here and throughout we denote by S^n the n-sphere and by $\wedge$ the smash product, i.e., the quotient of the cartesian product by the one-point union of pointed spaces. The space ΣX is called the *suspension* of X. The exactness axiom and the suspension isomorphism axiom are usually replaced by long exact sequences for pairs of spaces and the excision axiom in the case of nonreduced homology theories. Passage from reduced to nonreduced and conversely can be done as explained in [33, Sects. 7.34 and 7.35] or in [34, Sect. 5]. Generalized cohomology theories are defined in the same way, but contravariantly.

The graded abelian group $h_*(S^0)$ is called the *coefficients* of h_*. A reduced homology theory $\{h_n\}_{n\in\mathbb{Z}}$ is *ordinary* if $h_n(S^0) = 0$ for $n \neq 0$. Otherwise it is called *extraordinary* or *generalized*. Examples include the following, among many others:

- Complex K-theory, for which $\tilde{K}_*(S^0) = \mathbb{Z}[t, t^{-1}]$ with t in degree 2.
- Complex cobordism, such that $\widetilde{MU}_*(S^0) = \mathbb{Z}[x_1, x_2, \ldots]$ with x_i in degree $2i$.
- Morava K-theories, with $\widetilde{K(n)}_*(S^0) = \mathbb{F}_p[v_n, v_n^{-1}]$ and v_n in degree $2(p^n - 1)$.

It follows from results in [12] that, if a homology theory is ordinary, then there is an abelian group G such that $h_n(X) \cong \tilde{H}_n(X; G)$ for all finite CW-complexes X and all n, where $\tilde{H}_n$ denotes reduced singular homology. This result was extended by Milnor in [21] to arbitrary CW-complexes, not necessarily finite, under the following additional assumption. A reduced homology theory $\{h_n\}_{n\in\mathbb{Z}}$ is called *additive* if it satisfies the *Milnor axiom* about preservation of coproducts:

$$h_n\left(\bigvee\nolimits_{i\in I} X_i\right) \cong \bigoplus\nolimits_{i\in I} h_n(X_i)$$

for every set of indices I and all n. This property is a consequence of the previous axioms if the set of indices I is finite, but it is not if I is infinite. If h_* is additive and ordinary, then the natural isomorphism $h_* \cong \tilde{H}_*(-; h_0(S^0))$ can be proved by comparing the respective cellular chain complexes, as in [13, Theorem 4.5.9]. A similar argument yields the following more general result, whose proof is given in [29, Proposition II.3.19] and [33, Theorem 7.55].

Proposition 2.1 *If a natural transformation $h'_* \to h_*$ of additive homology theories induces an isomorphism $h'_*(S^0) \cong h_*(S^0)$, then it also induces an isomorphism $h'_*(X) \cong h_*(X)$ for every CW-complex X.*

3 Spectra and Representability

There are several different models for the homotopy category of spectra. Here we consider CW-spectra for consistency with the rest of the article. A *CW-spectrum* is a sequence of pointed CW-complexes $E = \{E_n\}_{n\in\mathbb{Z}}$ together with subcomplex inclusions $\Sigma E_n \hookrightarrow E_{n+1}$ for all n. Each CW-complex X yields a CW-spectrum with $X_n = \Sigma^n X$ if $n \geq 0$ and $X_n = *$ (a single point) for $n < 0$. We will not distinguish notationally a CW-complex from the corresponding CW-spectrum, and will omit "CW" from now on for shortness.

Spectra can be suspended and desuspended:

$$(\Sigma^k E)_n = E_{n+k} \quad \text{for } k \in \mathbb{Z}.$$

A *stable cell* of a spectrum E is a cell $c \subset E_n$ for some n, which is identified with $\Sigma^k c \subset E_{n+k}$ for $k \geq 1$. If c is a d-cell in E_n then it represents a $(d-n)$-cell of E. A spectrum with only a finite number of distinct stable cells is called *finite*. More generally, the *cardinality* of a spectrum is the cardinality of its set of stable cells.

Maps between spectra are defined up to cofinality [3, 33], and homotopies between maps of spectra are defined similarly as for topological spaces. We denote by $[X, Y]$ the set of homotopy classes of maps $X \to Y$. Suspension induces bijections

$$[X, Y] \cong [\Sigma^k X, \Sigma^k Y] \tag{1}$$

for all k and all spectra X and Y. Moreover there is a natural homotopy equivalence $\Sigma E \simeq S^1 \wedge E$ for every spectrum E. Consequently, the homotopy category of spectra is additive, since $[X, Y] \cong [\Sigma^2 X, \Sigma^2 Y]$ and the latter has a natural abelian group structure for all X and Y, resulting from the pinch map $S^2 \to S^2 \vee S^2$ on the domain.

Moreover, the homotopy category of spectra is *triangulated*. This means that each map $f\colon X \to Y$ fits into a *cofibre sequence* $X \to Y \to C$ that expands into

$$\cdots \longrightarrow X \xrightarrow{\ f\ } Y \longrightarrow C \longrightarrow \Sigma X \xrightarrow{\ \Sigma f\ } \Sigma Y \longrightarrow \cdots \tag{2}$$

in such a way that certain axioms are satisfied [16, 20, 24]. Most notably, (2) yields long exact sequences of abelian groups by applying $[E, -]$ or $[-, E]$ to it, where E is any spectrum. Indeed, it is a feature of spectra that there is no distinction between fibre sequences and cofibre sequences, in contrast with spaces.

The *homotopy groups* of a spectrum $E = \{E_n\}_{n\in\mathbb{Z}}$ are defined as

$$\pi_k(E) = [\Sigma^k S^0, E] \cong \operatorname{colim}_n \pi_{k+n}(E_n) \quad \text{for } k \in \mathbb{Z}.$$

A map of spectra $X \to Y$ inducing isomorphisms $\pi_k(X) \cong \pi_k(Y)$ for all k is a homotopy equivalence [3, Corollary III.3.5]. It is also remarkable that

$$\pi_k(X \vee Y) \cong \pi_k(X) \oplus \pi_k(Y) \tag{3}$$

for all k, since $X \to X \vee Y \to Y$ is a split cofibre sequence.

The stable homotopy groups of the sphere spectrum are of utmost importance. If $k > 0$ then $\pi_k(S^0)$ is finite [27], while $\pi_0(S^0)$ is infinite cyclic, and if $k < 0$ then $\pi_k(S^0) = 0$. We state the following consequence for its use in Sect. 5.

Lemma 3.1 *The set of homotopy types of finite spectra is countable, and given any two finite spectra A and B the abelian group $[A, B]$ is finitely generated.*

Proof For every finite spectrum A there is a finite CW-complex X and an integer k such that $A \simeq \Sigma^k X$, and if two CW-complexes are homotopy equivalent then their suspension spectra are also homotopy equivalent. Hence our first claim follows from the fact that every finite CW-complex is homotopy equivalent to a finite polyhedron; cf. [29, Lemma II.3.16].

To prove the second claim, observe first that, for every finite spectrum B, each of its homotopy groups $\pi_k(B)$ is finitely generated since it is obtained by means of finitely many group extensions starting from homotopy groups of spheres and using cofibre sequences as in (2) corresponding to the cells of B. Arguing in the same way, if A is another finite spectrum then the abelian group $[A, B]$ is finitely generated since it is obtained in finitely many steps starting from homotopy groups of B and using cofibre sequences determined by the cells of A. □

As shown in [34, Theorem 5.2], every spectrum E defines a homology theory as

$$E_n(X) = \pi_n(E \wedge X) \tag{4}$$

and similarly E defines a cohomology theory as

$$E^n(X) = [\Sigma^{-n} X, E], \tag{5}$$

where X is any pointed CW-complex. In fact (4) and (5) make perfectly sense if X is a spectrum, with any version of a smash product for spectra [2]; for instance, $(E \wedge X)_{2n} = E_n \wedge X_n$ and $(E \wedge X)_{2n+1} = E_{n+1} \wedge X_n$.

Thus E_* defines a homology theory on spectra, meaning that it is a functor from spectra to graded abelian groups which is homotopy invariant and exact in the sense that every cofibre sequence $X \to Y \to C$ of spectra yields an exact sequence

$$E_n(X) \longrightarrow E_n(Y) \longrightarrow E_n(C)$$

for every n, and there is a natural isomorphism $E_n(X) \cong E_{n+1}(\Sigma X)$ for all X.

Similarly, E^* is a cohomology theory on spectra. It is clear from (5) that E^* sends coproducts to products, and it is also true that E_* preserves coproducts, by the following argument. Recall that a partially ordered set I is *filtered* if for every two elements i and j there is another element k such that $i \leq k$ and $j \leq k$.

Lemma 3.2 *For every spectrum E, the homology theory E_* preserves coproducts and sends filtered unions of subspectra to filtered colimits.*

Proof As shown, for instance, in [33, Lemma 8.34], if a spectrum X is a filtered union of subspectra X_i then the inclusions $X_i \hookrightarrow X$ induce an isomorphism

$\operatorname{colim}_i [F, X_i] \cong [F, X]$ for every finite spectrum F. Moreover, $E \wedge X$ is also a filtered union of its subspectra $E \wedge X_i$. Therefore, since $\Sigma^n S^0$ is a finite spectrum,

$$E_n(X) = \pi_n(E \wedge X) = [\Sigma^n S^0, E \wedge X] \cong \operatorname{colim}_i [\Sigma^n S^0, E \wedge X_i] = \operatorname{colim}_i E_n(X_i).$$

As a special case, E_n preserves coproducts because every coproduct of spectra is a filtered union of finite coproducts and E_n preserves these by (3). □

The homology theory E_* and the cohomology theory E^* given by (4) and (5) are said to be *represented* by the spectrum E. Singular (co)homology with coefficients in G is represented by the Eilenberg–Mac Lane spectrum $\{K(G, n) \mid n \geq 0\}$, and complex K-theory is represented by the spectrum consisting of the unitary group U in odd dimensions and $\mathbb{Z} \times BU$ in even dimensions (where BU is the classifying space of U), with structure maps given by Bott periodicity $\Omega^2 BU \simeq \mathbb{Z} \times BU$.

Brown's representability theorem [7, Theorem II] for cohomology theories with countable coefficients was extended by Adams in [2, Theorem 1.6] by showing that every cohomology theory defined on finite CW-complexes is represented by some spectrum (not necessarily finite). This leads to the following central result.

Theorem 3.3 (Adams) *Every additive homology theory on CW-complexes is represented by some spectrum.*

Proof As a consequence of Alexander duality, if X is a finite nonempty proper subcomplex of S^n then there is a finite subcomplex $D_n X$ of $S^n \setminus X$ such that

$$E^k(X) \cong E_{n-k-1}(D_n X) \tag{6}$$

for all k and every spectrum E; see [30, p. 199]. Hence each homology theory h_* defines by means of such duality a cohomology theory on finite CW-complexes, as shown in [34, Corollary 7.10], which is representable by Adams' extension of Brown's theorem. Then the representing spectrum E defines an additive homology theory E_* whose restriction to finite CW-complexes is naturally isomorphic to the restriction of h_*. Moreover, for every CW-complex X and every n the group $E_n(X)$ is the colimit of $E_n(X_i)$ where $\{X_i\}_{i \in I_X}$ is the filtered set of all finite subcomplexes of X; see [33, Corollary 8.35]. Hence there is a natural transformation $E_* \to h_*$ inducing an isomorphism $E_*(S^0) \cong h_*(S^0)$. If h_* is also additive, this implies that $E_*(X) \cong h_*(X)$ for all X, by Proposition 2.1. □

The stable analogue of (6) is as follows; cf. [3, Part III, § 5]. Each finite spectrum A admits a homotopy unique *Spanier–Whitehead dual* DA, which is also finite and is equipped with a map

$$DA \wedge A \longrightarrow S^0$$

inducing isomorphisms $[X, Y \wedge DA] \cong [X \wedge A, Y]$ and $[X, A \wedge Y] \cong [DA \wedge X, Y]$ for all spectra X and Y; cf. [33, Theorem 14.34]. Therefore $DDA \simeq A$ and

$$E^{-n}(A) \cong E_n(DA) \tag{7}$$

for all spectra E and all n. Using Spanier–Whitehead duality it follows with the same argument as in the proof of Theorem 3.3 that every additive homology theory on spectra is represented by some spectrum [20, Chapter 4, Theorem 16].

4 Bousfield Equivalence Classes of Spectra

Given two spectra E and X, the spectrum X is called *E_*-acyclic* if $E_*(X) = 0$, where E_* denotes the homology theory represented by E as in (4). Two spectra E and F are called *Bousfield equivalent* if the classes of E_*-acyclic spectra and F_*-acyclic spectra coincide. Since the statement that $E_n(X) = 0$ for all n is equivalent to the statement that $E \wedge X \simeq 0$, where 0 denotes here the one-point spectrum, two spectra E and F are Bousfield equivalent if and only if

$$\{X \mid E \wedge X \simeq 0\} = \{X \mid F \wedge X \simeq 0\}. \tag{8}$$

It is also true that E and F are Bousfield equivalent if and only if E_*-localization and F_*-localization are naturally isomorphic. Here *E_*-localization* is meant in the sense of [5], where it was proved that for every spectrum X and every representable homology theory E_* there is a map $l\colon X \to L_E X$ such that $E_n(l)$ is an isomorphism for all n and $L_E X$ is *E_*-local*, that is, for every map $f\colon A \to B$ such that $E_n(f)$ is an isomorphism for all n, the function $[B, L_E X] \to [A, L_E X]$ is bijective. Then L_E defines an exact endofunctor in the homotopy category of spectra such that a map $X \to Y$ induces a homotopy equivalence $L_E X \simeq L_E Y$ if and only if it induces isomorphisms $E_n(X) \cong E_n(Y)$ for all n. Hence $L_E X \simeq 0$ if and only if X is E_*-acyclic. Therefore, the collection of E_*-acyclic spectra determines L_E up to a natural isomorphism.

Bousfield equivalence classes have been studied since the decade of 1980 in connection with homological localizations [5, 6, 26]. The Bousfield equivalence class of a spectrum E is usually denoted by $\langle E \rangle$, and it is also common to view $\langle E \rangle$ as the collection of all E_*-acyclic spectra. There is a partial order on Bousfield classes, namely $\langle E \rangle \le \langle F \rangle$ if and only if the class of F_*-acyclics is contained in the class of E_*-acyclics, or, equivalently, if there is a natural transformation $L_F \to L_E$ of coaugmented functors.

Thanks to Ohkawa's theorem, the collection of Bousfield classes becomes in fact a complete lattice with least upper bounds (*joins*) given by the wedge sum, and greatest lower bounds (*meets*) obtained as wedges of all lower bounds, which exist since there is only a set of those. The smash product provides lower bounds, but not greatest lower bounds in general. This lattice and other related lattices have been studied by a number of authors [4, 11, 14, 15, 17, 18, 23, 36, 37].

Ohkawa's injective hulls [25] are closely related to homological localizations. For a homology theory E_* on spectra, a spectrum Y is *E_*-injective* if, for every map $f\colon A \to B$ such that $E_n(f)$ is a monomorphism for all n, the function $[B, Y] \to [A, Y]$ is surjective. A map $h\colon X \to Y$ is an *E_*-injective enveloping map* if Y is E_*-injective and $E_n(h)$ is a monomorphism for all n, and, moreover, for all maps

$g\colon Y \to Z$ and every n the homomorphism $E_n(g)$ is monic if $E_n(g \circ h)$ is monic. In [25, Theorem 1] it was shown that if a homology theory E_* is representable then every spectrum X admits an E_*-injective enveloping map $h\colon X \to Y$, which is unique up to homotopy. Then Y is called an *injective hull* of X.

5 Okhawa's Argument

Choose a set $\mathscr{F}$ of representatives of all homotopy types of finite spectra and a set $\mathscr{M}$ of representatives with domains and codomains in $\mathscr{F}$ of all isomorphism classes of maps between finite spectra in the stable homotopy category. Thus for each map $f\colon A \to B$ between finite spectra the set $\mathscr{M}$ contains a unique map $f_0\colon A_0 \to B_0$ where A_0 and B_0 are in $\mathscr{F}$ and there exist two homotopy equivalences $h_A\colon A_0 \to A$ and $h_B\colon B_0 \to B$ such that $f \circ h_A \simeq h_B \circ f_0$. By Lemma 3.1, $\mathscr{F}$ has cardinality $\aleph_0$ and $\mathscr{M}$ also has cardinality $\aleph_0$ since for every two finite spectra A and B the abelian group $[A, B]$ of homotopy classes of maps $A \to B$ is finitely generated.

Given two maps of spectra $g\colon X \to Y$ and $f\colon X \to E$, we say that f *extends to* Y if there exists a map $\tilde{f}\colon Y \to E$ such that $\tilde{f} \circ g \simeq f$. For a map $f\colon X \to E$ of spectra with $X \in \mathscr{F}$, we denote, as in [25],

$$t(f) = \{g\colon X \to Y \mid g \in \mathscr{M} \text{ and } f \text{ extends to } Y\}. \tag{9}$$

Hence $t(f) \in \mathscr{P}(\mathscr{M})$, where the latter denotes the set of subsets of $\mathscr{M}$. Next, for a spectrum E, let $t_E\colon \mathscr{F} \to \mathscr{P}(\mathscr{P}(\mathscr{M}))$ be the function defined as

$$t_E(X) = \{t(f) \mid f\colon X \to E\} \tag{10}$$

for each $X \in \mathscr{F}$, and call two spectra E and F *elementarily equivalent* if $t_E = t_F$, that is, if $t_E(X) = t_F(X)$ for every $X \in \mathscr{F}$.

For a spectrum E, we consider the homology theory E_* on spectra represented by E, namely $E_n(X) = \pi_n(E \wedge X)$ for $n \in \mathbb{Z}$ and every spectrum X. If $\{X_i\}_{i \in I_X}$ is the collection of all finite subspectra of X, then the inclusions $X_i \hookrightarrow X$ induce an isomorphism

$$\operatorname{colim}_{i \in I_X} E_n(X_i) \cong E_n(X) \tag{11}$$

for every n by Lemma 3.2, since I_X is filtered.

Theorem 5.1 (Ohkawa) *Suppose that two spectra E and F are elementarily equivalent, and let $f\colon X \to Y$ be any map of spectra. For each $n \in \mathbb{Z}$, the homomorphism $E_n(f)\colon E_n(X) \to E_n(Y)$ is monic if and only if $F_n(f)\colon F_n(X) \to F_n(Y)$ is monic.*

Proof Suppose that $E_n(f)$ is a monomorphism, and let $\phi \in \operatorname{Ker} F_n(f)$. Our aim is to prove that $\phi = 0$.

Since F_n satisfies (11), there is a finite subspectrum $A \subseteq X$ and a class $\alpha \in F_n(A)$ such that $F_n(i_{A,X})(\alpha) = \phi$, where $i_{A,X}\colon A \to X$ denotes the inclusion. Therefore $F_n(f \circ i_{A,X})(\alpha) = 0$ and, using again the fact that F_n commutes with filtered colimits, we infer that there is a finite subspectrum $B \subseteq Y$ that contains $f(A)$ and such that $F_n(f')(\alpha) = 0$ if $f' \colon A \to B$ denotes the restriction of f:

$$\begin{array}{ccc} X & \xrightarrow{f} & Y \\ {\scriptstyle i_{A,X}}\uparrow & & \uparrow{\scriptstyle i_{B,Y}} \\ A & \xrightarrow[f']{} & B \end{array}$$

Let DA denote a Spanier–Whitehead dual of A. Then, since $F_n(A) \cong F^{-n}(DA)$, the class α is represented by a map $a \colon \Sigma^n DA \to F$, and the fact that $F_n(f')(\alpha) = 0$ implies that $a \circ \Sigma^n Df' \simeq 0$, where $Df' \colon DB \to DA$ is dual to f'.

Now replace $\Sigma^n DA$ by a homotopy equivalent finite spectrum belonging to $\mathscr{F}$ and choose a map $p \colon \Sigma^n DA \to P$ in $\mathscr{M}$ such that the following is a cofibre sequence:

$$\Sigma^n DB \xrightarrow{\Sigma^n Df'} \Sigma^n DA \xrightarrow{p} P \longrightarrow \Sigma^{n+1} DB.$$

Here the map $a \colon \Sigma^n DA \to F$ extends to P since $a \circ \Sigma^n Df' \simeq 0$, and this means precisely that $p \in t(a)$ as defined in (9).

Now $t(a) \in t_F(\Sigma^n DA)$ and, since we are assuming that $t_E = t_F$, we infer that $t(a) \in t_E(\Sigma^n DA)$. Therefore there is a map $b \colon \Sigma^n DA \to E$ with $t(a) = t(b)$. Thus $p \in t(b)$ and this implies that b extends to P.

Let $\beta \in E_n(A)$ be the class represented by b. Since b extends to P, we have that $b \circ \Sigma^n Df' \simeq 0$ and consequently $E_n(f')(\beta) = 0$. Since $E_n(f)$ is injective and $f \circ i_{A,X} = i_{B,Y} \circ f'$, it follows that $E_n(i_{A,X})(\beta) = 0$. Since E_n commutes with filtered colimits, there is a finite subspectrum $C \subseteq X$ containing A such that $E_n(i_{A,C})(\beta) = 0$.

Hence $b \circ \Sigma^n Di_{A,C} \simeq 0$ and therefore b extends to a homotopy cofibre Q of $\Sigma^n Di_{A,C}$, which we may choose so that the map $q \colon \Sigma^n DA \to Q$ is in $\mathscr{M}$:

$$\Sigma^n DC \xrightarrow{\Sigma^n Di_{A,C}} \Sigma^n DA \xrightarrow{q} Q \longrightarrow \Sigma^{n+1} DC.$$

Thus $q \in t(b)$, and using again that $t(a) = t(b)$, we find that $q \in t(a)$, and this means that $F_n(i_{A,C})(\alpha) = 0$. Hence $\phi = F_n(i_{A,X})(\alpha) = F_n(i_{C,X})F_n(i_{A,C})(\alpha) = 0$, from which it follows that $F_n(f)$ is indeed a monomorphism. Exchanging the roles of E and F completes the proof. □

Corollary 5.2 *If two spectra E and F are elementarily equivalent, then E and F are Bousfield equivalent.*

Proof Suppose that E and F are elementarily equivalent, and suppose that a given spectrum X is E_*-acyclic. Then the map from X to the zero spectrum induces a monomorphism $E_n(X) \to 0$ for all n. According to Theorem 5.1, the homomorphism $F_n(X) \to 0$ is also a monomorphism for all n, which means that X is F_*-acyclic. □

Hence Bousfield equivalence classes of spectra form a set of cardinality smaller than or equal to the cardinality of the set of elementary equivalence classes, which is at most $2^{2^{\aleph_0}}$.

6 Other Proofs and Extensions of Ohkawa's Theorem

The argument given in Sect. 5 uses Spanier–Whitehead duality and the fact that representable homology theories commute with filtered colimits. An alternative proof not requiring the use of duality was published by Dwyer and Palmieri in [10]. Their argument works in every algebraic stable homotopy category, as shown in [19]. It can be summarized as follows in the case of spectra.

For a spectrum E and a homology class $c \in E_*(A)$, where A is a finite spectrum, define the *annihilator* of c as

$$\mathrm{ann}_A^E(c) = \{f\colon A \to B \mid B \text{ is finite and } E_*(f)(c) = 0\},$$

and let the *Ohkawa class* of E consist of all annihilators of all classes $c \in E_*(A)$ where A is a finite spectrum.

Theorem 6.1 (Dywer–Palmieri) *If two spectra E and F give rise to the same Ohkawa class, then they are in the same Bousfield equivalence class.*

Proof If the Ohkawa class of F is contained into the Ohkawa class of E then the class of E_*-acyclics is contained in the class of F_*-acyclics. To prove this fact, suppose that $E_*(X) = 0$ and write X as a union of its finite subspectra. Given any class $c \in F_n(X)$, there is a finite subspectrum A of X and a class $a \in F_n(A)$ such that $F_n(i)(a) = c$ where $i\colon A \to X$ is the inclusion. By assumption there is a class $a' \in E_n(A)$ such that $\mathrm{ann}_A^F(a) = \mathrm{ann}_A^E(a')$. Since $E_*(X) = 0$, there is a finite subspectrum B of X containing A such that the homomorphism induced by the inclusion $j\colon A \to B$ satisfies $E_n(j)(a') = 0$; therefore $j \in \mathrm{ann}_A^E(a')$. Hence j also belongs to $\mathrm{ann}_A^F(a)$ and this implies that $F_n(j)(a) = 0$, so $c = 0$ in $F_n(X)$, as claimed. □

If the definition of Ohkawa classes is restricted after a choice of a set $\mathscr{F}$ of representatives of all homotopy types of finite spectra and a set $\mathscr{M}$ of representatives with domains and codomains in $\mathscr{F}$ of all homotopy classes of maps between finite spectra, as in Sect. 5 and as in [10], then the cardinality of the set of Ohkawa classes is bounded by $2^{2^{\aleph_0}}$, and hence so is the cardinality of the set of Bousfield equivalence classes of spectra. It is still unknown if this bound can be lowered.

Upper and lower bounds for the cardinality of the set of Bousfield classes in an arbitrary algebraic stable homotopy category have been given in [19] in terms of a generating set of small objects.

Dwyer and Palmieri proved in [11] that in the derived category of a truncated polynomial ring on countably many generators there is also only a set of Bousfield equivalence classes, and asked whether this was in fact the case in the derived category of every commutative ring—Bousfield equivalence of chain complexes of modules over a ring is defined as in (8) with the smash product replaced by the derived tensor product of chain complexes. This was known to be true for countable rings as shown in [10] and also for Noetherian rings due to a result of Neeman [22]: for a Noetherian commutative ring R there is a bijection between the Bousfield lattice in the derived category $\mathscr{D}(R)$ and the lattice of subsets of the spectrum of R. However, it had already been observed in [23] that something very different happens for rings that fail to be Noetherian. Further results in this research direction have been obtained by Wolcott in [36].

Around 2010 Stevenson extended in an unpublished article the Dwyer–Palmieri argument to compactly generated triangulated categories equipped with a biexact and coproduct-preserving tensor product, hence proving that, indeed, the Bousfield lattice of the derived category $\mathscr{D}(R)$ is a set for every commutative ring R. Shorty after, Iyengar and Krause proved in [18] that the same result holds in any well generated tensor triangulated category. Their argument was based on a restricted Yoneda embedding of a triangulated category $\mathscr{T}$ with a set of α-compact generators for some cardinal α into the category of abelian presheaves over that set.

Still Okhawa's theorem remained a result about additive categories. Another step was made in [8] by showing that it holds in fact in the homotopy category of every combinatorial model category, not necessarily stable. A proof of this fact is presented in [9] using Rosický's result [28, Proposition 5.1] that, for a combinatorial model category $\mathscr{K}$, the composite

$$\mathscr{K} \longrightarrow \mathrm{Ho}\,\mathscr{K} \longrightarrow \mathbf{Set}^{(\mathrm{Ho}\,\mathscr{K}_\lambda)^{\mathrm{op}}}$$

of the canonical functor from $\mathscr{K}$ to its homotopy category followed by a restricted Yoneda embedding preserves λ-filtered colimits for a sufficiently large regular cardinal λ. Here $\mathscr{K}_\lambda$ is a set of representatives of isomorphism classes of λ-presentable objects in $\mathscr{K}$.

7 Nonrepresentable Homology Theories

If a homology theory is not representable, then it need not preserve colimits of any kind. Therefore its value on a spectrum need not be determined by its values on finite subspectra. For this reason, there is no hope that the argument used in the proof of Theorem 5.1 can be extended to non necessarily representable homology theories. In this section we show that, indeed, Ohkawa's theorem does not hold for nonrepresentable homology theories.

For an abelian group A and a cardinal α, we denote by A^α the cartesian product of α copies of A, that is, the abelian group of functions $\alpha \to A$. Moreover, we denote by SA^α the subgroup of A^α consisting of *shrinking* functions, that is, functions $\alpha \to A$ whose image has cardinality smaller than α.

Note that $F_\alpha A = A^\alpha / SA^\alpha$ defines an exact functor from the category of abelian groups to itself. This fact has the following consequence.

Theorem 7.1 *For every uncountable cardinal α there is a reduced homology theory h_*^α on pointed CW-complexes such that if X has less than α cells then $h_*^\alpha(X) = 0$ but there exists a CW-complex with α cells which is not h_*^α-acyclic.*

Proof Consider the exact endofunctor $F_\alpha A = A^\alpha / SA^\alpha$ on the category of abelian groups. If the cardinality of A is less than α then the image of every function $\alpha \to A$

has cardinality smaller than α. Hence $SA^\alpha = A^\alpha$ and $F_\alpha A = 0$. On the other hand, there is an injective function $\alpha \to \bigoplus_{i<\alpha} \mathbb{Z}$ and hence $F_\alpha(\bigoplus_{i<\alpha} \mathbb{Z}) \neq 0$.

Next, define $h_n^\alpha = F_\alpha \circ \tilde{H}_n$ for all n, where $\tilde{H}_*$ denotes reduced singular homology. Since F_α is exact, h_*^α is a reduced homology theory. If X has less than α cells then the cardinality of $\tilde{H}_*(X)$ is smaller than α and therefore $h_*^\alpha(X) = 0$. However, for a wedge of α circles we have $H_1(\bigvee_{i<\alpha} S^1) \cong \bigoplus_{i<\alpha} \mathbb{Z}$ and this implies that $h_1^\alpha(\bigvee_{i<\alpha} S^1)$ is nonzero. □

We say that two homology theories h_* and h'_* (defined on spaces or spectra) are *Bousfield equivalent* if they have the same acyclics.

Corollary 7.2 *There is a proper class of distinct Bousfield equivalence classes of nonrepresentable homology theories of spaces or spectra.*

Proof In the case of spaces, consider the collection $\{h_*^\alpha\}$ given by Theorem 7.1 where α runs through all uncountable cardinals. Then any two of them belong to distinct Bousfield equivalence classes since if $\beta > \alpha$ then there is a space X which is h_*^β-acyclic but not h_*^α-acyclic. The same argument is valid for spectra by defining similarly $h_*^\alpha = F_\alpha \circ H_*$ where H_* is ordinary homology with $\mathbb{Z}$ coefficients. □

However, if we fix an arbitrary regular cardinal λ then there is only a set of Bousfield equivalence classes of homology theories that preserve λ-filtered colimits. For a proof of this claim, see [8, Corollary 3.8].

Acknowledgements The author wishes to acknowledge the kind hospitality of the University of Nagoya during the memorial conference for Professor Ohkawa held in August 2015. The content of Sect. 7 is based on joint work of the author with Pau Casassas and Fernando Muro. The author was supported by the Agency for Management of University and Research Grants of Catalonia with references 2014 SGR 114 and 2017 SGR 585, and the Spanish Ministry of Economy and Competitiveness under AEI/FEDER research grants MTM2013-42178-P and MTM2016-76453-C2-2-P, as well as grant MDM-2014-0445 awarded to the Barcelona Graduate School of Mathematics.

References

1. Adámek, J., Rosický, J.: Locally Presentable and Accessible Categories. London Mathematical Society Lecture Note Series, vol. 189. Cambridge University Press, Cambridge (1994)
2. Adams, J.F.: A variant of E. H. Brown's representability theorem. Topology **10**, 185–198 (1971)
3. Adams, J.F.: Stable Homotopy and Generalised Homology. Chicago Lectures in Mathematics. The University of Chicago Press, Chicago (1974)
4. Beardsley, J.: The Bousfield lattice of p-local harmonic spectra. Preprint (2013)
5. Bousfield, A.K.: The localization of spectra with respect to homology. Topology **18**, 257–281 (1979)
6. Bousfield, A.K.: The Boolean algebra of spectra. Comment. Math. Helv. **54**, 368–377 (1979)
7. Brown, E.H.: Cohomology theories. Ann. Math. **75**(2), 467–484 (1962)
8. Casacuberta, C., Gutiérrez, J.J., Rosický, J.: A generalization of Ohkawa's theorem. Compos. Math. **150**, 893–902 (2014)
9. Casacuberta, C., Rosický, J.: Combinatorial homotopy categories. In this volume

10. Dwyer, W.G., Palmieri, J.H.: Ohkawa's theorem: there is a set of Bousfield classes. Proc. Am. Math. Soc. **129**, 881–886 (2000)
11. Dwyer, W.G., Palmieri, J.H.: The Bousfield lattice for truncated polynomial algebras. Homol. Homotopy Appl. **10**, 413–436 (2008)
12. Eilenberg, S., Steenrod, N.: Foundations of Algebraic Topology. Princeton Mathematical Series, vol. 15. Princeton University Press, Princeton (1952)
13. Hatcher, A.: Algebraic Topology. Cambridge University Press, Cambridge (2002)
14. Hovey, M.: Cohomological Bousfield classes. J. Pure Appl. Algebra **103**, 45–59 (1995)
15. Hovey, M., Palmieri, J.H.: The structure of the Bousfield lattice. In: Meyer, J.-P., Morava, J., Wilson, W.S. (eds.), Homotopy Invariant Algebraic Structures. Contemporary Mathematics, vol. 239. American Mathematical Society, Providence (1999)
16. Hovey, M., Palmieri, J.H., Strickland, N.P.: Axiomatic Stable Homotopy Theory. Memoirs of the American Mathematical Society, vol. 128, no. 610. American Mathematical Society, Providence (1997)
17. Hovey, M., Strickland, N.P.: Morava K-Theories and Localisation. Memoirs of the American Mathematical Society, vol. 139, no. 666. American Mathematical Society, Providence (1999)
18. Iyengar, S.B., Krause, H.: The Bousfield lattice of a triangulated category and stratification. Math. Z. **273**, 1215–1241 (2013)
19. Kato, R., Okajima, H., Shimomura, K.: Notes on an algebraic stable homotopy category. In this volume
20. Margolis, H.R.: Spectra and the Steenrod Algebra. North-Holland Mathematical Library, vol. 29. Elsevier, New York (1983)
21. Milnor, J.: On axiomatic homology theory. Pacific J. Math. **12**, 337–341 (1962)
22. Neeman, A.: The chromatic tower for $\mathrm{D}(R)$. Topology **31**, 519–532 (1992)
23. Neeman, A.: Oddball Bousfield classes. Topology **39**, 931–935 (2000)
24. Neeman, A.: Triangulated Categories. Annals of Mathematics Studies, vol. 148. Princeton University Press, Princeton (2001)
25. Ohkawa, T.: The injective hull of homotopy types with respect to generalized homology functors. Hiroshima Math. J. **19**, 631–639 (1989)
26. Ravenel, D.C.: Localization with respect to certain periodic homology theories. Am. J. Math. **106**, 351–414 (1984)
27. Ravenel, D.C.: Complex Cobordism and Stable Homotopy Groups of Spheres. Pure and Applied Mathematics, vol. 121. Academic, Orlando (1986)
28. Rosický, J.: Generalized Brown representability in homotopy categories. Theory Appl. Categ. **14**, 451–479 (2005)
29. Rudyak, Y.B.: On Thom Spectra, Orientability, and Cobordism. Springer Monographs in Mathematics. Springer, Berlin (1998)
30. Spanier, E.H.: Duality and S-theory. Bull. Am. Math. Soc. **62**, 194–203 (1956)
31. Stevenson, G.: An extension of Dwyer's and Palmieri's proof of Ohkawa's theorem on Bousfield classes (2011) (Unpublished manuscript)
32. Strickland, N.P.: Counting Bousfield classes (1997) (Unpublished manuscript)
33. Switzer, R.M.: Algebraic Topology – Homology and Homotopy. Classics in Mathematics. Springer, Berlin (2002)
34. Whitehead, G.W.: Generalized homology theories. Trans. Am. Math. Soc. **102**, 227–283 (1962)
35. Whitehead, J.H.C.: Combinatorial homotopy. Part I. Bull. Am. Math. Soc. **55**, 213–245 (1949)
36. Wolcott, L.: Bousfield lattices of non-Noetherian rings: some quotients and products. Homol. Homotopy Appl. **16**, 205–229 (2014)
37. Wolcott, L.: Variations of the telescope conjecture and Bousfield lattices for localized categories of spectra. Pacific J. Math. **276**, 483–509 (2015)

From Ohkawa to Strong Generation via Approximable Triangulated Categories—A Variation on the Theme of Amnon Neeman's Nagoya Lecture Series

Norihiko Minami

Abstract This survey stems from Amnon Neeman's lecture series at Ohakawa's memorial workshop. Starting with Ohakawa's theorem, this survey intends to supply enough motivation, background and technical details to read Neeman's recent papers on his "approximable triangulated categories" and his $\mathbf{D}^b_{\mathrm{coh}}(X)$ strong generation sufficient criterion via de Jong's regular alteration, even for non-experts.

Keywords Research exposition · Derived categories · Triangulated categories · Stable homotopy theory · Bouseld class · Motivic homotopy theory

2010 Mathematics Subject Classification 14-02 · 18-02 · 55-02 · 14F05 · 14F42 · 18E30 · 18G55 · 55P42 · 55N20 · 55U35

1 Introduction

This survey stems from Amnon Neeman's lecture series at Ohakawa's memorial workshop.[1] The original lecture series started and ended with Ohkawa's theorem on the stable homotopy category. In the beginning Ohkawa's theorem was presented in its lovely, original form. The lecture series then meandered through some–definitely not all–of the developments and generalizations made by others in the years following Ohkawa's paper. And at the end came what was then a recent result of Amnon Neeman's–and the relevance was that the Ohkawa set and its properties, as developed in the years following Ohkawa, turned out to be key to the proof of the recent theorem.

Here, our presentation significantly modifies Neeman's original presentation, partially fueled by other distinguished submissions to this proceedings, mostly to

[1]The author also would like to thank Professors Mitsunori Imaoka, Takao Matumoto, Takeo Ohsawa, Katsumi Shimomura, and Masayuki Yamasaki, for coorganizing the workshop.

N. Minami (✉)
Nagoya Institute of Technology, Gokiso, Showa-ku, Nagoya 466-8555, Japan
e-mail: nori@nitech.ac.jp

T. Ohsawa and N. Minami (eds.), *Bousfield Classes and Ohkawa's Theorem*, Springer Proceedings in Mathematics & Statistics 309,
https://doi.org/10.1007/978-981-15-1588-0_3

motivate topologists to get interested in this rich subject. For this purpose, we have reorganized and expanded the original framework of Amnon Neeman's lecture series.

Still, the underlying philosophy of Neeman's presentation to start with Ohkawas's theorem remains kept in this survey. And most significantly, following a strong request of Professor Neeman, we reviewed Neeman's recent proof of: $\underline{\mathbf{D}^b_{\mathrm{coh}}(X)\ \textit{strong generation sufficient criterion via de Jong's regular alteration}}$ with enough background and technical details, expanding and sometimes even modifying parts of the original proof so as to make this review beginner-friendly from a homotopy theorist's point of view. Actually, this proof of Neeman also makes critical use of, in addition to de Jong's regular alteration, a couple of Thomason's theorems:

- First, the fundamental theorem of Hopkins, Neeman, Thomason and others on the classification of thick tensor ideals of $\mathbf{D}^{\mathrm{perf}}(X)$, the $\mathbf{D}^{\mathrm{perf}}(X) = \mathbf{D}_{\mathrm{qc}}(X)^c$ analogue of the Hopkins–Smith thick subcategory theorem of $\mathcal{SH}^{fin} = \mathcal{SH}^c$ whose proof heavily depends upon the (Devinatz-)Hopkins–Smith nilpotency theorem.
- Second, Thomason's localization theorem on $\mathbf{D}^{\mathrm{perf}}(X \setminus Z)$, for which Neeman found a homotopy theoretical proof in the framework of Miller's finite localization.

Considering these circumstance, we have also explained the role of (Devinatz-) Hopkins–Smith nilpotency theorem in the proof of Hopkins–Smith thick subcategory theorem, as well as essentially all the details of Neeman's proof of Thomason's localization theorem.

Now the rest of this survey is organized as follows:

Section 2: The first goal of this section is to recall Ohkawa's theorem in stable homotopy theory. Ohkawa's theorem claims the Bousfield classes in the stable homotopy category $\mathcal{SH}$ form a set which is very mysterious and beyond our imagination. Then the second goal of this section is the fundamental theorem of Hopkins, Neeman, Thomason and others, which roughly states the analogue of the Bousfield classes in $\mathbf{D}_{\mathrm{qc}}(X)$, in contrast to the Ohakawa's case of $\mathcal{SH}$, form a set with a clear algebro-geometric description. For these purposes, standard facts about the Bousfield localization and triangulated categories are reviewed, including the existence of Bousfield localization for perfectly generated triangulated subcategories, Miller's finite localization for triangulated subcategories generated a set of compact objects, and the telescope conjecture.

Section 3: In reality, Hopkins was not motivated by Ohkawa's Theorem 2.25 for his influential paper in algebraic geometry [48] (Theorem 2.37). Instead, Hopkins was motivated by his own theorem with Smith [50] in the triangulated subcategory $\mathcal{SH}^c$ consisting of compact objects, whose validity was already known to them back around the time Hopkins wrote [48]. In this section, we review this theorem of Hopkins–Smith, emphasizing the way how (Devinatz-)Hopkins–Smith nilpotency theorem is used in its proof. In Theorem 3.7, we summarize the main stories in $\mathcal{SH}^c_{(p)} \subset \mathcal{SH}_{(p)}$ (the Ohkawa theorem, the Hopkins–Smith theorem, Miller's version of the Ravenel telescope conjecture $(C \circ I \stackrel{?}{=} Id_{\mathbb{T}(\mathcal{SH}^{fin}_{(p)})})$, and the conjectures of Hovey and Hovey–Palmieri) in the following succinct commutative diagram:

$$\begin{array}{ccccc}
\textit{mysterious}\ \text{set} & \xrightarrow[\quad]{\text{Ohkawa Th.}} \!\!\!\!\!\! = & \mathbb{B}(\mathcal{SH}_{(p)}) & \overset{\text{Hovey Conj.}}{\underset{?}{\hookrightarrow}} \ \overset{?}{=} & \mathbb{L}(\mathcal{SH}_{(p)}) \\
\big\uparrow & & & & \big\uparrow \\
\underset{\cdots\subsetneq C_{n+1}\cdots\subsetneq C_n\cdots}{\text{chromatic hierachy}} & \overset{\text{Hopkins–Smith Th.}}{=\!=\!=} & \mathbb{T}\left(\mathcal{SH}^{fin}_{(p)}\right) & \underset{C\ (\text{split surj.})}{\overset{I\ (\text{split inj.})}{\rightleftarrows}} & \mathbb{S}(\mathcal{SH}_{(p)})
\end{array} \tag{1}$$

We then review anlogues of the Hopkins–Smith theorem in the motivic setting by Joachimi and Kelly. Also, inspired by this influence of Hopkins–Smith theorem to algebra and algebraic geometry, we briefly reviewed the couple of most prominent conjectures in homotopy theory, the telescope conjecture and the chromatic splitting conjecture, following a suggestion of Professor Morava.

Section 4: From the previous two sections, we are naturally led to investigate $\mathbf{D}_{\mathrm{qc}}(X)^c$. However, the story is not so simple. Whereas there is a conceptually simple algebro-geometrical interpretation $\mathbf{D}_{\mathrm{qc}}(X)^c = \mathbf{D}^{\mathrm{perf}}(X)$, it is its close relative (actually equivalent if X is smooth over a field) $\mathbf{D}^b_{\mathrm{coh}}(X)$ which traditionally has been intensively studied because of its rich geometric and physical information. So, we wish to understand both $\mathbf{D}^b_{\mathrm{coh}}(X)$ and $\mathbf{D}^{\mathrm{perf}}(X)$. In this section, we start with brief, and so inevitably incomplete, summaries of $\mathbf{D}^b_{\mathrm{coh}}(X)$ and $\mathbf{D}^{\mathrm{perf}}(X)$, focusing on their usages. Still, we hope this would convince non-experts that $\mathbf{D}^b_{\mathrm{coh}}(X)$ and $\mathbf{D}^{\mathrm{perf}}(X)$ are very important objects to study. Amongst of all, we shall recall the fundamental theorem of Hopkins, Neeman, Thomason and others on the classification of thick tensor ideals of $\mathbf{D}^{\mathrm{perf}}(X)$ and the Thomason's localization theorem on $\mathbf{D}^{\mathrm{perf}}(X \setminus Z)$, both of which play critical roles in Neeman's proof of the strong generation of $\mathbf{D}^b_{\mathrm{coh}}(X)$ reviewed in Sect. 5. For the classification of thick tensor ideals of $\mathbf{D}^{\mathrm{perf}}(X)$, we shall establish the following commmutative diagram (39) in Theorem 4.15, which is the $\mathbf{D}^c_{\mathrm{qc}}(X) = \mathbf{D}^{\mathrm{perf}}(X)$ analogue of the Hopkins–Smith theorem, coupled with the fundamental theorem of Hopkins, Neeman, Thomason, and others, reviewed in Sect. 2, which is the $\mathbf{D}_{\mathrm{qc}}(X)$ analogue of he Ohkawa theorem:

$$\begin{array}{ccccc}
2^{|X|} & & \underset{\text{supp}}{\overset{\{Q\in\mathbf{D}_{\mathrm{qc}}(X)\ \mid\ \mathrm{supp}(Q)\subseteq -\}}{\rightleftarrows}} & & \mathbb{L}(\mathbf{D}_{\mathrm{qc}}(X)) \\
\big\uparrow & & & & \big\uparrow \\
\mathrm{Tho}(|X|) & \underset{\text{supp}}{\overset{\mathbf{D}^{\mathrm{perf}}_{-}(X)}{\rightleftarrows}} & \mathbb{T}\left(\mathbf{D}^{\mathrm{perf}}(X)\right) & \underset{C_X}{\overset{I_X}{\rightleftarrows}} & \mathbb{S}(\mathbf{D}_{\mathrm{qc}}(X))
\end{array} \tag{2}$$

This commutative diagram is very important because it encapsulates the story (of not only this article, but also of this proceedings!). In fact, this commutative diagram in $\mathbf{D}^c_{\mathrm{qc}} \subset \mathbf{D}_{\mathrm{qc}}$, which is the analogue of the commutative diagram in $\mathcal{SH}^c \subset \mathcal{SH}$ (introduced in Sect. 3), leads us to extend these commutative diagrams to other triangulated categories. Furthermore, the mutually inverse arrows at the bottom right of the diagram yield a positive solution to the telescope conjecture (see Theorem 4.15 and Remark 4.16 for more detail), unlike the original problematic telescope conjecture in $\mathcal{SH}_{(p)}$ which shows up in the commutative diagram (1) (see the paragraph after Theorem 3.3). Finally, to close this section, we shall review Neeman's recent result, which claims two close relatives $\mathbf{D}^b_{\mathrm{coh}}(X)$ and $\mathbf{D}^{\mathrm{perf}}(X)$ actually determine each other, and its main technical tool: approximable triangulated category whose principal example is $\mathbf{D}_{\mathrm{qc}}(X)$, as well as $\mathcal{SH}$.

Section 5: Having been convinced that $\mathbf{D}^b_{\mathrm{coh}}(X)$ and $\mathbf{D}^{\mathrm{perf}}(X)$ carry rich information and are intimately related to each other in the previous section, we review here Neeman's recent investigations of the important "strong generation" property, in the sense of Bondal and Van den Bergh [20], for $\mathbf{D}^b_{\mathrm{coh}}(X)$ and $\mathbf{D}^{\mathrm{perf}}(X)$. The focus here (and in this paper) is Neeman's $\mathbf{D}^b_{\mathrm{coh}}(X)$ strong generation sufficient criterion via de Jong's regular alteration, for which we give a substantial part of its proof, including some modifications.

- Start with the $\mathbf{D}_{\mathrm{qc}}(X)$ strong compact generation sufficient criterion Theorem 5.12, and give an outline of its proof, emphasizing where the approximability of $\mathbf{D}_{\mathrm{qc}}(X)$ is used.
- Applying both the fundamental theorem of Hopkins, Neeman, Thomason and others on the classification of thick tensor ideals of $\mathbf{D}^{\mathrm{perf}}(X)$ and the Thomason localization theorem on $\mathbf{D}^{\mathrm{perf}}(X \setminus Z)$, both of which were reviewed in Sect. 4, we shall show how the $\mathbf{D}_{\mathrm{qc}}(X)$ strong compact generation sufficient criterion Theorem 5.12, reviewed above, implies the $\mathbf{D}_{\mathrm{qc}}(X)$ strong bounded generation sufficient criterion via de Jong's regular alteration Theorem 5.12. Here, we extend and partially modify Neeman's proof in order to make this review beginner-friendly.
- Having the $\mathbf{D}_{\mathrm{qc}}(X)$ strong compact generation sufficient criterion available, we can prove our desired $\mathbf{D}^b_{\mathrm{coh}}(X)$ strong generation sufficient criterion via de Jong's regular alteration Theorem 5.6. However, this proof is rather involved, and requires, in addition to Christensen's theory of phantom maps, some algebro-geometric result which we had to put in a black box. We have located this black box in Lemma 5.7 (ii).

Neeman's own results presented in this survey are not exactly what he talked about at the workshop. For instance, although the "strong generation" of $\mathbf{D}^b_{\mathrm{coh}}(X)$ and $\mathbf{D}^{\mathrm{perf}}(X)$ was still a major issue in Neeman's lecture series, Neeman's theory of approximable triangulated category, which first appeared in Neeman's series of arxiv preprints in 2017, was not touched upon during 2015 lectures. Likewise, nothing was mentioned from Sects. 3 and 4 in this survey during 2015 lectures. In contrast, Neeman actually talked about other results of his own , but they have been omitted in this survey. All of these decisions were made in order to make this proceedings a "coherent story," with this survey at its philosophical core. In fact, the author, who happened to be both an organizer of the workshop and an editor of this follow-up proceedings, became confident that the mathematics presented by Neeman at the workshop vividly interacts with lots of other talks at the workshop and articles submitted to this proceedings. So, the author repeatedly mentioned such interactions whenever appropriate.

In spite of such an excitement, the first version of this paper was just a twenty page short list of results with no proof,[2] but it was the requests and the suggestions by Professor Neeman and Professor Morava, which prompted the author to revise this article repeatedly to contain lots of useful results, including many proofs!

The author would like to express his hearty thanks to Professor Amnon Neeman for his beautiful lecture series, his encouragement to write up his lecture series from the author's perspective as a non-expert, and his request to write a beginner-friendly survey of his proof of the $\mathbf{D}^b_{\mathrm{coh}}(X)$ strong generation sufficient criterion, in such a way that the roles of the homotopical ideas of Bousfield, Ohkawa, Hopkins–Smith and others in its proof become transparent. Not only that, Professor Neeman kindly read a preliminary version of this survey and offered the author many many useful suggestions including locating author's confusions.

The author's thanks also goes to Professor Jack Morava for his suggestion to emphasize the telescope conjecture and the chromatic splitting in this article, as well as many inspiring and useful comments, some of which emerged as footnotes of this paper.

The author also thanks Dr. Tobias Barthel for his help with the chromatic splitting conjecture, Professor Mike Hopkins for his historical comment on an earlier version of this paper, Professors Srikanth B. Iyengar and Ryo Takahashi for their information of their work, and Professor Peter May for his comments on the definition of the tensor triangulated category and supporting our emphasis of the conjecture(s) of Hovey and Hovey–Palmieri. The author also would like to thank Dr. Ryo Kanda for preparing a tex file of Professor Neeman's lecture series for us.

Still, the author is solely responsible for any left over mistakes and confusions, as a matter of course.

Professor Haynes Miller informed the author of interesting works of Ruth Joachimi and Tobias Barthel, both of which have been incorporated in this survey and our proceedings, As an editor of this proceedings, the author would like to thank Professor Miller for these information and other valuable information, all of which were so crucial in organizing this proceedings.

To conclude the introduction, the author dedicates this survey to Professor Tetsusuke Ohkawa, the author's former colleague at Hiroshima University. Probably the author should express his heartfelt gratitude to Professor Tetsusuke Ohkawa with rhetorical flourish... However, the author does not have such an ability, and, what is probably even more importantly, the author knows very well that Professor Ohkawa prefers interesting mathematics much more than such rhetorical flourish! So, the author would like to close this section with a homework on behalf of Professor Tetsusuke Ohkawa to be submitted to Professor Tetsusuke Ohkawa:

[2] Actually, the author thought even such a short list is exciting.

Homework 1.1 *Extend the commutative diagrams below to other triangulated categories:*

$$
\begin{array}{ccccc}
\textit{mysterious set} & \xlongequal{\textit{Ohkawa Th.}} & \mathbb{B}(\mathcal{SH}_{(p)}) & \xhookrightarrow[\overset{?}{=}]{\textit{Hovey Conj.}} & \mathbb{L}(\mathcal{SH}_{(p)}) \\
\uparrow & & & & \uparrow \\
\underset{\cdots\subsetneq C_{n+1}\cdots\subsetneq C_n\cdots}{\textit{chromatic hierachy}} & \xlongequal{\textit{Hopkins–Smith Th.}} & \mathbb{T}\left(\mathcal{SH}^{fin}_{(p)}\right) & \underset{C\ (\textit{split surj.})}{\overset{I\ (\textit{split inj.})}{\rightleftarrows}} & \mathbb{S}(\mathcal{SH}_{(p)})
\end{array}
$$

$$
\begin{array}{ccccc}
2^{|X|} & & \underset{\mathrm{supp}}{\overset{\{Q\in\mathbf{D}_{\mathrm{qc}}(X)\ \mid\ \mathrm{supp}(Q)\subseteq -\}}{\rightleftarrows}} & & \mathbb{L}(\mathbf{D}_{\mathrm{qc}}(X)) \\
\uparrow & & & & \uparrow \\
\mathrm{Tho}(|X|) & \underset{\mathrm{supp}}{\overset{\mathbf{D}^{\mathrm{perf}}_{-}(X)}{\rightleftarrows}} & \mathbb{T}\left(\mathbf{D}^{\mathrm{perf}}(X)\right) & \underset{C_X}{\overset{I_X}{\rightleftarrows}} & \mathbb{S}(\mathbf{D}_{\mathrm{qc}}(X))
\end{array}
$$

2 Ohkawa's Theorem on Bousfield Classes Forming a Set, and Its Shadows in Algebraic Geometry

The first goal of this section is to recall Ohkawa's theorem in stable homotopy theory. Ohkawa's theorem claims the Bousfield classes in the stable homotopy category $\mathcal{SH}$ form a set which is very mysterious and beyond our imagination.[3]

Then the second goal of this section is the fundamental theorem of Hopkins, Neeman, Thomason and others, which roughly states the analogue of the Bousfield classes in $\mathbf{D}_{\mathrm{qc}}(X)$, in contrast to the Ohakawa's case of $\mathcal{SH}$, form a set with a clear algebro-geometric description.

Since both $\mathcal{SH}$ and $\mathbf{D}_{\mathrm{qc}}(X)$ are triangulated categories, we start with recalling some basic terminologies of triangulated categories.

---

[3]Concerning this sentence, Professor Morava communicated the following thoughts to the author: "When I read it I was reminded of a quotation from the English writer Sir Thomas Browne (from 'Urn Burial', in 1658):

What song the Sirens sang, or what name Achilles assumed when he hid himself among women, though puzzling questions, are not beyond all conjecture...

I believe understanding the structure of Ohkawa's set (perhaps by defining something like a topology on it) is very important, not just for homotopy theory but for mathematics in general. An analogy occurs to me, to other very complicated objects (like the Stone-Čech compactification of the rationals or the reals, or maybe the Mandelbrot set) which are very mysterious but can approached as limits of more comprehensible objects. Indeed I wonder if this is what Neeman's theory of approximable triangulated categories points toward."

2.1 Bousfield Localizations

Let $\mathcal{T}$ be a triangulated category. The suspension functor is denoted by Σ. In this article all triangulated categories are assumed to have small Hom-sets, except Verdier quotients to be defined now.

In fact, to study highly rich objects like triangulated categories, we should "localize" at various stages. This is exactly what Verdier [135] did in the context of derived categories.[4]

---

[4]Let us briefly recall the localization in the abelian category setting: [41, III, 1] [43, p. 122, Exer. 9]. Just as we may start with thick triangulated categories for Verdier quotients, which we will see in Remark 2.3 (iii), to localize an abelian category $\mathcal{A}$ by its full subcategory $\mathcal{B}$, we start with assuming $\mathcal{B}$ is a Serre subcategory, i.e.

for any exact sequene $0 \to B' \to B \to B'' \to 0$ in $\mathcal{A}$, $\quad \big(B \in \mathcal{B} \iff (B' \in \mathcal{B} \text{ and } B'' \in \mathcal{B})\big)$

Then the *quotient category $\mathcal{A}/\mathcal{B}$, in the sense of Gabriel, Grothendieck, and Serre*, is of the following form:

$$\mathrm{Ob}\,\mathcal{A}/\mathcal{B} := \mathrm{Ob}\,\mathcal{A}; \qquad \text{"Hom"}_{\mathcal{A}/\mathcal{B}}(A, A') := \varinjlim_{\underline{A}, \underline{A'} \text{ s.t. } A/\underline{A} \in \mathcal{B}, \underline{A'} \in \mathcal{B}} \mathrm{Hom}_{\mathcal{A}}\big(\underline{A}, A'/\underline{A'}\big)$$

Thus, an element of $\mathrm{Hom}_{\mathcal{A}/\mathcal{B}}(A, A')$ is of the following form:

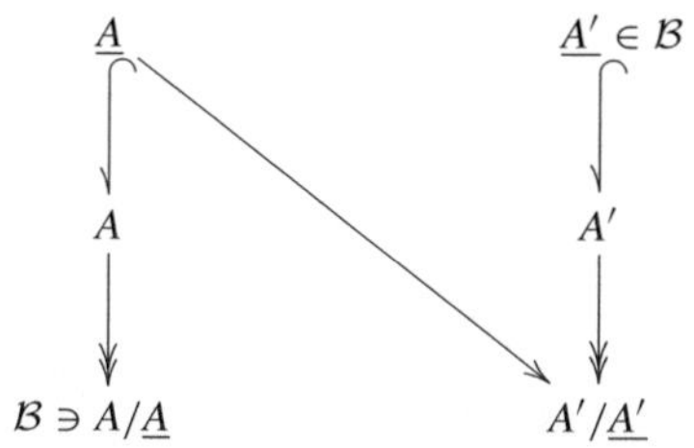

However, if we consider a similar diagram in the setting of derived categories, we may take the homotopy pullback $\widetilde{\underline{A}}$ as in the following diagram:

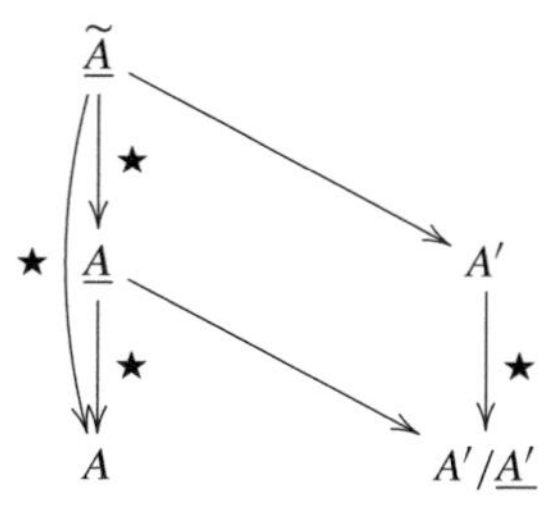

Here, arrows with ★ are local maps, and so, this gives a pair of maps $(A \overset{\bigstar}{\longleftarrow} \widetilde{\underline{A}} \to A')$, which is a typical element in the "Hom" class in the Verdier quotient.

Definition 2.1 *(Verdier quotient (a.k.a. Verdier localization))* [135] (see also [111, Chapter 2]) For a triangulated category $\mathcal{T}$ and its triangulated subcategory[5] $\mathcal{S}$, the <u>*Verdier quotient*</u> (a.k.a. <u>*Verdier localization*</u>) $\mathcal{T}/\mathcal{S}$ is a "triangulated category",[6] which are characterized by the following properties:

- $\mathrm{Ob}(\mathcal{T}/\mathcal{S}) = \mathrm{Ob}(\mathcal{T})$. For $X, Y \in \mathrm{Ob}(\mathcal{T}/\mathcal{S}) = \mathrm{Ob}(\mathcal{T})$, the <u>*class*</u> of morphisms, is given by

$$\text{“Hom”}_{\mathcal{T}/\mathcal{S}}(X,Y) = \frac{\text{diagrams of the form } (X \xleftarrow{l} Z \xrightarrow{f} Y) \text{ with } l, f \in \mathrm{Hom}_{\mathcal{T}}, \mathrm{Cone}(l) \in \mathrm{Ob}(\mathcal{S})}{(X \xleftarrow{l_1} Z_1 \xrightarrow{f_1} Y_1) \simeq (X \xleftarrow{l_2} Z_2 \xrightarrow{f_2} Y_2) \iff \begin{array}{ccccc} & & Z_1 & & \\ & \swarrow^{l_1} & \uparrow & \searrow^{f_1} & \\ X & \longleftarrow & \exists Z & \longrightarrow & Y \\ & \nwarrow_{l_2} & \downarrow & \nearrow_{f_2} & \\ & & Z_2 & & \end{array}}$$

- The *Verdier localization functor*

$$\begin{aligned} F_{univ} : \mathcal{T} &\to \mathcal{T}/\mathcal{S} \\ X &\mapsto X \\ (X \xrightarrow{f} Y) &\mapsto (X \xleftarrow{id_X} X \xrightarrow{f} Y) \end{aligned} \tag{3}$$

  is universal for all triangulated functors $F : \mathcal{T} \to \mathcal{T}$ which sends all morphisms $(Z \xrightarrow{l} X)$ with $\mathrm{Cone}(l) \in \mathrm{Ob}(\mathcal{S})$ to invertible morphisms.
- The triangulated structure of $\mathcal{T}/\mathcal{S}$ is induced from that of $\mathcal{T}$ via the Verdier localization functor F_{univ}:

– The suspension $\Sigma_{\mathcal{T}/\mathcal{S}}$ of $\mathcal{T}/\mathcal{S}$ is induced from the suspension $\Sigma_{\mathcal{T}}$ of $\mathcal{T}$:

$$\begin{aligned} \Sigma_{\mathcal{T}/\mathcal{S}} : \mathcal{T}/\mathcal{S} &\to \mathcal{T}/\mathcal{S} \\ X &\mapsto \Sigma_{\mathcal{T}} X \\ (X \xleftarrow{l} Z \xrightarrow{f} Y) &\mapsto (\Sigma_{\mathcal{T}} X \xleftarrow{\Sigma_{\mathcal{T}} l} \Sigma_{\mathcal{T}} Z \xrightarrow{\Sigma_{\mathcal{T}} f} \Sigma_{\mathcal{T}} Y) \end{aligned}$$

– A distinguished triangle in $\mathcal{T}/\mathcal{S}$ is isomorphics to the Verdier localization functor F_{univ} image of a distinguished triangle in $\mathcal{T}$.

As is always the case with such a localization procedure, the Verdier localization does not necessarily have small Hom-sets. It was Neeman's insight [105, 106, 111]

---

[5] <u>WARNING!:</u> In this article, we follow the convention of [111, Def. 1.5.1] [78, 4.5] for a <u>*triangulated subcategory*</u>, which is automatically full by this convention. On the other hand, it is not so in the convention of [131, p. 3, 1.1].

[6] Verdier quotient does not necessarily have small Hom-sets.

to make use of the Bousfield localization [21], which was introduced in the context of stable homotopy theory, to take case of this problem in general triangulated category theory.

To explain this theory of Neeman, we now prepare some definitions.

Definition 2.2 (WARNING!: A *triangulated subcategory* is by definition [111, Def. 1.5.1] [78, 4.5] automatically full.)

1. A triangulated subcategory $\mathcal{S}$ of a triangulated category $\mathcal{T}$ with small coproducts is called *localizing*, if it is closed under coproducts in $\mathcal{T}$.
2. A triangulated subcategory $\mathcal{S}$ of $\mathcal{T}$ is called *thick*, if it closed under direct summands in $\mathcal{T}$.
3. [111, p99, Rem. 2.1.39] The *thick closure* $\widehat{\mathcal{S}}$ of a triangulated subcategory $\mathcal{S}$ of a triangulated category $\mathcal{T}$ is the triangulated subcategory of $\mathcal{T}$ consisting of direct summands in $\mathcal{T}$ of objects in $\mathcal{S}$.
4. [131, 1.4] A triangulated subcategory $\mathcal{S}$ of a triangulated category $\mathcal{T}$ is called **dense**, if $\widehat{\mathcal{S}} = \mathcal{T}$.

Remark 2.3 (i) Every localizing triangulated subcategory is thick, for any direct summand decomposition in $\mathcal{T}$:

$$S \ni x = ex \oplus (1-e)x$$

can be realized using the cones in $\mathcal{S}$:

$$\begin{cases} ex & = \operatorname{Cone}(\oplus_{\mathbb{N}}x \to \oplus_{\mathbb{N}}x : (\xi_n)_{n\in\mathbb{N}} \mapsto (\xi_n - e\xi_{n-1})_{n\in\mathbb{N}}) \\ (1-e)x & = \operatorname{Cone}(\oplus_{\mathbb{N}}x \to \oplus_{\mathbb{N}}x : (\xi_n)_{n\in\mathbb{N}} \mapsto (\xi_n - (1-e)\xi_{n-1})_{n\in\mathbb{N}}) \end{cases}$$

(ii) A triangulated subcategory $\mathcal{S}$ of a triangulated category $\mathcal{T}$ is thick if and only if $\mathcal{S} = \widehat{\mathcal{S}}$.
(iii) [111, p99, Rem. 2.1.39] The thick closure is nothing but the kernel of the Verdier localization functor: For a triangulated subcategory $\mathcal{S}$ of a triangulated category $\mathcal{T}$, $\widehat{\mathcal{S}} = \operatorname{Ker}(F_{univ} : \mathcal{T} \to \mathcal{T}/\mathcal{S})$.
(iv) [111, p. 148, Cor. 4.5.12] If $\mathcal{S}$ is a dense triangulated subcategory of a triangulated category $\mathcal{T}$, then,

$$\forall x \in \mathcal{T}, \quad x \oplus \Sigma x \in \mathcal{S}. \tag{4}$$

To see this,[7] since $\exists y \in \mathcal{T}$ s.t. $x \oplus y \in \mathcal{S}$, form a triangle:

$$x \oplus 0 \oplus y \xrightarrow{0\oplus 0\oplus id_Y} 0 \oplus x \oplus y \xrightarrow{0\oplus id_X \oplus id_Y} \Sigma x \oplus x \oplus 0,$$

where the first and the second terms are contained in $\mathcal{S}$: $x \oplus 0 \oplus y \cong 0 \oplus x \oplus y \cong x \oplus y \in \mathcal{S}$, and so is the third term: $\Sigma x \oplus x \cong \Sigma x \oplus x \oplus 0 \in \mathcal{S}$, as desired.

[7] If $\mathcal{T}$ is essentially small, this result also follows immediately from a general result reviewed later in Proposition 4.3.

From Remark 2.3 (iii), to search for criteria which guarantee the Verdier quotient to have small Hom-sets, we may start with a thick triangulated subcategory $\mathcal{S}$ of $\mathcal{T}$. Also, while the original Bousfield localization [21] require $\mathcal{T}$ to have small coproducts, there are many cases where we wish Verdier quotients $\mathcal{T}/\mathcal{S}$ to have small Hom-sets, even when $\mathcal{T}$ does not have small coproducts, Now, Neeman [111] proposed the following general definition for Bousfield localization:

Definition 2.4 [111, Def. 9.1.1, Def. 9.1.3, Def. 9.1.4, Def. 9.1.10] [78] (i) Let $\mathcal{S}$ be a thick subcategory of a triangulated category $\mathcal{T}$.[8] Then the pair $\mathcal{S} \subset \mathcal{T}$ is said to possese a *Bousfield localization functor* when the Verdier localization functor $F_{univ} : \mathcal{T} \to \mathcal{T}/\mathcal{S}$ has a right adjoint $G : \mathcal{T}/\mathcal{S} \to \mathcal{T}$, which is called the *Bousfield localization functor*. The resulting composite

$$L := G \circ F_{univ} : \mathcal{T} \xrightarrow{F_{univ}} \mathcal{T}/\mathcal{S} \xrightarrow{G} \mathcal{T}$$

is also called the *Bousfield localization functor* by an abuse of terminology.
(ii) $\mathcal{S} \subset \mathcal{T}$ is, by definition, the full subcategory of $\mathcal{S}$ – colocal objects.
(iii) $\mathcal{S}^{\perp} \subset \mathcal{T}$ is, by definition, the full subcategory of $L - localobjects$ or $\mathcal{S} - localobjects$.

An adjoint functor between triangulated categories showed up in the above definition, but such an adjoint functor actually becomes a triangulated functor:

Lemma 2.5 [111, Lem. 5.3.6] *Suppose a pair of adjoint functors between triangulated categories are given:*

$$\mathcal{S} \underset{G}{\overset{F}{\rightleftarrows}} \mathcal{T}$$

If either one of F or G is a triangulated functor, then so is the other.

We shall freely use this useful fact for the rest of this article.

Still, readers might worry that the more existence of a right adjoint $G : \mathcal{T}/\mathcal{S} \to \mathcal{T}$ in the definition of the above Bousfield localization too weak. However, in this particular case, we have a very special property that the natural map from the category of fractions $\mathcal{T}\left[\Sigma(F_{univ})^{-1}\right]$ to the Verdier quotient $\mathcal{T}/\mathcal{S}$ becomes an equivalence:

$$\mathcal{T}\left[\Sigma(F_{univ})^{-1}\right] \xrightarrow{\cong} \mathcal{T}/\mathcal{S},$$

where $\Sigma(F_{univ})$ is the collection of morphisms in $\mathcal{T}$ whose image in $\mathcal{T}/\mathcal{S}$ is invertible, i.e. those maps in $\mathcal{T}$ whose mapping cone is in $\mathcal{S}$. And, using this useful fact, we can see any right adjoint $G : \mathcal{T}\mathcal{S} \to \mathcal{T}$ is fully faithful by applying the following useful fact:

[8] We dot not require $\mathcal{T}$ to have small coproducts in this definition.

Lemma 2.6 (see [42, I, Prop. 1.3] [78, Prop. 2.3.1]). *For an adjoint pair*[9]*:* $\mathcal{C} \underset{G}{\overset{F}{\rightleftarrows}} \mathcal{D}$ *, the following conditions are equivalent:*

- *The right adjoint G is fully faithful.*
- *The adjunction* $F \circ G \to \mathrm{Id}_{\mathcal{D}}$ *is an isomorphism.*
- *The functor* $\overline{F} : \mathcal{C}\left[\Sigma(F)^{-1}\right] \to \mathcal{D}$ *satisfying* $F = \overline{F} \circ Q_{\Sigma(F)}$ *is an equivalence, where* $\Sigma(F)$ *is the collection of morphisms in* $\mathcal{T}$ *whose images in* $\mathcal{T}'$ *by F becomes invertible, and* $Q_{\Sigma(F)} : \mathcal{C} \to \mathcal{C}\left[\Sigma(F)^{-1}\right]$ *is the canonical quotient functor to the category of fractions.*

Thus, from Lemma 2.6 and Lemma 2.5, we obtain the following:

Proposition 2.7 *Any right adjoint* $G : \mathcal{T}/\mathcal{S} \to \mathcal{T}$ *in Neeman's definition of the Bousfield localization Definition 2.4 is automatically a fully faithful triangulated functor.*

In fact, as is well known, if a triangulated functor $F : \mathcal{T} \to \mathcal{T}'$ enjoys good properties listed in Lemma 2.6, then we have the following very useful result[10,11]:

Proposition 2.8 (see e.g. [126, Lem. 3.4]) *If a triangulated functor* $F : \mathcal{T} \to \mathcal{T}'$ *has a fully faithful right adjoint G or a right adjoint G with its adjunction an isomorphism* $F \circ G \overset{\cong}{\to} \mathrm{Id}_{\mathcal{D}}$*, then* Ker F *becomes a thick triangulated subcategory of* $\mathcal{T}$*, and F induces the following equivalence of triangulated categories:*

$$\mathcal{T}/\,\mathrm{Ker}\,F \overset{\cong}{\to} \mathcal{T}'$$

Going back to Bousfield localization, we prepare some more definitions to state its basic properties.

Definition 2.9 1. (WARNING!: These conventions are those of [78, 4.8], which are the *opposite* of [111, Def. 9.1.10; Def. 9.1.11]!) For a full subcategory $\mathcal{A}$ of $\mathcal{T}$, define the full subcategory $\mathcal{A}^{\perp}$ of $\mathcal{T}$ by

$$\mathcal{A}^{\perp} = \{\, t \in \mathcal{T} \mid \mathrm{Hom}_{\mathcal{T}}(\mathcal{A}, t) = 0 \,\}.$$

Dually, ${}^{\perp}\mathcal{A}$ is defined by

---

[9]This is an adjoint pair of functors between ordinary categories, and we are not considering any triangulated structure.

[10]Goes back at least to Verdier.

[11]Let us recall the following precursor of this result in the setting of abelian categories, which goes back at least to Gabriel (see also [126, Lem. 3.2]): If an exact functor $F : \mathcal{A} \to \mathcal{B}$ between abelian categories has a fully faithful right adjoint G (i.e. the adjunction $F \circ G \to \mathrm{Id}_{\mathcal{B}}$ is an isomorphism, then Ker F is Serre subcategory of $\mathcal{A}$, and F induces the following equivalence of abelian categories: $\mathcal{A}/\,\mathrm{Ker}\,F \overset{\cong}{\to} \mathcal{B}$, where the left hand side is the abelian quotient category in the sense of Gabriel, Grothendieck, and Serre.

$$^{\perp}\mathcal{A} = \{t \in \mathcal{T} \mid \mathrm{Hom}_{\mathcal{T}}(t, \mathcal{A}) = 0\}.$$

2. For full subcategories $\mathcal{A}$ and $\mathcal{B}$ of $\mathcal{T}$, denote by $\mathcal{A} * \mathcal{B}$ the full subcategory of $\mathcal{T}$ consisting of all objects $y \in \mathcal{T}$ for which there exists a triangle $x \to y \to z \to \Sigma x$ with $x \in \mathcal{A}$ and $z \in \mathcal{B}$.

Proposition 2.10 [111, Prop. 9.1.18; Th. 9.1.16; Th. 9.1.13; Cor. 9.1.14] [78, Prop. 4.9.1] *Let $\mathcal{S}$ be a thick subcategory of a triangulated category $\mathcal{T}$.*[12] *Then the following assertions are equivalent.*

1. *The inclusion functor $I : \mathcal{S} \hookrightarrow \mathcal{T}$ has a right adjoint $\widetilde{\Gamma} : \mathcal{T} \to \mathcal{S}$.*
2. *$\mathcal{T} = \mathcal{S} * \mathcal{S}^{\perp}$.*
3. *$\mathcal{S} \subset \mathcal{T}$ posseses a Boundfield localization functor, i.e. the Verdier localization functor $F_{univ} : \mathcal{T} \to \mathcal{T}/\mathcal{S}$ has a right adjoint $G : \mathcal{T}/\mathcal{S} \to \mathcal{T}$.*
4. *The composite $E : \mathcal{S}^{\perp} \hookrightarrow \mathcal{T} \to \mathcal{T}/\mathcal{S}$ is an equivalence.*
5. *The inclusion $J : \mathcal{S}^{\perp} \hookrightarrow \mathcal{T}$ has a left adjoint $\mathcal{T} \to \mathcal{S}^{\perp}$ and $^{\perp}(\mathcal{S}^{\perp}) = \mathcal{S}$.*

These equivalent conditions can be succinctly expressed, via the standard adjoint functor notation,[13] *as follows:*

$$\mathcal{S} \underset{\widetilde{\Gamma}}{\overset{I}{\rightleftarrows}} \mathcal{T} \underset{G}{\overset{F_{univ}}{\rightleftarrows}} \mathcal{T}/\mathcal{S} \tag{5}$$

Remark 2.11 Assume that the inclusion $I \colon \mathcal{S} \hookrightarrow \mathcal{T}$ has a right adjoint $\widehat{\Gamma}$ as in Proposition 2.10(1). Then, for each $t \in \mathcal{T}$, embed the counit of adjunction $\Gamma(t) = I\widetilde{\Gamma}(t) \to t$, where $\Gamma : \mathcal{T} \to \mathcal{T}$ is called the Bousfield colocalization functor for the pair $\mathcal{T} \to \mathcal{T}/\mathcal{S}$,[14] into a triangle

$$\Gamma(t) \to t \to L(t) \to \Sigma\Gamma(t),$$

which yields a functor $L : \mathcal{T} \to \mathcal{T}$. Then we see $L(t) \in \mathcal{S}^{\perp}$, which

- implies $\mathcal{T} = \mathcal{S} * \mathcal{S}^{\perp}$ in Proposition 2.10(2);
- yields a left adjoint $\widetilde{L} : \mathcal{T} \to \mathcal{S}^{\perp}$ to the inclusion $J : \mathcal{S}^{\perp} \hookrightarrow \mathcal{T}$, stated in Proposition 2.10(5), and $\widetilde{L}$ yields the Bousfield localization functor, recovering the above functor L by the composition

$$L = J \circ \widetilde{L} : \mathcal{T} \to \mathcal{S}^{\perp} \to \mathcal{T}. \tag{6}$$

- yields a left adjoint $G : \mathcal{T}/\mathcal{S} \to \mathcal{T}$ to the Verdier localization functor $F_{univ}; \mathcal{T} \to \mathcal{T}/\mathcal{S}$ as the composition $G : \mathcal{T}/\mathcal{S} \xrightarrow{\widetilde{\widetilde{L}}} \mathcal{S}^{\perp} \hookrightarrow \mathcal{T}$ stated in Proposition 2.10(3), and,

---

[12] We do not require $\mathcal{T}$ to have small coproducts.

[13] An arrow above is left adjoint to the arrow below.

[14] A Bousfield colocalization functor means its opposite functor is a Bousfield localization functor [54, Def. 3.1.1] [78, 2.8]. *WARNING:* This terminology is not consistent with that of Bousfield [21] (see [54, Rem. 3.1.4]).

- assuming Proposition 2.10(4), $\widetilde{L}$ is equivalent to $E^{-1} \circ F_{univ} : \mathcal{T} \to \mathcal{T}/\mathcal{S} \to \mathcal{S}^{\perp}$.

Remark 2.12 Actually, the property in Proposition 2.10(2) is exactly what Bondal–Orlov [19, Def. 3.1] call *semiorthogonal decomposition* and denote by

$$\mathcal{T} = \langle \mathcal{S}^{\perp}, \mathcal{S} \rangle. \tag{7}$$

Of course, the fundamental question is when Bousfield location exists. Now, Neeman's insight [111, Th. 8.4.4] is to apply Brown representability to construct Bousfield localization. We now review this development following mostly Krause [77, 78].

Definition 2.13 Let $\mathcal{T}$ be a triangulated category with small coproducts.
(i) [111, Def. 6.2.8] A set G of objects in $\mathcal{T}$ is said to generate $\mathcal{T}$, if $(\bigcup_{n\in\mathbb{Z}} \Sigma^n G)^{\perp} = 0$, i.e.,
given $t \in \mathcal{T}$,

$$\forall g \in G, \forall n \in \mathbb{Z},\ \mathrm{Hom}_{\mathcal{T}}(\Sigma^n g, t) = 0 \qquad \Longrightarrow \qquad t = 0.$$

(ii) An element $t \in \mathcal{T}$ is called compact if, for every set of objects $\{t_\lambda\}_{\lambda\in\Lambda}$ in $\mathcal{T}$, the natural map

$$\oplus_{\lambda\in\Lambda} \mathrm{Hom}_{\mathcal{T}}(t, t_\lambda) \to \mathrm{Hom}_{\mathcal{T}}(t, \oplus_{\lambda\in\Lambda} t_\lambda)$$

is an isomorphism.
(iii) $\mathcal{T}$ is called compactly generated, if $\mathcal{T}$ is generated by a set of compact objects in $\mathcal{T}$.
(iii) (c.f. [77, Def. 1][78, 5.1][15] (see also [111, Def. 8.1.2])) A set of objects P in $\mathcal{T}$ is said to perfectly generate $\mathcal{T}$, if,

1. P generates $\mathcal{T}$,
2. for every countable set of morphisms $x_i \to y_i$ in $\mathcal{T}$ such that $\mathcal{T}(p, x_i) \to \mathcal{T}(p, y_i)$ is surjective for all $p \in P$ and i, the induced map

$$\mathcal{T}\left(p, \coprod_i x_i\right) \to \mathcal{T}\left(p, \coprod_i y_i\right)$$

is surjective.

$\mathcal{T}$ is called perfectly generated, if $\mathcal{T}$ is perfectly generated by a set P of objects in $\mathcal{T}$.

Remark 2.14 Any compactly generated triangulated category is perfectly generated.

Theorem 2.15 (Brown representability) [77, Th. A][78, Th. 5.1.1] [106, 111] *Suppose a triangulated category $\mathcal{T}$ is perfectly generated.*

[15] Strictly speaking, the definition here is slightly differently from Krause's, but essentially the same.

1. *A functor $F : \mathcal{T}^{op} \to Ab$, the category of abelian groups, is cohomological and sends coproducts in $\mathcal{T}$ to products in Ab if and only if*

$$F \cong \mathcal{T}(-, t)$$

for some object t in $\mathcal{T}$.

2. *A triangulated functor $\mathcal{T} \to \mathcal{U}$ preserves small coproducts if and only if it has a right adjoint.*

From the second part of this theorem and the second characterization of Bousfield localization in Proposition 2.10, we immediately obtain the following:

Corollary 2.16 (Existence of Bousfield localization) [78, Prop. 5.2.1] [111, Prop. 9.1.19] *Bousfield localization exists for any perfectly generated triangulated subcategory $\mathcal{S}$ of $\mathcal{T}$, a triangulated category with small coproducts.*

Corollary 2.17 *Bousfield localization exists for any compactly generated triangulated subcategory $\mathcal{S}$ of $\mathcal{T}$, a triangulated category with small coproducts.*

To be precise, the "compactly generated" assumption adapted in [105, Lem. 1.7] meant the smallest localizing triangulated subcategory containing the generating set is the entire triangulated category. But this can be reconciled by the following corollary of Corollary 2.16:

Corollary 2.18 [111, Th. 8.3.3; Prop. 8.4.1] *Suppose $\mathcal{T}$ is perfectly generated by a set P of objects in $\mathcal{T}$, then*

$$\mathcal{T} = \textit{the smallest localizing triangulated subcategory containing } P.$$

For a special case of Corollary 2.17, Neeman and Miller gave a simple explicit homotopy theoretical construction of Bousfield localization with a nice property:

Theorem 2.19 [93] [105, Lem. 1.7] *For any localizing triangulated subcategory $\mathcal{R}$ of a compactly generated triangulated category with small coproducts $\mathcal{T}$ such that $\mathcal{R}$ is the smallest localizing triangulated subcategory containing a set R consisting of compact objects in $\mathcal{T}$,*

1. *Bousfield localization exists,*[16] *given explicitly by Miller's finite localization [105, p. 554, Proof of Lem. 1.7] [93, From p. 384, -6th line to p. 385, 1st line]: for $x \in \mathcal{T}$, proceed inductively as follows:*

   - $x_0 := x$,
   - *Suppose x_n has been defined, then set*

$$x_{n+1} := \mathrm{Cone}\left(\oplus_{r\in R} \oplus_{f_r \in \mathrm{Hom}_{\mathcal{T}}(r,x_n)} r \xrightarrow{\oplus_{r\in R}\oplus_{f_r\in \mathrm{Hom}_{\mathcal{T}}(r,x_n)} f_r} x_n\right)$$

[16] This claim itself is a special case of Corollary 2.17.

- *Then Miller's finite localization of $x \in \mathcal{T}$ is simply given by the mapping telescope:*

$$x \;\to\; Lx := \operatorname{hocolim}(x_n).$$

2. *Miller's finite localization is smashing, i.e. L preserves arbitrary coproducts.*

Let us record the above definition of "smashing", because this definition of "smashing" without smash (tensor) product is not the traditional Ravenel's definition [122]:

Definition 2.20 [78, 5.5] A Bousfield localization $L : \mathcal{T} \to \mathcal{T}$ is *smashing* if L preserves arbitrary coproducts in L. Then, $\mathcal{S} = \operatorname{Ker} L$ is also called *smashing.*

We have the following equivalent characterizations of smashing Bousfield localization without smash (tensor) product:

Proposition 2.21 [78, Prop. 5.5.1] *For a thick subcategory $\mathcal{S}$ of a triangulated category with small coproducts, suppose there is a Bousfield localization $L = G \circ F_{univ} : \mathcal{T} \to \mathcal{T}$ for the pair $\mathcal{S} \to \mathcal{T}$ in the following set-up: (see* (5)*):*

$$\mathcal{S} \underset{\widetilde{\Gamma}}{\overset{I}{\rightleftarrows}} \mathcal{T} \underset{G}{\overset{F_{univ}}{\rightleftarrows}} \mathcal{T}/\mathcal{S} \tag{8}$$

Then the following conditions are equivalent:

1. *Bousfield localization $L = G \circ F_{univ}$ is smashing, i.e. $L = G \circ F_{univ} : \mathcal{T} \to \mathcal{T}$ preserves coproducts (see Definition 2.20).*
2. *Bousfield colocalization $\Gamma = I \circ \widetilde{\Gamma} : \mathcal{T} \to \mathcal{T}$ preserves coproducts.*
3. *The right adjoint $G : \mathcal{T}/\mathcal{S} \to \mathcal{T}$ of the Verdier quotient $F_{univ} : \mathcal{T} \to \mathcal{T}/\mathcal{S}$ preserves coproducts.*
4. *The right adjoint $\widetilde{\Gamma} : \mathcal{T} \to \mathcal{S}$ of the canonical inclusion $I : \mathcal{S} \to \mathcal{T}$ preserves coproducts.*
5. *The full subcategory $\mathcal{S}^{\perp}$ of all L-local ($\mathcal{S}$-local) objects is localizing.*

If $\mathcal{T}$ is perfectly generated,[17] *in addition the following is equivalent.*

6. *In the set-up* (8)*, both $\widetilde{\Gamma}$ and G have right adjoints and* (8) *is amplified to a recollement*[18] *of the following form:*

$$\begin{array}{ccccc} & \xrightarrow{\;I\;} & & \xrightarrow{\;F_{univ}\;} & \\ \mathcal{S} & \xleftarrow{\;\widetilde{\Gamma}\;} & \mathcal{T} & \xleftarrow{\;G\;} & \mathcal{T}/\mathcal{S} \\ & \longrightarrow & & \longrightarrow & \end{array} \tag{9}$$

---

[17] This "perfectly generated" condition is used to apply Brown representability (Theorem 2.15) to construct two right adjoints in the recollement.

[18] For the precise definition of recollement, consult [15, 1.4].

Later in Proposition 2.28, all of these conditions are shown to be equivalent to Ravenel's [122], when $\mathcal{T}$ is a rigidly compactly generated tensor triangulated category. Smashing localization is frequently referred in the context of the telescope conjecture, which asks whether the converse of the second claim in Theorem 2.19 holds or not[19]:

Conjecture 2.22 (Telescope conjecture without smash (tensor) product) [54, Def. 3.3.2, Def. 3.3.8] (see also Proposition 2.28) *In a rigidly compactly generated tensor triangulated category $\mathcal{T}$, a smashing localization $L : \mathcal{T} \to \mathcal{T}$ is a finite localization, i.e.* Ker L *is generated by a set of compact objects in $\mathcal{T}$.*

After we take into account the tensor product structure, we shall revisit the finite localization and the telescope conjecture in Theorem 3.3. For now, we record another easy consequence of Miller's finite localization construction presented in Theorem 2.19:

Proposition 2.23 (See [105, p. 556, from 7th to 10th lines])
Let R be a set of compact objects in a triangulated category with small coproducts $\mathcal{T}$, and $\mathcal{R}$ be the smallest localizing triangulated subcategory containing R.

Then, every element in $\mathcal{R}^c$ is isomorphic in $\mathcal{R}^c$ to a direct summand of a finite extensions of finite coproducts of elements in R. In particular, $\mathcal{R}^c$ is essentially small.

In fact, the Bousfield localization with respect to the pair $\langle R \rangle = \mathcal{R} \subset \mathcal{R}$, is trivial for any $x \in \mathcal{R}$:

$$x \to Lx := \mathrm{hocolim}(x_n) \simeq 0.$$

Then, if $x \in \mathcal{R}^c$, this map becomes trivial at some "finite" stage, which implies x is a direct summand of a finite extensions of finite coproducts of elements in R, as claimed.

2.2 Bousfield Classes and Ohkawa's Theorem

Now we focus on a special case: let $\mathcal{T} = \mathcal{SH}$ be the homotopy category of spectra. Then $\mathcal{T}$ is a triangulated category with coproducts. It has the smash product $\wedge : \mathcal{T} \times \mathcal{T} \to \mathcal{T}$ and the unit object $S^0 \in \mathcal{T}$ which make $\mathcal{T}$ a tensor triangulated category.[20]

The smash product preserves coproducts in each variable. $\mathcal{T}$ is generated by $\{S^0\}$, and $\mathcal{T}$ satisfies Brown representability.

For each $H \in \mathcal{T}$, put $H_* = H \wedge (-)$. We consider the localizing triangulated subcategory

[19] Strictly speaking, this is the telescope conjecture without smash (tensor) product, but coincides with the original Ravenel's telescope conjecture for $\mathcal{T} = \mathcal{SH}$, and more generally for rigidly compactly generated tensor triangulated categories [54, Def. 3.3.2, Def. 3.3.8] (see also Proposition 2.28).

[20] For a serious treatment of the definition of "tensor triangulated category," consult [92].

$$\operatorname{Ker} H_* = \{\, t \in \mathcal{T} \mid H \wedge t = 0 \,\},$$

which is called the *Bousfield class* of H.

Theorem 2.24 (Bousfield [21]) *Let $\mathcal{T} = \mathcal{SH}$ be the homotopy category of spectra.*

1. *If $\mathcal{S} \subset \mathcal{SH}$ is a localizing triangulated subcategory which is generated by a* set *of objects, then a Bousfield localization exists for $\mathcal{S}$.*
2. *For every $H \in \mathcal{SH}$, there exists a* set *of objects which generates* $\operatorname{Ker} H_*$. *Therefore a Bousfield localization exists for* $\operatorname{Ker} H_*$.

Somewhat surprisingly, Ohkawa's theorem had been elusive from researchers' attention for more than a decade. It was the paper of Dwyer and Palmieri [37] which drew researchers' attention to Ohkawa's surprising theorem[21]:

Theorem 2.25 (Ohkawa [115]) $\{\, \operatorname{Ker} H_* \mid H \in \mathcal{SH} \,\}$ *is a* set.

We note that no explicit structure of this set is known.

For more details, including a proof, of the Ohakawa theorem, see the survey [26] in this proceedings.

2.3 Casacuberta–Gutiérrez-Rosický Theorem, Motivic Analogue of Ohkawa's Theorem

Ohkawa's theorem is a statement in the stable homotopy category $\mathcal{SH}$, which is "a part" of the Morel–Voevodsky stable homotopy category[22] $\mathcal{SH}(k)$ when $k \subseteq \mathbb{C}$, via the retraction of the following form:

$$\underbrace{\mathcal{SH} \longrightarrow \mathcal{SH}(k) \xrightarrow{R_k} \mathcal{SH}}_{id} \tag{10}$$

So, a natural question here is whether there is a shadow of Ohkawa's theorem in this algebro-geometrical setting, i.e. whether there is a motivic analogue of Ohkawa's theorem or not.

Now, Casacuberta–Gutiérrez-Rosický [28] answered this question affirmatively under some very mild assumption.

Theorem 2.26 [28, Cor. 3.6] *For each Noetherian scheme S of finite Krull dimension, there is only a set of distinct Bousfield classes in the stable motivic homotopy category $\mathcal{SH}(S)$ with base scheme S.*

Once again, no explicit structure of this set is known.

For various generalizations of Ohkawa's theorem, see afore-quoted [28], also [66] and the review [27], both in this proceedings.

[21] For a concise summary of the academic life of Professor Tetsusuke Ohkawa, see [91] in this proceedings.

[22] For short reviews of the Morel-Voevodsky stable homotopy category, cosult [72, 94] for instance.

2.4 Localizing Tensor Ideals of Derived Categories and the Fundamental Theorem of Hopkins, Neeman, Thomason and Others

In both Ohkawa's Theorem 2.25 and its algebro-geometric shadow Theorem 2.26, the resulting sets are completely mysterious and beyond our imagination. However, if we take a look at the algebro-geometrical shadow of Ohkawa's theorem from a different angle, i.e. by considering $\mathbf{D}_{qc}(X)$ for a fixed Noetherian scheme instead of $\mathcal{SH}(k)$, then we see an explicit set representing clear algebro-geometric information. This is the fundamental theorem of Hopkins, Neeman, Thomason, and others, which has been the guiding principle of the area.

Now, the tensor structure is essential for this fundamental theorem, and we must start with some review of fundamental facts about general tensor triangulated categories and Bousfield localization from the tensor triangulated category point of view.

Definition 2.27 Let $\mathcal{T}$ be a tensor triangulated category.

1. A triangulated subcategory $\mathcal{I}$ of $\mathcal{T}$ is called a $\begin{cases}\underline{\text{tensor ideal}}\\ \underline{\text{prime}}\end{cases}$ if

$$\begin{cases}\mathcal{T}\otimes\mathcal{I}\subset\mathcal{I};\\ \text{it is a tensor ideal and } (\mathcal{T}\setminus\mathcal{I})\otimes(\mathcal{T}\setminus\mathcal{I})\subset(\mathcal{T}\setminus\mathcal{I})\neq\emptyset.\end{cases}$$

2. [81, Chapter III] (see also [54, App. A,] [7, p. 1163]) An element x in a closed symmetric monoidal triangulated category $(\mathcal{T},\otimes,\underline{\mathrm{Hom}})$ is called *strongly dualizable* or simply *rigid*,[23] if the natural map $Dx\otimes y\to\underline{\mathrm{Hom}}(x,y)$, where $Dx:=\underline{\mathrm{Hom}}(x,\mathbb{1})$, is an isomorphism for all $y\in\mathcal{T}$.
3. [54, Def. 1.1.4] (see also [7, Hyp. 1.1]) A closed symmetric monoidal triangulated category $(\mathcal{T}=\langle G\rangle,\otimes,\underline{\mathrm{Hom}})$ is called a *unital algebraic stable homotopy category* or a *rigidly compactly generated tensor triangulated category*, if $\mathbb{1}$ is compact and $\mathcal{T}=\langle G\rangle$ for a set G of rigid and compact objects.[24]

Now, we are ready to reconcile our previous definition (Definition 2.20) of smashing localization with Ravel's original definition in [122] for rigidly compactly generated tensor triangulated categories:

[23] If $x\in\mathcal{T}$ is strongly dualizable, i.e. rigid, the natural map $x\to D^2x$ is an isomorphism [81, Chapter III] [54, Th. A.2.5.(b)].

[24] In a rigidly compactly generated tensor triangulated category, any compact object is rigid, for, by Proposition 2.23, any compact object is seen to be isomorphic to a direct summand of a finite extensions of finite coproducts of rigid elements. In particular, in a rigidly compactly generated tensor triangulated category, $\mathbb{1}$ is both rigid and compact.

Proposition 2.28 [54, Def. 3.3.2] *For a thick subcategory $\mathcal{S}$ of a closed symmetric monoidal triangulated category with small coproducts* ($\mathcal{T} = \langle G \rangle, \otimes, \underline{\mathrm{Hom}}$),[25] *suppose there is a Bousfield localization* $L : \mathcal{T} \to \mathcal{T}$ *for the pair* $\mathcal{S} \to \mathcal{T}$. *Consider the following "smishing" conditions:*

(S): *(Ravenel's original definition of smashing localization [122]):*

$$L \cong L(\mathbb{1}) \otimes -, \text{ where } \mathbb{1} \text{ is the unit onject of } (\mathcal{T}, \otimes).$$

(C): *(The definition of smashing localization in Definition 2.20):*

$$L \text{ preserves arbitrary coproducts.}$$

Then, the implication (S) $\implies$ (C) *always holds. If* $\mathcal{T}$ *is also a rigidly compactly generated tensor triangulated category, the converse* (C) $\implies$ (S) *also holds, and so,* (C) *and* (S) *become equivalent.*

Proof The implication (S) $\implies$ (C) is easy:

$$L\left(\oplus_\lambda x_\lambda\right) \overset{(S)}{\cong} L(\mathbb{1}) \otimes \left(\oplus_\lambda x_\lambda\right) \cong \oplus_\lambda \left(L(\mathbb{1}) \otimes x_\lambda\right) \overset{(S)}{\cong} \oplus_\lambda L x_\lambda.$$

For the converse (C) $\implies$ (S), first note that (C) implies those $x \in \mathcal{T}$ which satisfies $L\mathbb{1} \otimes x \cong Lx$ form a localizing triangulated subcategory of $\mathcal{T}$, even without the rigidly compactly generated assumption. For instance, if $L\mathbb{1} \otimes x_\lambda \cong Lx_\lambda \ \forall \lambda \in \Lambda$, then

$$L(\mathbb{1}) \otimes \left(\oplus_\lambda x_\lambda\right) \cong \oplus_\lambda \left(L(\mathbb{1}) \otimes x_\lambda\right) \cong \oplus_\lambda L x_\lambda \overset{(C)}{\cong} L\left(\oplus_\lambda x_\lambda\right).$$

Now, we are reduced to showing $L\mathbb{1} \otimes g \cong Lg$ for any rigid element g. For this, we start with the tensor product of the localization distinguished sequence for $\mathbb{1}$ with g:

$$\Gamma(\mathbb{1}) \otimes g \to (g \cong \mathbb{1} \otimes g) \to L(\mathbb{1}) \otimes g,$$

and apply the Bousfield localization L to drive the equivalence $L(\mathbb{1}) \otimes g \cong Lg$ as follows:

$$\left(* \overset{(TI)}{\cong} L(\Gamma(\mathbb{1}) \otimes g)\right) \to$$

$$(Lg \cong L(\mathbb{1} \otimes g)) \xrightarrow[\because)(TI)]{\cong} \left(L(L(\mathbb{1}) \otimes g) \overset{(R)}{\cong} L\underline{\mathrm{Hom}}(Dg, L(\mathbb{1})) \overset{(L)}{\cong} \underline{\mathrm{Hom}}(Dg, L(\mathbb{1})) \overset{(R)}{\cong} L(\mathbb{1}) \otimes g\right),$$

where (TI) holds because $\operatorname{Ker} L$ is a tensor ideal, (R) holds because g is rigid, and (L) holds because $\underline{\mathrm{Hom}}(Dg, L(\mathbb{1}))$ is L-local. □

---

[25] Recall in this case $\mathcal{T}$ becomes distributive, because for any objects x_λ $(\lambda \in \Lambda), y, z$ in $\mathcal{T}$, $\mathrm{Hom}\left((\oplus_\lambda x_\lambda) \otimes y, z\right) \cong \mathrm{Hom}\left(\oplus_\lambda x_\lambda, \underline{\mathrm{Hom}}(y, z)\right) \cong \prod_\lambda \mathrm{Hom}\left(x_\lambda, \underline{\mathrm{Hom}}(y, z)\right) \cong \prod_\lambda \mathrm{Hom}(x_\lambda \otimes y, z) \cong \mathrm{Hom}\left(\oplus_\lambda x_\lambda \otimes y, z\right)$.

In general, when we talk about smashing Bousfield localization in tensor triangulated setting, we adopt the following equivalent conditions, where the localizing tensor ideal $\mathcal{I}$ is called a smashing ideal [7, Def. 2.15]:

Proposition 2.29 [See [7, Th. 2.13]] *Let $\mathcal{T}$ be a tensor triangulated category with coproducts, and let $\mathcal{I}$ be a localizing tensor ideal of $\mathcal{T}$ for which a Bousfield localization exists. Define the Bousfield localization functor $L\colon \mathcal{T} \to \mathcal{I}^{\perp}$ as in Remark 2.11. Then the following assertions are equivalent.*

(TI) $\mathcal{I}^{\perp}$ *is a tensor ideal. That is,* $\mathcal{T} \otimes \mathcal{I}^{\perp} \subset \mathcal{I}^{\perp}$.
(S) L *is smashing in Ravenel's sense:* $L \cong L(\mathbb{1}) \otimes -$.

Remark 2.30 (TI) is a tensor triangulated analogue of Proposition 2.21(5).

Proof *(Proof of Proposition)* 2.29 Now, for the implication (TI) $\Longrightarrow$ (S), consider the tensor product of the localization distinguished sequence for $\mathbb{1}$ with $x \in \mathcal{T}$:

$$\Gamma(\mathbb{1}) \otimes x \to (x \cong \mathbb{1} \otimes x) \to L(\mathbb{1}) \otimes x, \tag{11}$$

where $\Gamma(\mathbb{1}) \otimes x \in \mathcal{I}$ because $\mathcal{I}$ is a tensor ideal by assumption, and $L(\mathbb{1}) \otimes x \in \mathcal{I}^{\perp}$ because $\mathcal{I}^{\perp}$ is also a tensor ideal by (TI). Then, from the uniqueness of the localization distinguished sequence for $x \in \mathcal{T}$, we find $Lx \cong L(\mathbb{1}) \otimes x$, which implies (S).

The converse (S) $\Longrightarrow$ (TI) is easy; for, if $l = L(l) \in \mathcal{I}^{\perp}$ be a $\mathcal{I}$-local object and $x \in \mathcal{T}$, then

$$l \otimes x = L(l) \otimes x \overset{(S)}{=} (L(\mathbb{1}) \otimes l) \otimes x = L(\mathbb{1}) \otimes (l \otimes x) \overset{(S)}{=} L(l \otimes x) \in \mathcal{I}^{\perp}.$$

□

In the above proposition, we started with a localizing tensor ideal for which a Bousfield localization exists. However, we have the following example of a localizing tensor ideal for which an existence of the Bousfield localization is problematic:

Example 2.31 Let $\mathcal{T} = \mathcal{SH}$ be the homotopy category of spectra. For every $H \in \mathcal{T}$, its Bousfield class $\operatorname{Ker} H_*$ is a localizing tensor ideal. The subcategory

$$\operatorname{Ker} H^* = \{t \in \mathcal{T} \mid \operatorname{Hom}(t, \Sigma^i H) = 0 \text{ for all } i \in \mathbb{Z}\},$$

called the cohomological Bousfield class of H, is also a localizing tensor ideal. Actually, as was noticed by Hovey [52, Prop. 1.1], any Bousfield class is a cohomological Bousfield class:

$$\operatorname{Ker} H_* = \operatorname{Ker}(IH)^*,$$

where IH is the *Brown–Comenetz dual* of H, characterized by: $(IH)^*(t) = \operatorname{Hom}(H_*(t), \mathbb{Q}/\mathbb{Z})$, $\forall t \in \mathcal{T}$.

Here, Hovey [52] and Hovey–Palmieri [53] proposed the following conjectures, any one of which implies that an arbitrary localizing tensor ideal $\mathrm{Ker}\, H^*$ admits a Bousfield localization[26]:

Conjecture 2.32 (i) [52, Conj. 1.2] *Every cohomological Bousfield class is a Bousfield class.*
(ii) *Every localizing tensor ideal is a Bousfield class.*
(iii) [53, Conj. 9.1] *Every localizing triangulated subcategory is a Bousfield class.*

Of course, $(iii) \implies (ii) \implies (i)$, for we have an obvious inclusions of classes:

Bousfield-Ohkawa set := The class of Bousfield classes $\subseteq$ The class of cohomological Bousfield classes
$\subseteq$ The class of localizing tensor ideals $\subseteq$ The class of localizing triangulated subcategories,

where all the inclusings become $=$ if the above conjecture (iii) holds. However, even (i) is still open, and so it is still unknown even whether the class of cohomological Bousfield classes becomes a set or not. Similarly, it is still unknown even whether any cohomological Bousfield class admits a Bousfield localization or not. Here, we shall show an analogue of (ii) holds holds with an explicit geometric description of its set structure for $\mathbf{D}_{\mathrm{qc}}(X)$.

For a scheme X, $\mathbf{D}_{\mathrm{qc}}(X)$ is the derived category of complexes of arbitrary $\mathcal{O}_X$-modules on X whose cohomologies are quasi-coherent. If X is quasi-compact and separated, then $\mathbf{D}_{\mathrm{qc}}(X)$ is equivalent to $\mathbf{D}(\mathrm{QCoh}\, X)$, where $\mathbf{D}(\mathrm{QCoh}\, X)$ is the derived category of complexes of quasi-coherent sheaves on X ([17, Corollary 5.5]). Here we have the nice theorem of Gabriel [41] and Rosenberg [124]:

Theorem 2.33 *Any quasi-compact and separated scheme X can be reconstructed from* $\mathrm{QCoh}\, X$.

Glancing at this theorem of Gabriel and Rosenberg, we naturally hope $\mathbf{D}_{\mathrm{qc}}(X) \cong \mathbf{D}(\mathrm{QCoh}\, X)$ would carry rich information of X.

Now $\mathbf{D}_{\mathrm{qc}}(X) \cong \mathbf{D}(\mathrm{QCoh}\, X)$ is a tensor triangulated category with coproducts, with respect to the derived tensor product $- \otimes_X^{\mathbf{L}} -$, which is defined using flat resolutions (see e.g. [82, (2.5.7)]), and the unit object given by the structure sheaf $\mathcal{O}_X$. Let us also recall the following standard facts about derived functors:

Proposition 2.34 (i) (see e.g. [82, (2.1.1)(2.7.2)(3.1.3)(3.9.1)(3.6.4)*]) *For any map of schemes $f : X \to Y$, we can define the derived pullback triangulated functor*

$$\mathbf{L}f^* : \mathbf{D}_{\mathrm{qc}}(Y) \to \mathbf{D}_{\mathrm{qc}}(X),$$

via flat resolutions.

Furthermore, we have a natural functorial isomorphism

$$\mathbf{L}f^*\mathbf{L}g^* \xrightarrow{\sim} \mathbf{L}(gf)^*$$

[26] Conjecture 2.32 should be taken more seriously. In fact, Professor Peter May is very glad to see Conjecture 2.32 is advertised here.

(ii) (see e.g. [82, (2.1.1)(2.3.7)(3.1.2)(3.9.2)(3.6.4)$_*$]) *For any quasi-compact and quasi-separated map of schemes $f : X \to Y$, we can define the derived direct image (a.k.a. derived pushforward) triangulated functor*

$$\mathbf{R}f_* : \mathbf{D}_{\mathrm{qc}}(X) \to \mathbf{D}_{\mathrm{qc}}(Y),$$

via injective resolutions. Furthermore, we have a natural functorial isomorphism

$$\mathbf{R}(gj)_* \overset{\sim}{\to} \mathbf{R}g_*\mathbf{R}f_*,$$

when both f and g and quasi-compact and quasi-separated maps.
(iii) (see e.g. [82, (3.6.10)]) *For any quasi-compact and quasi-separated map of schemes $f : X \to Y$, $(\mathbf{L}f^*, \mathbf{R}f_*)$ gives an afjunction pair:*

$$\mathbf{D}_{\mathrm{qc}}(Y) \underset{\mathbf{R}f_*}{\overset{\mathbf{L}f^*}{\rightleftarrows}} \mathbf{D}_{\mathrm{qc}}(X)$$

(iv) (see e.g. [82, (3.2.1)(3.9.4)]) *For any quasi-compact and quasi-separated map of schemes $f : X \to Y$, the projection formula holds, i.e. we have natural isomorphisms for any $F \in \mathbf{D}_{\mathrm{qc}}(X)$, $G \in \mathbf{D}_{\mathrm{qc}}(Y)$:*

$$(\mathbf{R}f_*F) \otimes^{\mathbf{L}} G \overset{\cong}{\to} \mathbf{R}f_*\left(F \otimes^{\mathbf{L}} \mathbf{L}f^*G\right), \quad G \otimes^{\mathbf{L}} \mathbf{R}f_*F \overset{\cong}{\to} \mathbf{R}f_*\left(\mathbf{L}f^*G \otimes^{\mathbf{L}} F\right)$$

To investigate an analogue of Ohkawa's theorem for $\mathbf{D}_{\mathrm{qc}}(X)$, we must consider localizing tensor ideals of $\mathbf{D}_{\mathrm{qc}}(X)$. However, those smashing (localizing tensor ideals) are sometimes, more important. To study such (smashing) localizing tensor ideals of $\mathbf{D}_{\mathrm{qc}}(X)$, an appropriate concept of "stalk" becomes crucial:

Definition 2.35 (compare with [2, Proof of Th. 4.12] [61, App. A])[27] Let $x \in X$ be a point in a scheme. Then we have the following canonical maps involving the local ring $\mathcal{O}_{X,x}$ and the residue field k_x at $x \in X$:

$$\operatorname{Spec} k_x \xrightarrow[r_x]{} \operatorname{Spec} \mathcal{O}_{X,x} \xrightarrow[l_x]{\text{flat}} X,\ \ i_x = l_x r_x \quad \Longrightarrow \quad \mathbf{D}_{\mathrm{qc}}(\operatorname{Spec} k_x) \xleftarrow[\mathbf{L}(r_x)^*]{} \mathbf{D}_{\mathrm{qc}}\left(\operatorname{Spec} \mathcal{O}_{X,x}\right) \xleftarrow[\mathbf{L}(l_x)^*=(l_x)^*]{} \mathbf{D}_{\mathrm{qc}}(X),\ \ \mathbf{L}(i_x)^* \tag{12}$$

Then, for $E \in \mathbf{D}_{\mathrm{qc}}(X)$, we have four notions of "supports":

[27] Our presentation of "supports" in this definition and next proposition is somewhat different from those given in [2, Proof of Th. 4.12] [61, App. A], but the author hopes this would be more transparent to the reader.

$$
\begin{aligned}
&\mathrm{supp}(E) := \left\{x \in X \mid \mathbf{L}(i_x)^* E \neq 0 \in \mathbf{D}_{\mathrm{qc}}(\operatorname{Spec} k_x)\right\} \\
\subseteq\ &\mathrm{Supp}(E) := \left\{x \in X \mid (l_x)^* E \neq 0 \in \mathbf{D}_{\mathrm{qc}}(\operatorname{Spec} \mathcal{O}_{X,x})\right\}; \\
&\mathrm{supph}(E) := \left\{x \in X \mid \mathbf{L}(i_x)^* \left(\oplus_{\bullet\in\mathbb{Z}} \mathcal{H}^\bullet E\right) \neq 0 \in \mathbf{D}_{\mathrm{qc}}(\operatorname{Spec} k_x)\right\} \\
\subseteq\ &\mathrm{Supph}(E) := \left\{x \in X \mid \oplus_{\bullet\in\mathbb{Z}} (\mathcal{H}^\bullet E)_x = (l_x)^* \left(\oplus_{\bullet\in\mathbb{Z}} \mathcal{H}^\bullet E\right) \neq 0 \in \mathrm{QCoh}(\operatorname{Spec} \mathcal{O}_{X,x})\right\}
\end{aligned} \tag{13}
$$

where:

- $\mathcal{H}^\bullet E$ is the associated homology sheaves, regarded as a chain complex with trivial boundaries, of E.
- the inclusive relations follow from $\mathbf{L}(r_x)^*(l_x)^* = \mathbf{L}(r_x)^*\mathbf{L}(l_x)^* \xrightarrow{\sim} \mathbf{L}(l_x r_x) = \mathbf{L}(i_x)^*$, where the former equality follows from $\mathbf{L}(l_x)^* = (l_x)^*$, a consequence of the flatness of l_x, and the latter isomorphism is a direct consequence of Proposition 2.34(i).
- these inclusive relations become equalities when $E \in \mathbf{D}^b_{\mathrm{coh}}(X)$ because of Nakayama's lemma.
- If it becomes necessary to distinguish these four concepts, we call $\mathrm{supp}(E)$ the small support of E, $\mathrm{Supp}(E)$ the large support of E. $\mathrm{supph}(E)$ the small homology support of E, $\mathrm{Supph}(E)$ the large homology support of E. Otherwise, we simply call $\mathrm{supp}(E)$ the support of E, because this is the most essential object, and $\mathrm{Supph}(E)$ he homology support of E, because this is a tractible ordinary sheaf theoretical support for the associated homology sheaves $\oplus_{\bullet\in\mathbb{Z}}\mathcal{H}^\bullet E$.

Then the following useful fact will be used later:

Proposition 2.36 (i) *Given $E \in \mathbf{D}_{\mathrm{qc}}(X)$, we have for any $x \in X$ and $\bullet \in \mathbb{Z}$,*

$$
\mathcal{H}^\bullet\left((l_x)^* E\right) \cong (l_x)^* \left(\mathcal{H}^\bullet E\right) \in \mathrm{QCoh}\left(\operatorname{Spec} \mathcal{O}_{X,x}\right).
$$

Consequently, for any $E \in \mathbf{D}_{\mathrm{qc}}(X)$,

$$
\mathrm{Supp}\, E = \mathrm{Supph}\, E.
$$

(ii) *The commutative diagram of quasi-coherent sheaves in* (12) *restricts to coherent sheaves, and for any $E \in \mathbf{D}^b_{\mathrm{coh}}(X)$, all the four concepts of supports in Definition 2.35 coincide:*

$$
\mathrm{supp}\, E = \mathrm{Supp}\, E = \mathrm{Supph}\, E = \mathrm{supph}\, E.
$$

Proof In view of Definition 2.35, we only have to verify the first claim in (i): $\mathcal{H}^\bullet\left((l_x)^* E\right) \cong (l_x)^* \left(\mathcal{H}^\bullet E\right) \in \mathrm{QCoh}\left(\operatorname{Spec} \mathcal{O}_{X,x}\right)$. However, this follows immediately from the flatness of l_x which implies $(l_x)^*$ preserves exactness at the cochain level. □

Now, the fundamental theorem of Hopkins, Neeman, Thomason and others classify (smashing) localizing tensor ideals of $\mathbf{D}_{\mathrm{qc}}(X)$ under a mild assumption of X:

Theorem 2.37 ([48] [103, Th. 2.8, Th. 3.3] [131], [2, Cor4.6; Cor. 4.13; Th. 5.6] [7, Cor. 6.8] [34, Cor. 6.8; Ex. 6.9] [47, Th. B]) *Let X be a Noetherian scheme. Then every localizing tensor ideal of $\mathbf{D}_{\mathrm{qc}}(X)$ is of the form*

$$\mathrm{Ker}\, H_* = \{\, Q \in \mathbf{D}_{\mathrm{qc}}(X) \mid \mathrm{supp}\, Q \subseteq S \,\},$$

for some $S \subset X$.

The subcategory Ker H_* *is smashing if and only if the corresponding $S \subset X$ is closed under specialization.*

Note those S's with $S \subset X$ clearly form a set. So, we see an analogue of Ohkawa's theorem, however with a clear algebro-geometrical interpretation of "the Bousfield–Ohkawa set" in contrast to the case of Ohkawa's theorem. Furthermore, Theorem 2.37 solves Conjecture 2.32 (ii) affirmatively for the case $\mathbf{D}_{\mathrm{qc}}(X)$.

Also note that, in the special case when S in Theorem 2.37 is $Z = X \setminus U \subset X$, the complement of a quasi-compact Zariski open immersion $j : U \hookrightarrow X$, we have the following equivalence for not only noetherian, but also more general quasicompact, quasiseparated schemes (in which case, as $\mathbf{L}j^*$ has a right adjoint $\mathbf{R}j_*$ with $\epsilon : \mathbf{L}j^*\mathbf{R}j_* \to \mathrm{id}$ an isomorphism, we may apply Proposition 2.8)[28,29]:

$$\mathbf{D}_{\mathrm{qc}}(X) \big/ \left(\mathbf{D}_{\mathrm{qc}}\right)_Z(X) \xrightarrow[\cong]{\overline{\mathbf{L}j^*}} \mathbf{D}_{\mathrm{qc}}(U), \tag{14}$$

where $\left(\mathbf{D}_{\mathrm{qc}}\right)_Z(X) := \left\{ Y \in \mathbf{D}_{\mathrm{qc}}(X) \mid \mathrm{Supp}\, Y \subseteq Z \right\} = \mathrm{Ker}\, \mathbf{L}j^*$.[30] In this generality of quasicompact, separated schemes, Bousfield localization L is smashing (see Proposition 2.29), given explicitly as follows:

$$L \overset{(6)}{=} \mathbf{R}j_*\mathbf{L}j^* = (\mathbf{R}j_*\mathcal{O}_U) \otimes^{\mathbf{L}}_{\mathcal{O}_X} - : \mathbf{D}_{\mathrm{qc}}(X) \to \mathbf{D}_{\mathrm{qc}}(X) \big/ \left(\mathbf{D}_{\mathrm{qc}}\right)_Z(X) \xrightarrow[\cong]{\overline{\mathbf{L}j^*}} \mathbf{D}_{\mathrm{qc}}(U) \xrightarrow{\mathbf{R}j_*} \mathbf{D}_{\mathrm{qc}}(X). \tag{15}$$

[28] So, should had been known to Verdier.

[29] Let us recall the following precursor of this result in the setting of abelian category of quasi-coherent sheaves, which should go back at least to Gabriel (see e.g. [126, In the proof of Prop. 3.1]): $\mathrm{QCoh}(X) \big/ \mathrm{QCoh}_Z(X) \xrightarrow[\cong]{\overline{j^*}} \mathrm{QCoh}(U)$, where the left hand side is the abelian quotient category in the sense of Gabriel, Grothendieck, and Serre.

[30] Unlike Theorem 2.37 stated under the noetherian assumption, (14) is stated under more general quasicompact, quasiseparated assumption. Therefore, in this equality $\left(\mathbf{D}_{\mathrm{qc}}\right)_Z(X) := \left\{ Y \in \mathbf{D}_{\mathrm{qc}}(X) \mid \mathrm{Supp}\, Y \subseteq Z \right\} = \mathrm{Ker}\, \mathbf{L}j^*$, we may not replace Supp with supp. In fact, without the noetherian hypothesis, Theorem 2.37 becomes very bad as was shown in [107]. The author is grateful to Professor Neeman for this reference.

3 Hopkins–Smith Theorem and Its Motivic Analogue

In reality, Hopkins was not motivated by Ohkawa's Theorem 2.25 for his influential paper in algebraic geometry [48] (Theorem 2.37). Instead, Hopkins was motivated by his own theorem with Smith [50] in the sub stable homotopy category $\mathcal{SH}^c$, consisting of compact objects, whose validity was already known to them back around the time Hopkins wrote [48].

Theorem 3.1 [50] *For any prime p, any thick (épaisse) subcategories of the sub-triangulated category $\mathcal{SH}^c_{(p)}$ consisting of compact objects*

$$\mathcal{SH}^c_{(p)} = \mathcal{SH}^{fin}_{(p)} = \text{the homotopy category of } p\text{-local finite spectra}$$

is of the form

$$\begin{aligned} \mathcal{C}_n &:= \operatorname{Ker} E(n-1)_*|_{\mathcal{SH}^{fin}_{(p)}} = \left\{X \in \mathcal{SH}^{fin}_{(p)} \mid E(n-1) \wedge X = 0\right\} \\ &= \operatorname{Ker} K(n-1)_*|_{\mathcal{SH}^{fin}_{(p)}} = \left\{X \in \mathcal{SH}^{fin}_{(p)} \mid K(n-1) \wedge X = 0\right\}. \end{aligned} \tag{16}$$

Furthermore, these form a decreasing filtration of $\mathcal{F}_{(p)}$:

$$\{*\} \subsetneq \cdots \subsetneq \mathcal{C}_{n+1} \subsetneq \mathcal{C}_n \subsetneq \mathcal{C}_{n-1} \subsetneq \cdots \subsetneq \mathcal{C}_1 \subsetneq \mathcal{C}_0 = \mathcal{SH}^{fin}_{(p)}. \tag{17}$$

In this Hopkins–Smith classification of thick triangulated subcategories of $\mathcal{SH}^c$, the first step is an easy observation that any thick triangulated subcategory of $\mathcal{SH}^c$ is a thick (tensor) ideal,[31] furthermore, $E(n-1)$ and $K(n-1)$ are the $(n-1)$-st Johnson–Wilson spectrum and Morava K-theory, respectively, and the equality $\operatorname{Ker} E(n-1)_*|_{\mathcal{SH}^{fin}_{(p)}} = \operatorname{Ker} K(n-1)_*|_{\mathcal{SH}^{fin}_{(p)}}$ in (16) and the inclusions (17) are consequences of the following results found in Ravenel's paper [122]:

Theorem 3.2 (i) [122, Th. 2.1(d)] $\operatorname{Ker} E(n-1)_* = \operatorname{Ker}\left(\vee_{0\le i\le n-1} K(i)\right)_*$
(ii) [122, Th. 2.11] *For $X \in \mathcal{SH}^{fin}_{(p)}$, if $K(i)_*X = 0$, then $K(i-1)_*X = 0$.*

By the Hopkins–Smith work [50], the smashing conjecture for $E(n)$ [122] also holds [123], and so, $\operatorname{Ker} E(n-1)_*$ in (16) is a smashing tensor ideal. Actually, the first equality in (16) is a part of the following elegant reformulation of the telescope conjecture [122][54, Def. 3.3.8] (see also Conjecture 2.22) by Miller [93] [54, Th. 3.3.3] (here we follow more recent formulations of [7, Th. 4.1; Def. 4.2] [47, Cor. 2.1; Def. 3.1].):

Theorem 3.3 (Miller's finite localization and the Ravenel telescope conjecture) *Let $\mathcal{T}$ be a rigidly compactly generated tensor triangulated category. Let $\mathbb{S}(\mathcal{T})$ denote*

[31] Such a property is not usually satisfied for general triangulated categories. So, most effort to generalize the Hopkins–Smith theorem for a general triangulated category $\mathcal{T}$ aim at a classification of thick (tensor) ideals of $\mathcal{T}^c$.

the collection of all smashing localizing tensor ideals of $\mathcal{T}$, and let $\mathbb{T}(\mathcal{T}^c)$ denote the collection of all thick tensor ideals of $\mathcal{T}^c$.

(i) [93, Cor. 6; Prop. 9] [47, Th. 1.7] *For any $C \in \mathbb{T}(\mathcal{T}^c)$, the smallest localizing triangulated subcategory $\langle C \rangle$ containing C in $\mathcal{T}$ is smashing, i.e. $\in \mathbb{S}(\mathcal{T})$. Thus, we obtain the inflation map:*

$$I : \mathbb{T}(\mathcal{T}^c) \to \mathbb{S}(\mathcal{T}).$$

(ii) [54, Th. 3.3.3] *There is also the contraction map:*

$$C : \mathbb{S}(\mathcal{T}) \to \mathbb{T}(\mathcal{T}^c); \quad S \mapsto S \cap \mathcal{T}^c,$$

which enjoys:

$$C \circ I = id_{\mathbb{T}(\mathcal{T}^c)} : \mathbb{T}(\mathcal{T}^c) \underset{C}{\overset{I}{\rightleftarrows}} \mathbb{S}(\mathcal{T})$$

(iii) [93, Cor. 6; Prop. 9; Cor. 10] *The* telescope conjecture for $S \in \mathbb{S}(\mathcal{T})$ *holds if and only if, in addition to $C \circ I = id_{\mathbb{T}(\mathcal{T}^c)}$ stated in (ii), the following also holds:*

$$I \circ C(S) = S \in \mathbb{S}(\mathcal{T})$$

(iv) [93, Cor. 6; Prop. 9; Cor. 10] *The* telescope conjecture for $\mathcal{T}$[32] *holds if and only if I and C give mutually inverse equivalence:*

$$C \circ I = id_{\mathbb{T}(\mathcal{T}^c)} : \mathbb{T}(\mathcal{T}^c) \underset{C}{\overset{I}{\rightleftarrows}} \mathbb{S}(\mathcal{T}) \ : id_{\mathbb{S}(\mathcal{T})} = I \circ C.$$

However, the telescope conjecture of this generality has been shown to be false [69], and even the original telescope conjecture for $\mathcal{SH}$ is now believed to be false by many experts, including Ravenel himself [87]. Still, algebraicists have shown the validity of its various algebraic analogues (e.g. [7, 16, 80]) as we shall review an algebraic analogue of the Hopkins–Smith theorem, in conjunction with the above telescope conjecture, later in Theorem 4.15. Furthermore, Krause [76] showed the underlying philosophical message of the telescope conjecture that smashing tensor ideals are completely characterized by their restrictions to compact objects In fact, whereas the original telescope conjecture only concerns local compact objects, Krause proves his characterization of smashing tensor ideals via "local maps" between compacts objects. For more details, consult Krause's own paper [76].

Going back to the Hopkins–Smith theorem, a major part of its proof was to show:

Theorem 3.4 [50, Th. 7] *Any thick subcategory of $\mathcal{SH}^{fin}_{(p)}$ is of the form $\mathcal{C}_n$ for some $n \in \mathbb{Z}_{\geq 0}$.*

[32] This telescope conjecture is equivalent to the telescope conjecture without product Conjecture 2.22 via Propositions 2.28 and 2.29.

To show this, Hopkins–Smith prepared the following version of the nilpotency theorem [50, Cor. 2.5], building upon their earlier collaboration work with Devinatz [35]:

Theorem 3.5 [50, Cor. 2.5.(ii)] *For a map $f : F \to A$ between finite p-local spectra and another finite P-local spectra Y, the following conditions are equivalent:*

- $\exists m \gg 0$ *such that* $f^{\wedge m} \wedge I_Y : F^{\wedge m} \wedge Y \to A^{\wedge m} \wedge Y$ *is null.*
- $0 \leq \forall n < \infty, \quad K(n)_* (f \wedge I_Y) = 0.$

Now, to prove Theorem 3.4, it suffices to prove the following:

Lemma 3.6 [50, (2.9)] *Let $\mathcal{C}$ be a thick subcategory of $\mathcal{SH}_{(p)}$ and X, Y be p-local finite spectra. Then, if $X \in \mathcal{C}$ and $\{n \in \mathbb{Z}_{\geq 0} \mid K(n)_* Y \neq 0\} \subseteq \{n \in \mathbb{Z}_{\geq 0} \mid K(n)_* X \neq 0\}$, then $Y \in \mathcal{C}$.*

Actually, if Lemma 3.6 is shown to be correct, together with Ravenel's Theorem 3.2 (ii), it would imply

$$\mathcal{C} = \mathcal{C}_m, \text{ where } m = \min\{n \in \mathbb{Z}_{\geq 0} \mid \mathcal{C}_n \subseteq \mathcal{C}\}.$$

Then, the proof of Lemma 3.6 in [50] proceeds as follows (see also [123]:

- Starting with X, let $e : S^0 \to X \wedge DX$ be the S-dual of the identity map : $I_X : X \to X$, and extend it to a triangle with a map between p-local finite spectra $f : F \to S^0$ as the fiber as follows:

$$F \xrightarrow{f} S^0 \xrightarrow{e} X \wedge DX \simeq C_f, \text{ the cofiber of } f. \tag{18}$$

- Applying the smash product with Y to (18), we obtain:

$$F \wedge Y \xrightarrow{f \wedge I_Y} S^0 \wedge Y \cong Y \xrightarrow{e \wedge I_Y} X \wedge DX \wedge Y \simeq C_f \wedge Y, \tag{19}$$

  for which, we claim

$$0 \leq \forall n < \infty, \quad K(n)_*(f \wedge I_Y) = 0. \tag{20}$$

  - If $K(n)_* Y = 0$ then $K(n)_*(I_Y) = 0$, which implies the triviality of (20), by the Kunneth theorem for Morava K-theories:

$$K(n)_*(X \wedge Y) \cong K(n)_* X \otimes_{K(n)_*} K(n)_* Y \quad \text{for any } p\text{-local spectra } X, Y \tag{21}$$

  - If $K(n)_* Y \neq 0$ then $K(n)_* X \neq 0$ by the assumption of Lemma 3.6. Then, by the duality isomorphism for Morava K-theories:

$$\begin{gathered}\mathrm{Hom}_{K(n)_*}(K(n)_* X, K(n)_* Y) = \mathrm{Hom}_{K(n)_*}(K(n)_*, K(n)_*(Y \wedge DX)) = K(n)_*(Y \wedge DX) \\ \text{for any } p\text{-local spectra } X, Y,\end{gathered} \tag{22}$$

we also find the non-triviality: $K(n)_*(e \wedge I_Y) \neq 0$. But, this in turn implies the triviality: $K(n)_*(f \wedge I_Y) = 0$ from the Morava K-theory exact sequence associated to (19), making use of the Morava Kunneth isomorphism (21) again,

- Since (20), we may apply the Hopkins–Smith nilpotency Theorem 3.5 to $f \wedge I_Y$ in (19) to find $m \gg 0$ such that $f^{\wedge m} \wedge I_Y : F^{\wedge m} \wedge Y \to (S^0)^{\wedge m} \wedge Y \cong Y$ is null. This implies:
$$Y \text{ is a direct summand of } C_{f^{\wedge m} \wedge I_Y} \cong C_{f^{\wedge m}} \wedge Y. \tag{23}$$
- By the assumption, $X \in \mathcal{C}$, but as the thick subcategory $\mathcal{C}$ of $\mathcal{SH}^{fin}_{(p)}$ is also a thick ideal, this implies $C_f \overset{(19)}{\cong} X \wedge DX \in \mathcal{C}$.
- For any $n \in \mathbb{N}$, consider the commutative diagram:

$$\begin{array}{ccccc}
F^{\wedge n} \wedge F & \xrightarrow{f^{\wedge n} \wedge I_F} & (S^0)^{\wedge n} \wedge F \cong F & \longrightarrow & C_{f^{\wedge n}} \wedge F \\
\| & & \downarrow f & & \downarrow \\
F^{\wedge (n+1)} & \xrightarrow{f^{\wedge (n+1)}} & (S^0)^{\wedge (n+1)} \cong S^0 & \longrightarrow & C_{f^{n+1}} \\
\downarrow & & \downarrow & & \downarrow \\
\bullet & \longrightarrow & C_f & = & C_f
\end{array}$$

From this, we obtain a triangle
$$C_{f^{\wedge n}} \wedge F \to C_{f^{n+1}} \to C_f$$
Since $C_f \in \mathcal{C}$ and $\mathcal{C}$ is a tensor ideal, we see inductively from this triangle that
$$C_{f^m} \in \mathcal{C} \qquad (\forall m \in \mathbb{N}) \tag{24}$$
- Since $\mathcal{C}$ is a thick ideal, we conclude from (23) and (24) that $Y \in \mathcal{C}$. This complete the proof of Lemma 3.6.

□

Now, the basic philosophy underlying the above picture of Hopkins–Smith was already perceived by Morava much earlier (see the "exercises" in Sect. 0.5 of [96], whose preprint version was circulated nearly a decade ago before its publication). For a modern development of Morava K-theory, consult Morava's own paper [98] in this proceedings.

The author believes the Hopkins–Smith theorem (Theorem 3.1) and the Ohkawa theorem (Theorem 2.25) are best understood, when they are appreciated simultaneously in a single commutative diagram. Since this commutative diagram can be drawn for more general rigidly compactly generated tensor triangulated category $\mathcal{T}$, let us first set up our notations of our interests in this generality:

- $\mathbb{L}(\mathcal{T})$: the collection of *localizing tensor ideals* of $\mathcal{T}$.
- $\mathbb{S}(\mathcal{T})$: the collection of *smashing localizing tensor ideals* of $\mathcal{T}$.
- $\mathbb{T}(\mathcal{T}^c)$: the collection of *thick tensor ideals* of $\mathcal{T}^c$.
- $\mathbb{B}(\mathcal{T})$: the collection of *Bousfield classes*, i.e. those of the form $\mathrm{Ker}(h \otimes -) \subseteq \mathbb{L}(\mathcal{T})$ $(h \in \mathcal{T})$.

Now let us specialize to the case $\mathcal{T} = \mathcal{SH}_{(p)}$:

Theorem 3.7 *In $\mathcal{SH}_{(p)}$, the Ohkawa theorem, the Hopkins–Smith theorem, Miller's version of the Ravenel telescope conjecture ($C \circ I \overset{?}{=} Id_{\mathbb{T}\left(\mathcal{SH}^{fin}_{(p)}\right)}$), and the conjectures of Hovey and Hovey–Palmieri can be simultaneously expressed in the following succinct commutative diagram:*

$$\begin{array}{ccccc} \textit{mysterious set} & \overset{\textit{Ohkawa Th.}}{=\!=\!=\!=\!=} & \mathbb{B}(\mathcal{SH}_{(p)}) & \underset{\overset{?}{=}}{\overset{\textit{Hovey Conj.}}{\hookrightarrow}} & \mathbb{L}(\mathcal{SH}_{(p)}) \\ \uparrow & & & & \uparrow \\ \underset{\cdots\subsetneq C_{n+1}\cdots\subsetneq C_n\cdots}{\textit{chromatic hierachy}} & \overset{\textit{Hopkins–Smith Th.}}{=\!=\!=\!=\!=} & \mathbb{T}\left(\mathcal{SH}^{fin}_{(p)}\right) & \underset{C\ (\textit{split surj.})}{\overset{I\ (\textit{split inj.})}{\rightleftarrows}} & \mathbb{S}(\mathcal{SH}_{(p)}) \end{array} \qquad (25)$$

For more on the Hopkins–Smith theorem and related "chromatic mathematics," see [123] and, for some of the latest developments,[33] see [9, 13, 133] in this proceedings. Actually, Bartel's survey [9] focuses upon the telescope [122, 123] and chromatic splitting conjectures [51], which are major directions of research, not only in chromatic homotopy theory, but also in stable homotopy theory as a whole. Considering the traditional influence of stable homotopy theory, initiated by Hopkins, Rickard, Neeman, Thomason and others, to the representation theory of finite dimensional algebras and the derived category theory in algebraic, as is highlighted by Brown representability, Bousfield localization, Hopkins–Smith theorem, researchers in these areas might better to keep this fact in mind.

Comparing with the telescope conjecture, the chromatic splitting conjectre appears to be elusive for them. In short, the chromatic splitting conjecture predicts, for a p-completed finite spectrum F, the first map in the canonical cofiber sequence

$$\underline{\mathrm{Hom}}\left(L_{E(n-1)}S^0, L_{E(n)}F\right) \to L_{E(n-1)}F \to L_{E(n-1)}L_{K(n)}F \qquad (26)$$

is trivial; stated differently, the second map in (26) is split injective.[34]

In fact, Hopkins [51, Conj.4.2(iv)] further predicted, presumably hoping to provide a program to prove the triviality of the first map in (26), an explicit decomposition

---

[33] A trend here is to apply the higher algebra technique of Lurie [85, 86] to understand chromatic phenomena [13, 133], where the latter contains a concise review of higher algebra technology. Different kinds of applications of Lurie's higher algebra technique can be seen in [89, 90].

[34] This splitting conjecture implies, for any p-completed finite spectrum F and any infinite subset $\{n_i\}_{i=1}^{\infty} \subseteq \mathrm{N}$, the natural map $F \to \prod_{i=1}^{\infty} L_{K(n_i)}F$ is split injective. For this and much more, consult [9, 51].

of $\underline{\mathrm{Hom}}\left(L_{E(n-1)}S^0, L_{E(n)}F\right)$, inspired by Morava's old observation [96, Rem. 2.2.5]. The structure of $\underline{\mathrm{Hom}}\left(L_{E(n-1)}S^0, L_{E(n)}F\right)$ is highly reflected by its divisible homotopy group elements. In general, divisible homotopy group elements of a spectrum X can be isolated in the spectrum $\underline{\mathrm{Hom}}(L_0S^0, X)$, which is in the current case:

$$\underline{\mathrm{Hom}}\left(L_0S^0, \underline{\mathrm{Hom}}\left(L_{E(n-1)}S^0, L_{E(n)}F\right)\right) \cong \underline{\mathrm{Hom}}\left(L_0S^0 \wedge L_{E(n-1)}S^0, L_{E(n)}F\right) \cong \underline{\mathrm{Hom}}\left(L_0S^0, L_{E(n)}F\right)$$

To understand this, Morava [97] suggested to consider the following cohomology theory L_n^*:

$$X \mapsto L_n^*(X) := \mathrm{Hom}\left(\pi_{-*}\underline{\mathrm{Hom}}\left(L_0S^0, L_{E(n)}X\right), \mathbb{Q}\right)$$

Actually, Morava [97] noticed the validity of the Hopkins' prediction on the explicit structure of $\underline{\mathrm{Hom}}\left(L_{E(n-1)}S^0, L_{E(n)}F\right)$ would imply the cohomology theory L_n^* is represented by the p-adic rationalization of the spectrum[35]:

$$\Sigma^{2n}\left(\bigvee_{\{n_i\in\mathbb{Z}_{\geq 0}\}_{i=1}^{\infty};\ \sum_{i=1}^{\infty} in_i=n} \frac{\left(\sum_{i=1}^{\infty} n_i\right)!}{\prod_{i=1}^{\infty}(n_i!)}\left(\prod_{i=1}^{\infty} U(i-1)^{n_i}\right)_+\right) \tag{27}$$

While Hopkins' prediction [51] above of the explicit decomposition of $\underline{\mathrm{Hom}}\left(L_{E(n-1)}S^0, L_{E(n)}F\right)$, which the above work of Morava [97] is based upon, is known to hold for $n = 1$ or $n = 2$ and $p \geq 3$, Beaudry [11] has recently shown it to fail for the case $n = 2$ and $p = 2$. Still, as was pointed out to the author by Tobias Barthel, The above formula (27), which was derived from Morava's calculation, still holds even for this troublesome case of $n = 2$ and $p = 2$, because the discrepancy found by Beaudry [11] is p-torsion and so vanishes rationally. Thus, it could well be the case (27) holds for any pair of a prime p and a natural number n.

Furthermore, it could be the case that Hopkins' prediction of the explicit decomposition of $\underline{\mathrm{Hom}}\left(L_{E(n-1)}S^0, L_{E(n)}F\right)$ still holds, consequently so does Morava's deduction (27) above, when the base prime p is sufficiently large comparing with the height n.

It would be fantastic, if, as Professor Morava dreams of, there hold formulae analogous to the predicted Hopkins' and Morava's in algebraic examples like $\mathbf{D}_{\mathrm{qc}}(X)$, where the fundamental theorem of Hopkins, Neeman, Thomason and others gave us an explicit "Bousfield–Ohkawa set", not only for Bousfield classes, but also for localized tensor ideals, whereas the original Ohkawa's set for $\mathcal{SH}$ only takes into account Bousfield classes and is not explicit at all. Furthermore, as we mentioned before, while the telescope conjecture is now believed to be false by many experts, algebraicists have shown the validity of its various algebraic analogues. So, why not for the chromatic splitting conjecture, as Professor Morava dreams of!

Actually, restricting to the conjectured splitting of the second map in (26), recent effort of Beaudry-Goerss-Henn [12] has shown its validity even for the case $n = p = 2$, which is the case [11] showed Hopkins' conjectural decomposition of

---

[35] It appears that [97, p. 4, Corollary] should be modified as in (27).

$\underline{\text{Hom}}\left(L_{E(n-1)}S^0, L_{E(n)}F\right)$ is false. Furthermore, Barthel–Heard–Valenzuela [10] has recently proved an algebraic analogue of the conjectural splitting of the second map in (26). For this and much more, consult Bartel's survey [9].

Going back to the Hopkins–Smith theorem, it is natural to look after its motivic analogue (10) (This means efforts to classify thick (tensor) ideals of $\mathcal{SH}(k)^c$.).

In this regard, Ruth Joachimi [62] constructed some motivic thick ideals in $\mathcal{SH}(k)^c$ for $k \subseteq \mathbb{C}$:

Theorem 3.8 [62, Th. 13]

1. *If $k \subseteq \mathbb{C}$, then $(\mathcal{SH}(k)^c)_{(p)}$ contains at least an infinite chain of different thick ideals, given by $\overline{R}_k^{-1}(\mathcal{C}_n)$, $0 \leq n \leq \infty$, where $\overline{R}_k$ denotes the p-localisation of the restriction of R_k to $\mathcal{SH}(k)^c$:*

$$\begin{array}{ccccc} & & \xrightarrow{\qquad\qquad id \qquad\qquad} & & \\ (\mathcal{SH}^c)_{(p)} & \xrightarrow{c_k} & (\mathcal{SH}(k)^c)_{(p)} & \xrightarrow{\overline{R}_k} & (\mathcal{SH}^c)_{(p)} \\ \downarrow & & \downarrow & & \downarrow \\ (\mathcal{SH})_{(p)} & \longrightarrow & (\mathcal{SH}(k))_{(p)} & \xrightarrow{R_k} & (\mathcal{SH})_{(p)} \\ & & \xrightarrow[\qquad\qquad id \qquad\qquad]{} & & \end{array}$$

Here,

- *c_k is induced from the constant presheaf functor [62, Th. 10], which restricts to the compact objects [62, Rem. 53, Prop. 58, Prop. 61].*
- *The existence of $\overline{R}_k$ follows since R_k preserves compactness [62, Prop. 61].*

2. *If $k \subseteq \mathbb{R}$, then $(\mathcal{SH}(k)^c)_{(p)}$ contains at least a two-dimensional lattice of different thick ideals, given by $\left(\overline{R}'_k\right)^{-1}(\mathcal{C}_{m,n})$, for all $(m, n) \in \Gamma_p$ (see [62, Def. 35] for the definition of Γ_p and more detail):*

$$\begin{array}{ccccccc} & & \xrightarrow{\qquad\qquad id \qquad\qquad} & & & & \\ (\mathcal{SH}(\mathbb{Z}/2)^c)_{(p)} & \longrightarrow & (\mathcal{SH}(k)^c)_{(p)} & \xrightarrow{\overline{R}'_k} & (\mathcal{SH}(\mathbb{Z}/2)^c)_{(p)} & \overset{\phi^{\{1\}}}{\underset{\phi^{\mathbb{Z}/2}}{\rightrightarrows}} & (\mathcal{SH})^c)_{(p)} \\ \downarrow & & \downarrow & & \downarrow & & \downarrow \\ (\mathcal{SH}(\mathbb{Z}/2))_{(p)} & \xrightarrow{c'_k} & (\mathcal{SH}(k))_{(p)} & \xrightarrow{R'_k} & (\mathcal{SH}(\mathbb{Z}/2))_{(p)} & \overset{\phi^{\{1\}}}{\underset{\phi^{\mathbb{Z}/2}}{\rightrightarrows}} & (\mathcal{SH}))_{(p)} \\ & & \xrightarrow[\qquad\qquad id \qquad\qquad]{} & & & & \end{array}$$

Here,

- [62, Th. 11] $c'_k : (\mathcal{SH}(\mathbb{Z}/2))_{(p)} \to (\mathcal{SH}(k))_{(p)}$ *is induced by*

$$c' : sSet(\mathbb{Z}/2) \to sPre(\mathrm{Sm}/\mathbb{R})$$
$$M \mapsto \left(\coprod_{M^{\mathbb{Z}/2}}\right) \coprod \left(\coprod_{(M\setminus M^{\mathbb{Z}/2})/(\mathbb{Z}/2)} \operatorname{Spec}\mathbb{C}\right),$$

which restricts to the compact objects [62, Rem. 53, Prop. 58, Prop. 61].

- (Strickland's theorem [62, Cor. 34][36]) *Any thick ideal in the category* $(\mathcal{SH}(\mathbb{Z}/2)^c)_{(p)}$ *is of the form*

$$\mathcal{C}_{m,n} = \{X \mid \phi^{\{1\}}(X) \in \mathcal{C}_m \text{ and } \phi^{\mathbb{Z}/2}(X) \in \mathcal{C}_n\},$$

where $m, n \in [0, \infty]$.

Just like the nilpotency Theorem 3.5 was crucial in the proof of Hopkins–Smith Theorem 3.1, the above theorem of Strickland is shown by first proving an appropriate nilpotency theorem [62, Th. 3]. At the same time, Joachimi [62] explains various difficulties in proving an appropriate nilpotency theorem in the motivic setting. Furthermore, the above Joachimi's construction of motivic thick ideals in $\mathcal{SH}(k)^c$ for $k \subseteq \mathbb{C}$ is so far limited to importing the Hopkins–Smith stable homotopy thick ideals in $\mathcal{SH}^c$. Thus, constructions of motivic thick ideals of truly algebro-geometric origin is highly desired. For details and much more of Joachimi's work, construct her own exposition [62] in this proceeding.

For a case of $k \nsubseteq \mathbb{C}$, Kelly [72] obtained the following surprisingly simple description of the set of prime thick tensor ideals $\operatorname{Spc}\left(\mathcal{SH}(\mathbb{F}_q)^c_{\mathbb{Q}}\right)$,[37] up to a couple of widely believed conjectures:

Theorem 3.9 [72, Th. 1.1] *Let* $\mathbb{F}_q$ *be a field with a prime power,* q, *number of elements. Suppose that for all connected smooth projective varieties* X *we have:*

$$CH^i(X; j)_{\mathbb{Q}} = 0; \quad \forall j \neq 0; i \in \mathbb{Z} \quad \textit{(Beilinson-Parshin conjecture)},$$
$$CH^i(X)_{\mathbb{Q}} \otimes CH_i(X)_{\mathbb{Q}} \to CH_0(X)_{\mathbb{Q}} \text{ is non-degenerate.} \quad \textit{(Rat. and num. equiv. agree)}$$

Then

$$\operatorname{Spc}\left(\mathcal{SH}(\mathbb{F}_q)^c_{\mathbb{Q}}\right) \cong \operatorname{Spec}(\mathbb{Q}).$$

For details, consult Kelly's own exposition [72] in this proceeding.

[36] Strickland's theorem for $G = \mathbb{Z}/2$ has recently been generalized to arbitrary finite group G by Balmer–Sanders [8].

[37] See Defjnition 4.22 for this concept.

4 $\mathbf{D}^b_{\mathrm{coh}}(X)$ and $\mathbf{D}^{\mathrm{perf}}(X)$

In the last two sections, we reviewed:

- Ohkawa's theorem in $\mathcal{SH}$, which states the Bousfield classes form a somewhat mysterious set.
- Its analogue in $\mathbf{D}_{\mathrm{qc}}(X)$ is explicitly computable: the fundamental theorem of Hopkins, Neeman,..., identifies the set of Bousfield classes with the set of localizing tensor ideals, which turns out to have a concrete and algebro-geometric description.
- Hopkins' motivation of his fundamental theorem in $\mathbf{D}_{\mathrm{qc}}(X)$ was his own theorem with Smith in $\mathcal{SH}^c$.

Thus, we are naturally led to investigate $\mathbf{D}_{\mathrm{qc}}(X)^c$. However, the story is not so simple. Whereas there is a conceptually simple categorical interpretation $\mathbf{D}_{\mathrm{qc}}(X)^c = \mathbf{D}^{\mathrm{perf}}(X)$, it is its close relative (actually equivalent if X is smooth over a field) $\mathbf{D}^b_{\mathrm{coh}}(X)$ which traditionally has been intensively studied because of its rich geometric and physical information.[38]

So, we wish to understand both $\mathbf{D}^b_{\mathrm{coh}}(X)$ and $\mathbf{D}^{\mathrm{perf}}(X)$.

In this section, we start with brief, and so inevitably incomplete, summaries of $\mathbf{D}^b_{\mathrm{coh}}(X)$ and $\mathbf{D}^{\mathrm{perf}}(X)$, focusing on their usages. Still, we hope this would convince non-experts that $\mathbf{D}^b_{\mathrm{coh}}(X)$ and $\mathbf{D}^{\mathrm{perf}}(X)$ are very important objects to study.

Then, we shall review Neeman's recent result, which claims these two close relatives $\mathbf{D}^b_{\mathrm{coh}}(X)$ and $\mathbf{D}^{\mathrm{perf}}(X)$ actually determine each other, and its main technical tool: approximable triangulated category.

4.1 $\mathbf{D}^b_{\mathrm{coh}}(X)$

- There is a classical functoriality result of Grothendieck:

Theorem 4.1 [44, Th. 3.2.1] *Let $f : X \to Y$ be a proper morphism with Y locally noetherian. Then*

$$\mathbf{R}f_*\,\mathbf{D}^b_{\mathrm{coh}}(X) \subset \mathbf{D}^b_{\mathrm{coh}}(Y).$$

Actually, there is a sharp converse (i.e. we do not have to check $\mathbf{R}f_*\,\mathbf{D}^b_{\mathrm{coh}}(X) \subset \mathbf{D}^b_{\mathrm{coh}}(Y)$) to Theorem 4.1 [83, Cor. 4.3.2] [109, Lem. 0.20]:

Theorem 4.2 [109, Lem. 0.20] *Let $f : X \to Y$ be a separated, finite-type morphism of noetherian schemes such that*

$$\mathbf{R}f_*\,\mathbf{D}^{\mathrm{perf}}(X) \subset \mathbf{D}^b_{\mathrm{coh}}(Y).$$

[38] Or, researchers might prefer "♡-felt" $\mathbf{D}^b_{\mathrm{coh}}(X) \cong \mathcal{D}^b(\mathrm{Coh}(X))$ (although separated, not mere quasi-separated, assumption is needed for this equivalence) over simply formal $\mathbf{D}^{\mathrm{perf}}(X) \cong \mathbf{D}_{\mathrm{qc}}(X)^c$…

Then f is proper.

- For an essentially small triangulated category $\mathcal{T}$, its Grothendieck K_0-group $K_0(\mathcal{T})$ is defined by generators and relations as follows [111, Def. 4.5.8] [112, Def. 1]:

$$K_0(\mathcal{T}) := \frac{\mathbb{Z}\{[X] \mid [X] \text{ is an isomorphism class of } X \in \mathcal{T}\}}{\mathbb{Z}\{[X]-[Y]+[Z] \mid \text{there is a distiguished triangle } X \to Y \to Z \to \Sigma X\}} \tag{28}$$

  - Having defined $K_0(\mathcal{T})$, we should not be too optimistic to hope $K_0(\mathcal{T})$ always carries a rich information of $\mathcal{T}$, In fact, if $\mathcal{T}$ contains an arbitrary countable direct sum (coproduct),[39] then, for any $X \in \mathcal{T}$, we have a distinguished triangle of the following form:

$$\oplus_{n\in\mathbb{N}} X \xrightarrow{\text{index shift}} \oplus_{n\in\mathbb{N}} X \to X \to \Sigma\left(\oplus_{n\in\mathbb{N}} X\right)$$

  From the defining relation of $K_0(\mathcal{T})$ (28), this implies $[X] = 0 \in K_0(\mathcal{T})$ for any $X \in \mathcal{T}$. By the definition (28), this means $K_0(\mathcal{T}) = 0$ whenever $\mathcal{T}$ contains an arbitrary countable direct sum (coproduct). As a very important special case, we emphasize:

$$K_0(\mathbf{D}_{\mathrm{qc}}(X)) = 0.$$

  - Grothendieck K_0-group is useful to classify dense subcategories of an essentially small triangulated subcategory.

Proposition 4.3 [131, p. 5, Lem. 2.2, p. 6, Cor. 2.3] [111, Prop. 4.5.11] *Suppose a triangulated subcategory $\mathcal{S}$ of an essentially small triangulated category $\mathcal{T}$ is* ***dense****, i.e. $\widehat{\mathcal{S}} = \mathcal{T}$. Then,*

1. *The induced map $K_0(\mathcal{S}) \to K_0(\mathcal{T})$ is a monomorphism.*
2. *For any $X \in \mathcal{T}$,*

$$X \in \mathcal{S} \iff [X] \in \mathrm{Im}(K_0(\mathcal{S}) \to K_0(\mathcal{T})).$$

Theorem 4.4 [131, p. 5, Th. 2.1] *For an essentially small triangulated category $\mathcal{T}$, there is a one-to-one correspondence between the dense triangulated subcategories of $\mathcal{T}$ and the subgroups of $K_0(\mathcal{T})$:*

$$\begin{aligned} \{\textit{dense triangulated subcategories of } \mathcal{T}\} &\overset{\cong}{\rightleftarrows} \{\textit{subgroups of } K_0(\mathcal{T})\} \\ \mathcal{S} &\mapsto \mathrm{Im}\,(K_0(\mathcal{S}) \to K_0(\mathcal{T})) \\ \triangle\, \textit{subcategory consisting of } X \in \mathcal{T} \textit{ with } [X] \in H \subseteq K_0(\mathcal{T}) &\leftarrow\!\dashv H \end{aligned}$$

- For any small abelian category $\mathcal{A}$, the functor $\mathcal{D}^b$ comes with the canonical embedding $\mathcal{A} \to \mathcal{D}^b(\mathcal{A})$, which induces an equivalence of Grothendieck K-groups of an abelian category $\mathcal{T}$ and a triangulated category $\mathcal{D}^b(\mathcal{A})$:

[39] Having arbitrary small coproducts was an indispensable assumption for Brown representability and Bousfield localization (Theorem 2.15, Corollary 2.16).

$$K_0(\mathcal{A}) \overset{\cong}{\to} K_0\left(\mathcal{D}^b(\mathcal{A})\right), \tag{29}$$

– Whenever a bounded t-structure is given on $\mathcal{T}$, if we denote by $\mathcal{T}^\heartsuit$ its heart, then we have another isomorphism of K_0-groups of an abelian category and a triangulated category:

$$K_0(\mathcal{T}^\heartsuit) \overset{\cong}{\to} K_0(\mathcal{T}). \tag{30}$$

Applying (30) to $\mathcal{T} = \mathbf{D}^b_{\mathrm{coh}}(X)$, $\mathcal{T}^\heartsuit = \mathrm{Coh}(X)$,[40] we find the canonical isomorphism:

$$K(\mathrm{Coh}(X)) \overset{\cong}{\to} K\left(\mathbf{D}^b_{\mathrm{coh}}(X)\right) \tag{32}$$

- The sheaf theory has its origin in Oka-Cartan theory of complex functions of several variables (see e.g. [116] for a general picture, and [114] for a review of the L^2-technique in complex geometry, both by Ohsawa[41]). The pivotal achievement at the time was *Oka's Coherence Theorem*, which states that the structure sheaf $\mathcal{O}_M$ of a complex manifold M is coherent (for a proof, see e.g. [113]). From the viewpoint of algebraic geometry, interest of complex manifolds emerge through the GAGA theorem of Serre [128], which, for a proper scheme X over $\mathrm{Spec}\,\mathbb{C}$, can be stated as an equivalence of abelian categories of coherent modules [45, XII, Th. 4.4]:

$$\phi^* : \mathrm{Coh}(X) \overset{\cong}{\to} \mathrm{Coh}(X^{\mathrm{an}}),$$

where $\phi : X^{\mathrm{an}} \to X$ is the canonical morphism from the associated analytic space X^{an} of X [45, XII, 1.1], and ϕ^* consequently induces isomorphisms of resulting derived categories[42]:

$$\mathbf{D}_{\mathrm{Coh}}(X) \overset{\cong}{\to} \mathbf{D}_{\mathrm{Coh}}(X^{\mathrm{an}}); \quad \mathbf{D}^b_{\mathrm{coh}}(X) \overset{\cong}{\to} \mathbf{D}^b_{\mathrm{coh}}(X^{\mathrm{an}}); \; \cdots$$

---

[40] If we apply (29) in order to obtain the isomorphism (32), we must require the extra "separated" assumption, for then we should also use the isomorphism:

$$\mathbf{D}^b_{\mathrm{coh}}(X) = \mathcal{D}^b(\mathrm{Coh}(X)), \tag{31}$$

which requires the "separated" assumption of X. This fact, and the above approach to use (30) was communicated to the author by Professor Neeman.

[41] Professor Takeo Ohsawa is the AMS Stefan Bergman Prize 2014 recipient. His survey paper [114] in this proceedings is a concise summary of his work for which this prize was awarded. It was his Bergman Prize money which enabled us to invite distinguished lecturers to Ohkawa's memorial conference at Nagoya University in the summer of 2015. Takeo Ohsawa was also Tetsusuke Ohkawa's highschool classmate at Kanazawa University High School in Kanazawa, Japan.

[42] X being proper over $\mathrm{Spec}(\mathbb{C})$ implies (as part of the definition of properness) that it is separated, hence $\mathbf{D}^b(\mathrm{Coh}(X)) = \mathbf{D}^b_{\mathrm{coh}}(X)$. Hence, these two isomorphisms are trivial consequences of the isomorphism $\phi^* : \mathrm{Coh}(X) \overset{\cong}{\to} \mathrm{Coh}(X^{\mathrm{an}})$. These two isomorphism are supplied just for reader's information.

Recently, Jack Hall [46] proposed a unified treatment of "GAGA type theorems," in which, a prominent role of Oka's coherence theorem became transparent in his deduction of the classical GAGA theorem [46, Example 7.5] (also consult the updated version of [104, Remark 1.7 and Appendix A] to appreciate how short and simple the Jack Hall's new proof is.).

- Derived categories in the complex analytic setting shows up in the Kontsevich homological mirror symmetry [74][43] which in the Calabi–Yau setting is of the following form:

$$\mathbf{D}^b_{\mathrm{coh}}(X) \cong \mathbf{D}^b\, Fuk(X^\vee), \tag{33}$$

where X is expected to be a mirror of $X^\vee$, given by a sigma model:

$$(M, I, \omega, B),$$

where we only note I is the complex structure of M, and that whose category of D-branes of type B (B-model) is the left side of (33) :

$$DB(M, I, \omega, B) \;\cong\; \mathbf{D}^b_{\mathrm{coh}}(M, I) \;\cong\; \mathbf{D}^b_{\mathrm{coh}}(X).$$

On the other hand, $\mathbf{D}^b\, Fuk(X^\vee)$, the *derived Fukaya category* consisting of Lagrangian submanifolds of the mirror $X^\vee$, is not a derived category of an abelian category (but of an A_∞ category; see [38, 39] for more details).

- Recall that $\mathbf{D}^b_{\mathrm{coh}}(X)$ is given by the composite of functors:

$$\mathbf{D}^b_{\mathrm{coh}} : X \overset{\mathrm{Coh}}{\mapsto} \mathrm{Coh}(X) \overset{\mathcal{D}^b}{\mapsto} \mathcal{D}^b(\mathrm{Coh}(X)) = \mathbf{D}^b_{\mathrm{coh}}(X). \tag{34}$$

It is instructive to keep reconstruction problems arising from these functors in mind. For instance, Theorem 2.33 of Gabriel-Rosenberg can be specialized to the following (which is essentially the original theorem of Gabriel [41]) reconstruction theorem with respect to Coh[44]:

Theorem 4.5 *Any Noetherian and separated scheme X can be reconstructed from* $\mathrm{Coh}(X)$.

- Glancing at this theorem of Gabriel, we naturally hope $\mathbf{D}^b_{\mathrm{coh}}(X)$ would carries rich information of X. Concerning the reconstruction problem associated with (34), any smooth connected projective variety with either K_X ample or $-K_X$ ample can be reconstructed from $\mathbf{D}^b_{\mathrm{coh}}(X)$ (the Bondal–Orlov reconstruction theorem [18]).
- On the other hand, among those X with trivial K_X like an abelian variety or Calabi–Yau, many examples of so-called Fourier–Mukai partners, i.e. non-isomorphic

[43] Of course, there are many other mathematical approaches to physics. For instance, some of Costello's approach to quantum field theory via Lurie's higher algebra [85, 86] point of view are touched upon in Matsuoka's surveys [89, 90] in this proceedings.

[44] Theorem 4.5 is reduced to Theorem 2.33 for $\mathrm{QCoh}(X) \cong \mathrm{Ind}\,\mathrm{Coh}(X)$ under the Noetherian hypothesis [84, Lem. 3.9]. See also [25, p. 2] [121].

smooth projective varieties with equivalent $\mathbf{D}^b_{\mathrm{coh}}$, have been produced, starting with Mukai [101], Thus, the restruction for the composite $\mathbf{D}^b_{\mathrm{coh}} : X \mapsto \mathbf{D}^b_{\mathrm{coh}}(X)$ in (34) does not hold in general. Considering the Gabriel reconstruction Theorem 4.5, we find this failure results from that of the reconstruction of $\mathcal{D}^b$ among those X with trivial K_X. This suggests an existence of a of moduli of hearts of $\mathbf{D}^b_{\mathrm{coh}}(X)$ for these X.[45]

- If X is affine locally regular and finite-dimensional, then we have the following canonical equivalence (which is a local assertion):

$$\mathbf{D}^b_{\mathrm{coh}}(X) \overset{\simeq}{\to} \mathbf{D}^{\mathrm{perf}}(X)$$

This, in turn, suggests the Verdier quotient

$$\mathbf{D}_{Sg}(X) := \mathbf{D}^b_{\mathrm{coh}}(X)/\,\mathbf{D}^{\mathrm{perf}}(X)$$

reflects singular information of X, and is consequently called the derived category of singularities [118, Def. 1.8].

In the Kontsevich homological mirror symmetry, a mirror of varieties other than Calabi–Yau is not expected to be given by a sigma model. For a variety with either K_X ample or $-K_X$ ample, its mirror is expected to be given by a Landau–Ginzburg model

$$(Y, I, \omega, B, W),$$

where $W : Y \to \mathcal{A}^1$ is a regular function called the superpotential. In this case, the category of D-branes of type B is, via its identification with the category of matrix factorizations, shown to be of the following form [64, 118, 119] :

$$DB(Y, I, \omega, B, W) \cong \prod_{\lambda \in \mathcal{A}^1} \mathbf{D}_{Sg}\left(W^{-1}(\lambda)\right). \tag{35}$$

- This oracle of physics (35), which highlights essentially only the singular part, might appear surprising for mathematicians. However, in the development of the minimal model program in birational geometry, it has become clear that we should take into account singular information even if we are only interested in smooth ones [73, 88, 95].

Now, close relationship between $\mathbf{D}^b_{\mathrm{coh}}$ and birational geometry have been observed [19, 67]. A central problem here is the *Kawamata DK-hypothesis*:

Conjecture 4.6 [68, Conj.1.2] *For birationally equivalent smooth projective varieties X, Y, suppose there exists a smooth projective variety Z with birational morphisms $f : Z \to X,\ g : Z \to Y$.*

[45] As we shall briefly review later, Bridgeland's space of stability conditions is a kind of moduli space of "enriched hearts" of a triangulated category.

K-equivalence $\Longrightarrow$ D-equivalence:

$$\begin{array}{lll} & K\text{-equivalence} & \left(i.e.\ \ f^*K_X \sim g^*K_Y \quad (\textit{linearly equivalent})\right) \\ \textit{implies} & D\text{-equivalence} & \left(i.e.\ \ \mathbf{D}^b_{\mathrm{coh}}(X) \cong \mathbf{D}^b_{\mathrm{coh}}(Y)\right) \end{array}$$

K − inequality $\Longrightarrow$ fully faithful triangulated functor:

$$\begin{array}{ll} & K\text{-inequality} \left(\begin{array}{c} i.e.\quad \textit{there exists an effective divisor } E \textit{ on } Z \textit{ s.t.} \\ f^*K_X + E \sim g^*K_Y \quad (\textit{linearly equivalent}) \end{array}\right) \\ \textit{implies} & \left(\begin{array}{c} \textit{there is a fully faithful functor of triangulated categories} \\ \mathbf{D}^b_{\mathrm{coh}}(X) \to \mathbf{D}^b_{\mathrm{coh}}(Y). \end{array}\right) \end{array}$$

While the converse (D-equivalence $\Longrightarrow$ K-equivalence) does not hold in general [134], if there is a fully faithful functor $\Psi : \mathbf{D}^b_{\mathrm{coh}}(X) \to \mathbf{D}^b_{\mathrm{coh}}(Y)$, then we obtain a semi-orthogonal decomposition (7) [20]:

$$\mathbf{D}^b_{\mathrm{coh}}(Y) = \langle \Psi\left(\mathbf{D}^b_{\mathrm{coh}}(X)\right)^{\perp}, \Psi\left(\mathbf{D}^b_{\mathrm{coh}}(X)\right)\rangle \tag{36}$$

- Motivated by the Kontsevich homological mirror symmetry, some previously unexpected structures of $\mathbf{D}^b_{\mathrm{coh}}(X)$ have been discovered:

– Motivated by the generalized Dehn twist associated with the Lagrangian spheres of the (hypothetical) mirror $X^{\vee}$, Seidel–Thomas [127] constructed a braid group B_{m+1} action under the presence of the spherical A_m-configuration, i.e. there are $\mathcal{E}_i \in \mathbf{D}^b_{\mathrm{coh}}(X)$ $(1 \le i \le m)$ such that the following two conditions are satisfied:

(sphericality): For $1 \le i \le m$, $\mathcal{E}_i \otimes \omega_X \cong \mathcal{E}_i$ and

$$\mathrm{Hom}_{\mathbf{D}^b_{\mathrm{coh}}(X)}(\mathcal{E}_i, \mathcal{E}_i[r]) = \begin{cases} \mathbb{C} & \text{if } r = 0, \dim X \\ 0 & \text{if } r \neq 0, \dim X \end{cases}$$

(A_m-configuration):

$$\dim_{\mathbb{C}} \oplus_r \mathrm{Hom}_{\mathbf{D}^b_{\mathrm{coh}}(X)}\left(\mathcal{E}_i, \mathcal{E}_j[r]\right) = \begin{cases} 1 & |i-j| = 1 \\ 0 & |i-j| \ge 2. \end{cases}$$

– Going back to the reconstruction problem of $\mathcal{D}^b$ in (34), existence of Fourier–Mukai partners suggests an existence of a moduli of hearts of $\mathbf{D}^b_{\mathrm{coh}}(X) = \mathcal{D}^b(\mathrm{Coh}(X))$.

To begin with, we recall a related toy model for $\mathrm{Coh}(X)$, where we can construct moduli spaces, $M_{\mathcal{O}_X(1)}(P)$ for a fixed Hilbert polynomial, by restricting to (Gieseker–Maruyama–Simpson) (semi)-stable sheaves [57, Th. 4.3.4].

Thus, its not surprising that some kind of stability condition is needed to construct a moduli in of hearts of $\mathbf{D}^b_{\mathrm{coh}}(X) = \mathcal{D}^b(\mathrm{Coh}(X))$. In fact, axiomatizing Douglas' study [36] of the Π-stability of D-branes, Bridgeland [22] proposed a way of constructing a moduli space of "enriched hearts," space of stability conditions, out of certain triangulated categories. Bridgeland [22] defined a *stability condition* on a triangulated category $\mathcal{D}$ to be a data $(Z, \mathcal{A})$ such that:

⋆ $\mathcal{A} \subset \mathcal{D}$ is the heart of a bounded t-structure on $\mathcal{D}$.
⋆ $Z : K(\mathcal{A}) \to \mathbb{C}$ is a *stability function*, i.e.

· $Z : K(\mathcal{A}) \to \mathbb{C}$ is a group homomorphism.
· For any $E \in \mathcal{A} \setminus \{0\}$,

$$\begin{aligned} Z(E) &:= r(E) \exp(i\pi\phi(E)) \quad (r(E) > 0, 0 < \phi(E) \le 1) \\ &\in \overline{\mathbb{H}} := \{ r \exp(i\pi\phi) \mid r > 0, 0 < \phi \le 1 \}. \end{aligned}$$

⋆ This stability function $Z : K(\mathcal{A}) \to \mathbb{C}$ is furthermore a *stability condition*, i.e. any $E \in \mathcal{A}$ admits a *Harder–Narasimhan filtration*:

$$0 = E_0 \subset E_1 \subset \cdots \subset E_n = E,$$

such that

· each $F_i = E_i/E_{i-1}$ is Z- *semistable*, i.e. for all nonzero subobjects $F_i' \subset F_i$ we have

$$\phi(F_i') \le \phi(F_i).$$

· $\phi(F_1) > \phi(F_2) > \cdots > \phi(F_n)$.
Since Z is a homomorphism, we can easily verify:

$$E, F \,:\, Z - \text{semistable s.t.}\, \phi(E) > \phi(F) \quad \Longrightarrow \quad \mathrm{Hom}_{\mathcal{A}}(E, F) = 0.$$

Thus, topologists should recognize a similarity between the Harder–Narasimhan filtration and the (finite) Postnikov tower with the following analogy

$$\begin{aligned} &K(\pi_1, n_1), K(\pi_2, n_2) : \text{ Eilenberg-MacLane spectra s.t. } n_1 > n_2 \\ \Longrightarrow\ &\mathrm{Hom}_{\mathcal{SH}}(K(\pi_1, n_1), K(\pi_2, n_2)) = H^{n_2}(K(\pi_1, n_1), \pi_2) = 0. \end{aligned}$$

Here, we wish to vary the heart $\mathcal{A} = \mathcal{D}^{\heartsuit}$ while fixing the ambient triangulated category $\mathcal{D}$. For this purpose, in view of (29), we impose an extra structure on the stability function, i.e.

$$
\begin{array}{ccccc}
K\left(\mathcal{D}^{\heartsuit}\right) & \xrightarrow{\cong} & K(\mathcal{D}) & \xrightarrow{Z} & \mathbb{C} \\
 & & \Big\downarrow \mathrm{cl} & \nearrow \exists & \\
 & & \Gamma & &
\end{array}
, \text{ where } \begin{cases} \Gamma \text{ is a finitely generated free abelian group,} \\ \quad \text{s.t. } \Gamma \otimes_{\mathbb{Z}} \mathbb{R} \text{ is equipped with a norm} \\ \quad \text{(which allows us to define } \| \mathrm{cl}(E) \| \text{ for } E \in K(\mathcal{D})). \\ \mathrm{cl} : \Gamma \to \mathbb{C} \text{ is a homomorphism} \end{cases}
$$

We further impose the *support property* [75]:

$$\left\{ \frac{|Z(E)|}{\| \mathrm{cl}(E) \|} \;\middle|\; E \in \left(\cup_{i \in \mathbb{Z}} \mathcal{D}^{\heartsuit}[i] \right) \setminus 0 \right\} \text{ is bounded.}$$

When we fix $\mathcal{D}$ with such a homomorphism $K(\mathcal{D}) \to \Gamma$, Bridgeland [22] showed the set of such stability conditions can be topologized and becomes a complex manifold $\mathrm{Stab}_{\Gamma}(\mathcal{D})$.

However, for the case of our interest $\mathcal{D} = \mathbf{D}^b_{\mathrm{coh}}(X)$, as soon as $\dim X \geq 3$, there is no stability condition on $\mathcal{D} = \mathbf{D}^b_{\mathrm{coh}}(X)$ with $\mathcal{D}^{\heartsuit} = \mathrm{Coh}(X)$ [132, Lem. 2.7], and even the existence of such a stability condition is problematic, i.e. the possibility of $\mathrm{Stab}_{\Gamma}(\mathcal{D}) = \emptyset$ is yet to be excluded.

4.2 $\mathbf{D}^{\mathrm{perf}}(X)$

- The functoriality results for $\mathbf{D}^b_{\mathrm{coh}}$ reviewed in Theorems 4.1 and 4.2 have the following analogue for $\mathbf{D}^{\mathrm{perf}}$:

Theorem 4.7 [83, Th. 1.2] [109, Ill.0.19] *For a separated, finite-type morphism of noetherian schemes $f : X \to Y$,*

$$\mathbf{R}f_* \, \mathbf{D}^{\mathrm{perf}}(X) \subset \mathbf{D}^{\mathrm{perf}}(Y) \ (\textit{i.e. } \underline{\textit{perfect}}) \quad \Longleftrightarrow \quad f \textit{ is proper and of finite Tor-dimension}$$

- $\mathbf{D}^{\mathrm{perf}}(X)$ can be directly recovered from $\mathbf{D}_{\mathrm{qc}}(X)$:

Theorem 4.8 ([20, 106]) *The canonical functor*

$$\mathbf{D}^{\mathrm{perf}}(X) \to \mathbf{D}_{\mathrm{qc}}(X)$$

identifies $\mathbf{D}^{\mathrm{perf}}(X)$ as the full triangulated subcategory $\mathbf{D}_{\mathrm{qc}}(X)^c$ of compact objects in $\mathbf{D}_{\mathrm{qc}}(X)$:

$$\mathbf{D}^{\mathrm{perf}}(X) = \mathbf{D}_{\mathrm{qc}}(X)^c$$

- Thomason–Trobaugh [130, App. F] proved $\mathbf{D}^{\mathrm{perf}}(X) = \mathbf{D}_{\mathrm{qc}}(X)^c$ is essentially small (i.e. equivalent to a small category) for any quasi-compact and quasiseparated scheme X (e.g. for any noetherian scheme). Starting with this, Thomason [131, Th. 3.15] classified thick tensor triangulated ideals of $\mathbf{D}^{\mathrm{perf}}(X) = \mathbf{D}_{\mathrm{qc}}(X)^c$ for any quasi-compact and quasiseparated scheme X. Here, we review Paul Balmer's generalization [3] of such a classification to certain essentially small tensor triangulated categories.

Definition 4.9 For a tensor triangulated category $\mathcal{K}$,

- [3, Def. 4.1] [5, Def. 7] A thick tensor ideal $\mathcal{I} \subset \mathcal{K}$ is called *radical* if

$$\mathcal{I} = \sqrt{\mathcal{I}} := \left\{a \in \mathcal{K} \mid \exists n \geq 1 \text{ such that } a^{\otimes n} \in \mathcal{I}\right\}.$$

  The collection of radical thick tensor ideals of $\mathcal{K}$ is denoted by $\mathbb{R}(\mathcal{K})$.
- [3, Def. 2.1] [5, Con. 8], (see also Definition 2.27) A proper thick tensor ideal $\mathcal{P} \subsetneq \mathcal{K}$ is called *prime*, if

$$a \otimes b \in \mathcal{P} \quad \Longrightarrow \quad a \in \mathcal{P} \text{ or } b \in \mathcal{P}.$$

- [3, Def. 2.1] [5, Con. 8] If $\mathcal{K}$ is further essentially small, its spectrum $\mathrm{Spc}(\mathcal{K})$ is given by the following (set, by the "essentially small" assumption):

$$\mathrm{Spc}(\mathcal{K}) = \{\mathcal{P} \subsetneq \mathcal{K} \mid \mathcal{P} \text{ is a proper prime thick tensor ideal of } \mathcal{K}\},$$

  which is endowed with the topology whose open subsets are of the form

$$U(\mathcal{E}) := \{\mathcal{P} \in \mathrm{Spc}(\mathcal{K}) \mid \mathcal{E} \cap \mathcal{P} \neq \emptyset\} \quad (\mathcal{E} \subseteq \mathcal{K});$$

  in other words, given by the closed basis $\{\mathrm{supp}(a)\}_{a \in \mathcal{K}}$, where

$$\mathrm{supp}(a) = \{\mathcal{P} \in \mathrm{Spc}(\mathcal{K}) \mid a \notin \mathcal{P}\}$$

  is the *support* of $a \in \mathcal{K}$.[46]
- [5, Rem. 12] For a general topological space T (we are particularly interested in the case $T = \mathrm{Spc}(\mathcal{K})$), a subset $Y \subset T$ of the form

$$Y = \cup_{i \in I} Y_i \quad \text{with each complement } X \setminus Y_i \text{open and quasi-compact}$$

  is called a *Thomason subset* of T. The set of Thomason subsets of T is denoted by $\mathrm{Tho}(T)$.

Theorem 4.10 (i) [3, Th. 4.10] [5, Th. 14] [7, Th. 5.9] *For an essentially small tensor triangulated category $\mathcal{K}$, there are mutually inverse isomorphisms between radical thick tensor ideals of $\mathcal{K}$ and Thomason subsets of* $\mathrm{Spc}(\mathcal{K})$*:*

[46] WARNING! We had already introduced the same notation supp back in Definition 2.35. However, from Proposition 2.36, Theorem 4.11, these two usages of supp coincide for the most fundamental example of $\mathcal{K} = \mathbf{D}^{\mathrm{perf}}(X)$.

$$\begin{aligned} \mathcal{K}_{-} : \mathrm{Tho}\,(\mathrm{Spc}(\mathcal{K})) &\overset{\cong}{\rightleftarrows} \mathbb{R}(\mathcal{K}) : \mathrm{supp} \\ Y &\mapsto \mathcal{K}_Y := \{a \in \mathcal{K} \mid \mathrm{supp}(a) \subset Y\} \\ \mathrm{supp}(\mathcal{R}) := \cup_{a \in \mathcal{R}} \mathrm{supp}(a) &\leftarrow\!\mid \mathcal{R} \end{aligned} \tag{37}$$

(ii) [4, Prop. 2.4] *Suppose further $\mathcal{K}$ is rigid, then every thick tensor ideal is radical, and so, $\mathbb{R}(\mathcal{K}) = \mathbb{T}(\mathcal{K})$. Consequently, the mutually inverse isomorphisms in (i) becomes the following:*

$$\mathcal{K}_{-} : \mathrm{Tho}\,(\mathrm{Spc}(\mathcal{K})) \rightleftarrows \mathbb{T}(\mathcal{K}) : \mathrm{supp}$$

Theorem 4.11 [131] [3, Cor. 5.6] [23, Cor. 5.2] [5, Th. 16] *For a quasi-compact and quasi-separated scheme X, its underlying topological space $|X|$ is homeomorphic to the spectrum $\mathrm{Spc}(\mathbf{D}^{\mathrm{perf}}(X))$ via*

$$\begin{aligned} |X| &\overset{\cong}{\to} \mathrm{Spc}\left(\mathbf{D}^{\mathrm{perf}}(X)\right) \\ x &\mapsto \mathfrak{P}(x) := \left\{P \in \mathbf{D}^{\mathrm{perf}}(X) \ \mid \ P_x \cong 0\right\}. \end{aligned}$$

For any $P \in \mathbf{D}^{\mathrm{perf}}(X)$, this homeomorphism restricts to the homeomorphism

$$\mathrm{Supph}(P) \overset{\cong}{\to} \mathrm{supp}(P),$$

where $\mathrm{Supph}(P) \subseteq X$ is the homological support of $P \in \mathbf{D}^{\mathrm{perf}}(X)$, i.e. the usual sheaf theoretical support of the total homology of P given in Definition 2.35 and Proposition 2.36.

From Theorem 4.11, Theorem 4.10 (ii) yields the following theorem of Thomason, which is a $\mathbf{D}_{\mathrm{qc}}(X)$ analogue of the Hopkins–Smith Theorem 3.1:

Theorem 4.12 [131, Th. 3.15] *For a quasi-compact and quasi-separated scheme X, there are mutually inverse isomorphisms between thick tensor ideals of $\mathbf{D}^{\mathrm{perf}}(X)$ and Thomason subsets of $|X|$:*

$$\begin{aligned} \mathbf{D}^{\mathrm{perf}}_{-}(X) : \mathrm{Tho}\,(|X|) &\overset{\cong}{\rightleftarrows} \mathbb{T}\left(\mathbf{D}^{\mathrm{perf}}(X)\right) : \mathrm{supp} \\ Y &\mapsto \mathbf{D}^{\mathrm{perf}}_{Y}(X) := \{P \in \mathbf{D}^{\mathrm{perf}}(X) \mid \mathrm{Supph}(P) \subset Y\} \\ \mathrm{supp}(\mathcal{R}) := \cup_{a \in \mathcal{R}} \mathrm{supp}(a) &\leftarrow\!\mid \mathcal{R}. \end{aligned} \tag{38}$$

Remark 4.13 [109, Lem. 3.1] For an object H of a tensor triangulated category $\mathcal{T}$, denote by $\langle H \rangle_{\otimes}$ the thick tensor ideal (tensor) generated by H. Then we easily see:

$$\langle H \rangle_{\otimes} = \cup_{N \in \mathbb{N}, C \in \mathcal{T}} \langle C \otimes H \rangle_N,$$

where the notation $\langle - \rangle_N$ is recalled in Definition 4.28.

Many tensor triangulated categories $\mathcal{T}$ are (tensor) generated by a single element.

It should be mentioned that, just like the nilpotency Theorem 3.5 was crucial in the proof of Hopkins–Smith Theorem 3.1, some algebro-geometric analogue of (Devinatz-)Hopkins–Smith nilpotency is crucial to prove these algebro-geometric analogues of the Hopkins–Smith theorem (see e.g. [103, Th. 1.1] [131, Th. 3.6, Th. 3.8]). In this direction, Hovey–Palmieri–Strickland [54, 5] developed a general theory how nilpotence implies classifications of thick subcategories.

Now, the following simple consequence of the above theorem of Thomason will be used later:

Corollary 4.14 *For a quasi-compact and quasi-separated scheme X, any thick tensor ideal generated by a single $H \in \mathbf{D}^{\mathrm{perf}}(X)$ with* $\mathrm{Supph}(H) = |X|$ *is all of* $\mathbf{D}^{\mathrm{perf}}(X)$.

- In terms of $\mathbf{D}^{\mathrm{perf}}(X) = \mathbf{D}_{\mathrm{qc}}(X)^c$, we may refine the smashing part of the fundamental theorem of Hopkins, Neeman, Thomason and others (Theorem 2.37) to become an algebraic analogue of the Hopkins–Smith theorem (Theorem 3.1), with an extra bonus of the validity of an algebraic analogue of the telescope conjecture. We shall review it now, together with (a restatement of) Theorem 2.37. For the notations below, consult the list just before Theorem 3.7.

Theorem 4.15 ([48] [103, Th. 2.8, Th. 3.3] [131], [2, Cor 4.6; Cor. 4.13; Th. 5.6] [7, Cor. 6.8] [34, Cor. 6.8; Ex. 6.9] [47, Th. B]) *For a Noetherian scheme X, we have a commutative diagram consisting of mutually inverse horizontal arrows:*

$$\begin{array}{ccccc} 2^{|X|} & & \xrightleftharpoons[\mathrm{supp}]{\{Q\in\mathbf{D}_{\mathrm{qc}}(X)\ \mid\ \mathrm{supp}(Q)\subseteq -\}} & & \mathbb{L}(\mathbf{D}_{\mathrm{qc}}(X)) \\ \uparrow & & & & \uparrow \\ \mathrm{Tho}(|X|) & \xrightleftharpoons[\mathrm{supp}]{\mathbf{D}^{\mathrm{perf}}_{-}(X)} & \mathbb{T}\left(\mathbf{D}^{\mathrm{perf}}(X)\right) & \xrightleftharpoons[C_X]{I_X} & \mathbb{S}(\mathbf{D}_{\mathrm{qc}}(X)) \end{array} \tag{39}$$

Here,

- *The upper side mutually inverse arrows are those in Theorem 2.37, which is the analogue of the Ohkawa theorem and an affirmative solution of the Hovey Conjecture 2.32 (ii) for $\mathbf{D}_{\mathrm{qc}}(X)$.*
- *The lower left side mutually inverse arrows are those in Thomason's Theorem 4.12, which is a $\mathbf{D}_{\mathrm{qc}}(X)$ analogue of the Hopkins–Smith Theorem 3.1:*

Remark 4.16 The above commutative diagram (39) encapsulates our story; starting with Ohkawa's theorem in $\mathcal{SH}$, we then move on to the $\mathbf{D}_{\mathrm{qc}}$ analogue, encountering the fundamental theorem of Hopkins, Neeman, Thomason and others; then going back to $\mathcal{SH}^c$ to appreciate the Hopkins–Smith thick category theorem, and then, moving back again to the $\mathbf{D}^c_{\mathrm{qc}}$ analogue, we discover the above fantastic Theorem 4.15. In fact, the commutative diagram (39) is a $\mathbf{D}^c_{\mathrm{qc}} \subset \mathbf{D}_{\mathrm{qc}}$ analogue of the commutative diagram (25) for $\mathcal{SH}^c_{(p)} \subset \mathcal{SH}_{(p)}$.

Thus the underlying message here is to extend the commutative diagrams of (39) and (25) to other triangulated categories. There is a paper of Iyenger–Krause [58] in this direction, and this is exactly the theme of our Homework in the introduction.

– The mutually inverse arrows at the bottom right of the diagram yield a positive solution of the telescope conjecture (Theorem 3.3 (iv)) by [7, Cor. 6.8] [34, Cor. 6.8; Ex. 6.9] [47, Th. B]).

- However, the analogue of (14) for $\mathbf{D}^{\mathrm{perf}}$ does not hold in general, for $\mathbf{L}j^* : \mathbf{D}^{\mathrm{perf}}(X) \to \mathbf{D}^{\mathrm{perf}}(U)$ is not surjective in general. Still, as was noticed by Thomason–Trobaugh [130], there is a similar equivalence as soon as we apply the thick closure $\widehat{(-)}$:[47,48]

---

[47]Let us recall the following related result in the setting of abelian category of quasi-coherent sheaves, which should go back at least to Gabiriel (see e.g. [126, Prop. 3.1]): $\mathrm{Coh}(X)/\mathrm{Coh}_Z(X) \underset{\cong}{\overset{\overline{j^*}}{\to}} \mathrm{Coh}(U)$, where the left hand side is the abelian quotient category in the sense of Gabriel, Grothendieck, and Serre.

[48]The following interesting historical account on the difficulty of generalizing statements in $\mathbf{D}_{\mathrm{qc}}$ (14) (15):

$$\begin{cases} \mathbf{D}_{\mathrm{qc}}(X)/\left(\mathbf{D}_{\mathrm{qc}}\right)_Z(X) \underset{\cong}{\overset{\overline{\mathbf{L}j^*}}{\longrightarrow}} \mathbf{D}_{\mathrm{qc}}(U) \\ L = \mathbf{R}j_*\mathbf{L}j^* = (\mathbf{R}j_*\mathcal{O}_U) \otimes^{\mathbf{L}}_{\mathcal{O}_X} - : \mathbf{D}_{\mathrm{qc}}(X) \to \mathbf{D}_{\mathrm{qc}}(X)/\left(\mathbf{D}_{\mathrm{qc}}\right)_Z(X) \underset{\cong}{\overset{\overline{\mathbf{L}j^*}}{\longrightarrow}} \mathbf{D}_{\mathrm{qc}}(U) \overset{\mathbf{R}j_*}{\longrightarrow} \mathbf{D}_{\mathrm{qc}}(X) \end{cases}$$

and the precursor in the setting of abelian categories reviewed in footnote 27:

$$\mathrm{QCoh}(X)/\mathrm{QCoh}_Z(X) \underset{\cong}{\overset{\overline{j^*}}{\to}} \mathrm{QCoh}(U)$$

to the setting of $\mathbf{D}^{\mathrm{perf}}$, has been communicated to the author by Professor Neeman:

> *... But the right adjoints* $j_* : \mathrm{QCoh}(U) \to \mathrm{QCoh}(X)$ *and* $\mathbf{R}j_* : \mathbf{D}_{\mathrm{qc}}(U) \to \mathbf{D}_{\mathrm{qc}}(X)$ *fail to preserve the finite subcategories* $\mathrm{Coh}(-)$ *and* $\mathbf{D}^{\mathrm{perf}}(-)$. *For these categories some work is needed. Especially in the case of* $\mathbf{D}^{\mathrm{perf}}(-)$; *for a long time all that was known was that* $\mathbf{L}j^* : \mathbf{D}^{\mathrm{perf}}(X) \to \mathbf{D}^{\mathrm{perf}}(U)$ *isn't surjective on objects, hence the natural map*
>
> $$\frac{\mathbf{D}^{\mathrm{perf}}(X)}{\mathrm{Ker}(\mathbf{L}j^*)} \longrightarrow \mathbf{D}^{\mathrm{perf}}(U)$$
>
> *couldn't be an equivalence. So the assumption was that this map had to be worthless.*
>
> *Thomason's ingenious insight was that the old counterexamples were a red herring. Up to idempotent completion this map is an equivalence, and in particular induces an isomorphism in higher K-theory. This of course required proof. Thomason gave a rather involved proof, following SGA6, and I noticed that the proof simplifies and generalizes when one uses the methods of homotopy theory.*
>
> *It was an amusing role reversal: Thomason, the homotopy theorist, had the brilliant idea but gave a clumsy proof using the techniques of algebraic geometry, while I, the algebraic geometer, simplified the argument with the techniques of homotopy theory.*

Theorem 4.17 (Thomason's localization theorem) *Under the situation of* (14), *i.e. let X be a quasicompact and quasiseparated scheme, $Z = X \setminus U \subset X$, the complement of a quasi-compact Zariski open immersion $j : U \hookrightarrow X$, we have a triangulated embedding*

$$\mathbf{D}^{\mathrm{perf}}(X) / \left(\mathbf{D}^{\mathrm{perf}}\right)_Z (X) \subset \mathbf{D}^{\mathrm{perf}}(U),$$

which yields an equivalence upon applying the thick closure:

$$\left(\mathbf{D}^{\mathrm{perf}}(X) / \left(\mathbf{D}^{\mathrm{perf}}\right)_Z (X)\right)^{\widehat{}} \underset{\cong}{\xrightarrow{\mathbf{L}j^*}} \mathbf{D}^{\mathrm{perf}}(U). \tag{40}$$

In applications, we sometime have to take care of elements in $\left(\mathbf{D}^{\mathrm{perf}}\right)_Z (X)$. Then we wonder if they are in the image of $\mathbf{R}i_* \, \mathbf{D}^{\mathrm{perf}}(Z)$ or not. Now, Rouquier [125] gave an affirmative answer for a weaker question in the coherent setting:

Theorem 4.18 [125, Lem. 7.40] *Let X be a separated noetherian scheme and Z be its closed subscheme given by the ideal sheaf $\mathcal{I}$ of $\mathcal{O}_X$. For $n \in \mathbb{N}$, let Z_n be the closed subscheme of X with ideal sheaf $\mathcal{I}^n$ and $i_n : Z_n \to X$ the corresponding immersion. Then,*

$$\forall Q \in \left(\mathbf{D}^b_{\mathrm{coh}}\right)_Z (X), \qquad \exists n \in \mathbb{N}, \exists P_n \in \mathbf{D}^b_{\mathrm{coh}}(Z_n) \ s.t. \ \ Q = \mathbf{R}i_{n*} P_n.$$

While the original proof of Theorem 4.17 given in [130] is purely algebro geometric in the spirit of SGA6, Neeman [105, Th. 2.1] gave a proof from a general triangulated category theoretical point of view, in the homotopy theoretical spirit of Bousfield, Ohakawa, and others, building upon Corollary 2.19 [105, Lem. 1.7]:

Theorem 4.19 (Neeman's generalization of Thomason's localization theorem) *Let $\mathcal{T}$ be a compactly generated triangulated category, generated by a set K consisting of compact objects in $\mathcal{T}$. For a subset $S \subseteq K$, set $\mathcal{S}$ be the smallest localizing triangulated subcategory containing S. Then, the canonical sequence of triangulated categories*

$$\mathcal{S} \to \mathcal{T} \to \mathcal{T}/\mathcal{S} \tag{41}$$

induces another sequence of triangulated categories of compact objects

$$\mathcal{S}^c \to \mathcal{T}^c \to (\mathcal{T}/\mathcal{S})^c, \tag{42}$$

which induces an equivalence

$$\mathcal{S}^c = \mathcal{S} \cap \mathcal{T}^c, \tag{43}$$

a fully faithful embedding

$$\mathcal{T}^c/\mathcal{S}^c \to (\mathcal{T}/\mathcal{S})^c, \tag{44}$$

and, although it may fail to induce an equivalence $\mathcal{T}^c/\mathcal{S}^c \overset{\cong}{\to} (\mathcal{T}/\mathcal{S})^c$, it does induce an equivalence upon applying the thick closure:

$$\left(\mathcal{T}^c/\mathcal{S}^c\right)^{\widehat{}} \overset{\cong}{\to} (\mathcal{T}/\mathcal{S})^c . \tag{45}$$

Proof (i) The first triangulated functor in (42) is an easy consequence of Proposition 2.23. The second triangulated functor in (42) is induced by the smashing Bousfield localization functor $\mathcal{T} \to \mathcal{T}/\mathcal{S}$, which preserves arbitrary coproducts Theorem 2.19. Then for $c \in \mathcal{T}^c$, $t_\lambda \in \mathcal{T}$ $(\lambda \in \Lambda)$, regarding $\mathcal{T}/\mathcal{S}$ as the full subcategory of L-local objects, we evaluate as follows:

$$\begin{aligned}&\mathrm{Hom}_{\mathcal{T}/\mathcal{S}}\left(Lc, \oplus_{\lambda\in\Lambda} Lt_\lambda\right) = \mathrm{Hom}_{\mathcal{T}}\left(Lc, \oplus_{\lambda\in\Lambda} Lt_\lambda\right) \overset{L:\text{ smashing}}{=} \mathrm{Hom}_{\mathcal{T}}\left(Lc, L(\oplus_{\lambda\in\Lambda} t_\lambda)\right)\\ &= \mathrm{Hom}_{\mathcal{T}}\left(c, L(\oplus_{\lambda\in\Lambda} t_\lambda)\right) \overset{L:\text{ smashing}}{=} \mathrm{Hom}_{\mathcal{T}}\left(c, \oplus_{\lambda\in\Lambda} Lt_\lambda\right) \overset{c:\text{ compact}}{=} \oplus_{\lambda\in\Lambda} \mathrm{Hom}_{\mathcal{T}}\left(c, Lt_\lambda\right)\\ &= \oplus_{\lambda\in\Lambda} \mathrm{Hom}_{\mathcal{T}}\left(Lc, Lt_\lambda\right) = \oplus_{\lambda\in\Lambda} \mathrm{Hom}_{\mathcal{T}/\mathcal{S}}\left(Lc, Lt_\lambda\right),\end{aligned}$$

which implies Lc is also compact.

On the other hand, Krause [78] gave a conceptually simple, though more involved, proof of the existence of (42), applying the following easy observation [78, Lem. 5.4.1.(1)], which goes back at least to [106, Th. 5.1] where the converse, i.e. compactness preservation of $F \implies$ small coproducts preservation of G, is also shown under the additional compact generation assumption of $\mathcal{T}$:

For any pair of adjoint triangulated functors $\mathcal{T} \underset{G}{\overset{F}{\rightleftarrows}} \mathcal{U}$ such that G preserves small coproducts, F preserves compactness.

$\because$) In fact, for any $c \in \mathcal{T}^c$, $u_\lambda \in \mathcal{U}$ $(\lambda \in \Lambda)$,

$$\begin{aligned}\mathrm{Hom}_{\mathcal{U}}(Fc, \oplus_\lambda u_\lambda) &= \mathrm{Hom}_{\mathcal{T}}(c, G(\oplus_\lambda u_\lambda)) = \mathrm{Hom}_{\mathcal{T}}(c, \oplus_\lambda G(u_\lambda)) = \oplus_\lambda \mathrm{Hom}_{\mathcal{T}}(c, G(u_\lambda))\\ &= \oplus_\lambda \mathrm{Hom}_{\mathcal{U}}(Fc, u_\lambda).\end{aligned}$$

Now, (42) is induced from (41) by applying this easy observation to the recollement given by Proposition 2.21(6).[49]
(ii) To see (43), first note $\mathcal{S}^c \supset \mathcal{S} \cap \mathcal{T}^c$ is trivial from the definition. Then (43) follows since converse $\mathcal{S}^c \subset \mathcal{S} \cap \mathcal{T}^c$ also follows from (42).
(iii) For (44), suffices to show the composite

$$\mathrm{Hom}_{\mathcal{T}^c/\mathcal{S}^c}(c, c') \to \mathrm{Hom}_{(\mathcal{T}/\mathcal{S})^c}(c, c') \overset{\cong}{\to} \mathrm{Hom}_{(\mathcal{T}/\mathcal{S})}(c, c') \overset{Th. (2.19)}{=} \mathrm{Hom}_{\mathcal{T}}(c, \mathrm{hocolim}(x_n))$$

is an isomorphism.

For the surjectivity, take $(f : c \to \mathrm{hocolim}(x_n)) \in \mathrm{Hom}_{\mathcal{T}}(c, \mathrm{hocolim}(x_n))$, then we can find its preimage $(c \overset{\bigstar}{\leftarrow} c \times^h_{c_n} c' \overset{\widetilde{f'_n}}{\to} c') \in \mathrm{Hom}_{\mathcal{T}^c/\mathcal{S}^c}(c, c')$ by a straightforward contemplation summarized in the following commutative diagram:

[49]This is the involved part of this proof, for the existence of recollement there requires Brown representability.

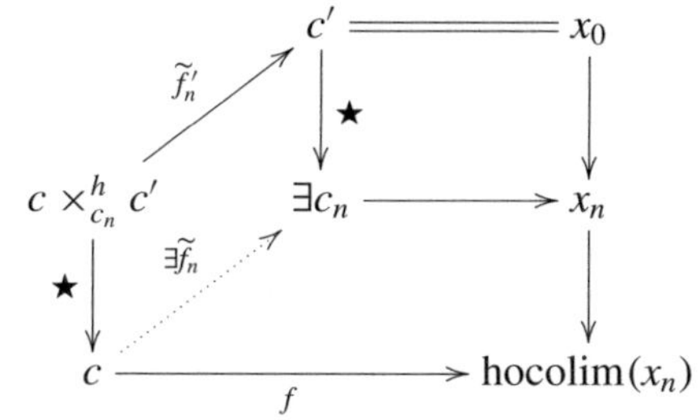

Here, c_n is some compact object so that arrows with $\bigstar$ have cones of the form finite extension of finite coproducts of elements in S, and $c \times^h_{c_n} c'$ is the homotopy pullback (see e.g. [130, p. 252, (1.1.2.5)]).

For the injectivity, suppose $(c \xleftarrow{\bigstar} \widetilde{c} \xrightarrow{f'} c') \in \mathrm{Hom}_{\mathcal{T}^c/\mathcal{S}^c}(c, c')$ is sent to $(c \xleftarrow{\bigstar} \widetilde{c} \xrightarrow{0} \mathrm{hocolim}(x_n)) = 0 \in \mathrm{Hom}_{\mathcal{T}}(c, \mathrm{hocolim}(x_n))$. Then we can see $(c \xleftarrow{\bigstar} \widetilde{c} \xrightarrow{f'} c') = (x \xleftarrow{\bigstar} \widetilde{c} \times^h_{c'_m} c' \xrightarrow{0} c') = 0 \in \mathrm{Hom}_{\mathcal{T}^c/\mathcal{S}^c}(c, c')$ by a straightforward contemplation summarized in the following commutative diagram:

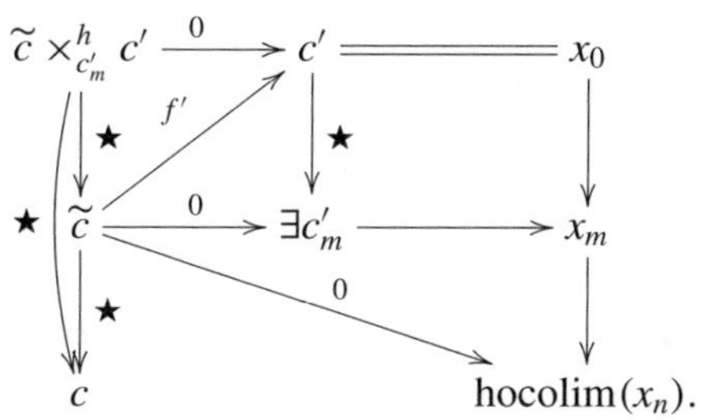

Here, c'_m is some compact object so that arrows with $\bigstar$ have cones of the form finite extension of finite coproducts of elements in S, and $c \times^h_{c'_m} c'$ is the homotopy pullback [130, p. 252, (1.1.2.5)].
(iv) To see (45), write $\mathcal{T} = \langle K \rangle$, and observe from the construction of the Verdier quotient $\mathcal{T} \xrightarrow{F_{univ}} \mathcal{T}/\mathcal{S}$ that $\mathcal{T}/\mathcal{S} = \langle F_{univ}(K) \rangle$, where $F_{univ}(K) \subseteq \mathcal{T}^c/\mathcal{S}^c \subseteq (\mathcal{T}/\mathcal{S})^c$ by (42) and (44). Now apply Proposition 2.23 to conclude any object y of $(\mathcal{T}/\mathcal{S})^c$ is a direct summand of a finite extension (in $(\mathcal{T}/\mathcal{S})^c$) of finite direct sums of objects in $F_{univ}(K) \subseteq \mathcal{T}^c/\mathcal{S}^c$, which is a full triangulated subcategory by (44). This implies the desired equivalence upon thick closure (45): $\widehat{(\mathcal{T}^c/\mathcal{S}^c)} \xrightarrow{\cong} (\mathcal{T}/\mathcal{S})^c$. □

The following consequence of Theorem 4.17 and Remark 2.3 (iv) will be used later:

Corollary 4.20 *Let X be a Noetherian scheme, and $Z = X \setminus U \subset X$, the complement of a quasi-compact Zariski open immersion $j : U \hookrightarrow X$. Then, for any $P \in \mathbf{D}^{\mathrm{perf}}(U)$, there exists $H \in \mathbf{D}^{\mathrm{perf}}(X)$ such that*

$$\mathbf{L}j^* H \cong P \oplus \Sigma P \in \mathbf{D}^{\mathrm{perf}}(U).$$

Now, to motivate Balmer's construction reviewed next, let us single out the following slight strengthening of Theorem 4.19 (and so also of Theorem 4.17):

Theorem 4.21 *Under the same assumption of Theorem 4.19, the extrinsic thick closure equivalence* (45) *can be upgraded to the intrinsic idempotent completion*[50] *equivalence:*

$$\left(\mathcal{T}^c/\mathcal{S}^c\right)^\sharp \overset{\cong}{\to} (\mathcal{T}/\mathcal{S})^c . \tag{46}$$

In particular, under the same assumption of Theorem 4.17, we have an equivalence upon applying the idempotent completion:

$$\left(\mathbf{D}^{\mathrm{perf}}(X)/\left(\mathbf{D}^{\mathrm{perf}}\right)_Z(X)\right)^\sharp \underset{\cong}{\overset{\mathbf{L}j^*}{\longrightarrow}} \mathbf{D}^{\mathrm{perf}}(U). \tag{47}$$

To show (46), it suffices to show $\widehat{(\mathcal{T}^c/\mathcal{S}^c)} \cong (\mathcal{T}^c/\mathcal{S}^c)^\sharp$ thanks to (45). For this, note from (44) a fully faithful embedding $\mathcal{T}^c/\mathcal{S}^c \to \mathcal{T}/\mathcal{S}$. Here, $\mathcal{T}/\mathcal{S}$ is idempotent complete, because $\mathcal{T}/\mathcal{S}$ is first seen to be equipped with arbitrary small coproducts by Theorem 2.19(2), Proposition 2.21(5), Proposition 2.10(5), and then we may apply Remark 2.3 (i) to find $\mathcal{T}/\mathcal{S}$ is idempotent complete. Thus, any added idempotent object of $(\mathcal{T}^c/\mathcal{S}^c)^\sharp$ shows up in $\mathcal{T}/\mathcal{S}$, but, because of $\mathcal{T}^c/\mathcal{S}^c \subseteq (\mathcal{T}/\mathcal{S})^c$ and any direct summand of a compact object is still compact, these added idempotent objects actually show up in $(\mathcal{T}/\mathcal{S})^c$. This implies the desired (46).

- In view of Theorem 4.11, we wonder whether the spectrum X is reconstructed from $(\mathbf{D}^{\mathrm{perf}}(X), \otimes^{\mathbb{L}})$. But, this is nothing but the theorem of Paul Balmer [3]:

Definition 4.22 For an essentially small tensor triangulated category $\mathcal{K}$, we defined in Definition 4.9 the spectrum (topological space) $\mathrm{Spc}(\mathcal{K})$.

- Here, motivated by (47), we can construct a presheaf of tensor triangulated categories by

$$U \mapsto \mathcal{K}(U) := \left(\mathcal{K}/\mathcal{K}_Z\right)^\sharp , \tag{48}$$

  where $\mathcal{K}_Z := \{a \in \mathcal{K} \mid \mathrm{supp}(a) \subseteq Z\}$ with $Z := X \setminus U$ and $\mathrm{supp}(a) := \mathrm{Spc}(\mathcal{K}) \setminus U(a) = \{\mathcal{P} \in \mathrm{Spc}(\mathcal{K}) \mid a \notin \mathcal{P}\}$.
- Finally, we obtain the ringed space

$$\mathrm{Spec}\,(\mathcal{K}) = (\mathrm{Spc}(\mathcal{K}), \mathcal{O}_{\mathcal{K}}) , \tag{49}$$

  as the sheafication of the presheaf of commutative rings

$$U \mapsto \mathrm{End}_{\mathcal{K}(U)}(\mathbb{1}), \tag{50}$$

  where $\mathbb{1}$ is the unit object of the tensor triangulated category $\mathcal{K}(U)$.

[50] For the fact that the idempotent completion of a triangulated category has a natural structure of a triangulated category, there is a proof in Balmer–Schlichting [6].

Now Balmer's reconstruction theorem [3] states:

Theorem 4.23 *For a quasi-compact and quasi-separated scheme X, we have an isomorphism of ringed spaces*[51]

$$\operatorname{Spec}\left(\mathbf{D}^{\mathrm{perf}}(X), \otimes^{\mathbf{L}}\right) \cong X.$$

4.3 $\mathbf{D}^b_{\mathbf{coh}}(X)$ *and* $\mathbf{D}^{\mathbf{perf}}(X)$ *Determine Each Other*

With the concepts "approximable", "noetherian approximable", "metric", "preferred t-structure", and "Cauchy sequence" in a black box, Amnon Neeman's strategy to prove this may be summarized as follows:

- [108, Ex. 8.4]:

  Out of an approximable triangulated category $\mathcal{T}$ with a preferred t-structure $(\mathcal{T}^{\le 0}, \mathcal{T}^{\ge 0})$, we can construct a couple of triangulated categories $\mathcal{S}$ with metrics:

  1. $\mathcal{S} = \mathcal{T}^c \subset \mathcal{T}$, and $\mathcal{M}_i = \mathcal{T}^c \cap \mathcal{T}^{\le -i}$.
  2. $\mathcal{S} = [\mathcal{T}_c^b]^{\mathrm{op}}$, and $\mathcal{M}_i^{\mathrm{op}} = \mathcal{T}_c^b \cap \mathcal{T}^{\le -i}$.

- [110, Def. 1.10] For an essentially small triangulated category $\mathcal{S}$ with a metric $\{\mathcal{M}_i\}$, we define three full subcategories $\mathfrak{L}(\mathcal{S})$, $\mathfrak{C}(\mathcal{S})$, $\mathfrak{S}(\mathcal{S})$ of the category

  $$\operatorname{Mod} -\mathcal{S} := \text{additive functors } \mathcal{S}^{\mathrm{op}} \to \mathbb{Z} - \operatorname{Mod}.$$

  With $Y : \mathcal{S} \to \operatorname{Mod} -\mathcal{S};\ A \mapsto Y(A) := \operatorname{Hom}(-, A)$ the *Yoneda functor*, we set

  $$\mathfrak{L}(\mathcal{S}) := \left\{ \operatorname*{colim}_{\longrightarrow} Y(E_i) \in \operatorname{Mod} -\mathcal{S} \;\middle|\; E_*, \text{ is a } \underline{\textit{Cauchy sequence}} \text{ in } \mathcal{S}. \right\}$$

  $$\mathfrak{C}(\mathcal{S}) := \left\{ A \in \operatorname{Mod} -\mathcal{S} \;\middle|\; \begin{matrix}\text{For every } j\in\mathbb{Z} \text{ there exists } i\in\mathbb{Z} \text{ with} \\ \operatorname{Hom}\left(Y(\mathcal{M}_i), \Sigma^{-j}A\right)=0.\end{matrix} \right\}$$

  $$\mathfrak{S}(\mathcal{S}) := \mathfrak{L}(\mathcal{S}) \cap \mathfrak{C}(\mathcal{S}).$$

  By construction, we see [110, Obs. 2.3]

  $$\mathfrak{S}(\mathcal{S}) = \bigcap_{j\in\mathbb{Z}} \bigcup_{i\in\mathbb{N}} \left[Y(\Sigma^j E_i)\right]^{\perp}.$$

  Intuitively, $\mathfrak{S}(\mathcal{S})$ consists of compactly supported objects (for contained in $\mathfrak{C}(\mathcal{S})$) of the Cauchy completion with respect to the given metric inside the Ind-completion given by the Yoneda embedding (for contained in $\mathfrak{L}(\mathcal{S})$).

[51] The weaker reconstruction just as a topological space was already shown by Thomason (see Theorem 4.11) in the course of his establishing a $\mathbf{D}_{\mathrm{qc}}(X)$ analogue of the Hopkins–Smith theorem (see Theorems 4.12 and 4.10).

Apriori, it is not clear whether $\mathfrak{S}(\mathcal{S})$ is triangulated or not. However, Neeman proves:

Theorem 4.24 [110, Def. 2.10, Th. 2.11] *$\mathfrak{S}(\mathcal{S})$ becomes a triangulated category with the distinguished triangles of the form* $\underrightarrow{\operatorname{colim}} Y(A_i \xrightarrow{f_i} B_i \xrightarrow{g_i} C_i \xrightarrow{h_i} \Sigma A_i)$, *where* $(A_* \xrightarrow{f_*} B_* \xrightarrow{g_*} C_* \xrightarrow{h_*} \Sigma A_*)$ *is a Cauchy sequence of triangles in* $\mathcal{S}$.

- [108, Th. 8.8] With the metrics as above, we have triangulated equivalences
  1. $\mathfrak{S}(\mathcal{T}^c) = \mathcal{T}^b_c$.
  2. If $\mathcal{T}$ is noetherian then $\mathfrak{S}\left([\mathcal{T}^b_c]^{\text{op}}\right) = [\mathcal{T}^c]^{\text{op}}$.
- [104, Ex. 3.6] The above theory works when X is separated and quasi-compact: If X is separated and quasi-compact, $\mathcal{T} = \mathbf{D}_{\text{qc}}(X)$ is approximable with the standard t-structure in the preferred equivalence class.
- Consequently, we obtain our desired result:

  When X is separated and quasi-compact, we have the following:
  1. $\mathfrak{S}(\mathbf{D}^{\text{perf}}(X)) = \mathbf{D}^b_{\text{coh}}(X)$.
  2. If X is further noetherian, $\mathfrak{S}\left(\left[\mathbf{D}^b_{\text{coh}}(X)\right]^{\text{op}}\right) = \left[\mathbf{D}^{\text{perf}}(X)\right]^{\text{op}}$.

For the rest of this section, we explain the concepts of "approximable", "noetherian approximable", "metric", "preferred t-structure", and "Cauchy sequence", which were put in a black box in the above summary. We urge readers to consult Neeman's own survey [108] for more details about the approximable triangulated categories.

Now, it is rather straightforward to define "metric" and "Cauchy sequence".

Definition 4.25 [110, Def. 1.2] [108, Def. 8.3] A *metric* on a triangulated category $\mathcal{S}$ is a sequence of additive subcategories $\{\mathcal{M}_i, i \in \mathbb{N}\}$, satisfying:

1. $\mathcal{M}_{i+1} \subset \mathcal{M}_i$ for every $i \in \mathbb{N}$.
2. Any $b \in \mathcal{S}$, with a distinguished triangle $a \to b \to c$ s.t. $a, c \in \mathcal{M}_i$, belongs to $\mathcal{M}_i$.

Definition 4.26 [110, Def. 1.6] [108, Def. 8.5] A *Cauchy sequence* in $\mathcal{S}$, a triangulated category with a metric $\{\mathcal{M}_i\}$, is a sequence

$$E_1 \to E_2 \to E_3 \to \cdots$$

such that, for any $i \in \mathbb{N}, j \in \mathbb{Z}$, there exists $M \in \mathbb{N}$ such that,

$$\operatorname{Cof}(E_m \to E_{m'}) \in \Sigma^{-j}\mathcal{M}_i$$

for any $m' > m \geq M$.

Next, we aim at "preferred t-structure", but we shall make a little detour for some later purpose.

Definition 4.27 [108, Rem. 3.1] Let $\mathcal{A}$ be a full subcategory of a category $\mathcal{T}$. Define the full subcategories add $\mathcal{A}$, Add $\mathcal{A}$, and smd $\mathcal{A}$ as follows.

1. Assume $\mathcal{T}$ has finite coproducts. add $\mathcal{A}$ consists of all finite coproducts of objects in $\mathcal{A}$.
2. Assume $\mathcal{T}$ has coproducts. Add $\mathcal{A}$ consists of all coproducts of objects in $\mathcal{A}$.
3. smd $\mathcal{A}$ consists of all direct summands in $\mathcal{T}$ of objects in $\mathcal{A}$.

The following construction will play major roles:

Definition 4.28 [108, Def. 3.3] [109, Rem. 0.1] Given $\mathcal{A} \subset \mathcal{T}$, a full subcategory of a triangulated category, and possibly infinite integers $m \leq n$, define the full subcategories:

1. $\mathcal{A}[m, n] = \cup_{i=m}^{n} \mathcal{A}[-i]$.
2. For $l \in \mathbb{N}$, define inductively the full subcategory $\langle \mathcal{A} \rangle_l^{[m,n]}$ (resp. $\overline{\langle \mathcal{A} \rangle}_l^{[m,n]}$ if $\mathcal{T}$ has coproducts) as follows.

   a. $\langle \mathcal{A} \rangle_1^{[m,n]} = \mathrm{smd}(\mathrm{add}\, \mathcal{A}[m, n])$ (resp. $\overline{\langle \mathcal{A} \rangle}_1^{[m,n]} = \mathrm{smd}(\mathrm{Add}\, \mathcal{A}[m, n])$)
   b. $\langle \mathcal{A} \rangle_{l+1}^{[m,n]} = \mathrm{smd}(\langle \mathcal{A} \rangle_1^{[m,n]} * \langle \mathcal{A} \rangle_l^{[m,n]})$ (resp. $\overline{\langle \mathcal{A} \rangle}_{l+1}^{[m,n]} = \mathrm{smd}(\overline{\langle \mathcal{A} \rangle}_1^{[m,n]} * \overline{\langle \mathcal{A} \rangle}_l^{[m,n]})$).

3. For the case $m = -\infty$, $n = \infty$ and $l \in \mathbb{N}$, following Bondal–Van den Bergh [20], we shall simply denote as follows[52]:

$$\langle \mathcal{A} \rangle_l := \langle \mathcal{A} \rangle_l^{[-\infty,\infty]} \quad (\textit{resp.}\ \overline{\langle \mathcal{A} \rangle}_l := \overline{\langle \mathcal{A} \rangle}_l^{[-\infty,\infty]})$$

Whereas the above definition might look complicated, its major part is reflected in the following simpler definition:

Definition 4.29 [109, Def. 1.3] Given $\mathcal{A} \subset \mathcal{T}$, a full subcategory of a triangulated category, and $l \in \mathbb{N}$, define inductively the full subcategory $\mathrm{coprod}_l(\mathcal{A})$ (resp. $\mathrm{Coprod}_l(\mathcal{A})$ if $\mathcal{T}$ has coproducts) as follows.

1. $\mathrm{coprod}_1(\mathcal{A}) = \mathrm{add}(\mathcal{A})$ (resp. $\mathrm{Coprod}_1(\mathcal{A}) = \mathrm{Add}(\mathcal{A})$,
2. $\mathrm{coprod}_{l+1}(\mathcal{A}) = \mathrm{coprod}_1(\mathcal{A}) * \mathrm{coprod}_l(\mathcal{A})$ (resp. $\mathrm{Coprod}_{l+1}(\mathcal{A}) = \mathrm{Coprod}_1(\mathcal{A}) * \mathrm{Coprod}_l(\mathcal{A})$.)

The key for Definition 4.29 to reflect a major part of Definition 4.28 is the following elementary observation of Bondal–Van den Bergh [20]:

Lemma 4.30 [20, Lem. 2.2.1] *Let $\mathcal{A}$ and $\mathcal{B}$ be full subcategories of a triangulated category with small coproducts. Then:*

[52] It was Neeman's insight to notice surprising usefullness of introducing related categories $\langle \mathcal{A} \rangle_l^{[m,n]}$ and $\overline{\langle \mathcal{A} \rangle}_l^{[m,n]}$ as well.

(1) $\mathrm{smd}(\mathcal{A}) * \mathcal{B} \subset \mathrm{smd}(\mathcal{A} * \mathcal{B}), \quad \mathcal{A} * \mathrm{smd}(\mathcal{B}) \subset \mathrm{smd}(\mathcal{A} * \mathcal{B})$;
(2) $\mathrm{smd}\,(\mathrm{smd}(\mathcal{A}) * \mathcal{B}) = \mathrm{smd}\,(\mathcal{A} * \mathrm{smd}(\mathcal{B})) = \mathrm{smd}(\mathcal{A} * \mathcal{B})$.

To show the first inclusion of (1): $\mathrm{smd}(\mathcal{A}) * \mathcal{B} \subset \mathrm{smd}(\mathcal{A} * \mathcal{B})$, pick $x \in \mathrm{smd}(\mathcal{A}) * \mathcal{B}$ fitting in a triangle:

$$s \to x \to b \quad (s \in \mathrm{smd}(\mathcal{A}), b \in \mathcal{B}),$$

for which we pick $s' \in \mathcal{T}$ with $s \oplus s' \in \mathcal{A}$ and form a new triangle:

$$s \oplus s' \to x \oplus s' \to b.$$

This shows the desired $x \in \mathrm{smd}(\mathcal{A} * \mathcal{B})$. The second inclusion of (1): $\mathcal{A} * \mathrm{smd}(\mathcal{B}) \subset \mathrm{smd}(\mathcal{S} * \mathcal{B})$ is shown similarly. Then (2) follows immediately from (1).

Using Lemma 4.30, we can easily prove, by induction on l, the following transparent expression relating Definition 4.28 with Definition 4.29.

Corollary 4.31 (c.f. [109, Cor. 1.11]) *Given $\mathcal{A} \subset \mathcal{T}$, a full subcategory of a triangulated category, a natural number $l \in \mathbb{N}$, and possibly infinite integers $m \leq n$,*

$$\langle \mathcal{A} \rangle_l^{[m,n]} = \mathrm{smd}\left(\mathrm{coprod}_l\, \mathcal{A}[m,n]\right), \qquad \overline{\langle \mathcal{A} \rangle}_l^{[m,n]} = \mathrm{smd}\left(\mathrm{Coprod}_l\, \mathcal{A}[m,n]\right).$$

The following Proposition 4.32 follows immediately by combining the second equality of Corollary 4.31 and Lemma 4.33 below. Philosophically Proposition 4.32 may be viewed as saying that $\overline{\langle - \rangle}_l$ and $\mathrm{Coprod}_l(-)$ are interchangeable.

Proposition 4.32 (c.f. [109, Cor. 1.11]) *Given $\mathcal{A} \subset \mathcal{T}$, a full subcategory of a triangulated category, a natural number $l \in \mathbb{N}$, and possibly infinite integers $m \leq n$,*

$$\mathrm{Coprod}_l\,(\mathcal{A}[m,n]) \subseteq \overline{\langle \mathcal{A} \rangle}_l^{[m,n]} \subseteq \mathrm{Coprod}_{2l}\,(\mathcal{A}[m-1,n]).$$

We include a proof of the following Lemma 4.33, to highlight the point at which infinite coproducts are used. Just in case the reader is wondering: the finite analogue of Proposition 4.32 is false. While the inclusion $\mathrm{coprod}_l\,(\mathcal{A}[m,n]) \subseteq \langle \mathcal{A} \rangle_l^{[m,n]}$ is true and easy, it isn't in general true that $\langle \mathcal{A} \rangle_l^{[m,n]} \subseteq \mathrm{coprod}_{2l}\,(\mathcal{A}[m-1,n])$.

Lemma 4.33 (c.f. [109, Lem. 1.9]) *Let $\mathcal{B}$ a subcategory of $\mathcal{T}$, a triangulated category with coproducts, and $l \in \mathbb{N}$. Then*

$$\mathrm{Coprod}_l(\mathcal{B}) \subseteq \mathrm{smd}\left(\mathrm{Coprod}_l(\mathcal{B})\right) \subseteq \mathrm{Coprod}_{2l}(\mathcal{B}[-1,0]).$$

Proof The first inclusion is obvious. For the second inclusion, recall from Remark 2.3 (i) that

$$\forall x \in \mathrm{smd}\left(\mathrm{Coprod}_l(\mathcal{B})\right), \quad \exists b \in \mathrm{Coprod}_l(\mathcal{B}) \text{ and an idempotent } \; e : b \to b, \text{ s.t. } x = eb = \mathrm{Cone}\,(\oplus_{\mathbb{N}} b \to \oplus_{\mathbb{N}} b).$$

From this, we obtain the following triangle:

$$\oplus_{\mathbb{N}} b \to \oplus_{\mathbb{N}} b \to x \to \Sigma\left(\oplus_{\mathbb{N}} b\right),$$

where $\oplus_{\mathbb{N}} b \in \text{Add}\left(\text{Coprod}_l(\mathcal{B})\right) = \text{Coprod}_l(\mathcal{B})$ and so $\Sigma\left(\oplus_{\mathbb{N}} b\right) \in \Sigma\, \text{Coprod}_l(\mathcal{B}) = \text{Coprod}_l(\Sigma\mathcal{B})$. Thus,

$$x \in \text{Coprod}_l(\mathcal{B}) * \text{Coprod}_l(\Sigma\mathcal{B}) \subseteq \text{Coprod}_l(\mathcal{B} \cup \Sigma\mathcal{B}) * \text{Coprod}_l(\mathcal{B} \cup \Sigma\mathcal{B}) \subseteq \text{Coprod}_{2l}(\mathcal{B}\Sigma\mathcal{B}).$$

□

The constructions $\langle - \rangle_l$ and $\overline{\langle - \rangle}_l$ are older than $\text{coprod}_l(-)$ and $\text{Coprod}_l(-)$, and for most purposes they work just fine. But there are results which become much easier to prove by working with $\text{coprod}_l(-)$ and $\text{Coprod}_l(-)$; for example the reader can look at the proof of [24, Lem. 4.4].[53] Thus one way to view the difference is to regard $\text{coprod}_l(-)$ and $\text{Coprod}_l(-)$ as technically more powerful than the older $\langle - \rangle_l$ and $\overline{\langle - \rangle}_l$.

Now, in practice, as their constructions suggest, coprod_l (resp. Coprod_l) are more tractible than $\langle \mathcal{A} \rangle_l^{[m,n]}$ (resp. $\overline{\langle \mathcal{A} \rangle}_l^{[m,n]}$). However, $\langle \mathcal{A} \rangle_l^{[m,n]}$ (resp. $\overline{\langle \mathcal{A} \rangle}_l^{[m,n]}$) occurs more frequently, for instance,

Theorem 4.34 [1, Th. A] (See also [104, Ex. 0.13]) *For a triangulated category $\mathcal{T}$ with coproducts and a compact generator $G \in \mathcal{T}$, there is a unique t-structure of the following form:*

$$\left(\mathcal{T}_G^{\leq 0}, \mathcal{T}_G^{\geq 0}\right) := \left(\overline{\langle G \rangle}^{[-\infty,0]}, \left(\overline{\langle G \rangle}^{[-\infty,0]}\right)^{\perp}[1]\right).$$

Definition 4.35 [108, Def. 7.3, Rem. 7.4]

1. Two t-structures $\left(\mathcal{T}_1^{\leq 0}, \mathcal{T}_1^{\geq 0}\right)$ and $\left(\mathcal{T}_2^{\leq 0}, \mathcal{T}_2^{\geq 0}\right)$ are called *equivalent*, if there exists $A \in \mathbb{N}$ with
$$\mathcal{T}_1^{\leq -A} \subset \mathcal{T}_2^{\leq 0} \subset \mathcal{T}_1^{\leq A}.$$
2. For a triangulated category $\mathcal{T}$ with coproducts and a compact generator, a t-structure $\left(\mathcal{T}^{\leq 0}, \mathcal{T}^{\geq 0}\right)$ is in the *preferred equivalence class* if it is equivalent to $\left(\mathcal{T}_G^{\leq 0}, \mathcal{T}_G^{\geq 0}\right)$ for some compact generator G (in fact, for every compact generator).

The importance of "preferred equivalence class" is that $\mathcal{T}^-$, $\mathcal{T}^+$, and $\mathcal{T}^b$, recalled in the next definition, are independent of the particular representative $(\mathcal{T}^{\leq 0}, \mathcal{T}^{\geq 0})$ in the preferred equivalence class [108, Fact. 0.5.(iii)]:

Definition 4.36 [108, Def. 7.5, Def. 7.6]

---

[53] The author is grateful to Professor Neeman for this reference.

1. Given a t-structure $(\mathcal{T}^{\le 0}, \mathcal{T}^{\ge 0})$, we have the usual subcategories:
$$\mathcal{T}^- = \cup_n \mathcal{T}^{\le n}, \quad \mathcal{T}^+ = \cup_n \mathcal{T}^{\ge n}, \quad \mathcal{T}^b = \mathcal{T}^- \cap \mathcal{T}^+.$$
2. For a triangulated category $\mathcal{T}$ with coproducts and a compact generator, choose a t-structure $(\mathcal{T}^{\le 0}, \mathcal{T}^{\ge 0})$ in the preferred equivalence class, define the full subcategories $\mathcal{T}_c^-$ and $\mathcal{T}_c^b$ as follows:
$$\mathcal{T}_c^- := \left\{ F \in \mathcal{T} \;\middle|\; \begin{array}{c}\text{For any } n\in\mathbb{N} \text{ there exists a triangle} \\ E\to F\to D\to E[1] \\ \text{with } E \text{ compact and } D\in\mathcal{T}^{\le -n-1}\end{array} \right\}, \qquad \mathcal{T}_c^b := \mathcal{T}^b \cap \mathcal{T}_c^-$$

Intuitively, $\mathcal{T}_c^-$ is the closure, with respect to the metric $\mathcal{M}_i = \mathcal{T}^{\le -i}$, of $\mathcal{T}^c$.

$\mathcal{T}_c^-$ and $\mathcal{T}_c^b$ in the above definition do not depend on the choice of compact generator G and are both intrinsic [108, Rem. 7.7, Fact. 0.5.(iv)].

Now we are ready to state the fundamental concepts of "approximable" and "noetherian (approximable)" triangulated categories:

Definition 4.37 [104, Def. 0.21] [108, Def. 4.1] A triangulated category $\mathcal{T}$ with coproducts is called approximable if there exits a compact generator $G \in \mathcal{T}$, a t-structure $(\mathcal{T}^{\le 0}, \mathcal{T}^{\ge 0})$, and $A \in \mathbb{N}$ such that

1. $G[A] \in \mathcal{T}^{\le 0}$ and $\mathrm{Hom}(G[-A], \mathcal{T}^{\le 0}) = 0$.
2. For every object $F \in \mathcal{T}^{\le 0}$, there exists a triangle
$$E \to F \to D \to E[1],$$
with $D \in \mathcal{T}^{\le -1}$ and $E \in \overline{\langle G \rangle}_A^{[-A,A]}$.

From the definition, we find for any approximable triangulated category $\mathcal{T}$, the closure, with respect to the metric $\mathcal{M}_i = \mathcal{T}^{\le -i}$, of $\bigcup_n \overline{\langle G \rangle}_n^{[-n,n]}$ is nothing but $\mathcal{T}^-$. Thus we may intuitively say every object in $\mathcal{T}^-$ may be "Taylor approximable" regarding $\langle G \rangle_n^{[-n,n]}$ as consisting of "Taylor polynomials of G of degree $\le n$." [108, Dis. 0.1, Rem. 02].

Definition 4.38 [110, Def. 5.1] [108, Not. 8.9] Suppose $\mathcal{T}$ is a triangulated category with coproducts, and assume it has a compact generator G with $\mathrm{Hom}(G, \Sigma^i G) = 0$ for $i \gg 0$. We declare $\mathcal{T}$ to be noetherian if there exists $N \in \mathbb{N}$ and a t-structure $(\mathcal{T}^{\le 0}, \mathcal{T}^{\ge 0})$ in the preferred equivalence class, s.t.
$$\forall X \in \mathcal{T}_c^-, \ \exists \text{ triangle } A \to X \to B \text{ s.t. } A \in \mathcal{T}_c^- \cap \mathcal{T}^{\le 0}, \ B \in \mathcal{T}_c^- \cap \mathcal{T}^{\ge -N} = \mathcal{T}_c^b \cap \mathcal{T}^{\ge -N}.$$

Remark 4.39 (i) *The noetherian hypothesis is somewhat weaker than the assumption that there exists a t-structure in the preferred equivalence class which restricts to a t-structure on* $\mathcal{T}_c^-$.
(ii) [104, Fac. 0.23, Exa. 3.6] *For a quasicompact and separated scheme X, the standard t-structure on $\mathcal{T} = \mathbf{D}_{\mathrm{qc}}(X)$ is in the preferred equivalence class. Suppose further*

that X is noetherian, then $\mathcal{T}_c^- = \mathbf{D}_{\text{coh}}^-$, *the category of bounded-above complexes of coherent sheaves, and so, the standard t-structure, which is in the preferred equivalence class, on* $\mathcal{T} = \mathbf{D}_{\text{qc}}(X)$ *restricts to a t-structure on* $\mathcal{T} = \mathbf{D}_{\text{qc}}(X)$. *This implies* $\mathbf{D}_{\text{qc}}(X)$ *becomes noetherian in the sense of Definition 4.38, provided X is noetherian and separated. This is the origin of the terminology "noetherian" of Definition 4.38.*
(iii) WARNING! *The "noetherian" triangulated category of Definition 4.38 is nothing to do with the "Noetherian" stable homotopy category of [54, Def. 6.0.1].*

For instance, for the case of $\mathcal{T} = \mathcal{SH}$, *the stable homotopy category of spectra, it is easy to see* $\mathcal{T}_c^-$ *consists of those spectra X whose homotopy group* $\pi_i(X)$ *is a finitely generated abelian groups for each i and vanishes for* $i \ll 0$. *Thus, the standard t-structure, which is obviously in the preferred equivalence class, restricts to a t-structure on* $\mathcal{T}_c^-$. *This implies* $\mathcal{SH}$ *is noetherian in the sense of Definition 4.38 [104, Fac. 0.23].*

On the other hand, $\mathcal{SH}$ *is clearly NOT a Noetherian stable homotopy category in the sense of [54, Def. 6.0.1], for the graded ring of the stable homotopy category of spheres* $\pi_* S^0$ *is not a Noetherian graded commutative ring, which can be easily seen by applying the Nishida nilpotency, the precursor of (Devinatz-)Hopkins–Smith nilpotency.*

Then we have the following somewhat straightforward result to produce examples of approximable triangulated categories:

Proposition 4.40 [104, Ex. 3.3] *If* $\mathcal{T}$ *has a compact generator G, such that* $\text{Hom}(G, \Sigma^i G) = 0$ *for all* $i > 0$, *then* $\mathcal{T}$ *is approximable. Just take the t-structure* $\left(\mathcal{T}_G^{\leq 0}, \mathcal{T}_G^{\geq 0}\right)$ *of Theorem 4.34 with* $A = 1$.

From this, we immediately see the stable homotopy category $\mathcal{SH}$ is approximable. (actually noetherian, as was remarked in Remark 4.39 (iii)).

Our principal example of approximable triangulated categories is supplied by the following theorem:

Theorem 4.41 [104, Ex. 3.6] *Let X be a quasicompact, separated*[54] *scheme. Then the category* $\mathbf{D}_{\text{qc}}(X)$ *is approximable. (actually noetherian if X is further noetherian, as was remarked in Remark 4.39 (ii)).*

The proof is very involved and we urge readers to consult Neeman's original paper [104].

For now, we shall record the following application of approximability:

Corollary 4.42 [108, Lem. 6.5] [109, Th. 0.18] *Let X be a quasicompact, separated scheme, let* $G \in \mathbf{D}_{\text{qc}}(X)$ *be a compact generator, and let* $u : U \to X$ *be an open immersion with U quasicompact. Then*

$$\exists n \in \mathbb{N} \text{ s.t. } \mathbf{R}u_* \mathcal{O}_U \in \overline{\langle G \rangle}_n^{[-n,n]} \subset \mathbf{D}_{\text{qc}}(X).$$

[54]Unlike (14) and Theorem 4.17, the general case (where X is quasicompact and quasiseparated) is still open—see [104, Just above Lem. 3.5].

Proof (*Outline of the proof of Corollary* 4.42 *using approximability presented in* [108])

Step 1: $\exists l \in \mathbb{N}$ s.t. $\mathrm{Hom}\left(\mathbf{R}u_*\mathcal{O}_U, \mathbf{D}_{\mathrm{qc}}(X)^{\le -l}\right) = 0$.
Step 2 (This is where the approximability of $\mathbf{D}_{\mathrm{qc}}(X)$ is used!) : By the approximability of $\mathbf{D}_{\mathrm{qc}}(X)$,[55] $\exists n \in \mathbb{N}$ and a triangle:

$$E \to \mathbf{R}u_*\mathcal{O}_U \to D$$

with $D \in \mathbf{D}_{\mathrm{qc}}(X)^{\le -l}$ and $E \in \overline{\langle G\rangle}_n^{[-n,n]}$.
Step 3: From Step 1 and Step 2, the map $\mathbf{R}u_*\mathcal{O}_U \to D$ in Step 2 is 0, which implies $\mathbf{R}u_*\mathcal{O}_U$ is a direct summand of $E \in \overline{\langle G\rangle}_n^{[-n,n]}$, as desired. □

For details about the approximable triangulated categories. Consult Neeman's own survey [108].

5 Strong Generation in Derived Categories of Schemes

In the previous section, we saw $\mathbf{D}^{\mathrm{perf}}(X)$ and $\mathbf{D}^b_{\mathrm{coh}}(X)$ carry rich information and are intimately related to each other. In this section, we would like to investigate the important "strong generation" property, in the sense of Bondal and Van den Bergh [20], for $\mathbf{D}^{\mathrm{perf}}(X)$ and $\mathbf{D}^b_{\mathrm{coh}}(X)$, via approximable triangulated category techniques.

For this purpose, we have to start with what we mean by a "generator" of $\mathbf{D}^{\mathrm{perf}}(X)$ and $\mathbf{D}^b_{\mathrm{coh}}(X)$, because our previous definition of a generator in Definition 2.13 only works for triangulated categories with small coproducts, which $\mathbf{D}^{\mathrm{perf}}(X)$ and $\mathbf{D}^b_{\mathrm{coh}}(X)$ are not.

Definition 5.1 [108, Expl. 5.4] Let G be an element of a triangulated category $\mathcal{S}$. Then, in the notation of Definition 4.28,

1. G is called a *classical generator* if $\mathcal{S} = \cup_n \langle G\rangle_n^{[-n,n]}$.
2. G is called a *strong generator* if there exists an integer $l > 0$ with $\mathcal{S} = \cup_n \langle G\rangle_l^{[-n,n]}$. In this case, $\mathcal{S}$ is called *strongly generated.*

With this opportunity, let us record the following important concept intimately related to the above definition:

Definition 5.2 [125, Def. 3.2] The *Rouquier dimension* of a triangulated category $\mathcal{S}$, denoted by $\dim \mathcal{S}$, is the smallest d for which there exists $G \in \mathcal{S}$ with $\mathcal{S} = \cup_n \langle G\rangle_{d+1}^{[-n,n]}$.

[55] There is some subtlety here. See e.g. [108, footnote 4 in Proof of Lem. 5; Sketch 7.19.(i)].

Remark 5.3 (i) Rouquier [125] proved the following properties of the Rouquier dimension of $\mathbf{D}^b_{\mathrm{coh}}(X)$:

- [125, Prop. 7.9] For a smooth quasiprojective scheme X over a field, we have $\dim \mathbf{D}^b_{\mathrm{coh}}(X) \leq 2 \dim X$.
- [125, Prop. 7.16] For a reduced separated scheme X of finite type over a field, $\dim \mathbf{D}^b_{\mathrm{coh}}(X) \geq \dim X$.
- [125, Th. 7.17] For a smooth affine scheme X of finite type over a field, $\dim \mathbf{D}^b_{\mathrm{coh}}(X) = \dim X$.

(ii) For a sample of examples of Rouquier dimension in affine case, see [30, 31, 59] for instance.

On the other hand, Neeman deduces strong generation of $\mathbf{D}^{\mathrm{perf}}(X)$ and $\mathbf{D}^b_{\mathrm{coh}}(X)$ from some properties of $\mathbf{D}_{\mathrm{qc}}(X)$:

Definition 5.4 Let X be a separated scheme.

1. $\mathbf{D}_{\mathrm{qc}}(X)$ is called *strongly compactly generated* if there exists $G \in \mathbf{D}^{\mathrm{perf}}(X)$ and integer $l > 0$ with $\mathbf{D}_{\mathrm{qc}}(X) = \overline{\langle G \rangle}_l^{(-\infty,\infty)}$.
2. $\mathbf{D}_{\mathrm{qc}}(X)$ is called *strongly boundedly generated* if there exists $G \in \mathbf{D}^b_{\mathrm{coh}}(X)$ and integer $l > 0$ with $\mathbf{D}_{\mathrm{qc}}(X) = \overline{\langle G \rangle}_l^{(-\infty,\infty)}$.

Remark 5.5 From Proposition 4.32, we may replace the required equality $\mathbf{D}_{\mathrm{qc}}(X) = \overline{\langle G \rangle}_l^{(-\infty,\infty)}$ showing up twice in Definition 5.4 with more tractible $\mathbf{D}_{\mathrm{qc}}(X) = \mathrm{Coprod}_l\,(G(-\infty,\infty))$ (of course, l here is a doubling of old l.).

Theorem 5.6 *Let X be a separated scheme.*

1. *[109, Proof of Lem. 2.2] If $\mathbf{D}_{\mathrm{qc}}(X)$ is strongly compactly generated, then $\mathbf{D}^{\mathrm{perf}}(X)$ is strongly generated.*
2. *[109, Proof of Lem. 2.7] Suppose X is noetherian. If $\mathbf{D}_{\mathrm{qc}}(X)$ is strongly boundedly generated, then $\mathbf{D}^b_{\mathrm{coh}}(X)$ is strongly generated.*

To prove these claims, the following observation is crucial:

Lemma 5.7 (i) *[109, Prop. 1.8.(i)] Let $\mathcal{T}$ be a triangulated category with coproducts, and let $\mathcal{B}$ be a subcategory of $\mathcal{T}^c$. Then, for any $l \in \mathbb{N}$,*

$$\mathcal{T}^c \cap \mathrm{Coprod}_l(\mathcal{B}) \subseteq \mathrm{smd}\left(\mathrm{coprod}_l(\mathcal{B})\right).$$

(ii) *[109, Lem. 2.6] Let X be a noetherian scheme, and let G be an object in $\mathbf{D}^b_{\mathrm{coh}}(X)$. Then, for any $l \in \mathbb{N}$,*

$$\mathbf{D}^b_{\mathrm{coh}}(X) \cap \mathrm{Coprod}_l(G(-\infty,\infty)) \subseteq \mathrm{smd}\left(\mathrm{coprod}_{2l}(G(-\infty,\infty))\right).$$

Of course, we are going to apply (i) with $\mathcal{T} = \mathbf{D}_{\mathrm{qc}}(X)$, $\mathcal{B} = G(-\infty, \infty) \subseteq \mathcal{T}^c = \mathbf{D}^{\mathrm{perf}}(X)$. Then (i) becomes

$$\mathbf{D}^{\mathrm{perf}}(X) \cap \mathrm{Coprod}_l(G(-\infty, \infty)) \subseteq \mathrm{smd}\left(\mathrm{coprod}_l(G(-\infty, \infty))\right),$$

a clear analogue of (ii).

However, the point is that we can not prove (ii) with a generality like (i). In fact, while the proof of (i) is somewhat straightforward, the proof of (ii) is more involved. For instance (see [109, Proof of Lem. 2.4]), the "phantom ideal" $\mathcal{I}$, consisting of those maps $f : x \to y$ such that any composite $\Sigma^i G \to x \xrightarrow{f} y$ vanishes for any $i \in \mathbb{Z}$ and any map $\Sigma^i G \to x$ is studied carefully, resorting Christensen's phantom map theory:

Theorem 5.8 [29, Th. 1.1] *Suppose $(\mathcal{P}, \mathcal{I})$ is a projective class of a triangulated category $\mathcal{T}$, i.e. $\mathcal{P}$ is a collection of objects in $\mathcal{T}$, $\mathcal{I}$ is a collection of maps in $\mathcal{T}$, such that*

- *$\mathcal{P} - null = \mathcal{I}$, where $\mathcal{P} -$ null is the collection of "$\mathcal{P}$-phantom maps", i.e. those maps $x \to y$ such that the composite $p \to x \to y$ is zero for all objects $p \in \mathcal{P}$ and all maps $p \to x$. (This condition makes $\mathcal{I}$ an ideal.)*
- *$\mathcal{I} - proj = \mathcal{P}$, where $\mathcal{I} -$ proj is the collection of all objects p such that the composite $p \to x \to y$ is zero for all maps $x \to y$ in $\mathcal{I}$ and all maps $p \to x$.*
- *For any object $x \in \mathcal{T}$, there exists a triangle $p \to x \to y$ with $p \in \mathcal{P}$ and $x \to y$ in $\mathcal{I}$.*

Then, for any $n \in \mathbb{N}$, $(\mathcal{P}_n, \mathcal{I}^n)$ is also a projective class, where $\mathcal{I}^n$ is the n-th power of the "phantom ideal" $\mathcal{I}$, and $\mathcal{P}_n = \langle \mathcal{P} \rangle_n$, which is by defined inductively analogous to Definition 4.28:

$$\langle \mathcal{P} \rangle_1 = \mathcal{P}, \quad \langle \mathcal{P} \rangle_{l+1} = \mathrm{smd}\left(\langle \mathcal{P} \rangle_1 * \langle \mathcal{P} \rangle_l\right).$$

But, we also need some algebro-geometric input also to prove (ii) (see [109, Lem. 2.5] [83, Th. 4.1]).

Anyway, assuming Lemma 5.7, the proof of Theorem 5.6 becomes straightforward:

Proof *(Proof of Theorem* 5.6 *assuming Lemma* 5.7) In both cases, assuming the respective assumption on $\mathbf{D}_{\mathrm{qc}}(X)$, together with Remark 5.5, the claims follow as follows:

$$\begin{aligned}
\mathbf{D}^{\mathrm{perf}}(X) &= \mathbf{D}^{\mathrm{perf}}(X) \cap \mathbf{D}_{\mathrm{qc}}(X) = \mathbf{D}^{\mathrm{perf}}(X) \cap \mathrm{Coprod}_l(G(-\infty, \infty)) \\
&\subseteq \mathrm{smd}\left(\mathrm{coprod}_l(G(-\infty, \infty))\right) \subseteq \cup_n \langle G \rangle_l^{[-n,n]}. \\
\mathbf{D}^b_{\mathrm{coh}}(X) &= \mathbf{D}^b_{\mathrm{coh}}(X) \cap \mathbf{D}_{\mathrm{qc}}(X) = \mathbf{D}^b_{\mathrm{coh}}(X) \cap \mathrm{Coprod}_l(G(-\infty, \infty)) \\
&\subseteq \mathrm{smd}\left(\mathrm{coprod}_{2l}(G(-\infty, \infty))\right) \subseteq \cup_n \langle G \rangle_{2l}^{[-n,n]}.
\end{aligned}$$

□

5.1 Strong Generation of $\mathbf{D}^{\mathrm{perf}}(X)$

From Theorem 5.6(1), we search for situations when $\mathbf{D}_{\mathrm{qc}}(X)$ becomes strongly compactly generated:

Theorem 5.9 (Max Kelly [70]) *Suppose $X = \operatorname{Spec} R$ is affine. Then $\mathbf{D}_{\mathrm{qc}}(X)$ is strongly compactly generated if and only if R is of finite global dimension.*

Theorem 5.10 (Bondal–Van den Bergh) *[20] Let X be smooth scheme of finite type over a field k. Then $\mathbf{D}_{\mathrm{qc}}(X)$ is strongly compactly generated.*

Theorem 5.10 has recently been improved by Orlov as a characterization of the strong generation of $\mathbf{D}^{\mathrm{perf}}(X)$:

Theorem 5.11 (Orlov [120, Th. 3,27]) *Let X be a separated noetherian scheme of finite Krull dimension over an arbitrary field k. Assume that the square $X \times X$ is noetherian too. Then the following conditions are equivalent:*

1. *X is regular;*
2. *$\mathbf{D}^{\mathrm{perf}}(X)$ is strongly generated.*

It was this paper of Orlov [120] which motivated Neeman to develop his theory of approximable triangulated category (see e.g. [109, p. 6, the paragraph before Rem. 0.10]).

In fact, the approximability of $\mathbf{D}_{\mathrm{qc}}(X)$ allowed Neeman to prove the following statement by reducing to the Kelly's old theorem in a straightforward way, i.e. by induction on the number of open affines covering X:

Theorem 5.12 (Neeman [109, Th. 2.1]) *Let X be a quasi-compact separated scheme. If X can be covered by open affines* Spec R_i *with R_i of finite global dimension, then $\mathbf{D}_{\mathrm{qc}}(X)$ is strongly compactly generated.*

Proof (*Outline of a proof of Theorem 5.12 following* [108, Sketch.6.6]) Proceed as follows:

- Write $X = \cup_{1 \le i \le r} U_i$ with $u_i : U_i = \operatorname{Spec}(R_i)$, by assumption.
- By induction on r unsing the Mayer Vietoris sequence [125, Prop. 5.10] (as in the proof given in [109, Proof of Theorem 2.1]), we find

$$\mathbf{D}_{\mathrm{qc}}(X) = \underbrace{\left(\operatorname{add}\left[\cup_{i=1}^{r} \mathbf{R}u_{i*}\,\mathbf{D}_{\mathrm{qc}}(U_i)\right]\right) * \left(\operatorname{add}\left[\cup_{i=1}^{r} \mathbf{R}u_{i*}\,\mathbf{D}_{\mathrm{qc}}(U_i)\right]\right) * \cdots * \left(\operatorname{add}\left[\cup_{i=1}^{r} \mathbf{R}u_{i*}\,\mathbf{D}_{\mathrm{qc}}(U_i)\right]\right)}_{r \text{ copies}}. \tag{51}$$

- By a minor variant of Max Kelly's Theorem 5.9,

$$\exists l \in \mathbb{N}, \text{ s.t. } \ 1 \le \forall i \le r, \quad \mathbf{D}_{\mathrm{qc}}(U_i) = \overline{\langle \mathcal{O}_{U_i} \rangle}_l^{(-\infty,\infty)}. \tag{52}$$

- From Corollary 4.42 (recall <u>*this is where the approximability of*</u> $\mathbf{D}_{qc}(X)$ <u>*was exploited*</u>),

$$\exists n \in \mathbb{N} \text{ s.t. } 1 \leq \forall i \leq r, \quad \mathbf{R}u_{i*}\mathcal{O}_{U_i} \in \overline{\langle G \rangle}_n^{[-n,n]} \subset \mathbf{D}_{qc}(X). \tag{53}$$

- From (52) and (53),

$$\mathbf{R}u_{i*}\,\mathbf{D}_{qc}(U_i) = \mathbf{R}u_{i*}\left[\overline{\langle \mathcal{O}_{U_i} \rangle}_l^{(-\infty,\infty)}\right] \subset \overline{\langle \mathbf{R}u_{i*}\mathcal{O}_{U_i} \rangle}_l^{(-\infty,\infty)} \subset \overline{\langle G \rangle}_{ln}^{[-\infty,\infty]},$$

  and so

$$\operatorname{add}\left[\cup_{i=1}^r \mathbf{R}u_{i*}\,\mathbf{D}_{qc}(U_i)\right] \subset \overline{\langle G \rangle}_{ln}^{[-\infty,\infty]}, \tag{54}$$

- From (51) and (54), we obtain the desired strong compact generation of $\mathbf{D}_{qc}(X)$:

$$\mathbf{D}_{qc}(X) = \overline{\langle G \rangle}_{lnr}^{[-\infty,\infty]},$$

□

Now, Neeman proves his main theorem on strong generation of $\mathbf{D}^{\text{perf}}(X)$:

Theorem 5.13 (Neeman [109, Th. 0.5] [108, Th. 6.1]) *Let X be a quasi-compact separated scheme. Then $\mathbf{D}^{\text{perf}}(X)$ is strongly generated if and only if X can be covered by open affines* Spec R_i *with R_i of finite global dimension.*

Proof <u>"if" part:</u> This is immediate from Theorem 5.12 and Theorem 5.6(1).
<u>"only if" part:</u> [109, Rem. 0.10] By Thomason–Trobaugh [130] recalled in Theorem 4.17 and (47), we have an equivalence upon idempotent completion:

$$\left(\mathbf{D}^{\text{perf}}(X) / \left(\mathbf{D}^{\text{perf}}\right)_Z(X)\right)^\sharp \underset{\cong}{\xrightarrow{\mathbf{L}j^*}} \mathbf{D}^{\text{perf}}(U).$$

Thus, if $G \in \mathbf{D}^{\text{perf}}(X)$ is a strong generator, then so is $\mathbf{L}j^*G \in \mathbf{D}^{\text{perf}}(U)$. Now the strong generation of an affine $U = \text{Spec}(R)$ forces R to be of finite global dimension, as is shown in [125, Prop. 7.25]. □

5.2 *Strong Generation of* $\mathbf{D}^b_{\text{coh}}(X)$

Here, we start with a nice theorem of Rouquier:

Theorem 5.14 (Rouquier [125, Th. 7.39]) *Let X be a scheme of finite type over a perfect field k. Then $\mathbf{D}_{qc}(X)$ is strongly boundedly generated, and $\mathbf{D}^b_{\text{coh}}(X)$ is strongly generated.*

To go further, let us recall:

- the canonical map $\mathbf{D}^{\mathrm{perf}}(X) \to \mathbf{D}^b_{\mathrm{coh}}(X)$ is an isomorphism when X is smooth over a field, and in this case, the strong generation of $\mathbf{D}^b_{\mathrm{coh}}(X) \cong \mathbf{D}^{\mathrm{perf}}(X)$ is already discussed in the previous subsection.
- the Verdier quotient $\mathbf{D}_{sg}(X) = \mathbf{D}^b_{\mathrm{coh}}(X)/\mathbf{D}^{\mathrm{perf}}(X)$ reflects singular information of X.

Thus, we must take care of singular property of X. However, while Theorem 5.13 is easy and classical in the case where X is affine, this problem is *neither easy nor classical for affine* X. See [108, H.S..6.12] for more on this point.[56]

Now, for this purpose, Neeman turned his attention to de Jong's alteration[57]:

Definition 5.15 [32, 33, 117] [109, Remi.0.13] Let X be a noetherian scheme. A *regular alteration* of X is a proper, surjective morphism $f : Y \to X$, so that

1. Y is regular and finite dimensional.
2. There is a dense open set $U \subset X$ over which f is finite.

Now, Neeman proves:

Theorem 5.16 (Neeman [109, Th. 2.3]) *Let X be a noetherian scheme, and assume every closed subscheme $Z \subset X$ admits a regular alteration. Then $\mathbf{D}_{\mathrm{qc}}(X)$ is strongly boundedly generated.*

Proof (*Outline of a proof of Theorem* 5.16 *following* [109, *Proof that Theorem 2.3 follows from Theorem 2.1*])[58]: This is proved in the following order:

- Suppose there is a counterexample X to Theorem 5.16 (SBG criterion). Since X is noetherian, we may choose a minimal closed subscheme $Z \subset X$ which does not satisfy Theorem 5.16 (SBG criterion).
- Replacing X by Z, may assume all proper closed subschemes $Z \subset X$ satisfy Theorem 5.16 (SBG criterion).
- To prove Theorem 5.16 (SBG criterion) for X, we may assume it is reduced: for, let $j : X_{\mathrm{red}} \to X$ be the inclusion of the reduced part of X, and let $\mathcal{J}$ be the corresponding ideal sheaf with $\mathcal{J}^n = 0$. Then, expressing any $C \in \mathbf{D}_{\mathrm{qc}}(X)$ by a complex of quasi-coherent sheaves, we obtain a filtration

$$0 = \mathcal{J}^n C \subset \mathcal{J}^{n-1} C \subset \cdots \subset \mathcal{J} C \subset C,$$

[56] In fact, when X is affine, strong generation of $\mathbf{D}_{\mathrm{qc}}(X)$ has been proved by Iyengar and Takahashi [60] under different hypotheses, and using quite different techniques, from Neeman's Theorem 5.16. And they give examples where strong generation fails; see [60] and references therein.

[57] (Gabber's strengthening [40] of) de Jong's alteration is now widely used in the Morel–Voevodsky motivic stable homotopy theory. See e.g. [55, 71]. For an introductory review of de Jong's alteration, consult Oort's [117] for instance.

[58] This proof does not directly use the approximability of $\mathbf{D}_{\mathrm{qc}}(X)$, the approximability enters only indirectly, when we appeal to Theorem 5.10. What we want to highlight here, following a strong suggestion of Professor Neeman, is "the pivotal role that the homotopy-theoretical ideas of Bousfield, Ohkawa, Hopkins–Smith and many others play in the reduction."

with $\mathcal{J}^j C/\mathcal{J}^{j+1} \in \mathbf{R}j_* \mathbf{D}_{\mathrm{qc}}(X_{\mathrm{red}})$ $(0 \le \forall j \le n-1)$. Then, as in [125, 7.3], we find:

$$C \in \left[\mathbf{R}j_* \mathbf{D}_{\mathrm{qc}}(X_{\mathrm{red}})\right]^{*n} = \underbrace{\left[\mathbf{R}j_* \mathbf{D}_{\mathrm{qc}}(X_{\mathrm{red}})\right] * \left[\mathbf{R}j_* \mathbf{D}_{\mathrm{qc}}(X_{\mathrm{red}})\right] * \cdots * \left[\mathbf{R}j_* \mathbf{D}_{\mathrm{qc}}(X_{\mathrm{red}})\right]}_{n}.$$

So, it suffices to prove the strong bounded generation $\mathbf{D}_{\mathrm{qc}}(X_{\mathrm{red}}) = \mathrm{Coprod}_{\widetilde{N}}\left(\widetilde{G}(-\infty, \infty)\right)$ for some $\widetilde{N} \in \mathbb{N}$ and some $\widetilde{G} \in \mathbf{D}^b_{\mathrm{coh}}(X_{\mathrm{red}})$, for then we would get:

$$\begin{aligned}\mathbf{D}^b_{\mathrm{coh}}(X) &\subseteq \left[\mathbf{R}j_* \mathbf{D}_{\mathrm{qc}}(X_{\mathrm{red}})\right]^{*n} = \left[\mathbf{R}j_* \mathrm{Coprod}_{\widetilde{N}}\left(\widetilde{G}(-\infty, \infty)\right)\right]^{*n} \\ &\subseteq \left[\mathrm{Coprod}_{\widetilde{N}}\left((\mathbf{R}j_*\widetilde{G})(-\infty, \infty)\right)\right]^{*n} = \mathrm{Coprod}_{\widetilde{N}n}\left((\mathbf{R}j_*\widetilde{G})(-\infty, \infty)\right),\end{aligned}$$

where $\mathbf{R}j_*\widetilde{G} \in \mathbf{D}^b_{\mathrm{coh}}(X)$ by Theorem 4.1. So, the strong bounded generation of $\mathbf{D}^b_{\mathrm{coh}}(X)$ would follow.

- Now that we may assume X is reduced, we may apply de Jong's regular alteration to X:

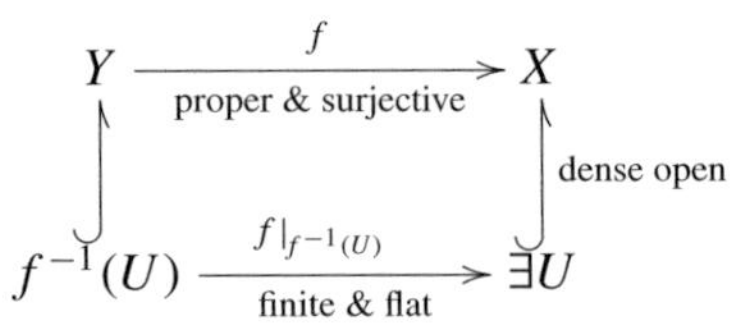

where we may apply Theorem 5.12 (SCG criterion) to Y, because Y is finite-dimensional, separated and regular: Here, let us consider $\mathbf{R}f_*(\mathcal{O}_Y \oplus \Sigma\mathcal{O}_Y) \in \mathbf{D}^b_{\mathrm{coh}}(X)$ (see Theorem 4.1). Then,

  - Since $f|_{f^{-1}(U)}$ is finite, flat and surjective, the restriction to U of the object $\mathbf{R}f_*\mathcal{O}_Y \in \mathbf{D}_{\mathrm{qc}}(X)$ is a nowhere vanishing vector bundle on U. In particular,

$$\left(\mathbf{L}j^*\mathbf{R}f_*\mathcal{O}_Y\right) \oplus \Sigma\left(\mathbf{L}j^*\mathbf{R}f_*\mathcal{O}_Y\right) = \mathbf{L}j^*\mathbf{R}f_*(\mathcal{O}_Y \oplus \Sigma\mathcal{O}_Y) \in \mathbf{D}^{\mathrm{perf}}(U). \quad (55)$$

  - Then, we can <u>*apply Corollary*</u> 4.20, <u>*a corollary of Thomason's localization theorem*</u> (Theorem 4.17), to (55) to find some $H \in \mathbf{D}^{\mathrm{perf}}(X)$ such that

$$\mathbf{L}j^*H \xrightarrow{\cong} \mathbf{L}j^*\mathbf{R}f_*(\mathcal{O}_Y \oplus \Sigma\mathcal{O}_Y) \in \mathbf{D}^{\mathrm{perf}}(U). \quad (56)$$

- To the local isomorphism (56), applying the ajoint isomorphism

$$\mathrm{Hom}_{\mathbf{D}_{\mathrm{qc}}(U)}\left(\mathbf{L}j^*H,\ \mathbf{L}j^*\mathbf{R}f_*(\mathcal{O}_Y \oplus \Sigma\mathcal{O}_Y)\right) \cong \mathrm{Hom}_{\mathbf{D}_{\mathrm{qc}}(X)}\left(H,\ \mathbf{R}j_*\mathbf{L}j^*\mathbf{R}f_*(\mathcal{O}_Y \oplus \Sigma\mathcal{O}_Y)\right),$$

we obtain a map[59]

[59] <u>WARNING!</u> In [109, Proof that Theorem 2.3 follows from Theorem 2.4], Neeman concluded the existence of an honest map $H \to \mathbf{R}f_*(\mathcal{O}_Y \oplus \Sigma\mathcal{O}_Y)$ corresponding to (56). However, this is quite problematic, and usually, such an honest map $H \to \mathbf{R}f_*\mathcal{O}_Y \oplus \Sigma\mathbf{R}f_*\mathcal{O}_Y$ does not exist. Thus,

$$\psi : H \to \mathbf{R}j_*\mathbf{L}j^*\mathbf{R}f_*\left(\mathcal{O}_Y \oplus \Sigma\mathcal{O}_Y\right). \tag{57}$$

- Recall, since $\left(\mathbf{D}_{\mathrm{qc}}\right)_Z(X)$ is compactly generated ([125, Th. 6.8]), we can apply Miller's finite localization Theorem 2.19 to form the Verdier quotient with the equivalence (14):

$$\mathbf{D}_{\mathrm{qc}}(X)/\left(\mathbf{D}_{\mathrm{qc}}\right)_Z(X) \xrightarrow[\cong]{\overline{\mathbf{L}j^*}} \mathbf{D}_{\mathrm{qc}}(U), \tag{58}$$

  and that $\mathbf{R}j_*\mathbf{L}j^*$ which shows up in the target of the ψ map (57) can be interpreted as the Bousfield localization, as in (15), which is consequently expressed by a mapping telescope **hocolim** as Miller's finite localization (Theorem 2.19). Then, consider the following pair of maps:

$$H \xrightarrow{\psi} \mathbf{R}j_*\mathbf{L}j^*\mathbf{R}f_*\left(\mathcal{O}_Y \oplus \Sigma\mathcal{O}_Y\right) = \mathrm{hocolim}(R_n) \xleftarrow[\text{canonical map}]{c} R_0 = \mathbf{R}f_*\left(\mathcal{O}_Y \oplus \Sigma\mathcal{O}_Y\right). \tag{59}$$

- The both maps in (59) are local isomorphism, i.e. isomorphisms when restricted to U. This is trivial for the canonical map (which is the Bousfield localization) and the claim for ψ follows from the local isomorphism (56).
- Since $H \in \mathbf{D}^{\mathrm{perf}}(X) = \mathbf{D}_{\mathrm{qc}}(X)^c$ is compact, arguing as in Proposition 2.23 and its comments below, we may factorize the pair of maps (59) as follows:

$$\begin{array}{ccccc} H & \xrightarrow{\psi} & \mathrm{hocolim}(R_n) & \xleftarrow{c} & \mathbf{R}f_*\left(\mathcal{O}_Y \oplus \Sigma\mathcal{O}_Y\right), \\ & \searrow^{\exists\widetilde{\psi}} & \uparrow{\scriptstyle \iota} & \swarrow_{\exists\widetilde{c}} & \\ & & \exists\widetilde{R} & & \end{array} \tag{60}$$

  where:
  - $\widetilde{R}$ is obtained from $\mathbf{R}f_*\left(\mathcal{O}_Y \oplus \Sigma\mathcal{O}_Y\right) \in \mathbf{D}^b_{\mathrm{coh}}(X)$ via $\widetilde{c}$ by a finite step extensions of finite coproducts of elements in $\mathbf{D}^{\mathrm{perf}}(X)$. Thus, we have a triangle of the following form:

$$\mathbf{R}f_*\left(\mathcal{O}_Y \oplus \Sigma\mathcal{O}_Y\right) \xrightarrow{\widetilde{c}} \widetilde{R} \to Q' \quad \left(Q' \in (\mathbf{D}^{\mathrm{perf}})_Z(X), \widetilde{R} \in \mathbf{D}^b_{\mathrm{coh}}(X)\right) \tag{61}$$

---

some sort of patch is needed. The "patch" presented above was communicated to the author by Professor Neeman, and the author replaced his own patch, which concentrates on $\widetilde{R}$ (see (63)), with Professor Neeman's "patch", which concentrates on $\widetilde{H}$ (see (63)), because Professor Neeman's patch delivers a simple message how to read [109, Proof that Theorem 2.3 follows from Theorem 2.4]: just replace H with $\widetilde{H}$ and pretend the map $\widetilde{\psi}' : \widetilde{H} \to \mathbf{R}f_*\left(\mathcal{O}_Y \oplus \Sigma\mathcal{O}_Y\right)$ obtained in (63) as our "honest map" $H \to \mathbf{R}f_*\left(\mathcal{O}_Y \oplus \Sigma\mathcal{O}_Y\right)$, and then, just proceed as is written in [109, Proof that Theorem 2.3 follows from Theorem 2.4].

According to Professor Neeman, this leap and omission of justification is standard. So, the reader is required to come up with this kind of patch spelled out in terms of elementary Bousfield (or Miller's finite) localization instantaneously at the top of his or her head. Thus, homotopy theoretical insight is prerequisite to read Professor Neeman's papers!

· From (61), we see $\widetilde{c}$ is a local isomorphism, then, since c is also a local isomorphism, ι is a local isomorphism as well from the right hand side commutative diagram of (60).
Then, since ϕ is also a local isomorphism, from the left hand side commutative diagram of (60), we find $\widetilde{\psi}$ is also a local isomorphism. Thus, we have a triangle of the following form:

$$Q'' \to H \xrightarrow{\widetilde{\psi}} \widetilde{R} \quad (Q'' \in (\mathbf{D}^b_{\mathrm{coh}})_Z(X)) \tag{62}$$

– Take the homotopy pullback $\widetilde{H}$ of the pair of maps $H \xrightarrow{\widetilde{\psi}} \widetilde{R} \xleftarrow{\widetilde{c}} \mathbf{R}f_* (\mathcal{O}_Y \oplus \Sigma\mathcal{O}_Y)$ obtained in (60):

$$\begin{array}{ccc} & \widetilde{H} := H \times^h_{\widetilde{R}} \mathbf{R}f_* (\mathcal{O}_Y \oplus \Sigma\mathcal{O}_Y) & \\ \swarrow^{\widetilde{c}'} & & \searrow^{\widetilde{\psi}'} \\ H & & \mathbf{R}f_* (\mathcal{O}_Y \oplus \Sigma\mathcal{O}_Y) \\ \searrow_{\widetilde{\psi}} & & \swarrow_{\widetilde{c}} \\ & \widetilde{R} & \end{array} \tag{63}$$

where:

· From (61), the homotopy pullback diagram (63) and $H \in \mathbf{D}^{\mathrm{perf}}(X)$, we have a triangle of the following form:

$$\widetilde{H} \xrightarrow{\widetilde{c}'} H \to Q' \quad (Q' \in (\mathbf{D}^{\mathrm{perf}})_Z(X),\ H, \widetilde{H} \in \mathbf{D}^{\mathrm{perf}}(X)) \tag{64}$$

· From (62) and the homotopy pullback diagram (63), we have a triangle of the following form:

$$Q'' \to \widetilde{H} \xrightarrow{\widetilde{\psi}'} \mathbf{R}f_* (\mathcal{O}_Y \oplus \Sigma\mathcal{O}_Y) \quad (Q'' \in (\mathbf{D}^b_{\mathrm{coh}})_Z(X)) \tag{65}$$

- Concerning the homological support $\mathrm{Supph}(\widetilde{H})$ of $\widetilde{H} \overset{(64)}{\in} \mathbf{D}^{\mathrm{perf}}(X)$, we see:
  – $\mathrm{Supph}(\widetilde{H})$ is closed, because $\widetilde{H} \in \mathbf{D}^{\mathrm{perf}}(X)$ implies $\mathcal{H}^\bullet \widetilde{H}$ is of finite type as an $\mathcal{O}_X$-module, and so we may apply [129, Lem. 17.9.6] for instance.

$$\begin{aligned} \mathrm{Supph}\,(\widetilde{H}) \bigcap U &\overset{(64)}{=} \mathrm{Supph}\,(H) \bigcap U \overset{(56)}{=} \mathrm{Supph}\,(\mathbf{R}f_* (\mathcal{O}_Y \oplus \Sigma\mathcal{O}_Y)) \bigcap U \\ &\overset{\text{direct summand}}{\supseteq} \mathrm{Supph}\,(\mathbf{R}f_*\mathcal{O}_Y) \bigcap U = U, \text{ a dense open of } X. \end{aligned}$$

where the last equality follows from the fact $\mathbf{R}f_*\mathcal{O}_Y$ restricted to U is a nowhere vanishing vector bundle.

Thus the homological support $\mathrm{Supph}\left(\widetilde{H}\right)$ is the whole X. Then, we can apply Corollary 4.14, a corollary of Thomason's theorem of Thomason sets (Theorem 4.12) to conclude that, $\langle \widetilde{H} \rangle_{\otimes}$, the tensor ideal generated by $\widetilde{H}$, is the whole $\mathbf{D}^{\mathrm{perf}}(X)$, which obviously contains $\mathcal{O}_X$. Then, applying Remark 4.13 and Proposition 4.32, we may pick some $C \in \mathbf{D}^{\mathrm{perf}}(X)$ and $L \in \mathbb{N}$ such that

$$\mathcal{O}_X \in \langle C \otimes \widetilde{H} \rangle_L \subseteq \mathrm{Coprod}_{2L}\left(\left(C \otimes \widetilde{H}\right)(-\infty, \infty)\right). \tag{66}$$

Consequently, for any $D \in \mathbf{D}_{\mathrm{qc}}(X)$,

$$D = D \otimes \mathcal{O}_X \in \langle D \otimes C \otimes \widetilde{H} \rangle_L \subseteq \mathrm{Coprod}_{2L}\left(\left(D \otimes C \otimes \widetilde{H}\right)(-\infty, \infty)\right). \tag{67}$$

- Having (67) in mind, we apply $D \otimes C \otimes -$ to (65) to obtain the following triangles:

$$D \otimes C \otimes Q'' \to D \otimes C \otimes \widetilde{H} \to D \otimes C \otimes \mathbf{R}f_* \left(\mathcal{O}_Y \oplus \Sigma \mathcal{O}_Y\right) \tag{68}$$

where $\mathbf{R}f_* \left(\mathcal{O}_Y \oplus \Sigma \mathcal{O}_Y\right) \in \mathbf{D}^b_{\mathrm{coh}}(X)$, $Q'' \in (\mathbf{D}^b_{\mathrm{coh}})_Z(X)$.
- For Y, obtained by de Jong's regular alteration, we may apply Theorem 5.12 to conclude its strong compact generation. Thus, $\exists G \in \mathbf{D}^{\mathrm{perf}}(X)$, $\exists N \in \mathbb{N}$, s.t. $\mathbf{D}_{\mathrm{qc}}(Y) = \mathrm{Coprod}_N\left(G(-\infty, \infty)\right)$. Hence,

$$\mathbf{L}f^*(D \otimes C) \otimes (\mathcal{O}_Y \oplus \Sigma \mathcal{O}_Y) \in \mathbf{D}_{\mathrm{qc}}(Y) = \mathrm{Coprod}_N\left(G(-\infty, \infty)\right) \qquad (G \in \mathbf{D}^{\mathrm{perf}}(X))$$

Consequently, by the projection formula,

$$\begin{aligned} D \otimes C \otimes \mathbf{R}f_* \left(\mathcal{O}_Y \oplus \Sigma \mathcal{O}_Y\right) &= \mathbf{R}f_*\left(\mathbf{L}f^*(D \otimes C) \otimes (\mathcal{O}_Y \oplus \Sigma \mathcal{O}_Y)\right) \\ &\in \mathbf{R}f_* \,\mathrm{Coprod}_N\left(G(-\infty, \infty)\right) \subseteq \mathrm{Coprod}_N\left((\mathbf{R}f_* G)(-\infty, \infty)\right) \end{aligned} \tag{69}$$

where $\mathbf{R}f_* G \in \mathbf{D}^b_{\mathrm{coh}}(X)$ by Theorem 4.1.
- For $Q'' \in (\mathbf{D}^b_{\mathrm{coh}})_Z(X)$ in (68), we may apply Rouquier's Theorem 4.18 to find $n \in \mathbb{N}$, $P_n \in \mathbf{D}^b_{\mathrm{coh}}(Z_n)$ s.t.

$$Q'' = \mathbf{R}i_{n*} P_n \qquad (P_n \in \mathbf{D}^b_{\mathrm{coh}}(Z_n)). \tag{70}$$

- For Z_n, whose underlying space is equal to that of the proper closed subscheme Z of X from their constructions in Theorem 4.18, we may apply Theorem 5.16 by inductive assumption to conclude their strong bounded generations. Thus, $\exists G'' \in \mathbf{D}^b_{\mathrm{coh}}(Z_n)$, $\exists M \in \mathbb{N}$ s.t. $\mathbf{D}_{\mathrm{qc}}(Z_n) = \mathrm{Coprod}_M\left(G''(-\infty, \infty)\right)$. Hence,

$$\mathbf{L}i_n^*(D \otimes C) \otimes P_n \in \mathbf{D}_{\mathrm{qc}}(Z_n) = \mathrm{Coprod}_M\left(G''(-\infty, \infty)\right) \qquad (G'' \in \mathbf{D}^b_{\mathrm{coh}}(Z_n)) \tag{71}$$

Consequently, by the projection formula,

$$D \otimes C \otimes Q'' = D \otimes C \otimes \mathbf{R}i_{n*}P_n = \mathbf{R}i_{n*}\Big(\mathbf{L}i_n^*(D \otimes C) \otimes P_n\Big)$$
$$\in \mathbf{R}i_{n*}\, \mathrm{Coprod}_M(G''(-\infty,\infty)) \subseteq \mathrm{Coprod}_M((\mathbf{R}i_{n*}G'')(-\infty,\infty)) \tag{72}$$

where $\mathbf{R}i_{n*}G'' \in \mathbf{D}^b_{\mathrm{coh}}(X)$ by Theorem 4.1.

- From (65), (69) and (72), we find[60]

$$\begin{aligned} D \otimes C \otimes \widetilde{H} &\in \mathrm{Coprod}_M((\mathbf{R}i_{n*}G'')(-\infty,\infty)) * \mathrm{Coprod}_N((\mathbf{R}f_*G)(-\infty,\infty)) \\ &\subseteq \mathrm{Coprod}_M((\mathbf{R}f_*G \oplus \mathbf{R}i_{n*}G'')(-\infty,\infty)) * \mathrm{Coprod}_N\big((\mathbf{R}f_*G \oplus \mathbf{R}i_{n*}G'')(-\infty,\infty)\big) \\ &\subseteq \mathrm{Coprod}_{M+N}\big((\mathbf{R}f_*G \oplus \mathbf{R}i_{n*}G'')(-\infty,\infty)\big), \end{aligned} \tag{73}$$

where $\mathbf{R}f_*G \oplus \mathbf{R}i_{n*}G'' \in \mathbf{D}^b_{\mathrm{coh}}(X)$.

- Finally, from (67) and (73) we see for any $D \in \mathbf{D}_{\mathrm{qc}}(X)$,

$$\begin{aligned} D &\overset{(67)}{\in} \mathrm{Coprod}_{2L}\big((D \otimes C \otimes \widetilde{H})(-\infty,\infty)\big) \\ &\overset{(73)}{\subseteq} \mathrm{Coprod}_{2L}\Big(\big(\mathrm{Coprod}_{M+N}\big((\mathbf{R}f_*G \oplus \mathbf{R}i_{n*}G'')(-\infty,\infty)\big)\big)(-\infty,\infty)\Big) \\ &\subseteq \mathrm{Coprod}_{2L(M+N)}\big((\mathbf{R}f_*G \oplus \mathbf{R}i_{n*}G'')(-\infty,\infty)\big), \end{aligned} \tag{74}$$

where $\mathbf{R}f_*G \oplus \mathbf{R}i_{n*}G'' \in \mathbf{D}^b_{\mathrm{coh}}(X)$. Thus, we have obtained the desired

$$\mathbf{D}_{\mathrm{qc}}(X) = \mathrm{Coprod}_{2L(M+N)}\big((\mathbf{R}f_*G \oplus \mathbf{R}i_{n*}G'')(-\infty,\infty)\big),$$

which shows the strong bounded generation of $\mathbf{D}_{\mathrm{qc}}(X)$ for $\mathbf{R}f_*G \oplus \mathbf{R}i_{n*}G'' \in \mathbf{D}^b_{\mathrm{coh}}(X)$. □

From Theorem 5.16 and Theorem 5.6 (2), we obtain Neeman's main theorem on strong generation of $\mathbf{D}^b_{\mathrm{coh}}(X)$:

Theorem 5.17 (Neeman [109, Th. 0.15] [108, Th. 6.11]) *Let X be a noetherian scheme, and assume every closed subscheme $Z \subset X$ admits a regular alteration. Then $\mathbf{D}^b_{\mathrm{coh}}(X)$ is strongly generated.*

From [32, 33, 102], we see any X, which is separated and essentially of finite type over a separated excellent scheme S of dimension ≤ 2, satisfies the assumptions of Theorems 5.16 and 5.17. Thus, Theorems 5.16 and 5.17 generalize Rouquier's Theorem 5.14.

For more details about strong generations of $\mathbf{D}^{\mathrm{perf}}(X)$ and $\mathbf{D}^b_{\mathrm{coh}}(X)$, consult Neeman's original article [109] and the survey [108].

[60] In Neeman's corresponding calculation [109, 1st paragraph in p. 24], the extension length of Coprod was doubled to be $2(M+N)$ rather than $M+N$ given in (73). However, the author does not see such a need, and so, the author opted to present as in (73).

References

1. Alonso Tarrío, L., Jeremías López, A., Souto Salorio, M.J.: Construction of t-structures and equivalences of derived categories. Trans. Am. Math. Society. **355**(6), 2523–2543 (2003) MR1974001 (2004c:18020)
2. Alonso Tarrío, L., Jeremías López, A., Souto Salorio, M.J.: Bousfield localization on formal schemes. J. Algebra **278**(2), 585–610 (2004). MR2071654 (2005g:14037)
3. Balmer, P.: The spectrum of prime ideals in tensor triangulated categories. J. Reine Angew. Math. **588**, 149–168 (2005). MR2196732
4. Balmer, P.: Supports and filtrations in algebraic geometry and modular representation theory. Am. J. Math. **129**(5), 1227–1250 (2007). MR2354319 (2009d:18017)
5. Balmer, P.: Tensor triangular geometry. In: Proceedings of the International Congress of Mathematicians, vol. II, pp. 85–112. Hindustan Book Agency, New Delhi (2010). MR2827786 (2012j:18016)
6. Balmer, P., Schlichting, M.: Idempotent completion of triangulated categories. J. Algebra **236**(2), 819–834 (2001). MR1813503 (2002a:18013)
7. Balmer, P., Favi, G.: Generalized tensor idempotents and the telescope conjecture. Proc. Lond. Math. Soc. (3) **102**(6), 1161–1185 (2011). MR2806103 (2012d:18010)
8. Balmer, P., Sanders B.: The spectrum of the equivariant stable homotopy category of a finite group. Invent. Math. **208**(1), 283–326 (2017). MR3621837
9. Barthel, T.: A short introduction to the telescope and chromatic splitting conjectures, in this proceedings
10. Barthel, T., Heard, D., Valenzuela, G.: The algebraic chromatic splitting conjecture for Noetherian ring spectra. Math. Z. **290**(3–4), 1359–1375 (2018). MR3856857
11. Beaudry, A.: The chromatic splitting conjecture at $n = p = 2$, Geom. Topol. **21**(6), 3213–3230 (2017). MR3692966
12. Beaudry, A., Goerss, P.G., Henn, H.-W.: Chromatic splitting for the $K(2)$-local sphere at $p = 2$. arXiv:1712.08182
13. Behrens, M., Rezk, C.: Spectral algebra models of unstable v_n-periodic homotopy theory, in this proceedings
14. Beĭlinson, A.A., Bernstein, J., Deligne, P.: Faisceaux pervers. Analysis and Topology on Singular Spaces, I (Luminy, 1981). Astérisque, vol. 100, pp. 5–171. Société Mathématique de France, Paris (1982). MR0751966 (86g:32015)
15. Benson, D. J., Carlson, J.F., Rickard, J., Complexity and varieties for infinitely generated modules. II, Math. Proc. Cambridge Philos. Soc. **120**(4), 597–615 (1996). MR1401950 (97f:20008)
16. Benson, D., Iyengar, S.B., Krause, H., Pevtsova, J.: Stratification for module categories of finite group schemes. J. Am. Math. Society. **31**(1), 265–302 (2018). MR3718455
17. Bökstedt, M., Neeman, A.: Homotopy limits in triangulated categories. Compos. Math. **86**(2), 209–234 (1993). MR1214458 (94f:18008)
18. Bondal, A., Orlov, D.: Reconstruction of a variety from the derived category and groups of autoequivalences. Compos. Math. **125**(3), 327–344 (2001). MR1818984 (2001m:18014)
19. Bondal, A., Orlov, D.: Derived categories of coherent sheaves. In: Proceedings of the International Congress of Mathematicians, Beijing, 2002, vol. II, pp. 47–56. Higher Education Press, Beijing (2002). MR1818984
20. Bondal, A., Van den Bergh, M.: Generators and representability of functors in commutative and noncommutative geometry. Mosc. Math. J. **3**(1), 1–36, 258 (2003). MR1996800 (2004h:18009)
21. Bousfield, A.K.: The localization of spectra with respect to homology. Topology, **18**(4), 257–281 (1979). MR551009 55N20 (55N15 55P60)
22. Bridgeland, T.: Stability conditions on triangulated categories. Ann. Math. (2) **166**(2), 317–345 (2007). MR2373143 (2009c:14026)
23. Buan, A.B., Krause, H., Solberg, Ø.: Support varieties: an ideal approach. Homol. Homotopy Appl. **9**(1), 45–74 (2007). MR2280286 (2008i:18007)

24. Burke, J., Neeman, A., Pauwels, B.: Gluing approximable triangulated categories. arxiv.1806.05342
25. Calabrese, J., Groechenig, M.: Moduli problems in abelian categories and the reconstruction theorem. Algebr. Geom. **2**(1), 1–18 (2015). MR3322195
26. Casacuberta, C.: Depth and simplicity of Ohkawa's argument, in this proceedings
27. Casacuberta, C., Rosický, J.: Combinatorial homotopy categories, in this proceedings
28. Casacuberta, C., Gutiérrez, J.J., Rosický, J.: A generalization of Ohkawa's theorem. Compos. Math. **150**(5), 893–902 (2014). MR3209799 55N20 18G55 55P42 55U40
29. Christensen, J.D.: Ideals in triangulated categories: phantoms, ghosts and skeleta. Adv. Math. **136**(2), 284–339 (1998). MR1626856 (99g:18007)
30. Dao, H., Takahashi, R.: The dimension of a subcategory of modules. Forum Math. Sigma **3**, e19 (2015) 31 pp. MR3482266
31. Dao, H., Takahashi, R.: Upper bounds for dimensions of singularity categories. Comptes Rendus Math. Acad. Sci. (Paris) **353**(4), 297–301 (2015). MR3319124
32. de Jong, A.J.: Smoothness, semi-stability and alterations. Inst. Ht. Études Sci. Publ. Math. (83), 51–93 (1996). MR1423020 (98e:14011)
33. de Jong, A.J.: Families of curves and alterations. Ann. Inst. Fourier (Grenoble) **47**(2), 599–621 (1997). MR1450427 (98f:14019)
34. Dell'Ambrogio, I., Stevenson, G.: On the derived category of a graded commutative Noetherian ring. J. Algebra **373**, 356–376 (2013). MR2995031
35. Devinatz, E.S., Hopkins, M.J., Smith, J.H.: Nilpotence and stable homotopy theory. I. Ann. Math. (2) **128**(2), 207–241 (1988). MR0960945 (89m:55009)
36. Douglas, M.R.: Dirichlet branes, homological mirror symmetry, and stability. In: Proceedings of the International Congress of Mathematicians, Beijing, 2002, vol. III, pp. 395–408. Higher Education Press, Beijing (2002). MR1957548
37. Dwyer, W.G., Palmieri, J.H., Ohkawa's theorem: there is a set of Bousfield classes, Proc. Amer. Math. Soc. **129**(3), 881-886 (2001). MR1712921 (2001f:55015)
38. Fukaya, K., Oh, Y.-G., Ohta, H., Ono, K.: Lagrangian Intersection Floer Theory: Anomaly and Obstruction. Part II. AMS/IP Studies in Advanced Mathematics, vol. 46.2. American Mathematical Society, Providence; International Press, Somerville (2009). xii+396 pp. MR2548482 (2011c:53218)
39. Fukaya, K., Oh, Y.-G., Ohta, H., Ono, K.: Lagrangian Intersection Floer Theory: Anomaly and Obstruction. Part I. AMS/IP Studies in Advanced Mathematics, vol. 46.1. American Mathematical Society, Providence; International Press, Somerville (2009). xii+396 pp. MR2553465 (2011c:53217)
40. Gabber, O.: Finiteness theorems for tale cohomology of excellent schemes. In: Conference in honor of P. Deligne on the occasion of his 61st birthday, IAS, Princeton, October 2005, p. 45 (2005)
41. Gabriel, P.: Des catégories abéliennes. Bull. Soc. Math. Fr. **90**, 323–448 (1962). MR0232821 (38 #1144)
42. Gabriel, P., Zisman, M.: Calculus of fractions and homotopy theory. Ergebnisse der Mathematik und ihrer Grenzgebiete, vol. 35. Springer, New York (1967) x+168 pp. MR0210125 (35 # 1019)
43. Gelfand, S.I., Manin, Y.I.: Methods of Homological Algebra. Springer Monographs in Mathematics, 2nd edn. Springer, Berlin (2003). xx+372 pp. MR1950475 (2003m:18001)
44. Grothendieck, A.: Éléments de géométrie algébrique. III. Étude cohomologique des faisceaux cohérents. I. Ins. Ht. Études Sci. Publ. Math. **11**, 5–167 (1961)
45. Grothendieck, A., Raynaud, M.: Revêtements étales et groupe fondamental, Séminaire de Géométrie Algébrique. I.H.E.S (1963)
46. Hall, J.: GAGA theorems. arXiv:1804.01976
47. Hall, J., Rydh, D.: The telescope conjecture for algebraic stacks. J. Topol. **10**(3), 776–794 (2017). MR3797596
48. Hopkins, M.: Global methods in homotopy theory. Homotopy Theory (Durham, 1985). London Mathematical Society Lecture Note Series, vol. 117, pp. 73–96. Cambridge University Press, Cambridge (1987). MR0932260 (89g:55022)

49. Hopkins, M.J., Gross, B.H.: The rigid analytic period mapping, Lubin-Tate space, and stable homotopy theory. Bull. Am. Math. Soc. (N.S.) **30**(1), 7686 (1994). MR1217353 (94k:55009)
50. Hopkins, M.J., Smith, J.H.: Nilpotence and stable homotopy theory. II. Ann. Math. (2) **148**(1), 1–49 (1998). MR1652975 (99h:55009)
51. Hovey, M.: Bousfield localization functors and Hopkins' chromatic splitting conjecture. The Čech centennial, Boston, MA, 1993. Contemporary Mathematics, vol. 181, pp. 225–250. American Mathematical Society, Providence, RI (1995). MR1320994 (96m:55010)
52. Hovey, M.: Cohomological Bousfield classes. J. Pure Appl. Algebra **103**(1), 45–59 (1995). MR1354066 (96g:55008)
53. Hovey, M., Palmieri, J.H.: The structure of the Bousfield lattice. Homotopy Invariant Algebraic Structures, Baltimore, MD, 1998, pp. 175–196. Contemporary Mathematics, vol. 239, 1999. MR1718080 (2000j:55033)
54. Hovey, M., Palmieri, J.H., Strickland, N.P.: Axiomatic Stable Homotopy Theory. Memoirs of the American Mathematical Society, vol. 128 (610), x+114 pp. (1997). MR1388895 (98a:55017)
55. Hoyois, M., Kelly, S., Østvær, P.A.: The motivic Steenrod algebra in positive characteristic. J. Eur. Math. Soc. (JEMS) **19**(12), 3813–3849 (2017). MR3730515
56. Huybrechts, D.: Fourier-Mukai transforms in algebraic geometry. Oxford Mathematical Monographs. The Clarendon Press, Oxford University Press, Oxford (2006). viii+307 pp. MR2244106 (2007f:14013)
57. Huybrechts, D., Lehn, M.: The geometry of moduli spaces of sheaves. Cambridge Mathematical Library, 2nd edn. Cambridge University Press, Cambridge (2010). xviii+325 pp. MR2665168 (2011e:14017)
58. Iyengar, S.B., Krause, H.: The Bousfield lattice of a triangulated category and stratification. Math. Z. **273**(3–4), 1215–1241 (2013). MR3030697
59. Iyengar, S.B., Takahashi, R.: Annihilation of cohomology and decompositions of derived categories. Homol. Homotopy Appl. **16**(2), 231–237 (2014). MR326389
60. Iyengar, S.B., Takahashi, R.: Annihilation of cohomology and strong generation of module categories. Int. Math. Res. Not. IMRN **2016**(2), 499–535 (2016). MR3493424
61. Iyengar, S.B., Lipman, J., Neeman, A.: Relation between two twisted inverse image pseudofunctors in duality theory. Compos. Math. **151**(4), 735–764 (2015). MR3334894
62. Joachimi, R.: Thick ideals in equivariant and motivic stable homotopy categories, in this proceedings
63. Jørgensen, P.: A new recollement for schemes. Houst. J. Math. **35**(4), 1071–1077 (2009). MR2577142 (2011c:14048)
64. Kapustin, A.N., Li, Y.: Topological correlators in Landau-Ginzburg models with boundaries. Adv. Theor. Math. Phys. **7**(4), 727–749 (2003). MR2039036 (2005b:81179a)
65. Kashiwara, M., Schapira, P.: Sheaves on manifolds. With a chapter in French by Christian Houzel. Corrected reprint of the 1990 original. Grundlehren der Mathematischen Wissenschaften [Fundamental Principles of Mathematical Sciences], vol. 292. Springer, Berlin (1994). x+512 pp. MR1299726 (95g:58222)
66. Kato, R., Okajima, H., Shimomura, K.: Notes on an algebraic stable homotopy category, in this proceedings
67. Kawamata, Y.: D-equivalence and K-equivalence. J. Differ. Geom. **61**(1), 147–171 (2002). MR1949787 (2004m:14025)
68. Kawamata, Y.: Birational geometry and derived categories. arXiv:1710.07370
69. Keller, B.: A remark on the generalized smashing conjecture. Manuscr. Math. **84**(2) (1994). 193198 MR1285956 (95h:18014)
70. Kelly, G.M.: Chain maps inducing zero homology maps. Proc. Camb. Philos. Soc. **61**, 847–854 (1965). MR0188273 (32 # 5712)
71. Kelly, S.: Triangulated categories of motives in positive characteristic. Ph.D. thesis, University of Paris 13; Australian National University (2013). arXiv:1305.5349v2
72. Kelly, S.: Some observations about motivic tensor triangulated geometry over a finite field, in this proceedings

73. Kollár, J., Mori, S.: Birational geometry of algebraic varieties, With the collaboration of C. H. Clemens and A. Corti; Translated from the 1998 Japanese original, Cambridge Tracts in Mathematics, vol. 134. Cambridge University Press, Cambridge, 1998, viii+254pp, MR1658959 (2000b:14018)
74. Kontsevich, M.: Homological algebra of mirror symmetry. In: Proceedings of the International Congress of Mathematicians, Zurich, 1994, vol. 1, 2, pp. 120–139. Birkhauser, Basel (1995). MR1403918 (97f:32040)
75. Kontsevich, M., Soibelman, Y.: Stability structures, motivic Donaldson-Thomas invariants and cluster transformations (2008). arXiv:0811.2435
76. Krause, H.: Smashing subcategories and the telescope conjecture–an algebraic approach. Invent. Math. **139**(1), 99–133 (2000). MR1728877 (2000k:55016)
77. Krause, H.: A Brown representability theorem via coherent functors. Topology **41**(4), 853–861 (2002). MR1905842 (2003c:18011)
78. Krause, H.: Localization theory for triangulated categories. Triangulated Categories. London Mathematical Society Lecture Note Series, vol. 375, pp. 161–235 (2010). MR2681709 (2012e:18026)
79. Krause, H.: Completing perfect complexes. arXiv:1805.10751
80. Krause, H., Šťovíček, J.: The telescope conjecture for hereditary rings via Ext-orthogonal pairs. Adv. Math. **225**(5), 2341–2364 (2010). MR2680168 (2011j:16013)
81. Lewis Jr., L.G., May, J.P., Steinberger, M.: Equivariant stable homotopy theory. With contributions by J. E. McClure. Lecture Notes in Mathematics, vol. 1213. Springer, Berlin (1986). x+538 pp. MR0866482 (88e:55002)
82. Lipman, J.: Notes on derived functors and Grothendieck duality. Foundations of Grothendieck duality for diagrams of schemes. Lecture Notes in Mathematics, vol. 1960, pp. 1–259. Springer, Berlin (2009). MR2490557 (2011d:14029)
83. Lipman, J., Neeman, A.: Quasi-of the twisted inverse image functor perfect scheme-maps and boundedness. Ill. J. Math. **51**(1), 209–236 (2007). MR2346195 (2008m:14004)
84. Lurie, J.: Tannaka duality for geometric stacks. arXiv:math/0412266v2
85. Lurie, J.: Higher Topos Theory. Annals of Mathematics Studies, vol. 170. Princeton University Press, Princeton (2009). MR2522659 (2010j:18001)
86. Lurie, J.: Higher algebra. www.math.harvard.edu/~lurie/ (2016)
87. Mahowald, M., Ravenel, D., Shick, P.: The triple loop space approach to the telescope conjecture. Homotopy Methods in Algebraic Topology, Boulder, CO, 1999. Contemporary Mathematics, vol. 271, pp. 217–284. American Mathematical Society, Providence, RI (2001). MR1831355 (2002g:55014)
88. Matsuki, K.: Introduction to the Mori Program. Universitext. Springer, New York (2002). xxiv+478, MR1875410 (2002m:14011)
89. Matsuoka, T.: Koszul duality for E_n-algebras in a filtered category, in this proceedings
90. Matsuoka, T.: Some technical aspects of factorization algebras on manifolds, in this proceedings
91. Matumoto, T.: Memories on Ohkawa's mathematical life in Hiroshima, in this proceedings
92. May, J.P.: The additivity of traces in triangulated categories. Adv. Math. **163**(1), 3473 (2001). MR1867203 (2002k:18019)
93. Miller, H.: Finite localizations. Papers in honor of Jos Adem Bol. Soc. Mat. Mex. (2) **37**(1–2), 383–389 (1992). MR1317588 (96h:55009)
94. Minami, N.: A topologist's introduction to the motivic homotopy theory for transformation group theorists–1, Geometry of transformation groups and combinatorics, RIMS Kôkyûroku Bessatsu. Res. Inst. Math. Sci. (RIMS), Kyoto, **B39**, 63–107 (2013). MR3156820
95. Miyaoka, Y., Peternell, T.: Geometry of Higher-Dimensional Algebraic Varieties. DMV Seminar, vol. 26. Birkhuser Verlag, Basel, 1997. vi+217 pp. MR1468476 (98g:14001)
96. Morava, J.: Noetherian localisations of categories cobordism comodules. Ann. Math. (2) **121**, 1–39 (1985). MR0782555 (86g:55004)
97. Morava, J.: A remark on Hopkins' chromatic splitting conjecture. arXiv:1406.3286,
98. Morava, J.: Operations on integral lifts of $K(n)$, in this proceedings

99. Morava, J.: Toward a fundamental groupoid for the stable homotopy category. In: Proceedings of the Nishida Fest, Kinosaki, 2003. Geometry and Topology Monographs, vol. 10, pp. 293–318. Geometry & Topology Publications, Coventry (2007). MR2402791 (2009e:55018)
100. Morel, F., Voevodsky, V.: $\mathbf{A}^1$-homotopy theory of schemes. Inst. Ht. Études Sci. Publ. Math. **90**, 1999, 45–143 (2001), MR1813224 (2002f:14029)
101. Mukai, S.: Duality between $D(X)$ and $D(\hat{X})$ with its application to Picard sheaves. Nagoya Math. J. **81**, 153–175 (1981). MR0607081 (82f:14036)
102. Nayak, S.: Compactification for essentially finite-type maps. Adv. Math. **222**(2), 527–546 (2009)
103. Neeman, A.: The chromatic tower for $D(R)$. Topology **31**(3), 519–532 (1992). With an appendix by Marcel Bökstedt. MR1174255 (93h:18018)
104. Neeman, A.: Triangulated categories with a single compact generator and a Brown representability theorem. arXiv:1804.02240
105. Neeman, A.: The connection between the K-theory localization theorem of Thomason, Trobaugh and Yao and the smashing subcategories of Bousfield and Ravenel. Ann. Sci. l'Ecole Norm. Supér. (4) **25**(5), 547–566 (1992). MR1191736 (93k:18015)
106. Neeman, A.: The Grothendieck duality theorem via Bousfield's techniques and Brown representability. J. Am. Math. Soc. **9**(1), 205–236 (1996). MRMR1308405
107. Neeman, A.: Oddball Bousfield classes. Topology **39**(5), 931–935 (2000). MR1763956 (2001c:18007)
108. Neeman, A.: Approximable triangulated categories. arXiv:1806.06995
109. Neeman, A.: Strong generators in $D^{perf}(X)$ and $D^b_{coh}(X)$. arXiv:1703.04484
110. Neeman, A.: The categories $\mathcal{T}^c$ and $\mathcal{T}^b_c$ determine each other. arXiv:1806.06471
111. Neeman, A.: Triangulated Categories. Annals of Mathematics Studies, vol. 148. Princeton University Press, Princeton (2001). MR1812507 (2001k:18010)
112. Neeman, A.: The K-theory of triangulated categories. Handbook of K-theory, vol. 1, 2, pp. 1011–1078 (2005). MR2181838 (2006g:19004)
113. Noguchi, J.: Analytic Function Theory of Several Variables. Elements of Oka's Coherence. Springer, Singapore (2016). xvi+397 pp. MR3526579
114. Ohsawa, T.: A role of the L^2 method in the study of analytic families, in this proceedings
115. Ohkawa, T.: The injective hull of homotopy types with respect to generalized homology functors. Hiroshima Math. J. **19**(3), 631–639 (1989). MR1035147
116. Ohsawa, T.: L^2 Approaches in Several Complex Variables. Development of Oka-Cartan Theory by L^2 Estimates for the $\bar{\partial}$ Operator. Springer Monographs in Mathematics. Springer, Tokyo (2015). MR3443603
117. Oort, F.: Alterations can remove singularities. Bull. Am. Math. Soc. (N.S.) **35**(4), 319–331 (1998). MR1638306 (99i:14021)
118. Orlov, D.O.: Triangulated categories of singularities and D-branes in Landau-Ginzburg models. (Russian. Russian summary) Tr. Mat. Inst. Steklova **246** (2004), Algebr. Geom. Metody, Svyazi i Prilozh, 240–262; translation in Proc. Steklov Inst. Math. 2004, no. 3(246), 227–248. MR2101296 (2006i:81173)
119. Orlov, D.O.: Matrix factorizations for nonaffine LG-models. Math. Ann. **353**(1), 95–108 (2012). MR2910782 14F05 18E30
120. Orlov, D.: Smooth and proper noncommutative schemes and gluing of DG categories. Adv. Math. **302**, 59–105 (2016). MR3545926 14F05 (16E45 18E30)
121. Perego, A.: A Gabriel theorem for coherent twisted sheaves. Math. Z. **262**(3), 571–583 (2009). MR2506308 (2011a:14032)
122. Ravenel, D.C.: Localization with respect to certain periodic homology theories. Am. J. Math. **106**(2), 351–414 (1984). MR0737778 (85k:55009)
123. Ravenel, D.C.: Nilpotence and periodicity in stable homotopy theory. Ann. Math. Stud. **128**, Appendix C by Jeff Smith (1992) xiv+209 MR1192553 (94b:55015)
124. Rosenberg, A.: Spectra of 'Spaces' Represented by Abelian Categories. MPIM Preprints, 2004-115

125. Rouquier, R.: Dimensions of triangulated categories. J. K-Theory **1**(2), 193–256 (2008). MR2434186
126. Rouquier, R.: Derived categories and algebraic geometry. Triangulated Categories. London Mathematical Society Lecture Note Series, vol. 375, pp. 351–370. Cambridge University Press, Cambridge (2010). MR2681712 (2011h:14022)
127. Seidel, P., Thomas, R.: Braid group actions on derived categories of coherent sheaves. Duke Math. J. **108**(1), 37–108 (2001). MR1831820 (2002e:14030)
128. Serre, J.P.: Géométrie algébrique et géométrie analytique. Ann. l'institut Fourier (Grenoble) **6**, 1–42 (1955–1956). MR0082175 (18,511a)
129. The Stacks Project, Part 1: Preliminaries, Chapter 17: Sheaves of modules, Section 17.9: Modules of finite type
130. Thomason, R.W., Trobaugh, T.: Higher algebraic K-theory of schemes and of derived categories. The Grothendieck Festschrift, Volume III. Progress in Mathematics, vol. 88, pp. 247–435. Birkhuser Boston, Boston (1990). MR1106918 (92f:19001)
131. Thomason, R.W.: The classification of triangulated subcategories. Compos. Math. **105**(1), 1–27 (1997). MR1436741 (98b:18017)
132. Toda, Y.: Limit stable objects on Calabi-Yau 3-folds. Duke Math. J. **149**(1), 157–208 (2009). MR2541209 (2011b:14043)
133. Torii, T.: On quasi-categories of comodules and Landweber exactness, in this proceedings
134. Uehara, H.: An example of Fourier-Mukai partners of minimal elliptic surfaces. Math. Res. Lett. **11**(2–3), 371–375 (2004). MR2067481 (2005g:14073)
135. Verdier, J.-L.: Catégories dérivées: quelques résultats (état 0). Cohomologie étale. Lecture Notes in Mathematics, vol. 569, pp. 262–311. Springer, Berlin (1977). MR3727440

Combinatorial Homotopy Categories

Carles Casacuberta and Jiří Rosický

Abstract A model category is called combinatorial if it is cofibrantly generated and its underlying category is locally presentable. As shown in recent years, homotopy categories of combinatorial model categories share useful properties, such as being well generated and satisfying a very general form of Ohkawa's theorem.

Keywords Combinatorial model category · Cofibrantly generated · Locally presentable · Well generated · Brown representability

1 Introduction

The term "combinatorial" in topology classically refers to discrete methods or, more specifically, to the use of polyhedra, simplicial complexes or cell complexes in order to deal with topological problems [17, 32].

In the context of Quillen model categories in homotopy theory [25], those called *combinatorial* are, by definition, the cofibrantly generated ones whose underlying category is locally presentable. For example, simplicial sets are combinatorial, but topological spaces are not. As a consequence of this fact, certain constructions involving homotopy colimits, such as Bousfield localizations, may seem intricate if one works with topological spaces while they have become standard technology in the presence of combinatorial models [2, 6, 12].

One key feature of combinatorial model categories is that they admit presentations in terms of generators and relations; in fact, as shown by Dugger in [11], they

---

C. Casacuberta
Facultat de Matemàtiques i Informàtica, Universitat de Barcelona (UB),
Gran Via de les Corts Catalanes 585, 08007 Barcelona, Spain
e-mail: carles.casacuberta@ub.edu

J. Rosický (✉)
Department of Mathematics and Statistics, Faculty of Sciences, Masaryk
University, Kotlářská 2, 611 37 Brno, Czech Republic
e-mail: rosicky@math.muni.cz

T. Ohsawa and N. Minami (eds.), *Bousfield Classes and Ohkawa's Theorem*,
Springer Proceedings in Mathematics & Statistics 309,
https://doi.org/10.1007/978-981-15-1588-0_4

are Quillen equivalent to localizations of categories of simplicial presheaves with respect to sets of maps. Moreover, for each combinatorial model category $\mathscr{K}$ there exist cardinals λ for which $\mathscr{K}$ admits fibrant and cofibrant replacement functors that preserve λ-filtered colimits and λ-presentable objects, and the class of weak equivalences is closed under λ-filtered colimits [4, 11, 28].

Cofibrantly generated model categories admit weak generators [13, 26]. Combinatorial model categories are, in addition, well generated in the sense of [18, 21]. This fact links the study of combinatorial model categories with the theory of triangulated categories in useful ways. For instance, it was shown in [8] that localizing subcategories of triangulated categories with combinatorial models are coreflective assuming a large-cardinal axiom (Vopěnka's principle), and similarly colocalizing subcategories are reflective.

In this article we show that a suitably restricted Yoneda embedding [1, 28] gives a way to implement Ohkawa's argument [24] in the homotopy category of any combinatorial model category, not necessarily stable. Ohkawa's original theorem becomes then a special case, since the homotopy category of spectra admits combinatorial models [15]. Thus we prove that, if $\mathscr{K}$ is a pointed strongly λ-combinatorial model category (see Sect. 3 below for details) then there is only a set of distinct kernels of endofunctors $H\colon \mathscr{K} \to \mathscr{K}$ preserving λ-filtered colimits and the zero object.

This statement (and our method of proof) is a variant of the main result in [9], where Ohkawa's theorem was broadly generalized. In independent work, Stevenson used abelian presheaves over compact objects to prove that Ohkawa's theorem holds in compactly generated tensor triangulated categories [31], and Iyengar and Krause extended this result to well generated tensor triangulated categories [16].

Our approach shows that Ohkawa's theorem is valid in the categories of motivic spaces and motivic spectra over any Noetherian base scheme of finite dimension [19], and also in categories of modules over (ordinary or motivic) ring spectra, since such categories have combinatorial models. Therefore, for example, Okhawa's theorem holds in the derived category of motives over any field k of characteristic zero, since these are modules over a motivic Eilenberg–Mac Lane spectrum [27].

2 Combinatorial Model Categories

The notion of a combinatorial model category was introduced by Jeff Smith in unpublished work made in the decade of 1990. The name refers to the fact that the underlying category and its model structure are both controlled by sets of sufficiently small objects and maps between them, in the precise sense that we next define. Further details and additional motivation can be found in [1, 4, 11–13].

For a regular cardinal λ, a small category $\mathscr{A}$ is *λ-filtered* if every diagram in $\mathscr{A}$ of cardinality smaller than λ has a cocone. An object A of a category $\mathscr{C}$ is called *λ-presentable* if the hom-functor $\mathscr{C}(A,-)\colon \mathscr{C} \to \mathbf{Set}$ preserves λ-filtered colimits. For example, a group (or a module over a ring) is λ-presentable if and only if it admits a presentation with less than λ generators and less than λ relations.

A category $\mathscr{C}$ is *locally λ-presentable* if it is cocomplete and has a set $\mathscr{A}$ of λ-presentable objects such that every object of $\mathscr{C}$ is a λ-filtered colimit of objects of $\mathscr{A}$. A category is *locally presentable* if it is locally λ-presentable for some regular cardinal λ. The category of sets is locally $\aleph_0$-presentable, since every set is the colimit of the inclusions of its finite subsets. As shown in [1, Corollary 3.7], every variety of finitary algebras is locally $\aleph_0$-presentable. Many more examples arise from the fact that every functor category from a small category to a locally presentable category is locally presentable [1, Corollary 1.54].

A model category $\mathscr{K}$ is *cofibrantly generated* if it has a set $\mathscr{I}$ of cofibrations such that the trivial fibrations of $\mathscr{K}$ are those morphisms having the right lifting property with respect to $\mathscr{I}$, and a set $\mathscr{J}$ of trivial cofibrations such that the fibrations of $\mathscr{K}$ are those morphisms having the right lifting property with respect to $\mathscr{J}$, and moreover $\mathscr{I}$ and $\mathscr{J}$ permit the small object argument, that is, their domains are small relative to transfinite compositions of pushouts of elements of $\mathscr{I}$ and $\mathscr{J}$ respectively. The category of simplicial sets is cofibrantly generated with $\mathscr{I}$ the set of inclusions $\partial\Delta[n] \hookrightarrow \Delta[n]$ for $n \geq 0$ and $\mathscr{J}$ the set of inclusions $\Lambda^k[n] \hookrightarrow \Delta[n]$ for $n \geq 0$ and $0 \leq k \leq n$; see [12, 13] for notation and a proof.

A model category is called *combinatorial* if it is locally presentable and cofibrantly generated. By a *combinatorial homotopy category* we mean a homotopy category of a combinatorial model category.

Every locally presentable category $\mathscr{C}$ can be viewed as a combinatorial homotopy category because the trivial model structure on $\mathscr{C}$ (that is, the one in which every morphism is both a cofibration and a fibration, and the weak equivalences are the isomorphisms) is cofibrantly generated by the argument given in [30, Example 4.6]. In general, combinatorial homotopy categories are far from being locally presentable themselves, but they behave in some sense like a homotopy-theoretical version of those.

A model category $\mathscr{K}$ is called *λ-combinatorial* for a regular cardinal λ if it is locally λ-presentable and cofibrantly generated by morphisms between λ-presentable objects. Then the functors giving factorizations of morphisms in $\mathscr{K}$ into cofibrations followed by trivial fibrations or into trivial cofibrations followed by fibrations can be chosen to be λ-accessible, that is, preserving λ-filtered colimits. Details are given in [28, Proposition 3.1].

3 Restricted Yoneda Embedding

Let $\mathscr{C}$ be a category and $\mathscr{A}$ a small full subcategory of $\mathscr{C}$. The *restricted Yoneda embedding*

$$E_{\mathscr{A}} : \mathscr{C} \longrightarrow \mathbf{Set}^{\mathscr{A}^{\mathrm{op}}}$$

sends every object X of $\mathscr{C}$ to the hom-set $\mathscr{C}(-, X)$ restricted to $\mathscr{A}$. Thus $E_{\mathscr{A}}$ is full and faithful on morphisms whose domain is an object of $\mathscr{A}$.

The subcategory $\mathscr{A}$ is called a *generator* of $\mathscr{C}$ if $E_{\mathscr{A}}$ is faithful, and a *strong generator* if $E_{\mathscr{A}}$ is faithful and conservative, that is, reflecting isomorphisms. We say that $\mathscr{A}$ is a *weak generator* if $E_{\mathscr{A}}$ reflects isomorphisms whose codomain is the terminal object of $\mathscr{C}$. This means that an object of $\mathscr{C}$ is terminal whenever its image under $E_{\mathscr{A}}$ is terminal; hence the objects in a weak generator of $\mathscr{C}$ form a *left weakly adequate* set in the sense of [26].

Recall from [13, 25] that if $\mathscr{K}$ is a model category then its homotopy category $\operatorname{Ho}\mathscr{K}$ is obtained by quotienting the full subcategory $\mathscr{K}_{cf}$ of objects that are fibrant and cofibrant by the homotopy relation on morphisms. Each choice of a fibrant replacement functor R_f and a cofibrant replacement functor R_c on $\mathscr{K}$ yields an essentially surjective functor

$$P\colon \mathscr{K} \longrightarrow \operatorname{Ho}\mathscr{K}, \tag{1}$$

namely the composite $R_c R_f \colon \mathscr{K} \to \mathscr{K}_{cf}$ followed by the projection $\mathscr{K}_{cf} \to \operatorname{Ho}\mathscr{K}$.

It was shown in [13, Theorem 7.3.1] that, if $\mathscr{I}$ is a set of generating cofibrations in a pointed cofibrantly generated model category $\mathscr{K}$, then the cofibres of morphisms in $\mathscr{I}$ form a weak generator of $\operatorname{Ho}\mathscr{K}$. The assumption that $\mathscr{K}$ be pointed can be removed if $\mathscr{K}$ has a set $\mathscr{I}$ of generating cofibrations between cofibrant objects, in which case the domains and codomains of morphisms in $\mathscr{I}$ form a weak generator of $\operatorname{Ho}\mathscr{K}$, as shown in [26, Theorem 1.2].

We also recall that a small full subcategory $\mathscr{A}$ of a category $\mathscr{C}$ is called *dense* if every object X in $\mathscr{C}$ is a colimit of its canonical diagram with respect to $\mathscr{A}$. This is equivalent to $E_{\mathscr{A}}$ being full and faithful; see [1, Proposition 1.26]. Correspondingly, $E_{\mathscr{A}}$ is full if and only if $\mathscr{A}$ is *weakly dense* in the sense that every object X is a weak colimit of its canonical diagram with respect to $\mathscr{A}$. Finally, $E_{\mathscr{A}}$ is full and conservative if and only if every X is a minimal weak colimit of its canonical diagram with respect to $\mathscr{A}$. Recall that a weak colimit $(\delta_d \colon Dd \to X)$ of a diagram $D\colon \mathscr{D} \to \mathscr{C}$ is called *minimal* if every morphism $f\colon X \to X$ such that $f \circ \delta_d = \delta_d$ for each $d \in \mathscr{D}$ is an isomorphism [10].

Theorem 3.1 *If $\mathscr{K}$ is a combinatorial model category, then there exist arbitrarily large regular cardinals λ such that $\mathscr{K}$ has the following properties:*

1. *$\mathscr{K}$ is locally λ-presentable.*
2. *There is a small weak generator of* $\operatorname{Ho}\mathscr{K}$ *consisting of λ-presentable objects.*
3. *There are fibrant and cofibrant replacement functors R_f and R_c on $\mathscr{K}$ that preserve λ-filtered colimits and λ-presentable objects.*

Proof If $\mathscr{K}$ is combinatorial, then, according to [11, Corollary 1.2], there is a zig-zag of Quillen equivalences into another combinatorial model category $\mathscr{M}$ where all objects are cofibrant. Consequently, the domains and codomains of morphisms in a set of generating cofibrations for $\mathscr{M}$ form a weak generator of the homotopy category $\operatorname{Ho}\mathscr{M}$ by [26, Theorem 1.2]. Since the latter is equivalent to $\operatorname{Ho}\mathscr{K}$, it follows that $\operatorname{Ho}\mathscr{K}$ also has a small weak generator $\mathscr{A}$.

As $\mathscr{K}$ is locally presentable, there are arbitrarily large regular cardinals μ such that $\mathscr{K}$ is locally μ-presentable, by [1, Theorem 1.20]. Thus we can choose μ big enough so that $\mathscr{K}$ is locally μ-presentable and cofibrantly generated by morphisms between μ-presentable objects, and, furthermore, the objects in the chosen weak generator $\mathscr{A}$ are μ-presentable. Then, as shown in the proof of [28, Proposition 3.1], there are μ-accessible functors giving factorizations of morphisms in $\mathscr{K}$ into cofibrations followed by trivial fibrations and into trivial cofibrations followed by fibrations. In particular we can pick a fibrant replacement functor R_f and a cofibrant replacement functor R_c that are μ-accessible. Moreover, using [1, Theorem 2.19] or [11, Proposition 7.2], we can pick a regular cardinal $\lambda \geq \mu$ such that R_f and R_c preserve both λ-filtered colimits and λ-presentable objects. □

Definition 3.2 We call a model category $\mathscr{K}$ *strongly λ-combinatorial* if it is combinatorial and λ satisfies the conditions stated in Theorem 3.1.

For a regular cardinal λ, let $\mathscr{K}$ be a strongly λ-combinatorial model category and denote by $\mathscr{K}_\lambda$ a small full subcategory of representatives of all isomorphism classes of λ-presentable objects. Here and in what follows we assume that fibrant and cofibrant replacement functors R_f and R_c have been chosen on $\mathscr{K}$ so that they preserve λ-filtered colimits and λ-presentable objects.

Let $\operatorname{Ho}\mathscr{K}_\lambda$ denote the full image of the composition

$$\mathscr{K}_\lambda \hookrightarrow \mathscr{K} \xrightarrow{P} \operatorname{Ho}\mathscr{K},$$

where P is the composite $R_c R_f$ followed by the canonical projection as in (1), and denote by $P_\lambda \colon \mathscr{K}_\lambda \to \operatorname{Ho}\mathscr{K}_\lambda$ the domain and codomain restriction of P.

Consider the restricted Yoneda embedding

$$E_\lambda \colon \operatorname{Ho}\mathscr{K} \longrightarrow \mathbf{Set}^{(\operatorname{Ho}\mathscr{K}_\lambda)^{\mathrm{op}}},$$

for which the composite $E_\lambda P$ preserves λ-presentable objects.

The next two results follow from [28, Proposition 5.1 and Corollary 5.2].

Theorem 3.3 *Let $\mathscr{K}$ be a strongly λ-combinatorial model category for a regular cardinal λ. Then the composite*

$$\mathscr{K} \xrightarrow{P} \operatorname{Ho}\mathscr{K} \xrightarrow{E_\lambda} \mathbf{Set}^{(\operatorname{Ho}\mathscr{K}_\lambda)^{\mathrm{op}}}$$

preserves λ-filtered colimits.

Corollary 3.4 *If $\mathscr{K}$ is strongly λ-combinatorial, then $E_\lambda P \cong \operatorname{Ind}_\lambda P_\lambda$.*

Here $\operatorname{Ind}_\lambda$ denotes free cocompletion with respect to λ-filtered colimits [1, Definition 2.25], so $\operatorname{Ind}_\lambda P_\lambda$ is a functor from $\mathscr{K}$ to $\operatorname{Ind}_\lambda \operatorname{Ho}\mathscr{K}_\lambda$. The statement of Corollary 3.4 means that E_λ factorizes through the inclusion

$$\operatorname{Ind}_\lambda \operatorname{Ho}\mathscr{K}_\lambda \subseteq \mathbf{Set}^{(\operatorname{Ho}\mathscr{K}_\lambda)^{\mathrm{op}}},$$

and its codomain restriction, which we keep denoting by E_λ, makes the composite $E_\lambda P$ isomorphic to $\operatorname{Ind}_\lambda P_\lambda$.

If the model category $\mathscr{K}$ is pointed, then $\operatorname{Ind}_\lambda \operatorname{Ho}\mathscr{K}_\lambda$ is also pointed and E_λ preserves the zero object 0, since $E_\lambda 0$ is terminal and it is also initial because 0 is λ-presentable and E_λ is full and faithful on morphisms with domain in $\operatorname{Ho}\mathscr{K}_\lambda$.

Corollary 3.5 *If $\mathscr{K}$ is a strongly λ-combinatorial model category, the codomain restriction $E_\lambda\colon \operatorname{Ho}\mathscr{K} \to \operatorname{Ind}_\lambda \operatorname{Ho}\mathscr{K}_\lambda$ preserves coproducts.*

Proof Pick a cofibrant replacement functor R_c preserving λ-filtered colimits and λ-presentable objects. Note that P preserves coproducts between cofibrant objects and every object in $\operatorname{Ho}\mathscr{K}$ is isomorphic to PX for some cofibrant object X in $\mathscr{K}$. Hence, using Corollary 3.4 it suffices to show that $\operatorname{Ind}_\lambda P_\lambda$ preserves coproducts between cofibrant objects. Since each coproduct is a λ-filtered colimit of λ-small coproducts and $\operatorname{Ind}_\lambda P_\lambda$ preserves λ-filtered colimits, we have to prove that $\operatorname{Ind}_\lambda P_\lambda$ preserves λ-small coproducts between cofibrant objects. Let $\coprod_{i\in I} K_i$ be such a coproduct, so that the cardinality of I is smaller than λ. Since the functor R_c preserves λ-filtered colimits and λ-presentable objects, each K_i is a λ-filtered colimit of cofibrant λ-presentable objects. Let $D_i\colon \mathscr{D}_i \to \mathscr{K}_\lambda$ denote the corresponding diagrams, so that $K_i \cong \operatorname{colim} D_i$. Then $\coprod_{i\in I} K_i$ is a colimit of a λ-filtered diagram whose values are coproducts $\coprod_{i\in I} D_i d_i$ with $d_i \in \mathscr{D}_i$, and each such coproduct $\coprod_{i\in I} D_i d_i$ is λ-presentable as the cardinality of I is smaller than λ. Since the functor $\operatorname{Ind}_\lambda P_\lambda$ preserves λ-filtered colimits and P_λ preserves λ-small coproducts of cofibrant objects, the result is proved. □

Definition 3.6 Let $\mathscr{C}$ be a category with coproducts and λ a cardinal. An object S of $\mathscr{C}$ is *λ-small* if for every morphism $f\colon S \to \coprod_{i\in I} X_i$ there is a subset J of I of cardinality less than λ such that f factorizes as

$$S \longrightarrow \textstyle\coprod_{j\in J} X_j \longrightarrow \coprod_{i\in I} X_i,$$

where the second morphism is the subcoproduct injection.

We also say that $\aleph_0$-small objects are *compact*. This terminology is due to Neeman [21], who found how compactness should be defined for uncountable cardinals in triangulated categories. His definition was simplified by Krause in [18]. They considered compactness in additive categories but the definition makes sense in general. Consider classes $\mathscr{S}$ of λ-small objects in a category $\mathscr{C}$ with coproducts such that for every morphism $f\colon S \to \coprod_{i\in I} X_i$ with $S \in \mathscr{S}$ there exist morphisms $g_i\colon S_i \to X_i$ for which $S_i \in \mathscr{S}$ for all $i \in I$ and f factorizes through

$$\textstyle\coprod_{i\in I} g_i\colon \coprod_{i\in I} S_i \longrightarrow \coprod_{i\in I} X_i.$$

Since the collection of such classes is closed under unions, there is a greatest class with this property. Its objects are called *λ-compact*.

Proposition 3.7 *If $\mathcal{K}$ is a strongly λ-combinatorial model category, then all objects in* $\operatorname{Ho}\mathcal{K}_\lambda$ *are λ-compact in* $\operatorname{Ho}\mathcal{K}$.

Proof Choose fibrant and cofibrant replacement functors R_f and R_c preserving λ-filtered colimits and λ-presentable objects, and let $P\colon \mathcal{K} \to \operatorname{Ho}\mathcal{K}$ be as in (1). Suppose given a morphism $g\colon PA \to \coprod_{i\in I} PK_i$ in $\operatorname{Ho}\mathcal{K}$ where A is in $\mathcal{K}_\lambda$. According to Corollary 3.5, we have

$$E_\lambda g\colon E_\lambda PA \longrightarrow \textstyle\coprod_{i\in I} E_\lambda PK_i.$$

Due to the fact that $E_\lambda P$ preserves λ-presentable objects, $E_\lambda PA$ is λ-presentable in $\operatorname{Ind}_\lambda \operatorname{Ho}\mathcal{K}_\lambda$. Since each coproduct is a λ-filtered colimit of λ-small subcoproducts, $E_\lambda g$ factorizes through some $\coprod_{j\in J} E_\lambda PK_j$ where J has cardinality smaller than λ. Since E_λ is full and faithful on morphisms with domain in $\operatorname{Ho}\mathcal{K}_\lambda$, we obtain a factorization of g through $\coprod_{j\in J} PK_j$ and therefore we conclude that PA is λ-small.

Moreover, the argument used in the proof of Corollary 3.5 shows in a similar way that $E_\lambda g$ factors through some coproduct $\coprod_{j\in J} E_\lambda PD_jd_j$ where J has cardinality smaller than λ and D_jd_j is in $\mathcal{K}_\lambda$ for all j. Using again the fact that E_λ is full and faithful on morphisms with domain in $\operatorname{Ho}\mathcal{K}_\lambda$, we find a factorization of g through $\coprod_{j\in J} PD_jd_j$. Hence PA is indeed λ-compact. □

Definition 3.8 A category with coproducts is called *well λ-generated* if it has a small weak generator consisting of λ-compact objects. It is called *well generated* if it is well λ-generated for some λ.

For example, every locally λ-presentable category is well λ-generated.

The following result was proved in [28, Proposition 6.10] with the additional assumption that $\mathcal{K}$ was stable, which is not necessary.

Theorem 3.9 *If $\mathcal{K}$ is a strongly λ-combinatorial model category, then* $\operatorname{Ho}\mathcal{K}$ *is well λ-generated.*

Proof Since, by assumption, there is a small weak generator of $\operatorname{Ho}\mathcal{K}$ whose objects are λ-presentable, $\operatorname{Ho}\mathcal{K}_\lambda$ weakly generates $\operatorname{Ho}\mathcal{K}$. The rest follows from Proposition 3.7. □

As a corollary one infers Neeman's result in [22] that, for any Grothendieck abelian category $\mathcal{A}$, the derived category $D(\mathcal{A})$ is well generated.

4 Ohkawa's Theorem

For an endofunctor $H\colon \mathcal{K} \to \mathcal{K}$ (not necessarily preserving weak equivalences) on a model category $\mathcal{K}$, we consider the composition

$$\mathcal{K} \xrightarrow{\;H\;} \mathcal{K} \xrightarrow{\;P\;} \operatorname{Ho}\mathcal{K},$$

where P is defined as in (1). The class of objects X in $\mathcal{K}$ such that PHX is the terminal object in $\operatorname{Ho}\mathcal{K}$ will be called the *kernel* of H and will be denoted by $\ker H$. Hence, if $\mathcal{K}$ is pointed and 0 denotes the zero object in $\mathcal{K}$ and also its image in $\operatorname{Ho}\mathcal{K}$, then $\ker H$ consists of objects X in $\mathcal{K}$ such that $PHX = 0$.

In this section we prove the following result.

Theorem 4.1 *Suppose that $\mathcal{K}$ is a pointed strongly λ-combinatorial model category. Then there is only a set of distinct kernels of endofunctors $H\colon \mathcal{K} \to \mathcal{K}$ preserving λ-filtered colimits and the zero object.*

Proof Consider the restricted Yoneda embedding as given by Corollary 3.4,

$$E_\lambda\colon \operatorname{Ho}\mathcal{K} \longrightarrow \operatorname{Ind}_\lambda \operatorname{Ho}\mathcal{K}_\lambda.$$

For a morphism $f\colon E_\lambda S \to E_\lambda PHA$ with $A \in \mathcal{K}_\lambda$ and $S \in \operatorname{Ho}\mathcal{K}_\lambda$, let us denote by $T_H(f)$ the set of all morphisms $t\colon A \to B$ in $\mathcal{K}_\lambda$ such that the composite

$$E_\lambda S \xrightarrow{\ f\ } E_\lambda PHA \xrightarrow{\ E_\lambda PHt\ } E_\lambda PHB$$

is the zero morphism, that is, $E_\lambda PHt \circ f$ factors through the zero object.

Next, we denote

$$J(H) = \{T_H(f) \mid f\colon E_\lambda S \to E_\lambda PHA \text{ with } A \in \mathcal{K}_\lambda \text{ and } S \in \operatorname{Ho}\mathcal{K}_\lambda\}.$$

We are going to prove that if $J(H_1) = J(H_2)$ then $\ker H_1 = \ker H_2$, assuming that H_1 and H_2 preserve λ-filtered colimits and the zero object. Thus suppose that $J(H_2) \subseteq J(H_1)$ and let $X \in \ker H_1$. In order to prove that $PH_2X = 0$, it is enough to show that every morphism $E_\lambda G \to E_\lambda PH_2X$ factors through the zero object if G is in $\operatorname{Ho}\mathcal{K}_\lambda$, since $\operatorname{Ho}\mathcal{K}_\lambda$ is a weak generator of $\operatorname{Ho}\mathcal{K}$ and E_λ is full and faithful on morphisms whose domain is in $\operatorname{Ho}\mathcal{K}_\lambda$.

Assume given such a morphism $f\colon E_\lambda G \to E_\lambda PH_2X$. Since the category $\mathcal{K}$ is locally λ-presentable, $X \cong \operatorname{colim}(D\colon \mathcal{D} \to \mathcal{K}_\lambda)$ for a certain λ-filtered diagram D. Since $E_\lambda PH_2$ preserves λ-filtered colimits by Theorem 3.3, we then have

$$E_\lambda PH_2X \cong \operatorname{colim}\left(\mathcal{D} \xrightarrow{\ D\ } \mathcal{K}_\lambda \xrightarrow{\ PH_2\ } \operatorname{Ho}\mathcal{K} \xrightarrow{\ E_\lambda\ } \operatorname{Ind}_\lambda \operatorname{Ho}\mathcal{K}_\lambda \right).$$

Since $E_\lambda G$ is λ-presentable, f factors through $\hat{f}\colon E_\lambda G \to E_\lambda PH_2Dd$ for some $d \in \mathcal{D}$. Note that the set $T_{H_2}(\hat{f})$ is nonempty, since the morphism $Dd \to 0$ is in it as H_2 preserves the zero object. Consequently, the assumption that $J(H_2) \subseteq J(H_1)$ implies that $T_{H_2}(\hat{f}) \in J(H_1)$. This means that there exist an object $V \in \operatorname{Ho}\mathcal{K}_\lambda$ and a morphism $g\colon E_\lambda V \to E_\lambda PH_1Dd$ such that $T_{H_2}(\hat{f}) = T_{H_1}(g)$.

Now, since $X \in \ker H_1$, we have $E_\lambda PH_1X = 0$. However,

$$E_\lambda PH_1X \cong \operatorname{colim}\left(\mathscr{D} \xrightarrow{D} \mathscr{K}_\lambda \xrightarrow{PH_1} \operatorname{Ho}\mathscr{K} \xrightarrow{E_\lambda} \operatorname{Ind}_\lambda \operatorname{Ho}\mathscr{K}_\lambda \right),$$

and, since $E_\lambda V$ is λ-presentable, there is a morphism $\delta\colon d \to d'$ in $\mathscr{D}$ such that

$$E_\lambda V \xrightarrow{g} E_\lambda PH_1Dd \xrightarrow{E_\lambda PH_1 D\delta} E_\lambda PH_1Dd'$$

factors through the zero object. Hence $D\delta \in T_{H_1}(g)$. Therefore $D\delta \in T_{H_2}(\hat{f})$ and this implies that $f\colon E_\lambda G \to E_\lambda PH_2X$ factors through the zero object, as we wanted to show.

Finally, since there is only a set of distinct sets $J(H)$, the theorem is proved. □

Ohkawa's theorem [24, Theorem 2] is a special case of Theorem 4.1. Recall that two (reduced) homology theories E_* and F_* on spectra are said to be *Bousfield equivalent* if the class of E_*-acyclic spectra coincides with the class of F_*-acyclic spectra. A spectrum X is called *E_*-acyclic* if $E_*(X) = 0$.

Corollary 4.2 *There is only a set of Bousfield equivalence classes of representable homology theories on spectra.*

Proof The homotopy category of spectra admits a combinatorial model category $\mathscr{K}$; for instance, symmetric spectra over simplicial sets [15]. For each cofibrant spectrum E we consider the endofunctor on $\mathscr{K}$ defined as $H_E X = E \wedge R_c X$ where R_c is a cofibrant replacement functor preserving filtered colimits. Since smashing with E has a right adjoint, H_E preserves filtered colimits. Moreover, a spectrum X is in $\ker H_E$ if and only if X is E_*-acyclic. Hence Theorem 4.1 implies that there is only a set of distinct kernels of endofunctors of the form H_E. □

5 Generalized Brown Representability

In this section we prove other properties of combinatorial homotopy categories related to results in [28]. Note that if $\mathscr{C}$ is a locally λ-presentable category with the trivial model structure then the functor $E_\lambda\colon \mathscr{C} \to \operatorname{Ind}_\lambda \mathscr{C}_\lambda$ is an isomorphism.

Definition 5.1 A strongly λ-combinatorial model category $\mathscr{K}$ is called *λ-Brown on morphisms* if $E_\lambda\colon \operatorname{Ho}\mathscr{K} \to \operatorname{Ind}_\lambda \operatorname{Ho}\mathscr{K}_\lambda$ is full. It is called *λ-Brown on objects* if E_λ is essentially surjective. Finally, $\mathscr{K}$ is called *λ-Brown* if it is λ-Brown both on objects and on morphisms.

Let us remark the following facts:

(i) A locally finitely presentable stable combinatorial model category is ω-Brown if it is Brown in the sense of [14], where ω denotes the first infinite ordinal.

(ii) Whenever $\mathscr{K}$ is strongly ω-combinatorial and E_ω is full then E_ω is essentially surjective as well. In fact, by Corollary 3.4, $\mathrm{Ind}_\omega P_\omega$ is full; since each object of $\mathrm{Ind}_\omega \mathscr{K}_\omega$ can be obtained by taking successive colimits of smooth chains [1] and P_ω is essentially surjective on objects, $\mathrm{Ind}_\omega P_\omega$ is essentially surjective on objects too. Hence $\mathscr{K}$ is ω-Brown on objects. This argument does not work for $\lambda > \omega$ because, in the proof, we need colimits of chains of cofinality ω.

(iii) E_λ is full if and only if $\mathrm{Ho}\,\mathscr{K}_\lambda$ is weakly dense in $\mathrm{Ho}\,\mathscr{K}$.

The homotopy category $\mathrm{Ho}\,\mathscr{K}$ of any model category $\mathscr{K}$ has weak colimits and weak limits. Weak colimits are constructed from coproducts and homotopy pushouts in the same way as colimits are constructed from coproducts and pushouts. A homotopy pushout of

$$PC \xleftarrow{\;Pg\;} PA \xrightarrow{\;Pf\;} PB$$

is a commutative diagram

$$\begin{array}{ccc} PA & \xrightarrow{\;Pf_1\;} & PB_1 \\ {\scriptstyle Pg_1}\downarrow & & \downarrow{\scriptstyle P\overline{g}} \\ PC_1 & \xrightarrow{\;P\overline{f}\;} & PE \end{array}$$

where $f = f_2 \circ f_1$ and $g = g_2 \circ g_1$ are factorizations of f and g, respectively, into a cofibration followed by a trivial fibration, and

$$\begin{array}{ccc} A & \xrightarrow{\;f_1\;} & B_1 \\ {\scriptstyle g_1}\downarrow & & \downarrow{\scriptstyle \overline{g}} \\ C_1 & \xrightarrow{\;\overline{f}\;} & E \end{array}$$

is a pushout in $\mathscr{K}$. The following definition is taken from [5].

Definition 5.2 A functor $H\colon \mathscr{C} \to \mathscr{D}$ will be called *nearly full* if for each commutative triangle

$$\begin{array}{ccccc} HA & & \xrightarrow{\quad Hh \quad} & & HC \\ & {\scriptstyle f}\searrow & & \nearrow{\scriptstyle Hg} & \\ & & HB & & \end{array}$$

there is a morphism $\overline{f}\colon A \to B$ in $\mathscr{C}$ such that $H\overline{f} = f$.

Proposition 5.3 *A strongly λ-combinatorial model category $\mathscr{K}$ is λ-Brown on morphisms if and only if the functor $E_\lambda\colon \operatorname{Ho}\mathscr{K} \to \operatorname{Ind}_\lambda \operatorname{Ho}\mathscr{K}_\lambda$ is nearly full.*

Proof Sufficiency is evident because any full functor is nearly full. Let $\mathscr{K}$ be a strongly λ-combinatorial model category and assume that E_λ is nearly full. Consider an object K in $\mathscr{K}$ and express it as a λ-filtered colimit $(\delta_d\colon Dd \to K)$ of its canonical diagram $D\colon \mathscr{D} \to \mathscr{K}_\lambda$. This means that we have

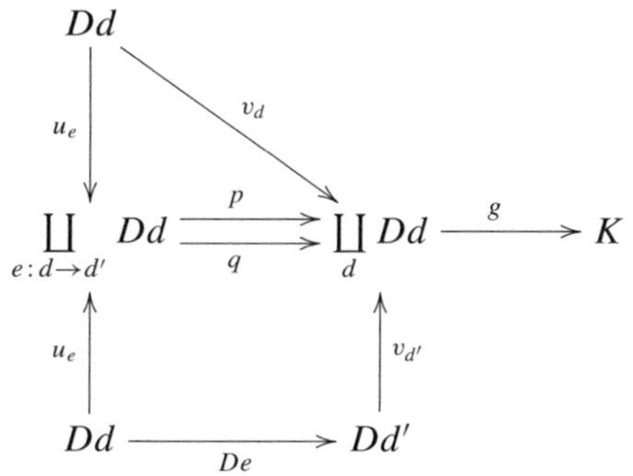

where g is given by a pushout

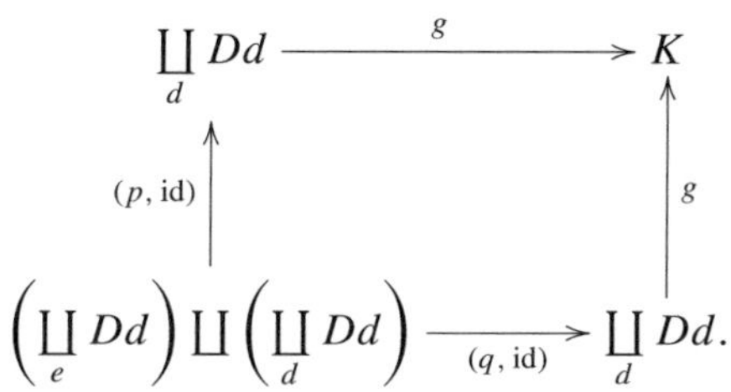

If we replace the pushout above by a homotopy pushout, we get $(\overline{\delta}_d\colon Dd \to \overline{K})$. It is not a cocone in $\mathscr{K}$ but $(P\overline{\delta}_d\colon PDd \to P\overline{K})$ is a standard weak colimit [10] in $\operatorname{Ho}\mathscr{K}$, and there is a comparison morphism $t\colon \overline{K} \to K$ such that $t \circ \overline{\delta}_d = \delta_d$ for each d. Since $H_\lambda = \operatorname{Ind}_\lambda P_\lambda$ preserves λ-filtered colimits, there is a morphism $u\colon H_\lambda K \to H_\lambda \overline{K}$ such that $u \circ H_\lambda \delta_d = H_\lambda \overline{\delta}_d$ for each d. Then $H_\lambda t \circ u = \mathrm{id}$ because

$$H_\lambda t \circ u \circ H_\lambda \delta_d = H_\lambda(t \circ \overline{\delta}_d) = H_\lambda \delta_d.$$

Now, since E_λ is nearly full, there is $\overline{u}\colon PK \to P\overline{K}$ such that $u = E_\lambda \overline{u}$.

Consider a morphism $h\colon H_\lambda K_1 \to H_\lambda K_2$. Let u_1, t_1, u_2, t_2 be as u, t above for K_1 and K_2. There is a cocone $(\gamma_d\colon PD_1 d \to P\overline{K}_2)$ from PD_1 such that

$$E_\lambda \gamma_d = u_2 \circ h \circ H_\lambda \delta_{1d}\colon H_\lambda D_1 d \longrightarrow H_\lambda \overline{K}_2$$

for each d in $\mathscr{D}_1$. Thus there is a morphism $\overline{h}\colon \overline{K}_1 \to \overline{K}_2$ such that $\overline{h} \circ P\overline{\delta}_{1d} = \gamma_d$ for each d in $\mathscr{D}_1$. Hence

$$E_\lambda \overline{h} \circ u_1 \circ H_\lambda \delta_{1d} = E_\lambda \overline{h} \circ H_\lambda \overline{\delta}_{1d} = E_\lambda \gamma_d = u_2 \circ h \circ H_\lambda \delta_{1d}$$

for each d in $\mathscr{D}_1$. Thus $E_\lambda \overline{h} \circ u_1 = u_2 \circ h$. Putting $h' = Pt_2 \circ \overline{h} \circ \overline{u}_1$, we obtain

$$E_\lambda h' = E_\lambda (Pt_2 \circ \overline{h}) \circ u_1 = H_\lambda t_2 \circ u_2 \circ h = h,$$

which proves that E_λ is full. □

Remark 5.4 In Proposition 5.3 it suffices to assume that E_λ is full on split monomorphisms. This means that $h = \mathrm{id}$ in Definition 5.2.

The proof of the following result is given in [28, Proposition 6.4].

Proposition 5.5 *If $\mathscr{K}$ is a combinatorial stable model category, then E_λ reflects isomorphisms for arbitrarily large regular cardinals λ.*

Remark 5.6 If E_λ is full and reflects isomorphisms then each object of $\mathrm{Ho}\,\mathscr{K}$ is a minimal weak colimit of its canonical diagram with respect to $\mathrm{Ho}\,\mathscr{K}_\lambda$.

One could ask if every combinatorial stable model category is λ-Brown for arbitrarily large regular cardinals λ, as discussed in [28, 29]. This fact would have important consequences [23], but it is unfortunately not true. The first counterexample was given in [7], and in [3] a large class was found of combinatorial stable model categories which are not λ-Brown for any λ. An obstruction theory for generalized Brown representability in triangulated categories was developed in [20], with special focus on derived categories of rings.

Acknowledgements This article has been written as a contribution to the proceedings of the memorial conference for Professor Tetsusuke Ohkawa held at the University of Nagoya in 2015. The content of Sect. 4 is based on previous joint work of the authors with Javier Gutiérrez published in [9]. We also appreciate useful discussions with George Raptis. The authors were supported by the Grant Agency of the Czech Republic under grant P201/12/G028, the Agency for Management of University and Research Grants of Catalonia with references 2014 SGR 114 and 2017 SGR 585, and the Spanish Ministry of Economy and Competitiveness under AEI/FEDER research grants MTM2013-42178-P and MTM2016-76453-C2-2-P, as well as grant MDM-2014-0445 awarded to the Barcelona Graduate School of Mathematics.

References

1. Adámek, J., Rosický, J.: Locally Presentable and Accessible Categories. London Mathematical Society Lecture Note Series, vol. 189. Cambridge University Press, Cambridge (1994)
2. Barwick, C.: On left and right model categories and left and right Bousfield localizations. Homol. Homotopy Appl. **12**, 245–320 (2010)

3. Bazzoni, S., Šťovíček, J.: On the abelianization of derived categories and a negative solution to Rosický's problem. Compos. Math. **149**, 125–147 (2013)
4. Beke, T.: Sheafifiable homotopy model categories. Math. Proc. Cambridge Philos. Soc. **129**, 447–475 (2000)
5. Beke, T., Rosický, J.: Abstract elementary classes and accessible categories. Ann. Pure Appl. Logic **163**, 2008–2017 (2012)
6. Bousfield, A.K.: Homotopical localizations of spaces. Am. J. Math. **119**, 1321–1354 (1997)
7. Braun, G., Göbel, R.: Splitting kernels into small summands. Israel J. Math. **188**, 221–230 (2012)
8. Casacuberta, C., Gutiérrez, J.J., Rosický, J.: Are all localizing subcategories of stable homotopy categories coreflective? Adv. Math. **252**, 158–184 (2014)
9. Casacuberta, C., Gutiérrez, J.J., Rosický, J.: A generalization of Ohkawa's theorem. Compos. Math. **150**, 893–902 (2014)
10. Christensen, J.D.: Ideals in triangulated categories: phantoms, ghosts and skeleta. Adv. Math. **136**, 284–339 (1998)
11. Dugger, D.: Combinatorial model categories have presentations. Adv. Math. **164**, 177–201 (2001)
12. Hirschhorn, P.S.: Model Categories and Their Localizations. Mathematical Surveys and Monographs, vol. 99. American Mathematical Society, Providence (2003)
13. Hovey, M.: Model Categories. Mathematical Surveys and Monographs, vol. 63. American Mathematical Society, Providence (1999)
14. Hovey, M., Palmieri, J.H., Strickland, N.P.: Axiomatic Stable Homotopy Theory. Memoirs of the American Mathematical Society, vol. 128, no. 610. American Mathematical Society, Providence (1997)
15. Hovey, M., Shipley, B., Smith, J.H.: Symmetric spectra. J. Am. Math. Soc. **13**, 149–208 (2000)
16. Iyengar, S.B., Krause, H.: The Bousfield lattice of a triangulated category and stratification. Math. Z. **273**, 1215–1241 (2013)
17. Kan, D.M.: On c.s.s. complexes. Am. J. Math. **79**, 449–476 (1957)
18. Krause, H.: On Neeman's well generated triangulated categories. Doc. Math. **6**, 121–126 (2001)
19. Morel, F., Voevodsky, V.: $\mathbb{A}^1$-homotopy theory of schemes. Publ. Math. IHÉS **90**, 45–143 (1999)
20. Muro, F., Raventós, O.: Transfinite Adams representability. Adv. Math. **292**, 111–180 (2016)
21. Neeman, A.: Triangulated Categories. Annals of Mathematics Studies, vol. 148. Princeton University Press, Princeton (2001)
22. Neeman, A.: On the derived category of sheaves on a manifold. Doc. Math. **6**, 483–488 (2001)
23. Neeman, A.: Brown representability follows from Rosický's theorem. J. Topol. **2**, 262–276 (2009)
24. Ohkawa, T.: The injective hull of homotopy types with respect to generalized homology functors. Hiroshima Math. J. **19**, 631–639 (1989)
25. Quillen, D.: Homotopical Algebra. Lecture Notes in Mathematics, vol. 43. Springer, Berlin (1967)
26. Raptis, G.: On the cofibrant generation of model categories. J. Homotopy Relat. Struct. **4**, 245–253 (2009)
27. Röndigs, O., Østvær, P.A.: Modules over motivic cohomology. Adv. Math. **219**, 689–727 (2008)
28. Rosický, J.: Generalized Brown representability in homotopy categories. Theory Appl. Categ. **14**, 451–479 (2005)
29. Rosický, J.: Generalized Brown representability in homotopy categories: Erratum. Theory Appl. Categ. **20**, 18–24 (2008)
30. Rosický, J.: On combinatorial model categories. Appl. Categ. Struct. **17**, 303–316 (2009)
31. Stevenson, G.: An extension of Dwyer's and Palmieri's proof of Ohkawa's theorem on Bousfield classes (2011) (Unpublished manuscript)
32. Whitehead, J.H.C.: Combinatorial homotopy; Part I. Bull. Am. Math. Soc. **55**, 213–245 (1949), Part II. Bull. Am. Math. Soc. **55**, 453–496 (1949)

Notes on an Algebraic Stable Homotopy Category

Ryo Kato, Hiroki Okajima and Katsumi Shimomura

Abstract Ohkawa showed that the collection of Bousfield classes of the stable homotopy category of spectra is a set (Ohkawa in Hiroshima Math. J. 19:631–639, [8]). Let $\mathcal{C}$ be an algebraic stable homotopy category in the sense of Hovey, Palmieri and Strickland (Axiomatic Stable Homotopy Theory, American Mathematical Society, Providence, RI, [6]). We here show that Bousfield classes of $\mathcal{C}$ form a set by introducing a homology theory based on the generators of $\mathcal{C}$, in a similar manner as Dwyer and Palmieri did in Dwyer and Palmieri (Proc. Am. Math. Soc. 129(3):881–886, [3]). We also consider a relation between Bousfield classes of finite objects and supports of them on a collection of objects.

Keywords Stable homotopy category · Bousfield lattice · Ohkawa theorem

1 Introduction

In the stable homotopy category $\mathcal{S}$ of spectra, the Bousfield class $\langle E \rangle$ of a spectrum E is the collection of spectra X with $E \wedge X = 0$. Ohkawa [8] showed that the Bousfield classes of $\mathcal{S}$ form a set (cf. [3]). Then, several authors generalized it to categories with some structure ([2, 4, 7, 9]). In this paper, we consider an algebraic stable homotopy category $\mathcal{C}$ in the sense of [6], which is a triangulated closed symmetric monoidal category $(\mathcal{C}, \wedge, S, F(-,-), \Sigma)$ with a set $\mathcal{G}$ of small objects of $\mathcal{C}$ such that $\operatorname{loc}\langle \mathcal{G} \rangle = \mathcal{C}$, satisfying that $\mathcal{C}$ admits arbitrary coproducts and that every cohomology

R. Kato (✉)
Faculty of Fundamental Science, National Institute of Technology, Niihama College, Niihama 792-8580, Japan
e-mail: ryo_kato_1128@yahoo.co.jp

H. Okajima · K. Shimomura
Department of Mathematics, Faculty of Science, Kochi University, Kochi 780-8520, Japan
e-mail: gg1122cc@gmail.com

K. Shimomura
e-mail: katsumi@kochi-u.ac.jp

T. Ohsawa and N. Minami (eds.), *Bousfield Classes and Ohkawa's Theorem*,
Springer Proceedings in Mathematics & Statistics 309,
https://doi.org/10.1007/978-981-15-1588-0_5

functor on $\mathcal{C}$ is representable. Here, $\mathrm{loc}\langle\mathcal{G}\rangle$ denotes the smallest localizing subcategory containing $\mathcal{G}$, and we call an object A *small* if $[A, \bigvee_\alpha X_\alpha]_* = \bigoplus_\alpha [A, X_\alpha]_*$, where $\bigvee_\alpha X_\alpha$ denotes the coproduct of $\{X_\alpha\}$ in $\mathcal{C}$. For examples of algebraic stable homotopy categories, see [6, 1.2. Examples]. The Bousfield class $\langle E\rangle$ of E in an algebraic stable homotopy category $\mathcal{C}$ is the collection $\{X \in \mathcal{C} \mid E \wedge X = 0\}$. Let $\mathfrak{a}$ denote the cardinal number of the set $\bigoplus_{F,F'\in \mathrm{thick}\langle\mathcal{G}\rangle}[F, F']_*$. Here, $\mathrm{thick}\langle\mathcal{G}\rangle$ denotes the smallest thick subcategory of $\mathcal{C}$ containing $\mathcal{G}$, whose objects we call $\mathcal{G}$*-finite*. Then, we have an analogous theorem to Ohkawa's:

Theorem 1.1 *Let $\mathcal{C}$ be an algebraic stable homotopy category. Then the Bousfield classes $\mathbb{B}(\mathcal{C})$ of $\mathcal{C}$ form a set, whose cardinal number is not greater than $2^{2^{\mathfrak{a}}}$.*

This follows from Lemma 2.4 and Corollary 2.6. We note that $\mathbb{B}(\mathcal{C})$ is a partially ordered set by setting $\langle E\rangle \geq \langle F\rangle$ if $\langle E\rangle \subset \langle F\rangle$. Consider a subset $\mathbb{DL}(\mathcal{C})$ of $\mathbb{B}(\mathcal{C})$ consisting of elements $x \in \mathbb{B}(\mathcal{C})$ satisfying $x \wedge x = x$. We call a non-zero element $a \in \mathbb{DL}(\mathcal{C})$ an *atom* if for any element $x \in \mathbb{B}(\mathcal{C})$, $a \wedge x = a$ or $a \wedge x = 0$. Consider the set $\mathbb{A}(\mathcal{C})$ of atoms of $\mathbb{B}(\mathcal{C})$, and let $\mathfrak{b}$ be the cardinal number of $\mathbb{A}(\mathcal{C})$. Then,

Proposition 1.2 *The cardinal number of $\mathbb{B}(\mathcal{C})$ is not less than $2^{\mathfrak{b}}$.*

Here, we show this by use of a surjection $\mathrm{supp}\colon \mathbb{B}(\mathcal{C}) \to 2^{\mathbb{A}(\mathcal{C})}$ defined by

$$\mathrm{supp}(\mathrm{b}) = \{\mathrm{a} \in \mathbb{A}(\mathcal{C}) \mid \mathrm{a} \wedge \mathrm{b} \neq 0\}. \tag{1.1}$$

In the stable homotopy category $\mathcal{S}_{(p)}$ of p-local spectra, finite spectra are classified by their types. A finite spectrum X has *type n* if $K(n)_*(X) \neq 0$ and $K(m)_*(X) = 0$ for $m < n$. Here, $K(n) \in \mathcal{S}_{(p)}$ denotes the nth Morava K-theory. It is well known that if E and F are finite spectra, then E and F have the same type if and only if $\langle E\rangle = \langle F\rangle$. We generalize this to an algebraic stable homotopy category. We say that $\mathbb{A}(\mathcal{C})$ *detects ring objects* if for any non-zero ring object R, there is an atom $a \in \mathbb{A}(\mathcal{C})$ such that $\langle R\rangle \wedge a \neq 0$.

Proposition 1.4 *Suppose that $\mathbb{A}(\mathcal{C})$ detects ring objects. Let E and F be $\mathcal{G}$-finite objects. Then, $\langle E\rangle = \langle F\rangle$ if and only if* $\mathrm{supp}(\langle \mathrm{E}\rangle) = \mathrm{supp}(\langle \mathrm{F}\rangle)$.

We prove this in section three.

2 Ohkawa Theorem

Let $\mathcal{C}$ denote an algebraic stable homotopy category with a set $\mathcal{G}$ of generators. We call a subcategory $\mathcal{D}$ *thick* if it is closed under cofibrations and retracts, and denote by $\mathrm{thick}\langle\mathcal{G}\rangle$ the smallest thick subcategory containing $\mathcal{G}$.

For $E \in \mathcal{C}$, put

$$E_*^{\mathcal{G}}(X) = \bigoplus_{G\in\mathcal{G}} [G, E \wedge X]_*. \tag{2.1}$$

Since $\mathcal{G} = \{S\}$ in the stable homotopy category of spectra, $E_*^{\mathcal{G}}(X) = [S, E \wedge X]_* = \pi_*(E \wedge X)$ is the homology theory represented by E in the usual sense. In this paper, a homology theory means a homology functor as defined in [6, Definition 1.1.3].

Lemma 2.2 *(1)* $E_*^{\mathcal{G}}(-)$ *is a homology theory.*
(2) ([6, Lemma 1.4.5 (b)]) If $E_*^{\mathcal{G}}(X) = 0$*, then* $E \wedge X = 0$.

For an object $X \in \mathcal{C}$, let $\Lambda(X)$ denote the category whose objects are morphisms $u\colon Z \to X$ of $\mathcal{C}$ for $Z \in \mathrm{thick}\langle\mathcal{G}\rangle$ and whose morphisms between objects $u\colon Z \to X$ and $u'\colon Z' \to X$ are morphisms $Z \xrightarrow{v} Z'$ of $\mathcal{C}$ such that $u'v = u$. Then, we read off the following from [6, Cor. 2.3.11]:

Lemma 2.3 *For any objects E and X of* $\mathcal{C}$, $E_*^{\mathcal{G}}(X) = \operatorname*{colim}_{\Lambda(X)} E_*^{\mathcal{G}}(X_\alpha)$, *where* $\{X_\alpha \to X\}$ *is the set of objects of* $\Lambda(X)$.

Consider the following subset of $A(X) = \bigoplus_{F\in\mathrm{thick}\langle\mathcal{G}\rangle}[X, F]_*$:

$$\mathrm{ann}_X^E(x) = \{f \in [X, F]_* \mid F \in \mathrm{thick}\langle\mathcal{G}\rangle,\ \mathrm{E}_*^{\mathcal{G}}(\mathrm{f})(\mathrm{x}) = 0\} \subset \mathrm{A(X)}$$

for $E \in \mathcal{C}$ and $x \in E_*^{\mathcal{G}}(X)$. Then the Ohkawa class of $E \in \mathcal{C}$ is the set

$$\langle\!\langle E\rangle\!\rangle = \{\mathrm{ann}_F^E(x) \mid F \in \mathrm{thick}\langle\mathcal{G}\rangle,\ \mathrm{x} \in \mathrm{E}_*^{\mathcal{G}}(\mathrm{F})\} \subset 2^{\bigoplus_{\mathrm{F}\in\mathrm{thick}\langle\mathcal{G}\rangle}\mathrm{A(F)}}.$$

Put

$$\mathbb{O} = \{\langle\!\langle E\rangle\!\rangle \mid E \in \mathcal{C}\}.$$

Lemma 2.4 $\mathbb{O}$ *is a set whose cardinal number is not greater than* $2^{2^{\mathfrak{a}}}$, *where* $\mathfrak{a}$ *denotes the cardinal number of* $\bigoplus_{F\in\mathrm{thick}\langle\mathcal{G}\rangle} A(F) = \bigoplus_{F,F'\in\mathrm{thick}\langle\mathcal{G}\rangle}[F, F']_*$.

For an object $E \in \mathcal{C}$, the Bousfield class of E is the collection

$$\langle E\rangle = \{X \in \mathcal{C} \mid E \wedge X = 0\}.$$

We denote the collection of all Bousfield classes of $\mathcal{C}$ by $\mathbb{B}$: $\mathbb{B} = \{\langle E\rangle \mid E \in \mathcal{C}\}$. We define a partial ordering on $\mathbb{B}$ and $\mathbb{O}$ as follows:

- $\langle E\rangle \geq \langle F\rangle$ if $E \wedge X = 0$ implies that $F \wedge X = 0$, and
- $\langle\!\langle E\rangle\!\rangle \geq \langle\!\langle F\rangle\!\rangle$ if for any $\mathrm{ann}_A^F(x) \in \langle\!\langle F\rangle\!\rangle$, there exists $y \in E_*^{\mathcal{G}}(A)$ such that $\mathrm{ann}_A^F(x) = \mathrm{ann}_A^E(y)$.

Then we have a similar lemma as [3, Lemma 1.7]:

Lemma 2.5 *If* $\langle\!\langle E\rangle\!\rangle \geq \langle\!\langle F\rangle\!\rangle$, *then* $\langle E\rangle \geq \langle F\rangle$.

Proof Suppose that $\langle\!\langle E\rangle\!\rangle \geq \langle\!\langle F\rangle\!\rangle$ and let X be an object such that $E \wedge X = 0$. Note that $F_*^{\mathcal{G}}(X) = \operatorname*{colim}_{\Lambda(X)} F_*^{\mathcal{G}}(X_\alpha)$ by Lemma 2.3. Take an element $x \in F_*^{\mathcal{G}}(X_\alpha)$. By hypothesis, for $\mathrm{ann}_{X_\alpha}^F(x) \in \langle\!\langle F\rangle\!\rangle$, there is an element $y \in E_*^{\mathcal{G}}(X_\alpha)$ such that $\mathrm{ann}_{X_\alpha}^F(x) =$

$\mathrm{ann}^E_{X_\alpha}(y)$. Since $E \wedge X = 0$, we have $0 = E^{\mathcal{G}}_*(X)$, which equals $\operatorname{colim}_{\Lambda(X)} E^{\mathcal{G}}_*(X_\alpha)$ by Lemma 2.3. It follows that there is a morphism $f_{\alpha\beta}\colon X_\alpha \to X_\beta \in \Lambda(X)$ for an object $f_\beta\colon X_\beta \to X \in \Lambda(X)$ such that $f_{\alpha\beta} \in \mathrm{ann}^E_{X_\alpha}(y) = \mathrm{ann}^F_{X_\alpha}(x)$. Therefore, $F^{\mathcal{G}}_*(f_{\alpha\beta})(x) = 0$, and so $F^{\mathcal{G}}_*(f_\alpha)(x) = F^{\mathcal{G}}_*(f_\beta)F^{\mathcal{G}}_*(f_{\alpha\beta})(x) = 0 \in F^{\mathcal{G}}_*(X)$. Since X_α and x are both arbitrary, we see that $F^{\mathcal{G}}_*(X) = 0$, and hence $F \wedge X = 0$ by Lemma 2.2. □

Corollary 2.6 *The map $f\colon \mathbb{O} \to \mathbb{B}$ defined by $f(\langle\langle E\rangle\rangle) = \langle E\rangle$ is well-defined. Furthermore, it is an order-preserving surjection.*

Let $\mathbb{DL}$ denote the subset of $\mathbb{B}$ consisting of elements x such that $x \wedge x = x$. Here, the pairing '$\wedge$' is inherited from $\mathcal{C}$, that is, if $x = \langle X\rangle$ and $y = \langle Y\rangle$ for objects X and $Y \in \mathcal{C}$, then $x \wedge y = \langle X \wedge Y\rangle$. We notice that '$\wedge$' is not always a meet in the lattice $\mathbb{B}$. The set $\mathbb{DL}$ is an ordered set bounded below. We call a non-zero element x of $\mathbb{DL}$ an *atom* if $x \wedge y = x$ or $x \wedge y = 0$ for any $y \in \mathbb{B}$. Let $\mathbb{A}$ denote the subset of $\mathbb{DL}$ consisting of atoms. Note that if both of x and y are atoms, then $x \wedge y = x$ if $x = y$ and $x \wedge y = 0$ otherwise. Consider the mapping $\mathrm{supp}\colon \mathbb{B} \to 2^{\mathbb{A}}$ defined by (1.1). We also consider the ordering on $2^{\mathbb{A}}$ by inclusion.

Lemma 2.7 *The mapping* supp *is an order-preserving surjection.*

Proof We see that supp is a surjection, since for a subset $S \subset \mathbb{A}$, we have $s = \bigvee_{a\in S} a \in \mathbb{B}$ satisfying supp(s) = S. Suppose that $e = \langle E\rangle \geq \langle F\rangle = f$. For an element $a = \langle A\rangle \notin$ supp(e), $A \wedge E = 0$, and so $A \wedge F = 0$. Thus, $a \notin$ supp(f), and supp(f) $\subset$ supp(e). □

Corollary 2.8 *The cardinal number of $\mathbb{B}$ is not less than $2^{\mathfrak{b}}$ for the cardinal number $\mathfrak{b}$ of $\mathbb{A}$.*

Remark 2.9 For the stable homotopy category $\mathcal{S}_{(p)}$ of p-local spectra, the role of $\mathbb{A}$ is played by $\{\langle K(n)\rangle \mid n \in \mathbb{N} \cup \{\infty\}\}$, whose cardinal number is $\aleph_0$. Here, $K(n)$ denotes the nth Morava K-theory if $n < \infty$, and the mod p Eilenberg-Mac Lane spectrum if $n = \infty$.

3 Bousfield Classes and Supports on $\mathcal{G}$-Finite Objects

In this section, we apply a thick subcategory theorem for the set $\mathbb{A}$ of atoms used in the previous section. Let $\mathbb{B}$ denote the set of Bousfield classes of a fixed algebraic stable homotopy category $\mathcal{C}$.

We call an object R a *ring object* if R admits an associative multiplication $\mu\colon R \wedge R \to R$ and a unit $\eta\colon S \to R$. Consider the following condition on the category $\mathcal{C}$:

$$\text{For any ring object } R \neq 0,\ \langle R\rangle \wedge \mathbb{A}^\vee \neq 0 \text{ for } \mathbb{A}^\vee = \bigvee_{a\in\mathbb{A}} a. \tag{3.1}$$

In this case, we say that $\mathbb{A}$ *detects ring objects*.

Remark 3.2 In the stable homotopy category $\mathcal{S}_{(p)}$ of p-local spectra, the nilpotence theorem [5, Th. 3 i)] of Hopkins and Smith says that an element α of a homotopy group of a ring spectrum R is nilpotent if and only if $K(n)_*(\alpha)$ is nilpotent for all $0 \le n \le \infty$. It follows that the set $\{\langle K(n)\rangle \mid n \in \mathbb{N} \cup \{\infty\}\} \subset \mathbb{A}$ detects ring objects.

We here call an object F *$\mathcal{G}$-finite* if $F \in \mathrm{thick}\langle\mathcal{G}\rangle$, that is, F belongs to the thick subcategory generated by $\mathcal{G}$, and a thick subcategory $\mathcal{D}$ a *$\mathcal{G}$-ideal* if $X \wedge G \in \mathcal{D}$ for any $X \in \mathcal{D}$ and $G \in \mathcal{G}$. We see that, under (3.1), the set $\mathbb{A}$ of atoms satisfies the conditions of [6, Th. 5.2.2], and so we have the following:

Proposition 3.3 ([6, Th. 5.2.2]) *Suppose that the condition* (3.1) *holds. Then, every $\mathcal{G}$-ideal $\mathcal{D}$ of small objects (= $\mathcal{G}$-finite objects) is expressed by*

$$\mathcal{D} = \{X \in \mathrm{thick}\langle\mathcal{G}\rangle \mid \mathrm{supp}(\langle \mathrm{X}\rangle) \subset \mathrm{supp}(\mathcal{D})\}.$$

Here $\mathrm{supp}(\mathcal{D}) = \bigcup_{\mathrm{X}\in\mathcal{D}} \mathrm{supp}(\langle \mathrm{X}\rangle)$.

Corollary 3.4 *Under the condition* (3.1)*, the class of $\mathcal{G}$-ideals of small objects is a set whose cardinal number is not greater than* $2^{\mathfrak{b}}$.

For an object E, consider the subcategories

$$\begin{aligned}\mathcal{T}_E &= \{X \in \mathrm{thick}\langle\mathcal{G}\rangle \mid \mathrm{supp}(\langle \mathrm{X}\rangle) \subset \mathrm{supp}(\langle \mathrm{E}\rangle)\} \text{ and}\\ \mathcal{T}_E^B &= \{X \in \mathrm{thick}\langle\mathcal{G}\rangle \mid \langle \mathrm{X}\rangle \le \langle \mathrm{E}\rangle\}.\end{aligned}$$

Lemma 3.5 *Both of $\mathcal{T}_E$ and $\mathcal{T}_E^B$ are $\mathcal{G}$-ideals and $\mathcal{T}_E^B \subset \mathcal{T}_E$.*

Proof The last statement follows from Lemma 2.7. By [6, Th. 2.1.3 (a)], it suffices to show that both of the categories are thick. If $X \vee Y \in \mathcal{T}_E$, then $\mathrm{supp}(\langle \mathrm{X}\rangle) \subset \mathrm{supp}(\langle \mathrm{X} \vee \mathrm{Y}\rangle) \subset \mathrm{supp}(\langle \mathrm{E}\rangle)$, and so $X \in \mathcal{T}_E$. Suppose that $X, Y \in \mathcal{T}_E$, and $X \to Y \to Z$ is a cofiber sequence. If $\langle A\rangle \notin \mathrm{supp}(\langle \mathrm{E}\rangle)$, then $\langle A\rangle \notin \mathrm{supp}(\langle \mathrm{X}\rangle)$ and $\langle A\rangle \notin \mathrm{supp}(\langle \mathrm{Y}\rangle)$, which implies that $A \wedge X = 0 = A \wedge Y$. It follows that $A \wedge Z = 0$. Therefore, $\mathrm{supp}(\langle \mathrm{Z}\rangle) \subset \mathrm{supp}(\langle \mathrm{E}\rangle)$. Thus, $\mathcal{T}_E$ is thick. For $\mathcal{T}_E^B$, a similar argument works. □

Corollary 3.6 *Let E be a $\mathcal{G}$-finite object. Then, $\mathcal{T}_E = \mathcal{T}_E^B$.*

Proof By Proposition 3.3 and Lemma 3.5, $\mathcal{T}_E^B = \{X \in \mathrm{thick}\langle\mathcal{G}\rangle \mid \mathrm{supp}(\langle \mathrm{X}\rangle) \subset \mathrm{supp}(\mathcal{T}_\mathrm{E}^\mathrm{B})\}$. For $X \in \mathcal{T}_E^B$, $\mathrm{supp}(\langle \mathrm{X}\rangle) \subset \mathrm{supp}(\langle \mathrm{E}\rangle)$ by Lemma 2.7. Since E is $\mathcal{G}$-finite, we see that $\mathrm{supp}(\mathcal{T}_\mathrm{E}^\mathrm{B}) = \mathrm{supp}(\langle \mathrm{E}\rangle)$. □

Corollary 3.7 *Let X and Y be $\mathcal{G}$-finite objects. Then, $\langle X\rangle = \langle Y\rangle$ if and only if* $\mathrm{supp}(\langle \mathrm{X}\rangle) = \mathrm{supp}(\langle \mathrm{Y}\rangle)$.

Proof The 'only if' part follows from Lemma 2.7. Suppose that $\mathrm{supp}(\langle \mathrm{X}\rangle) = \mathrm{supp}(\langle \mathrm{Y}\rangle)$. Then, $\mathcal{T}_X = \mathcal{T}_Y$, and so $\mathcal{T}_X^B = \mathcal{T}_Y^B$ by Corollary 3.6. Noticing that $X \in \mathcal{T}_X^B$, we see the 'if' part. □

References

1. Bousfield, A.K.: The Boolean algebra of spectra. Comment. Math. Helv. **54**, 368–377 (1979)
2. Casacuberta, C., Gutiérrez, J.J., Rosický, J.: A generalization of Ohkawa's theorem. Comp. Math. **150**, 893–902 (2014)
3. Dwyer, W.G., Palmieri, J.H.: Ohkawa's theorem: there is a set of Bousfield classes. Proc. Am. Math. Soc. **129**(3), 881–886 (2001)
4. Dwyer, W.G., Palmieri, J.H.: The Bousfield lattice for truncated polynomial algebras. Homol. Homotopy Appl. **10**, 413–436 (2008)
5. Hopkins, M.J., Smith, J.H.: Nilpotence and stable homotopy theory II. Ann. Math. **148**, 1–49 (1998)
6. Hovey, M., Palmieri, J.H., Strickland, N.P.: Axiomatic Stable Homotopy Theory. American Mathematical Society, vol. 128/610. Providence, RI (1997)
7. Iyengar, S.B., Krause, H.: The Bousfield lattice of a triangulated category and stratification. Math. Z. **273**, 1215–1241 (2013)
8. Ohkawa, T.: The injective hull of homotopy types with respect to generalized homology functors. Hiroshima Math. J. **19**, 631–639 (1989)
9. Stevenson, G.: An extension of Dwyer's and Palmieri's proof of Ohkawa's theorem on Bousfield classes (2011). https://www.math.uni-bielefeld.de/~gstevens/Ohkawa_thm_paper.pdf

Thick Ideals in Equivariant and Motivic Stable Homotopy Categories

Ruth Joachimi

Abstract We study thick ideals in the stable motivic homotopy category $\mathcal{SH}(k)$ and in its subcategories of compact and of finite cellular objects. If k is a subfield of the complex or even the real numbers, then using comparison functors we find thick ideals corresponding to thick ideals in classical or $\mathbb{Z}/2$-equivariant stable homotopy theory, respectively. We also study motivic Morava K-theories $AK(n)$, for which we prove the motivic analogue of the decomposition of the Bousfield class of $E(n)$ into Bousfield classes of $K(i)$'s over the complex numbers if $p > 2$. In that case we also prove that $AK(n)$-acyclicity implies $AK(n-1)$-acyclicity.

Keywords Motivic homotopy theory · Equivariant homotopy theory · Thick subcategories · Triangulated categories · Motivic Morava K-theories · Motivic Bousfield classes

1 Introduction

In a tensor triangulated category, a thick ideal is a full subcategory which is closed under exact triangles and retracts and under tensoring with arbitrary elements of the category. The classification of thick ideals in the stable homotopy category of p-local finite spectra, $\mathcal{SH}^{fin}_{(p)}$, is given by a famous theorem of Hopkins and Smith, [25, Theorem 7], see Sect. 2. It states that, in $\mathcal{SH}^{fin}_{(p)}$, the thick ideals are given as a chain

$$\mathcal{SH}^{fin}_{(p)} = \mathcal{C}_0 \supsetneq \mathcal{C}_1 \supsetneq \cdots \supsetneq \mathcal{C}_n \supsetneq \cdots \supsetneq \mathcal{C}_\infty = \{0\},$$

and each thick ideal is characterised by the vanishing of a particular Morava K-theory, that is, $\mathcal{C}_n = \{X \in \mathcal{SH}^{fin}_{(p)} \mid K(p, n-1)_*(X) = 0\}$ for $0 < n < \infty$. This theorem is a consequence of the nilpotence theorem [25, Theorem 3], the existence of type-n spectra for any $n \geq 0$ [58, Theorem 4.8] and the fact that $K(n)_*(X) = 0$ implies $K(n-1)_*(X) = 0$ for $X \in \mathcal{SH}^{fin}$ [73, Theorem 2.11].

R. Joachimi (✉)
Fachgruppe Mathematik und Informatik, Bergische Universität Wuppertal, Gaußstraße 20, 42119 Wuppertal, Germany
email: ruthjoachimi@aol.com

T. Ohsawa and N. Minami (eds.), *Bousfield Classes and Ohkawa's Theorem*, Springer Proceedings in Mathematics & Statistics 309,
https://doi.org/10.1007/978-981-15-1588-0_6

In equivariant stable homotopy theory for a finite group G, unpublished work of Strickland [86], see Sect. 3, contains a partial classification of thick ideals in the category $\mathcal{SH}(G)_f \subset \mathcal{SH}(G)$, which is the full subcategory of compact objects in the G-equivariant stable homotopy category. This is a generalisation of the above result, which concerns the special case $\mathcal{SH}(\{1\})_f = \mathcal{SH}^{fin}$. In $\mathcal{SH}(G)_f$, any thick ideal is characterised by the vanishing of particular equivariant Morava K-theories, which are indexed by a prime and a nonnegative integer (as the ordinary Morava K-theories) and, additionally, by a conjugacy class of subgroups of G. The set of thick ideals can be mapped to a lattice of such multi-indices and Strickland proves lower and upper bounds for a sublattice onto which this map is bijective.

In this article, we study thick ideals in $\mathcal{SH}(k)$, $k \subseteq \mathbb{C}$, and related motivic categories, like $(\mathcal{SH}(k)_f)_{(p)}$, the p-localisation of the full subcategory of all compact objects, and $\mathcal{SH}(k)^{fin}_{(p)}$, the category of p-localised finite cell spectra. We use different approaches, all of which are, in some sense, motivated by the results about thick ideals in $\mathcal{SH}^{fin}$.

One approach is to use the comparison functors,

$$\mathcal{SH} \xrightarrow{c_k} \mathcal{SH}(k) \xrightarrow{R_k} \mathcal{SH}$$

for $k \subseteq \mathbb{C}$, and also

$$\mathcal{SH}(\mathbb{Z}/2) \xrightarrow{c'_k} \mathcal{SH}(k) \xrightarrow{R'_k} \mathcal{SH}(\mathbb{Z}/2)$$

for $k \subseteq \mathbb{R}$. We show that, for $k \subseteq \mathbb{C}$, the preimages $R_k^{-1}(\mathcal{C}_n) \subseteq (\mathcal{SH}(k)_f)_{(p)}$, $n \geq 0$, form a chain of different thick ideals in $(\mathcal{SH}(k)_f)_{(p)}$ (Theorem 13). For $k \subseteq \mathbb{R}$, we also show that $(R'_k)^{-1}(\mathcal{C}) \subseteq (\mathcal{SH}(k)_f)_{(p)}$ are different thick ideals, where $\mathcal{C}$ runs over all thick ideals in $(\mathcal{SH}(\mathbb{Z}/2)_f)_{(p)}$, as studied in [86] and in Sect. 3, as well as in [8].

The second approach is to use methods of nilpotence theory. The thick subcategory theorem for $\mathcal{SH}$ is highly related to the nilpotence theorem, which states that the cobordism spectrum MU detects certain kinds of nilpotence. In this context, the Morava K-theories recover the information from MU, meaning that also the family $\{K(p,n) \mid p \text{ prime}, n \geq 0\}$ detects nilpotence. Since the Morava K-theories have a particularly easy structure (they are field theories satisfying the Künneth formula, see e.g. [75, p. 176]), this can be used to show that any thick ideal in $\mathcal{SH}^{fin}$ can be uniquely described in terms of the vanishing and non-vanishing of Morava K-theories. This is how [25, Theorem 7] is proven. Similar nilpotence arguments are used in [86] to classify thick ideals in $\mathcal{SH}(G)_f$ for finite groups G. Strickland shows that the equivariant Morava K-theories detect nilpotence (Theorem 3).

The motivic analog to MU is the algebraic cobordism spectrum MGL. But here, the situation is different, as MGL does not detect nilpotence. There is a notion of motivic Morava K-theories, which are more complicated than the topological ones, but they, too, describe a certain family of thick ideals. In Sect. 7, we show that they do not discover all thick ideals in $(\mathcal{SH}(k)_f)_{(p)}$, and also that not all thick ideals are of the form $R_k^{-1}(\mathcal{C}_n)$.

The third approach is to find different lifts of topological type-n spectra to the motivic world and to ask whether they generate the same thick ideals or different ones. We consider two explicit different such lifts to $(\mathcal{SH}(\mathbb{C})_f)_{(p)}$.

In $\mathcal{SH}_{(p)}^{fin}$, the thick ideals are ordered linearly by inclusion, due to the fact that $K(n+1)_*(X) = 0$ implies $K(n)_*(X) = 0$ for $X \in \mathcal{SH}^{fin}$. This raises the question whether this implication also holds in $\mathcal{SH}(k)_f$. For $p > 2$, we prove that the analog statement holds for motivic Morava K-theories over $\mathbb{C}$ if X is a finite cellular motivic spectrum, as studied in [16]. That is, it holds for $X \in \mathcal{SH}(\mathbb{C})^{fin} \subseteq \mathcal{SH}(\mathbb{C})_f$. On the way, we prove a couple of interesting facts concerning the motivic versions of BP, $K(n)$ and related theories. We prove that the analog of the decomposition of Bousfield classes $\langle E(n)\rangle = \bigvee_{i\leq n} \langle K(i)\rangle$ holds in $\mathcal{SH}(\mathbb{C})$ (for $p > 2$), as conjectured by Hornbostel in [27, Question 2.17] for arbitrary fields.

While we study the tensor triangulated spectrum of the Morel–Voevodsky stable homotopy category over fields $k \subseteq \mathbb{C}$, we refer to [42] for an account on the case of finite fields. Another paper to mention is [24], which studies this object for fields of characteristic different from 2 and proves the surjectivity of Balmer's comparison map.

Outline

In Sect. 2, we introduce basic notation concerning thick ideals and the stable homotopy category $\mathcal{SH}$. We recall the thick subcategory theorem of Hopkins and Smith (Theorem 1). In $\mathcal{SH}^{fin}$, there is no difference between thick subcategories and thick ideals (Lemma 1).

Section 3 is an account on Strickland's work [86]. It contains Strickland's main results on thick ideals in $\mathcal{SH}(G)_f$ and their proofs. The section begins with the necessary recollection from equivariant stable homotopy theory, such as compact objects in $\mathcal{SH}(G)$ and geometric fixed point functors. Equivariant Morava K-theories

$$K(n, H) = G/H_+ \wedge \tilde{E}[\not\supseteq H] \wedge K(n), \text{ for } H \subseteq G,$$

are introduced in Sect. 3.2. They are related to the classical Morava K-theories via the geometric fixed point functor (Proposition 14) and satisfy similar properties, such as the Künneth formula (Corollary 15). Section 3.3 introduces the terminology of lattices and contains the result of Strickland which establishes a general relation between thick ideals and the detection of nilpotence by some family of homology theories (Theorem 2). The equivariant analog of the nilpotence theorem [25, Theorem 3] is Theorem 3. We give a reformulation of the thick subcategory theorem [25, Theorem 7] in a non-p-localised way (Theorem 4) and prove a similar equivariant result,

Theorem 5, which describes an injective lattice homomorphism from the set of thick ideals in $\mathcal{SH}(G)_f$ to the lattice

$$GQ = \prod_{\mathrm{sub}(G)} \{u \in \prod_p \mathcal{Q}_p \mid u_p = 1 \,\forall\, p \text{ or } u_p \neq 1 \,\forall\, p\},$$

where $\mathcal{Q}_p = \{p^{-n} \mid 0 \leq n \leq \infty\}$, and gives a lower bound for its image:

Theorem 5 (Strickland) *The composition*

$$\tau : \mathrm{Idl}(\mathcal{SH}(G)_f) \xrightarrow{\mathrm{supp}} \mathcal{P}(GQ') \xrightarrow{\max} GQ$$

is injective. Its image contains all $u \in GQ$ which satisfy: if $H \subseteq H'$, then $u_H \geq u_{H'}$.

Here, $\mathcal{P}(GQ')$ denotes the power set of the set

$$GQ' = \{p^{-n} \mid p \text{ prime},\ 0 \leq n < \infty\} \times \mathrm{sub}(G).$$

An upper bound is given in Proposition 33. In Sect. 3.5, we apply Strickland's results to $\mathcal{SH}(\mathbb{Z}/2)_f$, which will be most interesting to us in our study of thick ideals in $\mathcal{SH}(k)_f$, $k \subseteq \mathbb{R}$. Any thick ideal in $(\mathcal{SH}(\mathbb{Z}/2)_f)_{(p)}$ is of the form

$$\mathcal{C}_{m,n} = \{X \mid \phi^{\{1\}}(X) \in \mathcal{C}_m \text{ and } \phi^{\mathbb{Z}/2}(X) \in \mathcal{C}_n\},$$

where $m, n \in [0, \infty]$ and $\phi^H : \mathcal{SH}(G) \to \mathcal{SH}$ is the geometric H-fixed point functor (Corollary 34). But not all $\mathcal{C}_{m,n}$ are different. Corollary 36 gives partial information on which ones are.

In Sect. 4, we introduce the comparison functors $\mathcal{SH} \xrightarrow{c_k} \mathcal{SH}(k) \xrightarrow{R_k} \mathcal{SH}$ for $k \subseteq \mathbb{C}$, and $\mathcal{SH}(\mathbb{Z}/2) \xrightarrow{c'_k} \mathcal{SH}(k) \xrightarrow{R'_k} \mathcal{SH}(\mathbb{Z}/2)$ for $k \subseteq \mathbb{R}$, which are symmetric monoidal and satisfy $R_k \circ c_k \cong \mathrm{id}$ and $R'_k \circ c'_k \cong \mathrm{id}$, respectively. This is mainly a recollection from various other sources. The same results are independently obtained in [23, Sect. 4].

In Sect. 5, we apply our knowledge concerning comparison functors to the study of thick ideals, proving the following theorem for any prime p.

Theorem 13 (Lower bound on the number of motivic thick ideals)

(1) If $k \subseteq \mathbb{C}$, the category $(\mathcal{SH}(k)_f)_{(p)}$ contains at least an infinite chain of different thick ideals given by $\overline{R}_k^{-1}(\mathcal{C}_n)$, $0 \leq n \leq \infty$, where $\overline{R}_k$ denotes the p-localisation of the restriction of R_k to $\mathcal{SH}(k)_f$ and $\mathcal{C}_n \subseteq \mathcal{SH}^{fin}_{(p)}$ is as defined in Sect. 1.

(2) If $k \subseteq \mathbb{R}$, then $(\mathcal{SH}(k)_f)_{(p)}$ contains at least a two-dimensional lattice of different thick ideals given by $(\overline{R}'_k)^{-1}(\mathcal{C}_{m,n})$, for all $(m, n) \in \Gamma_p$ as in Definition 35.

One ingredient of this theorem is Proposition 61, where we show that the realisation functors R_k and R'_k preserve compactness. In Sect. 5.3, we also prove a couple of additional results on the connection between motivic thick ideals and the comparison functors.

Section 6 begins with an account of homology and cohomology theories in the category of finite motivic cell spectra, $\mathcal{SH}(k)^{fin}$, as studied by Dugger and Isaksen in [16]. We show that, for a cellular ring spectrum E and a finite cellular spectrum X, $E_{**}X = 0$ is equivalent to $E^{**}X = 0$ (Proposition 69). For $k \subseteq \mathbb{R}$, we use a notion of cellular spectra which is more general than the notion from [16], see Definition 52. This yields another version of Proposition 69 (Corollary 72). In Sect. 6.2, we discuss different ways of defining thick ideals associated with a (ring) spectrum. This is applied to motivic Morava K-theories $AK(n)$ in Sect. 6.4. For example, we show that the thick ideal $\mathcal{C}_{AK(n)}$ associated with the n-th motivic Morava K-theory is contained in $R_k^{-1}(\mathcal{C}_{n+1})$ (Proposition 78). We recall the definition and some properties of the motivic Morava K-theories in Sect. 6.3. The motivic Atiyah Hirzebruch spectral sequence described in [32, Example 8.13], implies that the n-th motivic Morava K-theory over the field $\mathbb{C}$ has coefficient ring $H\mathbb{Z}/p_{**} \otimes K(n)_*$ (Lemma 5), as remarked in [96] below Corollary 3.9.

In Sect. 7, we study the thick ideal generated by the cofiber of the motivic Hopf map, $C\eta \cong \mathbb{P}^2_k$, and compare it to the thick ideals $R_k^{-1}(\mathcal{C}_n)$ and $\mathcal{C}_{AK(n)}$ for $k \subseteq \mathbb{C}$. We calculate the type of $R_k(C\eta_{(p)}) \in \mathcal{SH}^{fin}_{(p)}$, which is 1, and the equivariant type of $R'_k(C\eta_{(p)}) \in (\mathcal{SH}(\mathbb{Z}/2)_f)_{(p)}$, which is $(1, 2)$ for $p = 2$ and $(1, \infty)$ for odd p (Proposition 82). In Proposition 83, we show that $C\eta_{(p)}$ generates a thick ideal of $(\mathcal{SH}(k)_f)_{(p)}$ which is neither of the form $R_k^{-1}(\mathcal{C}_{n+1})$ or $\mathcal{C}_{AK(n)}$ for any $n \geq 0$, nor is it all of $(\mathcal{SH}(k)_f)_{(p)}$ (at least, if $p = 2$ or $k \subseteq \mathbb{R}$).

Proposition 83 *For $k \subseteq \mathbb{C}$, let* $\text{thickid}(C\eta_{(p)}) \subseteq (\mathcal{SH}(k)_f)_{(p)}$ *denote the thick ideal generated by the p-localised cofiber of the Hopf map. Then the following hold:*

(1) $\text{thickid}(C\eta_{(p)}) \nsubseteq \mathcal{C}_{AK(n)}$ *for any $n \geq 0$ and any prime p,*
(2) $\text{thickid}(C\eta_{(p)}) \nsubseteq R_k^{-1}(\mathcal{C}_n)$ *for any $n > 0$ and any prime p,*
(3) $\text{thickid}(C\eta_{(p)}) \subsetneq \text{thickid}(S^0_{(p)}) = (\mathcal{SH}(k)_f)_{(p)}$ *if $k \subseteq \mathbb{R}$ and p is any prime or $k \subseteq \mathbb{C}$ and $p = 2$.*
(4) For any prime p, the thick ideals $\text{thickid}(C\eta_{(p)}) \cap R_k^{-1}(\mathcal{C}_n)$ *are distinct for different $n \geq 0$ and in particular nonzero if $n < \infty$.*

This proves that $\mathcal{SH}(k)_f, k \subseteq \mathbb{C}$, really has "more" thick ideals than its topological counterpart. In Sect. 7.2, we compare our results to Balmer's work on prime ideals [7]. For the categories $\mathcal{SH}^{fin}$, $\mathcal{SH}(\mathbb{Z}/2)_f$, $\mathcal{SH}(\mathbb{C})_f$ and $\mathcal{SH}(\mathbb{R})_f$, we recover the information on prime ideals given in [7, Sect. 10] from a different point of view.

In Sect. 8, we study two preimages under $R_{\mathbb{C}}$ of a type-n spectrum $X_n \in \mathcal{SH}^{fin}_{(p)}$. One of them is $c_{\mathbb{C}}(X_n)$ and the other one, $\mathbb{X}_n$, is constructed by a motivic version of the construction of X_n, as given in [75, Appendix C]. In analogy to Mitchell's result [58, Theorem 4.8], we prove the following vanishing theorem for motivic Morava K-theory.

Theorem 14 (Vanishing criterion) *Let $s > 0$ and $X \in \mathcal{SH}(\mathbb{C})^{fin}$ be a finite motivic cell spectrum such that* $\mathrm{H}^{**}(X, \mathbb{Z}/p)$ *is free over the exterior algebra $\Lambda_{H\mathbb{Z}/p^{**}}(Q_s)$ as a module over the motivic Steenrod algebra. Then $AK(s)_{**}X = 0$.*

This is proven with the help of the motivic Adams spectral sequence for $Ak(s) \wedge X$, where $Ak(s)$ is the motivic analog of the connective Morava K-theory spectrum. This spectral sequence is studied in Sect. 8.3. In Sect. 8.5, we construct a spectrum $\mathbb{X}_n$ satisfying the assumption of the theorem, and we show that this spectrum is indeed of motivic type n (Theorem 16).

Section 9 is devoted to the study of the Bousfield classes of $AK(n)$ and related motivic spectra. The main goal in writing this section was to prove that $AK(n+1)_{**}(X) = 0$ implies $AK(n)_{**}(X) = 0$, which we show for $X \in \mathcal{SH}(\mathbb{C})^{fin}$ and $p > 2$ in Theorem 21.

Theorem 21 *Let $p > 2$. If $X \in \mathcal{SH}(\mathbb{C})^{fin}$ satisfies $AK(n+1)_{**}(X) = 0$, then it also satisfies $AK(n)_{**}(X) = 0$.*

A lot of results in Sect. 9 hold more generally. In Sect. 9.1, we prove that v_n-torsion in $ABP_{**}ABP$ is also v_{n-1}-torsion. This holds in any $\mathcal{SH}(k)$, $k \subseteq \mathbb{C}$ (Theorem 17). The proof uses methods similar to the topological version of the statement, [40, Theorem 0.1]. Another ingredient is the map of Hopf algebroids $(BP_*, BP_*BP) \to (ABP_{**}, ABP_{**}ABP)$, as studied in [65]. In Sect. 9.4, we construct certain operations on $AP(n)$ in $\mathcal{SH}(k)$, $k \subseteq \mathbb{C}$ (Theorem 18), similar to the operations on $P(n)$ constructed by Würgler in [94, Theorem 5.1]. These are used to prove the equality of Bousfield classes $\langle AK(n) \rangle = \langle AB(n) \rangle$ in $\mathcal{SH}(\mathbb{C})$ (Corollary 147) with methods similar to those of [39]. In the proof of Corollary 147, we assume $k = \mathbb{C}$ because we make use of the explicitly known coefficient rings $H\mathbb{Z}/p_{**}$ and $AK(n)_{**}$. The result is used to prove Theorem 20, which is the following decomposition of Bousfield classes in $\mathcal{SH}(\mathbb{C})$, as conjectured in [27, Question 2.17] for arbitrary fields.

Theorem 20 *For $p > 2$,*

$$\langle AE(n) \rangle = \bigvee_{i \leq n} \langle AK(i) \rangle \text{ in } \mathcal{SH}(\mathbb{C}).$$

Large parts of Sects. 8 and 9 also hold for $k = \mathbb{R}$, when p is odd, using Lemma 5 applied to $\mathrm{H}^{**}(\mathbb{R}, \mathbb{Z}/p) \cong \mathbb{Z}/p[\tau^2]$. Here $\deg(\tau^2) = (0, 2)$ and τ^2 is mapped to $\tau^2 \in \mathrm{H}^{**}(\mathbb{C}, \mathbb{Z}/p)$ by the base change functor.

2 Thick Ideals in Classical Stable Homotopy Theory

In this section, we introduce basic notation concerning thick ideals and the stable homotopy category $\mathcal{SH}$. In $\mathcal{SH}^{fin}$, a thick ideal is the same as a thick subcategory (Lemma 1). We recall the thick subcategory theorem of Hopkins and Smith in Theorem 1.

Definition 1 A tensor triangulated category is a triple $(\mathcal{T}, \wedge, S)$ consisting of a triangulated category $\mathcal{T}$ and a symmetric monoidal product $\wedge$ on $\mathcal{T}$ with unit S, such that for any $A \in \mathcal{T}$, $A \wedge -$ preserves exact triangles (see e.g. [6, Definition 1.1]).

Example 2 The stable homotopy category $(\mathcal{SH}, \wedge, S)$ with $S = S^0 = \Sigma^\infty S^0$ is a tensor triangulated category.

Definition 3 Let $(\mathcal{T}, \wedge, S)$ be a tensor triangulated category. A full triangulated subcategory $\emptyset \neq \mathcal{C} \subseteq \mathcal{T}$ is called a

(1) thick subcategory if it is closed under retracts.
(2) thick ideal if it is a thick subcategory and in addition satisfies:

$$\text{if } X \in \mathcal{T} \text{ and } Y \in \mathcal{C} \text{ then } X \wedge Y \in \mathcal{C}.$$

If $\mathcal{X}$ is an object or a set of objects, we denote the smallest thick ideal containing $\mathcal{X}$ by $\text{thickid}(\mathcal{X})$ and call it the thick ideal generated by $\mathcal{X}$. This is well defined because the intersection of thick ideals is again a thick ideal. If $\mathcal{X}$ is a finite set of objects, $\text{thickid}(\mathcal{X})$ is called finitely generated.

Remark 4 Any finitely generated thick ideal is generated by a single element, namely the direct sum of all generators.

Example 5 If $(\mathcal{T}, \wedge, S)$ is a tensor triangulated category, then

$$\text{thickid}(S) = \mathcal{T},$$

since for any $X \in \mathcal{T}$, $X \cong X \wedge S$.

More generally, if Z is in the Picard group $Pic(\mathcal{T})$, i.e., if there exists a Z' such that $Z' \wedge Z \cong S$, then $\text{thickid}(Z) = \mathcal{T}$. The Picard group of the stable homotopy category $\mathcal{SH}$ consists precisely of the spheres $\Sigma^n S^0$, $n \in \mathbb{Z}$ (see e.g. [26]), the Picard group of the equivariant stable homotopy category $\mathcal{SH}(G)$ is described in [19] and examples for elements in the Picard groups of motivic stable homotopy categories are given in [35].

Definition 6 The category $\mathcal{SH}^{fin}$ is the smallest full subcategory of $\mathcal{SH}$ that contains all finite desuspensions of suspension spectra of finite CW complexes and is closed under isomorphisms.

Remark 7 $\mathcal{SH}^{fin}$ is a tensor triangulated subcategory of $\mathcal{SH}$. It can equivalently be defined as the smallest thick subcategory of $\mathcal{SH}$ that contains S^0, or as the full subcategory of compact objects in $\mathcal{SH}$ (see, e.g. [78, Theorem II.7.4]).

Lemma 1 *In $\mathcal{SH}^{fin}$, any thick subcategory is already a thick ideal.*

Proof Let X be an element of the thick subcategory $\mathcal{C} \subseteq \mathcal{SH}^{fin}$ and let $Y \in \mathcal{SH}^{fin}$. By the definition of $\mathcal{SH}^{fin}$, there is a finite sequence of spectra $\{Y^k\}_{0 \le k \le n}$ such that $Y^0 = S^{n_0}$, $Y \cong Y^n$ and each Y^k is the cofiber of some map $S^{n_k} \to Y^{k-1}$, $n_k \in \mathbb{Z}$. Any thick subcategory is closed under suspensions and desuspensions because $\Sigma^{\pm 1}X$ lies in an exact triangle with $X \xrightarrow{1} X$. Hence, $X \wedge Y^0 \in \mathcal{C}$. Assume that $X \wedge Y^k \in \mathcal{C}$ for some k. Then $X \wedge S^{n_{k+1}} \to X \wedge Y^k \to X \wedge Y^{k+1}$ is an exact triangle whose first two objects are in $\mathcal{C}$. Since $\mathcal{C}$ is a thick subcategory, it follows that $X \wedge Y^{k+1} \in \mathcal{C}$, too, and inductively, $X \wedge Y^n \in \mathcal{C}$. Note further that thick subcategories are closed under isomorphisms as these are special cases of retractions. Hence, $X \wedge Y \in \mathcal{C}$, which proves that $\mathcal{C}$ is a thick ideal. □

Definition 8 Let p be a prime number. The p-local categories $\mathcal{SH}_{(p)}$ and $\mathcal{SH}_{(p)}^{fin}$ are defined as the Bousfield localisations of $\mathcal{SH}$ and $\mathcal{SH}^{fin}$ at the p-local Moore spectrum $M\mathbb{Z}_{(p)}$.

It is a common procedure to study spectra p-locally for each prime p, i.e. instead of $X \in \mathcal{SH}$ one studies its image $X_{(p)}$ under $\mathcal{SH} \to \mathcal{SH}_{(p)}$, and then fits the information together. For example, n-th Morava K-theory $K(n)$ is defined for any fixed prime p, where it satisfies $K(n)_*(X) = K(n)_*(X_{(p)})$. For the construction and properties of $K(n)$, see e.g. [39].

Now we are ready to state the thick subcategory theorem of Hopkins and Smith [25, Theorem 7], which was the main motivation for this paper. It gives a beautiful and complete description of the thick ideals in $\mathcal{SH}_{(p)}^{fin}$ in terms of Morava K-theories.

Theorem 1 (Hopkins, Smith) *In $\mathcal{SH}_{(p)}^{fin}$, the thick ideals are given as a chain*

$$\mathcal{SH}_{(p)}^{fin} = \mathcal{C}_0 \supsetneq \mathcal{C}_1 \supsetneq \cdots \supsetneq \mathcal{C}_n \supsetneq \cdots \supsetneq \mathcal{C}_\infty = \{0\},$$

with $\mathcal{C}_n = \{X \in \mathcal{SH}_{(p)}^{fin} \mid K(n-1)_*(X) = 0\}$ for $0 < n < \infty$.

Definition 9 A spectrum $X \in \mathcal{SH}_{(p)}^{fin}$ is said to be of type n if $K(n-1)_*(X) = 0$ and $K(n)_*(X) \neq 0$. We write $\text{type}(X) = n$.

For any fixed prime p, the type of a spectrum is well-defined by [73, Theorem 2.11]. The thick subcategory theorem implies that any spectrum X of type n generates $\mathcal{C}_n$ as a thick ideal.

3 Thick Ideals in Equivariant Stable Homotopy Theory

The contents of this section (except for the introductory section and some details) are due to Neil Strickland [86]. We state Strickland's main results on thick ideals in $\mathcal{SH}(G)_f$ and their proofs. We start with the necessary recollection from equivariant stable homotopy theory. Thick ideals in $\mathcal{SH}(G)_f$ are classified by equivariant Morava K-theories, $K(n, H) = G/H_+ \wedge \tilde{E}[\not\supseteq H] \wedge K(n)$, $H \subseteq G$, which are introduced in Sect. 3.2. They are related to the classical Morava K-theories via the geometric fixed point functor (Proposition 14) and satisfy similar properties, such as the Künneth formula (Corollary 15). As in the non-equivariant theory of Hopkins and Smith, the equivariant Morava K-theories detect nilpotence in $\mathcal{SH}(G)_f$ (Theorem 3). The general relation between the detection of nilpotence by a family of homology theories and thick ideals is described in Theorem 2. As a corollary, we reformulate the thick subcategory theorem [25, Theorem 7] in a non-p-localised way (Theorem 4) and prove a similar equivariant result, Theorem 5, which describes an injective lattice homomorphism from the set of thick ideals in $\mathcal{SH}(G)_f$ to a particular lattice $G\mathcal{Q}$ and gives a lower bound for its image. An upper bound is given in Proposition 33.

For our study of thick ideals in the motivic stable homotopy categories $\mathcal{SH}(k)_f$, $k \subseteq \mathbb{R}$, we will use the here given knowledge concerning thick ideals in the $\mathbb{Z}/2$-equivariant stable homotopy category. Therefore, the case $G = \mathbb{Z}/2$ is the interesting one for the rest of this paper and we will summarise all results on thick ideals in $\mathcal{SH}(\mathbb{Z}/2)_f$ in Sect. 3.5. Any thick ideal in $(\mathcal{SH}(\mathbb{Z}/2)_f)_{(p)}$ is of the form $\mathcal{C}_{m,n} = \{X \mid \phi^{\{1\}}(X) \in \mathcal{C}_m \text{ and } \phi^{\mathbb{Z}/2}(X) \in \mathcal{C}_n\}$, where $m, n \in [0, \infty]$ and ϕ^H is the geometric fixed point functor (Corollary 34). But not all $\mathcal{C}_{m,n}$ are different. Corollary 36 gives partial information on which ones are.

In the meanwhile, a full description of thick ideals in $\mathcal{SH}(\mathbb{Z}/2)_f$ has been given by [8].

3.1 Equivariant Stable Homotopy Theory

Let G be a finite group and $\mathcal{SH}(G)$ be the stable homotopy category of genuine G-spectra. This category has quite a few models. We switch between spectra of G-CW complexes and spectra of G-simplicial sets, depending on which is more convenient in the concrete situation. A good model for $\mathcal{SH}(G)$ as a tensor triangulated category is the category of orthogonal G-spectra, see e.g. [53] or [79]. In Sect. 4.2, we make use of two other models, namely symmetric G-spectra and $G\Sigma_G$-spectra. The following definition of finite G-spectra, for example, makes sense if we use the model of orthogonal G-spectra with G-CW complexes as the underlying category of spaces. The definition induces a notion of finiteness for any other model for $\mathcal{SH}(G)$.

Definition 10 For G a finite group, let $\mathcal{SH}(G)^{fin}$ be the smallest full subcategory of $\mathcal{SH}(G)$ that contains all finite desuspensions of suspension spectra of finite G-CW complexes and is closed under isomorphisms. The objects of $\mathcal{SH}(G)^{fin}$ are called finite G-CW spectra. We denote the closure of $\mathcal{SH}(G)^{fin}$ under retracts in $\mathcal{SH}(G)$ by $\mathcal{SH}(G)_f$.

Both $\mathcal{SH}(G)^{fin}$ and $\mathcal{SH}(G)_f$ are tensor triangulated subcategories of $\mathcal{SH}(G)$ because finite G-CW complexes are closed under cofiber sequences and under smash products and because retracts commute with smash products.

Definition 11 A spectrum $X \in \mathcal{SH}(G)$ is called dualisable, if the canonical map

$$F(X, S^0) \wedge Y \to F(X, Y)$$

is an isomorphism for any $Y \in \mathcal{SH}(G)$, where $F(-,-)$ denotes the derived function spectrum and $S^0 = S^0_G$ is the sphere spectrum in $\mathcal{SH}(G)$ (for possible definitions of S^0 and $F(-,-)$, see e.g. [79, Examples 2.10 and 5.12]). $DX = F(X, S^0)$ is called the dual of X. It satisfies $DDX \cong X$ ([51, Proposition III.1.3]). In the following, we also use the notation S for S^0_G, since it is the unit in $\mathcal{SH}(G)$.

$X \in \mathcal{SH}(G)$ is called compact if $[X, -]_{\mathcal{SH}(G)} = \mathrm{Hom}_{\mathcal{SH}(G)}(X, -)$ preserves arbitrary coproducts.

Proposition 12 *The subcategory $\mathcal{SH}(G)_f \subseteq \mathcal{SH}(G)$ has the following equivalent descriptions:*

(1) It is the full subcategory of retracts of finite G-CW spectra.
(2) It is the full subcategory of dualisable objects.
(3) It is the full subcategory of compact objects.

Proof Items (1) and (2) are equivalent by [54, Theorem XVI.7.4].

Furthermore, any dualisable object X is also compact, because

$$\left[X, \bigvee Y_i\right] = \left[S, F(X, \bigvee Y_i)\right] = \left[S, DX \wedge \bigvee Y_i\right] = \left[S, \bigvee (DX \wedge Y_i)\right]$$

$$= \bigoplus [S, DX \wedge Y_i] = \bigoplus [S, F(X, Y_i)] = \bigoplus [X, Y_i],$$

where we used that $F(X, -)$ is right adjoint to $X \wedge -$ and that the unit S is compact, see e.g. [79, Corollary 3.30(i)]. Since $\pi_n^H X = 0$ for all $H \subseteq G$ and $n \in \mathbb{Z}$ implies $X \cong 0$ in $\mathcal{SH}(G)$ by the definition of $\mathcal{SH}(G)$, $\{\Sigma^n \Sigma^\infty (G/H)_+ \mid H \subseteq G, n \in \mathbb{Z}\}$ is a detecting set and, hence, also a generating set by [55, Lemma 13.1.6]. That is, the smallest thick subcategory of $\mathcal{SH}(G)$ which is closed under infinite coproducts and contains $\{\Sigma^n \Sigma^\infty (G/H)_+ \mid H \subseteq G, n \in \mathbb{Z}\}$ is $\mathcal{SH}(G)$ itself. By [51, Corollary II.6.3], $\Sigma^\infty (G/H)_+$ is dualisable and, thus, compact. Hence, $\{\Sigma^\infty (G/H)_+ \mid H \subseteq G\}$ is a set of compact generators for $\mathcal{SH}(G)$. By general theory due to Neeman [66], see e.g. [55, Theorem 13.1.14], the full subcategory of compact objects in $\mathcal{SH}(G)$ is the thick subcategory generated by this set (i.e., the smallest thick subcategory of $\mathcal{SH}(G)$ containing this set). Therefore, (3) is also equivalent to (1) and (2). □

Let G be a finite group and $H \subseteq G$ a subgroup. There are functors $i : \mathcal{SH} \to \mathcal{SH}(G)$ and $\phi^H : \mathcal{SH}(G) \to \mathcal{SH}$, where i maps a nonequivariant spectrum to the corresponding G-spectrum with trivial G-action and ϕ^H is the geometric fixed point functor (as defined in [51, Definition 9.7], [53, Definition 4.3] or [79, Sect. 7.3]) concatenated with the forgetful functor from $\mathcal{SH}(W(H))$ to $\mathcal{SH}$, where $W(H)$ denotes the Weyl group of $H \subseteq G$. We will need the following properties [86, Proposition 12.1 and Theorem 12.4].

Proposition 13 *The geometric fixed point functor ϕ^H has the following properties:*

(1) In $\mathcal{SH}$, $\phi^H(\Sigma^\infty X) = \Sigma^\infty X^H$ for any suspension spectrum $\Sigma^\infty X \in \mathcal{SH}(G)$.
(2) In $\mathcal{SH}$, $\phi^H(X \wedge Y) = \phi^H(X) \wedge \phi^H(Y)$ for any $X, Y \in \mathcal{SH}(G)$.

(3) In $\mathcal{SH}$, $\phi^H(i(X)) = X$ for any $X \in \mathcal{SH}$.
(4) If $\phi^H(X) = 0$ in $\mathcal{SH}$ for all $H \subseteq G$, then $X = 0$ in $\mathcal{SH}(G)$.

Proof A proof of (1) can be found in [51, Corollary II.9.9], [53, Corollary 4.6], or in [79, Example 7.7]. (2) follows from [51, Theorem II.9.8(ii)] and [51, Proposition II.9.12(ii)]. (3) follows directly from the definition of ϕ^H, since H acts trivially on $i(X)$. (4) is proven in [79, Theorem 7.12] and in [86, Theorem 12.4]. □

3.2 Equivariant Morava K-Theories

For $H \subseteq G$, $\tilde{E}[\not\geq H]$ denotes a G-space which satisfies:

$$\tilde{E}[\not\geq H]^K \simeq \begin{cases} 0 \text{ if } K \not\geq_G H \\ S^0 \text{ if } K \geq_G H \end{cases},$$

where $K \geq_G H$ means that K contains a subgroup conjugate to H.

The existence of such a space $\tilde{E}[\not\geq H]$ follows from the theory of classifying spaces for families (see e.g. [51, Sect. II.2]), if one takes $\mathcal{F}$ as the family of all subgroups of G for which H is not subconjugate and then defines $\tilde{E}[\not\geq H]$ by the cofiber sequence

$$E\mathcal{F}_+ \to S^0 \to \tilde{E}[\not\geq H],$$

as in [53, Notations 4.14].

Fix a prime number p. Strickland [86, Definition 16.2] defines Morava K-theory spectra in $\mathcal{SH}(G)$, G a finite group, by

$$K(n, H) = G/H_+ \wedge \tilde{E}[\not\geq H] \wedge K(n)$$

for any subgroup $H \subseteq G$. He notes that, as a localisation of S^0, $\tilde{E}[\not\geq H]$ is a commutative ring spectrum, which together with the ring structure of $K(n)$ and the diagonal map on G/H induces a ring structure on $K(n, H)$, which is commutative for $p > 2$. We will only be interested in H up to conjugacy, because if H and H' are conjugate, then $K(n, H) \cong K(n, H')$.

The following proposition serves as motivation for this definition of equivariant Morava K-theories [86, Remark 16.4 ff].

Proposition 14

$$K(n, H)_*(X) = K(n)_*(\phi^H(X))$$

and

$$K(n, H)^*(X) = K(n)^*(\phi^H(X)).$$

Proof Here, we need the following formula for the geometric fixed point spectrum [51, Theorem II.9.8(ii)]:

$$\phi^H(X) \cong (\tilde{E}[\not\supseteq H] \wedge X)^H,$$

where $(-)^H$ is the spectrification of the levelwise fixed point functor. In the following, we abbreviate $\tilde{E}[\not\supseteq H]$ by $\tilde{E}$. The first equation follows from

$$K(n, H)_*(X) = \pi_*\left((X \wedge G/H_+ \wedge \tilde{E} \wedge K(n))^G\right)$$

$$= \pi_*\left((X \wedge \tilde{E} \wedge K(n))^H\right) = \pi_*\left(\phi^H(X \wedge K(n))\right)$$

$$= \pi_*\left(\phi^H(X) \wedge K(n)\right) = K(n)_*(\phi^H(X)).$$

For the second equation, we use the fact that G/H_+ is self-dual [51, Corollary II.6.3], hence

$$K(n, H)^*(X) = [X, F(G/H_+, \tilde{E} \wedge K(n))]^G_* = [X, \tilde{E} \wedge K(n)]^H_*.$$

We claim that this is isomorphic to $[\phi^H X, \phi^H K(n)]_* = K(n)^*(\phi^H X)$. To prove the claim, first note that because $\tilde{E} \wedge -$ is a Bousfield localisation functor,

$$[X, \tilde{E} \wedge K(n)]^H_* = [\tilde{E} \wedge X, \tilde{E} \wedge K(n)]^H_*.$$

From here, the H-fixed points yield a map

$$\alpha : [\tilde{E} \wedge X, \tilde{E} \wedge K(n)]^H_* \to [(\tilde{E} \wedge X)^H, (\tilde{E} \wedge K(n))^H]_*$$

and the latter group is isomorphic to $[\phi^H X, \phi^H K(n)]_*$. Assume that X^H is an orbit H/K_+ for $K \subseteq H$. If $K \neq H$, $\alpha : 0 \to 0$ is an isomorphism. If $K = H$, $[\tilde{E} \wedge X, \tilde{E} \wedge K(n)]^H_* = [S^0_G, \tilde{E} \wedge K(n)]^H_* = [S^0, \phi^H(K(n))]_* = [\phi^H X, \phi^H(K(n))]_*$. That is, α is an isomorphism on all orbit types and it follows that α is an isomorphism for any $X \in \mathcal{SH}(G)$. □

From this and the properties of nonequivariant Morava K-theories (see e.g. [25, Sect. 1]), it follows immediately that $K(n, H)$ has coefficients like $K(n)$ and satisfies the Künneth formula [86, Sect. 16].

Corollary 15 *The equivariant Morava K-theories satisfy the following properties:*

(1) $K(n, H)_* S^0_G = K(n)_* S^0 = \mathbb{F}_p[v_n^{\pm 1}]$ *for any* $n > 0$ *and any* $H \subseteq G$.
(2) $K(n, H)_*(X \wedge Y) \cong K(n, H)_*(X) \otimes_{K(n)_*} K(n, H)_*(Y)$ *for any* $X, Y \in \mathcal{SH}(G)$.
(3) If X *is dualisable, i.e., if* $X \in \mathcal{SH}(G)_f$*, then*

$$K(n, H)_*(DX) \cong \mathrm{Hom}_{K(n)_*}(K(n, H)_*(X), K(n)_*).$$

Furthermore, Strickland shows [86, Proposition 16.6]:

Proposition 16 *If $p \neq p'$ or $n \neq n'$ or $H \neq_G H'$ (i.e., not conjugate in G), then*

$$K(p, n, H) \wedge K(p', n', H') = 0.$$

Proof In the cases $p \neq p'$ and $n \neq n'$, this follows from $K(p, n) \wedge K(p', n') = 0$ (see [73, Theorem 2.1(i)]), as these appear as smash factors in $K(p, n, H) \wedge K(p', n', H')$. Therefore, assume $p = p'$, $n = n'$ and $H \neq_G H'$. Now it suffices to show $G/H_+ \wedge \tilde{E}[\not\geq H] \wedge G/H'_+ \wedge \tilde{E}[\not\geq H'] = 0$, which is easily checked on the level of K-fixed points for all $K \subseteq G$. □

3.3 Nilpotence and Lattices of Thick Ideals

For a convenient description of the collection of thick ideals in the equivariant homotopy category of dualisable spectra, $SH(G)_f$, [86] uses the language of lattices.

Definition 17 A lattice is a partially ordered set A for which any finite subset $F \subseteq A$ has a greatest lower bound (called meet) $\bigwedge F$ and a smallest upper bound (called join) $\bigvee F$. The largest element in A is $\bigwedge \emptyset$, which we denote by 1 and the smallest element is $0 = \bigvee \emptyset$. A lattice homomorphism is an order preserving map $f : A \to B$ which also preserves all joins and meets.

Example 18 (1) The collection of thick ideals $\mathcal{C}$ in a tensor triangulated category $\mathcal{T}$, partially ordered by inclusion, is a lattice. Meets are just intersections, whereas the join of a finite collection of thick ideals is the smallest thick ideal which contains all objects of the different ideals. We denote this lattice by $\mathrm{Idl}(\mathcal{T})$.
(2) The power set of any set is a lattice, partially ordered by inclusion, meets given by intersections and joins by unions.

We introduce a new notation, due to Strickland, which will be useful in the rest of this section.

Notation 19 For a prime p and a nonnegative integer n, let $K(p^{-n})$ denote the n-th Morava K-theory spectrum at the prime p (which above was denoted by $K(p, n)$ or just $K(n)$).

One advantage of this notation is that there is only one name for the zeroth Morava K-theory spectrum (which is independent of p): $K(1) = H\mathbb{Q}$.

Definition 20 Let

$$\mathcal{Q}_p = \{p^{-n} \mid 0 \leq n \leq \infty\}$$

and

$$\mathcal{Q} = \{u \in \prod_p \mathcal{Q}_p \mid u_p = 1\,\forall\, p \text{ or } u_p \neq 1\,\forall\, p\}.$$

The sets $\mathcal{Q}_p$ and $\mathcal{Q}$ are lattices, with the usual ordering of rational numbers and the componentwise partial ordering of products. We immediately see that Theorem 1 can be reformulated as follows.

Corollary 21 *Let* $\mathrm{Idl}(\mathcal{SH}^{fin}_{(p)})$ *be as in Example 18(1). The map*

$$\tau_p : \mathrm{Idl}(\mathcal{SH}^{fin}_{(p)}) \longrightarrow \mathcal{Q}_p,$$

$$\tau_p(\mathcal{C}) = \max\{p^{-n} \mid \mathrm{type}(X) = n \text{ for some } X \in \mathcal{C}\},$$

is a lattice isomorphism.

We will see in Theorem 4 how to merge the information for different p to a classification of finitely generated thick ideals of $\mathcal{SH}^{fin}$ using the lattice $\mathcal{Q}$. But before we are able to do so, we need some more theory on thick ideals and lattice homomorphisms.

Definition 22 For $X \in \mathcal{SH}(G)_f$, let $\mathrm{ann}(X)$ denote the fibre of the unit map $S \to F(X, X)$ and define

$$\mathcal{A}_{\mathrm{ann}(X)} = \{A \mid \mathrm{ann}(X)^{\wedge n} \wedge A \to A \text{ is null for some } n > 0\}.$$

The map here is the n-fold smash product of the map $\mathrm{ann}(X) \to S$ from the cofiber sequence, smashed with $A \xrightarrow{1} A$.

The following is [86, Proposition 15.6].

Proposition 23 *The smallest thick ideal containing X is*

$$\mathrm{thickid}(X) = \mathcal{A}_{\mathrm{ann}(X)}$$

and $\mathrm{thickid}(X) \subseteq \mathrm{thickid}(Y)$ *if and only if the map* $\mathrm{ann}(Y)^{\wedge n} \to S$ *factors through* $\mathrm{ann}(X) \to S$ *for some* $n > 0$.

Proof We first show that $\mathrm{thickid}(X) \subseteq \mathcal{A}_{\mathrm{ann}(X)}$. Since X is a module over $F(X, X) = DX \wedge X$, it is a retract of $DX \wedge X \wedge X$. It follows that $\mathrm{ann}(X) \wedge X$ is the fiber of a map $X \to DX \wedge X \wedge X$ which splits, so $\mathrm{ann}(X) \wedge X \to X$ is zero and hence $X \in \mathcal{A}_{\mathrm{ann}(X)}$. It is easy to see that $\mathcal{A}_{\mathrm{ann}(X)}$ is closed under exact triangles and retracts, as well as under smashing with arbitrary objects. Hence, $\mathcal{A}_{\mathrm{ann}(X)}$ is a thick ideal containing X, which proves $\mathrm{thickid}(X) \subseteq \mathcal{A}_{\mathrm{ann}(X)}$.

Now assume $A \in \mathcal{A}_{\mathrm{ann}(X)}$. We need to show $A \in \mathrm{thickid}(X)$. Consider the cofiber sequence

$$\operatorname{ann}(X)^{\wedge n} \wedge A \to A \to S/(\operatorname{ann}(X)^{\wedge n}) \wedge A.$$

By the assumption, we can choose n such that the first map is zero. Then it follows that A is a retract of $S/(\operatorname{ann}(X)^{\wedge n}) \wedge A$. Therefore, it suffices to show $S/(\operatorname{ann}(X)^{\wedge n}) \in \operatorname{thickid}(X)$. By the definition of $\operatorname{ann}(X)$, $S/\operatorname{ann}(X) = F(X, X) = DX \wedge X \in \operatorname{thickid}(X)$. There is a cofiber sequence

$$\operatorname{ann}(X) \wedge S/(\operatorname{ann}(X)^{\wedge j}) \to S/(\operatorname{ann}(X)^{\wedge j+1}) \to S/(\operatorname{ann}(X)^{\wedge j}),$$

which implies inductively that $S/(\operatorname{ann}(X)^{\wedge n}) \in \operatorname{thickid}(X)$.

The existence of this cofiber sequence follows from Verdier's axiom for triangulated categories, also known as octahedral axiom (see e.g. [67, Proposition 1.4.6]). Applied to the three cofiber sequences $I \wedge J \to I \wedge S \to I \wedge S/J$, as well as $I \wedge J \to S \to S/(I \wedge J)$ and $I \to S \to S/I$, the axiom yields a cofiber sequence $I \wedge S/J \to S/(I \wedge J) \to S/I$.

For the second claim, assume that $\operatorname{thickid}(X) \subseteq \operatorname{thickid}(Y)$, which is equivalent to $X \in \operatorname{thickid}(Y) = \mathcal{A}_{\operatorname{ann}(Y)}$. Let $n > 0$ be such that $\operatorname{ann}(Y)^{\wedge n} \wedge X \to X$ is the zero map. Consider the two cofiber sequences

$$\begin{array}{ccccc}
\operatorname{ann}(Y)^{\wedge n} & \longrightarrow & S & \longrightarrow & S/(\operatorname{ann}(Y)^{\wedge n}) \\
 & & \| & & \\
\operatorname{ann}(X) & \longrightarrow & S & \longrightarrow & F(X, X).
\end{array}$$

The smash product of the upper sequence with X is

$$\operatorname{ann}(Y)^{\wedge n} \wedge X \stackrel{0}{\to} X \to S/(\operatorname{ann}(Y)^{\wedge n}) \wedge X,$$

so there is a retraction $S/(\operatorname{ann}(Y)^{\wedge n}) \wedge X \to X$, which then induces a morphism $S/(\operatorname{ann}(Y)^{\wedge n}) \to F(X, X)$ making the diagram commutative. From the axioms for triangulated categories it follows that we can fill in the required map $\operatorname{ann}(Y)^{\wedge n} \to \operatorname{ann}(X)$, as claimed.

On the other hand, if $\operatorname{ann}(Y)^{\wedge n} \to S$ factors through $\operatorname{ann}(X) \to S$ and $\operatorname{ann}(X)^{\wedge m} \wedge A \to A$ is zero then also $\operatorname{ann}(Y)^{\wedge (nm)} \wedge A \to A$ is zero and it follows $\mathcal{A}_{\operatorname{ann}(X)} \subseteq \mathcal{A}_{\operatorname{ann}(Y)}$. □

Proposition 24 *Let $\mathcal{I}_0$ and $\mathcal{J}_0$ be collections of objects in $\mathcal{SH}(G)_f$ and let $\mathcal{I}$ and $\mathcal{J}$ be the thick ideals which they generate. Then*

$$\mathcal{I} \cap \mathcal{J} = \operatorname{thickid}\left(\{Y \wedge Z \mid Y \in \mathcal{I}_0,\ Z \in \mathcal{J}_0\}\right).$$

Proof This is [86, Proposition 15.8].

Let $\mathcal{K} = \operatorname{thickid}\left(\{Y \wedge Z \mid Y \in \mathcal{I}_0,\ Z \in \mathcal{J}_0\}\right)$. The intersection $\mathcal{I} \cap \mathcal{J}$ is a thick ideal which contains $Y \wedge Z$ for all $Y \in \mathcal{I}_0$ and $Z \in \mathcal{J}_0$. Therefore, $\mathcal{K} \subseteq \mathcal{I} \cap \mathcal{J}$. Now, let

$$\mathcal{I}' = \{Y \in \mathcal{I} \mid Y \wedge Z \in \mathcal{K} \,\forall Z \in \mathcal{J}_0\},$$

$$\mathcal{J}' = \{Z \in \mathcal{J} \mid Y \wedge Z \in \mathcal{K} \,\forall Y \in \mathcal{I}\}.$$

It is easy to check that $\mathcal{I}'$ is a thick subideal of $\mathcal{I}$ which contains $\mathcal{I}_0$. Hence, $\mathcal{I}' = \mathcal{I}$. It follows that the thick subideal $\mathcal{J}' \subseteq \mathcal{J}$ contains $\mathcal{J}_0$, so $\mathcal{J}' = \mathcal{J}$. That is, $Y \wedge Z \in \mathcal{K}$ for all $Y \in \mathcal{I}$, $Z \in \mathcal{J}$.

Now, let $X \in \mathcal{I} \cap \mathcal{J}$. Since X is an $F(X, X)$-module, X is a retract of $DX \wedge X \wedge X$. Consider $DX \wedge X$ as an object of $\mathcal{I}$ and the other X as an object of $\mathcal{J}$. It follows $DX \wedge X \wedge X \in \mathcal{K}$ and hence $X \in \mathcal{K}$. □

Remark 25 Note that Proposition 24 holds in any tensor triangulated category in which internal hom objects exist and all objects are dualisable. The following definition, theorem and corollary can also be formulated in such a general setting, given a suitable notion of homology theories.

Definition 26 Let $\{E_i \mid i \in I\}$ be a family of ring spectra in $\mathcal{SH}(G)$. For $X \in \mathcal{SH}(G)_f$, define

$$\mathrm{supp}(X) = \{i \in I \mid (E_i)_*(X) \neq 0\}.$$

If $\mathcal{C} \subseteq \mathcal{SH}(G)_f$ is a subcategory, let

$$\mathrm{supp}(\mathcal{C}) = \bigcup_{X \in \mathcal{C}} \mathrm{supp}(X).$$

For a map $f : X \to Y$, we also define

$$\mathrm{supp}(f) = \{i \in I \mid (E_i)_*(f) \neq 0\}.$$

Remark 27 If $(E_i)_*(X) = 0$, then $(E_i)_*(Y) = 0$ for any $Y \in \mathrm{thickid}(X)$ by the following arguments. If $A \to B \to C$ is a cofiber sequence and the $(E_i)_*$-homology of two of the three objects is zero, then, by the long exact $(E_i)_*$-sequence, $(E_i)_*(-)$ of the third object is zero, too. If $(E_i)_*(A) = \pi_*(E_i \wedge A) = 0$, then also $(E_i)_*(A \wedge B) = \pi_*(E_i \wedge A \wedge B) = 0$. And, finally, if $(E_i)_*(A) = 0$ and B is a retract of A, then $(E_i)_*(B) \to 0 \to (E_i)_*(B)$ is the identity map, hence, $(E_i)_*(B) = 0$.

This implies

$$\mathrm{supp}(\mathrm{thickid}(X)) = \mathrm{supp}(X).$$

The following theorem is one of the central results in [86], where it is Theorem 15.14.

Theorem 2 (Strickland) *Assume $\{E_i \mid i \in I\}$ is a family of ring spectra in $\mathcal{SH}(G)$ satisfying the following properties:*

(1) If $f : X \to Y$, with $X, Y \in \mathcal{SH}(G)_f$, and $(E_i)_*(f) = 0$ for all $i \in I$, then there exists $n > 0$ such that $f^{\wedge n} = 0$.

(2) For any $X, Y \in \mathcal{SH}(G)_f$ *and any* $i \in I$, $(E_i)_*(X \wedge Y) \cong (E_i)_*(X) \otimes_{(E_i)_*} (E_i)_*(Y)$.

(3) For any $i \in I$, $(E_i)_* = (E_i)_*(S^0)$ *is concentrated in even degrees and any nonzero homogeneous element in* $(E_i)_*$ *is invertible.*

Then, for any $X, Y \in \mathcal{SH}(G)_f$, $\text{thickid}(X) \subseteq \text{thickid}(Y)$ *if and only if* $\text{supp}(X) \subseteq \text{supp}(Y)$.

In other words, a family $\{E_i \mid i \in I\}$ of spectra detecting nilpotence (see Definition 28) and satisfying some additional properties can be used to distinguish any two different thick ideals with the help of the support functor $\text{supp}(-)$.

Proof Consider the cofiber sequence $\text{ann}(Y) \xrightarrow{v} S \xrightarrow{u} F(Y, Y)$. We first show that $\text{supp}(v) = I \setminus \text{supp}(Y)$. Consider the long exact sequence

$$(E_i)_*(F(Y, Y)) \to (E_i)_*(\text{ann}(Y)) \overset{(E_i)_*(v)}{\to} (E_i)_*(S) \overset{(E_i)_*(u)}{\to} (E_i)_*(F(Y, Y))$$

If $i \in I \setminus \text{supp}(Y)$, then $(E_i)_*(F(Y, Y)) \cong (E_i)_*(DY) \otimes_{(E_i)_*} (E_i)_*(Y) = 0$ and $(E_i)_*(v)$ is an isomorphism. Hence, $i \in \text{supp}(v)$. If, on the other hand, $(E_i)_*(v) \neq 0$, it already has to be surjective because $(E_i)_*(\text{ann}(Y))$ is an $(E_i)_*$-vector space (by property (3)). It follows that $(E_i)_*(u) = 0$. But u is the unit map of $F(Y, Y)$, so this implies $(E_i)_*(F(Y, Y)) = 0$. As Y is a retract of $F(Y, Y) \wedge Y$, it follows that $(E_i)_*(Y) = 0$. This proves $\text{supp}(v) = I \setminus \text{supp}(Y)$.

Now let $X, Y \in \mathcal{SH}(G)_f$ and $\text{supp}(X) \subseteq \text{supp}(Y)$. With v as above, we have $\text{supp}(v) = I \setminus \text{supp}(Y) \subseteq I \setminus \text{supp}(X)$. Hence,

$$\text{supp}\left(\text{ann}(Y) \xrightarrow{v} S \to F(X, X)\right) = \emptyset.$$

By property (1), this map is smash nilpotent, so there is some $m > 0$ such that

$$\text{ann}(Y)^{\wedge m} \to S \to F(X, X)^{\wedge m}$$

is the zero map. Concatenation defines a map $F(X, X)^{\wedge m} \to F(X, X)$, over which the unit map $S \to F(X, X)$ factors, so we get a diagram in which the lower row is a cofiber sequence, the composition of the two upper maps is zero and the square commutes:

$$\begin{array}{ccccc} \text{ann}(Y)^{\wedge m} & \longrightarrow & S & \longrightarrow & F(X, X)^{\wedge m} \\ \downarrow & & \| & & \downarrow \\ \text{ann}(X) & \longrightarrow & S & \longrightarrow & F(X, X), \end{array}$$

It follows that the map $\text{ann}(Y)^{\wedge m} \to S$ factors over $\text{ann}(X)$. By Proposition 23, this is equivalent to $\text{thickid}(X) \subseteq \text{thickid}(Y)$.

For the other direction, assume $\mathrm{thickid}(X) \subseteq \mathrm{thickid}(Y)$. Then by Remark 27,

$$\mathrm{supp}(X) = \mathrm{supp}(\mathrm{thickid}(X)) \subseteq \mathrm{supp}(\mathrm{thickid}(Y)) = \mathrm{supp}(Y).$$

□

Definition 28 We say that a family $\{E_i \mid i \in I\}$ detects nilpotence, if for any $f : X \to Y$ in $\mathcal{SH}(G)_f$, $\mathrm{supp}(f) = \emptyset$ implies $f^{\wedge n} = 0$ for some $n > 0$.

Corollary 29 (Strickland) *Under the assumptions of the above theorem, the map from the collection of thick ideals in $\mathcal{SH}(G)_f$ to the set of subsets of I,*

$$\mathrm{Idl}(\mathcal{SH}(G)_f) \longrightarrow \mathcal{P}(I),$$

$$\mathcal{C} \mapsto \mathrm{supp}(\mathcal{C}),$$

is a lattice homomorphism (see Definition 17). It is injective on the collection of finitely generated thick ideals, $\mathrm{FIdl}(\mathcal{SH}(G)_f)$ *(see Definition 3).*

Proof This is [86, Corollary 15.15]. It is clear from the definition of $\mathrm{supp}(\mathcal{C})$, that $\mathrm{supp}(-)$ is order preserving. The map preserves meets by Proposition 24 and by the Künneth formula for E_i. As $\mathrm{supp}(\mathcal{C})$ is the support of any set of generators for $\mathcal{C}$ and the join of thick ideals $\mathcal{C}_i$ is generated by the collection of generators of the individual thick ideals, it is also clear that $\mathrm{supp}(-)$ preserves joins. Recall that any finitely generated thick ideal is already generated by a single element (Remark 4). By the above theorem, $\mathrm{thickid}(X) = \mathrm{thickid}(Y)$ if and only if $\mathrm{supp}(X) = \mathrm{supp}(Y)$, which is the same as $\mathrm{supp}(\mathrm{thickid}(X)) = \mathrm{supp}(\mathrm{thickid}(Y))$. This proves the injectivity on FIdl. □

3.4 Thick Ideals and Equivariant Morava K-Theories

Definition 30 For a finite group G, let $\mathrm{sub}(G)$ denote the set of equivalence classes of conjugate subgroups of G. Let

$$\mathcal{Q}' = \{p^{-n} \mid p \text{ prime}, 0 \le n \le \infty\}$$

$$\text{and } \; G\mathcal{Q}' = \mathcal{Q}' \times \mathrm{sub}(G).$$

The following theorem shows that the family of equivariant Morava K-theories $\{K(p^{-n}, H) \mid (p^{-n}, H) \in G\mathcal{Q}'\}$ (see Sect. 3.2) detects nilpotence, as required in the assumptions of Theorem 2 and Corollary 29. As in [12, Theorem 1], there are different kinds of nilpotence, which are all detected by the Morava K-theories. Although we mainly work with smash nilpotence, the theorem, which is [86, Theorem 16.7], considers all three definitions.

Theorem 3 (Strickland)

(1) Let $R \in \mathcal{SH}(G)$ be a ring spectrum. Then $\alpha \in \pi_*^G R$ is nilpotent if and only if for all $v \in GQ'$, $K(v)_*(\alpha)$ is nilpotent as an element of $K(v)_* R$.
(2) A self-map $f : \Sigma^k W \to W$, $W \in \mathcal{SH}(G)_f$, is nilpotent if and only if for all $v \in GQ'$, $K(v)_*(f)$ is nilpotent.
(3) A map $f : W \to X$, with $W \in \mathcal{SH}(G)_f$ and $X \in \mathcal{SH}(G)$, is smash nilpotent (i.e., it exists an $n > 0$ such that $f^{\wedge n} = 0$) if and only if for all $v \in GQ'$, $K(v)_*(f)$ is nilpotent.

Proof (1) Let $\alpha : S_G^d \to R$ be such that $K(v)_*(\alpha)$ is nilpotent for all $v \in GQ'$. For any $H \subseteq G$, $\phi^H(S_G^d)$ is a non-equivariant sphere, so $\phi^H \alpha \in \pi_*(\phi^H R)$. By Proposition 14, $K(u)_*(\phi^H \alpha) = K(u, H)_*(\alpha)$, so this is nilpotent for all $u \in \mathcal{Q}'$. Furthermore, $\phi^H R$ is a ring spectrum and, by [25, Theorem 3(i)], it follows that $\phi^H(\alpha)$ is nilpotent, so $(\phi^H R)[\phi^H \alpha^{-1}] = 0$ (for the definition of this mapping telescope, see [12, p. 212]). By [79, Remark 7.15], ϕ^H preserves telescopes, hence, $\phi^H(R[\alpha^{-1}]) \cong (\phi^H R)[\phi^H \alpha^{-1}] = 0$. This holds for all H, which by Proposition 13(4) implies $R[\alpha^{-1}] = 0$, and, hence, $\pi_*^G R[\alpha^{-1}] = 0$. Thus, α is nilpotent.
(2) The adjoint of f is an element $\alpha \in \pi_d^G F(W, W)$, and $K(v)_*(\alpha)$ is nilpotent for all $v \in GQ'$. So, the claim follows from (1).
(3) Let $R = \bigvee_{k \geq 0} F(W, X)^{\wedge k} \in \mathcal{SH}(G)$ be the free associative ring spectrum generated by $F(W, X)$. The map f is adjoint to $\alpha \in \pi_d^G F(W, X) \subset \pi_d^G R$ such that $K(v)_*(\alpha)$ is nilpotent for all $v \in GQ'$, and the claim follows from (1). □

Definition 31 Let $GQ = \prod_{\mathrm{sub}(G)} \mathcal{Q}$, where $\mathcal{Q}$ is as in Definition 20 and let

$$\max : \mathcal{P}(GQ') \longrightarrow GQ,\ I \mapsto \left\{H \mapsto \max\{p^{-n} \mid (p^{-n}, H) \in I\}\right\},$$

with the convention $\max(\emptyset) = 0$. Let

$$\tau = \max \circ \operatorname{supp} : \operatorname{Idl}(\mathcal{SH}(G)_f) \longrightarrow GQ.$$

Furthermore, for $u \in GQ$, $H \subseteq G$ and p prime, we write $u_H \in Q$ for the projection of $u \in GQ$ onto the component of GQ corresponding to the equivalence class of H in sub(G), and we write $u_{H,p} \in Q_p$ for the projection of $u_H \in Q$ onto the component of $Q \subseteq \prod_p Q_p$ corresponding to p.

Corollary 32 *The functor*

$$\operatorname{Idl}(\mathcal{SH}(G)_f) \longrightarrow \mathcal{P}(GQ'),$$

$$\mathcal{C} \mapsto \operatorname{supp}(\mathcal{C}) = \left\{v \in GQ' \mid K(v)_*(X) \neq 0 \text{ for some } X \in \mathcal{C}\right\},$$

is a lattice homomorphism. Furthermore, it is injective.

Proof This is [86, Corollary 16.8]. It follows from Corollary 29 applied to the family of equivariant Morava K-theories. The assumptions are satisfied by Theorem 3(3) and Corollary 15.

Corollary 29 states that the restriction of supp to finitely generated thick ideals $\mathrm{FIdl}(\mathcal{SH}(G)_f)$ is injective. To show that supp is injective on arbitrary thick ideals, it suffices to show that τ is injective. Note that τ is injective on $\mathrm{FIdl}(\mathcal{SH}(G)_f)$ because, by [73, Theorem 2.11], $K(p^{-n})_*(\phi^H X) = 0$ implies $K(p^{-m})_*(\phi^H X) = 0$ for all $m \leq n$.

Now let $\tau(\mathcal{I}) = \tau(\mathcal{J})$ for some thick ideals $\mathcal{I}$ and $\mathcal{J}$ in $\mathcal{SH}(G)_f$. Thus, $\tau(\mathcal{I})_{H,p} = \tau(\mathcal{J})_{H,p}$ for all $H \subseteq G$ and all primes p. Assume $X \in \mathcal{I}$. Then $\tau(X)_{H,p} \leq \tau(\mathcal{I})_{H,p} = \tau(\mathcal{J})_{H,p}$. Let $I = \{(p, H) \mid \tau(X)_{H,p} \neq 0, 1\}$. We claim that I is finite. Since G is finite, there are only finitely many possibilities for H. If $(p, H) \in I$, then $K(p^{-0})_*(\phi^H X) = \mathrm{H}_*(\phi^H X, \mathbb{Q}) = 0$. Since $\phi^H X$ is a finite spectrum, this implies $\mathrm{H}_*(\phi^H X, \mathbb{F}_q) = 0$ for all but finitely many q, so I is finite. For any $u \in I$, there exists by assumption a $Y_u \in \mathcal{J}$ with $\tau(Y_u)_u \geq \tau(X)_u$. Similarly, let $I' = \{H \mid \tau(X)_H = 1\}$ and pick $Y_H \in \mathcal{J}$ with $\tau(Y_H) = 1$ for each $H \in I'$. Then, by the injectivity of τ on finitely generated thick ideals, X is contained in $\mathrm{thickid}(\{Y_u \mid u \in I\} \cup \{Y_H \mid H \in I'\})$, which is a finitely generated subideal of $\mathcal{J}$. It follows $\mathcal{I} \subseteq \mathcal{J}$ and, similarly, $\mathcal{J} \subseteq \mathcal{I}$. □

As promised, we now state a reformulation of the thick subcategory theorem from [25], which is no longer in p-local form. We do not claim that all the notation from above was necessary to state this. But it is helpful for generalising the result to $\mathcal{SH}(G)_f$ or maybe also other categories.

Theorem 4 *The composition*

$$\tau : \mathrm{Idl}(\mathcal{SH}^{fin}) \xrightarrow{\text{supp}} \mathcal{P}(\mathcal{Q}') \xrightarrow{\max} \mathcal{Q}$$

is bijective.

Its restriction to the finitely generated thick ideals maps $\mathrm{FIdl}(\mathcal{SH}^{fin})$ *bijectively to*

$$\mathrm{Fin}(\mathcal{Q}) = \{u \in \mathcal{Q} \mid u = 1 \text{ or } u_p = 0 \text{ for almost all } p\}.$$

Proof This is [86, Proposition 19.14]. The theorem states that supp maps injectively onto its image, which is isomorphic to $\mathcal{Q}$. The map supp is injective by the nonequivariant version of Corollary 32. The image of supp maps injectively to $\mathcal{Q}$ because $K(p^{-n})_* X = 0$ implies $K(p^{-m})_* X = 0$ for all $m \leq n$ by [73, Theorem 2.11]. We show that τ is surjective: By [58, Theorem B(b)] or [75, Sect. C.3], for any number $n > 0$ there is a spectrum $X_n \in \mathcal{SH}^{fin}_{(p)}$ of type n. Thus, the p-localisation of τ is surjective onto $\mathcal{Q}_p$ (see Definition 20). Let $u \in \mathcal{Q}$. If $u = 1$, then $\mathcal{C} = \mathrm{thickid}(S^0)$ satisfies $\tau(\mathcal{C}) = u$. Assume $u \neq 1$, so $u_p = p^{-n_p}$ with $0 < n_p \leq \infty$. Let X_{n_p} be a p-local spectrum of type n_p if $0 < n_p < \infty$ and $X_{n_p} = 0$ if $n_p = \infty$. Then $\mathcal{C} = \mathrm{thickid}(X_{n_2}, X_{n_3}, X_{n_5}, \cdots)$ satisfies $\tau(\mathcal{C}) = u$. Hence, τ is surjective. If $u = 1$ or $u_p = 0$ for almost all p, then $\mathcal{C}$ as above is finitely generated. As in the proof of Corollary 32, all finitely generated thick ideals are mapped to $\mathrm{Fin}(\mathcal{Q})$. □

This theorem identifies the thick ideals with a sublattice of $\mathcal{P}(\mathcal{Q}')$ that is bijective to $\mathcal{Q}$. Similarly, thick ideals in the equivariant category $\mathcal{SH}(G)_f$ are mapped injectively into $G\mathcal{Q}$. Unlike its nonequivariant version, the equivariant version of τ is not surjective. However, we can state the following:

Theorem 5 (Strickland) *The composition*

$$\tau : \mathrm{Idl}(\mathcal{SH}(G)_f) \xrightarrow{\text{supp}} \mathcal{P}(G\mathcal{Q}') \xrightarrow{\max} G\mathcal{Q}$$

is injective. Its image contains all $u \in G\mathcal{Q}$ which satisfy: If $H \subseteq H'$ for some $H, H' \in$ $\mathrm{sub}(G)$, then $u_H \geq u_{H'}$.

Proof We already know that supp is an injective lattice homomorphism. The injectivity of τ follows as in the nonequivariant case, by considering H-fixed points for each $H \in \mathrm{sub}(G)$ separately.

Now let $u \in G\mathcal{Q}$ be such that $H' \subseteq H$ implies $u_{H'} \geq u_H$. Let X_{u_H} be a finite spectrum in $\mathcal{SH}$ with $\tau(X_{u_H}) = u_H$. Recall that any nonequivariant spectrum maps to a G-spectrum with trivial G-action through the functor $i : \mathcal{SH} \to \mathcal{SH}(G)$. Let $Y_H = i(X_{u_H}) \wedge G/H_+$. This is a finite G-spectrum satisfying $\phi^{H'} Y_H = X_{u_H} \wedge G/H_+$ if H' is subconjugate to H and $\phi^{H'} Y_H = 0$ if H' is not subconjugate to H. Hence, $\tau_{H'}(Y_H) = u_H$ if $H' \subseteq H$ and $\tau_{H'}(Y_H) = 0$ if $H' \nsubseteq H$. Let $Y = \bigvee_{H \in \mathrm{sub}(G)} Y_H$. Then $\tau_H(Y) = \max\{u_{H'} \mid H \subseteq H'\} = u_H$ by the assumption. □

This theorem gives a lower bound on the set of thick ideals in $\mathcal{SH}(G)_f$. The whole set $G\mathcal{Q}$ is an upper bound. Strickland was able to show that $\tau(\mathrm{Idl}(\mathcal{SH}(G)_f))$ lies in a certain proper subset of $G\mathcal{Q}$ (if G is nontrivial). For a cyclic group $G = \mathbb{Z}/p$, his result is the following [86, Proposition 16.9]:

Proposition 33 (Strickland) *If $X \in \mathcal{SH}(\mathbb{Z}/p)_f$ and $K(p^{-(n+1)})_* \phi^{\{1\}} X = 0$, then $K(p^{-n})_* \phi^{\mathbb{Z}/p} X = 0$. (Note that the same prime p appears in two different roles.)*

Proof Again, the proof is taken from [86]. X is of the form $X = \Sigma^{\infty - V} T$, where T is a retract of a finite $\mathbb{Z}/p$-CW complex and V is a representation of $\mathbb{Z}/p$. Then, by Proposition 13, $\phi^{\mathbb{Z}/p} X = \Sigma^{\infty - \dim(V^{\mathbb{Z}/p})} T^{\mathbb{Z}/p}$. The assumption is equivalent to $K(p^{-(n+1)})_* T = 0$ and the claim is equivalent to $K(p^{-n})_* T^{\mathbb{Z}/p} = 0$. By [73, Theorem 2.11], we can formulate this in terms of Johnson-Wilson theories, namely: We know $E(n+1)_* T = 0$ and have to show $E(n)_* T^{\mathbb{Z}/p} = 0$. The proof uses the Greenlees–May theory of Tate spectra [20]. For any G-spectrum Y, let $t_G Y = F(EG_+, Y) \wedge \tilde{E}G$ and $P_G Y = (t_G Y)^G$. They have the following properties:

(1) The functors t_G and P_G preserve exact triangles.
(2) We have $t_G(X \wedge Y) = X \wedge t_G Y$ for finite G-spectra X, and $P_G(X \wedge Y) = X \wedge P_G Y$ for finite spectra X with trivial G-action.
(3) If Y is a free G-spectrum, then $t_G Y = 0$ and so $P_G Y = 0$.
(4) If Y is nonequivariantly contractible, then $t_G Y = 0$ and so $P_G Y = 0$.
(5) If p divides the order of G, then the spectrum $P_G E(n+1)$ has Bousfield class $\langle P_G E(n+1) \rangle = \langle E(n) \rangle$ [30, Theorem 1.1].

As $E(n+1)_*T = 0$, we see that the $\mathbb{Z}/p$-spectrum $E(n+1) \wedge T$ is nonequivariantly contractible, so $t_G(E(n+1) \wedge T) = 0$ by (4). We also have $t_G(E(n+1) \wedge T/T^{\mathbb{Z}/p}) = 0$ by (3), so (1) implies $t_G(E(n+1) \wedge T^{\mathbb{Z}/p}) = 0$. Hence, $(P_G E(n+1)) \wedge T^{\mathbb{Z}/p} = P_G(E(n+1) \wedge T^{\mathbb{Z}/p}) = 0$ and (5) gives $E(n) \wedge T^{\mathbb{Z}/p} = 0$, as required. □

This result on cyclic groups can be used to derive some restriction on the types of G-spectra that can occur for a general group G. This is done by applying the result to quotients of subgroups of G, see [86, Corollary 16.10]. We do not state this result here, because it needs additional notation and because we will only be interested in the case $G = \mathbb{Z}/2$ later on.

Further work on the topic has been done by Balmer and Sanders. In [8, Proposition 7.5], they show that $K(p^{-(n+1)})_* \phi^{\{1\}} X = 0$ in the above situation does not imply $K(p^{-(n+1)})_* \phi^{\mathbb{Z}/p} X = 0$, which completes the classification of thick ideals in $(\mathcal{SH}(\mathbb{Z}/p)_f)_{(p)}$.

3.5 Thick Ideals in $\mathcal{SH}(\mathbb{Z}/2)_f$

We apply the results of this section to $G = \mathbb{Z}/2$.

Recall that, for G a finite group, the thick ideals in $(\mathcal{SH}(G)_f)_{(p)}$ are characterised by their equivariant types, i.e., by the vanishing or non-vanishing of the different equivariant Morava K-theories.

Corollary 34 *By the injectivity of τ (Theorem 5), any thick ideal in the category $(\mathcal{SH}(\mathbb{Z}/2)_f)_{(p)}$ is of the form*

$$\mathcal{C}_{m,n} = \{X \mid \phi^{\{1\}}(X) \in \mathcal{C}_m \text{ and } \phi^{\mathbb{Z}/2}(X) \in \mathcal{C}_n\},$$

where $m, n \in [0, \infty]$. By Proposition 14, $X \in \mathcal{C}_{m,n}$ is equivalent to $K(m-1, \{1\})_*(X) = 0$ and $K(n-1, \mathbb{Z}/2)_*(X) = 0$ if $0 < m, n < \infty$.

Definition 35 We say that $X \in (\mathcal{SH}(\mathbb{Z}/2)_f)_{(p)}$ has type (m, n), $0 \leq m, n \leq \infty$, if $X \in \mathcal{C}_{m,n} \setminus (\mathcal{C}_{m+1,n} \bigcup \mathcal{C}_{m,n+1})$.

Let $\Gamma_p \subseteq (\mathbb{Z}_{\geq 0} \cup \{\infty\}) \times (\mathbb{Z}_{\geq 0} \cup \{\infty\})$ be the sublattice of all (m, n) such that a p-local spectrum of type (m, n) exists.

Corollary 36 *(1) Any type-(m, n) spectrum generates $\mathcal{C}_{m,n}$ as a thick ideal (by Theorem 5).*

(2) Not all pairs (m, n) occur as the type of some spectrum. For example, if $p = 2$, m cannot be greater than $n + 1$ (by Proposition 33).

(3) If $m \leq n$, then a type-(m, n) spectrum exists, namely $X_{m,n} = (X_m \wedge \mathbb{Z}/2_+) \vee X_n$ with $X_k \in \mathcal{SH}_{(p)}^{fin}$ a type-k spectrum and with $\mathbb{Z}/2$ acting nontrivially only on $\mathbb{Z}/2$ (by Theorem 5). Thus, $m \leq n$ implies $(m, n) \in \Gamma_p$.

These results of Strickland give upper and lower bounds for the set of thick ideals in $(\mathcal{SH}(\mathbb{Z}/2)_f)_{(p)}$ (which is bijective to Γ_p).

Remark 37 The results on thick ideals in $\mathcal{SH}(G)_f$ presented in this section also hold in the category of finite G-CW spectra, $\mathcal{SH}(G)^{fin} \subseteq \mathcal{SH}(G)_f$, as we did not use the closure under retracts in any argument.

4 Comparison Functors

Notation 38 For k a field, we write $\mathcal{SH}(k)$ for the motivic stable homotopy category over k. The standard spheres are denoted by $S^{p,q} = S_s^{\wedge(p-q)} \wedge \mathbb{G}_m^{\wedge q}$ and their $\mathbb{P}^1$-suspension spectra are also denoted by $S^{p,q}$ if no confusion can arise. $\mathcal{SH}(k)$ is a tensor triangulated category, whose unit is the sphere spectrum $S = \Sigma^\infty_{\mathbb{P}^1} S^{0,0}$, which we also denote by S^0. For any $E \in \mathcal{SH}(k)$, $\pi_{p,q}(E) = [S^{p,q}, E]_{\mathcal{SH}(k)}$ denotes the set of maps from $S^{p,q}$ to E in $\mathcal{SH}(k)$. For E and X in $\mathcal{SH}(k)$, let $E_{p,q}(X) = \pi_{p,q}(X \wedge E)$ and $E^{p,q}(X) = [X, S^{p,q} \wedge E]$.

For our study of thick ideals in motivic categories $\mathcal{SH}(k)$, $k \subseteq \mathbb{C}$, we want to use the given knowledge on thick ideals in the classical stable homotopy category and the $\mathbb{Z}/2$-equivariant stable homotopy category. For this purpose, we will make use of the functors $\mathcal{SH} \overset{c_k}{\to} \mathcal{SH}(k) \overset{R_k}{\to} \mathcal{SH}$ for $k \subseteq \mathbb{C}$ and $\mathcal{SH}(\mathbb{Z}/2) \overset{c'_k}{\to} \mathcal{SH}(k) \overset{R'_k}{\to} \mathcal{SH}(\mathbb{Z}/2)$ for $k \subseteq \mathbb{R}$. The functors c_k, $R_\mathbb{C}$ and $R'_\mathbb{R}$ appear in various places in the literature but none of the sources conveniently covers all of the constructions. Most details on $R_\mathbb{C}$ can be found in [70, Appendix] and the unstable functor $R'_\mathbb{R}$ is studied in [15, Sect. 5]. Other important references are [88, Sect. 3.4], [63, Sect. 3.3] and [5]. The stable functors R'_k and c'_k are constructed and studied by Heller and Ormsby in [23, Sect. 4], which was written independently and at the same time as this section. Since our approach is slightly different, we give another, mostly self-contained construction of all these functors.

We start with the construction of $R_\mathbb{C}$ and $R'_\mathbb{R}$.

4.1 Symmetric $\mathbb{C}P^1$-Spectra

Objects of the motivic stable homotopy category $\mathcal{SH}(\mathbb{C})$ are spectra with respect to suspension by $\mathbb{P}^1_\mathbb{C} \cong S^{2,1}$. The corresponding analytic space $\mathbb{P}^1_\mathbb{C}(\mathbb{C})$ is $\mathbb{C}P^1$. We want that the topological realisation $R(X)$ of a spectrum $X \in \mathcal{SH}(\mathbb{C})$ is a spectrum again. Therefore, we work with $\mathbb{C}P^1$-spectra. The category of symmetric $\mathbb{C}P^1$-spectra, $Sp^\Sigma_{\mathbb{C}P^1}$, is described in [70, Theorem A.44] and is a model for the stable homotopy category. A symmetric $\mathbb{C}P^1$-spectrum is defined in the same way as a usual symmetric spectrum with S^1 replaced by $\mathbb{C}P^1 \cong S^2$. The stable model structure is constructed analogously as for symmetric S^1-spectra in [31]. Hence, the following results also hold for symmetric $\mathbb{C}P^1$-spectra [31, Lemmas 3.4.5, 3.4.12, 3.4.13].

Proposition 39 *1. The stable trivial fibrations are the level trivial fibrations.*
2. *A map of $\mathbb{C}P^1$-spectra $f : X \to Y$ is a stable fibration if and only if it is a level fibration and*

$$\begin{array}{ccc} X_n & \xrightarrow{\tilde{\sigma}} & \Omega X_{n+1} \\ \downarrow & & \downarrow \\ Y_n & \xrightarrow{\tilde{\sigma}} & \Omega Y_{n+1} \end{array}$$

is homotopy cartesian for all n, where the horizontal maps are the adjoints of the structure maps.
3. *The fibrant objects are Ω-spectra, i.e., the adjoints of their structure maps are weak equivalences.*

4.2 $\mathbb{Z}/2$-Equivariant Symmetric Spectra

For the $\mathbb{Z}/2$-equivariant stable homotopy category we use the model constructed in [52]. We now recall Mandell's definitions and results. Let G be a finite group. We only need the case $G = \mathbb{Z}/2$.

Definition 40 Let $S(G)$ denote the based simplicial G-set obtained by smashing together copies of the simplicial circle S^1 indexed on the elements of G, where the G-action permutes the smash factors according to the multiplication in G.

For example, $S(\{1\}) = S^1$ and $S(\mathbb{Z}/2) \cong S^2$, where $\mathbb{Z}/2$ acts via an orientation reversing map of degree one.

Definition 41 A symmetric G-spectrum consists of a based $G \times \Sigma_n$-simplicial set $T(n)$ for each $n \in \mathbb{N}$ and structure maps $T(n) \wedge S(G) \to T(n+1)$ such that the m-th iterated structure maps are $G \times \Sigma_n \times \Sigma_m$-equivariant. Morphisms of spectra are defined in the usual way. We denote the category of symmetric G-spectra by $Sp^{\Sigma}(G)$.

Mandell replaces $Sp^{\Sigma}(G)$ by the isomorphic category $Sp(G\Sigma_G)$ of $G\Sigma_G$-spectra, which is defined as follows.

Definition 42 Let Σ be the category whose objects are nonnegative integers $\underline{n} = \{1, \cdots, n\}$ (with the convention $\underline{0} = \emptyset$) and whose morphisms are bijections. Let Σ_G be the category Σ together with the diagram S indexed on Σ taking $\underline{n}$ to the n-fold smash product of $S(G)$ and with arrows permuting these smash factors.

A $G\Sigma_G$-spectrum T is a functor from Σ to based simplicial G-sets together with natural transformations $\sigma_{\underline{n},\underline{m}} : T(\underline{n}) \wedge S(\underline{m}) \to T(\underline{n+m})$ satisfying certain associativity and unitality conditions [52, Definition 1.3]. A morphism of $G\Sigma_G$-spectra is a natural transformation commuting with the structure maps.

Mandell defines Ω-spectra in $Sp(G\Sigma_G)$ and constructs a projective level model structure with level equivalences, level fibrations and projective cofibrations. The resulting homotopy category is denoted by Ho^l.

Definition 43 A morphism of $G\Sigma_G$-spectra $f : T \to U$ is called a stable equivalence if it induces bijections $\mathrm{Ho}^l(U, E) \to \mathrm{Ho}^l(T, E)$ for all Ω-spectra E.

The following theorem is [52, Theorem 4.1].

Theorem 6 *The category $Sp(G\Sigma_G)$ has a symmetric monoidal model structure with stable equivalences as weak equivalences and projective cofibrations as cofibrations. A morphism $f : T \to U$ is a fibration if and only if it is a level fibration and*

$$\begin{array}{ccc} T(\underline{m}) & \xrightarrow{\tilde{\sigma}_{\underline{m},\underline{n}}} & \Omega_{\underline{n}} T(\underline{m+n}) \\ \downarrow & & \downarrow \\ U(\underline{m}) & \xrightarrow{\tilde{\sigma}_{\underline{m},\underline{n}}} & \Omega_{\underline{n}} U(\underline{m+n}), \end{array}$$

is homotopy cartesian for all $\underline{m}, \underline{n} \in \Sigma$, where the horizontal maps are adjoint to the structure maps.

Using this model structure, Mandell shows [52, Theorem 2]:

Theorem 7 *The category $Sp^{\Sigma}(G)$ is Quillen equivalent to the stable G-equivariant category indexed on a complete G-universe.*

4.3 Complex and Real Topological Realisation Functors

The aim of this section is to recall the construction of the stable topological realisation functors

$$R = R_{\mathbb{C}} : \mathcal{SH}(\mathbb{C}) \to \mathcal{SH},$$

$$R' = R'_{\mathbb{R}} : \mathcal{SH}(\mathbb{R}) \to \mathcal{SH}(\mathbb{Z}/2).$$

Various unstable and stable versions of these functors were constructed in [88, Sect. 3.4], [63, Sect. 3.3], [5, 15], [70, Appendix] and [23, Sect. 4].

For $k \subseteq \mathbb{C}$, let $\mathbf{Sm}\,/k$ be the category of smooth schemes of finite type over k and let $\mathbf{sPre}(\mathbf{Sm}\,/k)$ be the category of simplicial presheaves on the Nisnevich site $\mathbf{Sm}\,/k$, see e.g. [63] or [38, Appendix B].

We begin with the definition of $R_{\mathbb{C}}$. In the next section we will define R_k and R'_k also for subfields k of $\mathbb{C}$ and $\mathbb{R}$ respectively.

The functor

$$R : \mathbf{sPre}(\mathbf{Sm}\,/\mathbb{C}) \to \mathbf{sSet}$$

is defined in the following way: Any simplicial presheaf A can be written as

$$\operatorname*{colim}_{X\times\Delta^n\to A}(X\times\Delta^n)\xrightarrow{\cong} A,$$

where the colimit is taken over the over-category of A, in which X runs over representable presheaves, see [70, Sect. A.4]. We set

$$R(A)=\operatorname*{colim}_{X\times\Delta^n\to A}(X(\mathbb{C})\times\Delta^n)\in\mathbf{sSet}\,.$$

By $X(\mathbb{C})$ we actually mean the simplicial set $\mathrm{Sing}(X(\mathbb{C})^{an})$, where $X(\mathbb{C})^{an}$ denotes the set of complex points of X with the analytic topology.

Now let $k=\mathbb{R}$. Let $\mathbf{sSet}(\mathbb{Z}/2)$ denote the category of $\mathbb{Z}/2$-simplicial sets. The functor

$$R':\mathbf{sPre}(\mathbf{Sm}\,/\mathbb{R})\to\mathbf{sSet}(\mathbb{Z}/2)$$

is still defined by

$$R'(A)=\operatorname*{colim}_{X\times\Delta^n\to A}(X(\mathbb{C})\times\Delta^n),$$

but now $\mathbb{Z}/2$ acts on $X(\mathbb{C})$ by precomposing with conjugation. This induces an action of $\mathbb{Z}/2$ on $R'(A)$, see e.g. [15, Sect. 5].

If A is pointed, then so are $R(A)$ and $R'(A)$ respectively.

We equip $\mathbf{sPre}(\mathbf{Sm}\,/k)$ with the projective model structure defined in [15, Sect. 5.1], where this category is denoted by $\mathrm{Spc}'(k)_{\mathrm{Nis}}$. The model structure for **sSet** can be found in [28, Sect. 3.2] and the equivariant model structure on $\mathbf{sSet}(\mathbb{Z}/2)$ is, for example, described in [21, Example 4.2]:

- Weak equivalences are maps f that induce weak equivalences f^H on the fixed point sets for all $H\subseteq\mathbb{Z}/2$.
- The collection $\{(\mathbb{Z}/2)/H\times\partial\Delta^n\to(\mathbb{Z}/2)/H\times\Delta^n\mid H\subseteq\mathbb{Z}/2\}$ is a set of generating cofibrations.
- The collection $\{(\mathbb{Z}/2)/H\times\Lambda^n_i\to(\mathbb{Z}/2)/H\times\Delta^n\mid H\subseteq\mathbb{Z}/2\}$ is a set of generating trivial cofibrations.

Theorem 8 *The functors R and R' and their pointed versions are strict symmetric monoidal left Quillen functors.*

Proof This is [15, Theorems 5.2, 5.5] and [70, Theorem A.23]. □

Now we define the functor

$$\mathrm{Sing}:\mathbf{sSet}\to\mathbf{sPre}(\mathbf{Sm}\,/\mathbb{C}),$$

as in [70, Theorem A.23]. It maps a simplicial set Z to the simplicial presheaf sending $X\in\mathbf{Sm}\,/\mathbb{C}$ to the simplicial set which in degree n is the set of maps $\mathbf{sSet}(X(\mathbb{C})\times\Delta^n,Z)$, that is, $\mathrm{Sing}(Z)(X)$ is defined as an internal hom object in **sSet**.

Proposition 44 *The functor* Sing *is right adjoint to* R.

Proof We have to find a bijection

$$\Phi : \mathbf{sSet}(R(X), Y) \to \mathbf{sPre}(\mathbf{Sm}/\mathbb{C})(X, \mathrm{Sing}(Y))$$

for any $X \in \mathbf{sPre}(\mathbf{Sm}/\mathbb{C})$ and $Y \in \mathbf{sSet}$. The general case will follow by passage to colimits after we have shown this for the case that X is representable. So let $X = \mathbf{Sm}/\mathbb{C}(- \times \Delta^\bullet, X')$ with $X' \in \mathbf{Sm}/\mathbb{C}$. Let $f : R(X) \to Y$ be an element of the left hand side, which now is $\mathbf{sSet}(X'(\mathbb{C}) \times \Delta^\bullet, Y)$. We have to define a natural transformation

$$\Phi(f) : X = \mathbf{Sm}/\mathbb{C}(- \times \Delta^\bullet, X') \to \mathbf{sSet}(-(\mathbb{C}) \times \Delta^\bullet, Y) = \mathrm{Sing}(Y).$$

We do this by the following composition:

$$\Phi(f) : \mathbf{Sm}/\mathbb{C}(- \times \Delta^\bullet, X') \xrightarrow{R} \mathbf{sSet}(-(\mathbb{C}) \times \Delta^\bullet, X'(\mathbb{C}))$$

$$\xrightarrow{f_*} \mathbf{sSet}(-(\mathbb{C}) \times \Delta^\bullet, Y).$$

The map Φ is obviously injective. It is also surjective because any morphism from $\mathbf{Sm}/\mathbb{C}(- \times \Delta^\bullet, X')$ to $\mathbf{sSet}(-(\mathbb{C}) \times \Delta^\bullet, Y)$ factors through the realisation functor. □

The $\mathbb{Z}/2$-version of this functor,

$$\mathrm{Sing}' : \mathbf{sSet}(\mathbb{Z}/2) \to \mathbf{sPre}(\mathbf{Sm}/\mathbb{R}),$$

maps an equivariant simplicial set Z to the simplicial presheaf sending $X \in \mathbf{Sm}/\mathbb{R}$ to the internal hom $\mathbf{sSet}(\mathbb{Z}/2)(X(\mathbb{C}), Z)$.

Proposition 45 *The functor* Sing$'$ *is right adjoint to* R'.

Proof The proof is the same as in the complex case, except that we have to replace $\mathbf{Sm}/\mathbb{C}$ by $\mathbf{Sm}/\mathbb{R}$ and $\mathbf{sSet}$ by $\mathbf{sSet}(\mathbb{Z}/2)$. □

We want to define stable versions of the functors R, R' and Sing, Sing$'$. Therefore, we consider the category of symmetric $\mathbb{P}^1$-spectra on $\mathbf{sPre}(\mathbf{Sm}/k)$, denoted by $Sp^{\Sigma}_{\mathbb{P}^1}(k)$. The stable model structure on $Sp^{\Sigma}_{\mathbb{P}^1}(k)$ is constructed in the same way as in [38], except that we start with our different definitions of fibrations and cofibrations on $\mathbf{sPre}(\mathbf{Sm}/k)$. This construction is also described in [70, Sect. A.5]. We will only need the following information about this model structure.

Let J be a level fibrant replacement functor and let Q be the stabilisation functor defined in [38, Remark 2.4]. A map f in $Sp^{\Sigma}_{\mathbb{P}^1}(k)$ is called a stable equivalence if $QJ(f)$ is a level equivalence.

Proposition 46 *A map of symmetric $\mathbb{P}^1$-spectra, $f : X \to Y$, is a stable fibration if and only if it is a level fibration and*

$$\begin{array}{ccc} X_n & \longrightarrow & QJX_n \\ \downarrow & & \downarrow \\ Y_n & \longrightarrow & QJY_n \end{array}$$

is homotopy cartesian [38, Lemma 2.7].

A stable trivial fibration is the same as a levelwise trivial fibration [38, Theorem 2.9].

Since $R(\mathbb{P}^1_{\mathbb{C}}) \cong \mathbb{C}P^1$ and $R'(\mathbb{P}^1_{\mathbb{R}}) \cong R'(S^1_s \wedge \mathbb{G}_m) \cong S^1_+ \wedge S^1_- \cong S(\mathbb{Z}/2)$ (the 2-sphere with orientation reversing $\mathbb{Z}/2$-action) in the homotopy categories, we can define

$$R : Sp^{\Sigma}_{\mathbb{P}^1}(\mathbb{C}) \to Sp^{\Sigma}_{\mathbb{C}P^1}$$

and

$$R' : Sp^{\Sigma}_{\mathbb{P}^1}(\mathbb{R}) \to Sp^{\Sigma}(\mathbb{Z}/2)$$

levelwise.

We can also extend Sing and Sing$'$ to the categories of spectra, as follows.

Over $\mathbb{C}$, the simplicial presheaf $\mathbb{P}^1$ is equivalent to the simplicial presheaf sending $Z \in \mathbf{Sm}/\mathbb{C}$ to $\mathbf{Sm}/\mathbb{C}(Z \times \Delta^n, \mathbb{P}^1_{\mathbb{C}})$. realisation defines a map from $\mathbf{Sm}/\mathbb{C}(Z \times \Delta^n, \mathbb{P}^1_{\mathbb{C}})$ to $\mathbf{sSet}(Z(\mathbb{C}) \times \Delta^n, \mathbb{C}P^1)$. These can be assembled into a map of simplicial presheaves $\mathbb{P}^1 \to \mathrm{Sing}(\mathbb{C}P^1)$. For $X \in Sp^{\Sigma}_{\mathbb{C}P^1}$ we can, hence, define structure maps

$$\mathrm{Sing}(X_n) \wedge \mathbb{P}^1 \to \mathrm{Sing}(X_n) \wedge \mathrm{Sing}(\mathbb{C}P^1)$$

$$\cong \mathrm{Sing}(X_n \wedge \mathbb{C}P^1) \xrightarrow{\mathrm{Sing}(\sigma_n)} \mathrm{Sing}(X_{n+1}),$$

so that we get a spectrum $\mathrm{Sing}(X) \in Sp^{\Sigma}_{\mathbb{P}^1}(\mathbb{C})$ defined levelwise.

Over $\mathbb{R}$, the same argument holds if we consider $\mathbb{Z}/2$-equivariant maps of simplicial sets. We get a spectrum $\mathrm{Sing}'(X) \in Sp^{\Sigma}_{\mathbb{P}^1}(\mathbb{R})$ defined levelwise.

Corollary 47 *The functors* (R, Sing) *and* (R', Sing') *form adjoint pairs between the categories of symmetric spectra.*

Theorem 9 (The stable functors R, R') *The pairs* (R, Sing) *and* (R', Sing') *are Quillen adjunctions on the spectrum level and R, R' are strict symmetric monoidal.*

Proof The case of R is covered in [70, Theorem A.45] and the claim for R' is proven in [23, Proposition 4.8]. For completeness, we reprove the theorem in our own words.

To show that R and R' are Quillen functors, we only have to prove that their right adjoints preserve stable fibrations between stably fibrant objects and stable trivial fibrations [13, Corollary A.2]. In all model structures we are considering here, the stable trivial fibrations are the levelwise trivial fibrations. From the unstable version of this theorem it follows therefore that Sing and Sing′ preserve stable trivial fibrations.

Stable fibrations are, in all of these model structures, levelwise fibrations with some additional homotopy pullback properties and stably fibrant objects are always Ω-spectra. By Propositions 44 and 45, the unstable functors Sing and Sing′ are right Quillen functors. It follows that the levelwise-defined functors Sing and Sing′ preserve Ω-spectra and level fibrations. Let $f : X \to Y$ be a stable fibration between Ω-spectra in $Sp^{\Sigma}_{\mathbb{C}P^1}$ with the model structure from Proposition 39 (or in $Sp^{\Sigma}(\mathbb{Z}/2)$ with the model structure from Theorem 6). We have to show that

$$\begin{array}{ccc} \mathrm{Sing}(X)_n & \longrightarrow & QJ\,\mathrm{Sing}(X)_n \\ \downarrow & & \downarrow \\ \mathrm{Sing}(Y)_n & \longrightarrow & QJ\,\mathrm{Sing}(Y)_n \end{array}$$

is homotopy cartesian for all n. Since X and Y are in particular level fibrant and Sing preserves level fibrations, $J\,\mathrm{Sing}(X) \simeq \mathrm{Sing}(X)$ and similarly for Y. Since Q is defined using only the adjoint structure maps, $Q\,\mathrm{Sing}(X) \simeq \mathrm{Sing}(X)$ and $Q\,\mathrm{Sing}(Y) \simeq \mathrm{Sing}(Y)$ for the Ω-spectra $\mathrm{Sing}(X)$ and $\mathrm{Sing}(Y)$. It follows that the above square is in particular homotopy cartesian.

The functors R and R' are strict symmetric monoidal, since this holds unstably and the product of symmetric spectra is defined in the same way in all the categories considered here. □

4.4 Realisation Functors for Other Fields

For $k \subseteq K$ a subfield, the canonical map $\mathrm{Spec}\,K \to \mathrm{Spec}\,k$ induces a couple of base change functors between the corresponding motivic homotopy categories. These are studied in [63, Sect. 3.1] and also in [33, Sect. 2]. For the stable version, see [70, Sect. A.7]. For a more general approach, see also [4].

Let $f : k \hookrightarrow K$ be the inclusion of the subfield. On the level of unpointed schemes, f^* is given by

$$f^* : \mathbf{Sm}\,/k \to \mathbf{Sm}\,/K, \;\; f^*(X) = X \times_{\mathrm{Spec}\,k} \mathrm{Spec}\,K.$$

It induces a functor

$$f^* : \mathbf{sPre}(\mathbf{Sm}\,/k) \to \mathbf{sPre}(\mathbf{Sm}\,/K),$$

which has a right adjoint f_*. By [70, Proposition A.47], this adjunction induces a strict symmetric monoidal Quillen adjunction on the level of symmetric spectra, where f^* is given by $f^*(E)_n = f^*(E_n)$.

One can therefore define realisation functors

$$R_k : \mathcal{SH}(k) \xrightarrow{f^*} \mathcal{SH}(\mathbb{C}) \xrightarrow{R} \mathcal{SH} \text{ for } k \overset{f}{\hookrightarrow} \mathbb{C},$$

$$R'_k : \mathcal{SH}(k) \xrightarrow{f^*} \mathcal{SH}(\mathbb{R}) \xrightarrow{R'} \mathcal{SH}(\mathbb{Z}/2) \text{ for } k \overset{f}{\hookrightarrow} \mathbb{R},$$

which are strict symmetric monoidal.

4.5 Constant Presheaf Functors

The following construction of the constant presheaf functors $c_k : \mathcal{SH} \to \mathcal{SH}(k)$ for $k \subseteq \mathbb{C}$ and $c'_k : \mathcal{SH}(\mathbb{Z}/2) \to \mathcal{SH}(k)$ for $k \subseteq \mathbb{R}$ is close to the one given in [23, Sect. 4].

Let $k \subseteq \mathbb{C}$. For $X \in \mathbf{sSet}$ we define $c_k(X) \in \mathbf{sPre}(\mathbf{Sm}/k)$ by $c_k(X)(Z) = X$ for all $Z \in \mathbf{Sm}/k$. Using $c_k(S^1) = S^1_s$, we extend the functor $c_k : \mathbf{sSet} \to \mathbf{sPre}(\mathbf{Sm}/k)$ levelwise to symmetric S^1-spectra and get:

$$c_k : Sp^{\Sigma}_{S^1} \to Sp^{\Sigma}_{S^1_s}(k).$$

We postcompose this functor with the $\mathbb{P}^1$-suspension functor, yielding a functor to the category of symmetric $(S^1_s, \mathbb{P}^1)$-bispectra, $c_k : Sp^{\Sigma}_{S^1} \to Sp^{\Sigma}_{S^1_s,\mathbb{P}^1}(k)$. The homotopy category of $Sp^{\Sigma}_{S^1_s,\mathbb{P}^1}(k)$ is equivalent to $\mathcal{SH}(k)$ by [29, Theorem 9.1].

Theorem 10 (The stable functor c_k) *This induces a functor $c_k : \mathcal{SH} \to \mathcal{SH}(k)$, which is strict symmetric monoidal. It is right inverse to R_k and hence faithful. Furthermore, by [49, Theorem 1], c_k is full if k is algebraically closed.*

Proof We first show that the unstable functor, $c_k : \mathbf{sSet} \to \mathbf{sPre}(\mathbf{Sm}/k)$, is a left Quillen functor. The generating cofibrations of $\mathbf{sSet}$ are the maps $\partial\Delta^n \hookrightarrow \Delta^n$. The functor c_k takes these maps to the same maps considered as morphisms of constant simplicial presheaves. These are examples of generating cofibrations in the model structure for $\mathbf{sPre}(\mathbf{Sm}/k)$, as described in [70, Sect. A.3]. The same applies to the generating trivial cofibrations $\Lambda^n_r \hookrightarrow \Delta^n$. The functor c_k preserves colimits by definition, hence it is a left Quillen functor. We denote its right adjoint by r_0. It satisfies $r_0(S^1_s) = S^1$.

Now we show that $c_k : Sp^{\Sigma}_{S^1} \to Sp^{\Sigma}_{S^1_s}(k)$ is left Quillen, where the model structure on $Sp^{\Sigma}_{S^1_s}(k)$ is described in [38, Sect. 4.5] and satisfies the analogue of Proposition 46. The right adjoint, r, to c_k is defined by levelwise application of r_0. Since r_0, as a right Quillen functor, preserves fibrations and trivial fibrations, r preserves level fibrations and level trivial fibrations. Stable trivial fibrations are the same as level trivial fibrations, hence these are preserved by r. We have to show that r also preserves stable fibrations between stably fibrant objects. Let $f : X \to Y$ be a stable fibration in $Sp^{\Sigma}_{S^1_s}(k)$ with X and Y level fibrant Ω-spectra. We have to show that $r(f)$ is a level fibration—which we already know—and that the squares

$$\begin{array}{ccc} r(X)_n & \longrightarrow & \Omega r(X)_{n+1} \\ \downarrow & & \downarrow \\ r(Y)_n & \longrightarrow & \Omega r(Y)_{n+1} \end{array}$$

are homotopy pullbacks. This is trivial because r preserves Ω-spectra (it is defined levelwise and commutes with desuspension), so $r(X)$ and $r(Y)$ are Ω-spectra. This proves that $c_k : Sp^{\Sigma}_{S^1} \to Sp^{\Sigma}_{S^1_s}(k)$ is a left Quillen functor. It is symmetric monoidal by its pointset definition and by the definition of products of symmetric spectra.

The $\mathbb{P}^1$-suspension functor $Sp^{\Sigma}_{S^1_s}(k) \to Sp^{\Sigma}_{S^1_s,\mathbb{P}^1}(k)$ is also a symmetric monoidal left Quillen functor if the category of symmetric $\mathbb{P}^1$-spectra over $Sp^{\Sigma}_{S^1_s}(k)$ is endowed with the stable model structure (see [29, Theorems 5.1 and 9.1]). It follows that both functors induce a functor on the respective stable homotopy categories, and the concatenation of the induced functors, $c_k : \mathcal{SH} \to \mathcal{SH}(k)$, is also strict symmetric monoidal by [29, Theorem 8.11].

To show that c_k is right inverse to R_k, first note that, for $f : k \hookrightarrow \mathbb{C}$, we have $f^*(\mathrm{Spec}(k)) = \mathrm{Spec}(\mathbb{C})$, which implies $f^* \circ c_k = c_{\mathbb{C}}$. So, by definition of R_k, $R_k \circ c_k = R_{\mathbb{C}} \circ f^* \circ c_k = R_{\mathbb{C}} \circ c_{\mathbb{C}}$, and it suffices to consider the case $k = \mathbb{C}$. Unstably, for $A \in \mathbf{sSet}$,

$$(R \circ c)(A) = \operatorname*{colim}_{X \times \Delta^\bullet \to cA} (X(\mathbb{C}) \times \Delta^\bullet) = \operatorname*{colim}_{\Delta^\bullet \to A} \Delta^\bullet = A.$$

On the level of spectra, we have used different models for constructing R and c. We therefore have to check that the following diagram is commutative, where, by definition, the composition of the upper maps induces $c : \mathcal{SH} \to \mathcal{SH}(\mathbb{C})$ and the lower map induces $R : \mathcal{SH}(\mathbb{C}) \to \mathcal{SH}$.

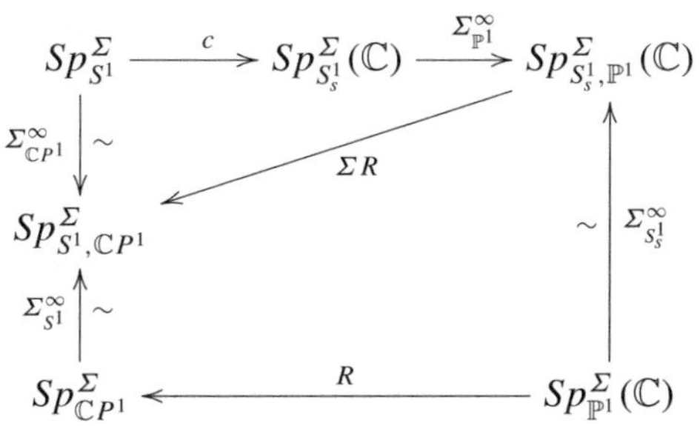

Since $R(S^1_s) = S^1$, the functor $R : Sp^{\Sigma}_{\mathbb{P}^1}(\mathbb{C}) \to Sp^{\Sigma}_{\mathbb{C}P^1}$ induces a functor ΣR on S^1_s-spectra on $Sp^{\Sigma}_{\mathbb{P}^1}(\mathbb{C})$ by the levelwise definition. This induced functor is drawn as a diagonal in the above diagram and it makes the lower subdiagram commutative by definition. Thus, it suffices to check that the upper diagram is commutative. Let $X = \{X_n\}_n \in Sp^{\Sigma}_{S^1}$. X is mapped to $\{cX_n\}_n$ in $Sp^{\Sigma}_{S^1_s}(\mathbb{C})$ and to $\{\mathbb{P}^m \wedge cX_n\}_{m,n}$ in $Sp^{\Sigma}_{S^1_s,\mathbb{P}^1}(\mathbb{C})$, which realises to $\{(\mathbb{C}P^1)^m \wedge X_n\}_{m,n}$ in $Sp^{\Sigma}_{S^1,\mathbb{C}P^1}$. This is the same as the image of X under the vertical map $Sp^{\Sigma}_{S^1} \to Sp^{\Sigma}_{S^1,\mathbb{C}P^1}$, which completes the proof that $R \circ c = \mathrm{id}$ on $\mathcal{SH}$. □

For subfields $k \subseteq \mathbb{R}$, we want to define functors $c'_k : \mathcal{SH}(\mathbb{Z}/2) \to \mathcal{SH}(k)$ which are right inverse to R'_k. For a better understanding, we first consider $k = \mathbb{R}$ and then generalise.

To define a functor $c' : \mathcal{SH}(\mathbb{Z}/2) \to \mathcal{SH}(\mathbb{R})$ which is right inverse to R', we first construct $R' : \mathbf{sSet}(\mathbb{Z}/2) \to \mathbf{sPre}(\mathbf{Sm}/\mathbb{R})$. Observe that $R'(\operatorname{Spec}\mathbb{R}) = *$ is the one-point set and $R'(\operatorname{Spec}\mathbb{C}) = \mathbb{Z}/2$ is the two-point set with non-trivial $\mathbb{Z}/2$-action. We let $c'(*) = \operatorname{Spec}\mathbb{R}$ and $c'(\mathbb{Z}/2) = \operatorname{Spec}\mathbb{C}$ and extend this to $\mathbb{Z}/2$-sets M by

$$c'(M) = \left(\coprod_{M^{\mathbb{Z}/2}} \operatorname{Spec}\mathbb{R}\right) \coprod \left(\coprod_{(M\setminus M^{\mathbb{Z}/2})/(\mathbb{Z}/2)} \operatorname{Spec}\mathbb{C}\right).$$

This can be done functoriality, just note that $-1 : \mathbb{Z}/2 \to \mathbb{Z}/2$ has to be mapped to the morphism $\operatorname{Spec}\mathbb{C} \to \operatorname{Spec}\mathbb{C}$ induced by complex conjugation. Furthermore, c' extends to simplicial $\mathbb{Z}/2$-sets by $c'(M \times \Delta^n) = c'(M) \times \Delta^n$. This defines the unstable, basepoint preserving functor $c' : \mathbf{sSet}(\mathbb{Z}/2) \to \mathbf{sPre}(\mathbf{Sm}/\mathbb{R})$. We extend this to a functor of spectra by postcomposing the levelwise defined functor

$$c' : Sp^{\Sigma}_{S(\mathbb{Z}/2)} \to Sp^{\Sigma}_{c'(S(\mathbb{Z}/2))}(\mathbb{R})$$

with the suspension spectrum functor

$$\Sigma^{\infty}_{\mathbb{P}^1} : Sp^{\Sigma}_{c'(S(\mathbb{Z}/2))}(\mathbb{R}) \to Sp^{\Sigma}_{c'(S(\mathbb{Z}/2)),\mathbb{P}^1}(\mathbb{R}).$$

Note that $c'(S(\mathbb{Z}/2)) \cong c'(S^1_+) \wedge c'(S^1_-) \cong S^1_s \wedge c'(S^1_-)$ and $c'(S^1_-) = F_{\mathbb{C}/\mathbb{R}}(S^V)$ in the notation of [33] (with V the sign representation), where it is shown that this is invertible in $\mathcal{SH}(\mathbb{R})$ [33, Theorem 3.5]. Thus, by [29, Theorem 9.1], $\Sigma^{\infty}_{c'(S(\mathbb{Z}/2))} : Sp^{\Sigma}_{\mathbb{P}^1}(\mathbb{R}) \to Sp^{\Sigma}_{c'(S(\mathbb{Z}/2)),\mathbb{P}^1}(\mathbb{R})$ is a Quillen equivalence.

Theorem 11 (The stable functor c') *This induces a functor $c' : \mathcal{SH}(\mathbb{Z}/2) \to \mathcal{SH}(\mathbb{R})$, which is strict symmetric monoidal and right inverse to R'. In particular, c' is faithful.*

Proof As in the previous proof, we start by considering the functor $c' : \mathbf{sSet}(\mathbb{Z}/2) \to \mathbf{sPre}(\mathbf{Sm}/\mathbb{R})$. It preserves colimits. The generating cofibrations of $\mathbf{sSet}(\mathbb{Z}/2)$ are the maps $(\mathbb{Z}/2)/H \times \partial\Delta^n \to (\mathbb{Z}/2)/H \times \Delta^n$, where $H \subseteq \mathbb{Z}/2$ is a subgroup. The images of these maps under c' can be written as pushout products:

$$c'(\mathbb{Z}/2 \times \partial\Delta^n \to \mathbb{Z}/2 \times \Delta^n) = (\emptyset \to \operatorname{Spec}\mathbb{C}) \,\square\, (\partial\Delta^n \to \Delta^n)$$

$$c'(\partial\Delta^n \to \Delta^n) = (\emptyset \to \operatorname{Spec}\mathbb{R}) \,\square\, (\partial\Delta^n \to \Delta^n).$$

These are examples of generating cofibrations for $\mathbf{sPre}(\mathbf{Sm}\,/\mathbb{R})$ as described in [70, Sect. A.3]. The same argument holds for the generating trivial cofibrations $(\mathbb{Z}/2)/H \times \Lambda^n_r \to (\mathbb{Z}/2)/H \times \Delta^n$. The passage to the spectrum level works similarly as in the previous proof. The induced functor $c' : \mathcal{SH}(\mathbb{Z}/2) \to \mathcal{SH}(\mathbb{R})$ is symmetric monoidal by the same arguments as before.

By its definition, c' is right inverse to R' on the level of simplicial $\mathbb{Z}/2$-sets. On the level of stable homotopy categories, $R' \circ c' = \mathrm{id}$ follows from the commutativity of the following diagram, similarly as in the previous proposition.

$$\begin{array}{ccccc}
Sp^{\Sigma}_{S(\mathbb{Z}/2)} & \xrightarrow{c'} & Sp^{\Sigma}_{c'(S(\mathbb{Z}/2))}(\mathbb{R}) & \xrightarrow{\Sigma^{\infty}_{\mathbb{P}^1}} & Sp^{\Sigma}_{c'(S(\mathbb{Z}/2)),\mathbb{P}^1}(\mathbb{R}) \\
\Sigma^{\infty}_{S(\mathbb{Z}/2)} \downarrow \sim & & \swarrow \Sigma R' & & \\
Sp^{\Sigma}_{S(\mathbb{Z}/2),S(\mathbb{Z}/2)} & & & & \sim \uparrow \Sigma^{\infty}_{c'(S(\mathbb{Z}/2))} \\
\Sigma^{\infty}_{S(\mathbb{Z}/2)} \uparrow \sim & & & & \\
Sp^{\Sigma}_{S(\mathbb{Z}/2)} & & \xleftarrow{\quad R' \quad} & & Sp^{\Sigma}_{\mathbb{P}^1}(\mathbb{R})
\end{array}$$

□

Now let $k \subseteq \mathbb{R}$. Then $R'_k(\operatorname{Spec} k) = *$ and $R'_k(\operatorname{Spec}(k[i])) = \mathbb{Z}/2$. Therefore, we let $c'_k(*) = \operatorname{Spec} k$ and $c'_k(\mathbb{Z}/2) = \operatorname{Spec}(k[i])$ and, for a $\mathbb{Z}/2$-set M,

$$c'_k(M) = \left(\coprod_{M^{\mathbb{Z}/2}} \operatorname{Spec} k\right) \coprod \left(\coprod_{(M\setminus M^{\mathbb{Z}/2})/(\mathbb{Z}/2)} \operatorname{Spec}(k[i])\right).$$

For functoriality, note that $-1 : \mathbb{Z}/2 \to \mathbb{Z}/2$ has to be mapped to $\operatorname{Spec}(k[i]) \to \operatorname{Spec}(k[i])$ induced by complex conjugation. As before, c'_k extends to $c'_k : \mathbf{sSet}(\mathbb{Z}/2) \to \mathbf{sPre}(\mathbf{Sm}\,/k)$ and then to

$$c'_k : Sp^{\Sigma}_{S(\mathbb{Z}/2)} \to Sp^{\Sigma}_{c'_k(S(\mathbb{Z}/2))}(k) \to Sp^{\Sigma}_{c'_k(S(\mathbb{Z}/2)),\mathbb{P}^1}(k),$$

where the first functor is defined levelwise by c'_k and the second one is the $\mathbb{P}^1_k$-suspension spectrum functor. Here, $c'_k(S(\mathbb{Z}/2)) \cong S^1_s \wedge c'_k(S^1_-)$ and $c'_k(S^1_-) = F_{k[i]/k}(S^V)$ in the notation of [33], which is invertible in $\mathcal{SH}(k)$ by [33, Theorem 3.5].

Theorem 12 (The stable functor c'_k) *This induces a functor* $c'_k : \mathcal{SH}(\mathbb{Z}/2) \to \mathcal{SH}(k)$ *which is strict symmetric monoidal and right inverse to* R'_k.

Proof The main claim follows exactly as in the case $k = \mathbb{R}$ considered above. It is also implied by [23, Theorem 4.6]. It remains to prove $R'_k \circ c'_k \cong \mathrm{id}$. Again, this follows from $f^* \circ c'_k \cong c'$ (for $k \overset{f}{\hookrightarrow} \mathbb{R}$) and $R' \circ c' \cong \mathrm{id}$, where $f^* \circ c'_k \cong c'$ holds because $f^*(\operatorname{Spec} k) \cong \operatorname{Spec} \mathbb{R}$ and $f^*(\operatorname{Spec}(k[i])) \cong \operatorname{Spec} \mathbb{R} \times_{\operatorname{Spec} k} \operatorname{Spec}(k[i]) \cong \operatorname{Spec} \mathbb{C}$. □

Remark 48 In [23, Theorem 1.1], Heller and Ormsby prove that if k is a real closed field, then c'_k is full after p-completion.

Remark 49 With similar methods as above, one can show that the functors $\mathbf{sSet} \overset{c_k}{\to} \mathbf{sPre}(\mathbf{Sm}/k) \overset{R_k}{\to} \mathbf{sSet}$ induce functors

$$\mathcal{SH} \overset{c_k}{\to} \mathcal{SH}_{S^1_s}(k) \overset{R_k}{\to} \mathcal{SH},$$

where $\mathcal{SH}_{S^1_s}(k)$ is the stable motivic homotopy category in which S^1_s got inverted but $\mathbb{G}_m$ did not.

For $k \subseteq \mathbb{R}$, the definition of the stable functor c'_k relied on the invertibility of $F_{\mathbb{C}/\mathbb{R}}(S^V)$ in $\mathcal{SH}(\mathbb{R}) = \mathcal{SH}_{\mathbb{P}^1}(\mathbb{R})$, as shown in [33, Theorem 3.5]. One can show that the functors $\mathbf{sSet}(\mathbb{Z}/2) \overset{c'_k}{\to} \mathbf{sPre}(\mathbf{Sm}/k) \overset{R'_k}{\to} \mathbf{sSet}(\mathbb{Z}/2)$ induce functors

$$\mathcal{SH}_{S^1}(\mathbb{Z}/2) \overset{c'_k}{\to} \mathcal{SH}_{S^1_s}(k) \overset{R'_k}{\to} \mathcal{SH}_{S^1}(\mathbb{Z}/2),$$

where $\mathcal{SH}_{S^1}(\mathbb{Z}/2)$ is the naive equivariant stable homotopy category, in which only the sphere with trivial action got inverted. The functor $R' : \mathcal{SH}_{S^1_s}(\mathbb{R}) \to \mathcal{SH}_{S^1}(\mathbb{Z}/2)$ sends $F_{\mathbb{C}/\mathbb{R}}(S^V)$ to $S^V = S(\mathbb{Z}/2)$, which is not invertible in $\mathcal{SH}_{S^1}(\mathbb{Z}/2)$. Therefore, $F_{\mathbb{C}/\mathbb{R}}(S^V)$ cannot be invertible in $\mathcal{SH}_{S^1_s}(\mathbb{R})$. This shows that it is not possible to extend $c'_k : \mathbf{sSet}(\mathbb{Z}/2) \to \mathbf{sPre}(\mathbf{Sm}/k)$ to a functor from $\mathcal{SH}(\mathbb{Z}/2)$ to $\mathcal{SH}_{S^1_s}(k)$.

5 Thick Ideals Discovered by Comparison Functors

The aim of this section is to draw conclusions concerning thick subcategories and thick ideals in $\mathcal{SH}(k)$, $k \subseteq \mathbb{C}$, and in finite, local versions of this category using the functors from the previous section. In the next section, we will study thick ideals that are described by motivic Morava K-theories.

5.1 Consequences of the Properties of R_k, R'_k, c_k and c'_k

Proposition 50 *(1) If $\mathcal{C} \subseteq \mathcal{SH}$ is a thick subcategory or a thick ideal, then, for $k \subseteq \mathbb{C}$, $R_k^{-1}(\mathcal{C}) \subseteq \mathcal{SH}(k)$ is a thick subcategory or a thick ideal, respectively.*

(2) If $\mathcal{C} \subseteq \mathcal{SH}(\mathbb{Z}/2)$ is a thick subcategory or a thick ideal, then, for $k \subseteq \mathbb{R}$, $(R'_k)^{-1}(\mathcal{C}) \subseteq \mathcal{SH}(k)$ is a thick subcategory or a thick ideal, respectively.

(3) If $f : k \hookrightarrow K$ and $\mathcal{C} \subseteq \mathcal{SH}(K)$ is a thick subcategory or a thick ideal, then $(f^*)^{-1}(\mathcal{C}) \subseteq \mathcal{SH}(k)$ is a thick subcategory or a thick ideal, respectively.

Proof Any functor preserves retracts, hence $R_k^{-1}(\mathcal{C})$ is closed under retracts whenever $\mathcal{C}$ is. We also have $R_k(S_s^1) = S^1$ and R_k preserves cofibers because it is a left adjoint, hence it preserves exact triangles. Therefore, $R_k^{-1}(\mathcal{C})$ is closed under triangles whenever $\mathcal{C}$ is. Hence, R_k^{-1} preserves thick subcategories. Since R_k is symmetric monoidal (see Sect. 4), $X \in R_k^{-1}(\mathcal{C})$ and $Y \in \mathcal{SH}(k)$ implies that $R_k(X \wedge Y) \cong R_k(X) \wedge R_k(Y)$ is in $\mathcal{C}$, if $\mathcal{C}$ is a thick ideal. Thus, $X \wedge Y \in R_k^{-1}(\mathcal{C})$. That is, R_k^{-1} preserves thick ideals, too.

The proofs for $(R_k')^{-1}$ and $(f^*)^{-1}$ are the same. □

Proposition 51 *For $k \subseteq \mathbb{C}$, c_k^{-1} preserves thick subcategories and thick ideals. Similarly, for $k \subseteq \mathbb{R}$, $(c_k')^{-1}$ preserves thick subcategories and thick ideals.*

Proof Since $S_s^1 = c_k(S^1)$ and c_k preserves mapping cones, c_k preserves exact triangles. It also preserves retracts and is strict symmetric monoidal, hence, c_k^{-1} preserves thick subcategories and thick ideals. □

5.2 *Finite Motivic Spectra*

The thick subcategory theorem of [25] concerns the category of finite spectra, $\mathcal{SH}^{fin}$, as defined in Sect. 1. The functors R_k and c_k can therefore only help us to understand subcategories of $\mathcal{SH}(k)$ which are at most as big as $R_k^{-1}(\mathcal{SH}^{fin})$. There are multiple equivalent possibilities to define $\mathcal{SH}^{fin}$, using the notions of finite CW-spectra, dualisable objects or compact objects. These notions are not equivalent in the motivic setting, therefore we obtain more than one possible category of "finite" objects in $\mathcal{SH}(k)$.

We will now discuss the various versions of finiteness, including non-standard notations which will be needed in Proposition 67. Let k be any field.

Definition 52 (1) The category $\mathcal{SH}(k)^{fin}$ of finite cellular motivic spectra over a field k is the smallest full subcategory of $\mathcal{SH}(k)$ that contains the spheres $S^{p,q}$ for all $p, q \in \mathbb{Z}$ and is closed under exact triangles [16, Definition 8.1].

(2) For $k \subseteq \mathbb{R}$, let $\mathcal{SH}(k)^{fin+}$ be the smallest full subcategory of $\mathcal{SH}(k)$ that contains $S^{p,q} \wedge (\operatorname{Spec} k[i])_+^{\wedge m}$ for all $p, q \in \mathbb{Z}, m \geq 0$ and is closed under exact triangles.

(3) The closures of $\mathcal{SH}(k)^{fin}$ and $\mathcal{SH}(k)^{fin+}$ under colimits are denoted by $\mathcal{SH}(k)^{cell}$ and $\mathcal{SH}(k)^{cell+}$. Their objects are called cellular, see [16, Definitions 2.1 and 2.10].

Remark 53 Note that these categories are closed under the bifunctor $\wedge$ because so are their sets of generators and because $\wedge$ preserves exact triangles and colimits, as it is a left adjoint.

With this definition, $\mathcal{SH}(k)^{fin}$ is the smallest tensor triangulated full subcategory of $\mathcal{SH}(k)$ that contains $c_k(\mathcal{SH}^{fin})$ and is closed under $- \wedge \mathbb{G}_m^{\pm 1}$, and $\mathcal{SH}(k)^{fin,+}$ is the smallest tensor triangulated full subcategory that contains $c'_k(\mathcal{SH}(\mathbb{Z}/2)^{fin})$ and is closed under $- \wedge \mathbb{G}_m^{\pm 1}$.

The following results can mostly be found in [65, Sect. 4].

Definition 54 (1) Let $\mathcal{D} \subseteq \mathcal{SH}(k)$ be the collection of all (strongly) dualisable objects. That is, all spectra X such that the canonical map $F(X, S) \wedge Y \to F(X, Y)$ is an isomorphism for all $Y \in \mathcal{SH}(k)$, where $F(-, -)$ denotes the derived internal hom in $\mathcal{SH}(k)$ and $S = S^{0,0}$ is the sphere spectrum. $F(X, S)$ is called the dual of X and is also denoted by DX (compare Definition 11).

(2) A motivic spectrum $F \in \mathcal{SH}(k)$ is called compact if $\mathrm{Hom}_{\mathcal{SH}(k)}(F, -)$ preserves arbitrary sums. Let $\mathcal{SH}(k)_f \subseteq \mathcal{SH}(k)$ denote the full subcategory of compact objects.

Remark 55 (1) $\mathcal{SH}(k)_f$ is a thick subcategory of $\mathcal{SH}(k)$ [65, Sect. 4].

(2) Any dualisable object is also compact, as shown in the proof of Proposition 12.

(3) The smash product of two dualisable objects X and Y is again dualisable, because $F(X \wedge Y, S) \wedge Z \cong F(X, F(Y, S)) \wedge Z \cong F(X, S) \wedge F(Y, S) \wedge Z \cong F(X, S) \wedge F(Y, Z) \cong F(X, F(Y, Z)) \cong F(X \wedge Y, Z)$.
Similarly, compact objects are closed under $\wedge$.

(4) By [33, Cor. 2.14 and Thm. 4.1], $F(\mathrm{Spec}(k[i])_+, E) \cong \mathrm{Spec}(k[i])_+ \wedge E$ in $\mathcal{SH}(k)$, $k \subseteq \mathbb{R}$. That is, $\Sigma^\infty \mathrm{Spec}(k[i])_+$ is self-dual and in particular compact.

Definition 56 For $\mathcal{R}$ a collection of objects in $\mathcal{SH}(k)_f$, let $\mathcal{SH}(k)_{\mathcal{R},f} \subseteq \mathcal{SH}(k)_f$ be the smallest thick subcategory containing $\mathcal{R}$.

Let $\mathcal{T}_k = \{S^{p,q} \mid p, q \in \mathbb{Z}\}$ be the collection of all motivic spheres in $\mathcal{SH}(k)$ and let $\mathcal{T}_k^+ = \{S^{p,q} \wedge (\mathrm{Spec}\, k[i])_+^{\wedge m} \mid p, q \in \mathbb{Z}, m \geq 0\}$ if $k \subseteq \mathbb{R}$. These are sets of compact objects in $\mathcal{SH}(k)$ by Remark 55, parts (2) and (4).

Comparing the definitions, we see that $\mathcal{SH}(k)_{\mathcal{T}_k,f}$ is the closure of $\mathcal{SH}(k)^{fin}$ under retracts and $\mathcal{SH}(k)_{\mathcal{T}_k^+,f}$ is the closure of $\mathcal{SH}(k)^{fin+}$ under retracts.

Proposition 57 *$\mathcal{SH}(k)_{\mathcal{D},f} \subseteq \mathcal{SH}(k)_f$ is the full subcategory of dualisable objects of $\mathcal{SH}(k)$.*

Proof This is [65, Lemma 4.2]. By Remark 55(2), $\mathcal{D}$ is a collection of compact objects. Furthermore, the full subcategory spanned by $\mathcal{D}$ is already a thick subcategory. □

Since $\mathcal{SH}(k)_{\mathcal{D},f} \subseteq \mathcal{SH}(k)_f$ is a thick subcategory and $\mathcal{SH}(k)_f \subseteq \mathcal{SH}(k)$ is a thick subcategory, it follows that the strongly dualisable objects form a thick subcategory of $\mathcal{SH}(k)$. Note also that all the categories mentioned above are closed under $\wedge$.

Proposition 58 *We have $\mathcal{SH}(k)^{fin} \subseteq \mathcal{SH}(k)_{\mathcal{T}_k,f} \subseteq \mathcal{SH}(k)_{\mathcal{D},f} \subseteq \mathcal{SH}(k)_f$ and for $k \subseteq \mathbb{R}$, $\mathcal{SH}(k)^{fin+} \subseteq \mathcal{SH}(k)_{\mathcal{T}_k^+,f} \subseteq \mathcal{SH}(k)_{\mathcal{D},f} \subseteq \mathcal{SH}(k)_f$. In particular, all objects in $\mathcal{SH}(k)^{fin}$ and $\mathcal{SH}(k)^{fin+}$ are strongly dualisable. Furthermore, $\mathcal{SH}(k)^{fin}$ and $\mathcal{SH}(k)^{fin+}$ are closed under taking duals.*

Proof The first line and the case $\mathcal{SH}(k)^{fin}$ are proven in [65, Sect. 4]. For $k \subseteq \mathbb{R}$, the only additional input is the self-duality of $\Sigma^\infty \operatorname{Spec}(k[i])_+$. □

Remark 59 ([65, Remark 8.2]) gives an example for an object in $\mathcal{SH}(S)$ that is compact but not dualisable, where S is the spectrum of a discrete valuation ring.

A stronger result holds if k is a field of characteristic 0. It is also proven in [76].

Proposition 60 *Let k be of characteristic* 0. *Then* $\mathcal{SH}(k)_{\mathcal{D},f} = \mathcal{SH}(k)_f$ *is the thick subcategory of* $\mathcal{SH}(k)$ *generated by* $\{\Sigma^{2n,n}\Sigma^\infty U_+ \mid U \in \mathbf{Sm}/k, n \in \mathbb{Z}\}$. *Hence, any object of* $\mathcal{SH}(k)$ *is dualisable if and only if it is compact.*

Proof By [16, Theorem 9.2], $\{\Sigma^{2n,n}\Sigma^\infty U_+ \mid U \in \mathbf{Sm}/k,\ n \in \mathbb{Z}\}$ is a set of compact generators for $\mathcal{SH}(k)$, which means two things: First, these objects are compact and second, the only full triangulated subcategory of $\mathcal{SH}(k)$ containing this set and being closed under infinite direct sums is $\mathcal{SH}(k)$ itself. Since schemes are locally affine, also $\{\Sigma^{2n,n}\Sigma^\infty U_+ \mid U \in \mathbf{Sm}/k$ quasi-projective$\}$ is a set of compact generators. General theory [55, Theorem 13.1.14] implies that $\mathcal{SH}(k)_f$ is the thick subcategory of $\mathcal{SH}(k)$ generated by $\{\Sigma^{2n,n}\Sigma^\infty U_+ \mid U \in \mathbf{Sm}/k$ quasi-projective$\}$. By [77, Theorem 4.9], $\Sigma^\infty U_+$ is dualisable for any such U. Since dualisability is preserved by exact triangles and retracts, the thick subcategory generated by $\{\Sigma^{2n,n}\Sigma^\infty U_+ \mid U \in \mathbf{Sm}/k$ quasi-projective$\}$ is contained in $\mathcal{SH}(k)_{\mathcal{D},f}$. Thus, $\mathcal{SH}(k)_f = \mathcal{SH}(k)_{\mathcal{D},f}$. □

For $k \subseteq \mathbb{C}$ ($k \subseteq \mathbb{R}$) all these categories are furthermore included in the preimage of compact topological spectra under R_k (R'_k):

Proposition 61 *For* $k \subseteq \mathbb{C}$,

$$\mathcal{SH}(k)_f \subseteq R_k^{-1}(\mathcal{SH}^{fin})$$

and for $k \subseteq \mathbb{R}$,

$$\mathcal{SH}(k)_f \subseteq R_k'^{-1}(\mathcal{SH}(\mathbb{Z}/2)_f).$$

Proof We have to show that R_k and R'_k preserve compact objects. Let $f : k \hookrightarrow \mathbb{C}$. Then f^* restricts to a functor between the categories of compact objects: $f^* : \mathcal{SH}(k)_f \to \mathcal{SH}(\mathbb{C})_f$, because the base change functor preserves smooth schemes and so f^* sends a compact generator $\Sigma^{(2n,n)}\Sigma^\infty U_+$, $U \in \mathbf{Sm}/k$, $n \in \mathbb{Z}$, of $\mathcal{SH}(k)$ to a compact generator of $\mathcal{SH}(\mathbb{C})$. Hence, for the first claim it suffices to prove that $R = R_\mathbb{C}$ preserves compact objects. Similarly, for the second claim it suffices to show that $R' = R'_\mathbb{R}$ preserves compact objects.

Let $X = \Sigma^{(2n,n)}\Sigma^\infty U_+$ be a compact generator of $\mathcal{SH}(\mathbb{C})$. As in the proof of the previous proposition, we can assume that U is a smooth quasi-projective scheme. By Jouanolou's trick [41, Lemma 1.5], there exists an affine vector bundle torsor over U. This is a vector bundle $E \to U$, together with a torsor $p : V \to U$ on E with V affine. By the definition of a torsor, V is locally isomorphic to E. This implies that for some m, $U \times \mathbb{A}^m$ is locally isomorphic to an affine smooth scheme V. In particular, U is $\mathbb{A}^1$-equivalent to V, and so $X \cong \Sigma^{(2n,n)}\Sigma^\infty V_+$ in $\mathcal{SH}(\mathbb{C})$. Now, since V is smooth and affine, its complex realisation has the homotopy type of a finite CW complex by [47, Example 3.1.9]. Hence, $R(X) \in \mathcal{SH}$ is isomorphic to $\Sigma^n\Sigma^\infty Y_+$ for some finite CW complex Y and, so, $R(X)$ is compact by Remark 7.

The proof that for any smooth and affine variety V, $V(\mathbb{C}) \subseteq \mathbb{C}^r$ has the homotopy type of a finite CW-complex, can be summarised as follows [47, Example 3.1.9]: $V(\mathbb{C})$ is a complex submanifold of $\mathbb{C}^r$ without boundary (because V is smooth, see e.g. [37, Sect. 3.1.2]), which is closed as a subset of $\mathbb{C}^r$ (because the zero locus of any polynomial is closed). For almost any $c \in \mathbb{C}^r$, the squared distance function $\phi_c : V(\mathbb{C}) \to \mathbb{R}, \phi_c(x) = ||x - c||^2$, has only non-degenerate critical points [57, Theorem 6.6]. Furthermore, ϕ_c being real algebraic implies that it has only finitely many critical points, as in [47, Example 3.1.9]. Using Morse theory [57, Theorem 3.5], it follows that $V(\mathbb{C})$ has the homotopy type of a CW complex with one cell of dimension n for each critical point of ϕ_c of index n.

For X a compact generator of $\mathcal{SH}(\mathbb{R})$, we also have $X \cong \Sigma^{(2n,n)}\Sigma^\infty V_+$ and now V is a smooth and affine real variety. Its realisation $R'(V) = V(\mathbb{C}) \subseteq \mathbb{C}^r$ is still a complex manifold, which is closed, but with the property that, for any $x \in V(\mathbb{C})$, also its complex conjugate $\bar{x}$ lies in $V(\mathbb{C})$. This was used for the definition of the $\mathbb{Z}/2$-action on $V(\mathbb{C})$: $\rho(x) = \bar{x}$ for $\rho \in \mathbb{Z}/2$ the generator. As before, we can choose $c \in \mathbb{C}^r$ such that ϕ_c has finitely many critical points, which are all non-degenerate. If $c \in \mathbb{R}^r$, ϕ_c is $\mathbb{Z}/2$-invariant and the claim follows from equivariant Morse theory: by the proof of [56, Theorem 2.2], any invariant Morse function can be turned into a special invariant Morse function and, by [56, Theorem 3.3], a manifold with a special invariant Morse function is equivariantly homotopy equivalent to an equivariant CW complex with one equivariant cell for each critical orbit. Thus, $V(\mathbb{C})$ is equivalent to a finite $\mathbb{Z}/2$-CW complex, whose suspension spectrum is compact in $\mathcal{SH}(\mathbb{Z}/2)$ by Proposition 12. For more information on equivariant Morse theory in English language, we refer the reader to [93].

Now, assume $c \in \mathbb{C}^r \setminus \mathbb{R}^r$. We would like to take $x \mapsto \min(\phi_c(x), \phi_c(\bar{x}))$ as a Morse function. It is continuous and $\mathbb{Z}/2$-invariant but it is not differentiable for $x \in \mathbb{R}^r$. Outside of $\mathbb{R}^r$, the critical points of $\min(\phi_c(x), \phi_c(\bar{x}))$ are a subset of the critical points of ϕ_c and their complex conjugates. The idea is to proceed in two steps: first, to take care of the real part $V(\mathbb{C}) \cap \mathbb{R}^r = V(\mathbb{R}) = V(\mathbb{C})^{\mathbb{Z}/2}$ and, second, to use $\min(\phi_c(x), \phi_c(\bar{x}))$ as a Morse function away from $\mathbb{R}^r$.

$V(\mathbb{R}) \subseteq \mathbb{R}^r$ is a manifold, which is closed and without boundary and for which there exists $d \in \mathbb{R}^r$ such that $\phi_d : V(\mathbb{R}) \to \mathbb{R}$ is a Morse function with finitely many critical points (analogously to $V(\mathbb{C}) \subseteq \mathbb{C}^r$). Let B be an open ball around $0 \in \mathbb{C}^r$ which contains a ball $D(d)$ around d containing all critical points of ϕ_d and a ball $D(c)$ around c containing all critical points of ϕ_c and their conjugates. $V(\mathbb{R})$ can be contracted inside $\mathbb{R}^r$ to $V(\mathbb{R}) \cap \overline{B}$, following the orthogonal trajectories of the hypersurfaces on which ϕ_d is constant, until $\overline{B}$ is reached (that is, we glue an infinite sequence of the diffeomorphisms $M^a \cong M^b$ constructed in the proof of [57, Theorem 3.1] to one homotopy equivalence). We extend this homotopy equivalence to a narrow open neighborhood $U \subset \mathbb{C}^r$ of $\mathbb{R}^r \setminus (B \cap \mathbb{R}^r)$ so that $\mathbb{C}^r \simeq \mathbb{C}^r \setminus U$ by a $\mathbb{Z}/2$-equivariant homotopy equivalence which does not add critical points to $\min(\phi_c(x), \phi_c(\bar{x}))$ outside of B. Using this homotopy equivalence, we can assume that $V(\mathbb{C}) \setminus B$ does not contain any points near the real subspace $\mathbb{R}^r$. Now we use ϕ_c to contract $V(\mathbb{C})$ to $V(\mathbb{C}) \cap \bar{B}$ with an equivariant homotopy equivalence: any $x \in V(\mathbb{C}) \setminus B$ with $\phi_c(x) < \phi_c(\bar{x})$ gets moved into B along the line of steepest descent of ϕ_c inside $V(\mathbb{C})$. If, however, $\phi_c(\bar{x}) < \phi_c(x)$, x gets moved along the line of steepest descent of $\phi_{\bar{c}}$. Note that $\phi_{\bar{c}}(x) = \phi_c(\bar{x})$ and that the homotopy equivalence is well defined because we removed $\mathbb{R}^r$ before and because all critical points of ϕ_c and $\phi_{\bar{c}}$ are inside B. Hence, this defines a $\mathbb{Z}/2$-equivariant homotopy equivalence between $V(\mathbb{C})$ and $V(\mathbb{C}) \cap \bar{B}$.

Thus, $V(\mathbb{C})$ is $\mathbb{Z}/2$-homotopy equivalent to a $\mathbb{Z}/2$-manifold (with boundary) which is closed and bounded and, hence, compact in $\mathbb{C}^r$. Using equivariant Morse theory, this compact manifold is $\mathbb{Z}/2$-homotopy equivalent to a finite $\mathbb{Z}/2$-CW complex (again, see [56, Theorems 2.2, 3.3] or [93]). Note that, although most of the literature only studies Morse theory for manifolds without boundary, it still covers this case for the following reason: Since $D(c) \subset B$, the Morse function ϕ_c evaluated on the boundary of $V(\mathbb{C}) \cap \overline{B}$ takes a higher value than on any critical point. Therefore, $V(\mathbb{C}) \cap \overline{B}$ appears at a finite step in the Morse theoretical construction of the CW complex associated with ϕ_c, i.e., $V(\mathbb{C}) \cap \overline{B} \simeq V(\mathbb{C})^m = \phi_c^{-1}([0, m])$ for some $m \in \mathbb{R}$, which is a finite $\mathbb{Z}/2$-CW complex. □

5.3 Motivic Thick Ideals

Let $k \subseteq \mathbb{C}$ and p be any prime. The following theorem identifies important families of thick ideals in $(\mathcal{SH}(k)_f)_{(p)}$. It is the main result in this section.

Theorem 13 (Lower bound on the number of motivic thick ideals)

(1) The category $(\mathcal{SH}(k)_f)_{(p)}$ contains at least an infinite chain of different thick ideals, given by $\overline{R}_k^{-1}(\mathcal{C}_n)$, $0 \leq n \leq \infty$, where $\overline{R}_k$ denotes the p-localisation of the restriction of R_k to $\mathcal{SH}(k)_f$ and $\mathcal{C}_n \subseteq \mathcal{SH}_{(p)}^{fin}$ is as defined in Sect. 1.

(2) If $k \subseteq \mathbb{R}$, then $(\mathcal{SH}(k)_f)_{(p)}$ contains at least a two-dimensional lattice of different thick ideals, given by $(\overline{R}_k^{\prime})^{-1}(\mathcal{C}_{m,n})$, for all $(m, n) \in \Gamma_p$ as in Definition 35.

Proof We prove the first part, the second is proven similarly. By Remark 53, Propositions 58 and 61, $c_k(\mathcal{SH}^{fin}) \subseteq \mathcal{SH}(k)_f \subseteq R_k^{-1}(\mathcal{SH}^{fin})$, hence c_k and R_k restrict to functors $\mathcal{SH}^{fin} \xrightarrow{c_k} \mathcal{SH}(k)_f \xrightarrow{\overline{R}_k} \mathcal{SH}^{fin}$. Since the motivic p-local Moore spectrum (the homotopy colimit of a diagram of sphere spectra whose arrows are multiplications by integers prime to p) is the image of the topological p-local Moore spectrum under c_k, we get induced functors between the localised categories,

$$\mathcal{SH}_{(p)}^{fin} \xrightarrow{c_k} (\mathcal{SH}(k)_f)_{(p)} \xrightarrow{\overline{R}_k} \mathcal{SH}_{(p)}^{fin},$$

which still have the properties that c_k and $\overline{R}_k$ preserve exact triangles and smash products and that $\overline{R}_k \circ c_k = \mathrm{id}$, as proven in Theorem 10. Similarly to Proposition 50, it follows that $\overline{R}_k^{-1}(\mathcal{C}_n)$ is a thick ideal. Now let $n < m$. Then $\mathcal{C}_m \subset \mathcal{C}_n$ in $\mathcal{SH}_{(p)}^{fin}$ and there is some $X \in \mathcal{C}_n \setminus \mathcal{C}_m$. It follows that $\overline{R}_k^{-1}(\mathcal{C}_m) \subseteq \overline{R}_k^{-1}(\mathcal{C}_n)$ and that $c_k(X) \in \overline{R}_k^{-1}(\mathcal{C}_n) \setminus \overline{R}_k^{-1}(\mathcal{C}_m)$. Thus, $\overline{R}_k^{-1}(\mathcal{C}_n)$, $0 \leq n \leq \infty$, form a chain of pairwise different thick ideals.

For the second part, one needs to consider the functors

$$(\mathcal{SH}(\mathbb{Z}/2)_f)_{(p)} \xrightarrow{c'_k} (\mathcal{SH}(k)_f)_{(p)} \xrightarrow{\overline{R}'_k} (\mathcal{SH}(\mathbb{Z}/2)_f)_{(p)}.$$

□

In the following, we will omit the overline and use the notation R_k, R'_k for the (p-localised) restricted functors, too.

Remark 62 In the above theorem, $\mathcal{SH}(k)_f$ can be replaced by any other tensor triangulated full subcategory $\mathcal{D} \subseteq \mathcal{SH}(k)$ satisfying

$$c_k(\mathcal{SH}^{fin}) \subseteq \mathcal{D} \subseteq R_k^{-1}(\mathcal{SH}^{fin}), \text{ or}$$

$$c'_k(\mathcal{SH}(\mathbb{Z}/2)^{fin+}) \subseteq \mathcal{D} \subseteq (R'_k)^{-1}(\mathcal{SH}(\mathbb{Z}/2)_f) \text{ respectively.}$$

In particular, the theorem applies to any

$$\mathcal{D} \in \{\mathcal{SH}(k)^{fin},\ \mathcal{SH}(k)^{fin+},\ \mathcal{SH}(k)_{\mathcal{T}_k,f},\ \mathcal{SH}(k)_{\mathcal{T}_k^+,f},$$

$$R_k^{-1}(\mathcal{SH}^{fin}),\ R_k'^{-1}(\mathcal{SH}(\mathbb{Z}/2)_f)\}$$

for the following reasons. Recall from Remark 37 that Strickland's characterisation of equivariant thick ideals works in $\mathcal{SH}(G)^{fin}$ as well as in its closure under retracts, $\mathcal{SH}(G)_f$. Therefore, we can here take $\mathcal{SH}(k)^{fin(+)}$ as well as its closure under retracts, $\mathcal{SH}(k)_{\mathcal{T}_k^{(+)},f}$. Note also that all categories mentioned here are closed under $\wedge$ because they are generated as thick or triangulated subcategories by classes of objects closed under $\wedge$: $\mathcal{SH}(k)^{fin(+)}$ and $\mathcal{SH}(k)_{\mathcal{T}_k^{(+)},f}$ are generated by $\{S^{p,q}(\wedge \operatorname{Spec}(k[i])_+^{\wedge m})\}$ and $\mathcal{SH}(k)_f$ is generated by smooth schemes, which are also closed under smash product. $R_k^{-1}(\mathcal{SH}^{fin})$ and $(R'_k)^{-1}(\mathcal{SH}(\mathbb{Z}/2)_f)$ are closed under $\wedge$ because $\wedge$ commutes with R_k, R'_k.

Remark 63 The construction of the functors R_k and c_k does not depend on the fact that $\mathbb{P}^1$ is invertible in $\mathcal{SH}(k)$. They can also be constructed for the category $\mathcal{SH}_{S_s^1}(k)$ in which only $S_s^1 = S^{1,0}$ got inverted and $\mathbb{G}_m$ did not. Therefore, part (1) of the theorem also holds for $(\mathcal{SH}_{S_s^1}(k)_f)_{(p)}$. The construction of c'_k, however, needed the invertibility of $\mathbb{P}^1$ (see Remark 49). Thus, part (2) cannot be applied to $(\mathcal{SH}_{S_s^1}(k)_f)_{(p)}$.

Definition 64 For a full subcategory $\mathcal{C}$ of a tensor triangulated category $\mathcal{T}$, let $\Delta(\mathcal{C})$ denote the smallest thick subcategory of $\mathcal{T}$ that contains $\mathcal{C}$.

Recall that $\operatorname{thickid}(\mathcal{C})$ denotes the smallest thick ideal that contains $\mathcal{C}$.

We state two more observations about thick ideals in $(\mathcal{SH}(k)_f)_{(p)}$. They also hold in any of the above categories $\mathcal{D}$.

Proposition 65 *Let $X_n \in \mathcal{SH}_{(p)}^{fin}$ be any spectrum of type n, i.e. $X_n \in \mathcal{C}_n \setminus \mathcal{C}_{n+1}$ (see Definition 9) and let $X_{m,n} \in \mathcal{SH}(\mathbb{Z}/2)_{(p)}^{fin}$ be any spectrum of type (m, n) (see Definition 35). Then*

$$\operatorname{thickid}(c_k X_n) = \operatorname{thickid}(c_k \mathcal{C}_n) \text{ in } (\mathcal{SH}(k)_f)_{(p)} \text{ if } k \subseteq \mathbb{C} \text{ and}$$

$$\operatorname{thickid}(c'_k X_{m,n}) = \operatorname{thickid}(c'_k \mathcal{C}_{m,n}) \text{ in } (\mathcal{SH}(k)_f)_{(p)} \text{ if } k \subseteq \mathbb{R}.$$

Proof It is clear that $\operatorname{thickid}(c_k X_n) \subseteq \operatorname{thickid}(c_k \mathcal{C}_n)$. Since c_k^{-1} preserves thick ideals by Proposition 51, $c_k^{-1}(\operatorname{thickid}(c_k X_n))$ is a thick ideal containing X_n. Since $\mathcal{C}_n$ is the smallest thick ideal containing X_n, we have $\mathcal{C}_n \subseteq c_k^{-1}(\operatorname{thickid}(c X_n))$ and hence $c_k \mathcal{C}_n \subseteq \operatorname{thickid}(c_k X_n)$, which implies the first claim. The same proof shows the second claim. □

Proposition 66 *If $X, Y \in (\mathcal{SH}(k)_f)_{(p)}$ with $\operatorname{type}(R_k X) \neq \operatorname{type}(R_k Y)$ (see Definition 9) or, if $k \subseteq \mathbb{R}$, $\operatorname{type}(R'_k X) \neq \operatorname{type}(R'_k Y)$ (see Definition 35), then*

$$\operatorname{thickid}(X) \neq \operatorname{thickid}(Y).$$

Proof Let $\operatorname{type}(R_k X) = n > \operatorname{type}(R_k Y)$. Then $\operatorname{thickid}(X) \subseteq R_k^{-1}(\mathcal{C}_n)$ but $Y \notin R_k^{-1}(\mathcal{C}_n)$. The case of R'_k is similar. □

The next proposition gives a description of thick ideals in the categories of finite cellular spectra, $\mathcal{SH}(k)^{fin}$, $k \subseteq \mathbb{C}$, and $\mathcal{SH}(k)^{fin+}$, $k \subseteq \mathbb{R}$, from Definition 52. In this case, a thick ideal is a thick subcategory that is closed under $- \wedge \mathbb{G}_m^{\pm 1}$ and under $- \wedge \operatorname{Spec}(k[i])_+$ if $k \in \mathbb{R}$.

Proposition 67 *Let $\mathcal{C} \subseteq \mathcal{SH}(k)^{fin}$ be a subcategory, $k \subseteq \mathbb{C}$. Then*

$$\operatorname{thickid}(\mathcal{C}) = \Delta\left(\bigcup_{n\in\mathbb{Z}} \mathcal{C} \wedge \mathbb{G}_m^{\wedge n}\right).$$

For a subcategory $\mathcal{C} \subseteq \mathcal{SH}(k)^{fin+}$, $k \subseteq \mathbb{R}$, we have

$$\operatorname{thickid}(\mathcal{C}) = \Delta\left(\bigcup_{n\in\mathbb{Z}} (\mathcal{C} \wedge \mathbb{G}_m^{\wedge n}) \cup \bigcup_{n\in\mathbb{Z}} (\mathcal{C} \wedge \mathbb{G}_m^{\wedge n} \wedge \operatorname{Spec}(k[i])_+)\right).$$

Proof We prove the second claim, the proof of the first claim is slightly shorter. Let $k \subseteq \mathbb{R}$. The subcategory thickid$(\mathcal{C})$ contains

$$\overline{\mathcal{C}} = \Delta\left(\bigcup_{n\in\mathbb{Z}}(\mathcal{C} \wedge \mathbb{G}_m^n) \cup \bigcup_{n\in\mathbb{Z}}(\mathcal{C} \wedge \mathbb{G}_m^n \wedge \mathrm{Spec}(k[i])_+)\right)$$

because thickid$(\mathcal{C})$ is closed under $- \wedge \mathbb{G}_m^n$ and $- \wedge \mathrm{Spec}(k[i])_+$ and is a thick subcategory. We have to show that the triangulated subcategory $\overline{\mathcal{C}}$ is already a thick ideal. Let $\mathcal{D} \subseteq \mathcal{SH}(k)^{fin+}$ be the full subcategory consisting of all objects X such that $\overline{\mathcal{C}}$ is closed under $- \wedge X$. We have to show $\mathcal{D} = \mathcal{SH}(k)^{fin+}$. First note that $\mathcal{D}$ contains all spheres $S^{p,q}$ because, as a triangulated subcategory, $\overline{\mathcal{C}}$ is closed under $- \wedge S^{p,0}$ and because we added $S^{q,q}$. If $X \to Y \to Z$ is a triangle with two objects in $\mathcal{D}$, then $A \wedge X \to A \wedge Y \to A \wedge Z$ is a triangle with two objects in $\overline{\mathcal{C}}$ for any $A \in \overline{\mathcal{C}}$, hence the third object is also in $\overline{\mathcal{C}}$. It follows that $\mathcal{D}$ is closed under exact triangles. Furthermore, $\mathcal{D}$ is closed under $\wedge$ by definition. It remains to show that $\mathcal{D}$ contains $\Sigma_T^\infty \mathrm{Spec}(k[i])_+$. By the equivalence

$$\begin{aligned}\mathrm{Spec}(k[i])_+ \wedge \mathrm{Spec}(k[i])_+ &\cong \mathrm{Spec}(k[i])_+ \vee \mathrm{Spec}(k[i])_+ \\ &\cong \mathrm{cof}\left(\mathrm{Spec}(k[i])_+ \overset{0}{\to} \mathrm{Spec}(k[i])_+\right),\end{aligned}$$

any triangulated subcategory of $\mathcal{SH}(k)$ containing $\Sigma_T^\infty \mathrm{Spec}(k[i])_+$ also contains $\Sigma_T^\infty \mathrm{Spec}(k[i])_+^{\wedge l}$, $l \geq 1$. It follows that $\overline{\mathcal{C}}$ contains $\mathcal{C} \wedge \mathbb{G}_m^n \wedge \mathrm{Spec}(k[i])_+^{\wedge l}$, $l \geq 0$. Hence, $\mathcal{D}$ contains $\Sigma_T^\infty \mathrm{Spec}(k[i])_+$ and is thus equal to $\mathcal{SH}(k)^{fin+}$. □

For $k \subseteq \mathbb{C}$, we have so far identified the following thick ideals in the category $(\mathcal{SH}(k)_f)_{(p)}$:

$$\begin{array}{ccccccc} R_k^{-1}(\mathcal{C}_0) & \supset & R_k^{-1}(\mathcal{C}_1) & \supset & R_k^{-1}(\mathcal{C}_2) & \supset & \cdots \\ \cup\!| & & \cup\!| & & \cup\!| & & \\ \mathrm{thickid}(c_k\mathcal{C}_0) & \supset & \mathrm{thickid}(c_k\mathcal{C}_1) & \supset & \mathrm{thickid}(c_k\mathcal{C}_2) & \supset & \cdots \end{array}$$

For $k \subseteq \mathbb{R}$, the picture has another dimension and depends on the classification of thick ideals in the equivariant category as described in Sect. 3. There is at least one spot where the inclusion from the lower row into the upper row is actually an equality.

Proposition 68

$$R_k^{-1}(\mathcal{C}_0) = \mathrm{thickid}(c_k(\mathcal{C}_0)) = (\mathcal{SH}(k)_f)_{(p)} \text{ and}$$

$$(R'_k)^{-1}(\mathcal{C}_{0,0}) = \mathrm{thickid}(c'_k(\mathcal{C}_{0,0})) = (\mathcal{SH}(k)_f)_{(p)}.$$

Proof This is because all these subcategories contain the sphere spectrum and are closed under smashing with arbitrary elements of $(\mathcal{SH}(k)_f)_{(p)}$. □

6 Thick Ideals Associated with Cohomology Theories

6.1 Equivalence of Homology and Cohomology Theories

We will now concentrate on the categories of finite cellular spectra $\mathcal{SH}(k)^{fin}$, $\mathcal{SH}(k)^{fin+}$, and on (co)homology theories represented by objects in $\mathcal{SH}(k)^{cell}$ or $\mathcal{SH}(k)^{cell+}$ respectively, because these satisfy some useful additional properties. A couple of these are proven in [16].

A consequence of the results in [16] is the following proposition, which states that cellular homology theories and cohomology theories for $\mathcal{SH}(k)^{fin}$ are exchangeable. We will state the analogous result for $\mathcal{SH}(k)^{fin+}$ below, in Corollary 72.

Proposition 69 *Let $E \in \mathcal{SH}(k)^{cell}$ (see Definition 52) be a ring spectrum and $X \in \mathcal{SH}(k)^{fin}$. Then $E_{**}(X) = 0$ if and only if $E^{**}(X) = 0$.*

Proof In the universal coefficient spectral sequence (see [16, Proposition 7.7] or Proposition 104),

$$E_2 = \mathrm{Ext}^{a,b,c}_{E_{**}}(M_{**}, N_{**}) \Rightarrow \pi_{-a-b,-c} F_E(M, N),$$

we set $M = X \wedge E$ and $N = E$:

$$E_2 = \mathrm{Ext}^{a,b,c}_{E_{**}}(E_{**}(X), E_{**}) \Rightarrow \pi_{-a-b,-c} F_E(X \wedge E, E).$$

Since $F_E(X \wedge E, E) = F(X, E)$, the spectral sequence converges conditionally to $E^{**}(X)$. If $E_{**}(X) = 0$, the spectral sequence collapses and thus converges strongly to $E^{**}(X)$, which, hence, is 0. For the other direction, we set $M = F(X, E)$ and $N = E$. Note that this M is cellular because X is dualisable and its dual is again a finite cell spectrum by Proposition 58. So we get

$$E_2 = \mathrm{Ext}^{a,b,c}_{E_{**}}(E^{**}(X), E_{**}) \Rightarrow \pi_{-a-b,-c} F_E(F(X, E), E).$$

Now $F_E(F(X, E), E) = F_E(D(X) \wedge E, E) = F(D(X), E) = X \wedge E$, so the sequence converges conditionally to $E_{**}(X)$. Hence, $E^{**}(X) = 0$ implies $E_{**}(X) = 0$. □

One important result of Dugger and Isaksen, [16, Proposition 7.1], states that for cellular objects $E \in \mathcal{SH}(k)^{cell}$, $\pi_{**}E = 0$ implies $E \cong 0$. We will show that an adjusted statement holds for $E \in \mathcal{SH}(k)^{cell+}$, $k \subseteq \mathbb{R}$.

We know from equivariant stable homotopy theory that a generalisation of homotopy groups is needed to obtain the corresponding result in $\mathcal{SH}(G)$. In $\mathcal{SH}(\mathbb{Z}/2)$, for example, we have the equivariant homotopy groups

$$\pi_n^{\mathbb{Z}/2}(X) = [S^n, X]_{\mathbb{Z}/2} = [S^n, X^{\mathbb{Z}/2}]_{\{1\}}$$

and

$$\pi_n^{\{1\}}(X) = [S^n \wedge \mathbb{Z}/2_+, X]_{\mathbb{Z}/2} = [S^n, X]_{\{1\}}.$$

We can similarly define homotopy groups in $\mathcal{SH}(k)$, $k \subseteq \mathbb{R}$ such that R_k' maps $\pi_{p,q}^k(X)$ to $\pi_p^{\mathbb{Z}/2}(R_k'X)$ and $\pi_{p,q}^{k[i]}(X)$ to $\pi_p^{\{1\}}(R_k'X)$.

Definition 70 For $k \subseteq \mathbb{R}$ and $X \in \mathcal{SH}(k)$, let

$$\pi_{p,q}^k(X) = [S^{p,q}, X]_{\mathcal{SH}(k)} \text{ and } \pi_{p,q}^{k[i]}(X) = [S^{p,q} \wedge \operatorname{Spec}(k[i])_+, X]_{\mathcal{SH}(k)}.$$

We write $\pi_{**}^+(X) = 0$ if $\pi_{**}^K(X) = 0$ for both $K = k$ and $K = k[i]$. Furthermore, $E_{**}^+(X) = 0$ will mean $\pi_{**}^+(E \wedge X) = 0$ and $(E^+)^{**}(X) = 0$ will mean $\pi_{**}^+ (F(X, E)) = 0$.

The same arguments as in [16, Proposition 7.1] now imply the following.

Proposition 71 *If $X \in \mathcal{SH}(k)^{cell+}$, $k \subseteq \mathbb{R}$ and $\pi_{**}^+(X) = 0$ then $X \cong 0$.*

Proof Assuming X is cofibrant and fibrant, one considers the class $\mathcal{D}$ of all Y such that $\operatorname{Map}(Y^{cof}, X)$ is contractible. By assumption, $\mathcal{D}$ contains $S^{p,q}$ and $S^{p,q} \wedge \operatorname{Spec}(k[i])_+$. Furthermore, $\mathcal{D}$ is closed under isomorphisms, (de-)suspensions and homotopy colimits. As in the proof of Proposition 67, it follows that it also contains $S^{p,q} \wedge (\operatorname{Spec}(k[i])_+)^{\wedge m}$, $m \geq 0$. As $\mathcal{SH}(k)^{cell+} \subseteq \mathcal{SH}(k)$ is generated by $\{S^{p,q} \wedge (\operatorname{Spec}(k[i])_+)^{\wedge m} \mid p, q \in \mathbb{Z}, m \geq 0\}$ under isomorphisms and homotopy colimits, it follows that $\mathcal{SH}(k)^{cell+} \subseteq \mathcal{D}$, in particular $X \in \mathcal{D}$. Hence, $\operatorname{Map}(X, X)$ is contractible, which implies $X \cong 0$. □

The following corollary is a $k \subseteq \mathbb{R}$-version of Proposition 69.

Corollary 72 *Let $E \in \mathcal{SH}(k)^{cell+}$ be a ring spectrum and $X \in \mathcal{SH}(k)^{fin+}$. Then $E_{**}^+(X) = 0$ if and only if $(E^+)^{**}(X) = 0$.*

Proof The spectral sequence from [16, Proposition 7.7] is derived similarly to the universal coefficient spectral sequence in [18, Sect. IV.5]. The crucial point is the convergence, where cellularity is needed to apply [16, Proposition 7.1]. The equivariant version of this spectral sequence (a spectral sequence of Mackey functors) is proven in [50]. Using [50] and Proposition 71, one can derive the universal coefficient spectral sequence for our motivic homotopy groups over $k \subseteq \mathbb{R}$. The new version of Proposition 69 can then be deduced from this spectral sequence as before. □

6.2 Thick Ideals

In $\mathcal{SH}^{fin}_{(p)}$, any thick ideal can be described by some cohomology theory (namely, n-th Morava K-theory). Conversely, any cohomology theory defines a thick ideal. We are therefore interested in thick ideals of $\mathcal{SH}(k)_f$, as well as $\mathcal{SH}(k)^{fin}$, $\mathcal{SH}(k)^{fin+}$, described by cohomology theories as follows.

Lemma 2 *Let k be any field and let $\mathcal{T} \subseteq \mathcal{SH}(k)$ be a tensor triangulated subcategory of $\mathcal{SH}(k)$. For any $E \in \mathcal{SH}(k)$, the full subcategory of $\mathcal{T}$ given by*

$$\mathcal{C}_E = \{X \in \mathcal{T} \mid X \wedge E \cong 0\}$$

is a thick ideal of $\mathcal{T}$.

Proof Since $- \wedge E$ preserves exact triangles, $\mathcal{C}_E$ is closed under these. If Y is a retract of X and $X \wedge E = 0$, then $Y \wedge E \to 0 \to Y \wedge E$ is the identity on $Y \wedge E$, hence $Y \wedge E \cong 0$. Let $X \in \mathcal{C}_E$ and $Y \in \mathcal{T}$. Then $X \wedge Y \wedge E \cong 0$, hence $X \wedge Y \in \mathcal{C}_E$. □

Proposition 73 *Let $\mathcal{C}_E$ be as defined in the above lemma.*
If $k \subseteq \mathbb{C}$, $E \in \mathcal{SH}(k)^{cell}$ and $\mathcal{T} = \mathcal{SH}(k)^{fin}$ or $\mathcal{T} = \mathcal{SH}(k)^{cell}$, then

$$\mathcal{C}_E = \{X \in \mathcal{T} \mid E_{**}(X) = 0\}.$$

If, furthermore, E is a ring spectrum and $\mathcal{T} = \mathcal{SH}(k)^{fin}$, then also

$$\mathcal{C}_E = \{X \in \mathcal{T} \mid E^{**}(X) = 0\}.$$

If $k \subseteq \mathbb{R}$, $E \in \mathcal{SH}(k)^{cell+}$ and $\mathcal{T} = \mathcal{SH}(k)^{fin+}$ or $\mathcal{T} = \mathcal{SH}(k)^{cell+}$, then

$$\mathcal{C}_E = \{X \in \mathcal{T} \mid (E^+)_{**}(X) = 0\}.$$

If, furthermore, E is a ring spectrum and $\mathcal{T} = \mathcal{SH}(k)^{fin+}$, then also

$$\mathcal{C}_E = \{X \in \mathcal{T} \mid (E^+)^{**}(X) = 0\}.$$

The same descriptions apply if $\mathcal{T}$ is the p-localisation of any of these categories for some prime p.

Proof For E and X as in the first claim, we have $E_{**}(X) = 0 \Leftrightarrow E \wedge X \cong 0$ by [16, Proposition 7.1] and for X finite, E a ring spectrum, $E^{**}(X) = 0 \Leftrightarrow E_{**}(X) = 0$ by Proposition 69. For $k \subseteq \mathbb{R}$, the same arguments hold by Proposition 71 and Corollary 72. □

6.3 Construction and Properties of $AK(n)$

Topology suggests that thick ideals described by Morava K-theories are particularly interesting. Therefore, we want to study motivic Morava K-theories and their properties.

Motivic Morava K-theories were introduced in [10]. The rough idea is as follows. One starts with MGL, the motivic analogue of MU. Since MU_* is a subring of MGL_{**}, elements in MU_* can be used to define maps on MGL and to construct motivic spectra which are analogous to certain other topological spectra constructed from MU.

Definition 74 Let $k \subseteq \mathbb{C}$. Let MGL be the algebraic cobordism spectrum as constructed in [88, Sect. 3.5], see also [89, Sect. 6.3] or [68, Sect. 6.5]. In [10, Theorem 10], elements $a_i \in MGL_{2i,i}$, $i \geq 1$, are defined, whose images under R_k are $a_i^{\text{top}} \in MU_{2i}$. If E is an MGL-module, then E/a_i and $a_i^{-1}E$ can be defined as in [27, Definition 2.10].

Note that the functor $MGL \wedge -$ from $\mathcal{SH}(k)$ to the category of MGL-modules in $\mathcal{SH}(k)$ has $F(MGL, -)$ as right adjoint and has the forgetful functor as left adjoint. As in [18, Lemma II.1.3] and [10, p. 99], it follows that the category of MGL-modules is complete and cocomplete. Thus, E/a_i and $a_i^{-1}E$ are again MGL-modules (see Lemma 18 for the proof that the action of a_i on E is a map of MGL-modules).

The motivic Brown–Peterson spectrum for a fixed prime p was first constructed in [87, Sect. 5]. By [32, Remark 6.20], it is equivalent as an MGL-module to $MGL_{(p)}/I$, where I is the image under $MU_* \to MGL_{**}$ of a regular sequence in the Lazard ring L that generates the vanishing ideal for p-typical formal group laws. We take this as the definition of the motivic Brown Peterson spectrum ABP at the prime p. That is,

$$ABP = MGL_{(p)}/(a_i \mid i \neq p^j - 1).$$

With $v_0 = p$ and $v_i = a_{p^i-1}$ for $i \geq 1$, we define:

$$AP(n) = ABP/(v_0, \cdots, v_{n-1}) \text{ and } AB(n) = v_n^{-1}AP(n),$$

$$Ak(n) = ABP/(v_0, \cdots, v_{n-1}, v_{n+1}, v_{n+2}, \cdots) \text{ and } AK(n) = v_n^{-1}Ak(n),$$

$$AE(n) = v_n^{-1}ABP/(v_{n+1}, v_{n+2}, \cdots).$$

We will make use of all these spectra in Sect. 9. For now, we are mostly interested in $AK(n)$.

Lemma 3 *Let Ah be one of the motivic spectra mentioned in the above definition, e.g. $Ah = AK(n)$ with $h = K(n)$. Then $R_k(Ah) = h$.*

Proof This follows from $R_k(MGL) = MU$ [88, Sect. 3.5] and $R_k(v_i) = v_i^{\text{top}}$ because R_k preserves colimits. □

Remark 75 (1) All spectra mentioned in the above definition are cellular: Theorem 6.4 in [16] shows that MGL is cellular. Since the spectra above are constructed from MGL by taking homotopy cofibers and homotopy colimits, they are cellular. The mod-p version of the main result in [32] states that $H\mathbb{Z}/p$ is the cofiber of $v_n : \Sigma^{2(p^n-1),p^n-1}Ak(n) \to Ak(n)$. Hence, $H\mathbb{Z}/p$ is cellular, too.

(2) ABP is a homotopy commutative ring spectrum by construction [87, Definition 5.3] and the orientation of MGL induces an orientation on ABP by [69, Theorem 1.1]. The ring structure of ABP induces an ABP-module structure on the quotients of ABP defined above.

(3) Since MGL is a ring spectrum, ABP is also an $MGL_{(p)}$-module spectrum and so is ABP/I for $I \subseteq MU_*$ a regular ideal.

(4) As remarked at the end of the introduction in [65], ABP is Landweber exact in the sense of [65] and their results for MGL also hold for ABP.

(5) $MGL_{2*,*}$ and $\mathrm{H}_{**}\, MGL$ (with coefficients in $\mathbb{Z}$ or $\mathbb{Z}/p$) are known (see e.g. [10, Theorem 5] and [32, Corollary 6.9]). From this, one can compute $Ah_{2*,*}$ and $\mathrm{H}_{**}\, Ah$ if Ah is one of the above spectra ([10, Theorem 12] and [32, Lemma 6.10]). More can be said about Ah_{**} if $\mathrm{H}_{**}(\mathrm{Spec}\, k)$ is known, see Lemma 5.

(6) In general, however, Ah_{**} is not known in degrees different from $(2i, i)$. It is also not known whether Ah can be given the structure of a ring spectrum (except for $Ah = ABP$). Another open question is whether $AK(n)$ satisfies the Künneth formula like $K(n)$.

We will make use of motivic Atiyah–Hirzebruch spectral sequences as discovered by Hopkins and Morel and worked out by Hoyois [32, Example 8.13]. See also [48, Sect. 11].

Proposition 76 *Let $h = MU_{(p)}/I$ and $Ah = MGL_{(p)}/I$ for some regular ideal $I \subseteq MU_*$. For $X \in \mathcal{SH}(k)^{fin}_{(p)}$, there are strongly convergent spectral sequences:*

$$E_2^{p,q,t} = \mathrm{H}^{p+2t,q+t}(X, h_t) \Rightarrow Ah^{p,q}(X),$$

$$E_2^{p,q,t} = \mathrm{H}^{p+2t,q+t}(X, (v_n^{-1}h)_t) \Rightarrow (v_n^{-1}Ah)^{p,q}(X).$$

Remark 77 (1) These are the spectral sequences associated with the slice filtrations of Ah and $v_n^{-1}Ah$. The n-th truncation of MGL in the slice filtration is described in the proof of [82, Theorem 4.6] as the colimit of a certain diagram $D_{\deg\geq n}$, meaning that $f_n MGL$ is constructed from MGL by quotienting out all monomials $a_1^{k_1}\cdots a_m^{k_m}$ with $a_i \in MGL_{2*,*}$ and $\sum_{i=1}^m ik_i \geq n$. The same construction with MU instead of MGL yields the Postnikov truncation of MU because $MU_* = \mathbb{Z}[a_1^{\mathrm{top}}, a_2^{\mathrm{top}}, \cdots]$ with $|a_i^{\mathrm{top}}| = 2i$. It follows that, for $k \subseteq \mathbb{C}$, $R_k : \mathcal{SH}(k) \to \mathcal{SH}$ maps the slice filtration of MGL to the Postnikov filtration of MU. By the construction of Ah and $v_n^{-1}Ah$ from MGL, this implies that their slice filtrations also realise to the Postnikov filtrations of h and $v_n^{-1}h$. For the associated spectral sequences, this means that R_k maps the above spectral sequences to the analogous topological Atiyah–Hirzebruch spectral sequences.

(2) In [32, Example 8.13], the convergence is stated for $X \in \mathbf{Sm}/S$ and follows from Ah, $v_n^{-1}Ah$ being convergent with respect to $[\Sigma^{0,q}\Sigma^{\infty}X_+, -]$, $X \in \mathbf{Sm}/S$, in the sense of [32, Sect. 8.5]. But this implies convergence with respect to $[X, -]$ for all finite cell spectra X by motivic cellular induction. Hence, the sequences converge for all finite cell spectra X. Note that [32, Example 8.13] shows the convergence for Landweber exact spectra, but his proof holds for quotients of MGL as well.

In Sects. 8 and 9, we will often assume $k = \mathbb{C}$ to be able to prove more results than for general k. The reason is that the coefficients of $H\mathbb{Z}/p$ are particularly simple in this case.

Lemma 4 *For* $k = \mathbb{C}$,

$$\mathrm{H}^{**}(\mathrm{Spec}(\mathbb{C}), \mathbb{F}_p) \cong \mathbb{F}_p[\tau]$$

with $\deg(\tau) = (0, 1)$.

Proof This is [92, Eq. (74)]. □

This, and the above spectral sequence, can be used to calculate the coefficients of theories like $Ak(n)$.

Lemma 5 *Let* $k = \mathbb{C}$ *and* $h = MU_{(p)}/I$ *with* $(p) \subseteq I \subseteq MU_*$ *a regular ideal. Let* $Ah = MGL_{(p)}/I$. *Then*

$$Ah_{**} \cong \mathrm{H}_{**}(\mathrm{Spec}\,\mathbb{C}, \mathbb{F}_p) \otimes_{\mathbb{F}_p} h_* \cong h_*[\tau].$$

Proof This is remarked in [96], below Corollary 3.9. The reason is the following: For $X = \mathrm{Spec}\,\mathbb{C}$, the motivic Atiyah–Hirzebruch spectral sequence is

$$\mathrm{H}^{p+2t,q+t}(\mathrm{Spec}\,\mathbb{C}, \mathbb{F}_p) \otimes h_t \Rightarrow Ah^{p,q}(\mathrm{Spec}\,\mathbb{C}).$$

By Lemma 4, $\mathrm{H}^{**}(\mathrm{Spec}\,\mathbb{C}, \mathbb{F}_p) \cong \mathbb{F}_p[\tau]$ with $\deg(\tau) = (0, 1)$. Thus, for fixed t, $E_2^{p,q,t} = \mathrm{H}^{p+2t,q+t}(\mathrm{Spec}\,\mathbb{C}, h_t)$ can only be nonzero in the column $p = -2t$, which implies that all differentials vanish. Therefore, the spectral sequence collapses immediately, proving $Ah_{**} \cong \mathrm{H}_{**} \otimes h_*$. □

6.4 Thick Ideals and Morava K-Theories

Let K be a cellular spectrum in $\mathcal{SH}(k)$, $k \subseteq \mathbb{C}$, such that $R_k(K) = K(n)$ is the n-th Morava K-theory with respect to a fixed prime p. We call such a K a motivic model for $K(n)$.

Let $\mathcal{C}_K = \{X \in (\mathcal{SH}(k)_f)_{(p)} \mid K \wedge X \cong 0\}$, which is a thick ideal in the p-localised category $(\mathcal{SH}(k)_f)_{(p)}$ by Lemma 2. If $K \wedge X \cong 0$, then $R_k(K \wedge X) \cong$

$K(n) \wedge R_k(X) \cong 0$. Consequently, $R_k(X) \in \mathcal{C}_{n+1}$ for any $X \in \mathcal{C}_K$. This is content of the following proposition.

Proposition 78 *For K a motivic model for $K(n)$, we have an inclusion of thick ideals in $(\mathcal{SH}(k)_f)_{(p)}$:*

$$\mathcal{C}_K \subseteq R_k^{-1}(\mathcal{C}_{n+1}).$$

The same is true in the category $\mathcal{SH}(k)^{fin}_{(p)}$.

Example 79 The motivic Morava K-theory spectra $AK(n)$, as defined in Definition 74, satisfy $R_k(AK(n)) = K(n)$ and are cellular. In particular, they satisfy the previous proposition.

Another possibility for such a spectrum K is the constant Morava K-theory spectrum $c_k(K(n))$. In Sects. 8.4–8.6 we will have a closer look at the thick ideals $\mathcal{C}_{AK(n)}$ and $\mathcal{C}_{c_k K(n)}$ for $k = \mathbb{C}$.

Remark 80 The spectrum $K(n)$ is not finite. One possibility to see this is by the equivalence [73, Theorem 2.1.(h) and (i)]:

$$K(m) \wedge K(n) \cong 0 \Leftrightarrow m \neq n$$

If $K(n)$ were finite, we would have $K(m)_* K(n) = 0 \Rightarrow K(m-1)_* K(n) = 0$ by [73, Theorem 2.11], which is wrong for $n = m - 1$. As a consequence, any spectrum K with $R_k(K) = K(n)$ can also not be finite cellular.

7 $\mathcal{SH}(k)_f$ Has More Thick Ideals than $\mathcal{SH}^{fin}$

7.1 *The Motivic Hopf Map*

In this section, we study the cofiber of the motivic Hopf map, which generates a thick ideal that is not of the form $R_k^{-1}(\mathcal{C}_n)$ or $\mathcal{C}_{AK(n)}$. The first part of this claim was proven by Balmer in [7, Proposition 10.4]. We will reprove and extend this result, as well as summarise and apply other results of Balmer's work on prime ideals in tensor triangulated categories.

For $k \subseteq \mathbb{R}$, we have already shown that $\mathcal{SH}(k)_f$ has more thick ideals than $\mathcal{SH}^{fin}$. This section is, therefore, in particular interesting for $k = \mathbb{C}$.

To understand why there cannot be a complete analogy between motivic thick ideals and topological thick ideals, recall the reasoning in the topological case: the thick subcategory (i.e., thick ideal) theorem of Hopkins and Smith [25] is derived from the fact that Morava K-theories detect nilpotence, which follows from the theorem that MU detects nilpotence. This is not the case in the motivic setting.

The Hopf map in $\mathcal{SH}(k)$ is given by $\eta : \mathbb{A}^2 \setminus \{0\} \to \mathbb{P}^1, (x, y) \mapsto [x : y]$, defining an element $\eta \in \pi_{1,1}(S)$. Unlike the topological Hopf element, η is not nilpotent [62,

Theorem 4.7]. The following lemma is due to Morel, see e.g. [60], but we could not find a proof in the literature. It shows that the spectrum MGL does not detect the non-nilpotence of η.

Lemma 6

$$MGL \wedge \eta \cong 0$$

Proof The unit map u of the ring spectrum MGL factors through $\Sigma^{-2,-1}C\eta$, as in the proof of [32, Theorem 3.8]. Consider the diagram:

$$\begin{array}{ccccc}
MGL \wedge S^{1,1} & \xrightarrow{1\wedge\Sigma^{-2,-1}\eta} & MGL \wedge S^{0,0} & \longrightarrow & MGL \wedge \Sigma^{-2,-1}C\eta \\
\downarrow \cong & & \downarrow \cong & \searrow^{1\wedge u} & \downarrow \\
MGL \wedge S^{1,1} & \longrightarrow & MGL & \xleftarrow[m]{} & MGL \wedge MGL
\end{array}$$

The upper row is a cofiber sequence and u, m are the structure maps of the ring spectrum MGL. The diagram is commutative. It follows that $1 \wedge \Sigma^{-2,-1}\eta$ factors through its own cofiber, hence it must be zero. Suspending by $S^{2,1}$, we get $MGL \wedge \eta \cong 0$. □

It seems likely that also $AK(n) \wedge \eta \cong 0$, but this does not immediately follow from the above lemma. For our interests, it will suffice to know that $AK(n) \wedge C\eta \ncong 0$, which we prove differently.

Remark 81 By [61, Lemma 6.2.1], $C\eta \cong \mathbb{P}^2$.

Proposition 82 *For any $k \subseteq \mathbb{C}$ and any prime p,*

$$\text{thickid}(R_k(C\eta_{(p)})) = \mathcal{C}_0 = \mathcal{SH}^{fin}_{(p)}.$$

For $k \subseteq \mathbb{R}$,

$$\text{thickid}(R'_k(C\eta_{(p)})) = \begin{cases} \mathcal{C}_{0,1} \subset (\mathcal{SH}(\mathbb{Z}/2)_f)_{(2)} & \text{if } p = 2 \\ \mathcal{C}_{0,\infty} \subset (\mathcal{SH}(\mathbb{Z}/2)_f)_{(p)} & \text{if } p \neq 2, \end{cases}$$

where $\mathcal{C}_n$ is the thick ideal defined in Theorem 1 and $\mathcal{C}_{m,n}$ is the thick ideal defined in Corollary 34.

Proof Since, by definition, $R_k = R_{\mathbb{C}} \circ f^*$ for $f : k \hookrightarrow \mathbb{C}$, $R'_k = R_{\mathbb{R}} \circ f^*$ for $f : k \hookrightarrow \mathbb{R}$, and $f^*(\eta_k) = \eta_K$ for $f : k \hookrightarrow K$, it suffices to prove the claims for $k = \mathbb{C}$ and $k = \mathbb{R}$.

For $k = \mathbb{C}$, we have $R(C\eta) = R(\mathbb{P}^2) = \mathbb{C}P^2$. Since $K(0)_*(\mathbb{C}P^2_{(p)}) = \text{H}_*(\mathbb{C}P^2_{(p)}, \mathbb{Q}) \neq 0$, $R(C\eta_{(p)})$ has type 0. By [25, Theorem 7], any spectrum of type 0 generates $\mathcal{SH}^{fin}_{(p)}$. So, thickid$(R(C\eta_{(p)})) = \mathcal{SH}^{fin}_{(p)}$.

For $k = \mathbb{R}$, $R'(C\eta)^{\{1\}} = C\eta(\mathbb{C}) = \mathbb{C}P^2$ as before. Therefore, if $R'(C\eta_{(p)}) \in \mathcal{C}_{m,n}$, then $m = 0$.

Furthermore, $R'(\eta)^{\mathbb{Z}/2} = \eta(\mathbb{R})$ is the quotient map $\mathbb{R}^2 \setminus \{0\} \to \mathbb{R}P^1$, which is isomorphic to $2 : S^1 \to S^1$. It follows that $R'(C\eta)^{\mathbb{Z}/2} = S/2$.

If $p \neq 2$, $S \xrightarrow{2} S$ is an isomorphism in $\mathcal{SH}^{fin}_{(p)}$ and, hence, $(S/2)_{(p)} = 0$, which has type ∞. This proves $\operatorname{thickid}(R'(C\eta_{(p)})) = \mathcal{C}_{0,\infty}$ if $p \neq 2$.

Now let $p = 2$. Since $S \xrightarrow{2} S$ is an isomorphism rationally, we have $K(0)_*(S/2) = H_*(S/2, \mathbb{Q}) = 0$. On the other hand, $K(1)$ is a direct summand of mod 2 topological K-theory, so $K(1)_*(S \xrightarrow{2} S) = 0$ and $K(1)_*(S/2) \neq 0$, as in the proof of [7, Proposition 9.4]. Hence, the type of $S/2 \in \mathcal{SH}^{fin}_{(2)}$ is 1 and, therefore, $\operatorname{thickid}(R'_k((C\eta)_{(2)})) = \mathcal{C}_{0,1}$. □

In terms of the motivic thick ideals $\operatorname{thickid}(C\eta_{(p)}) \subseteq (\mathcal{SH}(k)_f)_{(p)}$, the above proposition states the following: For $k \subseteq \mathbb{C}$, $\operatorname{thickid}(C\eta_{(p)}) \subseteq R_k^{-1}(\mathcal{C}_0)$ (which is a trivial statement) and $\operatorname{thickid}(C\eta_{(p)}) \nsubseteq R_k^{-1}(\mathcal{C}_n)$ for $n > 0$. For $k \subseteq \mathbb{R}$ and $p \neq 2$, $\operatorname{thickid}(C\eta_{(p)}) \subseteq (R'_k)^{-1}(\mathcal{C}_{0,\infty})$ and $\operatorname{thickid}(C\eta_{(p)}) \nsubseteq (R'_k)^{-1}(\mathcal{C}_{m,n})$ for any $m > 0$, n arbitrary. If $p = 2$, then $\operatorname{thickid}((C\eta)_{(2)}) \subseteq (R'_k)^{-1}(\mathcal{C}_{0,1})$ and $\operatorname{thickid}((C\eta)_{(2)}) \nsubseteq (R'_k)^{-1}(\mathcal{C}_{m,n})$ for any $m > 0$ or $n > 1$.

The next proposition contains two more results on $\operatorname{thickid}(C\eta_{(p)}) \subseteq (\mathcal{SH}(k)_f)_{(p)}$.

Proposition 83 *For $k \subseteq \mathbb{C}$, let $\operatorname{thickid}(C\eta_{(p)}) \subseteq (\mathcal{SH}(k)_f)_{(p)}$ denote the thick ideal generated by the p-localised cofiber of the Hopf map. Then the following hold:*

(1) $\operatorname{thickid}(C\eta_{(p)}) \nsubseteq \mathcal{C}_{AK(n)}$ *for any $n \geq 0$ and any prime p,*
(2) $\operatorname{thickid}(C\eta_{(p)}) \nsubseteq R_k^{-1}(\mathcal{C}_n)$ *for any $n > 0$ and any prime p,*
(3) $\operatorname{thickid}(C\eta_{(p)}) \subsetneq \operatorname{thickid}(S^0_{(p)}) = (\mathcal{SH}(k)_f)_{(p)}$ *if $k \subseteq \mathbb{R}$ and p is any prime or $k \subseteq \mathbb{C}$ and $p = 2$.*
(4) *For any prime p, the thick ideals* $\operatorname{thickid}(C\eta_{(p)}) \cap R_k^{-1}(\mathcal{C}_n)$ *are distinct for different $n \geq 0$ and in particular nonzero if $n < \infty$.*

Proof (1) By Proposition 78, $\mathcal{C}_{AK(n)} \subseteq R_k^{-1}(\mathcal{C}_{n+1})$.
Assuming $\operatorname{thickid}(C\eta_{(p)}) \subseteq \mathcal{C}_{AK(n)}$ therefore implies $\operatorname{thickid}(C\eta_{(p)}) \subseteq R_k^{-1}(\mathcal{C}_{n+1})$. Hence, (1) will follow from (2).

(2) This can either be derived from the previous proposition or can be seen by the following argument.
$R_k(\eta)$ is the topological Hopf map, which is nilpotent. Hence, for $n \geq 1$, $R_k(\eta)_*$ is not surjective in the sequence

$$\cdots \to K(n-1)_*(S^3) \xrightarrow{R_k(\eta)_*} K(n-1)_*(S^2) \to K(n-1)_*(C(R_k\eta)) \to \cdots .$$

It follows that $K(n-1)_*(C(R_k\eta)) \neq 0$ and, since $C(R_k\eta) \cong R_k(C\eta)$, $C\eta_{(p)} \notin R_k^{-1}(\mathcal{C}_n)$.

(3) Note that $C\eta \in \mathrm{thickid}(S^0) = \mathcal{SH}(k)_f$, so $\mathrm{thickid}(C\eta_{(p)}) \subseteq \mathrm{thickid}(S^0_{(p)})$ for any p. We have to show that $S^0_{(p)} \notin \mathrm{thickid}(C\eta_{(p)})$. We consider the sheaf cohomology theory $\mathrm{H}^*(-, \underline{K}^{MW}_*[\eta^{-1}])$, on which η induces an isomorphism as in the proof of [62, Theorem 4.7], where Morel concludes that η cannot be nilpotent. Localising at p, we get that $\eta_{(p)}$ induces an isomorphism on $\mathrm{H}^*(S^0_{(p)}, \underline{K}^{MW}_*[\eta^{-1}])$. Consider $\mathrm{H}^0(S^0_{(p)}, \underline{K}^{MW}_*[\eta^{-1}]) = K^{MW}_*(k)[\eta^{-1}]_{(p)}$. As in the proof of [62, Theorem 4.7], this is $K^W_0(k)[\eta, \eta^{-1}]_{(p)}$, and $K^W_0(k)$ is isomorphic to the Witt ring $W(k)$ by [62, Remark 4.2]. We have $W(\mathbb{C}) = \mathbb{Z}/2$ and $W(\mathbb{R}) = \mathbb{Z}$, see e.g. [43, p. 34].
Hence, for $k = \mathbb{C}$, $\mathrm{H}^0(S^0_{(p)}, \underline{K}^{MW}_*[\eta^{-1}]) = (\mathbb{Z}/2[\eta, \eta^{-1}])_{(p)}$, which is nonzero if $p = 2$. And, for $k = \mathbb{R}$, $\mathrm{H}^0(S^0_{(p)}, \underline{K}^{MW}_*[\eta^{-1}]) = (\mathbb{Z}[\eta, \eta^{-1}])_{(p)} \neq 0$ for any p. Therefore, the above mentioned isomorphism induced by $\eta_{(p)}$ is not the zero map, and, as in [62, Theorem 4.7], it follows that $\eta_{(p)}$ is not nilpotent in these cases.
Now we apply [7, Theorem 2.15], which states that, for a map $f : X \to Y$ between invertible objects X and Y in a tensor triangulated category $\mathcal{T}$, the thick ideal generated by Cf is equal to the full subcategory consisting of all objects A such that $f^{\wedge n} \wedge A = 0$ for some $n \geq 1$ (this is similar to Proposition 23). A corollary of this theorem is that $\mathrm{thickid}(Cf) = \mathcal{T}$ if and only if f is nilpotent.
It follows that, since $\eta_{(p)}$ is not nilpotent, $S^0_{(p)} \notin \mathrm{thickid}(C\eta_{(p)})$ and $\mathrm{thickid}(C\eta_{(p)}) \neq \mathrm{thickid}(S^0_{(p)})$ in the cases $k = \mathbb{C}$ and $p = 2$ or $k = \mathbb{R}$ and p any prime.
Now, let $f : k \hookrightarrow \mathbb{C}$. By Proposition 50, $(f^*)^{-1}$ preserves thick ideals. Since $f^*(C\eta_k) = C\eta_{\mathbb{C}}$, $\mathrm{thickid}((C\eta_k)_{(2)}) \subseteq (f^*)^{-1}(\mathrm{thickid}((C\eta_{\mathbb{C}})_{(2)}))$. As $f^*((S^0_k)_{(2)}) = (S^0_{\mathbb{C}})_{(2)} \notin \mathrm{thickid}((C\eta_{\mathbb{C}})_{(2)})$, it follows $(S^0_k)_{(2)} \notin (f^*)^{-1}(\mathrm{thickid}((C\eta_{\mathbb{C}})_{(2)}))$ and, hence, $(S^0_k)_{(2)} \notin \mathrm{thickid}((C\eta_k)_{(2)})$.

The same argument holds for arbitrary primes p if $f : k \hookrightarrow \mathbb{R}$. Alternatively, the statement for $k \subseteq \mathbb{R}$ can be derived from Proposition 82 and Theorem 13:

$$\begin{aligned}\mathrm{thickid}(C\eta_{(p)}) &\subseteq (R'_k)^{-1}(\mathrm{thickid}(R'_k(C\eta_{(p)})))\\ &\neq (R'_k)^{-1}(\mathcal{C}_{0,0}) = \mathrm{thickid}(S^0_{(p)}).\end{aligned}$$

(4) Let $X \in R_k^{-1}(\mathcal{C}_n)$ be such that $R_k(X)$ is of type n, i.e., $R_k(X) \in \mathcal{C}_n \setminus \mathcal{C}_{n+1}$. From the proof of (2), we know that $R_k(C\eta_{(p)})$ is of type 0. From the Künneth formulas for $K(n)$ and $K(n+1)$, it follows that $R_k(C\eta_{(p)} \wedge X) \cong R_k(C\eta_{(p)}) \wedge R_k(X)$ is of type n. Therefore,

$$C\eta_{(p)} \wedge X \in \left(\mathrm{thickid}(C\eta_{(p)}) \cap R_k^{-1}(\mathcal{C}_n)\right) \setminus \left(\mathrm{thickid}(C\eta_{(p)}) \cap R_k^{-1}(\mathcal{C}_{n+1})\right).$$

□

Corollary 84 *For* $k \subseteq \mathbb{C}$, $\operatorname{thickid}((C\eta)_{(2)})$ *is neither of the form* $\mathcal{C}_{AK(n)}$ *for any* $n \geq 0$ *(by Proposition 83(1)) nor of the form* $R_k^{-1}(\mathcal{C}_n)$ *for any* $n \geq 0$ *(by Proposition 83(2) and (3)).*

For $k \subseteq \mathbb{R}$, the statements analogous to (2) and (4) of Proposition 83 read as follows.

Proposition 85 *Let* η *denote the Hopf map in* $\mathcal{SH}(k)$, $k \subseteq \mathbb{R}$. *In the category* $(\mathcal{SH}(k)_f)_{(p)}$ *the following inequalities hold.*

(1) $\operatorname{thickid}(C\eta_{(p)}) \not\subseteq (R'_k)^{-1}(\mathcal{C}_{m,n})$ *for any* $m > 0$ *(*p, n *arbitrary) or, if* $p = 2$, $m = 0$ *and* $n > 1$.
(2) Let (m, n) *and* (m', n') *be pairs of integers* ≥ 0 *such that* $\mathbb{Z}/2$*-equivariant spectra of types* (m, n) *and* (m', n') *exist (see Sect. 3). If* p *is any prime and* $m \neq m'$ *or* $p = 2$ *and* $\max(n, 1) \neq \max(n', 1)$ *then*

$$\operatorname{thickid}(C\eta_{(p)}) \cap (R'_k)^{-1}(\mathcal{C}_{m,n}) \neq \operatorname{thickid}(C\eta_{(p)}) \cap (R'_k)^{-1}(\mathcal{C}_{m',n'}).$$

Otherwise, the two thick ideals are equal.

Proof (1) is a reformulation of the second part of Proposition 82. For (2), take $X = c'_k(X_{m,n})$ or any other spectrum whose realisation is a spectrum of type (m, n). Then $R'_k(C\eta_{(p)} \wedge X) \cong R'_k(C\eta_{(p)}) \wedge R'_k(X)$ is the smash product of a spectrum of type $(0, \infty)$ if p is odd ($(0, 1)$ if $p = 2$) with a spectrum of type (m, n). By the equivariant Künneth formula, Corollary 15, it follows that $R'_k(C\eta_{(p)} \wedge X)$ has type (m, ∞) if p is odd and $(m, \max(n, 1))$ if $p = 2$. Consequently, $\operatorname{thickid}(C\eta_{(p)}) \cap (R'_k)^{-1}(\mathcal{C}_{m,n})$ is equal to $\operatorname{thickid}(C\eta_{(p)}) \cap (R'_k)^{-1}(\mathcal{C}_{m,\infty})$ if p is odd and to $\operatorname{thickid}(C\eta_{(p)}) \cap (R'_k)^{-1}(\mathcal{C}_{m,\max(n,1)})$ if $p = 2$. The intersection is not contained in $(R'_k)^{-1}(\mathcal{C}_{m',n'})$ for any $m' > m$ or, if $p = 2$, for any $n' > \max(n, 1)$. □

Related to the previous propositions are the following conjectures.

Conjecture 86 *(1)* $R_k^{-1}(\mathcal{C}_n) \not\subseteq \operatorname{thickid}(C\eta_{(p)})$ *for all* $0 \leq n < \infty$.
(2) $\mathcal{C}_{AK(n)} \not\subseteq \operatorname{thickid}(C\eta_{(p)})$ *for all* $n \geq 0$.
(3) The thick ideals $\operatorname{thickid}(C\eta_{(p)}) \cap \mathcal{C}_{AK(n)}$ *are distinct for different* $n \geq 0$ *and in particular nonzero if* $n < \infty$.
(4) For $k \subseteq \mathbb{R}$, $\operatorname{thickid}(C\eta_{(p)}) \subsetneq (R'_k)^{-1}(\mathcal{C}_{0,\infty})$ *for odd primes* p *and, for* $p = 2$, $\operatorname{thickid}(C\eta_{(p)}) \subsetneq (R'_k)^{-1}(\mathcal{C}_{0,1})$.
(5) For $k \subseteq \mathbb{R}$, $(R'_k)^{-1}(\mathcal{C}_{m,n}) \not\subseteq \operatorname{thickid}(C\eta_{(p)})$ *for all* $0 \leq m, n < \infty$.

Remark 87 A possible approach for proving (1) or (2) might be to choose a motivic spectrum X whose realisation is of type n (e.g. $c_k(X_n)$) or a spectrum which is of motivic type n (e.g. as constructed in Sect. 8), respectively. If one can show that $\mathrm{H}^*(X, \underline{K}_*^{MW}[\eta^{-1}]) \neq 0$, this proves (1) respectively (2).

An idea to prove (3) is to choose a spectrum X of motivic type n and to show that $C\eta \wedge X$ is again of motivic type n for this particular choice.

For (4), p odd, one has to find a spectrum $X \in (R_k')^{-1}(\mathcal{C}_{0,\infty})$, which is not contained in thickid$(C\eta_{(p)})$. $X = (\Sigma^\infty \operatorname{Spec} \mathbb{C}_+)_{(p)}$ satisfies the first condition and it might be possible to show $\mathrm{H}^*(X, \underline{K}_*^{MW}[\eta^{-1}]) \neq 0$, which would imply the second. Since $\mathcal{C}_{0,\infty} \subseteq \mathcal{C}_{0,1}$, the same X could be used for the case $p = 2$.

Remark 88 If and only if the first conjecture is true, the ideals in Proposition 83(4) are different from the ideals $R_k^{-1}(\mathcal{C}_n)$ for all $n \geq 0$.

Remark 89 In [3], Andrews and Miller compute the ring $\pi_{**}(S)[\eta^{-1}]$ over $\mathbb{C}$. The computation implies the existence of a non nilpotent element μ_9. Using methods from this section, one can show a result similar to Proposition 83 for thickid$(C\mu_9)$.

7.2 *Prime Ideals*

Let us recall some results of [7], where Balmer studies prime ideals in tensor triangulated categories.

Definition 90 A prime ideal is a proper thick ideal $\mathcal{C}$ with the additional property that $X \wedge Y \in \mathcal{C}$ implies $X \in \mathcal{C}$ or $Y \in \mathcal{C}$. The set of prime ideals of a tensor triangulated category $\mathcal{T}$ is denoted by $\operatorname{Spc}(\mathcal{T})$.

In [7], the endomorphism ring of the unit object of $\mathcal{T}$ is denoted by $R_{\mathcal{T}}$. To avoid confusion with the realisation functors, we use the notation $\pi_0^{\mathcal{T}}$ instead of $R_{\mathcal{T}}$. In [7, Corollary 5.6], Balmer defines a functor

$$\rho_{\mathcal{T}} : \operatorname{Spc}(\mathcal{T}) \to \operatorname{Spec}(\pi_0^{\mathcal{T}}),$$

$$\rho_{\mathcal{T}}(\mathcal{P}) = \{f \in \pi_0^{\mathcal{T}} \mid C(f) \notin \mathcal{P}\},$$

which he proves to be surjective if $\mathcal{T}$ is connective [7, Theorem 7.13], where connectivity means that $\operatorname{Hom}_{\mathcal{T}}(S, \Sigma^i S) = 0$ for all $i > 0$ and S the unit object of $\mathcal{T}$. Furthermore, this functor is natural for tensor triangulated functors $f : \mathcal{T} \to \mathcal{T}'$ by [7, Theorem 5.3(c) and Corollary 5.6(b)].

7.3 *Prime Ideals in the Topological Categories $\mathcal{SH}^{fin}$ and $\mathcal{SH}(\mathbb{Z}/2)_f$*

Applied to $\mathcal{SH}_{(p)}^{fin}$, this yields the following [7, Proposition 9.4]:

All proper thick ideals $\mathcal{C}_n \subset \mathcal{SH}_{(p)}^{fin}$, $0 < n \leq \infty$ are prime ideals, as can be seen either from the Künneth formula for $K(n)$ or by the linear ordering of the $\mathcal{C}_n$. The functor

$$\rho_{\mathcal{SH}^{fin}_{(p)}} : \mathrm{Spc}(\mathcal{SH}^{fin}_{(p)}) \to \mathrm{Spec}(\mathbb{Z}_{(p)})$$

has the following values:

$$\rho_{\mathcal{SH}^{fin}_{(p)}}(\mathcal{C}_n) = \begin{cases} p\mathbb{Z}_{(p)} & \text{if } n > 1, \\ 0 & \text{if } n = 1. \end{cases}$$

Note that $\mathcal{C}_0$ is not proper, so it is not in $\mathrm{Spc}(\mathcal{SH}^{fin}_{(p)})$. Note also that [7] uses a different indexing convention, in which the indices are shifted by one.

The prime ideals in the non-localised category $\mathcal{SH}^{fin}$ are given as follows.

Proposition 91 *A subcategory $\mathcal{C} \subset \mathcal{SH}^{fin}$ is a prime ideal if and only if there exist a prime p and a number $1 \leq n \leq \infty$ such that*

$$\mathcal{C} = \mathcal{C}_{p,n} = \{X \in \mathcal{SH}^{fin} \mid K(p, n-1)_* X = 0\}$$

if $n < \infty$, or

$$\mathcal{C} = \mathcal{C}_{p,\infty} = \{X \in \mathcal{SH}^{fin} \mid X_{(p)} = 0\}$$

if $n = \infty$. Here, $\mathcal{C}_{p,1} = \mathcal{C}_{q,1} = \mathcal{C}_1$ for any primes p and q. Except for $n = 1$, the $\mathcal{C}_{p,n}$ are pairwise different.

The functor $\rho_{\mathcal{SH}^{fin}} : \mathrm{Spc}(\mathcal{SH}^{fin}) \to \mathrm{Spec}(\mathbb{Z})$ maps $\mathcal{C}_{p,n}$ to $p\mathbb{Z}$ for any $n > 1$ and it maps $\mathcal{C}_1$ to 0.

Proof This is [7, Corollary 9.5]. We give another proof of the first statement. By Theorem 4, any thick ideal of $\mathcal{SH}^{fin}$ is an intersection

$$\mathcal{C} = \bigcap_{p \in P} \{X \in \mathcal{SH}^{fin} \mid X_{(p)} \in \mathcal{C}_{n_p} \subseteq \mathcal{SH}^{fin}_{(p)}\}$$

for some set of primes P and some numbers $1 \leq n_p \leq \infty$. If $|P| = 0$, then $\mathcal{C} = \mathcal{SH}^{fin}$, which is not proper, and, thus, is no prime ideal. Assume $|P| \geq 1$. By Proposition 24, the intersection of two thick ideals $\mathcal{I}$ and $\mathcal{J}$ satisfying $\mathcal{I} \not\subseteq \mathcal{J}$ and $\mathcal{J} \not\subseteq \mathcal{I}$ is never a prime ideal, because it contains all $X \wedge Y$ with $X \in \mathcal{I}$ and $Y \in \mathcal{J}$. It follows that, if $\mathcal{C}$ is a prime ideal, then $|P| = 1$. This proves that any prime ideal of $\mathcal{SH}^{fin}$ is of the form $\mathcal{C} = \{X \in \mathcal{SH}^{fin} \mid X_{(p)} \in \mathcal{C}_{n_p}\}$ for some prime p and some $1 \leq n_p \leq \infty$.

On the other hand, the Künneth formula implies that $K(p, n)_*(X \wedge Y)$ can only be zero if $K(p, n)_* X = 0$ or $K(p, n)_* Y = 0$, proving that $\mathcal{C}_{p,n}$ as above is indeed a prime ideal. Note that, for $n = \infty$, we have $X \in \mathcal{C}_{p,n}$ if and only if $K(m)_*(X) = 0$ for all $m \geq 0$, so $\mathcal{C}_{p,\infty}$ is also a prime ideal by the Künneth formula (using that $K(m+1)_*(X) = 0$ implies $K(m)_*(X) = 0$). □

Now we turn to the $\mathbb{Z}/2$-equivariant category, $\mathcal{T} = \mathcal{SH}(\mathbb{Z}/2)_f$. By [81, Corollary 1], the map

$$\pi_0^{\mathcal{T}} \to A(\mathbb{Z}/2),\ [f] \mapsto (\deg(f^{\{1\}}), \deg(f^{\mathbb{Z}/2}))$$

is an isomorphism to the Burnside ring $A(\mathbb{Z}/2) \cong \mathbb{Z} \oplus \mathbb{Z}$. Hence, [7, Theorem 7.13] yields surjective functors

$$\rho_{\mathcal{SH}(\mathbb{Z}/2)_f} : \mathrm{Spc}(\mathcal{SH}(\mathbb{Z}/2)_f) \to \mathrm{Spec}(\mathbb{Z} \oplus \mathbb{Z})$$

and

$$\rho_{(\mathcal{SH}(\mathbb{Z}/2)_f)_{(p)}} : \mathrm{Spc}((\mathcal{SH}(\mathbb{Z}/2)_f)_{(p)}) \to \mathrm{Spec}(\mathbb{Z}_{(p)} \oplus \mathbb{Z}_{(p)}).$$

Proposition 92 *A thick ideal $\mathcal{C}_{m,n} \subseteq (\mathcal{SH}(\mathbb{Z}/2)_f)_{(p)}$ is a prime ideal if and only if $\mathcal{C}_{m,n} = \mathcal{C}_{m,0}$, $0 < m \leq \infty$, or $\mathcal{C}_{m,n} = \mathcal{C}_{0,n}$, $0 < n \leq \infty$.*

Proof From Sect. 3.5, we know that any thick ideal in $(\mathcal{SH}(\mathbb{Z}/2)_f)_{(p)}$ is of the form

$$\mathcal{C}_{m,n} = \{X \mid \phi^{\{1\}}(X) \in \mathcal{C}_m \text{ and } \phi^{\mathbb{Z}/2}(X) \in \mathcal{C}_n\} = \mathcal{C}_{m,0} \cap \mathcal{C}_{0,n}.$$

By Proposition 24, this is equal to $\{X \wedge Y \mid X \in \mathcal{C}_{m,0}, Y \in \mathcal{C}_{0,n}\}$. It follows that $\mathcal{C}_{m,n}$ being a prime ideal implies $\mathcal{C}_{m,n} = \mathcal{C}_{m,0}$ or $\mathcal{C}_{m,n} = \mathcal{C}_{0,n}$. On the other hand, if either m or n is 0 then $\mathcal{C}_{m,n}$ is a prime ideal by the Künneth formula for $K(m, \{1\})$ or $K(n, \mathbb{Z}/2)$. Note that $\mathcal{C}_{0,0} = (\mathcal{SH}(\mathbb{Z}/2)_f)_{(p)}$ is not a prime ideal since it is no proper subcategory. □

Remark 93 As a consequence of Strickland's results [86], $\mathcal{C}_{m,n} = \mathcal{C}_{0,n}$ implies $m = 0$ but $\mathcal{C}_{m,n} = \mathcal{C}_{m,0}$ does not always imply that $n = 0$. This follows from Corollary 36: for $m \leq n$, a type (m, n)-spectrum $X_{m,n}$ always exists, and $X_{0,n} \in \mathcal{C}_{0,n} \setminus \mathcal{C}_{m,n}$ for all $m > 0$. On the other hand, Proposition 33 implies that, for $p = 2$, $\mathcal{C}_{m,0} = \mathcal{C}_{m,n}$ for any $0 \leq n \leq m - 1$.

Recall from Definition 35 that the thick ideals in $(\mathcal{SH}(\mathbb{Z}/2)_f)_{(p)}$ are in bijection with a lattice $\Gamma_p \subseteq (\mathbb{Z}_{\geq 0} \cup \{\infty\}) \times (\mathbb{Z}_{\geq 0} \cup \{\infty\})$ containing all (m, n) such that a spectrum of type (m, n) exists. In the case just considered, $(m, 0)$ would not be in Γ_2 (at least for $m > 1$). In terms of the bijection to Γ_p, the above proposition instead reads as follows:

Corollary 94 *For any $(m, n) \in \Gamma_p$, the thick ideal $\mathcal{C}_{m,n}$ is a prime ideal if and only if one of the following conditions holds:*

(1) $(m, n) = (0, n)$, with $0 < n \leq \infty$,
(2) $0 < m \leq \infty$, $0 \leq n \leq \infty$ and $(m, n - 1) \notin \Gamma_p$.

For evaluating the functor $\rho_{(\mathcal{SH}(\mathbb{Z}/2)_f)_{(p)}}$, we use the first description of the prime ideals in $(\mathcal{SH}(\mathbb{Z}/2)_f)_{(p)}$, given by Proposition 92.

Proposition 95 *The functor*

$$\rho_{(\mathcal{SH}(\mathbb{Z}/2)_f)_{(p)}} : \mathrm{Spc}((\mathcal{SH}(\mathbb{Z}/2)_f)_{(p)}) \to \mathrm{Spec}(\mathbb{Z}_{(p)} \oplus \mathbb{Z}_{(p)})$$

has the following values:

$$\rho_{(\mathcal{SH}(\mathbb{Z}/2)_f)_{(p)}}(\mathcal{C}_{m,0}) = \begin{cases} 0 \oplus \mathbb{Z}_{(p)} & \textit{if } m = 1 \\ p\mathbb{Z}_{(p)} \oplus \mathbb{Z}_{(p)} & \textit{if } m > 1, \end{cases}$$

$$\rho_{(\mathcal{SH}(\mathbb{Z}/2)_f)_{(p)}}(\mathcal{C}_{0,n}) = \begin{cases} \mathbb{Z}_{(p)} \oplus 0 & \textit{if } n = 1 \\ \mathbb{Z}_{(p)} \oplus p\mathbb{Z}_{(p)} & \textit{if } n > 1. \end{cases}$$

Proof By [7, Corollary 5.6(b)], $\rho_{\mathcal{T}}$ is natural in $\mathcal{T}$. Since $X \in \mathcal{C}_{m,0}$ is only a condition on $\phi^{\{1\}}X$ and $X \in \mathcal{C}_{0,n}$ is only a condition on $\phi^{\mathbb{Z}/2}X$, the fixed point functors can be used to derive this result from the nonequivariant case. Note that the induced functor between the endomorphism rings,

$$\pi_0(\phi^H) : \pi_0^{(\mathcal{SH}(\mathbb{Z}/2)_f)_{(p)}} \to \pi_0^{\mathcal{SH}^{fin}_{(p)}}$$

is the projection onto the first summand of $\mathbb{Z} \oplus \mathbb{Z}$ if $H = \{1\}$ and onto the second summand of $\mathbb{Z} \oplus \mathbb{Z}$ if $H = \mathbb{Z}/2$ [81, Corollary 1]. □

The generalisation to the non-localised category $\mathcal{SH}(\mathbb{Z}/2)_f$ works similarly as in the non-equivariant case. In particular, any prime ideal in $\mathcal{SH}(\mathbb{Z}/2)_f$ is the preimage under the p-localisation functor of a prime ideal in $(\mathcal{SH}(\mathbb{Z}/2)_f)_{(p)}$ for some prime p.

Summarising the result, we can say that all information on thick ideals in $\mathcal{SH}(\mathbb{Z}/2)_f$ given in [7]—namely, that they map surjectively to $\mathrm{Spec}(\mathbb{Z} \oplus \mathbb{Z})$—is recovered in the results of [86] as presented in Sect. 3. Furthermore, [86] does not only specify a preimage of any element in $\mathrm{Spec}(\mathbb{Z} \oplus \mathbb{Z})$ but gives a complete list of all possible such preimages.

7.4 Prime Ideals in the Motivic Category $\mathcal{SH}(k)_f$

Reference [7, Sect. 10] studies $\mathcal{T} = \mathcal{SH}(F)_f$ for F a perfect field, in which case the map $\rho_{\mathcal{T}} : \mathrm{Spc}(\mathcal{T}) \to \mathrm{Spec}(GW(F))$ is surjective. That is, $\mathrm{Spec}(GW(F))$ gives a lower bound on the thick ideals of $\mathcal{SH}(F)_f$.

For $F = \mathbb{C}$, the naturality of $\rho_{\mathcal{T}}$ yields a commutative diagram:

$$\begin{array}{ccc} \mathrm{Spc}(\mathcal{SH}(\mathbb{C})_f) & \xrightarrow{\ \rho\ } & \mathrm{Spec}(\pi_0^{\mathcal{SH}(\mathbb{C})_f}) \\ {\scriptstyle \mathrm{Spc}(R)} \uparrow & & \uparrow {\scriptstyle \mathrm{Spec}(\pi_0(R))} \\ \mathrm{Spc}(\mathcal{SH}^{fin}) & \xrightarrow[\ \rho\ ]{} & \mathrm{Spec}(\pi_0^{\mathcal{SH}^{fin}}). \end{array}$$

The map $\mathrm{Spc}(R)$ takes a prime ideal $\mathcal{C}$ of $\mathcal{SH}^{fin}$ to its preimage under $R = R_{\mathbb{C}}$. The endomorphism ring $\pi_0^{\mathcal{SH}^{fin}}$ is isomorphic to $\mathbb{Z}$, generated by the identity $S^0 \to S^0$. By [62, Corollary 4.11], the same holds for $\pi_0^{\mathcal{SH}(\mathbb{C})_f} \cong \mathbb{Z}$. Since $R(\mathrm{id} : S^0_{\mathbb{C}} \to S^0_{\mathbb{C}}) = (\mathrm{id} : S^0 \to S^0)$, it follows that the map $\mathrm{Spec}(\pi_0(R))$ is isomorphic to the identity map $\mathrm{Spec}(\mathbb{Z}) \to \mathrm{Spec}(\mathbb{Z})$. Thus:

Corollary 96 *For any* $0 < n \leq \infty$, *the thick ideal* $R^{-1}(\mathcal{C}_n) \subseteq (\mathcal{SH}(\mathbb{C})_f)_{(p)}$ *is a prime ideal, and*

$$\rho_{(\mathcal{SH}(\mathbb{C})_f)_{(p)}}(R^{-1}(\mathcal{C}_n)) = \begin{cases} 0 & \text{if } n = 1, \\ p\mathbb{Z}_{(p)} & \text{if } n > 1. \end{cases}$$

Remark 97 From Proposition 83(2) or [7, Proposition 10.4], we know that $\mathrm{Spc}(R)$ is not surjective, since $C\eta$ lies in some prime ideal which is not of the form $R^{-1}(\mathcal{C}_n)$.

For $F = \mathbb{R}$, there is also a commutative diagram:

$$\begin{array}{ccc} \mathrm{Spc}(\mathcal{SH}(\mathbb{R})_f) & \xrightarrow{\ \rho\ } & \mathrm{Spec}(\pi_0^{\mathcal{SH}(\mathbb{R})_f}) \\ {\scriptstyle \mathrm{Spc}(R')}\uparrow & & \uparrow{\scriptstyle \mathrm{Spec}(\pi_0(R'))} \\ \mathrm{Spc}(\mathcal{SH}(\mathbb{Z}/2)_f) & \xrightarrow[\ \rho\ ]{} & \mathrm{Spec}(\pi_0^{\mathcal{SH}(\mathbb{Z}/2)_f}). \end{array}$$

Lemma 7 *In the above diagram, the right map* $\mathrm{Spec}(\pi_0(R'))$ *is an isomorphism.*

Proof The realisation functor R' maps generators of $\pi_0^{\mathcal{SH}(\mathbb{R})} \cong GW(\mathbb{R}) \cong \mathbb{Z} \oplus \mathbb{Z}$ to generators of $\pi_0^{\mathcal{SH}(\mathbb{Z}/2)}$. This follows from [62, Sect. 4] and is explained on [14, Slides 15–16 and 22–24]: $\pi_0^{\mathcal{SH}(\mathbb{R})}$ is generated by 1 and $\epsilon = -1 - \rho_{-1}\eta$ (in $GW(\mathbb{R})$, ϵ corresponds to the quadratic form $q(x) = -x^2$). The element $\epsilon \in \pi_0^{\mathcal{SH}(\mathbb{R})}$ is also represented by the twist map $S^{1,1} \wedge S^{1,1} \to S^{1,1} \wedge S^{1,1}$ [14, Slide 22]. Now, $S^{1,1} = \mathbb{G}_m = \mathbb{A}^1 \setminus \{0\}$ is mapped by R' to the circle with $\mathbb{Z}/2$ acting by involution, which is also denoted by $S^{1,1}$. Hence, $R'(\epsilon) = \epsilon : S^{1,1} \wedge S^{1,1} \to S^{1,1} \wedge S^{1,1}$, the twist map in $\mathcal{SH}(\mathbb{Z}/2)$. By [14, Slide 16], 1 and ϵ are generators for $\pi_0^{\mathcal{SH}(\mathbb{Z}/2)}$. Thus, $\pi_0(R') : \pi_0^{\mathcal{SH}(\mathbb{R})} \to \pi_0^{\mathcal{SH}(\mathbb{Z}/2)}$ is an isomorphism. □

Corollary 98 *For any prime ideal* $\mathcal{C}_{m,n} \subseteq (\mathcal{SH}(\mathbb{Z}/2)_f)_{(p)}$, $(R')^{-1}(\mathcal{C}_{m,n})$ *is a prime ideal of* $(\mathcal{SH}(\mathbb{R})_f)_{(p)}$ *and* $\rho_{(\mathcal{SH}(\mathbb{R})_f)_{(p)}}((R')^{-1}(\mathcal{C}_{m,n}))$ *can be identified with the same prime ideal of* $\mathbb{Z}_{(p)} \oplus \mathbb{Z}_{(p)}$ *as* $\rho_{(\mathcal{SH}(\mathbb{Z}/2)_f)_{(p)}}(\mathcal{C}_{m,n})$.

Remark 99 Conjecture 86(4) would imply that in the above diagram, $(R')^{-1}$ is not surjective, as $\mathrm{thickid}(C\eta_{(p)})$ would have to lie in some prime ideal which is not of the form $(R')^{-1}(\mathcal{C}_{m,n})$.

Remark 100 Since the Künneth formula might not hold for $AK(n)$, we do not know whether $\mathcal{C}_{AK(n)}$ or $\mathcal{C}_{AK(m,n)}$ are prime ideals.

Remark 101 If $k \subset \mathbb{C}$ or $k \subset \mathbb{R}$ is a field such that $\mathrm{Spec}(GW(\mathbb{C})) \to \mathrm{Spec}(GW(k))$ is not surjective or $\mathrm{Spec}(GW(\mathbb{R})) \to \mathrm{Spec}(GW(k))$ is not surjective respectively then [7, Corollary 10.1] already implies that there is an infinite family of thick ideals in $\mathcal{SH}(k)_f$ which are not of the form $R_k^{-1}(\mathcal{C})$ or $(R_k')^{-1}(\mathcal{C})$ respectively.

8 Motivic Type-n Spectra

Definition 102 Let $AK(n)$ be the motivic Morava K-theory spectrum as defined in Definition 74. We say that $X \in (\mathcal{SH}(k)_f)_{(p)}$ has motivic type n if $AK(n-1)_{**}(X) = 0$ and $AK(n)_{**}(X) \neq 0$.

A priori, the motivic type of X might not be unique, as we do not know whether $\mathcal{C}_{AK(n-1)} \subseteq \mathcal{C}_{AK(s)}$ for all $s < n$. In Sect. 9.6, we will prove that any $X \in \mathcal{SH}(\mathbb{C})^{fin}_{(p)}$, $p > 2$, has a unique motivic type. For any prime p, the motivic type-n spectra that we are going to consider in this section satisfy $AK(s)_{**}(X) = 0$ for all $s < n$.

Remark 103 In the topological category $\mathcal{SH}^{fin}_{(p)}$, the notion of types is equivalent to the notion of thick ideals by the thick subcategory theorem [25, Theorem 7]. In Sect. 3, we have seen that, in equivariant homotopy theory, a more general notion of types is required. Also in the motivic world, not every thick ideal can be described in the language of types, as defined above. For example, as shown in Sect. 7, the motivic Morava K-theories $AK(n)$ do not distinguish between the nonequal thick ideals $\mathrm{thickid}(C\eta_{(2)})$ and $\mathrm{thickid}(S^0_{(2)})$, as both are generated by a spectrum of motivic type 0.

In this section, we will often assume $k = \mathbb{C}$, because we will need explicit knowledge of H^{**} and of $AK(n)_{**}$ for some of our arguments. The results might hold in greater generality, but this seems to require different methods of proof.

If $X_n \in \mathcal{SH}^{fin}_{(p)}$ has type n (see Definition 9), then $c(X_n) \in \mathcal{SH}(\mathbb{C})^{fin}_{(p)}$ has motivic type n (Sect. 8.6). That is, $c(X_n) \in \mathcal{C}_{AK(n-1)}$ and $c(X_n) \notin \mathcal{C}_{AK(n)}$. For any given n, we show how to construct a spectrum $\mathbb{X}_n \in (\mathcal{SH}(\mathbb{C})_f)_{(p)}$ with motivic type n that is not in the image of the functor c. The construction will be similar to topological constructions given by [58, Sect. 4] and [75, Appendix C]. We stick to the approach in [75] but we believe that a motivic version of Mitchell's spectrum would give another spectrum of motivic type n.

This section is organised as follows:

We first discuss some foundations that are needed to apply the motivic Adams spectral sequence and obtain a vanishing result for motivic Morava K-theory (Theorem 14). Afterwards, we construct a spectrum satisfying the conditions of the theorem and we prove that it has indeed motivic type n (Sect. 8.5). Finally, we compare our findings to a constant type-n spectrum, $c(X_n)$ (Sect. 8.6), realising that motivic Morava K-theory does not distinguish the thick ideal generated by $\mathbb{X}_n$ from the one generated by cX_n, meaning that both spectra have motivic type n. We partly calculate their types with respect to the cohomology theories $c(K(s))$, as well. However, we do not have the answer to the question whether $\mathrm{thickid}(\mathbb{X}_n)$ equals $\mathrm{thickid}(c(X_n))$.

8.1 Universal Coefficient and Künneth Theorems

This section states some general results, which hold in $\mathcal{SH}(k)$, k any field, and which will be used later on. In the whole section, we use the notation $H = H\mathbb{Z}/p$. Recall from Remark 75(1) that H is cellular.

Proposition 7.7 in [16] describes the following universal coefficient spectral sequences (one of which we already encountered in the proof of Proposition 69):

Proposition 104 *Let $E \in \mathcal{SH}(k)$ be a motivic ring spectrum, M a right E-module and N a left E-module. Furthermore, assume that $E, M \in \mathcal{SH}(k)^{cell}$ (see Definition 52).*

(1) There is a strongly convergent spectral sequence

$$E^2 = \mathrm{Tor}^{E_{**}}_{a,b,c}(M_{**}, N_{**}) \Rightarrow \pi_{a+b,c}(M \wedge_E N).$$

(2) There is a conditionally convergent spectral sequence

$$E_2 = \mathrm{Ext}^{a,b,c}_{E_{**}}(M_{**}, N_{**}) \Rightarrow \pi_{-a-b,-c} F_E(M, N),$$

where $F_E(-,-)$ denotes the E-function spectrum.

We apply (2) to the case $E = H = H\mathbb{Z}/p$, $M = A \wedge H$ and $N = H$ for A a cell spectrum and get:

$$\mathrm{Ext}_{\mathrm{H}_{**}}(\mathrm{H}_{**}\, A, \mathrm{H}_{**}) \Rightarrow \pi_{**} F_H(A \wedge H, H) = \mathrm{H}^{**}\, A.$$

If A is a finite cell spectrum, we can also apply the spectral sequence to the case $E = H$, $M = F(A, H)$, $N = H$:

$$\mathrm{Ext}_{\mathrm{H}^{**}}(\mathrm{H}^{**}\, A, \mathrm{H}^{**}) \Rightarrow \pi_{**} F_H(F(A, H), H) = \pi_{**} F(F(A, S^0), H) = \mathrm{H}_{**}\, A.$$

The first equality holds because $F(A, H) = F(A, S^0) \wedge H$ for finite A and the second holds because taking the dual of A twice gives A again (see [51, Proposition III.1.3]).

In the case of vanishing higher Ext-groups, the spectral sequence collapses and we get the following result:

Corollary 105 *If $A \in \mathcal{SH}(k)^{cell}$ is any cell spectrum such that $\mathrm{H}_{**}\, A$ is free over H_{**}, then $\mathrm{H}^{**}\, A \cong \mathrm{Hom}_{\mathrm{H}_{**}}(\mathrm{H}_{**}\, A, \mathrm{H}_{**})$. If A is a finite cell spectrum with $\mathrm{H}^{**}\, A$ free over H^{**}, then $\mathrm{H}_{**}\, A \cong \mathrm{Hom}_{\mathrm{H}^{**}}(\mathrm{H}^{**}\, A, \mathrm{H}^{**})$.*

The universal coefficient spectral sequence also implies the following Künneth theorem [16, Remark 8.7]:

Proposition 106 *Let A and B be motivic spectra such that A is a finite cell spectrum. If $\mathrm{H}^{**} A$ is free over H^{**}, then*

$$\mathrm{H}^{**}(A) \otimes_{\mathrm{H}^{**}} \mathrm{H}^{**}(B) \cong \mathrm{H}^{**}(A \wedge B).$$

8.2 The Motivic Steenrod Algebra

In this section, too, we let $H = H\mathbb{Z}/p$. Let $k \subseteq \mathbb{C}$.

The motivic mod-p Steenrod algebra, $\mathcal{A} = \mathcal{A}^{**}$ has first been defined in [91, Sect. 11] as the algebra of certain bistable natural transformations $\mathrm{H}^{**}(-) \to \mathrm{H}^{**}(-)$. By [92, Theorem 3.49] and [32, Lemma 5.7], $\mathcal{A}$ is the algebra of all such operations and, as an H^{**}-module,

$$\mathcal{A} \cong \mathrm{H}^{**} \otimes_{\mathbb{F}_p} \mathcal{A}_{\mathrm{top}},$$

where $\mathcal{A}_{\mathrm{top}}$ is the topological mod-p Steenrod algebra. Thus, as H^{**}-modules,

$$\mathcal{A} \cong \mathrm{H}^{**} \otimes_{\mathbb{F}_p} RP \otimes_{\mathbb{F}_p} \Lambda_{\mathbb{F}_p}(Q_0, Q_1, \cdots),$$

where RP is the $\mathbb{F}_p$-module generated by certain products of reduced powers $P^i : \mathrm{H}^{**}(-) \to \mathrm{H}^{*+2i(p-1),*+i(p-1)}(-), i \geq 0$, and $\Lambda_{\mathbb{F}_p}(Q_0, Q_1, \cdots)$ denotes the exterior algebra over $\mathbb{F}_p$ generated by $Q_i : \mathrm{H}^{**}(-) \to \mathrm{H}^{*+2p^i-1,*+p^i-1}$, $i \geq 0$, as defined in [91, Sect. 13]. See also [10, Corollaries 3 and 4] or [96, Eq. (2.18)].

Borghesi [10, Theorem 12] computes the cohomology of the motivic connective Morava K-theory spectrum, which can be expressed by the same formula as in topology.

Proposition 107

$$\mathrm{H}^{**}(Ak(s)) = \mathcal{A}/\mathcal{A}Q_s = \mathrm{H}^{**} \otimes RP \otimes \Lambda_{\mathbb{F}_p}(Q_0, \cdots, Q_{s-1}, Q_{s+1}, \cdots).$$

If X is a finite cell spectrum such that $\mathrm{H}^{**}(X)$ is free over H^{**}, we can apply the Künneth theorem to $Ak(s) \wedge X$.

Corollary 108 *Let $X \in \mathcal{SH}(k)^{fin}$ with $\mathrm{H}^{**}(X)$ free over H^{**}. Then*

$$\mathrm{H}^{**}(Ak(s) \wedge X) = \mathcal{A}/\mathcal{A}Q_s \otimes_{\mathrm{H}^{**}} \mathrm{H}^{**}(X).$$

Writing $\Lambda(Q_s)$ for the exterior algebra over H^{**} generated by Q_s, we have $\mathcal{A} = \mathcal{A}/\mathcal{A}Q_s \otimes_{\mathrm{H}^{**}} \Lambda(Q_s)$. Any resolution of $\mathrm{H}^{**} X$ by projective $\Lambda(Q_s)$-modules P_i yields a resolution $\mathcal{A}/\mathcal{A}Q_s \otimes_{\mathrm{H}^{**}} P_i$ of $\mathcal{A}/\mathcal{A}Q_s \otimes_{\mathrm{H}^{**}} \mathrm{H}^{**} X$ by projective $\mathcal{A}$-modules. This implies the following change of rings isomorphism.

Corollary 109 *For any finite cell spectrum X with $\mathrm{H}^{**}(X)$ free over H^{**}, we have*

$$\mathrm{Ext}_{\mathcal{A}}(\mathrm{H}^{**}(Ak(s) \wedge X), \mathrm{H}^{**}) \cong \mathrm{Ext}_{A(Q_s)}(\mathrm{H}^{**}(X), \mathrm{H}^{**}).$$

This isomorphism will later be applied to the motivic Adams spectral sequence for $Ak(s)_{**}(X)$.

8.3 The Motivic Adams Spectral Sequence

Recall the notation $H = H\mathbb{Z}/p$. Let $k \subseteq \mathbb{C}$.

Our aim in this section is to show the existence and convergence of the following Adams spectral sequence:

$$E_2^{s,t,u} = \mathrm{Ext}_{\mathcal{A}}^{s,t,u}(\mathrm{H}^{**}(Ak(s) \wedge X), \mathrm{H}^{**}) \Rightarrow Ak(s)_{**}(X)$$

for finite cell spectra $X \in \mathcal{SH}(k)^{fin}$ (see Definition 52), with $\mathrm{H}^{**}\,X$ free over H^{**}.

Motivic Adams spectral sequences were first described in [59]. Corollary 3 of [36] shows that over fields of characteristic 0, the Adams spectral sequence for a motivic cell spectrum X of finite type converges to the homotopy groups of the completion $X^{\wedge}_{p,\eta}$. In [17], calculations are made for the case $p = 2$, $X = S$. More details and explanations can be found in [83]. The convergence of the Adams spectral sequence for $Ak(s) \wedge X$ will follow from [36]. However, we have to start again from the Adams resolution, to get the limit term $Ak(s)_{**}(X)$ using arguments from [74, Sect. 2.1], and to get a module structure on the sequence.

Definition 110 An Adams resolution $(Y_s, g_s, K_s, f_s)_{s\geq 0}$ for a motivic cell spectrum $Y \in \mathcal{SH}(k)^{cell}$ is a diagram

$$\begin{array}{ccccccc}
Y = Y_0 & \xleftarrow{g_0} & Y_1 & \xleftarrow{g_2} & Y_2 & \xleftarrow{g_3} & \cdots, \\
\downarrow{\scriptstyle f_0} & & \downarrow{\scriptstyle f_1} & & \downarrow{\scriptstyle f_2} & & \\
K_0 & & K_1 & & K_2 & &
\end{array}$$

where each K_s is a wedge of suspensions of H, f_s is surjective on motivic mod-p cohomology and Y_{s+1} is the fiber of f_s.

Such an Adams resolution exists whenever $\mathrm{H}^{**}\,Y$ is a free module over H^{**} of motivically finite type [36], [83, Sect. 2.5.4]. The finite type condition is defined in [17, Definition 2.12]. For our purposes, it suffices to know that, if the generators of $\mathrm{H}^{**}\,Y$ are located in degrees (i_α, j_α) with i_α bounded below and, for each α, there are only finitely many β's with $i_\alpha = i_\beta$ and $j_\alpha \geq j_\beta$, then $\mathrm{H}^{**}\,Y$ is of motivically finite type. This holds, for example, if $Y \in \mathcal{SH}(k)^{fin}$.

Corollary 111 *Let X be a finite cell spectrum such that $\mathrm{H}^{**}\,X$ is free over H^{**} and let $Ak(s)$ be the motivic connective Morava K-theory spectrum (see Definition 74). Then $Y = Ak(s) \wedge X$ satisfies the finite type condition as described above.*

Proof By Corollary 108, $\mathrm{H}^{**}(Y) \cong \mathcal{A}/\mathcal{A}Q_s \otimes_{\mathrm{H}^{**}} \mathrm{H}^{**} X$. The motivic Steenrod algebra $\mathcal{A}$ is of motivically finite type, since, for the bidegrees (i_α, j_α) of its H^{**}-generators, i_α and j_α are nonnegative and there are only finitely many generators of a fixed bidegree. The same holds for $\mathcal{A}/\mathcal{A}Q_s$. Furthermore, $\mathrm{H}^{**} X$ is of motivically finite type, since X is finite. It follows that the tensor product $\mathcal{A}/\mathcal{A}Q_s \otimes_{\mathrm{H}^{**}} \mathrm{H}^{**} X$ is also of motivically finite type. □

Corollary 112 (Convergence of the ASS) *Let $k = \mathbb{C}$ (or any other field satisfying the assumptions of [36, Theorem 1]) and let X be a finite cell spectrum such that $\mathrm{H}^{**}(X)$ is a free H^{**}-module. Then the Adams spectral sequence for $Y = Ak(s) \wedge X$, $s > 0$, strongly converges to $Ak(s)_{**}(X)$.*

Proof Since Y is of motivically finite type by the previous corollary, strong convergence follows from [36, Corollary 3]. By [36, Theorem 1], the spectral sequence converges to the p-completion of $\pi_{**}Y$. Since $Ak(s)$ is a quotient of $ABP/(p)$ by definition, $Y_p^\wedge = Y$, whence the limit term is $Ak(s)_{**}(X)$. □

Remark 113 In Corollary 112, X does not necessarily need to be finite. All we need to know is that $Ak(s) \wedge X$ is of motivically finite type.

In Theorem 14, it will be crucial that this particular spectral sequence is a spectral sequence of $(MGL_{(p)})_{**}$-modules.

Since $Ak(s)$ is an $MGL_{(p)}$-module spectrum by Remark 75(3), $Ak(s)_{**}X$ and $\mathrm{H}^{**}(Ak(s) \wedge X)$ are $(MGL_{(p)})_{**}$-modules. Before we can prove compatibility of the Adams spectral sequence with the module structure, we need to determine $\mathrm{H}_{**}\, Ak(s)$. Borghesi [10, Theorem 5 and Remark 2.2] shows:

Lemma 8

$$\mathrm{H}_{**}(MGL) \cong \mathrm{H}_{**}[m_1, m_2, \cdots] \cong \mathrm{H}_{**} \otimes_{\mathbb{F}_p} \mathrm{H}_*(MU, \mathbb{Z}/p).$$

Since $\pi_{**}(H \wedge MGL_{(p)}) = \pi_{**}(H_{(p)} \wedge MGL)$ and $H_{(p)} = H$, it follows that also $\mathrm{H}_{**}(MGL_{(p)})$ is a free H_{**}-module with basis elements $m_I = m_{i_1}^{k_1} m_{i_2}^{k_2} \cdots m_{i_t}^{k_t}$. Here, m_i is defined as the Hurewicz image of $a_i \in MGL_{2i,i}$ and a_i is the image of a polynomial generator of MU_*. The motivic connective Morava K-theory spectra are defined as homotopy colimits of spectra E_i, which are defined by successively taking cofibers of maps $\Sigma^{2(i-1),i-1} E_{i-1} \xrightarrow{a_{i-1}} E_{i-1}$, starting from $MGL_{(p)}$ (see Definition 74). Passing from $\mathrm{H}_{**}\, E_{i-1}$ to $\mathrm{H}_{**}\, E_i$, the H_{**}-basis changes but the fact that the homology is a free H_{**}-module remains true for each i.

Corollary 114 $\mathrm{H}_{**}(Ak(s))$ *is a free* H_{**}*-module. Hence,*

$$\mathrm{H}_{**}(Ak(s)) \cong \mathrm{Hom}_{\mathrm{H}^{**}}(\mathrm{H}^{**}(Ak(s)), \mathrm{H}^{**}).$$

Proof The second statement follows from Corollary 105. For the cellularity of $Ak(s)$, see Remark 75(1). □

Proposition 115 *Let k and X be as in Corollary 112. The Adams spectral sequence for $Ak(s) \wedge X$ is a spectral sequence of $(MGL_{(p)})_{**}$-modules.*

Proof We show that there is an Adams resolution by $MGL_{(p)}$-modules. The construction of the spectral sequence from the Adams resolution preserves such a module structure in every step and the claim will follow.

We use similar arguments as in Corollary 112, but starting with an Adams resolution $\{X_t, g_t, K_t, f_t\}$ for X. Then we take the smash product of such a resolution with $Ak(s)$ and show that this yields an Adams resolution for $Y = Ak(s) \wedge X$. Since $Ak(s)$ is an $MGL_{(p)}$-module, this will be a resolution by $MGL_{(p)}$-modules. By definition, f_t is surjective on cohomology, which is equivalent to g_t inducing zero in cohomology. It follows that $Ak(s) \wedge g_t$ induces zero in cohomology, hence, $Ak(s) \wedge f_t$ is surjective on cohomology by the long exact fiber sequence.

It remains to show that $Ak(s) \wedge K_t$ is a wedge of suspensions of H. Consider one wedge summand H of K_t. By Corollary 114, $\pi_{**}(Ak(s) \wedge H)$ is free over $\pi_{**}(H)$. Furthermore, $Ak(s) \wedge H$ is a cellular H-module. Therefore, [32, Lemma 5.3] implies that $Ak(s) \wedge H$ is split, that is, $Ak(s) \wedge H \cong \bigvee \Sigma^{**} H$, as we wanted to show. □

8.4 Vanishing Criterion for Motivic Morava K-Theory

Now we can prove the following motivic version of the vanishing result [58, Theorem 4.8].

Theorem 14 (Vanishing criterion) *Let p be any prime, $s > 0$ and k be as in Corollary 112. Let $X \in \mathcal{SH}(k)^{fin}$ be a finite motivic cell spectrum such that $\mathrm{H}^{**} X$ is free over $\Lambda(Q_s)$ (the exterior algebra over H^{**}) as a module over the Steenrod algebra. Then*

$$AK(s)_{**}X = 0.$$

Proof With the preparations made so far, the rest of the proof is exactly as in [58, Theorem 4.8]. Since $AK(s) = v_s^{-1}Ak(s)$, it suffices to show that $Ak(s)_{**}X$ is v_s-torsion. We apply the change of rings isomorphism, Corollary 109, to the Adams spectral sequence for $Ak(s) \wedge X$ and get:

$$E_2 \cong \mathrm{Ext}_{\Lambda(Q_s)}(\mathrm{H}^{**} X, \mathrm{H}^{**}) \Rightarrow Ak(s)_{**}X.$$

By the assumption on X, this collapses to

$$\mathrm{Hom}_{\Lambda(Q_s)}(\mathrm{H}^{**} X, \mathrm{H}^{**}) \cong Ak(s)_{**}X,$$

and, by the previous proposition, this is an isomorphism of $(MGL_{(p)})_{**}$-modules. Since $(MGL_{(p)})_{**}$ acts trivially on the left hand side of this isomorphism, v_s acts trivially on $Ak(s)_{**}X$, too. Hence, $AK(s)_{**}X = 0$. □

8.5 Construction of Motivic Type-n Spectra

In this section, we assume $k = \mathbb{C}$ because we will work explicitly with $\mathrm{H}^{**} \cong \mathbb{F}_p[\tau]$, $\deg(\tau) = (0, 1)$ (see Lemma 4). The construction of a spectrum X with the properties required in the previous theorem can be done similarly as in [75, Appendix C]. That is, one starts with a so-called weakly type-n spectrum ($n > s$) and then uses a particular idempotent to split off a (strongly) type-n spectrum of a certain smash power of it.

In the following, we study idempotents for free H^{**}-modules.

For V^{**} an Adams graded graded abelian group (i.e., V^{**} has a sign rule in the first grading but not in the second one, see e.g. [65, Sect. 3]) which is a free H^{**}-bimodule, let $V^+ = \bigoplus_{p \text{ even}} V^{p,q}$ and $V^- = \bigoplus_{p \text{ odd}} V^{p,q}$ be the even and odd dimensional parts of V. That is, commuting with an element of V^+ does not change the sign but commuting two elements of V^- does. For a vector space V^* over $\mathbb{F}_p$, Ravenel defines a number k_V and an idempotent $e_V \in \mathbb{Z}_{(p)}[\Sigma_{k_V}]$, which only depend on $\dim_{\mathbb{F}_p} V^+$ and $\dim_{\mathbb{F}_p} V^-$ [75, Appendix C.2]. The analogous definition can be formulated using $\dim_{\mathrm{H}^{**}} V^+$ and $\dim_{\mathrm{H}^{**}} V^-$ for our bigraded H^{**}-modules. The symmetric group Σ_{k_V} acts on $V^{\otimes k_V}$ by permuting the factors. As V is an $\mathbb{F}_p$-module, this induces an action of $\mathbb{Z}_{(p)}[\Sigma_{k_V}]$ on $V^{\otimes k_V}$. For our purposes, it will not be important to know the precise definitions of k_V and e_V. We just need to know that they are defined in such a way that the following analogue of [75, Theorem C.2.1] holds.

Proposition 116 *Let k_V and $e_V \in \mathbb{Z}_{(p)}[\Sigma_{k_V}]$ be the number and idempotent defined in [75, Appendix C.2] and let $W = V^{\otimes k_V}$. Then $e_V W \neq 0$. If $U \subset V$ has $\dim U^+ \leq \dim V^+ - 1$ or $\dim U^- \leq \dim V^- - (p-1)$, then $e_V U^{\otimes k_V} = 0$. Here, $\dim$ denotes the H^{**}-dimension and $\otimes$ denotes the tensor product over H^{**}.*

Proof The proof of [75, Theorem C.2.1] applies to our setting without changes. □

We will need the following lemma.

Lemma 9 *Let M be a module over $\mathbb{F}_p[\tau, Q]/Q^2$ which is free as a module over $\mathbb{F}_p[\tau]$ and free as a module over $\mathbb{F}_p[Q]/Q^2$. Then M is a free $\mathbb{F}_p[\tau, Q]/Q^2$-module.*

Proof Let $\{m_i\}_{i\in I}$ be a basis of M over $\mathbb{F}_p[\tau]$ and $\{n_j\}_{j\in J}$ a basis over $\mathbb{F}_p[Q]/Q^2$. Then M is a free $\mathbb{F}_p$-module with bases $\{\tau^k m_i\}_{i\in I, k\in\mathbb{N}}$ and $\{n_j, Qn_j\}_{j\in J}$. As an $\mathbb{F}_p[Q]/Q^2$-module, M decomposes as $M \cong M' \oplus QM'$ with $M' \cong QM' \cong QM$ as $\mathbb{F}_p$-modules. Hence, the elements m_i can be written as $m_i = a_i + Qb_i$ with $a_i, b_i \in M'$. For any $i \in I$ such that both a_i and b_i are nonzero, we replace the basis element m_i by the two elements a_i and Qb_i. Then we still have a set of generators for M over $\mathbb{F}_p[\tau]$, which can be turned into a basis by removing elements. Hence, we can assume that all m_i are of the form $m_i = a_i$ or $m_i = Qb_i$. Let $I' = \{i \in I \mid m_i = a_i \in M'\}$. Then $M' \cong \mathbb{F}_p\{\tau^k a_i\}_{i\in I', \tau\in\mathbb{N}}$ as an $\mathbb{F}_p$-module. Hence, $QM' \cong \mathbb{F}_p\{Q\tau^k a_i\}_{i\in I', \tau\in\mathbb{N}}$ and $M \cong \mathbb{F}_p\{\tau^k a_i, Q\tau^k a_i\}$. It follows that $\{a_i\}_{i\in I'}$ is a basis of M as a free $\mathbb{F}_p[\tau, Q]/Q^2$-module. □

Theorem C.2.2 of [75] explains how to split off a free module over the exterior $\mathbb{F}_p$-algebra generated by Q_s from a module with nontrivial Q_s-action. In our setting, this also works for the exterior H^{**}-algebra $\Lambda(Q_s)$.

Proposition 117 *Let $V = U \oplus F$ be a splitting of $\Lambda(Q_s)$-modules which are free over H^{**}, and $F \neq 0$ be free over $\Lambda(Q_s)$. Then $e_V V^{\otimes k_V}$ is a free $\Lambda(Q_s)$-module.*

Proof We write the tensor product as $V^{\otimes k_V} = U^{\otimes k_V} \oplus F'$, where $F' = \bigoplus_{\substack{a+b=k_V \\ b\geq 1}} U^{\otimes a} \otimes F^{\otimes b}$. By the proof of [75, Theorem C.2.2], F' is free over the Hopf algebra $\mathbb{F}_p[Q_s]/Q_s^2$, which we abbreviate by E. Let us give the reason for this statement. We show that if the E-module U has basis $\{u_i\}_I$ over $\mathbb{F}_p$ and F has basis $\{f_j\}_J$ over E, then $U \otimes_{\mathbb{F}_p} F$ has $\mathbb{F}_p$-basis $\{u_i \otimes f_j, Q_s(u_i \otimes f_j)\}_{I,J}$ and hence is a free module over E. The module structure of $U \otimes F$ is defined by $Q_s(u \otimes f) = Q_s u \otimes f + u \otimes Q_s f$. Since $\{f_j, Q_s f_j\}_J$ defines an $\mathbb{F}_p$-basis of F, an $\mathbb{F}_p$-basis of $U \otimes F$ can be given by $\{u_i \otimes f_j, u_i \otimes Q_s f_j\}$. In the formula $Q_s(u_i \otimes f_j) = Q_s u_i \otimes f_j + u_i \otimes Q_s f_j$, $Q_s u_i = \Sigma r_k u_k$ can be expressed by the basis elements of U, hence $Q_s u_i \otimes f_j \in \mathbb{F}_p\{u_k \otimes f_j\}$ and the basis elements $u_i \otimes Q_s f_j$ of $U \otimes F$ can be replaced by $Q_s(u_i \otimes f_j)$. We obtain $U \otimes F = \mathbb{F}_p\{u_i \otimes f_j, Q_s(u_i \otimes f_j)\}$. Inductively, it follows that all mixed summands $U^{\otimes a} \otimes F^{\otimes b}$ in $V^{\otimes k_V}$ are free and hence F' is free over E.

The analogue holds if we consider U and F as free modules over $\mathbb{F}_p[\tau]$ instead of $\mathbb{F}_p$ and use $\otimes_{\mathbb{F}_p[\tau]}$. It follows that F' is free over $\Lambda(Q_s)$. The direct summands are invariant under the Σ_{k_V}-action. Hence, we have a short exact sequence

$$0 \to e_V U^{\otimes k_V} \to e_V V^{\otimes k_V} \to e_V F' \to 0.$$

Since $\deg(Q_s) = (2p^i - 1, p^i - 1)$, multiplication by Q_s sends V^+ to V^- and vice versa. It follows that $\dim F^+ > 0$ (and $\dim F^- > 0$) and, hence, $\dim U^+ < \dim V^+$. By the previous proposition, this implies $e_V U^{\otimes k_V} = 0$. It follows that $e_V V^{\otimes k_V} = e_V F'$. We have to show that $e_V F'$ is a free $\Lambda(Q_s)$-module. As a module over the exterior $\mathbb{F}_p$-algebra over Q_s, this is a direct summand of a free module over a local ring. Hence, $e_V F'$ is free over $\mathbb{F}_p[Q_s]/(Q_s^2)$. Since $e_V F'$ is also a free H^{**}-module, it is free over $\Lambda(Q_s) = \mathbb{F}_p[\tau, Q_s]/Q_s^2$ by Lemma 9. □

We can apply this to motivic cohomology in the following way:

Theorem 15 (Splitting off free $\Lambda(Q_s)$-modules) *Let $X \in \mathcal{SH}(\mathbb{C})_{(p)}^{fin}$ be a p-local finite cell spectrum such that Q_s acts nontrivially on $\mathrm{H}^{**}(X)$ as an element of the Steenrod algebra. Assume that $V = \mathrm{H}^{**}(X)$ is a free H^{**}-module and let $Y = e_V(X^{\wedge k_V})$. Then $\mathrm{H}^{**}(Y)$ is free over $\Lambda(Q_s)$.*

Proof This is analogous to a statement in [75, Theorem C.3.2]. Since Q_s acts nontrivially, $\mathrm{H}^{**}(X)$ contains a nontrivial summand which is free over $\Lambda(Q_s)$. The previous proposition yields the claim. Note that $\mathrm{H}^{**}(e_V X^{\wedge k_V}) = e_V\, \mathrm{H}^{**}(X)^{\otimes k_V}$ by the Künneth theorem (Proposition 106) and by the way $\mathbb{Z}_{(p)}[\Sigma_{k_V}]$ acts. The Künneth theorem also holds for p-local finite spectra because p-localisation commutes with $\mathrm{H}^{**}(-)$. □

Y is a retract of the p-local finite cell spectrum $X^{\wedge k_V}$, but maybe it is not finite itself. Therefore, we need an additional argument which shows that Theorem 14 holds for Y.

Corollary 118 *For $s > 0$ and Y as in Theorem 15, $AK(s)_{**}(Y) = 0$.*

Proof We have to show $\mathrm{H}^{**}(Ak(s) \wedge Y) \cong \mathcal{A}/\mathcal{A}Q_s \otimes_{\mathrm{H}^{**}} \mathrm{H}^{**}(Y)$. Since $\mathrm{H}^{**}\, Y \cong e_V\, \mathrm{H}^{**}(X)^{\otimes k_V}$, Y is of motivically finite type. Remark 113 applies and the claim follows as in the proof of Theorem 14. The left hand side of the claimed isomorphism can be rewritten as

$$\begin{aligned} \mathrm{H}^{**}(Ak(s) \wedge e_V X^{\wedge k_V}) &\cong \mathrm{H}^{**}((1 \wedge e_V)(Ak(s) \wedge X^{\wedge k_V})) \\ &\cong (1 \otimes e_V)\, \mathrm{H}^{**}(Ak(s) \wedge X^{\wedge k_V}). \end{aligned}$$

Now we can apply the Künneth isomorphism and get $(1 \otimes e_V)(\mathcal{A}/\mathcal{A}Q_s \otimes_{\mathrm{H}^{**}} \mathrm{H}^{**}(X)^{\wedge k_V})$. This is isomorphic to

$$\mathcal{A}/\mathcal{A}Q_s \otimes_{\mathrm{H}^{**}} e_V\, \mathrm{H}^{**}(X)^{\wedge k_V} \cong \mathcal{A}/\mathcal{A}Q_s \otimes_{\mathrm{H}^{**}} \mathrm{H}^{**}(Y),$$

which is the right hand side. □

This result tells us that, given a nontrivial Q_s-action on $\mathrm{H}^{**}(X)$, $X \in \mathcal{SH}(\mathbb{C})^{fin}_{(p)}$, we can construct a spectrum Y for which $AK(s)_{**}(Y) = 0$. So, let's construct such an X.

Next, we will construct a finite cell spectrum with nontrivial Q_s-action and trivial Q_n-action.

Let $k = \mathbb{C}$. We combine ideas of Ravenel [75] with computations by Voevodsky [91]. In [75, Lemma 6.2.6], the given example of a spectrum with nontrivial Q_s^{top}-action, $s < n$, and trivial Q_n^{top}-action on $\mathrm{H}^*(X)$ is $X = (B\mathbb{Z}/p)_2^{2p^n}$, that is, the suspension spectrum of the $2p^n$-skeleton of the classifying space $B\mathbb{Z}/p$ modulo its 1-skeleton. Cutting off higher dimensional cells leads to a trivial Q_n^{top}-action, which is needed for nontrivial n-th Morava K-theory. In [91, Sect. 6], the algebraic analogue to $B\mathbb{Z}/p$ is defined as $B\mu_p = \mathrm{colim}_n\, \tilde{V}_n/\mu_p$, where $\tilde{V}_n = \mathbb{A}^n \setminus \{0\}$ (see the proof of [91, Lemma 6.3]) and μ_p acts by multiplication with a p-th root of unity in each of the n coordinates. Under $R = R_{\mathbb{C}}$, this action realises to the $\mathbb{Z}/p$-action on $S^{2n-1} \subset \mathbb{C}^n$ rotating each $\mathbb{C}$ factor by a p-th root of unity.

Lemma 10

$$R(\tilde{V}_n/\mu_p) \cong S^{2n-1}/(\mathbb{Z}/p)$$

is the $(2n-1)$-skeleton of $B\mathbb{Z}/p$ in the CW-structure having one cell in each dimension.

Proof $B\mathbb{Z}/p$ is the infinite dimensional lens space, as studied for example in [22, Example 2.43]. There, it is explained that the $(2n-1)$-skeleton is precisely the $(2n-1)$-dimensional lens space, which is defined as the orbit space $S^{2n-1}/(\mathbb{Z}/p)$. □

By [22, Example 2.43, p. 146], the attaching map of the $2k$-cell of $B\mathbb{Z}/p$ is the quotient map $S^{2k-1} \to S^{2k-1}/(\mathbb{Z}/p)$. We define $V_n \in \mathcal{SH}(\mathbb{C})$ to be the cofiber of the quotient map of suspension spectra $\tilde{V}_{p^n} \to \tilde{V}_{p^n}/\mu_p$, so that the following lemma holds.

Lemma 11

$$R(V_n) = (B\mathbb{Z}/p)^{2p^n}.$$

Let $\mathbb{B}$ be the cofiber of the composite map $\tilde{V}_1/\mu_p \to \tilde{V}_{p^n}/\mu_p \to V_n$. Then $\mathbb{B}$ is a finite cell spectrum and satisfies the following corollary.

Corollary 119

$$R(\mathbb{B}) = (B\mathbb{Z}/p)_2^{2p^n}.$$

The following is a special case of [91, Proposition 6.10] (as explained on [91, p. 20]).

Proposition 120 $\mathrm{H}^{**}(B\mu_p)$ *is a free* H^{**}*-module with basis* $\{v^i, uv^i \mid i \geq 0\}$, *where* $v \in \mathrm{H}^{2,1}(B\mu_p)$ *and* $u \in \mathrm{H}^{1,1}(B\mu_p)$.

From this, it follows for dimensional reasons:

Proposition 121 *The cohomology* $\mathrm{H}^{**}\,\mathbb{B}$ *is the free* H^{**}*-module with basis* $\{v^i \mid 1 \leq i \leq p^n\} \cup \{uv^i \mid 1 \leq i \leq p^n - 1\}$.

Furthermore, Voevodsky shows [91, Lemmas 11.2 and 11.3]:

Lemma 12 *On*

$$\mathrm{H}^{**}(B\mu_p)/\,\mathrm{H}^{*,>0}\,\mathrm{H}^{**}(B\mu_p) \cong \mathbb{F}_p[u, v]/(u^2 = 0),$$

$\mathcal{A}/(\mathrm{H}^{*,>0}\mathcal{A})$ *acts by* $\beta(u) = v$, $P^i(u) = 0$ *for* $i > 0$, $\beta(v^k) = 0$ *and* $P^i(v^k) = \binom{k}{i} v^{k+i(p-1)}$.

Over $\mathbb{C}$, the action of Q_i can be defined by $Q_0 = \beta$ and $Q_{i+1} = P^{p^i} Q_i - Q_i P^{p^i}$ [88, Proposition 3.1]. From this, we can inductively compute the action of Q_s on $uv^k \in \mathrm{H}^{**}(\mathbb{B})/\,\mathrm{H}^{*,>0}\,\mathrm{H}^{**}(\mathbb{B})$. We get $Q_s(uv^k) = cv^{k+p^s}$ with $c \equiv 1 \mod p$, which is nontrivial for $s < n$ and $k < p^n - p^s$. It follows that Q_s acts nontrivially on $\mathrm{H}^{**}(\mathbb{B})$ for $s < n$.

Now we have all ingredients for the motivic type n spectrum in $\mathcal{SH}(\mathbb{C})$.

Theorem 16 (A spectrum of motivic type n) *For a fixed* $n > 0$, *let* $V = \mathrm{H}^{**}(\mathbb{B}_{(p)})$ *and* $X = e_V(\mathbb{B}_{(p)})^{\wedge k_V}$, *then* $AK(s)_{**}(X) = 0$ *for all* $s < n$ *and* $AK(n)_{**}(X) \neq 0$.

Proof For $s > 0$, $AK(s)_{**}(X) = 0$ follows from Corollary 118, whose assumptions are satisfied by the above considerations.

For $s = 0$, note that $AK(0) = p^{-1}MGL_{(p)}/(a_1, a_2, \cdots)$ by Definition 74. The main result of [32] implies $AK(0) \cong p^{-1}(H\mathbb{Z})_{(p)}$. It follows that

$$AK(0)_{**}X \cong \pi_{**}(p^{-1}(H\mathbb{Z} \wedge X)_{(p)}).$$

But p acts trivially on $\tilde{V}_m/\mu_p$, which implies that X is p-torsion. Therefore, $p^{-1}X \cong 0$ and $AK(0)_{**}X = 0$. It remains to show that $AK(n)_{**}(X) \neq 0$. This can either be done analogously to [58, Theorem 4.8], using the motivic Atiyah Hirzebruch spectral sequence from Proposition 76, or by considering the topological realisation $R_{\mathbb{C}}(X)$. In Proposition 124, we show that $R(X)$ is of type n. It follows that $X \in R^{-1}(\mathcal{C}_n \setminus \mathcal{C}_{n+1})$. In particular, $X \notin R^{-1}(\mathcal{C}_{n+1})$. By Proposition 78, $\mathcal{C}_{AK(n)} \subseteq R^{-1}(\mathcal{C}_{n+1})$. This proves $X \notin \mathcal{C}_{AK(n)}$. □

Remark 122 The spectrum X is a retract of the p-local finite cell spectrum $\mathbb{B}_{(p)}^{\wedge k_V}$ and it follows by Remark 55(1) that X is compact, i.e., $X \in (\mathcal{SH}(\mathbb{C})_f)_{(p)}$.

Remark 123 $X = e_V(\mathbb{B}_{(p)})^{\wedge k_V}$ is an example of a motivic spectrum with vanishing Margolis homology groups $MH_s^{p,q}(X)$ for all $s < n$, $p, q \in \mathbb{Z}$, as defined in [90, Sect. 3]: we have shown that $\mathrm{H}^{**}(X)$ is free over $\Lambda(Q_s)$, which implies that $\ker Q_s = \operatorname{im} Q_s$ for $Q_s : \mathrm{H}^{**}(X) \to \mathrm{H}^{**}(X)$.

Proposition 124 *The topological realisation $R(X)$ of the spectrum X constructed above is of type n.*

Proof Let $k = \mathbb{C}$ and $B = (B\mathbb{Z}/p)_2^{2p^n}$. We already know that for $R : \mathcal{SH}(\mathbb{C}) \to \mathcal{SH}$, $R(\mathbb{B}) = B$. Since R preserves $\wedge$-products, it follows

$$R(X) = R(e_V(\mathbb{B}_{(p)})^{\wedge k_V}) = e_V(R(\mathbb{B}_{(p)}))^{\wedge k_V} = e_V(B_{(p)})^{\wedge k_V}.$$

The $\mathbb{F}_p$-vector space $\mathrm{H}^*(B_{(p)})$ is generated by similar elements as the $\mathbb{F}_p[\tau]$-vector space $V = \mathrm{H}^{**}(\mathbb{B}_{(p)})$ (compare Proposition 121 with [75, Lemma 6.2.6]). In particular,

$$\dim_{\mathbb{F}_p}(\mathrm{H}^*(B_{(p)}))^+ = \dim_{\mathrm{H}^{**}} V^+$$

and

$$\dim_{\mathbb{F}_p}(\mathrm{H}^*(B_{(p)}))^- = \dim_{\mathrm{H}^{**}} V^-.$$

It follows that $R(X)$ is the type-n spectrum defined in [75, Theorem C.3.2]. □

8.6 The Constant Type-n Spectrum

In this section, let $k = \mathbb{C}$ and let $X_n = e_V B^{\wedge k_V}$ be the type-n spectrum defined by Ravenel and let $c : \mathcal{SH} \to \mathcal{SH}(\mathbb{C})$ as in Sect. 4. X_n is constructed via an idempotent from the finite cell spectrum $B = (B\mathbb{Z}/p)_2^{2p^n}$ [75, Lemma 6.2.6]. We calculate $\mathrm{H}^{**}(cB)$.

Proposition 125 *Let $X = \Sigma^\infty Y$ be the suspension spectrum of a finite CW complex. Then $\mathrm{H}^{**}(cX) \cong \mathrm{H}^*(X)[\tau]$ as $\mathbb{F}_p$-modules, where a generator in degree i from the right hand side maps to bidegree $(i, 0)$ on the left hand side.*

Proof For any $F \in \mathcal{SH}(\mathbb{C})$, we have $\mathrm{H}^{**}(F) \cong \mathrm{H}^{*+n,*}(F \wedge S_s^n)$. For $F = S^0$, we get $\mathrm{H}^{**}(S_s^n) \cong \mathrm{H}^{*-n,*}(S^0) \cong \mathrm{H}^{*-n,*} \cong \mathrm{H}^*(S^n)[\tau]$. Now let $(Y^k)_k$ be a CW decomposition of Y, that is, $\Sigma^\infty Y^k$ is the cofiber of some $\Sigma^\infty \alpha_k : S^{n_k} \to \Sigma^\infty Y^{k-1}$. We write X^k for $\Sigma^\infty Y^k$. Since c preserves cofiber sequences, we get a cofiber sequence of suspension spectra $S_s^{n_k} \overset{c\alpha}{\to} cX^{k-1} \to cX^k$. It induces a long exact sequence

$$\cdots \to \mathrm{H}^{*-1,*}(cX^{k-1}) \overset{(c\alpha)^*}{\to} \mathrm{H}^{*-1,*}(S_s^{n_k}) \to \mathrm{H}^{**}(cX^k) \to \mathrm{H}^{**}(cX^{k-1})$$

$$\overset{(c\alpha)^*}{\to} \mathrm{H}^{**}(S_s^{n_k}) \to \cdots$$

We assume inductively that $\mathrm{H}^{i,j}(cX^{k-1}) \cong \mathrm{H}^i(X^{k-1})\{\tau^j\}$.

Let $x\tau^j \in \mathrm{H}^{i,j}(cX^{k-1})$. Since $R(\tau) = 1$ and $R(c\alpha) = \alpha$, we have

$$R((c\alpha)^*(x\tau^j)) = \alpha^*(R(x\tau^j)) = \alpha^*(x).$$

The only element in $\mathrm{H}^{i,j}(S_s^{n_k})$ which is mapped to $\alpha^*(x)$ by R is $\alpha^*(x)\tau^j$. This proves $(c\alpha)^* \cong \alpha^*[\tau]$. By the five lemma, it follows that the map $\mathrm{H}^{**}(cX^k) \to \mathrm{H}^*(X^k)[\tau]$, given by sending $x \in \mathrm{H}^{i,j}(cX^k)$ to $R(x)\tau^j$, is an isomorphism and, inductively, $\mathrm{H}^{**}(cX) \cong \mathrm{H}^*(X)[\tau]$. □

Corollary 126 *As $\mathbb{F}_p$-modules,*

$$\mathrm{H}^{**}(cB) \cong \mathrm{H}^*(B)[\tau] \cong \mathrm{H}^{**}\{x^k \mid 1 \le k \le p^n\} \cup \{yx^k \mid 1 \le k \le p^n - 1\},$$

with $\deg(x) = (2, 0)$ *and* $\deg(y) = (1, 0)$.

Proof The cohomology of B is described in the proof of [75, Lemma 6.2.6]. We only have to add the polynomial generator $\tau \in \mathrm{H}^{**}$. □

Recall that on the one hand, $\mathrm{thickid}(\mathbb{X}_n) \subseteq \mathcal{C}_{AK(n-1)}$. On the other hand, thickid $(cX_n) \subseteq \mathcal{C}_{cK(n-1)}$, as the following proposition shows.

Proposition 127 *$cK(s) \wedge c(X_n) \cong 0$ if and only if $s < n$. Hence,*

$$\mathrm{thickid}(cX_n) \subseteq \mathcal{C}_{cK(n-1)} \subseteq R^{-1}(\mathcal{C}_n)$$

in $(\mathcal{SH}(\mathbb{C})_f)_{(p)}$. Furthermore, $\mathrm{thickid}(cX_n) \nsubseteq R^{-1}(\mathcal{C}_{n+1})$.

Proof We have $c(K(s) \wedge X_n) \cong 0$ if and only if X_n is in $\mathcal{C}_{s+1}$, since c is fully faithful by [49, Theorem 1]. Since $X_n \in \mathcal{C}_n \setminus \mathcal{C}_{n+1}$, the first claim follows. The second claim holds because $R(cX_n) = X_n$. □

One may also ask whether $\mathrm{thickid}(cX_n) \subseteq \mathcal{C}_{AK(n-1)}$ or $\mathrm{thickid}(\mathbb{X}_n) \subseteq \mathcal{C}_{cK(n-1)}$. In [84, Sect. 3.6], Stahn constructs a counterexample to the first inclusion for $n = 2$ and shows

$$\mathrm{thickid}(cX_n) \not\subseteq \mathcal{C}_{AK(n-1)}$$

and

$$\mathcal{C}_{AK(n)} \subsetneq R^{-1}(\mathcal{C}_{n+1}).$$

Summary
We have constructed two different lifts of a topological type-n spectrum to the motivic category $(\mathcal{SH}(\mathbb{C})_f)_{(p)}$, one of them is in $c(\mathcal{SH}_{(p)}^{fin})$ and the other one is not. These are candidates for generators of different thick sub-ideals of $R^{-1}(\mathcal{C}_n)$ inside $(\mathcal{SH}(\mathbb{C})_f)_{(p)}$. We proved that $\mathbb{X}_n$ has motivic type n. By [84], the thick ideals generated by the two lifts are different and can be distinguished using motivic Morava K-theories $AK(s)$.

9 Bousfield Classes

So far, we have seen that the thick ideals $R_k^{-1}\mathcal{C}_n$ form a descending chain and that $\mathcal{C}_{AK(n-1)}$ is a thick ideal contained in $R_k^{-1}\mathcal{C}_n$. However, we have not seen that the thick ideals $\mathcal{C}_{AK(n)}$ form a descending chain themselves. The aim of this section is to prove this, at least for $k = \mathbb{C}$ and finite cell spectra. That is, we prove that $AK(n)_{**}X = 0$ implies $AK(n-1)_{**}X = 0$ for $X \in \mathcal{SH}(\mathbb{C})^{fin}$, where $n \geq 1$ and $p > 2$ (Theorem 21). As is done in topology [73, Theorem 2.11], we will work in terms of Bousfield classes (Definition 128).

We proceed as follows. In the first two sections, k can be any subfield of $\mathbb{C}$. In Sect. 9.1, we show that v_n-torsion is also v_{n-1}-torsion (Theorem 17). Then we show that some basic results on Bousfield classes also apply to the motivic setting (Lemma 17). In Sect. 9.3, we construct a product on $AP(n)$. Here, we need to know that $AP(n)_{**}$ vanishes in certain degrees, which we do if we assume $k = \mathbb{C}$, $p > 2$ and $n > 0$. We continue by showing that, for $p > 2$ and $n > 0$, $\langle AK(n)\rangle = \langle AB(n)\rangle$ in $\mathcal{SH}(\mathbb{C})$ (Corollary 147), passing from $AK(n)$ to the cohomology theory $AB(n) = v_n^{-1}AP(n)$, which is slightly easier to understand. On the way, we need to compute a couple of things like $AP(m)^{**}AP(n)$, to construct stable operations $AP(n)^{**}(-) \to AP(n)^{**}(-)$ (Theorem 18). Here, the assumption $k = \mathbb{C}$ is also helpful, as we make explicit use of the formula $AP(n)_{**} \cong P(n)_*[\tau]$ (Lemma 5). An application of all these results is Theorem 20, where, for $p > 2$ and $k = \mathbb{C}$, we prove

$$\langle AE(n)\rangle = \bigvee_{0 \leq i \leq n} \langle AK(i)\rangle.$$

For the definitions of $AK(n)$, $AB(n)$, $AP(n)$ and $AE(n)$, see Definition 74.

Let's start with a definition of Bousfield classes.

Definition 128 Let k be a field and let $\mathcal{T} = \mathcal{SH}(k), \mathcal{SH}(k)^{cell}$ or $\mathcal{SH}(k)^{fin}$. For any $E \in \mathcal{SH}(k)$, the class of all spectra $X \in \mathcal{T}$ satisfying $E_{**}X \neq 0$ is denoted by $\langle E \rangle$. We write $\langle E \rangle \leq \langle F \rangle$ if $\langle E \rangle$ is a subclass of $\langle F \rangle$. Meet and join of Bousfield classes are given by $\langle E \rangle \wedge \langle F \rangle = \langle E \wedge F \rangle$ and $\langle E \rangle \vee \langle F \rangle = \langle E \vee F \rangle$ [73, Definition 1.20].

Remark 129 In [73, Definition 1.19], $\langle E \rangle$ is defined to be the equivalence class consisting of all F such that, for all X, $E_*X = 0$ if and only if $F_*X = 0$. Thus, $\langle E \rangle$ is determined by the collection of all X such that $E_*X \neq 0$, as in the definition above. For $\mathcal{T} = \mathcal{SH}(k)^{cell}$ or $\mathcal{SH}(k)^{fin}$ and $E \in \mathcal{SH}(k)^{cell}$, $\langle E \rangle = \{X \in \mathcal{T} \mid E \wedge X \not\simeq 0\} = \mathcal{T} \setminus \mathcal{C}_E$ by Proposition 73.

9.1 v_n-Torsion

In this section, we work in the category $\mathcal{SH}(k)$, $k \subseteq \mathbb{C}$. We prove the following theorem, refining [40, Theorem 0.1].

Theorem 17 *Any v_n-torsion element in an $ABP_{**}ABP$-comodule is also a v_{n-1}-torsion element.*

Recall that $BP_*BP \cong BP_*\{t^E_{\text{top}}\}$ for some generators t^E_{top}, where $E = (e_1, e_2, \cdots)$ runs over all finite sequences of non-negative integers and $\deg(t^E_{\text{top}}) = \sum e_i(2p^i - 2)$, as is shown in [2, Theorem II.16.1(ii)].

From [87, Definition 5.3], we know that ABP is a homotopy commutative ring spectrum. It follows that $ABP_{**}ABP$ is an ABP_{**}-module. We describe its structure:

Lemma 13 *(1) As a left ABP_{**}-module,*

$$ABP_{**}ABP \cong ABP_{**}\{t^E\},$$

where E runs over all finite sequences of non-negative integers,

$$\deg\left(t^{(e_1,e_2,\cdots)}\right) = \left(\sum_i e_i(2p^i - 2), \sum_i e_i(p^i - 1)\right),$$

and $R_k(t^E) = t^E_{\text{top}}$ Consequently, as a right ABP_{**}-module,

$$ABP_{**}ABP \cong ABP_{**}\{c(t^E)\},$$

where $c : ABP_{**}ABP \to ABP_{**}ABP$ is the conjugation, induced by the twist map $ABP \wedge ABP \to ABP \wedge ABP$.

(2) As a left ABP^{**}-module,

$$ABP^{**}ABP \cong ABP^{**}[[s^E]],$$

which is the completion of $ABP^{**}\{s^E\}$ under infinite sums.
Here, $\deg(s^E) = \deg(t^E)$ *and the s^E are the dual basis elements to t^E.*

In particular, $ABP_{**}ABP$ is a flat ABP_{**}-module.

Proof By Remark 75(4), [65, Proposition 9.1.(i)] applies to ABP, so we have $ABP_{**}ABP \cong ABP_{**} \otimes_{BP_*} BP_*BP$. Since $BP_*BP \cong BP_*\{t^E_{\text{top}}\}$, the first claim follows. As BP_*BP is projective over BP_*, [65, Proposition 9.7(i)] implies $ABP^{**}ABP \cong \text{Hom}_{BP_*}(BP_*BP, ABP_{**})$. Since the analogue holds for BP^*BP, this is the same as $ABP^{**} \otimes_{BP^*} BP^*BP$, which is $ABP^{**}[[s^E]]$ by [40, Lemma 5.12]. □

For any finite motivic cell spectrum X, the morphism

$$m_* : ABP_{**}(ABP) \otimes_{ABP_{**}} ABP_{**}(X) \to ABP_{**}(ABP \wedge X)$$

induced by the ABP-module structure of $ABP \wedge X$ is an isomorphism: this holds for $X = S^0$, since $ABP_{**}ABP$ is free over ABP_{**} and, hence, for any finite X by cellular induction via the five lemma, see also [1, Lecture 3, Lemma 1]. More precisely, one has to check that a cofiber sequence $X \to Y \to Z$ induces a long exact sequence on both ends of m_*. On the right hand side, this is the long exact ABP_{**}-sequence induced by $ABP \wedge X \to ABP \wedge Y \to ABP \wedge Z$ and on the left hand side, we get a long exact ABP_{**}-sequence tensored over the field $\mathbb{F}_p$ with $\mathbb{F}_p\{t^E\}$, which is still exact. Actually, the above map is an isomorphism for any motivic spectrum X by [65, Lemma 5.1(i)].

This can be used to define elementary ABP-operations as in [40, Sect. 1]:

Definition 130 Let

$$\psi_X : ABP_{**}X \to ABP_{**}ABP \otimes_{ABP_{**}} ABP_{**}X$$

be the map induced by $1 \wedge i \wedge 1 : ABP \wedge S^0 \wedge X \to ABP \wedge ABP \wedge X$ (where i is the unit of the ring spectrum ABP) followed by $(m_*)^{-1}$. Then the elementary ABP-operation $s_E : ABP_{**}X \to ABP_{**}X$ is defined by

$$\psi_X(x) = \sum_E c(t^E) \otimes s_E(x).$$

The $s^E \in ABP^{**}ABP$ from the above lemma are special cases of these operations (see [40, Lemma 5.12]). The ABP-operations satisfy a Cartan formula similar to [40, Formula (1.7)]:

Lemma 14 *If $y \in ABP_{**}$ and $x \in ABP_{**}X$, then*

$$s_E(yx) = \sum_{F+G=E} s_F(y)s_G(x).$$

Proof We have to show that

$$\psi_X(yx) = \sum_E \left(c(t^E) \otimes \sum_{F+G=E} s_F(y)s_G(x) \right).$$

Since ψ_X is a map of ABP_{**}-modules, $\psi_X(yx) = \psi_{S^0}(y)\psi_X(x)$, which is equal to $\sum_E \sum_{F+G=E}(c(t^F)c(t^G) \otimes s_F(y)s_G(x))$. As F and G are exponent sequences, $t^F t^G = t^E$ and it follows that $\psi_X(yx) = \sum_E(c(t^E) \otimes \sum_{F+G=E} s_F(y)s_G(x))$. □

The next lemma compares the Hopf algebroid structures of (BP_*, BP_*BP) and $(ABP_{**}, ABP_{**}ABP)$ and is closely related to [65, Sect. 5].

Lemma 15 *$(ABP_{**}, ABP_{**}ABP)$ is a flat Hopf algebroid, and there is a map of Hopf algebroids $(BP_*, BP_*BP) \to (ABP_{**}, ABP_{**}ABP)$ such that the following hold:*

(1) $BP_* \to ABP_{**}$ is the inclusion into $\bigoplus_i ABP_{(2i,i)}$, mapping v_i^{top} to v_i.

(2) $BP_*BP \to ABP_{**}ABP$ is the map $BP_*\{t_{\text{top}}^E\} \to ABP_{**}\{t^E\}$ given by (1) on BP_* and mapping t_{top}^E to t^E.

(3) The map $\psi = \psi_{S^0}$ from Definition 130 is the coaction map of ABP_{**} as a left $(ABP_{**}, ABP_{**}ABP)$-comodule and, similarly, for the map ψ^{top} from [40]. Furthermore, the map of Hopf algebroids preserves the comodule structure in the sense that

$$\begin{array}{ccc} BP_* & \xrightarrow{\psi^{\text{top}}} & BP_*BP \otimes_{BP_*} BP_* \\ \downarrow & & \downarrow \\ ABP_{**} & \xrightarrow{\psi} & ABP_{**}ABP \otimes_{ABP_{**}} ABP_{**} \end{array}$$

commutes.

Proof In [65, Corollary 5.2(i)], it is shown that $(E_{**}, E_{**}E)$ is a flat Hopf algebroid whenever E is a cellular ring spectrum and $E_{**}E$ is a flat E_{**}-module. This is the case for $E = ABP$ by Remark 75 and Lemma 13(1). Furthermore, an orientation on E induces a map of Hopf algebroids $(MU_*, MU_*MU) \to (E_{**}, E_{**}E)$ by [65, Corollary 6.7], where $MU_* \to E_{**}$ is the map classifying the formal group law (FGL) given by the orientation on E and $MU_*MU \to E_{**}E$ classifies the strict isomorphism of formal group laws induced on $E \wedge E$ by the left and right units $E \to E \wedge E$.

If the FGL associated with the orientation of E is p-typical, the map of Hopf algebroids factors through (BP_*, BP_*BP) because BP_* and BP_*BP classify p-typical group laws and strict isomorphisms of p-typical group laws, respectively (see [74, Appendix 2]). Recall that ABP is oriented by $MU_* \to MGL_{**} \to ABP_{**}$ and its FGL is p-typical because this factors as $MU_* \to BP_* \to ABP_{**}$ by the construction of ABP, where the latter map is as claimed in (1). Hence, we get a map of Hopf algebroids $(BP_*, BP_*BP) \to (ABP_{**}, ABP_{**}ABP)$ satisfying (1).

Before we prove (2), we will show the analogous statement for MGL. By [65, Corollary 6.7], as above, there is a map of Hopf algebroids

$$(MU_*, MU_*MU) \to (MGL_{**}, MGL_{**}MGL),$$

determined by the complex orientation on MGL. Let x be the orientation on $MGL \wedge MGL$ induced by the left unit $MGL \to MGL \wedge MGL$ and x' be the orientation induced by the right unit. By [65, Lemma 6.4.(ii)], $x' = \sum_{i \geq 0} b_i x^{i+1}$. The orientations x and x' correspond to formal group laws F_L and F_R and the formula implies that the b_i are the coefficients of the power series of the strict isomorphism φ between F_L and F_R (as in the proof of [74, Theorem 4.1.11]). The same formula holds for the orientations on $MU \wedge MU$ by [74, Lemma 4.1.8] and the strict isomorphism between the FGLs F_L^{top} and F_R^{top} therefore has coefficients b_i^{top}. By definition, b_i is the image of b_i^{top} under $MU_*[b_i^{\mathrm{top}}] \cong MU_*MU \to MGL_{**} \otimes_{MU_*} MU_*MU \cong MGL_{**}MGL$, as in [65, Lemma 6.4.(i)].

MU_*MU classifies strict isomorphisms $F \xrightarrow{f} G$ of FGLs in the following way: $MU_*MU \cong MU_*[b_i^{\mathrm{top}}]$, where MU_* classifies F, f is given by a power series in b_i^{top} and G is determined by F and f.

In our setting, the map $MU_*MU \to MGL_{**}MGL$ is the map corresponding to the strict isomorphism $F_L \xrightarrow{\varphi} F_R$. Furthermore, F_L is the FGL associated with the orientation of $MGL_{**}MGL$ given by $MU_* \to MGL_{**} \to MGL_{**}[b_i] \cong MGL_{**}MGL$ (because the isomorphism herein is an isomorphism of left MGL_{**}-modules), and, similarly, F_L^{top} is the FGL associated with the orientation $MU_* \to MU_*[b_i^{\mathrm{top}}] \cong MU_*MU$. This implies that the following square commutes, where the left horizontal maps are the obvious inclusions, and the vertical maps are the maps from [65, Corollary 6.7].

$$\begin{array}{ccccc} MU_* & \longrightarrow & MU_*[b_i^{\mathrm{top}}] & \xrightarrow{\cong} & MU_*MU \\ \downarrow & & & & \downarrow \\ MGL_{**} & \longrightarrow & MGL_{**}[b_i] & \xrightarrow{\cong} & MGL_{**}MGL. \end{array}$$

In terms of group laws, the right hand map sends φ^{top} to φ. Since these are the power series described above, b_i^{top} is sent to b_i. This proves the MGL-version of (2).

For the ABP-version, one has to show that the following diagram commutes, where the right map is the Hopf algebroid morphism and the left map is as described in (2).

$$\begin{array}{ccc} BP_*[t_i^{\mathrm{top}}] & \xrightarrow{\cong} & BP_*BP \\ \downarrow & & \downarrow \\ ABP_{**}[t_i] & \xrightarrow{\cong} & ABP_{**}ABP. \end{array}$$

The proof is exactly the same as in the case of MGL. One simply has to replace each FGL by the corresponding p-typical FGL.

For (3), note that, by its definition, ψ is the coaction map that comes naturally with any flat Hopf algebroid $(E_{**}, E_{**}E)$ (as in [65, Corollary 5.2(i)]), meaning in particular that the diagram in (3) commutes. □

Definition 131 For an exponent sequence $E = (e_1, e_2, \cdots)$ as in Lemma 13(1), we set $|E| = \sum_i e_i(2p^i - 2)$. Let $I_m = (p, v_1, \cdots, v_{m-1}) \subset ABP_{**}$ be the usual prime ideal.

Corollary 132 *Consider $s_E : ABP_{**} \to ABP_{**}$ as in Definition 130 and assume that $|E| \geq 2kp^s(p^n - p^m)$ for $n \geq m$, $s \geq 0$ and $k \geq 1$. Then*

$$s_E(v_n^{kp^s}) = \begin{cases} v_m^{kp^s} & \text{mod } I_m^{s+1} \text{ if } e_{n-m} = kp^{s+m} \text{ and } e_i = 0 \text{ for } i \neq n-m \\ 0 & \text{mod } I_m^{s+1} \text{ otherwise.} \end{cases}$$

Proof By the above lemma, the following diagram commutes:

$$\begin{array}{ccc} BP_* & \xrightarrow{\psi^{\mathrm{top}}} & BP_*BP \otimes_{BP_*} BP_* \\ \downarrow & & \downarrow \\ ABP_{**} & \xrightarrow{\psi} & ABP_{**}ABP \otimes_{ABP_{**}} ABP_{**}, \end{array}$$

where the vertical arrows send v_n^{top} to v_n and t_{top}^E to t^E. We consider the element $(v_n^{\mathrm{top}})^{kp^s} \in BP_*$. It is mapped horizontally to

$$\psi^{\mathrm{top}}((v_n^{\mathrm{top}})^{kp^s}) = \sum_E c^{\mathrm{top}}(t_{\mathrm{top}}^E) \otimes s_E^{\mathrm{top}}((v_n^{\mathrm{top}})^{kp^s}).$$

By [40, Lemma 2.1], $s_E^{\mathrm{top}}((v_n^{\mathrm{top}})^{kp^s})$ satisfies the formula we want to prove. Since all the elements from the topological case map to the corresponding elements in the lower row, the formula has to hold there, too. □

The rest of the proof of Theorem 17 is exactly the same as [40, Lemmas 2.2 and 2.3], relying mainly on the above lemma and the Cartan formula. Theorem 17 implies the following corollary. The analogous topological statement can be found in the proof of [73, Theorem 2.1(d)].

Corollary 133 *Let $k \subseteq \mathbb{C}$. If $AE(n)_{**}X = 0$, then also $AE(i)_{**}X = 0$ for all $i \leq n$. In terms of Bousfield classes in $\mathcal{SH}(k)$:*

$$\langle AE(n)\rangle \geq \langle AE(i)\rangle \text{ for all } n \geq i.$$

Proof Since $E(n)$ is Landweber exact (see [46] or [74, Sect. 4.2]), the ABP-version of [65, Theorem 8.7] applies to $AE(n)_{**}(X)$, yielding

$$AE(n)_{**}(X) \cong ABP_{**}(X) \otimes_{BP_*} E(n)_*,$$

which is an $ABP_{**}ABP$-comodule via the map ψ_X from Definition 130.

As $E(n) = (v_n^{\text{top}})^{-1} BP/(v_{n+1}^{\text{top}}, v_{n+2}^{\text{top}}, \cdots)$, the condition $AE(n)_{**}(X) = 0$ is equivalent to $ABP_{**}(X) \otimes_{BP_*} BP/(v_{n+1}^{\text{top}}, v_{n+2}^{\text{top}}, \cdots)$ being v_n-torsion. By Theorem 17, it follows that $ABP_{**}(X) \otimes_{BP_*} BP/(v_{n+1}^{\text{top}}, v_{n+2}^{\text{top}}, \cdots)$ is v_i-torsion for any $i \leq n$. This implies that also $ABP_{**}(X) \otimes_{BP_*} BP/(v_{i+1}^{\text{top}}, v_{i+2}^{\text{top}}, \cdots)$ is v_i-torsion, which is equivalent to $AE(i)_{**}(X) = 0$. □

9.2 Properties of Bousfield Classes

Ravenel has shown the following properties of Bousfield classes [73, Sect. 1]. They hold in any tensor triangulated category $(\mathcal{T}, \wedge)$.

Lemma 16 *(1) In an exact triangle, each Bousfield class is less or equal to the wedge of the other two [73, Proposition 1.23].*

(2) If M is a module spectrum over the ring spectrum E, then $\langle M\rangle \leq \langle E\rangle$ [73, Proposition 1.24].

(3) Let Σ be an auto-equivalence in $\mathcal{T}$. If Y is the homotopy cofiber of $\Sigma^d X \xrightarrow{f} X$ and $\hat{X} = \operatorname{colim}_f(\Sigma^{-kd}X)$, then $\langle X\rangle = \langle \hat{X}\rangle \vee \langle Y\rangle$ [73, Lemma 1.34].

Furthermore, the following relations from [73, Sect. 2] also hold in $\mathcal{SH}(k)$:

Lemma 17 *(1) $\langle AE(n)\rangle \geq \langle AK(n)\rangle$,*

(2) $\langle AE(n)\rangle \wedge \langle AP(n+1)\rangle = \langle v_n^{-1}ABP\rangle \wedge \langle AP(n+1)\rangle = \langle 0\rangle$,

(3) $\langle AP(n)\rangle = \langle AB(n)\rangle \vee \langle AP(n+1)\rangle$.

Proof Constructing $AK(n)$ from $AE(n)$, (1) follows from Lemma 16(1). (3) is a direct application of Lemma 16(3) (see also [73, Theorem 2.1(c)]). For the first part of (2), note that $\langle AE(n)\rangle \leq \langle v_n^{-1}ABP\rangle$ since $AE(n) = v_n^{-1}ABP/(v_{n+1}, v_{n+2}, \cdots)$. When we prove the second equation in (2), the first one will follow from this inequality because $\langle 0\rangle$ is the empty set.

It remains to show that $AP(n+1) \wedge v_n^{-1}ABP \cong 0$, as proven in the topological setting in [73, Lemma 2.3]. This spectrum is the homotopy cofiber of the map

$$v_n \wedge 1 : AP(n) \wedge v_n^{-1}ABP \to AP(n) \wedge v_n^{-1}ABP.$$

We claim that $(v_n \wedge 1)_* = (1 \wedge v_n)_*$ on $\pi_{**}(AP(n) \wedge ABP)$. Since $(1 \wedge v_n)_*$ is an isomorphism on $\pi_{**}(AP(n) \wedge v_n^{-1}ABP)$, this will imply that the homotopy cofiber is contractible.

To prove this claim, note that $(v_n \wedge 1)_*$ and $(1 \wedge v_n)_*$ are induced by the respective maps on $\pi_{**}(ABP \wedge ABP)$, where they are given by applying the left respectively right unit $ABP_{**} \to ABP_{**}ABP$ to v_n. In the topological case, the left and right units applied to v_n are the same modulo I_n by [2, II.16.1 (ii)]. By Lemma 13(1) and the inclusion of BP_* in ABP_{**}, this also holds motivically. Hence, $(v_n \wedge 1)_*$ and $(1 \wedge v_n)_*$ are the same modulo I_n. It remains to show that $I_n \subseteq ABP_{**}ABP$ maps to 0 under $ABP_{**}ABP \to AP(n)_{**}ABP$. For $n = 0$, there is nothing to show. Assume that I_n is mapped to zero in $AP(n)_{**}ABP$ for some n. Consider the map $AP(n)_{**}ABP \xrightarrow{i_*} AP(n+1)_{**}ABP$ induced by the map to the cofiber in $AP(n) \xrightarrow{v_n} AP(n) \xrightarrow{i} AP(n+1)$. The inductive assumption implies that I_n is still zero in $AP(n+1)_{**}ABP$. Recall that $I_{n+1} \subseteq ABP_{**}ABP$ is the ideal generated by I_n and v_n. Since $i_* \circ (v_n)_* = 0$ in the long exact sequence

$$\cdots \to AP(n)_{**}ABP \xrightarrow{(v_n)_*} AP(n)_{**}ABP \xrightarrow{i_*} AP(n+1)_{**}ABP \to \cdots,$$

i_* maps $v_n = (v_n)_*(1)$ to 0. Hence, $I_{n+1} = 0$ in $AP(n+1)_{**}ABP$. □

9.3 The Action of v_i on $AP(n)$

In [39, Appendix], a geometric proof using the Baas–Sullivan construction of $P(n)$ shows that the action of v_i^{top} on $P(n)_*(X)$ is zero for any $0 \leq i < n$. A non-geometric proof of this result is given by [64, Satz 1.3.4], which was motivated by [95]. Nassau's proof can be simplified using the language of triangulated categories of modules, which is basically done in [18, Lemma V.2.4], as well as in [85, Lemma 3.2]. These proofs rely on the fact that the v_i are non-zero divisors of MU_* and that BP_* vanishes in certain degrees, which is not known in the motivic case. In the following, we will use ideas from [64] and [18] to give a proof which also works in $\mathcal{SH}(\mathbb{C})$. The main difference is that we only know coefficients after passing to $MGL_{(p)}/(p)$ (see Lemma 5), which is why we have to work with $R/(x, y)$, while [18, Chap. V] only works with R/x for some ring spectrum R.

In this section, we will prove that v_i acts trivially on $AP(n)$ if $k = \mathbb{C}$. Furthermore, we will show that if p is odd, then $AP(n)_{**}(X)$ and $AP(n)^{**}(X)$ are $AP(n)_{**}$-modules for any $X \in \mathcal{SH}(\mathbb{C})$.

Recall that MGL can be constructed as an E_∞-ring spectrum [34, Theorem 14.2], which is equivalent to a strictly commutative ring spectrum by the motivic version of [18, Corollary II.3.6].

Let $R \in \mathcal{SH}(k)$ be a strictly commutative ring spectrum with multiplication $m : R \wedge R \to R$ and unit $i : S^{0,0} \to R$. Let $x : S^{k,l} \to R$ for some $k, l \in \mathbb{Z}$. In our application, we will have $R = MGL_{(p)}$. Note that $MGL_{(p)}$ is the homotopy colimit of the diagram of maps $MGL \xrightarrow{n} MGL$ for all positive integers relatively prime to p (see [34, end of Sect. 14]). As these are maps of strictly commutative ring spectra, $MGL_{(p)}$ is also a strictly commutative ring spectrum ([80, Theorem 4.1(3)] applied to $S^{0,0}$-algebras implies that the category of strictly commutative ring spectra is cocomplete).

Let M be an R-module with action map $\nu_M : R \wedge M \to M$. Let

$$\phi = \nu_M \circ (x \wedge 1_M) : S^{k,l} \wedge M \to R \wedge M \to M.$$

The map ϕ is the action of x on M.

The R-module structure on $S^{k,l} \wedge M$ is given by

$$\nu_{S^{k,l}\wedge M} : R \wedge S^{k,l} \wedge M \xrightarrow{\tau \wedge 1_M} S^{k,l} \wedge R \wedge M \xrightarrow{1_{S^{k,l}} \wedge \nu_M} S^{k,l} \wedge M.$$

Lemma 18 *The map ϕ is an R-module map.*

Proof We have to check the commutativity of the following diagram:

$$\begin{array}{ccccccc} R \wedge S^{k,l} \wedge M & & \xrightarrow{1\wedge x \wedge 1} & & R \wedge R \wedge M & \xrightarrow{1\wedge \nu_M} & R \wedge M \\ \downarrow{\scriptstyle \tau\wedge 1} & & & & & & \downarrow{\scriptstyle \nu_M} \\ S^{k,l} \wedge R \wedge M & \xrightarrow{1\wedge \nu_M} & S^{k,l} \wedge M & \xrightarrow{x \wedge 1} & R \wedge M & \xrightarrow{\nu_M} & M. \end{array}$$

In this diagram, we can replace $(x \wedge 1) \circ (1 \wedge \nu_M)$ by

$$S^{k,l} \wedge R \wedge M \xrightarrow{x\wedge 1 \wedge 1} R \wedge R \wedge M \xrightarrow{1 \wedge \nu_M} R \wedge M$$

and we can fill in a diagonal across the upper left corner,

$$S^{k,l} \wedge R \wedge M \xrightarrow{(\tau\wedge 1)\circ(x \wedge 1 \wedge 1)} R \wedge R \wedge M.$$

It follows that the above diagram commutes if and only if the following diagram commutes:

$$\begin{array}{ccccccc} S^{k,l} \wedge R \wedge M & \xrightarrow{x\wedge 1\wedge 1} & R \wedge R \wedge M & \xrightarrow{\tau\wedge 1} & R \wedge R \wedge M & \xrightarrow{1\wedge \nu_M} & R \wedge M \\ \downarrow{\scriptstyle x\wedge 1\wedge 1} & & & & & & \downarrow{\scriptstyle \nu_M} \\ R \wedge R \wedge M & & \xrightarrow{\qquad 1\wedge\nu_M \qquad} & & R \wedge M & \xrightarrow{\nu_M} & M. \end{array}$$

Since R is commutative, we have $m \circ \tau = m$ and, hence,

$$\nu_M \circ (m \wedge 1) \circ (\tau \wedge 1) = \nu_M \circ (m \wedge 1).$$

Since M is an R-module, this is the same as

$$\nu_M \circ (1 \wedge \nu_M) \circ (\tau \wedge 1) = \nu_M \circ (1 \wedge \nu_M),$$

proving the commutativity of the above diagram. □

In the following, we denote the homotopy category of R-modules by R-Mod. A stable model structure on R-modules is given by [80, Theorem 4.1] applied to the motivic stable model structure from [38], so that R-Mod is a triangulated category (compare [65, p. 554]). Since ϕ from above is a map of R-modules (Lemma 18), there is an exact triangle in R-Mod,

$$S^{k,l} \wedge M \xrightarrow{\phi} M \xrightarrow{\eta} N \xrightarrow{\partial} S^{k+1,l} \wedge M.$$

The cofiber N is also denoted M/x. Application of $[-, N]_{R\text{-Mod}}$ to this exact triangle yields a long exact sequence

$$\cdots \to [S^{2k+1,2l} \wedge M, N]_{R\text{-Mod}} \xrightarrow{\partial^*} [S^{k,l} \wedge N, N]_{R\text{-Mod}} \xrightarrow{\eta^*} [S^{k,l} \wedge M, N]_{R\text{-Mod}} \to$$

Let $\psi = \nu_N \circ (x \wedge 1_N) : S^{k,l} \wedge N \to R \wedge N \to N$. This map is the action of x on M/x, and a map of R-modules by Lemma 18. We want to show that, under certain assumptions, $\psi = 0$, meaning that x acts trivially on M/x.

First, we consider

$$\eta^*\psi = \psi \circ (1_{S^{k,l}} \wedge \eta) : S^{k,l} \wedge M \xrightarrow{1\wedge\eta} S^{k,l} \wedge N \xrightarrow{\psi} N.$$

By the definition of ψ, this is the map $\nu_N \circ (x \wedge 1_N) \circ (1_{S^{k,l}} \wedge \eta) = \nu_N \circ (1_R \wedge \eta) \circ (x \wedge 1_M)$. Since η is a map of R-modules, this is the same as $\eta \circ \nu_M \circ (x \wedge 1_M)$, which, by definition of ϕ, is the map $\eta \circ \phi$. By the above exact triangle, it follows that $\eta^*\psi = \eta \circ \phi = 0$. The long exact sequence implies that there is a map in R-Mod,

$$\overline{\psi} : S^{2k+1,2l} \wedge M \to N,$$

such that $\psi = \partial^*\overline{\psi}$.

Now, we assume that either $M = R$ or that $M = R/y$ for some $y \in R_{**}$. Furthermore, we assume that $\pi_{2k+1,2l}N = 0$.

Note that Case 1 is a special instance of Case 2 (with $y = 0$), so the reader may skip the following paragraph and continue reading at Case 2. However, Case 1 is easier, for which reason it might still be a good idea to read it, anyway.

Case 1. $M = R$.

We have $\overline{\psi} : S^{2k+1,2l} \wedge R \to N$. Since R is a ring spectrum, the unit $i : S^{0,0} \to R$ satisfies $1_R = m \circ (1_R \wedge i) : R \wedge S^{0,0} \to R \wedge R \to R$. Hence,

$$\overline{\psi} = \overline{\psi} \circ (1_{S^{2k+1,2l}} \wedge m) \circ (1_{S^{2k+1,2l} \wedge R} \wedge i).$$

By the definition of the R-module structure $\nu_{S^{2k+1,2l} \wedge R}$ on $S^{2k+1,2l} \wedge R$,

$$(1_{S^{2k+1,2l}} \wedge m) = \nu_{S^{2k+1,2l} \wedge R} \circ (\tau \wedge 1_R) :$$

$$S^{2k+1,2l} \wedge R \wedge R \to R \wedge S^{2k+1,2l} \wedge R \to S^{2k+1,2l} \wedge R.$$

Hence,

$$\overline{\psi} = \overline{\psi} \circ \nu_{S^{2k+1,2l} \wedge R}(\tau \wedge 1_R)(1_{S^{2k+1,2l} \wedge R} \wedge i)$$

$$= \overline{\psi} \circ \nu_{S^{2k+1,2l} \wedge R}(1_R \wedge 1_{S^{2k+1,2l}} \wedge i)(\tau \wedge 1_R).$$

Since $\overline{\psi}$ is an R-module map, $\overline{\psi}\nu_{S^{2k+1,2l} \wedge R} = \nu_N(1_R \wedge \overline{\psi})$, and therefore

$$\overline{\psi} = \nu_N(1_R \wedge \overline{\psi})(1_R \wedge 1_{S^{2k+1,2l}} \wedge i)(\tau \wedge 1_R)$$

$$= \nu_N(1_R \wedge (\overline{\psi}(1_{S^{2k+1,2l}} \wedge i)))(\tau \wedge 1_R).$$

Now, $\overline{\psi}(1_{S^{2k+1,2l}} \wedge i) : S^{2k+1,2l} \wedge S^{0,0} \to N$ is in $\pi_{2k+1,2l}N$, which we assumed to be zero. Thus, $\overline{\psi} = 0$ and it follows that also $\psi = \partial^*\overline{\psi}$, which is the action of x on M/x, is zero, as we wanted to show.

Thus, we have shown:

Proposition 134 *Let $R \in \mathcal{SH}(k)$ be a strictly commutative ring spectrum and $x \in \pi_{k,l}R$. Assume that $\pi_{2k+1,2l}(R/x) = 0$. Then x acts trivially on R/x, i.e., the map ψ from above is zero. The same holds if $x \in \pi_k R$ for a strictly commutative ring spectrum $R \in \mathcal{SH}$ such that $\pi_{2k+1}(R/x) = 0$.*

Note that part of the above argument can be formulated more generally:

Lemma 19 *Let R be a (homotopy) ring spectrum, M a left R-module, and $\pi_{k,l}M = 0$. Then any R-module map $\psi : S^{k,l} \wedge R \to M$ is homotopically trivial.*

Proof Let $i : S^{0,0} \to R$ be the unit of R. It satisfies $1_R = m(1_R \wedge i)$. Thus,

$$\psi = \psi \circ m(1_R \wedge i) = \nu_M(1_R \wedge \psi)(1_R \wedge i) = \nu_M(1_R \wedge \psi i),$$

with $\psi i \in \pi_{k,l}M = 0$. It follows $\psi = \nu_M(1_R \wedge 0) = 0$. □

Now we pass on to case 2.

Case 2. $M = R/y$.

Let $y : S^{k',l'} \to R$ and let $\phi' = m(y \wedge 1_R) : S^{k',l'} \wedge R \to R \wedge R \to R$ be the action of y on R. We have an exact triangle in R- Mod,

$$S^{k',l'} \wedge R \overset{\phi'}{\to} R \overset{\eta'}{\to} M \overset{\partial'}{\to} S^{k'+1,l'} \wedge R,$$

and, again, an exact sequence

$$\cdots \to [S^{2k+k'+2,2l+l'} \wedge R, N]_{R\text{- Mod}} \overset{\partial'^*}{\to} [S^{2k+1,2l} \wedge M, N]_{R\text{- Mod}}$$

$$\overset{\eta'^*}{\to} [S^{2k+1,2l} \wedge R, N]_{R\text{- Mod}} \to \cdots$$

We consider

$$\eta'^*\overline{\psi} = \overline{\psi} \circ (1_{S^{2k+1,2l}} \wedge \eta') : S^{2k+1,2l} \wedge R \to S^{2k+1,2l} \wedge M \to N.$$

Let $i : S^{0,0} \to R$ be the unit of R, as before. Since $\eta' : R \to M$ is a map of R-modules (using Lemma 18), $\eta' \circ m = \nu_M(1_R \wedge \eta')$, and, hence,

$$\nu_M(1_R \wedge \eta')(1_R \wedge i) = \eta' \circ m(1_R \wedge i) = \eta'.$$

Thus,

$$\overline{\psi}(1_{S^{2k+1,2l}} \wedge \eta') = \overline{\psi}(1_{S^{2k+1,2l}} \wedge \nu_M(1_R \wedge \eta')(1_R \wedge i))$$

$$= \overline{\psi}(1_{S^{2k+1,2l}} \wedge \nu_M(1_R \wedge \eta' i)).$$

Since $\overline{\psi}$ is a map of R-modules, $\overline{\psi} \circ \nu_{S^{2k+1,2l} \wedge M} = \nu_N \circ (1_R \wedge \overline{\psi})$, where, by definition, $\nu_{S^{2k+1,2l} \wedge M} = (1_{S^{2k+1,2l}} \wedge \nu_M)(\tau \wedge 1_M)$. Hence, $\overline{\psi}(1_{S^{2k+1,2l}} \wedge \nu_M) = \nu_N(1_R \wedge \overline{\psi})(\tau \wedge 1_M)$, and, therefore,

$$\eta'^*\overline{\psi} = \nu_N(1_R \wedge \overline{\psi})(\tau \wedge 1_M)(1_{S^{2k+1,2l}} \wedge 1_R \wedge \eta' i)$$

$$= \nu_N(1_R \wedge \overline{\psi})(1_R \wedge 1_{S^{2k+1,2l}} \wedge \eta' i)(\tau \wedge 1_{S^{0,0}})$$

$$= \nu_N(1_R \wedge (\overline{\psi}(1_{S^{2k+1,2l}} \wedge \eta' i)))(\tau \wedge 1_{S^{0,0}}).$$

Now, $\overline{\psi}(1_{S^{2k+1,2l}} \wedge \eta' i) : S^{2k+1,2l} \wedge S^{0,0} \to N$ lies in $\pi_{2k+1,2l}N$, which we assumed to be zero. Hence, $\eta'^*\overline{\psi} = 0$. By the long exact sequence from above, it follows that $\overline{\psi} = \partial'^*\overline{\overline{\psi}}$ for some R-module map $\overline{\overline{\psi}} : S^{2k+k'+2,2l+l'} \wedge R \to N$. Thus, $\psi = \partial^*\overline{\psi} = \partial^*\partial'^*\overline{\overline{\psi}}$.

Consider the following commutative diagram. The map ψ is the precomposition of $\overline{\overline{\psi}}$ with the diagonal of the righthand square.

$$
\begin{array}{ccccccc}
S^{2k,2l} \wedge M & \xrightarrow{\phi} & S^{k,l} \wedge M & \xrightarrow{\eta} & S^{k,l} \wedge N & \xrightarrow{\partial} & S^{2k+1,2l} \wedge M \\
\downarrow{\scriptstyle \partial'} & & \downarrow{\scriptstyle \zeta} & \swarrow{\scriptstyle \xi} & \downarrow{\scriptstyle \partial} & & \downarrow{\scriptstyle \partial'} \\
S^{2k+k'+1,2l+l'} \wedge R & \xrightarrow{\phi'} & S^{2k+1,2l} \wedge R & \xrightarrow{\eta'} & S^{2k+1,2l} \wedge M & \xrightarrow{\partial'} & S^{2k+k'+2,2l+l'} \wedge R.
\end{array}
$$

Since both rows are exact triangles, we can fill in a map $\zeta : S^{k,l} \wedge M \to S^{2k+1,2l} \wedge R$. We have $\phi' \circ \partial' = 0$, as both of these are maps in the lower triangle. Thus, the diagonal in the first square is zero and the map ζ lifts to a map $\xi : S^{k,l} \wedge N \to S^{2k+1,2l} \wedge R$. It follows that $\partial' \circ \partial = \partial' \circ \eta' \circ \xi = 0$, and, hence, $\psi = \overline{\overline{\psi}} \circ \partial' \circ \partial = 0$.

We have proven:

Proposition 135 *Let $R \in \mathcal{SH}(k)$ be a commutative ring spectrum, $x \in \pi_{k,l} R$ and $y \in \pi_{k',l'} R$. Assume that $\pi_{2k+1,2l}(R/(y,x)) = 0$. Then x acts trivially on $R/(y,x)$.*

The same holds if $x \in \pi_k R$ and $y \in \pi_{k'} R$ for a commutative ring spectrum $R \in \mathcal{SH}$ such that $\pi_{2k+1}(R/(y,x)) = 0$.

This result can be applied to the action of v_i on $AP(n)$ for $0 \leq i < n$, at least for $k = \mathbb{C}$.

Corollary 136 *Let $k = \mathbb{C}$ and $n \geq 1$. Then v_i acts trivially on $AP(n)$ for any $0 \leq i < n$ and v_i acts trivially on $Ak(n)$ for any $i \neq n$.*

Proof First, we consider $MGL_{(p)}/(p, v_i)$ for some $0 < i < n$ (thus, $n \geq 2$). Since MU_* is concentrated in even degrees and Lemma 5 holds for $k = \mathbb{C}$ and quotients of $MGL_{(p)}/p$, we get $\pi_{2k+1,2l}(MGL_{(p)}/(p, v_i)) = 0$ for any k, l. By Proposition 135 it follows that $v_0 = p$ and v_i act trivially on $MGL_{(p)}/(p, v_i)$.

By [18, Lemma V.1.10],

$$MGL_{(p)}/(p, v_i) \cong MGL_{(p)}/p \wedge_{MGL_{(p)}} MGL_{(p)}/v_i,$$

and, by [32, Remark 6.20], $AP(n) \cong MGL_{(p)}/J$, where J contains $a_i \in MGL_{**}$, $i \neq 2p^i - 2$, as well as v_i, $0 \leq i \leq n-1$. From [18, Lemma V.1.10], it follows that

$$AP(n) \cong MGL_{(p)}/(p, v_i) \wedge_{MGL_{(p)}} MGL_{(p)}/(J \setminus \{p, v_i\}).$$

Now v_i acts trivially on $MGL_{(p)}/(p, v_i)$, i.e., the respective map ϕ_i on $MGL_{(p)}/(p, v_i)$ is zero. It follows that also the map

$$\phi_i \wedge_{MGL_{(p)}} 1_{MGL_{(p)}/(J \setminus \{p, v_i\})}$$

is zero, meaning that v_i acts trivially on $AP(n)$. Similarly, p acts trivially on $AP(n)$. This proves that all v_i, $0 \leq i < n$, act trivially on $AP(n)$ if $n \geq 2$.

If $n = 1$, one has to replace $MGL_{(p)}/(p, v_i)$ by $MGL_{(p)}/(p)$ in the above argument.

Furthermore,

$$Ak(n) \cong AP(n) \wedge_{MGL_{(p)}} MGL_{(p)}/(p, v_{n+1}, v_{n+2}, \cdots),$$

so v_i acts trivially on $Ak(n)$, too, for $0 \leq i < n$. For $i > n$, the claim follows analogously to the above argument. □

Remark 137 Working with modules over $MGL_{(p)}$ has the advantage that the E_∞-structure allows us to use the isomorphism from [18, Lemma V.1.10], as well as the results from [80] (below Lemma 18). For ABP, we only know of a commutative ring structure in the weak sense (see [87, Definition 5.3]). Note that for BP, an E_4-structure is constructed in [9].

The ABP-module structure on $AP(n)$ is the action of ABP on itself in $ABP \wedge_{MGL_{(p)}} MGL_{(p)}/(v_0, \cdots, v_{n-1}) \cong AP(n)$ and it is (by its construction) compatible with the $MGL_{(p)}$-action on $AP(n)$.

Recall that $\nu : BP \wedge P(n) \to P(n)$ induces a BP_*-module structure on $P(n)_*(X)$ and $P(n)^*(X)$ for any $X \in \mathcal{SH}$ and that $P(n)_* = BP_*/(v_0, \cdots, v_{n-1})$. Therefore, the classical version of the above corollary immediately implies that the BP_*-module structure on $P(n)_*(X)$ and $P(n)^*(X)$ induces a $P(n)_*$-module structure, as also concluded in [39, Remark 2.5(a)].

Our next aim is to show that also for $X \in \mathcal{SH}(\mathbb{C})$, the ABP_{**}-module structure on $AP(n)_{**}(X)$ and $AP(n)^{**}(X)$ induces a structure of $AP(n)_{**}$-modules, where the ring structure on $AP(n)_{**}$ is defined via the isomorphism $AP(n)_{**} \cong \mathrm{H}_{**} \otimes_{\mathbb{F}_p} P(n)_*$ (Lemma 5). We will show in Lemma 22 that this is the right choice of ring structure on $AP(n)_{**}$.

Let R be a strictly commutative ring spectrum, let $M = R/y$ satisfy $\pi_{2k'+1,2l'}M = 0$ (where (k', l') is the degree of y, as in Case 2 above), and let $N = M/x$ satisfy $\pi_{2k+1,2l}N = 0$.

In the commutative diagram,

$$\begin{array}{ccccc} S^{k',l'} \wedge R \wedge M & \xrightarrow{y\wedge 1\wedge 1} & R \wedge R \wedge M & \xrightarrow{m\wedge 1} & R \wedge M \\ \downarrow{\scriptstyle 1\wedge\nu_M} & & \downarrow{\scriptstyle 1\wedge\nu_M} & & \downarrow{\scriptstyle \nu_M} \\ S^{k',l'} \wedge M & \xrightarrow[y\wedge 1]{} & R \wedge M & \xrightarrow[\nu_M]{} & M, \end{array}$$

the composition $\nu_M(y \wedge 1_M)$ is zero by Proposition 134. Furthermore, $(m \wedge 1_M) \circ (y \wedge 1_R \wedge 1_M) = \phi' \wedge 1_M$, where ϕ' is, as before, the map whose cofiber is M. Thus, we have

$$\begin{array}{ccccc} S^{k',l'} \wedge R \wedge M & \xrightarrow{\phi'\wedge 1} & R \wedge M & \xrightarrow{\eta'\wedge 1} & M \wedge M \\ & {\scriptstyle 0}\searrow & \downarrow{\scriptstyle \nu_M} & \swarrow{\scriptstyle \mu_M} & \\ & & M & & \end{array},$$

and there exists a map $\mu_M : M \wedge M \to M$ in the homotopy category R-Mod such that $\mu_M \circ (\eta' \wedge 1_M) = \nu_M$.

Next, we define a map $\nu_{M,N} : M \wedge N \to N$ by $\nu_{M,N} = \mu_M \wedge_R 1_{R/x}$, using $N \cong M \wedge_R R/x$ [18, Lemma V.1.10]. It satisfies $\nu_{M,N}(\eta' \wedge 1_N) = \nu_N$ (by applying $- \wedge_R R/x$ to the analogous equation for μ_M) and $\nu_{M,N}(1 \wedge \eta) = \eta \circ \mu_M$ (because $\eta : M \to N$ is the canonical map $M \wedge_R R \to M \wedge_R R/x$).

In the commutative diagram (where the right square commutes because $\nu_{M,N}$ is a map of R-modules)

$$
\begin{array}{ccccc}
S^{k,l} \wedge M \wedge N & \xrightarrow{x \wedge 1 \wedge 1} & R \wedge M \wedge N & \xrightarrow{\nu_M \wedge 1} & M \wedge N \\
\downarrow{\scriptstyle 1 \wedge \nu_{M,N}} & & \downarrow{\scriptstyle 1 \wedge \nu_{M,N}} & & \downarrow{\scriptstyle \nu_{M,N}} \\
S^{k,l} \wedge N & \xrightarrow[x \wedge 1]{} & R \wedge N & \xrightarrow[\nu_N]{} & N,
\end{array}
$$

the lower composition is the action of x on N, which is trivial by Proposition 135. Thus,

$$\nu_{M,N}(\phi \wedge 1_N) = \nu_{M,N}(\nu_M \wedge 1_N)(x \wedge 1_M \wedge 1_N) = 0.$$

Hence, there exists a map $\mu_N : N \wedge N \to N$ in R-Mod making the following diagram commutative.

$$
\begin{array}{ccccc}
S^{k,l} \wedge M \wedge N & \xrightarrow{\phi \wedge 1} & M \wedge N & \xrightarrow{\eta \wedge 1} & N \wedge N \\
& \searrow^{0} & \downarrow{\scriptstyle \nu_{M,N}} & \swarrow_{\mu_N} & \\
& & N & &
\end{array}
$$

In particular, this applies to $N = MGL_{(p)}/(p, x)$ as in Corollary 136, yielding an $MGL_{(p)}$-module map

$$\mu_x : MGL_{(p)}/(p, x) \wedge MGL_{(p)}/(p, x) \to MGL_{(p)}/(p, x).$$

Lemma 20 *$AP(n)$ is isomorphic as an $MGL_{(p)}$-module to the $\wedge_{MGL_{(p)}}$-product of all $MGL_{(p)}/(p, x)$, $x \in J$, where J is as in the proof of Corollary 136.*

Proof By Proposition 134, p acts trivially on $MGL_{(p)}/p$, which proves $MGL_{(p)}/(p, p) = MGL_{(p)}/p$. By [18, Lemma V.1.10], it follows that

$$MGL_{(p)}/(p, x) \wedge_{MGL_{(p)}} MGL_{(p)}/(p, y) \cong MGL_{(p)}/(p, x) \wedge_{MGL_{(p)}} MGL_{(p)}/y$$

for any $x, y \in J$. With $p = v_0 \in J$, this implies that the $\wedge_{MGL_{(p)}}$-product of all $MGL_{(p)}/(p, x)$ is isomorphic to the $\wedge_{MGL_{(p)}}$-product of all $MGL_{(p)}/x$, which is the quotient $MGL_{(p)}/J \cong AP(n)$ (as in Corollary 136). □

We can, therefore, define a map of $MGL_{(p)}$-modules,

$$\mu_{AP(n)} : AP(n) \wedge AP(n) \to AP(n)$$

by applying the maps μ_x on each factor $MGL_{(p)}/(p, x)$.

Lemma 21 *If, in the above setting,*

$$\pi_{k'+1,l'} M = \pi_{2k'+2,2l'} M = \pi_{3k'+3,3l'} M = 0,$$

then μ_M is homotopy associative.

If, furthermore,

$$\pi_{k+1,l} N = \pi_{2k+2,2l} N = \pi_{3k+3,3l} N = 0,$$

then μ_N is also homotopy associative.

Proof Let $\mu = \mu_M$ and $\nu = \nu_N$. We have to show that

$$\delta = \mu(\mu \wedge 1_M - 1_M \wedge \mu) : M \wedge M \wedge M \to M$$

is zero. Let $\delta' = \delta(\eta' \wedge 1_M \wedge 1_M) : R \wedge M \wedge M \to M$, $\delta'' = \delta'(1_R \wedge \eta' \wedge 1_M) : R \wedge R \wedge M \to M$ and $\delta''' = \delta''(1_R \wedge 1_R \wedge \eta') = \delta \circ (\eta')^{\wedge 3} : R \wedge R \wedge R \to M$. Consider

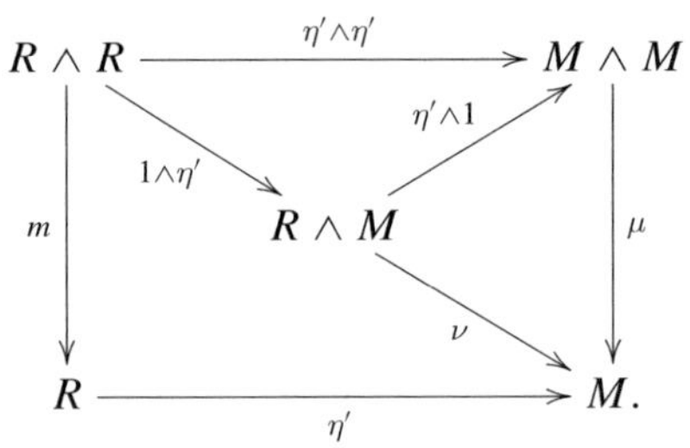

The top triangle obviously commutes, the right triangle commutes (up to homotopy) by the definition of μ, and the large triangle commutes because η' is an R-module map. Thus, $\mu \circ (\eta')^{\wedge 2} = \eta' \circ m$, and it follows

$$\mu(\mu \wedge 1 - 1 \wedge \mu)(\eta')^{\wedge 3} = \mu(\eta' m \wedge \eta' - \eta' \wedge \eta' m)$$

$$= \mu \circ (\eta')^{\wedge 2}(m \wedge 1 - 1 \wedge m) = \eta' m(m \wedge 1 - 1 \wedge m),$$

which vanishes by the associativity of m. Thus, $\delta''' = \delta''(1 \wedge 1 \wedge \eta') = 0$.

This implies that δ'' factors through an R-module map $\zeta : S^{k'+1,l'} \wedge R \wedge R \wedge R \to M$, as in the following diagram:

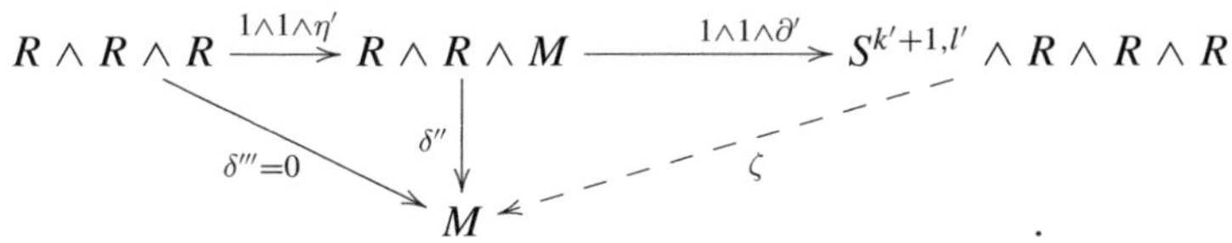

Now, $R \wedge R \wedge R$ is a ring spectrum and ζ can be considered as a map of $R \wedge R \wedge R$-modules. By Lemma 19 and the assumption on $\pi_{k'+1,l'}M$, ζ must be trivial. Thus, $\delta'' = 0$. Again, this implies $\delta' = \zeta'(1_R \wedge \partial' \wedge 1_M)$ for some $\zeta' : S^{k'+1,l'} \wedge R \wedge R \wedge M \to M$. The $R \wedge R \wedge R$-module map $\zeta'(1_{S^{k'+1,l'}} \wedge 1_R \wedge 1_R \wedge \eta')$ is a degree $(k'+1, l')$-map from the ring spectrum $R \wedge R \wedge R$ to M, and therefore vanishes by Lemma 19. It follows that $\zeta' = \zeta''(1_{S^{k'+1,l'}} \wedge 1_R \wedge 1_R \wedge \partial')$ for some map $\zeta'' : S^{2k'+2,2l'} \wedge R \wedge R \wedge R \to M$.

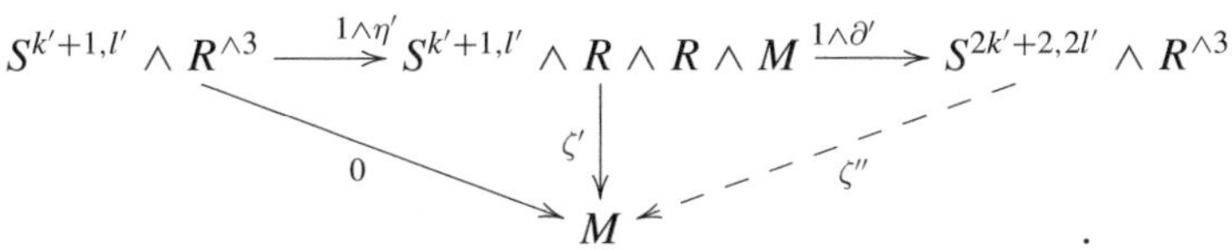

By the second assumption on $\pi_{**}M$, $\zeta'' = 0$. It follows that $\delta' = \zeta'(1 \wedge \partial' \wedge 1) = \zeta''(1 \wedge \partial' \wedge \partial') = 0$. That is, $\delta(\eta' \wedge 1_M \wedge 1_M) = \delta' = 0$, which implies $\delta = \zeta'''(\partial' \wedge 1_M \wedge 1_M)$ for some $\zeta''' : S^{k'+1,l'} \wedge R \wedge M \wedge M \to M$.

Now, $\zeta'''(1 \wedge \eta' \wedge 1)(1 \wedge 1 \wedge \eta')$ is a map from $R \wedge R \wedge R$ to M of degree $(k'+1, l')$, and therefore trivial. Thus, $\zeta'''(1 \wedge \eta' \wedge 1) = \zeta^{(4)}(1 \wedge 1 \wedge \partial')$ with $\zeta^{(4)} : S^{2k'+2,2l'} \wedge R \wedge R \wedge R \to M$, which is also trivial. It follows that $\zeta''' = \zeta^{(5)}(1 \wedge \partial' \wedge 1)$ for some $\zeta^{(5)} : S^{2k'+2,2l'} \wedge R \wedge R \wedge M \to M$, and $\delta = \zeta'''(\partial' \wedge 1 \wedge 1) = \zeta^{(5)}(\partial' \wedge \partial' \wedge 1)$. The map $\zeta^{(5)}(1 \wedge 1 \wedge \eta) : S^{2k'+2,2l'} \wedge R \wedge R \wedge R \to M$ is again zero, which implies $\zeta^{(5)} = \zeta^{(6)}(1 \wedge 1 \wedge \partial')$ for some $\zeta^{(6)} : S^{3k'+3,3l'} \wedge R \wedge R \wedge R \to M$. By the third condition on $\pi_{**}M$, this $\zeta^{(6)}$ is zero. Finally, we have

$$\delta = \zeta^{(6)}(\partial' \wedge \partial' \wedge \partial') = 0.$$

The same line of proof can be used to derive the associativity of μ_N from that of μ_M. It only needs to be checked that

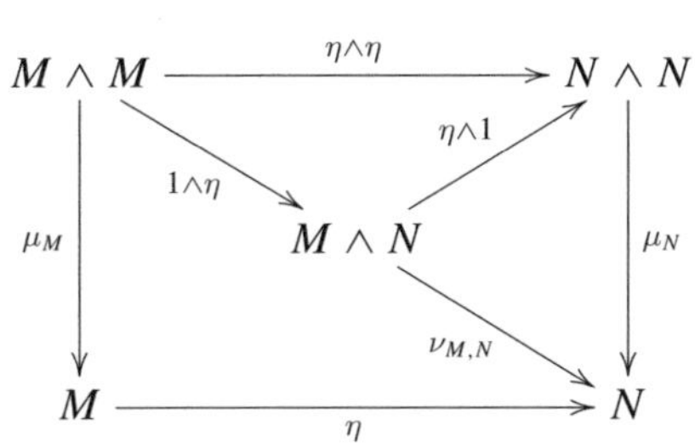

is commutative, but this was part of the definitions of $\nu_{M,N}$ and μ_N. $\square$

Proposition 138 *If $p > 2$, then $\mu_{AP(n)}$ is homotopy associative.*

Proof Note that this does not follow immediately from the above lemma, as $\pi_{2k+2,2l}N \neq 0$ for $N = MGL_{(p)}/(p, x)$. However, we can use a trick from [18, Theorem V.3.1], where the topological analogue of this statement is proven.

Let $N = R/(x, y)$ and $A = R/(J - \{x, y\})$ for some set J of elements in $\pi_{**}R$, and assume that we already know that A is equipped with a (homotopy) associative map $\mu_A : A \wedge A \to A$ as above. The product on $R/J \cong A/(x, y) \cong N \wedge_R A$ is given by

$$(N \wedge_R A) \wedge (N \wedge_R A) \xrightarrow{\tau} (N \wedge N) \wedge_R (A \wedge A) \xrightarrow{1 \wedge_R \mu_A} (N \wedge N) \wedge_R A \xrightarrow{\mu_N \wedge_R 1} N \wedge_R A.$$

Therefore, to prove associativity for $\mu_{N \wedge_R A}$, it suffices to prove that the associativity diagram for μ_N commutes after applying $- \wedge_R 1_A$ to it. Applying $- \wedge_R 1_A$ to all diagrams appearing in the above lemma yields the following result: If $\pi_{i,j}(M \wedge_R A) = 0$ for $(i, j) \in \{(k' + 1, l'), (2k' + 2, 2l'), (3k' + 3, 3l')\}$ and $\pi_{i,j}(N \wedge_R A) = 0$ for $(i, j) \in \{(k + 1, l), (2k + 2, 2l), (3k + 3, 3l)\}$, then $\mu_{N \wedge_R A}$ is associative.

Now, let $R = MGL_{(p)}$, $A = ABP$, and furthermore $M = MGL_{(p)}/p$ and $N = MGL_{(p)}/(p, x)$. From [87, Definition 5.3], we know that μ_{ABP} is associative. The assumptions on the homotopy groups are satisfied by Lemma 5, by which $\pi_{**}(ABP/p) \cong \mathrm{H}_{**} \otimes \pi_*(BP/p)$ and $\pi_{**}(ABP/(p, x)) \cong \mathrm{H}_{**} \otimes \pi_*(BP/(p, x))$. Note that these homotopy groups vanish in degrees $(k + 1, l)$ and $(3k + 3, 3l)$ for any p because $\pi_* BP$ is concentrated in even degrees and k is even. However, for $\pi_{(2k+2,2l)}$ to vanish, we need to assume that p is odd.

This proves that $\mu_{ABP/(p,x)}$ is associative. Inductively, we can apply this argument to $A = ABP/J'$ for some $(p) \subset J' \subset J$, where J is as in the proof of Corollary 136, using $ABP/(J' \cup \{y\}) = MGL_{(p)}/(p, y) \wedge_{MGL_{(p)}} ABP/J'$ (compare Lemma 20). □

Recall that, for $k = \mathbb{C}$ and $n > 0$, $AP(n)_{**} \cong \mathrm{H}_{**} \otimes_{\mathbb{F}_p} P(n)_*$ (Lemma 5), which is a ring. Hence, we can speak of $AP(n)_{**}$-modules. Note that $AP(0)_{**}$ is a ring, anyway, since $AP(0) = ABP$ is a ring spectrum.

Lemma 22 *Let $k = \mathbb{C}$. On coefficients, the map*

$$\mu_{AP(n)} : AP(n) \wedge AP(n) \to AP(n)$$

induces the multiplication on $AP(n)_{**}$ given by $AP(n)_{**} \cong \mathrm{H}_{**} \otimes_{\mathbb{F}_p} P(n)_*$.

Proof By [71, Theorem 3.6.16], the motivic Atiyah Hirzebruch spectral sequence in Lemma 5 is multiplicative, yielding an isomorphism of rings between $AP(n)_{**}$ with multiplication induced from $\mu_{AP(n)}$ and $E_2 = \mathrm{H}_{**} \otimes_{\mathbb{F}_p} P(n)_*$ with multiplication induced from the ring structure of H_{**} and from $R_{\mathbb{C}}(\mu_{AP(n)}) : P(n) \wedge P(n) \to P(n)$. It therefore suffices to show that $R_{\mathbb{C}}(\mu_{AP(n)})$ induces the ring structure on $P(n)_*$. Now, $R_{\mathbb{C}}$ carries all the above diagrams to the analogous topological diagrams and, therefore, $BP \wedge P(n) \to P(n)$ (inducing the action of BP_* on $P(n)_*$) factors through $R_{\mathbb{C}}(\mu_{AP(n)})$, which, hence, induces the induced action of $P(n)_*$ on $P(n)_*$. □

Corollary 139 *In $\mathcal{SH}(\mathbb{C})$, the action $\nu_n : ABP \wedge AP(n) \to AP(n)$ factors through a map $\mu_{AP(n)} : AP(n) \wedge AP(n) \to AP(n)$. If $p > 2$, the induced action of ABP_{**} on $AP(n)_{**}(X)$ and $AP(n)^{**}(X)$ gives $AP(n)_{**}(X)$ and $AP(n)^{**}(X)$ the structure of left $AP(n)_{**}$-modules for any $n \geq 0$ and $X \in \mathcal{SH}(\mathbb{C})$.*

Proof We have to check that the action of $AP(n)_{**}$ on $AP(n)_{**}(X)$ and $AP(n)^{**}(X)$ is unital and associative. This is equivalent to $\mu_{AP(n)}$ being homotopy left unital and associative. Unitality follows from the maps $\nu_M : R \wedge M \to M$ being unital, since, by definition of μ_M, $\nu_M = \mu_M \circ (\eta' \wedge 1) : R \wedge M \to M \wedge M \to M$. Associativity is proven in Proposition 138 for $p > 2$. □

9.4 Bousfield Classes of $AK(n)$ and $AB(n)$

Recall $AB(n) = v_n^{-1} AP(n)$.

We want to use methods of [39, 94] to prove that

$$AK(n)_{**}X = 0 \Leftrightarrow AB(n)_{**}X = 0$$

for $X \in \mathcal{SH}(\mathbb{C})$. For this, we need the following two results, which hold in $\mathcal{SH}(k)$ for any $k \subseteq \mathbb{C}$ and are analogous to [94, Formula (2.8)]. The reader may skip the first proposition, as it is a special case of the second one. The proof of the first one is maybe more illustrative.

Let $\mu_n : ABP^{**}(-) \to AP(n)^{**}(-)$ be induced from

$$ABP \overset{1\wedge i}{\to} ABP \wedge AP(n) \overset{\nu_n}{\to} AP(n),$$

where $i : S \to AP(n)$ is induced from the unit map $S \to ABP$ and ν_n is the structure map of the ABP-module $AP(n)$. For $s^E \in ABP^{**}ABP$ as in Lemma 13(2), we have $\mu_n(s^E) \in AP(n)^{**}(ABP)$.

Proposition 140 *For any $n \geq 0$, the map*

$$h_n : AP(n)^{**}[[\mu_n(s^E)]] \to AP(n)^{**}ABP,$$

given by $h_n(\sum_E x_E \cdot \mu_n(s^E)) = \sum_E \nu_n(s^E \wedge x_E)$, is an isomorphism of ABP^{**}-modules. If $k = \mathbb{C}$ and $p > 2$, it is an isomorphism of $AP(n)^{**}$-modules by Corollary 139.

Proof We proceed by induction. For $AP(0) = ABP$, μ_0 is the identity and the claim holds by Lemma 13(2). Now, assume h_n is an isomorphism for some $n \geq$

0. Consider the following diagram, consisting of two exact sequences induced by $AP(n) \overset{v_n}{\to} AP(n) \to AP(n+1)$.

$$\begin{array}{ccccccccc}
AP(n)^{**}ABP & \xrightarrow{v_n^*} & AP(n)^{**}ABP & \xrightarrow{\lambda_n^*} & AP(n+1)^{**}ABP & \xrightarrow{\delta^*} & AP(n)^{**}ABP & \xrightarrow{v_n^*} & AP(n)^{**}ABP \\
\uparrow h_n & & \uparrow h_n & & \uparrow h_{n+1} & & \uparrow h_n & & \uparrow h_n \\
AP(n)^{**}[[\mu_n s^E]] & \to & AP(n)^{**}[[\mu_n s^E]] & \to & AP(n+1)^{**}[[\mu_{n+1} s^E]] & \to & AP(n)^{**}[[\mu_n s^E]] & \to & AP(n)^{**}[[\mu_n s^E]]
\end{array}$$

The lower sequence is the exact sequence

$$\cdots \to AP(n)^{**} \to AP(n)^{**} \to AP(n+1)^{**} \to AP(n)^{**} \to AP(n)^{**} \cdots,$$

tensored over $\mathbb{F}_p$ with $\mathbb{F}_p[[s^E]]$. We show that the diagram commutes.

In the first square, the upper composition takes

$$x \cdot \mu_n(s^E) \in AP(n)^{**}[[\mu_n s^E]]$$

to

$$v_n^*(h_n(x\mu_n s^E)) : ABP \overset{s^E \wedge x}{\longrightarrow} ABP \wedge AP(n) \overset{\nu_n}{\longrightarrow} AP(n) \overset{v_n}{\longrightarrow} AP(n),$$

and the lower composition takes the same element to

$$h_n((v_n \cdot x) \cdot \mu_n(s^E)) : ABP \overset{s^E \wedge x}{\longrightarrow} ABP \wedge AP(n) \overset{1 \wedge v_n}{\longrightarrow} ABP \wedge AP(n) \overset{\nu_n}{\longrightarrow} AP(n).$$

Therefore, the commutativity of the first square is equivalent to the commutativity of the square

$$\begin{array}{ccc}
ABP \wedge AP(n) & \xrightarrow{1 \wedge v_n} & ABP \wedge AP(n) \\
\downarrow \nu_n & & \downarrow \nu_n \\
AP(n) & \xrightarrow[v_n]{} & AP(n).
\end{array}$$

This square commutes because v_n is a map of ABP-modules (compare Lemma 18).

In the second square, the upper composition takes

$$x \cdot \mu_n(s^E) \in AP(n)^{**}[[\mu_n s^E]]$$

to

$$(\lambda^* \circ h_n)(x\mu_n s^E) : ABP \overset{s^E \wedge x}{\longrightarrow} ABP \wedge AP(n) \overset{\nu_n}{\longrightarrow} AP(n) \overset{\lambda_n}{\longrightarrow} AP(n+1)$$

and the lower composition takes $x \cdot \mu_n(s^E)$ to

$$h_{n+1}(\lambda_n x \cdot \mu_n s^E) : ABP \xrightarrow{s^E \wedge x} ABP \wedge AP(n)$$

$$\xrightarrow{1\wedge\lambda_n} ABP \wedge AP(n+1) \xrightarrow{\nu_{n+1}} AP(n+1).$$

Thus, the commutativity of the second square is equivalent to the commutativity of

$$\begin{array}{ccc} ABP \wedge AP(n) & \xrightarrow{1\wedge\lambda_n} & ABP \wedge AP(n+1) \\ \downarrow{\scriptstyle \nu_n} & & \downarrow{\scriptstyle \nu_{n+1}} \\ AP(n) & \xrightarrow[\lambda_n]{} & AP(n+1), \end{array}$$

which holds because λ_n is, by definition, a map in the category of ABP-modules (see Definition 74).

In the third square, the upper composition takes $x \cdot \mu_{n+1}(s^E) \in AP(n+1)^{**}[[\mu_{n+1}s^E]]$ to $(\delta \circ h_{n+1})(x\mu_{n+1}s^E) = \delta(\nu_{n+1}(s^E \wedge x))$ and the lower composition takes $x \cdot \mu_{n+1}(s^E)$ to $\nu_n(s^E \wedge \delta(x))$. Thus, the commutativity of the third square follows from the commutativity of

$$\begin{array}{ccc} ABP \wedge AP(n+1) & \xrightarrow{1\wedge\delta} & ABP \wedge AP(n) \\ \downarrow{\scriptstyle \nu_{n+1}} & & \downarrow{\scriptstyle \nu_n} \\ AP(n+1) & \xrightarrow[\delta]{} & AP(n). \end{array}$$

Finally, the five lemma implies that h_{n+1} is an isomorphism. □

The above proposition holds more generally:

Proposition 141 *Let $h \in \mathcal{SH}(k)^{cell}$ be any cellular ABP-module spectrum. Then*

$$h^{**}ABP \cong h^{**}[[s^E]]$$

as ABP^{**}-modules. In particular, this also holds for $h = Ak(n)$.

Proof We apply the universal coefficient spectral sequence from [16, Proposition 7.7] (see also Proposition 104) to $E = ABP$ (which is a ring spectrum by [87, Definition 5.3] and is cellular by Remark 75(1)), $M = ABP \wedge ABP$ and $N = h$.

$$\mathrm{Ext}^{***}_{ABP_{**}}(ABP_{**}ABP, h_{**}) \Rightarrow \pi_{**}F_{ABP}(ABP \wedge ABP, h),$$

converging conditionally to $\pi_{**}F_{ABP}(ABP \wedge ABP, h) \cong h^{**}(ABP)$. As $ABP_{**}ABP \cong ABP_{**}\{t^E\}$ is free over ABP_{**} (Lemma 13(1)), the higher Ext-groups vanish and the sequence collapses to

$$h^{**}ABP \cong \mathrm{Hom}_{ABP_{**}}(ABP_{**}\{t^E\}, h_{**}),$$

which is isomorphic to $h^{**}[[s^E]]$, as in Lemma 13(2). □

Würgler constructs operations

$$s_Q^E : MUQ^i(-) \to MUQ^{i+|E|}(-)$$

for regular sequences Q [94, Theorem 5.1]. For $MUQ = P(n)$, these operations specify a choice of the operations $(r_E)_n : P(n)^*(-) \to P(n)^{*+|E|}(-)$ considered in [39, Sect. 4]. They are needed in the proof of the isomorphism $B(n)_*(X) \cong K(n)_*(X) \otimes \mathbb{F}_p[v_{n+1}, v_{n+2}, \cdots]$ in [39, Proposition 4.14]. We will now state motivic analogues of some of Würgler's lemmas for $MUQ = P(n)$.

In $\mathcal{SH}(k)^{cell}$, let $AP(n)[\mathbf{t}]$ represent the cohomology theory

$$AP(n)^{**}(-)[\mathbf{t}] = ABP^{**}[t_1, t_2, \cdots] \otimes_{ABP^{**}} AP(n)^{**}(-)$$

with $\deg(t_i) = -(2(p^i - 1), p^i - 1)$.

Remark 142 Würgler defines $h^*(-)[\mathbf{t}] = h^*(-) \otimes_{h^*} h^*[t_1, t_2, \cdots]$, where the t_i take any even degree, i.e., $\deg(t_i) = -2i$ in [94, Sect. 5]. He also considers operations s^E of any degree $\sum_i 2ie_i$. However, Johnson and Wilson [39, Sects. 1 and 4] are only interested in operations of degree $|E| = \sum_i 2(p^i - 1)e_i$. We will see that it suffices to restrict to those degrees. Most of Würgler's results we refer to in the following are formulated in a much greater generality and do not depend on any degrees of particular elements. The only point where the degrees of the t_i are important is in [94, Theorem 5.1] and we will comment on that below Theorem 18.

Lemma 23 *For* $X \in \mathcal{SH}(k)^{fin}$,

$$(ABP \wedge AP(n))^{**}(X) \cong AP(n)^{**}(X)[\mathbf{t}].$$

As any cohomology theory on $\mathcal{SH}(k)^{fin}$ *extends uniquely to* $\mathcal{SH}(k)^{cell}$ *by [65, Lemma 4.10], it follows that we can take*

$$AP(n)[\mathbf{t}] = ABP \wedge AP(n).$$

Proof We consider the map

$$(ABP \wedge ABP)^{**} \otimes_{ABP^{**}} AP(n)^{**}(X) \to (ABP \wedge AP(n))^{**}(X),$$

induced by the ABP-module structure on $AP(n)$. First, we see that

$$(ABP \wedge ABP)^{**} \cong (ABP \wedge ABP)_{-*,-*} \cong ABP_{-*,-*}[\mathbf{t}] \cong ABP^{**}[\mathbf{t}]$$

by Lemma 13(1). Note that the t_i appearing here are dual to the t_i in Lemma 13(1), hence $\deg(t_i) = -(2(p^i - 1), p^i - 1)$. To prove the claim, it suffices to show that the above map is an isomorphism.

For $n = 0$ and $X = S^0$, this is clear. Induction on n shows

$$(ABP \wedge ABP)^{**} \otimes_{ABP^{**}} AP(n)^{**} \cong (ABP \wedge AP(n))^{**}$$

for any n, since $AP(n) \to AP(n) \to AP(n+1)$ induces long exact sequences on both sides and the diagram commutes because $AP(n) \to AP(n) \to AP(n+1)$ are maps of ABP-modules. As also any cofiber sequence $X \to Y \to Z$ induces long exact sequences on both sides, cellular induction proves the claim for any finite spectrum X. □

We state an analogue of [94, Lemma 3.14].

Lemma 24 *Let $h \in \mathcal{SH}(k)^{cell}$, $k \subseteq \mathbb{C}$, be an ABP-module. The multiplication $ABP \wedge h \to h$ induces an isomorphism of ABP^{**}-modules,*

$$ABP^{**}ABP \otimes_{ABP^{**}} h^{**}AP(n) \cong h^{**}(ABP \wedge AP(n)).$$

Proof We apply the spectral sequence from [16, Proposition 7.7] to $E = ABP$, $M = ABP \wedge ABP \wedge ABP$ and $N = h$:

$$\mathrm{Ext}^{***}_{ABP_{**}}(ABP_{**}(ABP \wedge ABP), h_{**}) \Rightarrow \pi_{**}F_{ABP}(ABP \wedge ABP \wedge ABP, h),$$

converging conditionally to $h^{**}(ABP \wedge ABP)$. By [65, Lemma 5.1(i)],

$$ABP_{**}(ABP \wedge ABP) \cong ABP_{**}ABP \otimes_{ABP_{**}} ABP_{**}ABP,$$

which is free over ABP_{**}. Thus, the spectral sequence collapses and

$$h^{**}(ABP \wedge ABP) \cong \mathrm{Hom}_{ABP_{**}}(ABP_{**}ABP \otimes_{ABP_{**}} ABP_{**}ABP, h_{**}).$$

Now, $ABP_{**}ABP \otimes_{ABP_{**}} ABP_{**}ABP \cong ABP_{**}\{t_1^E t_2^F\}$, and, therefore,

$$\mathrm{Hom}_{ABP_{**}}(ABP_{**}ABP \otimes_{ABP_{**}} ABP_{**}ABP, h_{**}) \cong h^{**}[[s_1^E s_2^F]]$$

$$\cong ABP^{**}[[s_1^E]] \otimes_{ABP_{**}} h^{**}[[s_2^F]] \cong ABP^{**}ABP \otimes_{ABP_{**}} h^{**}ABP,$$

using Proposition 141. This proves the claim for $n = 0$.

Assume we have shown that $ABP \wedge h \to h$ induces an isomorphism

$$ABP^{**}ABP \otimes_{ABP^{**}} h^{**}AP(n) \cong h^{**}(ABP \wedge AP(n))$$

for some $n \geq 0$. The cofiber sequence $AP(n) \to AP(n) \to AP(n+1)$ induces long exact sequences on both sides of this isomorphism, which form a commutative diagram because the maps $AP(n) \to AP(n) \to AP(n+1)$ are maps of ABP-modules. Thus, the five lemma implies the claim for $n+1$. $\square$

Combining the above isomorphism with the one from Proposition 141, we get:

Lemma 25 *Let* $k = \mathbb{C}$ *and* $p > 2$. *Let* $X = AP(n)$ *or* $X = AP(n)[\mathbf{t}]$. *The action of* ABP *on* $AP(m)$, $ABP \wedge AP(m) \to AP(m)$, *induces an isomorphism of* $AP(m)^{**}$-*modules*

$$AP(m)^{**}(ABP) \otimes_{AP(m)^{**}} AP(m)^{**}(X) \xrightarrow{\cong} AP(m)^{**}(ABP \wedge X).$$

Proof From Proposition 140, we know that

$$AP(m)^{**}(ABP) \cong AP(m)^{**} \otimes_{ABP^{**}} ABP^{**}ABP$$

as $AP(m)^{**}$-modules. With $h = AP(m)$, the above lemma immediately implies

$$AP(m)^{**}(ABP) \otimes_{AP(m)^{**}} AP(m)^{**}(AP(n)) \xrightarrow{\cong} AP(m)^{**}(ABP \wedge AP(n))$$

as ABP^{**}-modules and, by Corollary 139, also as $AP(m)^{**}$-modules.

For $X = AP(n)[\mathbf{t}] = ABP \wedge AP(n)$, we set $M = ABP^{\wedge 4}$ in the proof of the previous lemma. Using $\pi_{**}ABP^{\wedge 4} \cong ABP_{**}ABP \otimes_{ABP_{**}} \pi_{**}ABP^{\wedge 3}$ [65, Proposition 5.1(i)], the proof proceeds with exactly the same arguments as the proof of Lemma 24. $\square$

As a consequence of Corollary 136, we show the following (see [39, Lemma 2.8(b)]):

Corollary 143 *Let* $k = \mathbb{C}$ *and* $p > 2$. *The cofibration* $S^{2p^n-2,p^n-1} \wedge AP(n) \xrightarrow{v_n} AP(n) \to AP(n+1)$ *induces short exact sequences*

$$0 \to Ak(j)^{**}(S^{2p^n-1,p^n-1} \wedge AP(n)) \to Ak(j)^{**}(AP(n+1))$$
$$\to Ak(j)^{**}(AP(n)) \to 0$$

for every $j > n$.

Proof We have to show that, in the $Ak(j)^{**}$-long exact sequence, the map induced by v_n is zero. This map is defined as the composition of the two left arrows in the following commutative diagram:

$$\begin{array}{ccc}
Ak(j)^{**}(S^{2p^n-2,p^n-1} \wedge AP(n)) & \xleftarrow{\cong} & ABP^{**}(S^{2p^n-2,p^n-1}) \otimes_{ABP^{**}} Ak(j)^{**}(AP(n)) \\
\uparrow{\scriptstyle Ak(j)^{**}(v_n \wedge 1)} & & \uparrow{\scriptstyle ABP^{**}(v_n)\otimes 1} \\
Ak(j)^{**}(ABP \wedge AP(n)) & \xleftarrow{\cong} & ABP^{**}(ABP) \otimes_{ABP^{**}} Ak(j)^{**}(AP(n)) \\
\uparrow{\scriptstyle Ak(j)^{**}(v_n)} & & \\
Ak(j)^{**}(AP(n)) & &
\end{array}$$

The horizontal maps are induced by the ABP-module structure on $Ak(j)$. The lower horizontal map is an isomorphism by the previous lemma. The upper isomorphism is proven similarly, setting $M = ABP \wedge ABP$ in the above proof. We show that the right hand map is zero. Let $\sum s^E \otimes x_E$ be an element of $ABP^{**}(ABP) \otimes_{ABP^{**}} Ak(j)^{**}(AP(n))$. It maps to $\sum s^E(v_n) \otimes x_E$. In Corollary 132, we set $k = 1$, $s = 0$ and $m = n$, to see that for all $|E| \geq 0$, either $s^E(v_n) \equiv v_n \mod I_n$ or $s^E(v_n) \equiv 0 \mod I_n$. Thus, $s^E(v_n) \in I_{n+1} = (v_0, \cdots, v_n)$. By Corollary 136, I_{n+1} acts trivially on $Ak(j)^{**}(AP(n))$ for $j > n$, hence, the right hand map is zero.

It follows that the left map factors through zero, proving the claim. □

The isomorphism from Lemma 25 is now used to define an $AP(m)^{**}(ABP)$-comodule structure on $AP(m)^{**}(AP(n))$ by

$$AP(m)^{**}AP(n) \xrightarrow{\nu_n^*} AP(m)^{**}(ABP \wedge AP(n))$$

$$\xleftarrow{\cong} AP(m)^{**}ABP \otimes_{AP(m)^{**}} AP(m)^{**}AP(n),$$

where ν_n is the ABP-module structure map on $AP(n)$. Note that, for $k \neq \mathbb{C}$ or $p = 2$, we do not know if these groups are $AP(m)^{**}$-modules (see Corollary 139). In this case, we might only get an $ABP^{**}ABP$-comodule structure on $AP(m)^{**}AP(n)$.

Lemma 26 *Let $k = \mathbb{C}$, $m > n \geq 0$ and E be an exponent sequence. Let $s^E : ABP \to ABP$ be as in Lemma 13 and $\mu_m : ABP \to AP(m)$ be as defined in the beginning of this section. Then*

$$\mu_m \circ s^E \circ v_n : S \to ABP \to ABP \to AP(m)$$

is the zero map.

Proof The realisation functor $R_{\mathbb{C}}$ takes the composition $\mu_m \circ s^E \circ v_n \in AP(m)_{**}$ to $\mu_m^{\text{top}} \circ s^E_{\text{top}} \circ v_n^{\text{top}} \in P(m)_*$. By the BP-version of [94, Lemma 2.2], $\mu_m^{\text{top}} : BP \to P(m)$ is the canonical projection. By the invariant prime ideal theorem, $I_{n+1}^{\text{top}} = (v_0^{\text{top}}, v_1^{\text{top}}, \cdots, v_n^{\text{top}}) \subset BP_*$ is invariant under the action of $s^E_{\text{top}} \in BP^*BP$, hence, $s^E_{\text{top}}(v_n^{\text{top}}) \in I_{n+1}^{\text{top}}$. Thus, $\mu_m^{\text{top}}(s^E_{\text{top}} \circ v_n^{\text{top}}) \in P(m)_*$ lies in the image of I_{n+1} under the projection $BP_* \to P(m)_* = BP_*/(v_0^{\text{top}}, \cdots, v_{m-1}^{\text{top}})$, which is zero, since $m > n$. This implies $R_{\mathbb{C}}(\mu_m \circ s^E \circ v_n) = 0 \in P(m)_*$.

For $k = \mathbb{C}$, $AP(m)_{**} \cong P(m)_*[\tau]$ by Lemma 5, and any homogeneous element in $AP(m)_{**}$ that realises to zero in $P(m)_*$ already has to be zero in $AP(m)_{**}$. Hence, $\mu_m \circ s^E \circ v_n = 0$ in $AP(m)_{**}$. □

Now we present an analogue of a special case of [94, Proposition 4.12].

Lemma 27 *Let $k = \mathbb{C}$, $p > 2$ and $m \geq n$. As $AP(m)^{**}ABP$-comodules,*

$$AP(m)^{**}AP(n) \cong AP(m)^{**}ABP \otimes_{AP(n)^{**}} \Lambda_{AP(n)^{**}}[[\beta_0, \cdots, \beta_{n-1}]]$$

with $\deg(\beta_i) = (2p^i - 1, p^i - 1)$.

Proof For $n = 0$, the statement is trivial. Assume the proposition holds for some pair (m, n) with $m > n$. We show that it also holds for $(m, n+1)$.

Consider $AP(n) \xrightarrow{v_n} AP(n)$. We show that the induced map

$$v_n^* : AP(m)^{**}AP(n) \to AP(m)^{**}AP(n)$$

is trivial. This works similar to [94, Lemma 3.15]. Let $\phi : S \to ABP$ represent $v_n \in ABP_{**}$ and let $\phi^* : AP(m)^{**}ABP \to AP(m)^{**}$ be the map induced on $AP(m)^{**}(-)$. For $x \in AP(m)^{**}ABP = AP(m)^{**}[[\mu_m s^E]]$ (see Proposition 140), we have $x = \sum_E \lambda_E \mu_m s^E$ and

$$\phi^*(x) = \phi^*(\sum_E \lambda_E \mu_m s^E) = \sum_E \lambda_E \mu_m (s^E(v_n)),$$

because ϕ^* is precomposition with $v_n \in ABP_{**}$. By Lemma 26, $\mu_m(s^E(v_n)) = 0$ in $AP(m)^{**}$, hence, $\phi^* = 0$. Now consider the following commutative square:

$$\begin{array}{ccc}
AP(m)^{**}(ABP \wedge AP(n)) & \xrightarrow{(\phi\wedge 1)^*} & AP(m)^{**}(AP(n)) \\
\uparrow \cong & & \uparrow \cong \\
AP(m)^{**}ABP \otimes_{AP(m)^{**}} AP(m)^{**}AP(n) & \xrightarrow[\phi^*\otimes 1]{} & AP(m)^{**} \otimes_{AP(m)^{**}} AP(m)^{**}AP(n)
\end{array}$$

The left map is an isomorphism by Lemma 25. Since $\phi^* = 0$, it follows that $(\phi \wedge 1)^* = 0$. Since v_n^* is, by definition, the precomposition of $(\phi \wedge 1)^*$ with the map $AP(m)^{**}AP(n) \to AP(m)^{**}(ABP \wedge AP(n))$, $v_n^* = 0$, as claimed.

It follows that the long exact $AP(m)^{**}(-)$-sequence induced by

$$S^{2p^n-2, p^n-1} \wedge AP(n) \xrightarrow{v_n} AP(n) \to AP(n+1)$$

splits into short exact sequences of $AP(m)^{**}ABP$-comodules

$$0 \to AP(m)^{**}(S^{2p^n-1, p^n-1} \wedge AP(n)) \to AP(m)^{**}AP(n+1)$$
$$\to AP(m)^{**}AP(n) \to 0.$$

Analogously to [94, Proposition 4.12], it follows inductively that

$$AP(m)^{**}AP(n+1) \cong AP(m)^{**}AP(n) \otimes_{AP(m)^{**}} \Lambda_{AP(m)^{**}}(\beta_n)$$

and the degree of β_n is determined by the degrees appearing in the exact sequence. □

If M is an $AP(n)^{**}(ABP)[\mathbf{t}]$-comodule with structure map ψ, an element $a \in M$ is called primitive if $\psi(a) = 1 \otimes a$. Similarly to [94, Lemma 4.13], the following holds:

Lemma 28 *Let $k = \mathbb{C}$ and $p > 2$. Let $g : AP(n) \to AP(n)[\mathbf{t}]$ be a map of spectra. Then g is a map of ABP-module spectra if and only if it is a primitive element of the $AP(n)^{**}(ABP)[\mathbf{t}]$-comodule $AP(n)^{**}(AP(n))[\mathbf{t}]$.*

Proof This follows from Lemma 25 in the same manner as [94, Lemma 4.13] follows from [94, Lemma 3.14] for $X = P(n)$ and $X = P(n)[\mathbf{t}]$. □

Now we state a result which is directly used for the construction of the operation we are aiming at. This corresponds to [94, Theorem 4.17].

Proposition 144 *Let $k = \mathbb{C}$ and $p > 2$. There is a degree-preserving group isomorphism*

$$\mathrm{Hom}^{**}_{ABP}(AP(n), AP(n)[\mathbf{t}]) \cong \Lambda_{AP(n)^{**}[\mathbf{t}]}[[\beta_0, \beta_1, \cdots]],$$

where the left hand side is the bigraded abelian group of maps of ABP-module spectra and the right hand side is an exterior algebra over $AP(n)^{**}[\mathbf{t}]$.

Proof Würgler derives this from [94, Proposition 4.12] and [94, Lemma 4.13] using that inverse limits of primitive elements are primitive [94, Lemma 4.16]. The same line of proof proves this proposition using Proposition 27 and Lemma 28. Denoting the set of primitive elements in M by $\mathrm{Pr}\{M\}$, the proof can be summarised by the following sequence of isomorphisms:

$$\mathrm{Hom}^{**}_{ABP}(AP(n), AP(n)[\mathbf{t}]) \cong \mathrm{Pr}\left\{AP(n)^{**}(AP(n))[\mathbf{t}]\right\}$$

$$\cong \mathrm{Pr}\left\{AP(n)^{**}(ABP)[\mathbf{t}] \otimes_{AP(n)^{**}[\mathbf{t}]} \Lambda_{AP(n)^{**}[\mathbf{t}]}[[\beta_0, \beta_1, \cdots]]\right\}$$

$$\cong \mathrm{Pr}\left\{AP(n)^{**}(ABP)[\mathbf{t}]\right\} \otimes_{AP(n)^{**}[\mathbf{t}]} \Lambda_{AP(n)^{**}[\mathbf{t}]}[[\beta_0, \beta_1, \cdots]]$$

$$\cong AP(n)^{**}[\mathbf{t}] \otimes_{AP(n)^{**}[\mathbf{t}]} \Lambda_{AP(n)^{**}[\mathbf{t}]}[[\beta_0, \beta_1, \cdots]] \cong \Lambda_{AP(n)^{**}[\mathbf{t}]}[[\beta_0, \beta_1, \cdots]].$$

□

This completes the preparation Würgler needs for [94, Theorem 5.1]. We state our version of this theorem, now using the notation from [39].

Theorem 18 *Let $p > 2$. In $\mathcal{SH}(\mathbb{C})^{cell}$, there exists a family $(r_E)_n$, $n \geq 0$, of natural stable operations*

$$(r_E)_n : AP(n)^{**}(-) \to AP(n)^{**}(-)$$

of degree $(|E|, |E|/2)$, such that

$$(r_E)_n(ux) = \sum_{F+G=E} s_F(u)(r_G)_n(x)$$

for $u \in ABP^{**}(X)$ *and* $x \in AP(n)^{**}(X)$ *(see Definition 130 for the definition of* s_F*) and such that*

$$\begin{array}{ccc} ABP^{**}(-) & \xrightarrow{s_E} & ABP^{**}(-) \\ \downarrow & & \downarrow \\ AP(n)^{**}(-) & \xrightarrow[(r_E)_n]{} & AP(n)^{**}(-) \end{array}$$

commutes.

Proof Let $s_t : ABP^{**}(-) \to ABP^{**}(-)[\mathbf{t}]$ be given by

$$s_t(x) = \sum_E s_E(x) \otimes t^E.$$

As in [94, Theorem 5.1], the square

$$\begin{array}{ccc} ABP^{**}(-) & \xrightarrow{s_t} & ABP^{**}(-)[\mathbf{t}] \\ \downarrow & & \downarrow \\ AP(n)^{**}(-) & \overset{s_{t,n}}{\dashrightarrow} & AP(n)^{**}(-)[\mathbf{t}] \end{array}$$

can be completed with the help of Proposition 144 and we define $(r_E)_n$ by $s_{t,n}(x) = \sum_E (r_E)_n(x) \otimes t^E$.

Since Proposition 144 gives us a map of ABP-module spectra, the operation $s_{t,n}$ satisfies $s_{t,n}(ux) = s_t(u)s_{t,n}(x)$. It follows that

$$\sum_E (r_E)_n(ux) \otimes t^E = \sum_{F+G=E} s_F(u)(r_G)_n(x) \otimes t^F t^G$$

and, hence, $(r_E)_n(ux) = \sum_{F+G=E} s_F(u)(r_G)_n(x)$, which proves the first property claimed. The second property holds because the above commutative square has to commute on the level of each t^E. □

Remark 145 Originally (see e.g. [72, Formula (2.4)]), one first defines the operation $s_t : BP^*(-) \to BP^*(-)[t_1, t_2, \cdots]$, $\deg(t_i) = -2i$, and then uses it to construct operations s_E of degree $\sum_i 2ie_i$. The s_t used in the above proof contains only those summands $s_E \otimes t^E$ with $|E| = \sum_i 2(p^i - 1)$, because we work with ABP-modules instead of MGL-modules. Hence, the $s_{t,n}$ also consists of less summands than the corresponding operation in [94, Theorem 5.1], but this suffices to define exactly those $(r_E)_n$ with the degrees needed in [39, Sect. 4].

In [39, Proposition 4.14], Johnson and Wilson show that, for any $X \in \mathcal{SH}^{fin}$, there is a natural isomorphism

$$B(n)_*(X) \cong K(n)_*(X) \otimes \mathbb{F}_p[v_{n+1}, v_{n+2}, \cdots]$$

for $n < 2p-2$. Johnson and Wilson needed the condition on n because they were only able to canonically define the operations $(r_E)_n$ in this range. As stated in [94, Remark 6.19], the condition becomes redundant if Würgler's operations are used.

With the above operations at hand, we can proceed exactly as in [39, Sect. 4]:

Definition 146 Let n be fixed and let $\mathcal{E}$ be the set of all exponent sequences of the form $E = (0, \cdots, 0, e_{n+1}, e_{n+2}, \cdots)$. Recall the definition of $|E|$ from Definition 131. For $E \in \mathcal{E}$, let $q = |E| = 2(p^n - 1)b + a$ with $0 \leq a < 2(p^n - 1)$. With $\sigma^n E = (p^n e_{n+1}, p^n e_{n+2}, \cdots)$ and $c = b - (e_{n+1} + e_{n+2} + \cdots)$, it follows $|\sigma^n E| = c2(p^n - 1) + a$, exactly as in [39]. Note that q and, hence, a are even. We define $\bar{s}_E \in Ak(n)^{a,a/2}(AP(n))$ by

$$\bar{s}_E : AP(n) \xrightarrow{(r_{\sigma^n E})_n} \Sigma^{c2(p^n-1)+a, c(p^n-1)+a/2} AP(n)$$

$$\xrightarrow{v_n^c} \Sigma^{a,a/2} AP(n) \xrightarrow{\lambda_n} \Sigma^{a,a/2} Ak(n),$$

where λ_n is the quotient map from $AP(n)$ to $Ak(n) = AP(n)/(v_{n+1}, \cdots)$.

Furthermore, we assume that the set $\{E \in \mathcal{E} \mid |E| = q\}$ is ordered as $\{E_1, \cdots, E_v\}$, where v is the $\mathbb{F}_p$-dimension of $(\mathbb{F}_p[v_{n+1}, v_{n+2}, \cdots])_q$, and we denote $\bar{s}_{E_u}$ by s_u.

Lemma 29 *Let $k = \mathbb{C}$, $p > 2$ and $\bar{s}_E : AP(n)_{**} \to Ak(n)_{**}$ be induced by the above map. Assume $n > 0$ and let $\tau \in AP(n)_{**}$ be as in Lemma 5. Then $\bar{s}_E(\tau) = 0$ for $E \neq 0$ and $\bar{s}_0(\tau) = \tau$. If, furthermore, $x \in AP(n)_{**}X$ with $X \in \mathcal{SH}(\mathbb{C})^{cell}$, then $\bar{s}_E(x\tau) = \bar{s}_E(x)\tau$.*

Proof Since $\bar{s}_E(\tau) = (\lambda_n \circ v_n^c \circ (r_{\sigma^n E})_n)(\tau)$ with $\lambda_n(\tau) = \tau$, and $c = 0$ for $E = 0$, the first claim is equivalent to $(r_F)_n(\tau) = 0$ for $F \neq 0$ and $(r_0)_n(\tau) = \tau$. The realisation functor $R_{\mathbb{C}}$ maps τ to $1 \in P(n)^0$ and it takes $(r_F)_n : AP(n)^{**} \to AP(n)^{**}$ to $(r_F^{\text{top}})_n : P(n)^* \to P(n)^*$, which, by [94, Theorem 5.1], is compatible with s_F^{top} in the sense that the following diagram commutes:

$$\begin{array}{ccc} BP^* & \xrightarrow{s_F^{\text{top}}} & BP^* \\ \downarrow & & \downarrow \\ P(n)^* & \xrightarrow{(r_F^{\text{top}})_n} & P(n)^*. \end{array}$$

The map s_F^{top} is defined via the coaction map

$$\psi^{\text{top}} : BP_* \xrightarrow{(1\wedge i)_*} BP_*BP \xrightarrow{m_*^{-1}} BP_*BP \otimes_{BP_*} BP_*,$$

which clearly takes 1 to $1 \otimes 1$. Therefore, in the formula

$$\psi^{\text{top}}(1) = \sum_F c(t_{\text{top}}^F) \otimes s_F^{\text{top}}(1),$$

all $s_F^{\mathrm{top}}(1)$ have to be 0, except for $s_0^{\mathrm{top}}(1)$, which is 1. It follows that also $(r_F^{\mathrm{top}})_n(1) = 0$ for $F \neq 0$ and $(r_0^{\mathrm{top}})_n(1) = 1$. Hence, $R_{\mathbb{C}}((r_F)_n(\tau))$ is 0 for $F \neq 0$ and it is 1 for $F = 0$.

The preimage of $x \in P(n)^*$ under $R_{\mathbb{C}} : AP(n)^{**} \to P(n)^*$ is

$$\left\{ \sum_{k \geq 0} x_k \cdot \tau^k \mid \sum_{k \geq 0} x_k = x \right\}.$$

Since $(r_F)_n$ is a map of degree $(|F|, |F|/2)$, the only possible preimage of $R_{\mathbb{C}}((r_F)_n(\tau))$ is $(r_F)_n(\tau) = 0$ for $F \neq 0$ and $(r_0)_n(\tau) = \tau$.

The claim on $\bar{s}_E(x\tau)$ now follows from the Cartan formula, Lemma 14. □

The following is a motivic analogue to [39, Proposition 4.14].

Theorem 19 *For any $X \in \mathcal{SH}(\mathbb{C})^{fin}$ and any $n > 0$, $p > 2$, there is a natural isomorphism*

$$AB(n)_{**}(X) \to AK(n)_{**}(X) \otimes \mathbb{F}_p[v_{n+1}, v_{n+2}, \cdots].$$

Furthermore, for any $X \in \mathcal{SH}(\mathbb{C})$,

$$AK(n)_{**}(X) = 0 \text{ if and only if } AB(n)_{**}(X) = 0.$$

Proof For a given exponent sequence E, consider the composition

$$AP(n) \xrightarrow{\bar{s}_E} \Sigma^{a,a/2} Ak(n) \to \Sigma^{a,a/2} AK(n) \xrightarrow{v_n^{-b}} \Sigma^{q,q/2} AK(n),$$

where $a = a_E$ and $b = b_E$ are as in Definition 146. These induce a natural homomorphism

$$\hat{\Lambda} : AP(n)_{**}(X) \to AK(n)_{**}(X) \otimes \mathbb{F}_p[v_{n+1}, v_{n+2}, \cdots]$$

$$\text{by } \hat{\Lambda}(y) = \sum_{E \in \mathcal{E}} v_n^{-b_E} \bar{s}_E(y) \otimes v^E.$$

By Corollary 132 and the Cartan formula, $r_E(yv_n) \equiv r_E(y)v_n \bmod I_n$. As $I_n = (v_0, \cdots, v_{n-1})$ acts trivially on $AP(n)$ and $Ak(n)$ (Proposition 136), it follows that $\bar{s}_E : AP(n)_{**}(X) \to Ak(n)_{**}(X)$ is an $\mathbb{F}_p[v_n]$-homomorphism. Therefore, $\hat{\Lambda}$ can be extended to $AB(n)_{**}(X) = v_n^{-1} AP(n)_{**} X$. This yields a map

$$\Lambda : AB(n)_{**}(X) \to AK(n)_{**}(X) \otimes \mathbb{F}_p[v_{n+1}, v_{n+2}, \cdots].$$

We show that Λ is an isomorphism for $X = S^0$, and, thus, for all $X = S^{p,q}$. By Lemma 5, we are considering a map $B(n)_*[\tau] \to K(n)_*[\tau] \otimes \mathbb{F}_p[v_{n+1}, v_{n+2}, \cdots]$ which, under $R_{\mathbb{C}}$, realises to the isomorphism $B(n)_* \xrightarrow{\cong} K(n)_* \otimes \mathbb{F}_p[v_{n+1}, v_{n+2}, \cdots]$ in

[39, Proposition 4.14]. Thus, it suffices to calculate $\Lambda(\tau)$, or, equivalently, $\hat{\Lambda}(\tau) = \sum_{E\in\mathcal{E}} v_n^{-b_E}\bar{s}_E(\tau)\otimes v^E$. By the previous lemma, this is equal to $v_n^{-b_0}\bar{s}_0(\tau)\otimes v^0 = \tau$. Thus, Λ is an isomorphism if X is a sphere.

Cellular induction via the five lemma shows that Λ is an isomorphism for all $X \in \mathcal{SH}(\mathbb{C})^{fin}$.

By Definition 52(3), any cell spectrum is the colimit of a diagram of finite cell spectra. As in [73, Theorem 2.1(a)], it follows that, for any $X \in \mathcal{SH}(\mathbb{C})^{cell}$, $AB(n)_{**}X = 0$ if and only if $AK(n)_{**}X = 0$. Here, we use that $\pi_{**}(-)$ commutes with filtered colimits by [16, Proposition 9.3].

Furthermore, by [16, Proposition 7.3], any $X \in \mathcal{SH}(\mathbb{C})$ has a cellular approximation $X' \in \mathcal{SH}(\mathbb{C})^{cell}$, $f : X' \to X^{fib}$ (X^{fib} a fibrant replacement of X), such that $\pi_{**}(f) : \pi_{**}X' \xrightarrow{\cong} \pi_{**}X$. Hence, also $\pi_{**}(1_E \wedge f) : E_{**}X' \to E_{**}X$ is an isomorphism for any $E \in \mathcal{SH}(\mathbb{C})$. In particular,

$$AB(n)_{**}X = 0 \Leftrightarrow AB(n)_{**}X' = 0 \Leftrightarrow AK(n)_{**}X' = 0 \Leftrightarrow AK(n)_{**}X = 0.$$

□

Corollary 147 *For $p > 2$ and $n > 0$,*

$$\langle AK(n)\rangle = \langle AB(n)\rangle \text{ in } \mathcal{SH}(\mathbb{C}).$$

This also holds for $n = 0$, in which case we do not need to assume $k = \mathbb{C}$ or $p > 2$.

Proposition 148 *For any $k \subseteq \mathbb{C}$ and any prime p,*

$$\langle AK(0)\rangle = \langle AB(0)\rangle \text{ in } \mathcal{SH}(k).$$

Proof By definition, $AB(0) = p^{-1}ABP$ and

$$AK(0) = p^{-1}ABP/(v_1, v_2, \cdots) = p^{-1}MGL_{(p)}/(a_1, a_2, \cdots)$$

$$= MGL_{\mathbb{Q}}/(a_1, a_2, \cdots).$$

By the main result in [32], $MGL/(a_1, a_2, \cdots) \cong H\mathbb{Z}$, which implies

$$p^{-1}MGL_{(p)}/(a_1, a_2, \cdots) \cong p^{-1}H\mathbb{Z}_{(p)} = H\mathbb{Q}.$$

Hence, $AK(0) \cong H\mathbb{Q}$. From Lemma 16(1), we already know

$$\langle AK(0)\rangle \le \langle AB(0)\rangle.$$

To prove

$$\langle AK(0)\rangle \ge \langle AB(0)\rangle,$$

let $X \in \mathcal{SH}(k)^{fin}$ satisfy $AK(0)_{**}(X) = 0$. We have to show $AB(0)_{**}(X) = 0$. We use [32, Lemma 7.10], in which we rationalise $H\mathbb{Z}$ and MGL and set $X = MGL_{\mathbb{Q}}$. Here, rationalising a spectrum E means forming the homotopy colimit $E_{\mathbb{Q}} = p^{-1}E_{(p)}$. The lemma then states the following.

If $F \in \mathcal{SH}(k)$ satisfies $H\mathbb{Q} \wedge F = 0$, then $[F, MGL_{\mathbb{Q}}] = 0$.

When rationalising the proof of [32, Lemma 7.10], one uses that $\kappa_0(MGL_{\mathbb{Q}}) \cong (MGL_{\mathbb{Q}})_{\leq 0}$ and $\kappa_0(H\mathbb{Q}) \cong H\mathbb{Q}_{\leq 0}$, which hold because $MGL_{\mathbb{Q}}$ and $H\mathbb{Q}$ are connective (which follows from the connectivity of MGL [32, Corollary 3.9] and of $H\mathbb{Z}$ [32, Lemma 7.3]). Furthermore, one uses that

$$(MGL_{\mathbb{Q}})_{\leq 0} \cong (MGL_{\leq 0})_{\mathbb{Q}} \cong (H\mathbb{Z}_{\leq 0})_{\mathbb{Q}} \cong H\mathbb{Q}_{\leq 0},$$

where the first and third isomorphisms follow from the fact that $(-)_{\leq d}$ preserves filtered homotopy colimits [32, Lemma 2.1] and the middle isomorphism is by [32, Lemma 7.5].

We apply this version of [32, Lemma 7.10] to see that $(MGL_{\mathbb{Q}})^{**}X = 0$ for X as above, which is equivalent to $(MGL_{\mathbb{Q}})_{**}X = 0$, since X is finite (Proposition 73). Using Lemma 16(1) again, it follows that $(p^{-1}ABP)_{**}(X) = 0$, as we wanted to show.

As in the previous proof, the equivalence

$$AK(0)_{**}X = 0 \Leftrightarrow AB(0)_{**}X = 0$$

passes from finite spectra to cellular spectra and then to arbitrary spectra $X \in \mathcal{SH}(k)$. □

As a corollary of the above results, we can prove the following analogue of [73, Theorem 2.1(i)].

Corollary 149 *In $\mathcal{SH}(\mathbb{C})$,*

$$\langle AK(n)\rangle \wedge \langle AK(m)\rangle = 0$$

for any $m \neq n$.

Proof Assume $m > n$. By (1) and (2) of Lemma 17,

$$\langle AK(n)\rangle \wedge \langle AP(n+1)\rangle \leq \langle AE(n)\rangle \wedge \langle AP(n+1)\rangle = \langle 0\rangle.$$

Furthermore, by Lemma 17(3) and the above result,

$$\langle AP(m)\rangle = \langle AB(m)\rangle \vee \langle AP(m+1)\rangle = \langle AK(m)\rangle \vee \langle AP(m+1)\rangle.$$

This implies $\langle AK(m)\rangle \le \langle AP(m)\rangle$. Since $m > n$, $\langle AP(m)\rangle \le \langle AP(n+1)\rangle$ by Lemma 16(1). Hence,

$$\langle AK(n)\rangle \wedge \langle AK(m)\rangle \le \langle AK(n)\rangle \wedge \langle AP(n+1)\rangle = \langle 0\rangle.$$

□

9.5 Decomposition of $\langle AE(n)\rangle$

Recall from Definition 74 that

$$AE(n) = v_n^{-1} ABP/(v_{n+1}, v_{n+2}, \cdots).$$

With the above preparations, we are ready to prove an analogue of the decomposition of Bousfield classes given in [73, Theorem 2.1(d)]. This answers a special case of [27, Question 2.17].

Theorem 20 *For $p > 2$,*

$$\langle AE(n)\rangle = \bigvee_{0\le i\le n} \langle AK(i)\rangle \text{ in } \mathcal{SH}(\mathbb{C}).$$

Proof These are the same arguments as for [73, Theorem 2.1(d)]: By Lemma 17(3) and Corollary 147, $\langle AP(n)\rangle = \langle AB(n)\rangle \vee \langle AP(n+1)\rangle = \langle AK(n)\rangle \vee \langle AP(n+1)\rangle$. Since $AP(0) = ABP$, it follows inductively:

$$\langle ABP\rangle = \langle AK(0)\rangle \vee \langle AK(1)\rangle \vee \cdots \vee \langle AK(n)\rangle \vee \langle AP(n+1)\rangle.$$

Since $AE(n)$ is an ABP-module spectrum, $\langle AE(n)\rangle \le \langle ABP\rangle$ by Lemma 16(2). By Lemma 17(2), $\langle AE(n)\rangle \wedge \langle AP(n+1)\rangle = \langle 0\rangle$. It follows that

$$\langle AE(n)\rangle \le \langle AK(0)\rangle \vee \langle AK(1)\rangle \vee \cdots \langle AK(n)\rangle.$$

By Corollary 133 and Lemma 17(1), $\langle AE(n)\rangle \ge \langle AE(i)\rangle \ge \langle AK(i)\rangle$ for $i \le n$ and hence also $\langle AE(n)\rangle \ge \bigvee_{i\le n} \langle AK(i)\rangle$. □

Remark 150 The restriction $p > 2$ originates from Sect. 9.3, where it was needed to prove homotopy associativity for the map $\mu_{AP(n)} : AP(n) \wedge AP(n) \to AP(n)$. There might be a different way to show that the ABP_{**}-action on $AP(n)_{**}(X)$ induces an $AP(n)_{**}$-action, in which case the condition $p > 2$ could be removed in the previous section and in the theorem.

9.6 $AK(n)$ and $AK(n+1)$

Lemma 30 *Let p be any fixed prime. For $n \geq 1$ and $X \in \mathcal{SH}(\mathbb{C})^{fin}$, $AP(n)_{**}(X)$ is an $AP(1)_{**}AP(1)$-comodule and a coherent module over $AP(1)_{**}$.*

Proof Note that $AP(1) = ABP/p$ is a ring spectrum because $p : ABP \to ABP$ is a map of ring spectra and the category of ring spectra is cocomplete. The coaction on $AP(n)_{**}(X)$ is defined via

$$AP(n)_{**}X \to AP(n)_{**}(AP(1) \wedge X) \leftarrow AP(1)_{**}AP(1) \otimes_{AP(1)_{**}} AP(n)_{**}X,$$

where the left map is induced by the unit of $AP(1)$ and we need to show that the right map (induced by $AP(1) \wedge AP(n) \to AP(n)$, compare Corollary 139) is an isomorphism. To prove this isomorphism, it suffices to show that, in the spectral sequence from [16, Proposition 7.7],

$$\mathrm{Tor}^{AP(1)_{**}}(AP(1)_{**}AP(1), AP(n)_{**}X) \Rightarrow AP(n)_{**}(AP(1) \wedge X),$$

$AP(1)_{**}AP(1)$ is free over $AP(1)_{**}$, so that the spectral sequence collapses immediately. We want to apply Lemma 5 to $AP(1) \wedge AP(1)$. Recall that the slice spectral sequence considered in Lemma 5 converges for quotients of Landweber exact spectra by [32, Theorem 8.12 and Example 8.13]. Now, $ABP \wedge ABP$ is a product of Landweber exact spectra and is therefore Landweber exact, see e.g. [65, Remark 9.2], and $AP(1) \wedge AP(1)$ is a quotient of $ABP \wedge ABP$. Hence, the slice spectral sequence converges strongly, and it collapses for the same degree reasons as in Lemma 5. Thus, $AP(1)_{**}AP(1) \cong P(1)_*P(1)[\tau]$. By [40, Sect. 1], this is isomorphic to $P(1)_*[\tau, z^{E,A}]$ for certain $z^{E,A}$. In particular, it is free over $AP(1)_{**} \cong P(1)_*[\tau]$, as we wanted to show.

Now we show that $AP(n)_{**}(X)$ is coherent over $AP(1)_{**}$. Recall that $P(1)_* \cong \mathbb{F}_p[v_1^{\mathrm{top}}, v_2^{\mathrm{top}}, \cdots]$ is a coherent ring (see [11, Sect. 1]). The same holds for $AP(1)_{**} \cong \mathbb{F}_p[\tau, v_1, v_2, \cdots]$. By [11, Proposition 1.2], coherence of modules satisfies the two out of three property for exact triangles of graded modules (i.e., long exact sequences). It follows that $AP(n)_{**}$ is a coherent $AP(1)_{**}$-module, and cellular induction implies that $AP(n)_{**}(X)$ is a coherent $AP(1)_{**}$-module, too. □

Setting $n = 1$ and $X = S^0$, the above lemma tells us that $AP(1)_{**}$ is a coherent $AP(1)_{**}AP(1)$-comodule, where coherent means coherent as an $AP(1)_{**}$-module.

Lemma 31 (Invariant prime ideals) *For $k = \mathbb{C}$ and p any prime, the invariant prime ideals of $AP(1)_{**}$ (that is, prime ideals which are also sub-comodules) are given by $I_m = (v_1, \cdots, v_{m-1})$ and $\overline{I}_m = (\tau, v_1, \cdots, v_{m-1})$.*

Proof We have $AP(1)_{**} \cong P(1)_*[\tau]$ and $AP(1)_{**}AP(1) \cong P(1)_*P(1)[\tau]$, as in the proof of the previous lemma. Under the functor $R_{\mathbb{C}}$, the coaction

$$AP(1)_{**} \to AP(1)_{**}AP(1) \otimes_{AP(1)_{**}} AP(1)_{**}$$

realises to

$$P(1)_* \to P(1)_*P(1) \otimes_{P(1)_*} P(1)_*$$

(similarly as in the proof of Corollary 132). By the classical invariant prime ideal theorem (see [44, Theorem 2.7] or compare [40, Theorem 1.16]), the invariant prime ideals of $P(1)_*$ are given by $I_m^{\text{top}} = (v_1^{\text{top}}, \cdots, v_{m-1}^{\text{top}})$. The isomorphism

$$AP(1)_{**}AP(1) \otimes_{AP(1)_{**}} AP(1)_{**} \cong P(1)_*P(1) \otimes_{P(1)_*} AP(1)_{**}$$

implies that I_m is an invariant prime ideal in $AP(1)_{**}$. By $AP(1)_{**} \cong P(1)_*[\tau]$, it follows also that $\overline{I}_m$ is an invariant prime ideal of $AP(1)_{**}$, too.

It remains to show that there cannot be any further invariant prime ideals. As in [44], this follows from the fact that the only primitive elements in $AP(1)_{**}/\overline{I}_m \cong P(1)_*/I_m^{\text{top}}$ are multiples of powers of v_m (compare [44, Proposition 2.11]). □

Corollary 151 (Motivic Landweber filtration theorem) *Let p be any prime and $n \geq 1$. For $X \in \mathcal{SH}(\mathbb{C})^{fin}$, $AP(n)_{**}(X)$ can be filtered by $AP(1)_{**}$-modules*

$$AP(n)_{**}(X) = M_0 \supset \cdots \supset M_k = 0$$

such that $M_i/M_{i+1} \cong AP(1)_{**}/I_m$ or $AP(1)_{**}/\overline{I}_m$ for some I_m and $\overline{I}_m$ $(m \geq n)$ as above.

Proof By Lemma 30, $AP(n)_{**}(X)$ is a coherent $AP(1)_{**}AP(1)$-comodule. Landweber's filtration theorem [45, Theorem 3.3] (see also [40, Theorem 1.16]) implies that $AP(n)_{**}(X)$ has a filtration

$$AP(n)_{**}(X) = M_0 \supset \cdots \supset M_k = 0$$

such that $M_i/M_{i+1} \cong AP(1)_{**}/I$ for some I which is invariant under the comodule action. Thus, the claim follows from the previous lemma. □

The following is a motivic version of one statement in [73, Theorem 2.11]. In the proof, we use ideas from Ravenel's proof.

Theorem 21 *Let $p > 2$. If $X \in \mathcal{SH}(\mathbb{C})^{fin}$ satisfies $AK(n+1)_{**}(X) = 0$, then also $AK(n)_{**}(X) = 0$. That is, $\langle AK(n+1)\rangle \geq \langle AK(n)\rangle$ in $\mathcal{SH}(\mathbb{C})^{fin}$.*

Proof Assume $n > 0$. The case $n = 0$ will be considered at the end of the proof. Let $E_{**}(-)$ be defined by

$$E_{**}X = AE(n+1)_{**} \otimes_{ABP_{**}} AP(n)_{**}(X).$$

As in [40, Lemma 3.5], the above Landweber filtration theorem and the fact that $AE(n+1)$ is Landweber exact (see [65, Theorem 8.7]) yield

$$\mathrm{Tor}_1^{AP(n)_{**}}(AE(n+1)_{**} \otimes_{ABP_{**}} AP(n)_{**}, AP(n)_{**}/I) = 0$$

for all invariant prime ideals $I \subseteq AP(n)_{**}$ as above. (Alternatively, this can be derived from the topological analogue and Lemma 5, using [65, Theorem 8.7].) As in [73, Theorem 2.11], this implies that $E_{**}(-)$ is an exact functor.

In analogy to [73], we show that there is an injective pairing

$$AK(n+1)_{**} \otimes_{E_{**}} E_{**}X \hookrightarrow AK(n+1)_{**}X.$$

To construct this pairing, note that the ABP-action on $AK(n+1)$ factors through a map $AP(n) \wedge AK(n+1) \to AK(n+1)$ by methods from Sect. 9.3, since v_i, $i < n+1$, acts trivially on $AK(n+1)$ by Corollary 136. This induces a map

$$AK(n+1)_{**} \otimes_{ABP_{**}} AP(n)_{**}X \to AK(n+1)_{**}X.$$

As in [73, Theorem 2.11], this map factors through a pairing

$$AK(n+1)_{**} \otimes_{E_{**}} E_{**}X \to AK(n+1)_{**}X,$$

the reason being again that the relevant elements act trivially. Such a pairing induces a universal coefficient spectral sequence, whose motivic version is constructed in [16, Propositions 7.7 and 7.10],

$$\mathrm{Tor}_i^{E_{**}}(E_{**}(X), AK(n+1)_{**}) \Rightarrow AK(n+1)_{**}(X).$$

As in [73], to prove the injectivity of the pairing, it suffices to prove the vanishing of

$$\mathrm{Tor}_i^{ABP_{**}}(AP(n)_{**}(X), AK(n+1)_{**})$$

for $i > 1$. Since $n \geq 1$, p acts trivially on both of these modules, and we can replace ABP by $AP(1) = ABP/p$. Hence, we have to show the vanishing of

$$\mathrm{Tor}_i^{AP(1)_{**}}(AP(n)_{**}(X), AK(n+1)_{**})$$

and, by Corollary 151, the question reduces to the vanishing of

$$\mathrm{Tor}_i^{AP(1)_{**}}(AP(1)_{**}/I, AK(n+1)_{**})$$

for $i > 1$ and $I = I_m$ or $\overline{I}_m$, $m \geq n$. By Lemma 5, this equals

$$\mathrm{Tor}_i^{P(1)_*[\tau]}(P(1)_*/I_m[\tau], K(n+1)_*[\tau]) \text{ or}$$

$$\mathrm{Tor}_i^{P(1)_*[\tau]}(P(1)_*/I_m, K(n+1)_*[\tau]), \quad \text{respectively.}$$

A projective resolution of $K(n+1)_*$ over $P(1)_*$ yields a projective resolution of $K(n+1)_*[\tau]$ over $P(1)_*[\tau]$ by applying $- \otimes \mathbb{F}_p[\tau]$. It follows that both of the above torsion terms vanish if

$$\mathrm{Tor}_i^{P(1)_*}(P(1)_*/I_m, K(n+1)_*) = 0.$$

For $i > 1$, this follows from

$$\mathrm{Tor}_i^{BP_*}(BP_*/I_m, K(n+1)_*) = 0,$$

as in the proof of [73, Theorem 2.11]. This proves the injectivity of the above pairing.

Now, assume that $AK(n+1)_{**}X = 0$. The injectivity of the pairing implies that $AK(n+1)_{**} \otimes_{E_{**}} E_{**}X = 0$. Recall $AK(n+1)_{**} \cong \mathrm{H}_{**}[v_{n+1}^{\pm 1}]$ and note that

$$E_{**} \cong AE(n+1)_{**} \otimes_{ABP_{**}} AP(n)_{**} \cong E(n+1)_* \otimes_{BP_*} P(n)_*[\tau] \cong \mathrm{H}_{**}[v_n, v_{n+1}^{\pm 1}]$$

by [65, Theorem 8.7] and Lemma 5. Therefore, $AK(n+1)_{**} \otimes_{E_{**}} E_{**}X = 0$ implies $v_n^{-1}E_{**}X = 0$. Now,

$$0 = v_n^{-1}E_{**}X = AE(n+1)_{**} \otimes_{ABP_{**}} AB(n)_{**}X,$$

by the definitions of $E_{**}(-)$ and of $AB(n)$. By Theorem 19, since X is finite, this is equal to

$$AE(n+1)_{**} \otimes_{ABP_{**}} AK(n)_{**}(X) \otimes \mathbb{F}_p[v_{n+1}, v_{n+2}, \cdots].$$

By [65, Theorem 8.7], $AE(n+1)_{**} \cong E(n+1)_* \otimes_{BP_*} ABP_{**}$. It follows

$$0 = \mathbb{Z}_{(p)}[v_1, \cdots, v_n, v_{n+1}^{\pm 1}] \otimes_{\mathbb{Z}_{(p)}[v_1, \cdots]} AK(n)_{**}X \otimes \mathbb{F}_p[v_{n+1}, \cdots].$$

As v_m acts trivially on $AK(n)_{**}X$ for all $m \neq n$ by Corollary 136, this implies that $AK(n)_{**}X = 0$.

It remains to show that $AK(1)_{**}X = 0$ implies $AK(0)_{**}X = 0$. Recall the definitions $AE(1) = v_1^{-1}ABP/(v_2, \cdots)$ and $AK(1) = v_1^{-1}ABP/(p, v_2, \cdots) = AE(1)/p$. Since multiplication by p commutes with $\pi_{**}(-)$, it follows that $AK(1)_{**}X = 0$ implies $p^{-1}AE(1)_{**}X = 0$. (Note that $p^{-1}(AE(1)_{**}X) \cong (p^{-1}AE(1))_{**}X$ since $\pi_{**}(-)$ commutes with filtered colimits by [16, Proposition 9.3].) The following argument is closely related to Corollary 133. Since $p^{-1}AE(1)_{**}X = 0$, $p^{-1}(ABP/(v_2, \cdots))_{**}X$ is v_1-torsion. By Theorem 17, this implies that $p^{-1}(ABP/(v_2, \cdots))_{**}X$ is p-torsion. But this can only be the case if $p^{-1}(ABP/(v_2, \cdots))_{**}X = 0$. It follows that

$$AE(0)_{**}X = p^{-1}(ABP/(v_1, v_2, \cdots))_{**}X = 0.$$

Now, by Theorem 20, $\langle AE(0)\rangle = \langle AK(0)\rangle$. Hence, $AK(0)_{**}X = 0$. $\square$

In terms of thick ideals in $\mathcal{SH}(\mathbb{C})^{fin}_{(p)}$, $p > 2$, we have proven that the motivic Morava K-theory spectra $AK(n)$ indeed describe a descending chain of thick ideals, similarly to the chain of thick subcategories in $\mathcal{SH}^{fin}_{(p)}$. The inclusions are proper by Sect. 8.

Corollary 152 *For $p > 2$,*

$$\mathcal{SH}(\mathbb{C})^{fin}_{(p)} \supsetneq \mathcal{C}_{AK(0)} \supsetneq \mathcal{C}_{AK(1)} \supsetneq \cdots .$$

To sum up, we have identified three sequences of thick ideals:

$$\begin{array}{ccccccc}
\mathcal{SH}(\mathbb{C})^{fin}_{(p)} & \supset & R^{-1}(\mathcal{C}_1) & \supset & R^{-1}(\mathcal{C}_2) & \supset & \cdots \\
\| & & \cup & & \cup & & \\
\mathcal{SH}(\mathbb{C})^{fin}_{(p)} & \supset & \mathcal{C}_{AK(0)} & \supset & \mathcal{C}_{AK(1)} & \supset & \cdots \\
\| & & \not\cup & & \not\cup & & \\
\mathcal{SH}(\mathbb{C})^{fin}_{(p)} & \supset & \mathrm{thickid}(c\mathcal{C}_1) & \supset & \mathrm{thickid}(c\mathcal{C}_2) & \supset & \cdots .
\end{array}$$

The lower vertical inequalities are shown in [84, Sect. 3.6].

Acknowledgements This paper is the outcome of my dissertation which has been supervised by Jens Hornbostel. I am grateful for his steady support and for all the helpful discussions. Thanks to Neil Strickland for sharing his preprint on thick ideals in equivariant stable homotopy categories with me and for allowing me to incorporate parts of it in this work. I also thank my colleagues Marcus Zibrowius and Jeremiah Heller, who helped me a lot with their comments, and Sven Stahn, who pointed out a mistake in a previous version of this work. Furthermore, I thank Marc Hoyois for answering some of my questions. This work was financed by the Deutsche Forschungsgemeinschaft.

References

1. Adams, J.F.: Lectures on generalised cohomology. Category Theory, Homology Theory and Their Applications III. Springer, Berlin (1969)
2. Adams, J.F.: Stable Homotopy and Generalised Homology. University of Chicago Press, Chicago (1974)
3. Andrews, M., Miller, H.: Inverting the Hopf map. J. Topol. **10**(4), 1145–1168 (2017)
4. Ayoub, J.: Les six oprations de Grothendieck et le formalism des cycles vanescents dans le monde motivique. Astérisque, pp. 314–315 (2007)
5. Ayoub, J.: Note sur les oprations de Grothendieck et la ralisation de Betti. J. Inst. Math. Jussieu **9**(2), 225–263 (2010)
6. Balmer, P.: The spectrum of prime ideals in tensor triangulated categories. J. Reine Angew. Math. **588**, 149–168 (2005)
7. Balmer, P.: Spectra, spectra, spectra - tensor triangular spectra versus Zariski spectra of endomorphism rings. Algebr. Geom. Topol. **10**(3), 1521–1563 (2010)

8. Balmer, P., Sanders, B.: The spectrum of the equivariant stable homotopy category of a finite group. Invent. Math. **208**(1), 283–326 (2017)
9. Basterra, M., Mandell, M.A.: The multiplication on BP. J. Topol. **6**, 285–310 (2013)
10. Borghesi, S.: Algebraic Morava K-theories. Invent. Math. **151**, 381–413 (2003)
11. Conner, P.E., Smith, L.: On the complex bordism of finite complexes. Publ. Math. Inst. Hautes Etudes Sci. **37**, 117–221 (1969)
12. Devinatz, E.S., Hopkins, M.J., Smith, J.H.: Nilpotence and stable homotopy theory I. Ann. Math. **128**, 207–241 (1988)
13. Dugger, D.: Replacing model categories with simplicial ones. Trans. Am. Math. Soc. **353**, 5003–5027 (2001)
14. Dugger, D.: Motivic stable homotopy groups of spheres (2012). http://sma.epfl.ch/~hessbell/arolla/slides12/Dugger.pdf
15. Dugger, D., Isaksen, D.: Topological hypercovers and A1-realizations. Math. Z. **246**, 667–689 (2004)
16. Dugger, D., Isaksen, D.: Motivic cell structures. Algebr. Geom. Topol. **5**, 615–652 (2005)
17. Dugger, D., Isaksen, D.: The motivic Adams spectral sequence. Geom. Topol. **14**, 967–1014 (2010)
18. Elmendorf, A.D., et al.: Rings, Modules and Algebras in Stable Homotopy Theory. American Mathematical Society, Providence (1997)
19. Fausk, H., Lewis Jr., L.G., May, J.P.: The Picard group of equivariant stable homotopy theory. Adv. Math. **163**, 17–33 (2001)
20. Greenlees, J.P.C., May, J.P.: Generalized Tate cohomology. Mem. AMS (1995)
21. Guillou, B.: A short note on models for equivariant homotopy theory (2006). http://www.ms.uky.edu/~guillou/EquivModels.pdf
22. Hatcher, A.: Algebraic Topology. Cambridge University Press, Cambridge (2001)
23. Heller, J., Ormsby, K.: Galois equivariance and stable motivic homotopy theory. Trans. Am. Math. Soc. **368**, 8047–8077 (2016)
24. Heller, J., Ormsby, K.: Primes and fields in stable motivic homotopy theory. Geom. Topol. **22**(4), 2187–2218 (2018)
25. Hopkins, M.J., Smith, J.H.: Nilpotence and stable homotopy theory II. Ann. Math. **148**, 1–49 (1998)
26. Hopkins, M.J., Mahowald, M., Sadofsky, H.: Constructions of elements in Picard groups. Contemp. Math. **158**, 89–126 (1994)
27. Hornbostel, J.: Localizations in motivic homotopy theory. Math. Proc. Camb. Philos. Soc. **140**, 95–114 (2006)
28. Hovey, M.: Model Categories. AMS, Providence (1999)
29. Hovey, M.: Spectra and symmetric spectra in general homotopy categories. J. Pure Appl. Algebra **165**, 63–127 (2001)
30. Hovey, M., Sadofsky, H.: Tate cohomology lowers chromatic Bousfield classes. Proc. Am. Math. Soc. **124**, 3579–3585 (1996)
31. Hovey, M., Shipley, B., Smith, J.: Symmetric spectra. J. Am. Math. Soc. **13**, 149–208 (1999)
32. Hoyois, M.: From algebraic cobordism to motivic cohomology. Journal für die reine und angewandte Mathematik (Crelles J.) (2013)
33. Hu, P.: Base change functors in the A1-stable homotopy category. Homol. Homotopy Appl. **3**, 417–451 (2001)
34. Hu, P.: S-modules in the category of schemes. Mem. Am. Math. Soc. **767** (2003), 125 pp
35. Hu, P.: On the Picard group of the A1-stable homotopy category. Topology **44**(3), 609–640 (2005)
36. Hu, P., Kriz, I., Ormsby, K.: Convergence of the motivic Adams spectral sequence. J. K-theory **7**(03), 573–596 (2011)
37. Hulek, K.: Elementare Algebraische Geometrie. Springer, Berlin (2012)
38. Jardine, J.F.: Motivic symmetric spectra. Doc. Math. **5**, 445–552 (2000)
39. Johnson, D.C., Wilson, W.S.: BP operations and Morava's extraordinary K-theories. Math. Z. **144**, 55–75 (1975)

40. Johnson, D.C., Yosimura, Z.: Torsion in Brown-Peterson homology and Hurewicz homomorphisms. Osaka J. Math. **17**, 117–136 (1980)
41. Jouanolou, J.P.: Une Suite exacte de Mayer-Vietoris en K-Theorie algebrique. Higher K-Theories (1973)
42. Kelly, S.: Some observations about motivic tensor triangulated geometry over a finite field. To appear in this proceedings (2019)
43. Lam, T.Y.: Introduction to Quadratic Forms over Fields. Graduate Studies in Mathematics, vol. 67. AMS, Providence (2005)
44. Landweber, P.S.: Annihilator ideals and primitive elements in complex cobordism. Ill. J. Math. **17**, 273–284 (1973)
45. Landweber, P.S.: Associated prime ideals and Hopf algebras. J. Pure Appl. Algebra **3**, 43–58 (1973)
46. Landweber, P.S.: Homological properties of comodules over MU*MU and BP*BP. Am. J. Math. **98**, 591–610 (1976)
47. Lazarsfeld, R.K.: Positivity in Algebraic Geometry I. Springer, Berlin (2004)
48. Levine, M.: The homotopy coniveau tower. J. Topol. **1**, 217–267 (2008)
49. Levine, M.: A comparison of motivic and classical stable homotopy theories. J. Topol. **7**, 327–362 (2014)
50. Lewis, L.G., Mandell, M.A.: Equivariant universal coefficient and Küunneth spectral sequences. Proc. Lond. Math. Soc. **92**, 505–544 (2006)
51. Lewis, L.G., May, J.P., Steinberger, M.: Equivariant Stable Homotopy Theory. Springer, Berlin (1986)
52. Mandell, M.A.: Equivariant symmetric spectra. Contemporary Mathematics, vol. 346, pp. 399–452. AMS, Providence (2002)
53. Mandell, M.A., May, J.P.: Equivariant orthogonal spectra and S-modules. Mem. AMS (2002)
54. May, J.P.: Equivariant Homotopy and Cohomology Theory. AMS and CBMS, Providence (1996)
55. May, J.P., Sigurdsson, J.: Parametrized Homotopy Theory. Mathematical Surveys and Monographs, vol. 132. American Mathematical Society, Providence (2006)
56. Mayer, K.H.: G-invariante Morse Funktionen. Manuscr. Math. **63**, 99–114 (1989)
57. Milnor, J.: Morse Theory. Princeton University Press, Princeton (1963)
58. Mitchell, S.A.: Finite complexes with A(n)-free cohomology. Topology **24**, 227–248 (1985)
59. Morel, F.: Suite spectrale d'Adams et invariants cohomologiques des formes quadratiques. C. R. Acad. Sci. Paris Ser. I **328**, 963–968 (1999)
60. Morel, F.: On the motivic π_0 of the sphere spectrum. In: Greenlees, J.P.C. (ed.) Axiomatic, Enriched and Motivic Homotopy Theory. NATO Science Series, pp. 219–260 (2002)
61. Morel, F.: An introduction to A1-homotopy theory. Contemporary Developments in Algebraic K-Theory, I.C.T.P. Lecture Notes, vol. 15, pp. 357–441 (2003)
62. Morel, F.: A1-algebraic topology. In: Proceedings of the International Congress of Mathematicians Madrid (2006)
63. Morel, F., Voevodsky, V.: A1-homotopy theory of schemes. Publ. Math. l'IHS **90**, 45–143 (1999)
64. Nassau, C.: Eine nichtgeometrische Konstruktion der Spektren P(n) und multiplikative Automorphismen von K(n). Diplomarbeit. Goethe Universität Frankfurt am Main (1995). http://publikationen.ub.uni-frankfurt.de/frontdoor/index/index/docId/4311
65. Naumann, N., Spitzweck, M., Østvœr, P.A.: Motivic Landweber exactness. Doc. Math. **14**, 551–593 (2009)
66. Neeman, A.: The Grothendieck duality theorem via Bousfield's techniques and Brown representability. J. AMS **9**, 205–236 (1996)
67. Neeman, A.: Triangulated Categories. Princeton University Press, Princeton (2001)
68. Panin, I., Yagunov, S.: Rigidity for orientable functors. J. Pure Appl. Algebra **172**, 49–77 (2002)
69. Panin, I., Pimenov, K., Röndigs, O.: A universality theorem for Voevodsky's algebraic cobordism spectrum. Homol. Homotopy Appl. **10**(2), 211–226 (2008)

70. Panin, I., Pimenov, K., Röndigs, O.: On Voevodsky's algebraic K-theory spectrum. In: Baas, N., et al. (eds.) Algebraic Topology. Abel Symposia, vol. 4, pp. 279–330. Springer, Berlin (2009)
71. Pelaez, P.: Multiplicative properties of the slice filtration. Asterisque **335**, 1–289 (2011)
72. Quillen, D.: Elementary proofs of some results of cobordism theory using Steenrod operations. Adv. Math. **7**, 29–56 (1971)
73. Ravenel, D.C.: Localization with respect to certain periodic homology theories. Am. J. Math. **106**, 351–414 (1984)
74. Ravenel, D.C.: Complex Cobordism and Stable Homotopy Groups of Spheres. Academic, New York (1986)
75. Ravenel, D.C.: Nilpotence and Periodicity in Stable Homotopy Theory. Princeton University Press, Princeton (1992)
76. Riou, J.: Dualité de Spanier-Whitehead en géométrie algébrique. C. R. Acad. Sci. Paris I **340**, 431–436 (2005)
77. Röndigs, O., Østvær, P.A.: Modules over motivic cohomology. Adv. Math. **219**(2), 689–727 (2008)
78. Schwede, S.: Symmetric spectra (2012). http://www.math.uni-bonn.de/people/schwede/SymSpec-v3.pdf
79. Schwede, S.: Lectures on equivariant stable homotopy theory (2013). http://www.math.uni-bonn.de/people/schwede/equivariant.pdf
80. Schwede, S., Shipley, B.: Algebras and modules in monoidal model categories. Proc. Lond. Math. Soc. **80**, 491–511 (2000)
81. Segal, G.B.: Equivariant stable homotopy theory. Actes Congrés intern. Math. **2**, 59–63 (1970)
82. Spitzweck, M.: Relations between slices and quotients of the algebraic cobordism spectrum. Homol. Homotopy Appl. **12**(2), 335–351 (2010)
83. Stahn, S.: Die motivische Adams-Spektralsequenz (2012). http://www2.math.uni-wuppertal.de/~hornbost/DiplomarbeitSvenStahn.pdf
84. Stahn, S.: Stable motivic homotopy groups and periodic self maps at odd primes. Dissertation. Bergische Universität Wuppertal (2018)
85. Strickland, N.: Products on MU-modules. Trans. Am. Math. Soc. **351**, 2569–2606 (1999)
86. Strickland, N.P.: Thick ideals of finite G-spectra. Unpublished (2010)
87. Vezzosi, G.: Brown-Peterson spectra in stable A1-homotopy theory. Rend. Sem. Mat. Univ. Padova **106**, 47–64 (2001)
88. Voevodsky, V.: The Milnor conjecture. Preprint (1996)
89. Voevodsky, V.: A1-homotopy theory. Doc. Math. Extra Vol. ICM 579–604 (1998)
90. Voevodsky, V.: Motivic cohomology with Z/2 coeffcients. Publ. Math. Inst. Hautes Etudes Sci. **98**, 1–57 (2003)
91. Voevodsky, V.: Reduced power operations in motivic cohomology. Publ. Math. l'IHS **98**, 1–57 (2003)
92. Voevodsky, V.: Motivic Eilenberg-MacLane spaces. Publ. Math. l'IHES **112**, 1–99 (2010)
93. Wasserman, A.G.: Equivariant differential topology. Topology **8**, 127–150 (1969)
94. Würgler, U.: Cobordism theories of unitary manifolds with singularities and formal group laws. Math. Z. **150**, 239–260 (1976)
95. Würgler, U.: On products in a family of cohomology theories associated to the invariant prime ideals of $\pi_*(BP)$. Comment. Math. Helv. **52**, 457–481 (1977)
96. Yagita, N.: Applications of Atiyah-Hirzebruch spectral sequences for motivic cobordisms. Proc. Lond. Math. Soc. **90**, 783–816 (2005)

Some Observations About Motivic Tensor Triangulated Geometry over a Finite Field

Shane Kelly

Abstract We give a brief introduction to tensor triangulated geometry, a brief introduction to various motivic categories, and then make some observations about the conjectural structure of the tensor triangulated spectrum of the Morel–Voevodsky stable homotopy category over a finite field.

Keywords Tensor triangulated categories · Motivic cohomology · Finite fields · Milnor-Witt K-theory

1 Introduction

These are notes based on three lectures I gave at the workshop "Bousfield classes form a set: a workshop in memory of Tetsusuke Ohkawa" at Nagoya University in August 2015.

The goal of the lectures was to give a brief sketch of the Morel–Voevodsky stable homotopy category $SH(S)$ and motivic stable homotopy groups of spheres, aimed at someone with no previous experience with motives, and then in the last lecture see if anything could be said about the tensor triangulated spectrum $\mathrm{Spc}(SH(S)^c)$ of $SH(S)$. I expected, perhaps naïvely, $\mathrm{Spc}(SH(S)^c)$ to be completely intractable, but to my surprise, it is possible to give a conjectural description of $\mathrm{Spc}(SH(\mathbb{F}_q)^c_{\mathbb{Q}})$ (cf. [32, Conj. 51] for Beilinson–Parshin and [34, pg. 17] for Rat = Num.).

Theorem 1.1 (Theorem 4.1) *Let $\mathbb{F}_q$ be a field with a prime power, q, number of elements. Suppose that for all connected smooth projective varieties X we have:*

$$\begin{array}{lll} CH^i(X, j)_{\mathbb{Q}} = 0, & \forall j \neq 0, i \in \mathbb{Z} & \textit{(Beilinson–Parshin conjecture)}, \\ CH^i(X)_{\mathbb{Q}} \otimes CH_i(X)_{\mathbb{Q}} \to CH_0(X)_{\mathbb{Q}} \textit{ is non-degenerate.} & & \textit{(Rat. and num. equiv. agree)} \end{array} \quad (1)$$

Then

S. Kelly (✉)
Tokyo Institute of Technology, 2-12-1 Ookayama, Meguro-ku, Tokyo 152-8551, Japan
e-mail: shanekelly@math.titech.ac.jp

T. Ohsawa and N. Minami (eds.), *Bousfield Classes and Ohkawa's Theorem*,
Springer Proceedings in Mathematics & Statistics 309,
https://doi.org/10.1007/978-981-15-1588-0_7

$$Spc(SH(\mathbb{F}_q)^c_{\mathbb{Q}}) \cong Spec(\mathbb{Q}). \tag{2}$$

Later, I found out there are a considerable number of results about $\mathrm{Spc}(SH(k)^c)$ for $k \subseteq \mathbb{C}$ in [31], using methods adapted from the study of the classical stable homotopy category. Another paper studying this object is [25], where it is proven that Balmer's comparison map is surjective (for any field of non-even characteristic!). They also address what is possibly one of the most important questions in this area of study—the production of field spectra.

For a speculative discussion about the structure of $\mathrm{Spc}(SH(\mathbb{F}_q)^c)$ see Sect. 4.

In this last section we also prove that Balmer's comparison map

$$\mathrm{Spc}(SH(\mathbb{F}_q)^c) \to \mathrm{Spec}^h(K^{MW}(\mathbb{F}_q)) \tag{3}$$

is surjective in the special case of finite fields of prime characteristic, Corollary 4.6. To do this we exploit the fact that we can describe $\mathrm{Spec}^h(K^{MW}(\mathbb{F}_q))$ completely—an approach due to Ormsby.

Outline. The first section contains the basic definitions of tensor triangulated geometry, and references to the literature. It is aimed at the motivic reader who has had minimal to no exposure to these things.

The second section contains basic definitions of various motivic categories, and references to the literature. It is aimed at the tensor triangulated geometer who has had minimal to no exposure to these things.

The last section contains some observations and guesses about the structure of $\mathrm{Spc}(SH(\mathbb{F}_q)^c)$. In particular, it contains Theorem 4.1 and its proof.

2 Tensor Triangulated Geometry

In this section we recall some basic definitions from tensor triangulated geometry. For a much more readable exposition of this material the reader is encouraged to consult [10].

Example 2.1 The most enlightening example to keep in mind when reading the following definitions is the bounded derived category of coherent sheaves on smooth variety X over a field, or equivalently, the derived category $D^{\mathrm{perf}}(X)$ of perfect complexes on X. Of course, there are other important examples coming from scheme theory, stable homotopy theory, modular representation theory, noncommutative topology, and, as we shall see below, the theory of motives, [10, §1].

Warning 2.2 Whereas the elements of a ring R behave like functions, the objects in a $\otimes$-triangulated $\mathscr{K}$ behave like (bounded complexes of) sheaves: a closed subset of $\mathrm{Spec}(R)$ corresponds to the set of functions vanishing on it, whereas a closed subset of $D^{\mathrm{perf}}(\mathrm{Spec}(R))$ corresponds to the set of (perfect complexes of) sheaves supported on it. Consequently, the correspondence $\{$ ideals $\} \leftrightarrow \{$ closed subsets $\}$ is inclusion preserving in the $\otimes$-triangulated world, where it is inclusion reversing in the world of commutative algebra. Cf. also Proposition 2.7(2) and (3) below. For a more surprising example of this phenomenon see [10, Rmk. 27]. Cf. also [35].

Definition 2.3 Let $(\mathscr{K}, \otimes, \mathbb{1})$ be an essentially small $\otimes$-triangulated category, i.e., a triangulated category equipped with a monoidal structure $\mathscr{K} \times \mathscr{K} \xrightarrow{\otimes} \mathscr{K}$ with a unit object $\mathbb{1}$, such that $\otimes$ is exact in each variable, [10, Def. 3]. Let $\mathscr{J}$ be a non-empty full triangulated subcategory.

1. $\mathscr{J}$ is called *thick* if it is stable under direct summands; $a \oplus b \in \mathscr{J} \Rightarrow a \in \mathscr{J}$ or $b \in \mathscr{J}$, [10, Def. 7]. For any closed subvariety Z of a smooth variety X, the subcategory of objects supported on Z, i.e., objects sent to zero by the canonical functor $D^{\mathsf{perf}}(X) \to D^{\mathsf{perf}}(X-Z)$, is thick.
2. $\mathscr{J}$ is said to be a $\otimes$-*ideal* if $\mathscr{K} \otimes \mathscr{J} \subseteq \mathscr{J}$, [10, Def. 7]. The subcategories $ker(D^{\mathsf{perf}}(X) \to D^{\mathsf{perf}}(X-Z))$ just mentioned are tensor ideals. For an example of a thick subcategory which is not a tensor ideal, the reader could consider using the Fourier-Mukai transform [27, Prop. 9.19] between the derived category of an abelian variety and its dual, as this preserves thick subcategories, but not the tensor structure.
3. A *prime* of $\mathscr{K}$ is a thick $\otimes$-ideal $\mathscr{P}$ such that $\mathbb{1} \notin \mathscr{P}$ and $a \otimes b \in \mathscr{P} \Rightarrow a \in \mathscr{P}$ or $b \in \mathscr{P}$, [10, Constr. 8]. The set of primes is denoted by
$$\mathrm{Spc}(\mathscr{K}) = \{\text{ primes of } \mathscr{K}\}. \tag{4}$$
The ideals $ker(D^{\mathsf{perf}}(X) \to D^{\mathsf{perf}}(X-Z))$ mentioned above are prime if and only if Z is irreducible.
4. The *support*, denoted by $\mathrm{supp}(a)$, of an object $a \in \mathscr{K}$ is the set of primes not containing it,
$$\mathrm{supp}(a) = \{\mathscr{P} \in \mathrm{Spc}(\mathscr{K}) : a \notin \mathscr{P}\}. \tag{5}$$
The complement of $\mathrm{supp}(a)$ is denoted by, [10, Constr. 8],
$$U(a) = \{\mathscr{P} \in \mathrm{Spc}(\mathscr{K}) : a \in \mathscr{P}\}. \tag{6}$$
5. The set $\mathrm{Spc}(\mathscr{K})$ has a canonical topology with basis the sets $U(a)$ as a ranges over all objects in $\mathscr{K}$, [10, Constr. 8].
6. To a subset Y of $\mathrm{Spc}(\mathscr{K})$, we associate the full subcategory
$$\mathscr{K}_Y = \{a \in \mathscr{K} : \mathrm{supp}(a) \subseteq Y\}. \tag{7}$$
If Y is a union $Y = \cup_{i \in I} Y_i$ of subsets Y_i whose complement $\mathrm{Spc}(\mathscr{K}) - Y_i$ is open and *quasi-compact* (in the sense that every open cover admits a finite subcover), then $\mathscr{K}_Y$ is a thick $\otimes$-ideal of $\mathscr{K}$, [10, Rmk. 12, Thm. 14].
7. If U is a quasi-compact open of $\mathrm{Spc}(\mathscr{K})$, with closed complement $Z = \mathrm{Spc}(\mathscr{K}) - U$, one defines the tensor triangulated category

$$\mathscr{K}(U) = (\mathscr{K}/\mathscr{K}_Z)^{\natural} \tag{8}$$

as the idempotent completion of the Verdier localisation of $\mathscr{K}$ by $\mathscr{K}_Z$. An inclusion $V \subseteq U$ of such opens comes with a canonical $\otimes$-triangulated functor $\mathscr{K}(U) \to \mathscr{K}(V)$, [10, Constr. 24].

8. The sheafification of the assignment sending a quasi-compact open U to the ring $\hom_{\mathscr{K}(U)}(\mathbb{1}, \mathbb{1})$ is denoted by $\mathscr{O}_{\mathscr{K}}$. This gives $\mathrm{Spc}(\mathscr{K})$ the structure of a *ringed space*, and, at least when $\mathscr{K}$ is rigid[1] and idempotent complete, $(\mathrm{Spc}(\mathscr{K}), \mathscr{O}_K)$ is a *locally ringed space*, [9, Cor. 6.6]. This ringed space is referred to as the *spectrum* of $\mathscr{K}$, [10, Constr. 29].

Example 2.4 1. Let X be a quasi-compact quasi-separated scheme (for example a variety over a field). Then there is an isomorphism of locally ringed spaces $X \cong \mathrm{Spc}(D^{\mathrm{perf}}(X))$ where $D^{\mathrm{perf}}(X)$ is the derived category of perfect complexes on X, [10, Thm. 54].

2. Let SH_{top} be the classical stable homotopy category. Then $\mathrm{Spc}(SH^c_{\mathrm{top}})$ is

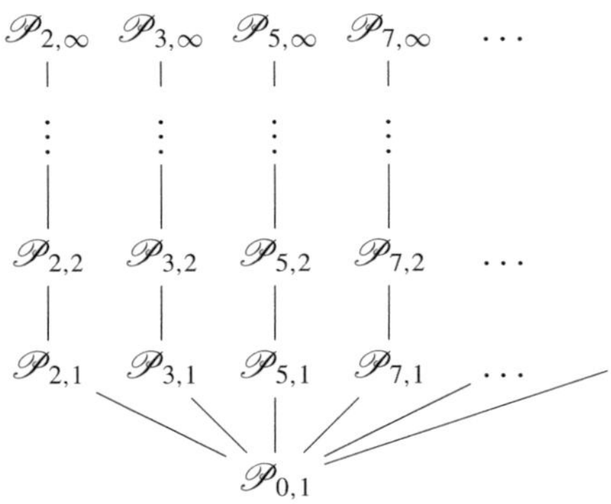

(9)

The lines indicate that the higher prime is in the closure of the lower one. For every prime number p and every $n \geq 1$, the prime $\mathscr{P}_{p,n}$ of SH^c_{top} is the kernel of the nth Morava K-theory (composed with localisation at p) and $\mathscr{P}_{p,\infty} = \cap_{n\geq 1}\mathscr{P}_{p,n}$ is the kernel of localisation at p. The generic point $\mathscr{P}_{0,1} = (SH^c_{\mathrm{top}})_{\mathrm{tor}} = \ker(H(-,\mathbb{Q}))$ is the kernel of singular cohomology with $\mathbb{Q}$-coefficients, Hopkins–Smith [26], [9, Cor. 9.5], [10, Thm. 51].

3. Let G be a finite group, and let SH_G be the G-equivariant stable homotopy category. For a subgroup $H \subseteq G$ let $\Phi^H : SH^c_G \to SH^c_{\mathrm{top}}$ denote the geometric H-fixed points functor. Then, *as a set*,

$$\mathrm{Spc}(SH^c_G) = \left\{ \mathscr{P}(H, p, n) \overset{def}{=} (\Phi^H)^{-1}\mathscr{P}_{p,n} : H \leq G, \mathscr{P}_{p,n} \in \mathrm{Spc}(SH^c_{\mathrm{top}}) \right\}. \tag{10}$$

---

[1] An object a in a $\otimes$-category is called *strongly dualisable* if there exists an object Da such that $a \otimes -$ is left adjoint to $(Da) \otimes -$. A $\otimes$-category is called *rigid* if every object is strongly dualisable.

Furthermore, $\mathscr{P}(H, p, n) = \mathscr{P}(H', p', n')$ if and only if H and H' are conjugate in G, and $p = p', n = n'$. For more details see [14].
In the case of a cyclic group $G = \mathbb{Z}/n$, the space $\mathrm{Spc}(SH_G^c)$ contains a copy of $\mathrm{Spc}(SH_{\mathrm{top}}^c)$ for every m dividing n, including 1 and n. Over $\mathrm{Spec}(\mathbb{Z}[1/n])$, the copies are disjoint, but there are some specialisation-generisation relations between the points lying over $\mathrm{Spec}(\mathbb{Z}/n)$, cf. [14, Eq. 1.3], [31, Sect. 3, Proof of Neil's theorem].

Just as schemes admit a canonical comparison morphism to the spectrum of their global sections, there are canonical comparison morphisms from $\mathrm{Spc}(\mathscr{K})$ to the spectrum of the ring of endomorphisms of the unit object.

Theorem 2.5 ([9, Thm. 5.3, Cor. 5.6, Thm. 7.13, Not. 3.1]) *Let $\mathscr{K}$ be an essentially small $\otimes$-triangulated category and $u \in \mathscr{K}$ an invertible object. There are two continuous maps of topological spaces*

$$\rho_{\mathscr{K}} : Spc(\mathscr{K}) \to Spec\Big(\hom_{\mathscr{K}}(\mathbb{1}, \mathbb{1})\Big), \qquad \rho^{\bullet}_{\mathscr{K}} : Spc(\mathscr{K}) \to Spec^h\Big(\oplus_{n\in\mathbb{Z}} \hom_{\mathscr{K}}(\mathbb{1}, u^n)\Big). \tag{11}$$

Here, $Spec^h$ indicates the set of proper homogeneous ideals which are prime.[2] *It is equipped with the topology whose closed sets are of the form $V(I^\bullet) = \{\mathfrak{p}^\bullet \in Spec^h : I^\bullet \subseteq \mathfrak{p}^\bullet\}$ for homogeneous ideals $I^\bullet$, [9, Rmk. 3.4].*

Futhermore, if $\hom_{\mathscr{K}}(\mathbb{1}[i], \mathbb{1}) = 0$ *for $i < 0$, then $\rho_{\mathscr{K}}$ is surjective. In the case $u = \mathbb{1}[1]$, and the graded endomorphism ring is coherent (e.g., noetherian), the map $\rho^{\bullet}_{\mathscr{K}}$ is surjective.*

For any morphism $s : \mathbb{1} \to u^n$, the premiage of the principle open $D(s)$ of homogeneous primes not containing s is the open $U(Cone(s))$ of primes of $\mathscr{K}$ containing $Cone(s)$.

Remark 2.6 The maps of Theorem 2.5 are as follows. The first one is defined on primes $\mathscr{P}$ by $\rho_{\mathscr{K}}(\mathscr{P}) = \{\mathbb{1} \xrightarrow{f} \mathbb{1} : \mathrm{Cone}(f) \notin \mathscr{P}\}$. The second one takes a prime $\mathscr{P}$ to the homogeneous ideal generated by those $\mathbb{1} \xrightarrow{f} u^n$ such that $\mathrm{Cone}(f) \notin \mathscr{P}$ as n ranges over all integers.

Later on we will use the following facts.

Proposition 2.7 (Balmer) *Let $\mathscr{K}$ be an essentially small $\otimes$-triangulated category.*

1. *If $F : \mathscr{K} \to \mathscr{L}$ is a $\otimes$-exact functor to another essentially small $\otimes$-triangulated category $\mathscr{L}$, then the assignment $\mathscr{P} \mapsto F^{-1}\mathscr{P}$ defines a continuous map of topological spaces $Spc(F) : Spc(\mathscr{L}) \to Spc(\mathscr{K})$. For every $a \in \mathscr{K}$, we have $supp(F(a)) = Spc(F)^{-1}(supp(a))$, [10, Prop. 11(c)].*

[2] Recall that Proj of a non-negatively graded ring is the set of those proper homogeneous prime ideals *which don't contain all elements of positive degree*. There is no such exclusion in the definition of Spec^h.

2. *Let $\mathscr{J} \subset \mathscr{K}$ be a thick $\otimes$-ideal. Then the Verdier localisation $\mathscr{K} \to \mathscr{K} / \mathscr{J}$ induces a homeomorphism from Spc$(\mathscr{K} / \mathscr{J})$ onto the subspace $\{\mathscr{P} : \mathscr{J} \subseteq \mathscr{P}\}$ of Spc$(\mathscr{K})$. This subspace is not always* open *(although it is if, for example, $\mathscr{J}$ is generated by a single element), but it is always* closed under specialisation, *[10, Thm. 18(a)].*
3. *Let $u \in \mathscr{K}$ be an object such that the cyclic permutation $u^{\otimes 3} \xrightarrow{\sim} u^{\otimes 3}$ is the identity. Then the localisation $\mathscr{K} \to \mathscr{K}[u^{\otimes -1}]$ induces a homeomorphism from Spc$(\mathscr{K}[u^{\otimes -1}])$ onto the* closed *subspace supp$(u) = \{\mathscr{P} : u \notin \mathscr{P}\}$ of Spc$(\mathscr{K})$, [10, Thm. 18(c)].*
4. *Let f be a morphism between tensor invertible objects of $\mathscr{K}$. Then there exists a prime containing Cone(f) if and only if $f^{\otimes n} \neq 0$ for all $n \geq 0$, [9, Thm. 2.15], [7, Cor. 2.5].*

The following notion of prime ideal of an abelian category which is analogous to primes of a $\otimes$-triangulated category was developed in [51].

Definition 2.8 (cf. [51, Def. 4.2]) Let $\mathscr{A}$ be an abelian category. Recall that a full[3] subcategory of $\mathscr{A}$ is called *thick*, [36, Def. 8.3.21.(iv)], if it is closed under extensions, kernels, and cokernels.[4] Suppose now that $\mathscr{A}$ is a tensor abelian category. A *(thick) tensor ideal* of $\mathscr{A}$ is a (full) thick subcategory $\mathscr{M} \subseteq \mathscr{A}$ such that $\mathscr{M} \otimes \mathscr{A} \subseteq \mathscr{M}$. A proper $\otimes$-ideal $\mathscr{M} \subset \mathscr{A}$ is called a *(thick) prime ideal* if $a \otimes b \in \mathscr{M}$ implies $A \in \mathscr{M}$ or $B \in \mathscr{M}$.

Lemma 2.9 *Suppose that $(\mathscr{A}, \otimes, \mathbb{1})$ is a rigid semisimple[5] tensor abelian category. If $\mathbb{1}$ is simple, then both $K^b(\mathscr{A})$ and $\mathscr{A}$ possess a unique prime: $\{0\}$.*

Proof Since $\mathscr{A}$ is semisimple, every object of $K^b(\mathscr{A})$ is isomorphic to the sum of its shifted cohomology objects, i.e., $K^b(\mathscr{A}) \cong \bigoplus_{i \in \mathbb{Z}} \mathscr{A}$ as a triangulated category; the triangulated structure on $\bigoplus_{i \in \mathbb{Z}} \mathscr{A}$ is given by shifting the indices, and Cone$(a \xrightarrow{f} b) = \ker(f)[1] \oplus \operatorname{coker}(f)$. Inspecting the definitions, we see that $\mathscr{M} \mapsto \bigoplus_{i \in I} \mathscr{M}$ is a bijection from the set of prime ideals of the tensor abelian category $\mathscr{A}$ to the set of primes of the $\otimes$-triangulated category $\bigoplus_{i \in \mathbb{Z}} \mathscr{A}$. So it suffices to treat the case of $\mathscr{A}$.

Let a be a nonzero object of some prime ideal $\mathscr{M}$. Since it is a tensor ideal, it must also contain $(Da) \otimes a$ where Da is the strong dual of a, which exists by the assumption that $\mathscr{A}$ is rigid. By definition, $(Da) \otimes -$ is right adjoint to $a \otimes -$, and so we have canonical morphisms $\mathbb{1} \xrightarrow{\epsilon} (Da) \otimes a$ and $a \otimes (Da) \xrightarrow{\eta} \mathbb{1}$ such that the composition $a \xrightarrow{\mathrm{id}_a \otimes \epsilon} a \otimes (Da) \otimes a \xrightarrow{\eta \otimes \mathrm{id}_a} a$ is id_a. If ϵ were to be zero, then id_a would be zero, contradicting the assumption that a is nonzero. So ϵ is nonzero, and by the

---

[3] This assumption does not appear in [51, Def. 4.2], but it seems it should be there.

[4] In [51, Def. 4.2], the definition uses *coherent* abelian subcategories, which, as Oliver Braunling pointed out to me, are just thick subcategories containing a zero object.

[5] An object a in an abelian category is called *simple* if every monomorphism $b \to a$ (in the categorical sense) is either zero or an isomorphism. An object is *semisimple* if it is a sum of simple objects. An abelian category is *semisimple* if all of its objects are semisimple.

assumption that $\mathbb{1}$ is simple, and $\mathscr{A}$ is semisimple, ϵ must be the inclusion of a direct summand. Since prime ideals are closed under direct summand, it follows that $\mathscr{M}$ contains $\mathbb{1}$, and therefore $\mathscr{M} = \mathscr{A}$.

3 Motivic Categories

In this section we rapidly review the motivic categories that we will discuss. Specifically, the Morel–Voevodsky stable homotopy category $SH(S)$, Voevodsky's triangulated category of motives $DM(S)$, Grothendieck's classical categories of motives with respect to an adequate equivalence relation $\mathscr{M}_{\sim}(k)$, and some of the relationships between these. This section will be too basic for the experts, and too terse for the non-experts, but we hope that it will at least serve to set notation for the experts, and provide references to the literature for the non-experts.

Remark 3.1 Philosophically, categories of motives should be defined by universal properties. Consequently, all the constructions have a "generators and relations" feel to them, cf. Remarks 3.4, 3.11, and Definition 3.3.

3.1 *Grothendieck Motives*

A nice introduction to classical Grothendieck motives over a field is [54]. For the extension to a smooth base and a beautiful application of this extension see [20].

Definition 3.2 *Cycle groups*. Let k be a field, and S a smooth k-scheme of pure dimension d_S. For a smooth projective S-scheme X, let $\mathscr{Z}^i(X)$ denote the free abelian group generated by the closed integral subscheme of X of codimension i. If $\sim$ is an adequate equivalence relation[6] such as rational equivalence,[7] homological equivalence,[8] or numerical equivalence,[9] denoted rat, hom, and num respectively, we will write $A^i_{\sim}(X) = \mathscr{Z}^i(X)/\sim$.

---

[6]An *adequate equivalence relation* is a family of equivalence relations $\sim_X$ on the $\mathscr{Z}^*(X)$ which satisfy three properties, which essentially require that composition as defined above is well-defined, [53]. In short, pullback, pushforward, and intersection are well-defined.

[7]Two cycles $\alpha, \alpha' \in \mathscr{Z}^i(X)$ are rationally equivalent if there is a cycle $\beta \in \mathscr{Z}^i(\mathbb{P}^1_X)$ such that $\beta \cdot [\{0\} \times X] = \alpha$ and $\beta \cdot [\{\infty\} \times X] = \alpha'$. Rational equivalence is the coarsest equivalence relation.

[8]A cycle α is homologically equivalent to zero if its image under the cycle class map $\mathscr{Z}^i(X) \to H^{2i}(X)$ is zero, for some prechosen Weil cohomology theory, such as étale cohomology $H^i(X) = H^{2i}_{et}(X, \mathbb{Q}_l(i))$ for some $l \nmid \mathrm{char}\ p$. In other words, $A^i_{\mathrm{hom}}(X)$ is the image of the cycle class map $A^i_{\mathrm{hom}}(X) = \mathrm{im}(\mathscr{Z}^i(X) \to H^{2i}(X))$.

[9]Numerical equivalence is the coarsest equivalence relation which makes the intersection product $A^i_{\mathrm{num}}(X) \otimes A^{d-i}_{\mathrm{num}}(X) \to A^d_{\mathrm{num}}(X)$ nondegenerate, where $d = \dim X$. That is, $\alpha \in A^i_{\mathrm{rat}}(X)$ is numerically equivalent to zero if and only if $\alpha \cdot \beta = 0$ for all $\beta \in A^{d-i}_{\mathrm{rat}}(X)$.

Cycle categories. For any triple X, Y, Z of smooth projective S-schemes, and cycles $\alpha \in A^i_\sim(X \times_S Y)$, and $\beta \in A^j_\sim(Y \times_S Z)$, the composition $\beta \circ \alpha \in A^{i+j}_\sim(X \times_S Z)$ is defined by pulling α and β back to the triple product $X \times_S Y \times_S Z$ along the canonical projections, intersecting them, and then obtaining a cycle on $X \times_S Z$ by pushing forward along the canonical projections. In this way we obtain a category, whose objects are smooth projective S-schemes, and $\hom(X, Y) = A^{\dim X/S}_\sim(X \times_S Y)$. Here $\dim X/S$ denotes the relative dimension of the morphism $X \to S$. Identity morphisms are given by the cycles associated to the diagonals $\Delta_X \subseteq X \times_S X$. Fibre product of S-schemes induces a tensor product on this category.

Motivic categories. The objects of $\mathscr{M}_\sim(S)$ are triples (X, p, n) where X is a smooth projective S-scheme, $p \in A^{\dim X/S}_\sim(X \times_S X)$ satisfies $p \circ p = p$, and $n \in \mathbb{Z}$. We set

$$\hom_{\mathscr{M}_\sim(S)}((X, p, n), (Y, q, m)) = \{\alpha \in A^{\dim X/S+n-m}_\sim(X \times_S Y) : \alpha = p \circ \alpha \circ q\}. \tag{12}$$

This category is a tensor additive category with sum induced by disjoint union of S-schemes, and tensor product induced by fibre product. There is a canonical functor

$$M : \mathrm{SmProj}(S) \to \mathscr{M}_\sim(S); \qquad X \mapsto (X, [\Delta_X], 0) \tag{13}$$

from smooth projective S-schemes which sends a morphism $f : X \to Y$ to the cycle associated to its graph $\Gamma_f \subseteq X \times_S Y$.

Remark 3.3 Any section $s : S \to X$ of an S-scheme $f : X \to S$, for example the section at infinity $\infty : S \to \mathbb{P}^1_S$, gives rise to any idempotent endomorphism $s \circ f$, and consequently a decomposition $M(X) \cong M(S) \oplus (\ker M(s \circ f))$. The Lefschetz motive $\mathbf{L}$ is defined as the kernel of the projection to infinity $M(\mathbb{P}^1_S) \cong M(S) \oplus \mathbf{L}$. We then have a canonical isomorphism

$$(X, p, n) \cong \left(\mathrm{im}\left(M(X) \xrightarrow{p} M(X)\right) \right) \otimes \mathbf{L}^{\otimes(-n)}. \tag{14}$$

Remark 3.4 This definition can be seen as a generators and relations construction of a category. The generators are smooth projective varieties and correspondences, and the relations we have forced are the equivalence relation $\sim$, the existence of kernels and images of idempotent endomorphisms, and the tensor inverse of $\mathbf{L}$.

Definition 3.5 For an abelian group A and Λ a flat $\mathbb{Z}$-algebra, we write $A_\Lambda = A \otimes \Lambda$. If $\mathscr{A}$ is an additive category, we write $\mathscr{A}_\Lambda$ for the category which has the same objects as $\mathscr{A}$ and $\hom_{\mathscr{A}_\Lambda}(a, b) = \hom_{\mathscr{A}}(a, b)_\Lambda$.

Remark 3.6 Since $\mathbb{Q}$ is a flat $\mathbb{Z}$-algebra, the categories $\mathscr{M}_\sim(k)_\mathbb{Q}$ are the same as the ones constructed as above, using $A^*_\sim(X)_\mathbb{Q}$ instead of $A^*_\sim(X)$.

Theorem 3.7 ([29, Thm. 1]) *Let k be a field and $\sim$ an adequate equivalence relation. The category $\mathscr{M}_\sim(k)_\mathbb{Q}$ is semisimple if and only if $\sim$ is numerical equivalence.*

Remark 3.8 ([54, 1.15]) There is a canonical functor $^\vee : \mathscr{M}_\sim(k)^{op} \to \mathscr{M}_\sim(k)$ defined by $(X, p, m)^\vee = (X, {}^t, d - m)$ when X is purely d dimensional. Here ${}^t : A^d_\sim(X \times_S X) \to A^d_\sim(X \times_S X)$ corresponds to swapping the Xs.

One can verify by hand that $M^{\vee\vee} = M$ and that for any three objects M, N, P one has

$$\hom(M \otimes N, P) = \hom(M, N^\vee \otimes P). \tag{15}$$

Consequently, the categories $\mathscr{M}_\sim(k)$ are rigid.

3.2 Voevodsky Motives

Definition 3.9 *Correspondences.* Let S be a regular[10] noetherian separated scheme, such as the spectrum of a field. The category SmCor(S) has as objects smooth S-schemes, and $\hom_{\mathrm{SmCor}(S)}(X, Y)$ is the free abelian group generated by closed integral subschemes $Z \subseteq X \times_S Y$ such that the map $Z \to X$ induced by projection is finite and surjective. Composition of two morphisms $\alpha \in \hom_{\mathrm{SmCor}(S)}(X, Y)$ and $\beta \in \hom_{\mathrm{SmCor}(S)}(Y, W)$ is defined as above: by pulling back to the triple product $X \times_S Y \times_S W$, intersecting, and then pushing forward to $X \times_S W$. The condition that the generators of the hom groups be finite and surjective over the first component ensures that the two pullbacks to $X \times_S Y \times_S W$ intersect properly. The category SmCor(S) is a tensor additive category, with direct sum induced by disjoint union of schemes, and tensor product induced by fibre product. As above, there is a canonical functor

$$[-] : \mathrm{Sm}(S) \to \mathrm{SmCor}(S) \tag{16}$$

which sends a morphism to the cycle induced by its graph.

Effective geometric motives. Since SmCor(S) is additive, we can consider its bounded homotopy category $K^b(\mathrm{SmCor}(S))$. Define HI to be the set of complexes of the form

$$(\dots \to 0 \to [\mathbb{A}^1_X] \to [X] \to 0 \to \dots) \tag{17}$$

where X ranges over all smooth S-schemes. Define NMV to be the set of complexes of the form

$$(\dots \to 0 \to [U \times_X V] \stackrel{[k]+[l]}{\to} [U] \oplus [V] \stackrel{[i]-[j]}{\to} [X] \to 0 \to \dots) \tag{18}$$

---

[10] The category SmCor(S) can be defined for any noetherian separated scheme S, but we have not mentioned this construction because branch points make composition more subtle. For a more general S, the group $\hom_{\mathrm{SmCor}(S)}(X, Y)$ is only a proper subgroup of the free abelian group we describe, since branches of X introduce an ambiguity in the composition of some cycles. In fact, it can be defined as the largest subgroup for which composition is well-defined, as soon as the notion of composition has been formalised appropriately, [33, Chap. 2].

ranging over all Nisnevich distinguished squares.[11,12] The category of *effective geometric motives* is defined as

$$DM_{gm}^{eff}(S) = \left(\frac{K^b(\mathrm{SmCor}(S))}{\langle \mathrm{HI} \cup \mathrm{NMV} \rangle} \right)^{\natural} \tag{19}$$

where $(-)^{\natural}$ denotes idempotent completion, $\langle \mathrm{HI} \cup \mathrm{NMV} \rangle$ is the smallest thick triangulated subcategory generated by the sets of objects HI and NMV, and the fraction denotes the Verdier quotient.

Noneffective geometric motives. The tensor structure on $\mathrm{SmCor}(S)$ induces a canonical tensor structure on $DM_{gm}^{eff}(S)$, and there is a canonical monoidal functor

$$M : \mathrm{Sm}(S) \to DM_{gm}^{eff}(S) \tag{20}$$

induced by $[-] : \mathrm{Sm}(S) \to \mathrm{SmCor}(S)$. As above, projection to infinity defines an idempotent endomorphism of $M(\mathbb{P}^1_S)$ the Tate motive $\mathbb{Z}(1)[2]$ to be the kernel of this endomorphism: $M(\mathbb{P}^1_S) \cong M(S) \oplus \mathbb{Z}(1)[2]$. Since we are working with complexes, we have an explicit model for this:

$$\mathbb{Z}(1) \stackrel{def}{=} (\cdots \to 0 \to \underset{2}{[\mathbb{P}^1_S]} \to \underset{3}{[S]} \to 0 \to \dots) \tag{21}$$

as a complex concentrated in degrees 2 and 3. We obtain the category of *noneffective geometric motives* is defined by forcing $\mathbb{Z}(1)$ to be tensor invertible.

$$DM_{gm}(S) = DM_{gm}^{eff}(S)[\mathbb{Z}(1)^{-1}]. \tag{22}$$

Formally, the objects of $DM_{gm}(S)$ are pairs (M, m) with M an object of $DM_{gm}^{eff}(S)$ and $m \in \mathbb{Z}$, and morphisms are $\hom((M, m), (N, n)) = \varinjlim_{k \geq m,n}(M \otimes \mathbb{Z}(1)^{k-m}, N \otimes \mathbb{Z}(1)^{k-n})$.

Remark 3.10 1. To simplify notation, one usually writes $\mathbb{Z}(m) = \mathbb{Z}(1)^{\otimes m}$ and $M(m) = M \otimes \mathbb{Z}(m)$.
2. To show that the tensor structure on $DM_{gm}(S)$ is well defined on morphisms, one needs to show that the cyclic permutation $\mathbb{Z}(1)^{\otimes 3} \xrightarrow{\sim} \mathbb{Z}(1)^{\otimes 3}$ is the identity. This is [58, Cor. 2.1.5].

---

[11] A *Nisnevich distinguished square* is a cartesian square $\begin{array}{ccc} U \times_X V & \to & V \\ \downarrow & & \downarrow \\ U & \to & X \end{array}$ such that $U \to X$ is an open immersion, $V \to X$ is étale, and $(X - U)_{red} \times_X V \to (X - U)_{red}$ is an isomorphism.

[12] In [58] Voevodsky only uses Zariski distinguished squares, i.e., those squares for which j is also an open immersion. However, [58, Thm. 3.1.12] implies that, at least when the base is a perfect field, the Zariski and Nisnevich versions produce the same category. On the other hand, it is the Nisnevich descent property which is often used in most of the proofs in [58]. Nisnevich locally, closed immersions of smooth schemes look like zero sections of trivial affine bundles, cf. [47, Proof of Lemma 2.28].

3. The transition morphisms in the colimit defining the hom groups in $DM_{gm}(S)$ are actually all isomorphisms, at least when the base is a perfect field: Voevodsky shows in the "incredibly short and absolutely ingenious"[13] article [59] that $- \otimes \mathbb{Z}(1)$ is a fully faithful functor on $DM_{gm}^{eff}(S)$ when S is the spectrum of a perfect field.

Remark 3.11 Again, this definition is clearly of the generators and relations form. One starts with smooth schemes and correspondences as the generators, and then forces $\mathbb{A}^1$-invariance, Nisnevich descent, kernels and cokernels of idempotents to exist, and $\mathbb{Z}(1)$ to be tensor invertible.

The most important facts we will need about $DM_{gm}(S)$ are the following.

Theorem 3.12 *Let k be a perfect field of exponential characteristic p.*

1. *The category $DM_{gm}(k)_{\mathbb{Z}[1/p]}$ is rigid. If X is a smooth projective k variety of pure dimension d, then the dual of $M(X)$ is $M(X)(-d)[-2d]$, [33, Thm. 5.5.14], [58, Thm. 4.3.7].*
2. *For any smooth k-variety X, $n \in \mathbb{Z}$, $i \geq 0$, there are canonical isomorphisms*

$$\hom_{DM_{gm}(k)}(M(X), \mathbb{Z}(i)[n]) \cong CH^i(X, 2i - n) \tag{23}$$

   towards Bloch's higher Chow groups, defined in [11]. In particular, for any smooth k-variety X and smooth projective k-variety Y of pure dimension d, there are canonical isomorphisms

$$\hom_{DM_{gm}(k)_{\mathbb{Z}[1/p]}}(M(X), M(Y)(i)[n]) \cong CH^{i+d}(X \times_k Y, 2i - n)_{\mathbb{Z}[1/p]}, \tag{24}$$

   [33, Thm. 5.6.4], [48, Thm. 19.1], [55, Thm. 3.2].
3. *Consequently, there is a canonical fully faithful tensor additive embedding*

$$\mathscr{M}_{rat}(k)_{\mathbb{Z}[1/p]} \to DM_{gm}(k)_{\mathbb{Z}[1/p]}. \tag{25}$$

4. *The category $DM_{gm}(k)_{\mathbb{Z}[1/p]}$ is generated (as an idempotent complete tensor triangulated category) by motives of smooth projective k-varieties. [13], [58, Cor. 3.5.5].*

Remark 3.13 If one believes in strong resolution of singularities, in the sense of [22, Def. 3.4], then one doesn't have to invert p in the above theorem.

Using the fact that étale cohomology has a structure of transfers, is homotopy invariant, satisfies Nisnevich descent, and is $\mathbb{P}^1$-stable, one can construct a canonical functor to l-adic sheaves.

---

[13] In his MathSciNet review, Röndigs attributes this quote to Suslin.

Theorem 3.14 ([28, Thm. 3.1], cf. also [37, Appendix A]) *For any noetherian separated scheme S, and l invertible on S, and any $n > 0$ there are canonical tensor triangulated functors*

$$DM_{gm}(S) \to D_{et}(S, \mathbb{Z}/l^n), \qquad DM_{gm}(S)_{\mathbb{Q}} \to D_{et}(S, \mathbb{Q}_l). \tag{26}$$

Here, the target categories are the derived categories associated to the small étale site of S. If $f : X \to S$ is a smooth morphism, then the image of $M(X)$ is the pushforward $Rf_*(\mathbb{Z}/l^n)_X$, resp. $Rf_*(\mathbb{Q}_l)_X$, of the constant sheaf on the small étale site of X. If S is the spectrum of a perfect field, then this also holds for non-smooth X.

3.3 Morel–Voevodsky's Stable Homotopy Category

The definition of $SH(S)$ was sketched in [57] using the unstable theory of [47]. A more explicit construction, which includes the use of symmetric spectra is in [30], and a more modern treatment which incorporates advances in the theory of model categories which happened in the meantime (many of which were motivated precisely for the study of $SH(S)$) appears in [3, Def. 4.5.52]. Even more recently, the universal property which $SH(S)$ satisfies was made formal in [52, Cor. 2.39], using the language of infinity categories. Below we sketch the construction of [16], which we find to be the most accessible.

Heuristic "Definition" 3.15. Let S be a noetherian scheme. The *Morel–Voevodsky stable homotopy category* $SH(S)$ is the universal tensor triangulated category such that:

1. There is a monoidal functor $\Sigma^\infty(-)_+ : \mathrm{Sm}(S) \to SH(S)$.
2. ($\mathbb{A}^1$-invariance) $\Sigma^\infty(\mathbb{A}^1_S)_+ \to \Sigma^\infty(S)_+$ is an isomorphism.
3. (Nisnevich descent) $\Sigma^\infty(-)_+$ sends Nisnevich distinguished squares (see footnote 11) to homotopy cocartesian squares. That is, in the notation of footnote 11, the Mayer-Vietoris style triangle

$$\Sigma^\infty(U \times_X V)_+ \longrightarrow \Sigma^\infty U_+ \oplus \Sigma^\infty V_+ \longrightarrow \Sigma^\infty X_+ \tag{27}$$

   fits into a distinguished triangle.
4. (($\mathbb{P}^1, \infty$)-Stability) The cofibre of the image of the section at infinity $\infty : S \to \mathbb{P}^1_S$ is tensor invertible. That is, $\mathrm{Cone}\big(\Sigma^\infty S_+ \to \Sigma^\infty(\mathbb{P}^1_S)_+\big)$ is tensor invertible.

   The tensor triangulated category $SH(S)$ can be constructed as follows.

Construction 3.16 [16, 1.2, 2.15] Let $Sp_{S^1}(S)$ denote the category of presheaves of symmetric S^1-spectra on $\mathrm{Sm}(S)$. This category is equipped with the projective model structure.[14] Via Yoneda, every smooth S-scheme equipped with a section, such as $\mathbb{P}^1_S$ equipped with the section at infinity, gives rise to an object of $Sp_{S^1}(S)$.[15] Let T be an cofibrant replacement[16] for the pointed scheme $\mathbb{P}^1_S$ in $Sp_{S^1}(S)$, and let $Sp_T Sp_{S^1}(S)$ denote the category of symmetric T-spectra in $Sp_{S^1}(S)$. It comes equipped with a canonical "Yoneda" functor[17]

$$\Sigma^\infty(-)_+ : \mathrm{Sm}(S) \to Sp_T Sp_{S^1}(S), \tag{28}$$

and a "constant sheaf" functor[18]

$$Sp_{S^1} \to Sp_T Sp_{S^1}(S). \tag{29}$$

Let HI be the set of images in the homotopy category $\mathrm{Ho}(Sp_T Sp_{S^1}(S))$ of the morphisms $\Sigma^\infty(\mathbb{A}^1_X)_+ \to \Sigma^\infty X_+$ as X ranges over all smooth S-schemes. Let NMV be the set of images in $\mathrm{Ho}(Sp_T Sp_{S^1}(S))$ of the morphisms $\mathrm{Cone}(\Sigma^\infty(U \times_X V)_+ \to \Sigma^\infty U_+ \oplus \Sigma^\infty V_+) \to \Sigma^\infty X_+$ ranging over all distinguished Nisnevich squares (see footnote 11) in $\mathrm{Sm}(S)$. We define $SH(S)$ as the Verdier quotient

$$SH(S) = \frac{\mathrm{Ho}(Sp_T Sp_{S^1}(S))}{\langle\langle \mathrm{HI} \cup \mathrm{NMV} \rangle\rangle}. \tag{30}$$

Here the double angle brackets $\langle\langle - \rangle\rangle$ indicate the localising category, i.e., the smallest triangulated category closed under direct summands and arbitrary small sums, containing the objects of HI and NMV.

Its subcategory of compact objects is the smallest thick triangulated category containing the objects $\Sigma^\infty X_+$ for all $X \in \mathrm{Sm}(S)$. This is denoted by

---

[14]The projective model structure is the model structure for which a morphism is a fibration (resp. weak equivalence) if and only if it is a fibration (resp. weak equivalence) of symmetric S^1-spectra after evaluation on every $X \in \mathrm{Sm}(S)$.

[15]The Yoneda functor produces a presheaf of pointed sets $\hom_{\mathrm{Sm}(S)}(-, \mathbb{P}^1_S)$, and then working schemewise, we associated to every pointed set its induced pointed simplicial set, and from there its associated symmetric S^1-spectrum.

[16]The projective model structure has the nice property that representable presheaves are cofibrant, however, by representable we mean the image of a scheme with a disjoint base point, cf. Footnote 17. Presheaves which are the image of pointed schemes whose basepoint is not disjoint are not in general cofibrant. Hence, we need to take some cofibrant model. For example, the pushout of $S_+ \wedge \Delta^1_+ \stackrel{0}{\leftarrow} S_+ \stackrel{\infty}{\to} \mathbb{P}^1_+$ is a cofibrant model for $(\mathbb{P}^1_S, \infty)$, where Δ^1_+ is the constant presheaf corresponding to the simplicial interval. This is exactly the analogue of the complex $([S] \stackrel{\infty}{\to} [\mathbb{P}^1_S])$ which we could have used to define $\mathbb{Z}(1)$ in $DM^{eff}_{gm}(S)$.

[17]We first equip any S-scheme X with a base point by replacing it with $X_+ = X \sqcup S$. Then using the procedure described in Footnote 15 we get a functor $\mathrm{Sm}(S) \to Sp_{S^1}(S)$, which we compose with the canonical functor $Sp_{S^1}(S) \to Sp_T Sp_{S^1}(S)$.

[18]This is actually the composition of the constant presheaf functor $Sp_{S^1} \to Sp_{S^1}(S)$ from symmetric S^1-spectra to presheaves of symmetric S^1-spectra, and the canonical functor $Sp_{S^1}(S) \to Sp_T Sp_{S^1}(S)$.

$$SH(S)^c = \langle \Sigma^\infty X_+ : X \in \mathrm{Sm}(S) \rangle \subset SH(S) \tag{31}$$

Remark 3.17 There are also constructions using $S^1 \wedge \mathbb{G}_m$ (where $\mathbb{G}_m$ is pointed at the identity) instead of $\mathbb{P}^1_S$, and constructions which take the HI $\cup$ NMV localisations as Bousfield localisations before passing to T-spectra. These all produce equivalent categories, [16, 2.15].

Remark 3.18 The canonical inclusion (see footnote 10) $\mathrm{Sm}(S) \to \mathrm{SmCor}(S)$, and the canonical functor $Sp_{S^1} \to \mathrm{Cpx(Ab)}$ from symmetric S^1-spectra to (unbounded) complexes of abelian groups induces a canonical tensor triangulated functor

$$SH(S) \to DM(S). \tag{32}$$

Here, $DM(S)$ can defined in the same way as we have defined $SH(S)$, but using $\mathrm{SmCor}(S)$ instead of $\mathrm{Sm}(S)$, and $\mathrm{Cpx(Ab)}$ instead of Sp_{S^1}. As $DM_{gm}(S)$ can be identified with the thick subcategory of $DM(S)$ generated by the motives of smooth schemes, this functor restricts to a functor

$$SH(S)^c \to DM_{gm}(S). \tag{33}$$

Theorem 3.19 (Morel, [45], cf. also [15, §A.3, §C.3]) *If S is the spectrum of a field such that -1 is a sum of squares (such as a finite field, or the field of complex numbers), the canonical functor, cf. Remark 3.18,*

$$SH(S)^c_{\mathbb{Q}} \to DM_{gm}(S)_{\mathbb{Q}} \tag{34}$$

is a $\otimes$-triangulated equivalence of categories.

Remark 3.20 Morel proved this theorem is true for the rational versions of the big categories $SH(S)$ and $DM(S)$, but Definition 3.5 does not work properly for non-compact objects, so some care must be taken with the term "rational version". Either one can localise at all morphisms $n \cdot \mathrm{id}_M$ for all objects M and all integers $n \neq 0$, or since the rational versions of Sp^1_S and $\mathrm{Cpx(Ab)}$ are Quillen equivalent to the category of unbounded complexes of $\mathbb{Q}$-vector spaces, one could just use these in the constructions instead.

The study of the motivic stable homotopy groups of spheres—the abelian groups $\hom_{SH(S)}(\mathbb{1}[p+q], T^q)$ for $p, q \in \mathbb{Z}$—is one of the central problems in motivic homotopy theory.

Theorem 3.21 *Suppose that k is the spectrum of a perfect field.*

1. $\hom_{SH(k)}(\mathbb{1}[p+q], T^q) = 0$ *if* $p < 0$, *[42, Thm. 4.2.10, §6 Intro.].*
2. *The graded ring* $\oplus_{q\in\mathbb{Z}} \hom_{SH(k)}(\mathbb{1}[q], T^q)$ *is canonically isomorphic to the graded associative ring* $K^{MW}_\bullet(k)$ *generated by symbols* $[a], a \in k^*$, *of degree* 1, *and one symbol* η *of degree* -1, *subject to the following relations.*

a. $[a][1-a]=0$.
b. $[ab]=[a]+[b]+\eta[a][b]$.
c. $\eta[a]=[a]\eta$.
d. $\eta \cdot h=0$, *where* $h=1+(\eta[-1]+1)$, *[46, Def. 12, Cor. 25].*

In degree zero, we have the Grothendieck-Witt group $GW(k)\cong K_0^{MW}(k)$, *[46, Cor. 24]. If the characteristic of k is not 2, then for every* $n\in\mathbb{Z}$, *there is a canonical short exact sequence of abelian groups*

$$0\to I(k)^{n+1}\to K_n^{MW}(k)\to \underbrace{K_n^M(k)}_{\cong K_n^{MW}(k)/\eta}\to 0 \tag{35}$$

where $I(k)=\ker(W(k)\to\mathbb{Z}/2)$ *is the augmentation ideal in the Witt ring* $W(k)$ *of the field k, we set* $I(k)^n=W(k)$ *for* $n<0$ *by convention, and* $K_n^M(k)$ *is the Milnor K-theory of the field k, [44], [23, Def. 3.7, Thm. 3.8, Thm. 5.4].*[19]

3. *If k is algebraically closed of characteristic zero then* $\hom_{SH(k)}(\mathbb{1}[n],\mathbb{1})\cong\pi_n^s$, *the classical stable homotopy groups, [39, Cor. 2]. If k is an algebraic closure of a finite field with p elements, and l is a prime different from p, then the we have the same result after l-completion* $\hom_{SH(k)}(\mathbb{1}[n],\mathbb{1})_l^\wedge\cong(\pi_n^s)_l^\wedge$, *[60, Thm. A]. In fact, for any subfield* $k\subseteq\mathbb{C}$, *the* $\otimes$*-triangulated functor* $SH_{top}\to SH(k)$ *induced by the constant presheaf funtor, cf. Eq.* (29), *is fully faithful.*
4. *In fact, let* $k\subseteq\mathbb{C}$ *be a subfield. Then sending a smooth k-scheme X to its complex valued points* $X(\mathbb{C})$ *considered as a topological space in the obvious way induces a* $\otimes$*-triangulated functor* $SH(k)\to SH_{top}$ *towards the classical stable homotopy category, [4], which is a retraction of the* $\otimes$*-triangulated functor* $SH_{top}\to SH(k)$ *induced by the constant presheaf funtor, cf. Eq.* (29).
5. *There are canonically defined "Hopf" elements* $\eta\in\hom(\mathbb{1}[-1],T^{-1})$, $\nu\in\hom(\mathbb{1}[-1],T^{-2})$, $\sigma\in\hom(\mathbb{1}[-1],T^{-4})$, *satisfying the relations* $(1-\epsilon)\eta=\eta\nu=\nu\sigma=0$, *where* $\epsilon:\mathbb{1}[-1]\to T^{-1}$ *corresponds to the map* $\mathbb{G}_m\otimes\mathbb{G}_m\to\mathbb{G}_m\otimes\mathbb{G}_m$ *swapping the factors, [19, Def. 4.7].*

Remark 3.22 For more calculations about motivic stable homotopy groups of spheres, see, for example, [18, 24, 49, 50], and the references therein.

Remark 3.23 In the $K_\bullet^{MW}(k)$ description of $\oplus_{q\in\mathbb{Z}}\hom_{SH(k)}(\mathbb{1}[q],T^q)$, the symbol $[a]$ corresponds to the section $k\to\mathbb{G}_m$ associated to the rational point $a\in\mathbb{G}_m(k)$, and the symbol η corresponds to the Hopf map $\mathbb{A}^2-\{0\}\to\mathbb{P}^1$, under the isomorphisms [57, Lem. 4.1]

$$(\mathbb{A}^n-\{0\},1)\cong T^n[-1],\qquad \mathbb{P}^n/\mathbb{P}^{n-1}\cong\mathbb{A}^n/(\mathbb{A}^n-\{0\})\cong T^n. \tag{36}$$

[19]There are some weird sign and bracket conventions in [23]. In [23], $\hat{\eta}$ and $\{a\}$ are used to denote what [46] writes as η and $[a]$. On the other hand, [23] use η and $[a]$ for the elements in $K^W\overset{def}{=}K^{MW}/h$ corresponding to our η and $-[a]$.

The element h corresponds to the class of the hyperbolic plane in $GW(k) \cong K_0^{MW}(k)$, [42, §6.2, §6.3].

Remark 3.24 The maps $I^{n+1} \to K_n^{MW}$ in the short exact sequence (35) are as follows. First, there is an isomorphism of graded rings $\oplus_{n\in\mathbb{Z}} I^n \cong K_\bullet^W \stackrel{def}{=} (\oplus_n K_n^{MW})/h$ where $[a] \in K_1^W$ corresponds to the Pfister form $\langle 1, -a\rangle \in I$ and $1 + \eta[a] \in K_0^{MW}$ corresponds to the one dimensional quadratic space $\langle a\rangle \in W$. For $n \leq 0$, multiplication by η induces the isomorphism $K_n^W \to K_{n-1}^W$ corresponding to $W^n = W^{n-1}$ [23, p. 13]. Next, since $\eta \cdot h = 0$, multiplication by η induces a map $K_\bullet^W = K_\bullet^{MW}/h \to K_{\bullet-1}^{MW}$. The map in question is the composition

$$I^{n+1} \cong K_{n+1}^W \stackrel{-\cdot\eta}{\to} K_n^{MW}. \tag{37}$$

Remark 3.25 The prime homogeneous ideals of $K^{MW}(k)$ for any field k of characteristic not 2 are classified in [56].

Example 3.26 Let k be a finite field with q elements, and q odd. Then

$$K_{\geq 0}^{MW}(\mathbb{F}_q) \cong \left(\mathbb{Z} \oplus \mathbb{Z}/2\right) \oplus \mathbb{F}_q^* \oplus 0 \oplus 0 \oplus \dots, \tag{38}$$

and for $n > 0$ we have

$$K_{-n}^{MW}(\mathbb{F}_q) = \begin{cases} \mathbb{Z}/4 & q \equiv 3 \mod 4, \\ \mathbb{Z}/2[\epsilon]/\epsilon^2 & q \equiv 1 \mod 4, \end{cases} \tag{39}$$

[38, p. 37], [38, p. 36, Thm. 3.5], [41, Exam. 1.5].

Let ω be a multiplicative generator for $\mathbb{F}_q^*$. Studying Remark 3.24 we find that K_1^{MW} is additively generated by $[\omega]$, the copy of $\mathbb{Z}/2$ in K_0^{MW} is additively generated by $\eta \cdot [\omega]$, and for $n > 0$, K_{-n}^{MW} is additively generated by η^n if $q \equiv 3 \mod 4$, or by η^n and $\eta^{n+1}[\omega]$ if $q \equiv 1 \mod 4$.

If $\mathbb{F}$ is the algebraic closure of $\mathbb{F}_q$, then we have

$$K_n^{MW}(\mathbb{F}) \cong \dots \oplus \underset{-3}{\mathbb{Z}/2} \oplus \underset{-2}{\mathbb{Z}/2} \oplus \underset{-1}{\mathbb{Z}/2} \oplus \underset{0}{\mathbb{Z}} \oplus \underset{1}{\overset{\cong(\mathbb{Q}/\mathbb{Z})[1/p]}{\mathbb{F}^*}} \oplus \underset{2}{0} \oplus \underset{3}{0} \oplus \underset{4}{0} \oplus \dots. \tag{40}$$

4 Observations

In this section we make some observations and guesses about the structure of $\mathrm{Spc}(SH(\mathbb{F}_q)^c)$.

4.1 Rational Coefficients

Theorem 4.1 *Let $\mathbb{F}_q$ be a field with a prime power, q, number of elements. Suppose that for all connected smooth projective varieties X we have:*

$$\begin{array}{lll} CH^i(X, j)_{\mathbb{Q}} = 0, & \forall j \neq 0, i \in \mathbb{Z} & \textit{(Beilinson–Parshin conjecture)}, \\ CH^i(X)_{\mathbb{Q}} \otimes CH_i(X)_{\mathbb{Q}} \to CH_0(X)_{\mathbb{Q}} & \textit{is non-degenerate.} & \textit{(Rat. and num. equiv. agree)} \end{array} \tag{41}$$

Then

$$Spc(SH(\mathbb{F}_q)^c_{\mathbb{Q}}) \cong Spec(\mathbb{Q}). \tag{42}$$

Proof First we observe that by Theorem 3.19 of Morel, the canonical functor $SH(\mathbb{F}_q)^c_{\mathbb{Q}} \to DM_{gm}(\mathbb{F}_q)_{\mathbb{Q}}$ is an equivalence. So we are reduced to studying $DM_{gm}(\mathbb{F}_q)_{\mathbb{Q}}$.

On the other hand, since we are assuming that rational and numerical equivalence agree, the canonical functor $K^b(\mathcal{M}_{\text{rat}}(\mathbb{F}_q)_{\mathbb{Q}}) \to K^b(\mathcal{M}_{\text{num}}(\mathbb{F}_q)_{\mathbb{Q}})$ is an equality. Let us write $\mathcal{M}(\mathbb{F}_q)_{\mathbb{Q}}$ for the category $\mathcal{M}_{\text{rat}}(\mathbb{F}_q)_{\mathbb{Q}} = \mathcal{M}_{\text{num}}(\mathbb{F}_q)_{\mathbb{Q}}$. Jannsen's semisimplicity theorem, Theorem 3.7, says that $\mathcal{M}(\mathbb{F}_q)_{\mathbb{Q}}$ is semisimple, and therefore $K^b(\mathcal{M}(\mathbb{F}_q)_{\mathbb{Q}}) \cong \bigoplus_{i \in \mathbb{Z}} \mathcal{M}(\mathbb{F}_q)_{\mathbb{Q}}$ as a tensor triangulated category, cf. the proof of Lemma 2.9. It follows that the canonical (tensor) functor $\mathcal{M}_{\text{rat}}(\mathbb{F}_q)_{\mathbb{Q}} \to DM_{gm}(\mathbb{F}_q)_{\mathbb{Q}}$ of Theorem 3.12(3) extends to a (tensor) triangulated functor

$$\bigoplus_{i \in \mathbb{Z}} \mathcal{M}(\mathbb{F}_q)_{\mathbb{Q}} \to DM_{gm}(\mathbb{F}_q)_{\mathbb{Q}}. \tag{43}$$

For smooth projective varieties X and Y, the Beilinson–Parshin conjecture says that

$$\hom_{DM_{gm}(\mathbb{F}_q)_{\mathbb{Q}}}(M(X), M(Y)[i]) \overset{Thm.3.12(2)}{\cong} CH^{\dim Y}(X \times Y, -i) \tag{44}$$

vanishes unless $i = 0$. It follows that the functor (43) is fully faithful. Since $DM_{gm}(\mathbb{F}_q)_{\mathbb{Q}}$ is generated by motives of smooth projective varieties, Theorem 3.12(4), the functor (43) is also essentially surjective. That is, it is an equivalence of tensor triangulated categories. Finally we apply Lemma 2.9 and Remark 3.8 to notice that $\mathcal{M}(\mathbb{F}_q)_{\mathbb{Q}}$ has a unique prime: the zero prime. Consequently, the same is true for $DM_{gm}(\mathbb{F}_q)_{\mathbb{Q}}$. For the structure sheaf: we have $\hom_{\mathcal{M}_{\text{rat}}(\mathbb{F}_q)_{\mathbb{Q}}}(\mathbb{1}, \mathbb{1}) \cong \mathbb{Q}$ by definition.

Remark 4.2 If one is willing to accept that Bondarko's weight complex functor $DM_{gm}(\mathbb{F}_q)_{\mathbb{Q}} \xrightarrow{t} K^b(\mathcal{M}_{\text{rat}}(\mathbb{F}_q)_{\mathbb{Q}})$ is monoïdal, there is a much more conceptual approach to the above proof. One considers the sequence of monoïdal functors

$$SH(\mathbb{F}_q)^c_{\mathbb{Q}} \to DM_{gm}(\mathbb{F}_q)_{\mathbb{Q}} \xrightarrow{t} K^b(\mathcal{M}_{\text{rat}}(\mathbb{F}_q)_{\mathbb{Q}}) \to K^b(\mathcal{M}_{\text{num}}(\mathbb{F}_q)_{\mathbb{Q}}). \tag{45}$$

The first one is an equivalence by Morel, the second one is an equivalence if and only if the Beilinson–Parshin conjecture holds, [12], and the third one is an equivalence if and only if rat = num. We didn't use this because at the time of writing there was no written proof that t is monoïdal. Since then, the dg-statement has appeared in [,], and a detailed proof in the generality of ∞-categories is available in [2].

Observation 4.3 *Conversely, if* $Spc(SH(\mathbb{F}_q)^c_{\mathbb{Q}}) \cong Spec(\mathbb{Q})$*, then the étale realisation functor of Theorem 3.14*

$$R : DM_{gm}(\mathbb{F}_q)^{op}_{\mathbb{Q}} \to D^b_{et}(\mathbb{F}_q, \mathbb{Q}_l) \tag{46}$$

is conservative for any $l \nmid q$*. That is, for any object* M*, we have* $R(M) \cong 0$ *if and only if* $M \cong 0$*.*

Proof First note that by Morel's Theorem 3.19, we have $SH(\mathbb{F}_q)^c_{\mathbb{Q}} \cong DM_{gm}(\mathbb{F}_q)_{\mathbb{Q}}$ so by assumption, $\mathrm{Spec}(DM_{gm}(\mathbb{F}_q)_{\mathbb{Q}}) \cong \mathrm{Spec}(\mathbb{Q})$. In particular, we are assuming that the only prime of $DM_{gm}(\mathbb{F}_q)_{\mathbb{Q}}$ is the zero ideal [9, Prop. 4.2(vi)].

Next note that the rigid category $D^b_{et}(\mathbb{F}_q, \mathbb{Q}_l)$ is *local*, in the sense that for any two objects $E, F \in D^b_{et}(\mathbb{F}_q, \mathbb{Q}_l)$, if $E \otimes F \cong 0$ then $E \cong 0$ or $F \cong 0$, [9, Sect.4]. From this and the fact that R is an exact monoïdal functor, it follows that the kernel of R is a prime. Since zero is the only prime, it follows that R is conservative.

Observation 4.3 is true if the base is the complex numbers, and we use the Betti realisation instead. Consequences of this conservativity are explored in [5, §2].[20]

Conjecture 4.4 (Cisinski, [17]) *If* $Spc(SH(\mathbb{F}_q)^c_{\mathbb{Q}}) \cong Spec(\mathbb{Q})$*, then the Beilinson–Parshin conjecture is true, and rational and numerical equivalence agree.*

4.2 The Structural Morphism

Proposition 4.5 *Let* q *be an odd prime power and consider the open-closed decomposition of the topological space* $Spc(SH(\mathbb{F}_q)^c) = U(\mathbb{P}^2) \cup supp(\mathbb{P}^2)$*, cf. Definition 2.3(4), where* $\mathbb{P}^2$ *is pointed at any rational point. The canonical morphism of* topological spaces

$$supp(\mathbb{P}^2) \to Spec(\mathbb{Z}), \qquad resp.\ U(\mathbb{P}^2) \to Spec(\mathbb{Z}) \tag{47}$$

is surjective, resp. has image $(2) \in Spec(\mathbb{Z})$*. Moreover, the closure of* $U(\mathbb{P}^2)$ *in* $Spc(SH(\mathbb{F}_q)^c)$ *intersects* $supp(\mathbb{P}^2)$ *nontrivially.*

Proof We begin with $U(\mathbb{P}^2)$. First we claim that $U(\mathbb{P}^2)$ is non-empty, cf. [9, Proof of Prop. 10.4]. Recall that $\eta : \mathbb{1}[-1] \to T^{-1}$ is the morphism induced by the canonical morphism $\mathbb{A}^2 - \{0\} \to \mathbb{P}^1$, and note that $\mathbb{P}^2 \cong \mathrm{Cone}(\eta)$ by the Mayer-Vietoris distinguished triangle associated to $\{\mathbb{P}^2 - \{x\} \to \mathbb{P}^2, \mathbb{P}^2 - \mathbb{P}^1 \to \mathbb{P}^2\}$ (and $\mathbb{A}^1$-invariance)

[20] In fact, the Conservativity Conjecture described in [5] is one of the major problems in the area. It is a kind of triangulated analogue of the Hodge conjecture.

where x is any rational point not in $\mathbb{P}^1$. So by Proposition 2.7(4) the set $U(\mathbb{P}^2)$ is non-empty if and only if $\eta^{\otimes n}$ is nonzero for all $n > 0$. But this latter follows from the fact that η is not nilpotent in $K^{MW}(\mathbb{F}_q)$, Theorem 3.21, Example 3.26, (this nonnilpotence is true for any field [43, Cor. 6.4.5,p. 258]).

By Proposition 2.7(2) the subspace $U(\mathbb{P}^2)$ is isomorphic to $\mathrm{Spc}\left(SH(\mathbb{F}_q)^c/\langle\mathbb{P}^2\rangle\right)$. Since $\mathbb{P}^2 \cong \mathrm{Cone}(\eta)$, the morphism η is invertible in $SH(\mathbb{F}_q)^c/\langle\mathbb{P}^2\rangle$, and therefore the canonical morphism of graded rings

$$\oplus_{n\in\mathbb{Z}} \hom_{SH(\mathbb{F}_q)}(\mathbb{1}[n], T^n) \to \oplus_{n\in\mathbb{Z}} \hom_{SH(\mathbb{F}_q)/\langle\mathbb{P}^2\rangle}(\mathbb{1}[n], T^n) \tag{48}$$

factors through the localised ring $\oplus_{n\in\mathbb{Z}} \hom_{SH(\mathbb{F}_q)}(\mathbb{1}[n], T^n)[\eta^{-1}]$. It follows from the description of Example 3.26 that 4 is zero in this latter. Hence $4 \cdot \mathrm{id}_{\mathbb{1}}$ is zero in $SH(\mathbb{F}_q)/\langle\mathbb{P}^2\rangle$ and therefore the canonical morphism of topological spaces $U(\mathbb{P}^2) \cong \mathrm{Spc}(\left(SH(\mathbb{F}_q)/\langle\mathbb{P}^2\rangle\right)^c) \to \mathrm{Spec}(\mathbb{Z})$ given by Theorem 2.5 factors through $\mathrm{Spec}(\mathbb{Z}/2) \subset \mathrm{Spec}(\mathbb{Z})$.

Now consider $\mathrm{supp}(\mathbb{P}^2)$. Using Theorems 2.5, 3.21 again, we find a surjective morphism of topological spaces

$$\mathrm{Spc}(SH(\mathbb{F}_q)^c) \to \mathrm{Spec}(GW(\mathbb{F}_q)) \overset{\text{Exm.3.26}}{\cong} \mathrm{Spec}(\mathbb{Z}). \tag{49}$$

We have seen that the open subspace $U(\mathbb{P}^2)$ is sent to the closed subspace (2). So to see that $\mathrm{supp}(\mathbb{P}^2) \to \mathrm{Spec}(\mathbb{Z})$ is still surjective, it suffices to show that $\mathrm{supp}(\mathbb{P}^2)$ intersects the closure of $U(\mathbb{P}^2)$ nontrivially. That is, it suffices to show that the topological space $\mathrm{Spc}(SH(\mathbb{F}_q)^c)$ is not the disjoint union of the topological spaces $U(\mathbb{P}^2)$ and $\mathrm{supp}(\mathbb{P}^2)$. Such a decomposition would induce a decomposition of the ring $End_{SH(\mathbb{F}_q)^c_{\mathbb{Z}[1/p]}}(\mathbb{1})$ into the direct product of the non-zero rings $End_{U(\mathbb{P}^2)}(\mathbb{1})[1/p]$ and $End_{\mathrm{supp}(\mathbb{P}^2)}(\mathbb{1})[1/p]$, [8, Theorem 2.11], [40, Riou's Appendix B]. Since $1 \in GW(\mathbb{F}_q)[1/p]$ is not a sum of two non-zero idempotents, this is not possible.

Corollary 4.6 *Let q be an odd prime power. The canonical morphism*

$$Spc(SH(\mathbb{F}_q)^c) \to Spec^h(K^{MW}(\mathbb{F}_q)) \tag{50}$$

of Theorem 2.5 and Theorem 3.21 is surjective.

Remark 4.7 This surjectivity is proven for a general field by Heller and Ormsby in [25] using the classification of the prime ideals of K^{MW} from [56]. It was Ormsby's idea of using this classification which lead to our proof.

Proof One can classify the homogenous prime ideals of $K^{MW}(\mathbb{F}_q)$ by hand[21] using the description in Example 3.26, or consult [56], and find that they are: $([\omega], \eta, p)$ with $p \in \mathbb{Z}$ an odd prime, $([\omega], 2)$, and $([\omega], \eta)$.

[21]They are in bijection with the homogeneous prime ideals of $K^{MW}(\mathbb{F}_q)_{red} \cong \mathbb{Z}[\eta]/(2 \cdot \eta)$. Then we have $\mathrm{Spec}(\mathbb{Z}[\eta]/(2 \cdot \eta)) = \mathrm{Spec}(\mathbb{Z}) \cup \mathrm{Spec}(\mathbb{Z}/2[\eta])$. Clearly, apart from $\mathrm{Spec}(\mathbb{Z}/2) = \mathrm{Spec}(\mathbb{Z}) \cap \mathrm{Spec}(\mathbb{Z}/2[\eta])$, the graded ring $\mathbb{Z}/2[\eta]$ has exactly one other homogeneous prime: (η).

It follows from $U(\mathbb{P}^2) = U(\mathrm{Cone}(\eta))$ being non-empty that $([\omega], 2)$ is in the image of $\mathrm{Spc}(SH(\mathbb{F}_q)^c)$. The primes $([\omega], \eta, p)$ with $p \in \mathbb{Z}$ an odd prime are in the image by the surjectivity of $\mathrm{Spc}(SH(\mathbb{F}_q)^c) \to \mathrm{Spec}(\mathbb{Z})$. Finally, $([\omega], \eta)$ is in the image by the claim that $\mathrm{supp}(\mathbb{P}^2)$ intersects the closure of $U(\mathbb{P}^2)$ nontrivially.

4.3 Equivariant Stable Homotopy Theory

Much information about $SH(\mathbb{R})$ comes from the realisation functor, $SH(\mathbb{R}) \to SH_{\mathbb{Z}/2}$, towards the $\mathbb{Z}/2$-equivariant stable homotopy category, cf. Theorem 3.21(4). Here, $\mathbb{Z}/2$ appears because $Gal(\mathbb{R}) \cong \mathbb{Z}/2$. The étale homotopy type, [1, 21], provides a functor from $\mathrm{Sm}(S)$ to the homotopy category Ho(Pro-SS) of pro-finite simplicials sets. We have $Gal(\mathbb{F}_q) \cong \hat{\mathbb{Z}}$, and we hope that the étale homotopy type induces a tensor triangulated functor towards an appropriate $\hat{\mathbb{Z}}$-equivariant stable homotopy category. The structure (as a set) of the spectrum of the equivariant stable homotopy category, $\mathrm{Spc}(SH_G^c)$, of a finite group G has recently been completely determined by Balmer–Sanders, [14]. They also describe much about its topology. This suggests that the $\hat{\mathbb{Z}}$-equivariant stable homotopy category mentioned above has a good chance of being accessible, and providing information about the structure of $\mathrm{Spc}(SH(\mathbb{F}_q)^c_{\mathbb{Z}_{(l)}})$.

4.4 Final Observations

Recall the following.

1. It seems highly likely that

$$\mathrm{Spc}(SH(\mathbb{F}_q)^c_{\mathbb{Q}}) \cong \mathrm{Spec}(\mathbb{Q}), \tag{51}$$

   since this is implied by conjectures which are widely believed to be true, cf. Theorem 4.1.
2. Let $\mathbb{F}$ be the algebraic closure of the finite field with an odd number of elements. In Example 3.26 above, we have seen that the endomorphism ring of the unit $GW(\mathbb{F}) \cong End_{SH(\mathbb{F})}(\mathbb{1})$ and the graded endomorphism ring $K^{MW}(\mathbb{F}) \cong \oplus_{n\in\mathbb{Z}} \hom_{SH(\mathbb{F})}(\mathbb{1}[n], T^n)$ are extremely simple, especially if we invert 2, and ignore nilpotents. Indeed,

$$K^{MW}(\mathbb{F})[1/2]_{red} \cong \mathbb{Z}[1/2]. \tag{52}$$

3. Due to the realisation functor, for a subfield $k \subseteq \mathbb{C}$, there is a retraction

$$\mathrm{Spc}(SH^c_{\mathrm{top}}) \longrightarrow \mathrm{Spc}(SH(k)^c) \longrightarrow \mathrm{Spc}(SH^c_{\mathrm{top}}) \; . \qquad \text{(composite: } \cong\text{)} \tag{53}$$

It seems likely that the étale realisation should give rise to a similar phenomenon for all fields.

Based on these observations, lets make the following wild speculation. We have taken the algebraic closure of the base field to avoid equivariant phenomena which might appear, as discussed in Sect. 4.3, but one could also ask if the case of a finite field base is just the appropriate combination of the $\mathbb{F}$-base case together with $\mathrm{Gal}(\mathbb{F}/\mathbb{F}_p)$-equivariant phenomena, cf. Example 2.4(3).

Guess 4.8 *Let $\mathbb{F}$ be an algebraic closure of a finite field with p elements. Then the canonical constant presheaf functor $SH_{top} \to SH(\mathbb{F})$ from the classical stable homotopy category, cf. Eq.* (29)*, induces an isomorphism*

$$Spc(SH(\mathbb{F})^c_{\mathbb{Z}_{(l)}}) \to Spc((SH^c_{top})_{\mathbb{Z}_{(l)}}) \tag{54}$$

whenever l and p are odd.

Note that if $\mathbb{Z}_{(l)}$ is replaced by $\mathbb{Q}$ in the above guess, we recover the isomorphism $\mathrm{Spec}(SH(\mathbb{F})^c_{\mathbb{Q}}) \cong \mathrm{Spec}(\mathbb{Q})$ implied by Theorem 4.1.

Acknowledgements I thank the organisers of the conference “Bousfield classes form a set: a workshop in memory of Tetsusuke Ohkawa” for the invitation to speak which led me to think about these things, and also for having organised such an interesting conference. I also thank Paul Balmer, Jens Hornbostel, and Denis-Charles Cisinski for interesting discussions about potential future work, Marc Hoyois for discussions about the étale homotopy type, and Jeremiah Heller and Kyle Ormsby for pointing out that an “elementary fact” I was using in the proof of Proposition 4.5 is actually a combination of theorems of Ayoub, Balmer, Gabber, and Riou.

References

1. Artin, M., Mazur, B.: Etale homotopy. Lecture Notes in Mathematics, vol. 100. Springer, Berlin (1986). Reprint of the 1969 original
2. Aoki, K.: The weight complex functor is symmetric monoidal (2019). Arxiv preprint arXiv:1904.01384
3. Ayoub, J.: Les six opérations de Grothendieck et le formalisme des cycles évanescents dans le monde motivique. II. Astérisque **315**(2008), vi+364 (2007)
4. Ayoub, J.: Note sur les opérations de Grothendieck et la réalisation de Betti. J. Inst. Math. Jussieu **9**(2), 225–263 (2010)
5. Ayoub, J.: Motive and algebraic cycles: a selection of conjectures and open questions (2015)
6. Bachmann, T.: On the invertibility of motives of affine quadrics. Doc. Math. **22**, 363–395 (2017)
7. Balmer, P.: The spectrum of prime ideals in tensor triangulated categories. J. für Die Reine Und Angew. Math. **2005**(588), 149–168 (2005)

8. Balmer, P.: Supports and filtrations in algebraic geometry and modular representation theory. Am. J. Math. 1227–1250 (2007)
9. Balmer, P.: Spectra, spectra, spectra-tensor triangular spectra versus Zariski spectra of endomorphism rings. Algebr. Geom. Topol. **10**(3), 1521–1563 (2010)
10. Balmer, P.: Tensor triangular geometry. In: Proceedings of the International Congress of Mathematicians, vol. II, pp. 85–112. Hindustan Book Agency, New Delhi (2010)
11. Bloch, S.: Algebraic cycles and higher K-theory. Adv. Math. **61**(3), 267–304 (1986)
12. Bondarko, M.V.: Differential graded motives: weight complex, weight filtrations and spectral sequences for realizations; voevodsky versus hanamura. J. Inst. Math. Jussieu **8**(01), 39–97 (2009)
13. Bondarko, M.V.: $\mathbb{Z}[1/p]$-motivic resolution of singularities. Compos. Math. **147**(5), 1434–1446 (2011)
14. Balmer, P., Sanders, B.: The spectrum of the equivariant stable homotopy category of a finite group (2015)
15. Cisinski, D.C., Déglise, F.: Triangulated categories of mixed motives (2012). Arxiv preprint arXiv:0912.2110
16. Cisinski, D.C.: Descente par éclatements en K-théorie invariante par homotopie. Ann. Math. (2), **177**(2), 425–448 (2013)
17. Cisinski, D.C.: Conversation (2016)
18. Dugger, D., Isaksen, D.C.: The motivic Adams spectral sequence. Geom. Topol. **14**(2), 967–1014 (2010)
19. Dugger, D., Isaksen, D.C.: Motivic Hopf elements and relations. New York J. Math. **19**, 823–871 (2013)
20. Deninger, C., Murre, J.: Motivic decomposition of abelian schemes and the Fourier transform. J. Reine Angew. Math. **422**, 201–219 (1991)
21. Friedlander, E.M.: Fibrations in etale homotopy theory. Inst. Hautes Études Sci. Publ. Math. **42**, 5–46 (1973)
22. Friedlander, E.M., Voevodsky, V.: Bivariant cycle cohomology. In: Cycles, Transfers, and Motivic uomology Theories. Annals of Mathematics Studies, vol. 143, pp. 138–187. Princeton University Press, Princeton, NJ (2000)
23. Gille, S., Scully, S., Zhong, C.: Milnor-Witt K-groups of local rings. Adv. Math. **286**, 729–753 (2016)
24. Hu, P., Kriz, I., Ormsby, K.: Remarks on motivic homotopy theory over algebraically closed fields. J. K-Theory **7**(1), 55–89 (2011)
25. Heller, J., Ormsby, K.: Primes and fields in stable motivic homotopy theory (2016)
26. Hopkins, M.J., Smith, J.H.: Nilpotence and stable homotopy theory II. Ann. Math. **148**(1), 1–49 (1998)
27. Huybrechts, D.: Fourier-Mukai Transforms in Algebraic Geometry. Oxford University Press, Demand (2006)
28. Ivorra, F.: Réalisation l-adique des motifs mixtes. Comptes Rendus Math. Acad. Sci. Paris **342**(7), 505–510 (2006)
29. Jannsen, U.: Motives, numerical equivalence, and semi-simplicity. Invent. Math. **107**(3), 447–452 (1992)
30. Jardine, J.F.: Motivic symmetric spectra. Doc. Math. **5**, 445–553 (electronic) (2000)
31. Joachimi, R.: Thick ideals in equivariant and motivic stable homotopy categories. To appear in this proceedings
32. Kahn, B.: Algebraic K-theory, algebraic cycles and arithmetic geometry. Handbook of K-Theory, pp. 351–428. Springer, Berlin (2005)
33. Kelly, S.: Triangulated categories of motives in positive characteristic (2013)
34. Kleiman, S.: The standard conjectures. Motives (Seattle, WA, 1991). Proceedings of Symposia in Pure Mathematics Part 1, vol. 55, pp. 3–20. A. M. S., Providence, RI
35. Kock, J., Pitsch, W.: Hochster duality in derived categories and point-free reconstruction of schemes (2013)

36. Kashiwara, M., Schapira, P.: Categories and Sheaves. vol. 332. Springer Science & Business Media (2005)
37. Kelly, S., Saito, S.: Weight homology of motives. Int. Math. Res. Not. **13**, 3938–3984 (2017)
38. Lam, T.Y.: Introduction to quadratic forms over fields. Graduate Studies in Mathematics, vol. 67. American Mathematical Society, Providence, RI (2005)
39. Levine, M.: A comparison of motivic and classical stable homotopy theories. J. Topol. **7**(2), 327–362 (2014)
40. Levine, M., Yang, Y., Gufang, Z.: Algebraic elliptic cohomology theory and flops, I (2013)
41. Milnor, J.: Algebraic K-theory and quadratic forms. Invent. Math. **9**, 318–344 (1969/1970)
42. Morel, F.: An introduction to $\mathbf{A}^1$-homotopy theory. In: Contemporary Developments in Algebraic K-Theory. ICTP Lecture Notes, vol. XV, pp. 357–441 (electronic). Abdus Salam International Centre for Theoretical Physics, Trieste (2004)
43. Morel, F.: On the motivic π_0 of the sphere spectrum. In: Axiomatic, Enriched and Motivic Homotopy Theory. NATO Science Series II: Mathematics, Physics and Chemistry, vol. 131, pp. 219–260. Kluwer Academic Publishers, Dordrecht (2004)
44. Morel, F.: Sur les puissances de l'idéal fondamental de l'anneau de Witt. Comment. Math. Helv. **79**(4), 689–703 (2004)
45. Morel, F.: Rationalized motivic sphere spectrum (2006)
46. Fabien, M.: $\mathbb{A}^1$- algebraic topology over a field. Lecture Notes in Mathematics, vol. 2052. Springer, Heidelberg (2012)
47. Morel, F., Voevodsky, V.: $\mathbf{A}^1$-homotopy theory of schemes. Inst. Hautes Études Sci. Publ. Math. **90**(2001), 45–143 (1999)
48. Mazza, C., Voevodsky, V., Weibel, C.: Lecture notes on motivic cohomology. Clay Mathematics Monographs, vol. 2 American Mathematical Society, Providence, RI; Clay Mathematics Institute, Cambridge, MA (2006)
49. Ormsby, K.M., Østvær, P.A.: Motivic Brown-Peterson invariants of the rationals. Geom. Topol. **17**(3), 1671–1706 (2013)
50. Ormsby, K.M., Østvær, P.A.: Stable motivic π_1 of low-dimensional fields. Adv. Math. **265**, 97–131 (2014)
51. Peter, T.J.: Prime ideals of mixed Artin-Tate motives. J. K-Theory **11**(2), 331–349 (2013)
52. Robalo, M.: K-theory and the bridge from motives to noncommutative motives. Adv. Math. **269**, 399–550 (2015)
53. Samuel, P.: Relations d'équivalence en géométrie algébrique. In: Proceedings of the International Congress of Mathematicians 1958, pp. 470–487. Cambridge University Press, New York (1960)
54. Scholl, A.J.: Classical motives. In: Motives (Seattle, WA, 1991). Proceedings of Symposia in Pure Mathematics, vol. 55, pp. 163–187. American Mathematical Society, Providence, RI (1994)
55. Suslin, A.A.: Higher Chow groups and etale cohomology. In: Cycles, Transfers, and Motivic Homology Theories. Annals of Mathematics Studies, vol. 143, pp. 239–254. Princeton University Press, Princeton, NJ (2000)
56. Thornton, R.: The homogeneous spectrum of Milnor-Witt K-theory (2015)
57. Voevodsky, V.: $\mathbf{A}^1$-homotopy theory. In: Proceedings of the International Congress of Mathematicians, vol. I. number Extra Vol. I, pp. 579–604 (electronic) Berlin (1998)
58. Voevodsky, V.: Triangulated categories of motives over a field. In: Cycles, Transfers, and Motivic Homology Theories. Annals of Mathematics Studies, vol. 143, pp. 188–238. Princeton University Press, Princeton, NJ (2000)
59. Voevodsky, V.: Cancellation theorem. Doc. Math. (Extra volume: Andrei A. Suslin sixtieth birthday):671–685, 2010
60. Wilson, G.M., Østvær, P.A.: Two-complete stable motivic stems over finite fields (2016)

Operations on Integral Lifts of $K(n)$

Jack Morava

Abstract This very rough sketch is a sequel to [27, 28]; it presents evidence that operations on lifts of the functors $K(n)$ to cohomology theories with values in modules over valuation rings $\mathfrak{o}_L$ of local number fields, indexed by Lubin–Tate groups of such fields, are extensions of the groups of automorphisms of the associated group laws, by the exterior algebras on the normal bundle to the orbit of the group law in the space of lifts.

Keywords Stable homotopy · Perfectoid fields · Koszul construction · Lubin-Tate theory · Morava K-theory

1 Introduction

1.1 In a symmetric monoidal category, e.g. of schemes or structured spectra, the morphisms defining an action of a monoid M on an object X can be presented as a cosimplicial object; for example [24] if $M = M\mathrm{U}$ is the Thom spectrum for complex cobordism (i.e. the universal complex-oriented S^0-algebra), then

$$(S^0 \dashrightarrow)\ M\mathrm{U} \rightrightarrows M\mathrm{U} \wedge_{S^0} M\mathrm{U} \mathrel{\substack{\textstyle\rightarrow\\[-0.6ex]\textstyle\rightarrow\\[-0.6ex]\textstyle\rightarrow}} \cdots$$

is a kind of $M\mathrm{U}$-free Adams-Mahowald-Novikov resolution of S^0. Its homotopy groups define a cosimplicial commutative algebra resolution

$$\pi_* S^0 \dashrightarrow \pi_* M\mathrm{U} = M\mathrm{U}_* \rightrightarrows \pi_*(M\mathrm{U} \wedge_{S^0} M\mathrm{U}) = M\mathrm{U}_* M\mathrm{U} \mathrel{\substack{\textstyle\rightarrow\\[-0.6ex]\textstyle\rightarrow\\[-0.6ex]\textstyle\rightarrow}} \cdots$$

J. Morava (✉)
Department of Mathematics, The Johns Hopkins University, Baltimore, MD 21218, USA
e-mail: jack@math.jhu.edu

T. Ohsawa and N. Minami (eds.), *Bousfield Classes and Ohkawa's Theorem*,
Springer Proceedings in Mathematics & Statistics 309,
https://doi.org/10.1007/978-981-15-1588-0_8

of the stable homotopy algebra.[1] Regarding these algebras as affine schemes over Spec $\mathbb{Z}$, this diagram becomes a presentation for a groupoid-scheme

$$\operatorname{Spec} MU_*MU \rightrightarrows \operatorname{Spec} MU_*$$

which, by work of Quillen [29], can be identified with a moduli stack for one-dimensional commutative formal groups.

Using a great deal of work on Lubin–Tate spectra by others, we construct in Sect. 3.3.2 below, certain (p-adically complete, where $p > 3$) A_∞ periodic MU-algebra spectra $\mathsf{K}(L)$, indexed by Lubin–Tate formal group laws LT_L for local number fields $L \supset \mathbb{Q}_p$, Galois of degree $[L : \mathbb{Q}_p] = n$ with valuation rings $\mathfrak{o}_L$. These spectra have homotopy groups

$$\pi_* \mathsf{K}(L) = \mathsf{K}(L)_* \cong \mathfrak{o}_{L*}[v^{\pm 1}]$$

($|v| = 2$), and in Sect. 4 we present a conjectural description of an associated groupoid-scheme

$$\operatorname{Spec} \mathsf{K}(L)_*\mathsf{K}(L) \rightrightarrows \operatorname{Spec} \mathsf{K}(L)_*$$

of homological co-operations in terms of the isotropy or stabilizer groups of LT_L, as objects in the Quillen–Lazard moduli stack. These automorphism groups are by now well-understood, almost classical in local arithmetic geometry, and the first section below summarizes some of that knowledge; it will serve as a model for our applications to algebraic topology.

Perhaps the point of this paper is to explain that, in spite of the notation, our construction of the spectra $\mathsf{K}(L)$ is **not** functorial in L; this note is instead a plea for a natural construction. The third section below contains preliminary results toward an identification of the endomorphisms or (co)operations of their associated cohomology theories, and argues that these have close connections with the Weil group of L [25]: or, more precisely, with the Galois group of a maximal totally ramified abelian extension L^{trab} of L, over $\mathbb{Q}_p$. Our partial results can perhaps be read as evidence toward an interpretation of the spectra $\mathsf{K}(L)$ as something like a K-theory spectrum associated to the (topological, perfectoid) completion L^∞ of L^{trab} [28]. For example, our $\mathsf{K}(\mathbb{Q}_p)$ can be (non-canonically) identified with the p-adically completed algebraic K-theory spectrum of the completion ${\mathbb{Q}_p}^\infty$ of the field of p-power roots of unity over $\mathbb{Q}_p$, and thus with the p-adic completion of Atiyah's topological K-theory of $\mathbb{C}$.

To return to the organization of this paper: its second section uses the theory of highly structured spectra to define, following the original work of Sullivan and Baas [4, 38, 39], the spectra $\mathsf{K}(L)$ as Koszul quotients of spectra $\mathsf{E}(\Phi_L)$ associated to Lubin–Tate formal group laws [14, 31]. The resulting constructions are integral lifts of the 'extraordinary' spectra $K(n)$ [44], in that smashing with a mod p Moore

[1] The terms in this display are graded, but it is convenient to regard them as $\mathbb{Z}_2$-graded comodules over the multiplicative groupscheme $\mathbb{G}_m = \mathbb{Z}[t_0^{\pm 1}]$, with coaction $M_{2k} \ni x \mapsto x \otimes t_0^k$, thus providing an excuse for often suppressing this grading.

spectrum defines natural isomorphisms

$$\mathsf{K}(L)_*(X \wedge M(p,0)) \cong K(n)_*(X, \mathbb{F}_p) \otimes \mathfrak{o}_L/p\mathfrak{o}_L ,$$

where $\mathfrak{o}_L/p\mathfrak{o}_L \cong \mathbb{F}_q[\pi]/(\pi^e)$ (with $n = ef$ and $q = p^f$, see Sect. 2.5). For example, if L is unramified then $e = 1,\ q = p^n$, and the mod p reduction of $\mathsf{K}(L)$ agrees with $K(n) \otimes \mathbb{F}_q$. In some sense the $K(n)$ are indexed by the finite fields, while the $\mathsf{K}(L)$ are indexed by finite Galois extensions of $\mathbb{Q}_p$.

2 Notation and Recollections

2.1 If A is a commutative ring, let $\mathsf{FG}(A) \subset A[[X,Y]]$ be the set of power series $F(X,Y) = X + Y + \ldots$ satisfying the standard axioms for a commutative formal group law over A, and let $\Gamma(A) \subset A[[T]]$ be the group of invertible power series $t(T) = t_0 T + \ldots$ (i.e. with $t_0 \in A^\times$) under composition; then the group Γ acts on the set FG by

$$\Gamma(A) \times \mathsf{FG}(A) \ni t, F \mapsto F^t(X,Y) = t^{-1}(F(t(X), t(Y))) \in \mathsf{FG}(A) .$$

Both Γ and FG are co-representable functors: $\mathsf{FG}(A) \cong \mathrm{Hom}_{\mathrm{alg}}(\mathbb{L}, A)$, where Lazard's ring $\mathbb{L}$ is polynomial over $\mathbb{Z}$, and $\Gamma(A) \cong \mathrm{Hom}_{\mathrm{alg}}(S, A)$, where $S = t_0^{-1}\mathbb{Z}[t_i]_{i \geq 0}$ is a Hopf algebra with coproduct

$$(\Delta t)(T) = (t \otimes 1)((1 \otimes t)(T)) \in (S \otimes_{\mathbb{Z}} S)[[T]] .$$

Yoneda's lemma then implies the existence of a coproduct homomorphism

$$\psi : \mathbb{L} \to \mathbb{L} \otimes_{\mathbb{Z}} S$$

of rings, corepresenting the group action. These rings are implicitly graded by the coaction of the multiplicative subgroup $\mathbb{G}_m \subset \mathrm{Spec}\ S$.

2.2 A group action $\alpha : G \times X \to X$ in (Sets) defines a groupoid

$$[X//G] : G \times X \underset{t}{\overset{s}{\rightrightarrows}} X$$

with X as set of objects, $G \times X$ as set of morphisms, and $s(g,x) = x,\ t(g,x) = \alpha(g,x)$ as source and target maps. The usual convention in algebraic topology regards $\mathbb{L} \otimes_{\mathbb{Z}} S$ as a two-sided $\mathbb{L}$-algebra, with the obvious structure on the left, and a right $\mathbb{L}$-algebra structure

$$(\mathbb{L} \otimes_{\mathbb{Z}} S) \otimes_{\mathbb{Z}} \mathbb{L} \xrightarrow{1 \otimes \psi} (\mathbb{L} \otimes_{\mathbb{Z}} S) \otimes_{\mathbb{L}} (\mathbb{L} \otimes_{\mathbb{Z}} S) \longrightarrow (\mathbb{L} \otimes_{\mathbb{Z}} S) \; ;$$

this is what's meant by saying that

$$\mathbb{L} \underset{\eta_R}{\overset{\eta_L}{\rightrightarrows}} \mathbb{L} \otimes_{\mathbb{Z}} S$$

is a Hopf algebroid.

Following Grothendieck and Segal, a category $\mathcal{C}$ with set $\mathcal{C}[0]$ of objects and $\mathcal{C}[1]$ of morphisms can be presented as a simplicial set

$$\mathcal{C}[0] \leftleftarrows \mathcal{C}[1] \;\overset{\leftarrow}{\leftleftarrows}\; \mathcal{C}[1] \times_{\mathcal{C}[0]} \mathcal{C}[1] \qquad \cdots$$

(where $X \times_Z Y$ denotes the fiber product or equalizer of two maps $X, Y \to Z$). In the case of a group action as above, this is isomorphic to a simplicial object

$$X \leftleftarrows G \times X \;\overset{\leftarrow}{\leftleftarrows}\; G \times G \times X \qquad \cdots$$

which can alternatively be regarded as a bar construction. The functor $A \mapsto [\mathsf{FG}(A)//\Gamma(A)]$ thus defines a simplicial scheme: the moduli stack of one-dimensional formal groups.

2.3 A homomorphism $A \to B$ of commutative rings defines an extension of scalars map

$$\mathsf{FG}(A) \ni F \mapsto F \otimes_A B \in \mathsf{FG}(B) \; .$$

Definition $[\mathsf{iso}(F)](B)$ is the groupoid with the orbit

$$\mathcal{O}_\Gamma(B)(F) = \{(F \otimes_A B)^g \mid g \in \Gamma(B)\}$$

(of $F \otimes_A B$ under coordinate changes) as its set of objects, and

$$\mathrm{mor}_{\mathsf{iso}_F(B)}(G, G') = \{h \in \Gamma(B) \mid G^h = G'\}$$

as (iso)morphisms of G with G'; thus

$$[\mathsf{iso}(F)](B) = [\mathcal{O}_\Gamma(B)(F)//\Gamma(B)] \; .$$

This groupoid maps fully and faithfully to its skeleton (which has one object) and the group $\mathrm{Aut}_B(F) \subset \Gamma(B)$ (of automorphisms of $F \otimes_A B$ as a formal group law over B) as its morphisms. The homomorphism $F : \mathbb{L} \to A$ classifying F thus defines a Hopf A - algebroid

$$[\mathsf{iso}(F)] : A \rightrightarrows A \otimes_{\mathbb{L}} (\mathbb{L} \otimes_{\mathbb{Z}} S) \otimes_{\mathbb{L}} A$$

equivalent to a simplicial groupoid-scheme $(A-\mathrm{alg}) \ni B \mapsto [\mathsf{iso}(F)](B)$ over Spec A.

2.4 It is nonstandard, but it will be convenient below to write $q = p^n$ and let $\mathbb{Q}_q$ denote the quotient field of the ring $W(\mathbb{F}_q)$ of Witt vectors, i.e. the degree n unramified extension of $\mathbb{Q}_p$. Following Ravenel [30, §5.1.13], a Lubin–Tate group law $\mathsf{LT}_{\mathbb{Q}_q}$ for this field can be defined over the p-adic integers $\mathbb{Z}_p$ by Honda's logarithm

$$\log_{\mathbb{Q}_q}(T) = \sum_{k\geq 0} p^{-k} T^{p^{nk}} \;;$$

this has, as its mod p reduction, a formal group law $\mathrm{H}(n)$ of height n over $\mathbb{F}_p$, associated to the cohomology theory $K(n)$. The resulting left and right $\mathbb{F}_p$-algebra structures on

$$\mathbb{F}_p \otimes_{\mathrm{H}(n)} (\mathbb{L} \otimes_{\mathbb{Z}} S) \otimes_{\mathrm{H}(n)} \mathbb{F}_p = C(\mathfrak{o}_D{}^{\times}, \mathbb{F}_q)^{\mathrm{Gal}(\mathbb{F}_q/\mathbb{F}_p)-\mathrm{inv}} = \Sigma(n)$$

coincide, representing $[\mathsf{iso}(\mathrm{H}(n))]$ by the algebra of functions on a certain pro-algebraic group scheme over Spec $\mathbb{F}_p$.

In more detail [24], a finite field $k = \mathbb{F}_q$ has a local domain $W(k)$ of Witt vectors, with maximal ideal generated by p and a canonical isomorphism $W(k)/pW(k) \to k$; its quotient field $W(k) \otimes_{\mathbb{Z}} \mathbb{Q} = \mathbb{Q}_q$ is the extension of $\mathbb{Q}_p$ obtained by lifting the roots $\mathbb{F}_q^{\times}$ of unity to $\mathbb{Q}_p$. This construction is functorial, and a generator σ of the cyclic group $\mathrm{Gal}(\mathbb{Q}_q/\mathbb{Q}_p) \cong \mathrm{Gal}(\mathbb{F}_q/\mathbb{F}_p)$ sends a root ω of unity to $\sigma(\omega) = \omega^p$. Let

$$D = \mathbb{Q}_q\langle \mathsf{F}\rangle/(\mathsf{F}^n = p)$$

be the noncommutative division algebra obtained from $\mathbb{Q}_q$ by adjoining an nth root F of p satisfying, for any $a \in \mathbb{Q}_q$, the relation $\sigma(a) \cdot \mathsf{F} = \mathsf{F} \cdot a$. The valuation on $\mathbb{Q}_q$ (normalized so $\mathrm{ord}(p) = 1$) extends to D to define a semidirect product extension

$$1 \longrightarrow \mathfrak{o}_D{}^{\times} \longrightarrow D^{\times} \xrightarrow{\mathrm{ord}} \tfrac{1}{n}\mathbb{Z} \longrightarrow 0$$

with a generator of the infinite cyclic group on the right acting on an element u of the compact kernel $\mathfrak{o}_D{}^{\times}$ as F-conjugation. This kernel thus acquires an action of the cyclic group of order n, which may be identified with $\mathrm{Gal}(\mathbb{F}_q/\mathbb{F}_p)$, making $\mathfrak{o}_D{}^{\times}$ the group of points of a pro-étale groupscheme over $\mathbb{F}_p$. It is represented by the $\mathbb{F}_p$-algebra of (Galois equivariant, continous) $\mathbb{F}_q$-valued functions h on $\mathfrak{o}_D{}^{\times}$ satisfying $\sigma(h(u)) = h(\mathsf{F}u\mathsf{F}^{-1})$. More concisely,

$$[\mathsf{iso}(\mathrm{H}(n))] \simeq [*//\mathfrak{o}_D{}^{\times}]$$

as groupoid-valued functors.

2.5 Similar results [21] hold for Lubin–Tate groups of local number fields (i.e. extensions L of $\mathbb{Q}_p$ with $[L : \mathbb{Q}_p] = n < \infty$); I will assume here that this extension is Galois. Such a field has a local valuation ring $\mathfrak{o}_L$ with finite residue field $k_L \cong \mathbb{F}_q$, where now $q = p^f$; moreover L contains a maximal unramified extension $L_0 = W(k_L) \otimes_{\mathbb{Z}} \mathbb{Q} \supset \mathbb{Q}_p$, such that $[L : L_0] = e = f^{-1}n$. The maximal ideal $\mathfrak{m}_L = (\pi) \subset \mathfrak{o}_L$ is principal, and we will choose a generator π; it satisfies some Eisenstein equation

$$E_L(\pi) = \pi^e + \sum_{0 \leq i < e} e_i \pi^i = 0$$

with $\mathrm{ord}(e_i) > 0$ and $\mathrm{ord}(e_0) = 1$. Lubin and Tate construct from this data, a formal group law LT_L over L (with logarithm $\log_L$ and exponential $\exp_L$) such that

$$\mathfrak{o}_L \ni a \mapsto [a]_L(T) = \exp_L(a \cdot \log_L(T)) \in \mathrm{End}\mathfrak{o}_L(\mathsf{LT}_L)$$

is a ring isomorphism. The reduction Φ_L over the residue field of LT_L is independent of the choices.

We will sometimes write $X +_L Y = \mathsf{LT}_L(X, Y)$. There is some degree of choice in the construction of the lift LT_L of Φ_L:

$$\log_L(T) = \sum_{i \geq 0} \pi^{-k} T^{q^k}$$

is one possibility, and $[\pi]_L(T) = \pi T +_L T^q$ defines another[2] but all constructions are isomorphic. Reduction modulo π defines a monomorphism

$$\mathfrak{o}_L \cong \mathrm{End}_{\mathfrak{o}_L}(\mathsf{LT}_L) \to \mathrm{End}_{\bar{k}_L}(\Phi_L)$$

of rings, which embeds the units $\mathfrak{o}_L{}^\times$ as a maximal commutative subgroup (a torus of some sort) in $\mathfrak{o}_D{}^\times$.

The associated simplicial scheme $[\mathsf{iso}(\mathsf{LT}_L)]$ over Spec $\mathfrak{o}_L$ can then be defined, as above, by the Hopf $\mathfrak{o}_L$-algebroid

$$\mathfrak{o}_L \rightrightarrows \mathfrak{o}_L \otimes_{\mathsf{LT}_L} (\mathbb{L} \otimes_{\mathbb{Z}} S) \otimes_{\mathsf{LT}_L} \mathfrak{o}_L \; ;$$

over the generic point of Spec $\mathfrak{o}_L$ its fiber is the groupoid $[*//\mathfrak{o}_L{}^\times]$, while over the closed point it is $[*//\mathfrak{o}_D{}^\times]$. Note that any degree n extension of $\mathbb{Q}_p$ embeds in D as a maximal commutative subfield, so the maximal toruses of $D^\times$ in some sense parametrize Lubin–Tate groups of degree n extensions of $\mathbb{Q}_p$. The Weyl groups of these toruses are then Galois groups $\mathrm{Gal}(L/\mathbb{Q}_p)$, and the normalizers of these toruses are essentially the Weil groups $W(L^{\mathrm{trab}}/\mathbb{Q}_p)$ associated to maximal totally ramified abelian extensions of L [41, 43]; the (cohomology classes of the) group extensions defining them are the 'fundamental classes' of local classfield theory.

When $L = \mathbb{Q}_q$ is unramified [27] we can assume that

[2] The author believes Serre's account [36] of the Lubin–Tate construction to be effectively optimal.

$$[p]_L(T) = pT +_L v^{q-1}T^q$$

i.e. that (a graded version of) $\mathsf{LT}_{\mathbb{Q}_q}$ is p-typical, defined by a homomorphism

$$\mathrm{BP}_* = \mathbb{Z}_p[v_i]_{i\geq 1} \to \mathbb{Z}_p[v^{\pm 1}]$$

sending Araki's [2] generators v_i to 0 when $i \neq n$, and v_n to v^{q-1}.

3 Some Koszul Constructions

3.1 To construct the spectra $\mathsf{K}(L)$ we work[3] at a prime away from 6, in a symmetric monoidal category of p-adically complete spectra, e.g. $S^0_{\hat{p}}$ - modules. We will be concerned below with $K(n)$-local spectra, and we will $K(n)$-localize their smash products [17]. Recall [33] that the Gaussian integer spectrum $S^0[(-1)^{1/2}]$ is not E_∞: the behavior of the stable homotopy category under arithmetic ramification seems potentially very interesting.

Lurie's étale topology on the category of spectra [23, Def 7.5.1.4] defines commutative ring-spectra $S^0_{W(\mathbb{F}_q)}$ étale over $S^0_{\hat{p}}$ (roughly, $W(\mathbb{F}_q) \otimes_{\mathbb{Z}_p} S^0_{\hat{p}}$). Schwede's Moore spectra (functorial away from 6 [34, §II Rem. 6.44]) can then be used to define, for a valuation ring $\mathfrak{o}_L$ (free of rank e over $W(k_L)$) a p-adic A_∞ ring spectrum $S^0_{\mathfrak{o}_L}$ (roughly, $M(\mathfrak{o}_L, 0) \otimes_{W(k_L)} S^0_{W(\mathbb{F}_q)}$) with

$$\pi_* S^0_{\mathfrak{o}_L} \cong \pi_* S^0 \otimes_{\mathbb{Z}} \mathfrak{o}_L \ .$$

Following Sect. 2.5, the commutative W(k)-algebra

$$\mathfrak{o}_L = \oplus_{0\leq i \leq e-1} W(k) \cdot \pi^i$$

(where $k = k_L$ for simplicity) is defined by classical structure constants $m_l^{i,j} \in W(k)$, $0 \leq i, j, l \leq e-1$, such that

$$\pi^i \cdot \pi^j = \sum_{0\leq l \leq e-1} m_l^{i,j} \pi^l \ .$$

Let $S^0_{\mathfrak{o}_L}$ denote the wedge sum $\bigvee_{0\leq i\leq e-1} S^0_{W(k)} \cdot t^i$ (with t^i a book-keeping indeterminate), and let

$$S^0_{\mathfrak{o}_L} \times S^0_{\mathfrak{o}_L} \to S^0_{\mathfrak{o}_L} \wedge_{S^0_{W(k)}} S^0_{\mathfrak{o}_L} \to S^0_{\mathfrak{o}_L}$$

[3]The typeface is intended to distinguish these constructions from Quillen's K-theory

be the morphism of $S^0_{W(k)}$-module spectra defined component-wise, as the composition

$$S^0_{W(k)} \times S^0_{W(k)} \cdot t^i \times t^j \longrightarrow S^0_{W(k)} \wedge_{S^0_{W(k)}} S^0_{W(k)} \cdot t^l \xrightarrow{\cdot m_l^{i,j}} S^0_{W(k)} t^l \;,$$

(where the final map is multiplication by the structure constant). This is the product map for a weak $S^0_{W(k)}$-algebra structure on $S^0_{\mathfrak{o}_L}$, i.e. a kind of H_∞ structure making

$$\pi_* S^0_{\mathfrak{o}_L} = \pi_0 \mathrm{Hom}_{S^0_{W(k)}}(S^*_{W(k)}, S^0_{\mathfrak{o}_L}) \cong \pi_*(S^0_{\hat{p}}) \otimes_{\mathbb{Z}_p} \mathfrak{o}_L$$

as algebras. In particular we have a morphism

$$\mathfrak{o}_L \times S^0_{\mathfrak{o}_L} \to S^0_{\mathfrak{o}_L}$$

of $S^0_{W(k)}$-algebras, representing $\mathfrak{o}_L$-multiplication on $\pi_* S^0_{\mathfrak{o}_L}$, used in Sect. 3.3.2 below.

Remark If G is the Galois group of a finite extension L of $\mathbb{Q}_p$, a theorem of Noether implies that its valuation ring $\mathfrak{o}_L$ is projective over the group ring $\mathbb{Z}_p G$ iff L is tamely ramified; but work of Swan [40] implies, more generally, that the class of $\mathfrak{o}_L$ in the Grothendieck group $G_0(\mathbb{Z}_p G)$ (defined by splitting short exact sequences) is the image of a (not necessarily unique) class $[P]$ in $K_0(\mathbb{Z}_p G)$, perhaps analogous to Wall's finiteness obstruction for CW complexes. Such Swan elements suggest constructing analogs of Moore spectra for $\mathfrak{o}_L$ as representing objects for functors such as

$$X \mapsto \pi_*(X \wedge G_+) \otimes_{\mathbb{Z}_p G} P := \pi_*(X; P) \; \ldots$$

3.1.1 Work several mathematical generations deep [10, 14, 31, 32]... associates to a one-dimensional formal group law Φ, of finite height n over a perfect field k of characteristic $p > 0$, an E_∞ p-adic complex oriented $S^0_{W(k)}$-algebra spectrum $\mathsf{E}(\Phi)$ with homotopy algebra

$$\pi * \mathsf{E}(\Phi) \cong E(\Phi)_* \cong W(k)[[u_1, \ldots, u_{n-1}]][v^{\pm 1}]]$$

of formal power series, representing Lubin and Tate's functor [22] which sends a complete noetherian local ring A with residue field k to the set (modulo isomorphisms which reduce to the identity over k) of lifts of Φ to A. We will sometimes take $v = 1$ to suppress the grading, and to simplify notation we may write E_F for $\mathsf{E}(\Phi)$ for a chosen lift F of Φ to a local ring (e.g. $W(k)$ or $\mathfrak{o}_L$) with residue field k; we may even write E_{L_0} for the $S^0_{W(k_L)} = S^0_{\mathfrak{o}_{L0}}$-module spectrum $\mathsf{E}(\mathsf{LT}_{L_0})$. Similarly $+_{E\Phi}$ or $+_{EF}$ may denote the associated formal group sum, and $\mathfrak{m}_F = [[u_*]]$ may signify the 'maximal ideal' of E_{F*} over $W(k)$ or $\mathfrak{o}_L$.

Lubin and Tate show that the (proétale) group $\mathrm{Aut}_{\bar{k}}(\Phi) \cong \mathfrak{o}_D{}^\times$ of automorphisms of $\Phi \otimes_k \bar{k}$, with its natural $\mathrm{Gal}(\bar{k}/k)$-action, lifts to a (continuous but not smooth) action on $W(\bar{k}) \otimes_{W(k)} E(\Phi)_*$; in particular, their Theorem 3.1 shows that this action takes $W(\bar{k}) \otimes_{W(k)} \mathfrak{m}_{E(\Phi)}$ to itself. In the formalism of Sects. 2.3–2.4, this defines a formal groupoid-scheme of equivalence classes of lifts to Artin local rings, represented by a Hopf algebroid

$$[\mathrm{Spf}\ E(\Phi)_*//\mathrm{Aut}_{\bar{k}}(\Phi)] :\ E(\Phi)_* \underset{\eta_R}{\overset{\eta_L}{\rightrightarrows}} E(\Phi)_*\mathsf{E}(\Phi) \cong H_{\mathrm{Aut}(\Phi)}\ ;$$

where $H_{\mathrm{Aut}(\Phi)}$ [17] is a Hopf algebra of Galois-equivariant continuous functions from $\mathrm{Aut}(\Phi)$ to $E(\Phi)$. Note that the two (left and right) unit homomorphisms send $\mathfrak{m}_E$ to $\mathfrak{m}_E \hat{\otimes} H_{\mathrm{Aut}(\Phi)}$, and that this Hopf algebroid is equivalent to

$$E(\Phi)_* \underset{\eta_R}{\overset{\eta_L}{\rightrightarrows}} E(\Phi)_* \otimes_{\mathbb{L}} (\mathbb{L}\otimes_{\mathbb{Z}}) \otimes_{\mathbb{L}} E(\Phi)_*\ .$$

3.1.2 If, for example, $\Phi_L/\mathbb{F}_p$ is the Lubin–Tate group law for $\mathsf{LT}_{\mathbb{Q}_q}$ as in Sect. 2.3, we can take the u_i to be Araki generators satisfying

$$[p]_E(T) = \sum_{E,i\geq o} v_i T^{p^i}$$

(with $v_0 = p$); the classifying homomorphism from $M\mathrm{U}_*$ then sends $\mathbb{C}P_{q^k-1}$ to

$$\prod_{1\leq i\leq k} (1 - p^{q^i-1})^{-1} \cdot (p^{-1}q^k)v^{q^k-1}$$

and the remaining $\mathbb{C}P_l$ to 0 [27].

3.2.1 A parallel (but even more venerable) line of research, leading to the modern theory of highly structured spectra, allows us to associate to the (by definition, regular) sequence $v_* = v_1, \ldots, v_{n-1}$ of elements of $E_{\mathbb{Q}_q*}$, a choice

$$v_i : S^{2(p^i-1)} \to \mathsf{E}_{\mathbb{Q}_q}$$

of representatives defining, by the construction of [12, V §1, §3.4], p-adic complex-oriented A_∞ ring-spectra

$$\mathsf{K}(\mathbb{Q}_q) = \mathsf{E}_{\mathbb{Q}_q}/(v_*)$$

with $\pi_*\mathsf{K}(\mathbb{Q}_q) \cong W(\mathbb{F}_q)[v^{\pm 1}]$, having $\mathsf{LT}_{\mathbb{Q}_q}$ as formal group law.

Remark When $n = 1$ this construction recovers a model for Atiyah's p-adic completion [3] of complex topological K-theory, and when $n = 2$ it defines a p-adic lift of Baker's supersingular elliptic cohomology [5]. Away from the prime 6, elliptic cohomology [13, §4.1 ex 4.2] has as coefficients, the polynomial ring of modular forms (generated [37, III §5.6.2] by the Eisenstein series E_2, E_3). A theorem of

Deligne [19] identifies the modular form defined by the Eisenstein series E_{p-1} and the (Hasse) parameter v_1, modulo p. This suggests a close relation between p-adic elliptic cohomology, mod E_{p-1}, with $\mathsf{K}(\mathbb{Q}_{p^2})$; but understanding that would require an understanding of E_{p-1} as a polynomial in E_2, E_3, which evidently depends on the prime p.

3.2.2 More generally, a Lubin–Tate group law for a ramified local field of degree n over $\mathbb{Q}_p$ lifts its mod π reduction to a homomorphism

$$u_i \mapsto u_i^* : E(\Phi_L)_* = W(k_L)[[u_*]] \to \mathfrak{o}_L$$

of local $W(k_L)$-algebras.

Lemma *Let $u_i^* \in \mathfrak{m}_L = (\pi) \subset \mathfrak{o}_L$, $i \leq i \leq n-1$ be a sequence of elements in the maximal ideal $\mathfrak{m}_L$: then $u_i \mapsto u_i - u_i^* = v_i$ defines an isomorphism $\mathfrak{o}_L[[u_*]] \cong \mathfrak{o}_L[[v_*]]$ of local $\mathfrak{o}_L$-algebras.*

[For if w$\in \mathfrak{m}_L$ and $\mathfrak{o}_L[[x]] \ni a(x) = \sum_{i \geq 1} a_i x^i$, then

$$a(x) = \sum_{k \geq 1} (\sum_{l \geq 1} \binom{k+l}{l} a_{k+l} w^l) \tilde{x}^k = \sum_{k \geq 1} \tilde{a}_k \tilde{x}^k ,$$

where $\tilde{x} = x - w$. The argument for multiple variables is similar.] □

Definition The A_∞ complex-oriented $S^0_{\mathfrak{o}_L}$-algebra spectrum

$$\mathsf{E}_L = S^0_L \wedge_{S^0_{W(k_L)}} \mathsf{E}(\Phi_{\mathsf{LT}_L})$$

has

$$\pi_* \mathsf{E}_L = E_{L*} = \mathfrak{o}_L[[v_*]][v^{\pm 1}]$$

as algebra of homotopy groups, generated by a regular sequence of elements $v_i = u_i - u_i^*$ such that $v_i \mapsto 0$ specializes the modular lift to the chosen Lubin–Tate group law of L.

3.3.1 The definition in Sect. 3.2.1 of $\mathsf{K}(\mathbb{Q}_q)$ uses the E_∞ structure on $\mathsf{E}(\Phi_{\mathsf{LT}_{\mathbb{Q}_q}})$, which is not available for nontrivially ramified fields. This issue can be avoided by reorganizing the induction in [12] (which follows Sullivan and Baas, based on iterated cofibrations) as a computation of the spectral sequence for the homotopy groups of the geometric realization of a suitable simplicial (Koszul) spectrum [11, §17.5], ex 18.2; [35]:

An element a of a commutative k-algebra A defines[4] an (elementary) differential graded A-algebra

---

[4]or, more generally, a homogeneous element of a graded commutative A_*; however this will be largely suppressed from our notation.

$$\mathrm{Ksz}_A(a) = (A[e]/(e^2),\ d_a(e) = a)\ .$$

More generally, a k-module homomorphism $a_\star : k^m \to A$ defines the classical differential graded commutative algebra

$$\mathrm{Ksz}_A(a_\star) = \bigotimes_{A, 1 \le i \le m} \mathrm{Ksz}_A(a_i) \cong A \otimes_k \Lambda_k(e_i \mid 1 \le i \le m)$$

(with an exterior algebra denoted by Λ, to reduce the multiplicity of things called E), and differential

$$de_I = \sum_{1 \le i \le m} (-1)^{i+1} a_\star(e_i) \cdot e_{\hat{I}(i)}\ ,$$

where $e_I = \wedge_{l \in I} e_l$ is indexed by (totally ordered) subsets I of $\{m\} = \{1, \dots, m\}$, $\hat{I}(i)$ is obtained from I by omitting its ith element, $e_I \wedge e_K$ is 0 if $I \cap K$ is nonempty and equals $\pm e_{I,K}$ if they are disjoint, with sign equal to that of the permutation putting $\{I, K\}$ in proper order.

In the elementary case, if a is not a 0-divisor in A, this defines an A-free resolution of the quotient algebra $A/(a)$, i.e. of the cofiber of the map of A to itself defined by a-multiplication. More generally, if a_i is not a 0-divisor in the quotient ring $A/(a_1, \dots, a_{i-1})$ (i.e. a_* is a **regular** sequence), this construction defines an A-free resolution of $A/(a_\star)$.

In the context of commutative S^0-algebras or ring spectra, we can associate to morphisms

$$a_i : S^{|a_i|} \to A$$

a semi-simplicial (i.e. without degeneracy operators [42, §8.2.2]) A-algebra

$$k \mapsto \mathsf{Ksz}_A(a_\star)[k] = \bigvee_{I \subset \{m\}, |I| = k} A \cdot e_I$$

with A-module face operators $\partial_i e_I = \mu(a_i) \cdot e_{\hat{I}(i)}$, where

$$\mu(a) : S^{|a|} A \to A$$

is defined by multiplication by a. The fat realization [11, §4.8, §18.2] $|\mathsf{Ksz}_A(a_\star)|$ of such a semisimplicial object is canonically filtered, leading to the construction of a spectral sequence computing its homotopy groups, with E_1 page

$$\pi_* \mathsf{Ksz}_A(a_\star) = \mathrm{Ksz}_{A_*}(a_\star) \Rightarrow |\mathsf{Ksz}_A(a_\star)|_*$$

3.3.2 When $a_\star$ is a regular sequence this complex is a resolution, and the spectral sequence collapses to an isomorphism

$$|\mathsf{Ksz}_A(a_\star)|_* \cong A_*/(a_\star) \ .$$

Applying this as above yields a definition for $\mathsf{K}(\mathbb{Q}_q) = |\mathsf{Ksz}_{\Phi(\mathbb{Q}_q)}(v_\star)|$ as an $S^0_{W(\mathbb{F}_q)}$-algebra spectrum, essentially equivalent to the construction of [12]. The ramified case is more delicate, because its building blocks are not E_∞; we need a

Lemma *The morphisms*

$$\tilde{v}_i, \tilde{v}_j : S^*_{\mathfrak{o}_L} \mathsf{E}_L \to \mathsf{E}_L$$

($1 \leq i, j \leq n-1$) defined by multiplication with

$$\tilde{v}_i = u_i^* \wedge_W 1_{\mathsf{E}} - 1_L \wedge_W u_i : S^*_{\mathfrak{o}_L} \to S^0_{\mathfrak{o}_L} \wedge_{W(k_L)} \mathsf{E}(\Phi_L) \ (= \mathsf{E}_L)$$

commute.

Proof Define a twist isomorphism

$$\mathsf{E}(\Phi_L) \wedge_{W(k_L)} \mathfrak{o}_L \to \mathfrak{o}_L \wedge_{W(k_L)} \mathsf{E}(\Phi_L)$$

adjoint to the composition

$$\mathsf{E}(\Phi_L) \to \mathrm{Hom}_{W(k_L)}(\mathfrak{o}_L, \mathfrak{o}_L) \wedge_{W(k_L)} \mathsf{E}(\Phi_L) \to \mathrm{Hom}_{W(k_L)}(\mathfrak{o}_L, \mathfrak{o}_L \wedge_{W(k_L)} \mathsf{E}(\Phi_L))$$

of $S^0_{W(k_L)}$-module morphisms. Since both S^0_L and $\mathsf{E}(\Phi_L)$ are commutative $S^0_{W(k_L)}$-modules, we have (with some abbreviation)

$$\tilde{v}_i \wedge_W \tilde{v}_j = (u_i^* \wedge_W 1_{\mathsf{E}} - 1_L \wedge_W u_i) \wedge_W (u_j^* \wedge_W 1_{\mathsf{E}} - 1_L \wedge_W u_i) = \cdots = \tilde{v}_j \wedge \tilde{v}_i \ .$$

Proposition

$$\mathsf{K}(L) = |\mathsf{Ksz}_{\mathsf{E}_L}(\tilde{v}_\star)|$$

is an A_∞ $S^0_{\mathfrak{o}_L}$-algebra spectrum with $\mathsf{K}(L)_* \cong \mathfrak{o}_L[v^{\pm 1}]$, complex-oriented by the morphism $M\mathrm{U}_* \to \mathsf{K}(L)_*$classifying the chosen Lubin-Tate group law for L. □

Remark The kernel of

$$\mathfrak{o}_L \otimes_{\mathbb{Z}} M\mathrm{U}_* \to E_{L*} \to \mathsf{K}(L)_*$$

is generated by an (infinite) regular sequence (which can be chosen to belong to $M\mathrm{U}_*$ in degree greater than $2(p^n - 1)$). The Koszul construction above, together with [12, V §4.2], leads to a construction for a connective spectrum $\mathsf{k}(L)$ with $\mathsf{K}(L)$ as v_n-localization.

4 Some Trivial Spectral Sequences

4.1 The spectral sequence of a geometric realization, together with the Eilenberg–Moore/Künneth spectral sequence for the smash product of module spectra provide some understanding of the bialgebra

$$(\mathsf{K}(L) \wedge_{S^0_L} \mathsf{K}(L)_* = \mathsf{K}(L)_*\mathsf{K}(L) \ .$$

To begin, note that

$$(\mathsf{K}(L) \wedge_{S^0_L} \mathsf{E}_L)_* = |\mathrm{Ksz}_{\mathsf{E}_L}(\tilde{v}_\star)|_*(\mathsf{E}_L)$$

is the $\mathsf{E}(\Phi_L)_*$ homology of a filtered $\mathsf{E}(\Phi_L)$-module spectrum, and that the E_1 page of the associated spectral sequence is the Koszul algebra

$$\mathrm{Ksz}_{(\mathsf{E}_L \wedge_{S^0_L} \mathsf{E}_L)_*}(\tilde{v}_\star) \ ;$$

but by [17], as in Sect. 3.1, $(\mathsf{E}_L \wedge_{S^0_L} \mathsf{E}_L)_* \cong H_{\mathrm{Aut}(\Phi_L)}$, with deformation parameters acting as left $\tilde{v}_*$-multiplication. This sequence is regular, so this is a resolution, and the spectral sequence collapses to an isomorphism

$$(\mathsf{K}(L) \wedge_{S^0_L} \mathsf{E}_L)_* \cong \mathsf{K}(L)_* \otimes_{E_{L*}} H_{\mathrm{Aut}(\Phi_L)} \cong \mathfrak{o}_L \otimes_{W(k_L)} H_{\mathrm{Aut}(\Phi_L)}$$

of $\mathfrak{o}_L$-algebras. Now observe that

$$\mathsf{K}(L) \wedge_{S^0_L} \mathsf{K}(L) \simeq (\mathsf{K}(L) \wedge_{S^0_L} \mathsf{E}_L) \wedge_{\mathsf{E}_L} \mathsf{K}(L)$$

which is accessible via [12, IV Thm 6.4].

Proposition *The Künneth spectral sequence collapses at E_2 to an isomorphism*

$$\mathsf{K}(L)_*\mathsf{K}(L) \cong \big(\mathfrak{o}_L \otimes_{E_{L*}} H_{\mathrm{Aut}(\Phi_L)}\big) \otimes_{\mathfrak{o}_L} \Lambda^*_{\mathfrak{o}_L}(\mathfrak{m}_L/\mathfrak{m}_L^2) \ ,$$

where the term on the right is the exterior algebra on the (free, of rank $n-1$) tangent $\mathfrak{o}_L$-module to the space of deformations of LT_L.

Proof The E_1-page of this spectral sequence is again a Koszul algebra, now of the form

$$\mathrm{Ksz}_{(\mathsf{K}(L)_* \otimes_{\mathsf{E}_{L*}} H_{\mathrm{Aut}(\Phi_L)})}(\eta(v_\star)) \ ,$$

where the images

$$\eta(v_i) = \sum_\alpha v_{i,\alpha} \otimes g_{i,\alpha} \in \mathfrak{m}_L H_{\mathrm{Aut}(\Phi_L)}$$

of the generators v_i under the right unit have coefficients $v_{i,\alpha}$ in the ideal $\mathfrak{m}_L$ ([22, Thm 3.1], see Sect. 3.1), and thus map to zero under $\mathsf{E}_{L*} \to \mathsf{K}(L)_*$. The homology of this DGA is therefore just its underlying graded algebra, which can be identified with the algebra of Galois-equivariant functions from $\mathrm{Aut}(\Phi_L)$ to the exterior algebra on $\mathfrak{m}_L/\mathfrak{m}_L^2$. [Note that $\mathsf{E}_{L*}\mathsf{E}_L$ is analogous [by Kodaira–Spencer theory, cf. [15, ex 2.8.1], [20]] to an algebra of functions from $\mathrm{Aut}(\Phi_L)$ to the symmetric algebra on $\mathfrak{m}_L/\mathfrak{m}_L^2$.] □

4.2 It seems reasonable to conjecture that this spectral sequence collapse implies an interpretation of $\mathsf{K}(L)_*\mathsf{K}(L)$ as a Hopf algebroid of functions on a (super, ie nontrivially $\mathbb{Z}_2$-graded) groupoid scheme, an extension of the automorphism group of LT_L by an exterior algebra of deformations parametrized by its tangent space as a point in Spf E_{L*}. However, the author feels that this and related questions (e.g. the possible nontriviality of such extensions, the action of $\mathrm{Gal}(L/\mathbb{Q}_p)$ and its previously mentioned relation to Weil groups, Massey product structures [1, §3.3], relations [16] with Azumaya algebras ...) are best left to younger, more vigorous and reliable researchers.

In particular: recent advances in the algebra of non-discretely valued fields suggest that the topological Hochschild homology of the perfectoid completion L^∞ of the fields L^{trab} (Sect. 2.5) associates to the p-adic completion of $B\mathbb{Q}/\mathbb{Z}$ (regarded as an analog of $\mathbb{C}P^\infty$), a rigid analytic analog [28] of a Lubin–Tate group for L. If the spectra $\mathsf{K}(L)$ have a natural construction in terms of fields like L^∞, one might hope for the existence of a generalized Chern character or cyclotomic-like trace, mapping $\mathsf{k}(L)$ Galois-equivariantly to $\mathrm{THH}(\mathcal{O}_{L^\infty}, \mathbb{Z}_p)$.

References

1. Angeltveit, V.: Uniqueness of Morava K-theory. Compos. Math. **147**, 633 – 648 (2011). arxiv.org/0810.5032
2. Araki, S.: Typical Formal Groups in Complex Cobordism and K-Theory. Lectures in Mathematics, Kyoto University, No. 6. Kinokuniya Book-Store Co., Tokyo (1973)
3. Atiyah, M.F., Tall, D.O.: Group representations, Λ-rings and the J-homomorphism. Topology **8**, 253–297 (1969)
4. Baas, N.A.: On bordism theory of manifolds with singularities. Math. Scand. **33**(1973), 279–302 (1974)
5. Baker, A.: Isogenies of supersingular elliptic curves over finite fields and operations in elliptic cohomology. arXiv:0712.2052
6. Behrens, M., Lawson, T.: Topological automorphic forms. Mem. AM **204**, 958 (2010). arXiv:math/0702719
7. Bousfield, A.K., Kan, D.: Homotopy limits, completions and localizations. Springer LNM **304** (1972)
8. Cartier, P.: Relévements des groupes formels commutatifs, Seminaire Bourbaki 359. Springer LNM **179**, 217–230 (1971)
9. Clarke, F.: p-adic Analysis and Operations in K-Theory. Groupe de Travail d'analyse Ultramétrique tome. **14**(15), 1–12 (1986–1987). www.numdam.org/article/GAU_1986-198_14_A7_0.pdf
10. Devinatz, E., Hopkins, M.: The action of the Morava stabilizer group on the Lubin-Tate moduli space of lifts. Am. J. Math. **117**, 669–710 (1995)
11. Dugger, D.: A primer on homotopy colimits. www.pages.uoregon.edu/ddugger/hocolim.pdf

12. Elmendorf, A., Kriz, I., Mandell, M., May, J.P.: Rings, modules, and algebras in stable homotopy theory. Mathematical Surveys and Monographs 47, AMS (1997)
13. Goerss, P.: Topological modular forms [after Hopkins, Miller and Lurie], Séminaire Bourbaki 2008/2009. Exp. 1005, Astérisque **332**, 221–255 (2010)
14. Goerss, P., Hopkins, M.J.: Moduli spaces of commutative ring spectra In: Structured Ring Spectra. LMS Lecture Notes, vol. 315, pp. 151–200. CUP (2004)
15. Hopkins, M.J.: Lectures on Lubin-Tate spaces, Arizona Winter School (2019). www.swc.math.arizona.edu/aws/2019/2019HopkinsNotes.pdf
16. Hopkins, M.J., Lurie, J.: On Brauer groups of Lubin-Tate spectra I. www.math.harvard.edu/~lurie/papers/Brauer.pdf
17. Hovey, M.: Operations and co-operations in Morava E-theory. Homol. Homotopy Appl. **6**, 201–236 (2004)
18. Hovey, M., Strickland, N.: Morava K-theories and localisation. Memoirs AMS, vol. 139 no. 666 (1999)
19. Katz, N.M.: p-adic properties of modular schemes and modular forms. In.: Modular Functions of One Variable III, vol. 350, pp. 69–190. Springer LNM (1973)
20. Katz, N.M., Oda, T.: On the differentiation of de Rham cohomology classes with respect to parameters. J. Math. Kyoto Univ. **8**, 199–213 (1968)
21. Lang, S.: Cyclotomic fields I and II Springer Graduate Texts in Mathematics. vol. 121 (1990)
22. Lubin, J., Tate, J.: Formal moduli for one-parameter formal Lie groups. Bull. Soc. Math. France **94**, 49–59 (1966)
23. Lurie, J.: Higher algebra. http://www.math.harvard.edu/~lurie/
24. Morava, J.: Noetherian localisations of categories of cobordism comodules. Ann. Math. **121**, 1–39 (1985)
25. Morava, J.: The Weil group as automorphisms of the Lubin-Tate group. Journées de Géométrie Algébrique de Rennes I, Astérisque, **63**, pp. 169–177. Societe mathematique de Franc, Paris (1979)
26. Morava, J.: Stable homotopy and local number theory. In: Algebraic Analysis, Geometry, and Number Theory, pp. 291–305. JHU Press, Baltimore (1989)
27. Morava, J.: Local fields and extraordinary K-theory. arXiv:1207.4011
28. Morava, J.: Complex orientations for THH of some perfectoid fields. In: Homotopy theory: tools and applications. Contemporary Mathematics, vol. 729, pp. 221–237. AMS (2019). arxiv.org/1608.04702
29. Quillen, D.: Elementary proofs of some results of cobordism theory using Steenrod operations. Adv. Math. **7**, 29–56 (1971)
30. Ravenel, D.C.: Complex cobordism and stable homotopy groups of spheres. Pure and Applied Mathematics, vol. 121. Academic Press (1986)
31. Rezk, C.: Notes on the Hopkins-Miller theorem. In: Homotopy Theory via Algebraic Geometry and Group Representations. Contemporary Mathematics, vol. 220, pp. 313–366. AMS (1998)
32. Rezk, C.: The congruence criterion for power operations in Morava E-theory. Homol. Homotopy Appl. **11**, 327–379 (2009). arXiv:0902.2499
33. Schwänzl, R., Vogt, R.M., Waldhausen, F.: Adjoining roots of unity to E_∞ ring spectra. In: Homotopy Invariant Algebraic Structures. Contemporary Mathematics, vol. 239, pp. 245–249. AMS (1999)
34. Schwede, S.: Symmetric spectra. www.math.uni-bonn.de/people/schwede/SymSpec-v3.pdf
35. Segal, G.: Classifying spaces and spectral sequences. IHES Publ. Math. **34**, 105–112 (1968)
36. Serre, J.P.: Local class field theory. In: Algebraic Number Theory, pp. 128–161. Thompson, Brighton (1987)
37. Serre, J.P.: A course in arithmetic. Springer Graduate Texts. vol. 7 (1973)
38. Shimada, N., Yagita, N.: Multiplications in the complex bordism theory with singularities. Publ. Res. Inst. Math. Sci. **12**, 259–293 (1976/77)
39. Sullivan, D.: Singularities in spaces. In: Proceedings of Liverpool Singularities Symposium II, vol. 209, pp. 196–206. Springer LNM (1971)
40. Swan, R.: The Grothendieck ring of a finite group. Topology **2**, 85–110 (1963)

41. Tate, J.: Number theoretic background. In: Automorphic Forms, Representations and L-Functions. Proceedings of Symposia in Pure Mathematics, vol. XXXIII, pp. 3–26. AMS (1979)
42. Weibel, C.: An introduction to homological algebra. Cambridge Studies in Advanced Mathematics. vol. 38 (1994)
43. Weil, A.: Basic Number Theory, vol. 144, 3rd edn. Springer Grundlehren der Mathematischen Wissenschaften (1974)
44. Würgler, U.: Morava K-theories: a survey. In: Algebraic topology(Poznan'), vol. 1474, pp. 111–138. Springer LNM (1991)

A Short Introduction to the Telescope and Chromatic Splitting Conjectures

Tobias Barthel

Abstract In this note, we give a brief overview of the telescope conjecture and the chromatic splitting conjecture in stable homotopy theory. In particular, we provide a proof of the folklore result that Ravenel's telescope conjecture for all heights combined is equivalent to the generalized telescope conjecture for the stable homotopy category, and explain some similarities with modular representation theory.

Keywords Bousfield localization · Telescope conjecture · Chromatic splitting conjecture

This document contains a slightly expanded and updated version of an overview talk, delivered at the Talbot Workshop 2013 on chromatic homotopy theory, on two of the major open conjectures in stable homotopy theory: the telescope conjecture and the chromatic splitting conjecture. As such, these notes are entirely expositional and are not aimed to give a comprehensive account; rather, we hope some might see them as an invitation to the subject.

We have augmented the original content of the talk by some material which is well-known to the experts but difficult to trace in the literature. In particular, we prove the folklore result that the telescope conjecture for all heights combined is equivalent to the classification of smashing Bousfield localizations of the stable homotopy category. In the final section, we discuss algebraic incarnations of chromatic structures in modular representation theory.

We will assume some familiarity with basic notions from stable homotopy theory, and refer the interested reader to [44] as well as [5] for a more thorough discussion of chromatic homotopy theory.

T. Barthel (✉)
Max Planck Institute for Mathematics, Vivatsgass 7, 53111 Bonn, Germany
e-mail: tbarthel@mpim-bonn.mpg.de

T. Ohsawa and N. Minami (eds.), *Bousfield Classes and Ohkawa's Theorem*, Springer Proceedings in Mathematics & Statistics 309,
https://doi.org/10.1007/978-981-15-1588-0_9

1 Motivation: Freyd's Generating Hypothesis

In 1966, Freyd [22] proposed one of the most fundamental conjectures in stable homotopy theory:

Conjecture 1.1 (Generating hypothesis) *Let $f\colon X \to Y$ be a map of finite spectra with $\pi_* f = 0$, then f is nullhomotopic.*

As of today, this hypothesis is completely open—since the computation of stable homotopy groups of finite complexes is notoriously difficult, there is essentially no evidence supporting either conclusion. However, one important statement that would follow if the hypothesis was true is that the map

$$\pi_*\colon [X, Y]_* \longrightarrow \mathrm{Hom}_{\pi_* S^0}(\pi_* X, \pi_* Y)$$

is an isomorphism for all finite spectra X and Y, the target being the group of graded homomorphisms of $\pi_* S^0$-modules. The generating hypothesis also has a number of other curious consequences, see for example [29].

In the early 1990s, Devinatz and Hopkins described a chromatic approach to the generating hypothesis in the special case when $Y = S^0$ is the sphere spectrum [17]. We explain their idea first in the global setting. Suppose $f\colon X \to S^0$ is not null and write $f^\vee\colon S^0 \to DX$ for its Spanier–Whitehead dual; we have to show that $\pi_* f \neq 0$. If f is of infinite order, then the question reduces to a rational statement, so suppose f has finite order d. Recall that by Brown representability there exists a spectrum I with

$$[W, I] \cong \mathrm{Hom}_{\mathbb{Z}}(\pi_0 W, \mathbb{Q}/\mathbb{Z})$$

for all W, the so-called Brown–Comenetz dual of the sphere spectrum.

This can be used to reduce the generating hypothesis with target S^0 to a set of universal examples, a strategy reminiscent of the proof of the nilpotence theorem [21]. Indeed, there is a map $f_d\colon [X, S^0] \to \mathbb{Q}/\mathbb{Z}$ sending f to $1/d$. By construction of I, f_d corresponds to a map $f_d\colon DX \to I$. Writing I as a directed colimit of finite spectra I^α, we see that f_d factors through some $f_d^\alpha\colon DX \to I^\alpha$, i.e., there is a commutative diagram

$$\begin{array}{ccccc} S^0 & \xrightarrow{f^\vee} & DX & & \\ & \searrow^{1/d} & \downarrow{\scriptstyle f_d} & \searrow^{f_d^\alpha} & \\ & & I & \longleftarrow & I^\alpha. \end{array}$$

Spanier–Whitehead duality gives a map $(f_d^\alpha)^\vee\colon DI^\alpha \to X$ such that the composite

$$DI^\alpha \xrightarrow{(f_d^\alpha)^\vee} X \xrightarrow{f} S^0$$

is not nullhomotopic and depends only on α and d. Therefore, it suffices to prove the claim for these universal examples $DI^{\alpha} \to S^0$.

In order to deal with them, we need to construct suitable models for the I^{α} and then prove the generating hypothesis for these examples. Instead of running this programme for S^0 directly, Devinatz and Hopkins propose to use the chromatic convergence theorem [3, 44], which says that, p-locally, S^0 is equivalent to the limit of the chromatic tower

$$\ldots \longrightarrow L_n S^0 \longrightarrow L_{n-1} S^0 \longrightarrow \ldots \longrightarrow L_1 S^0 \longrightarrow L_0 S^0 \simeq S^0_{\mathbb{Q}}, \quad (1)$$

where L_n denotes $E(n)$-localization (reviewed below). It consequently suffices to prove an analogue of the generating hypothesis for the $E(n)$-local analogues of the universal examples considered above, for each height $n \geq 0$. The filtration steps of the chromatic tower are built out of the monochromatic layers $M_n S^0 = \mathrm{fib}(L_n S^0 \to L_{n-1} S^0)$, which leads to the study of $IM_n S^0$ via Gross–Hopkins duality [25] and the $K(n)$-local Picard group [26]. The original approach relied on the telescope conjecture as well as the chromatic splitting conjecture in order to control the universal examples, and it has been carried out successfully at height 1 [17]:

Theorem 1.2 (Devinatz) *If $p > 2$ and $f\colon X \to S^0$ a map between p-local finite spectra with $\pi_* f = 0$, then $L_1 f$ is nullhomotopic.*

In response to subsequent progress on the telescope conjecture and the chromatic splitting conjecture as outlined in the next sections, Devinatz describes a modified approach in [19], which appears to be the current state of the art.

2 Recollections on Bousfield Localization

Throughout this section, we will implicitly work locally at a fixed prime p. Let E be a spectrum. A spectrum X is called E-acyclic if $E \wedge X \simeq 0$ and X is called E-local if any map from an E-acyclic spectrum into X is nullhomotopic. Moreover, a map $f\colon X \to Y$ is called an E-equivalence if $E \wedge f$ is an equivalence or, equivalently, if the fiber of f is E-acyclic. A localization functor is an endofunctor L of the stable homotopy category together with a natural transformation $\eta\colon \mathrm{id} \to L$ such that $L\eta\colon L \to L^2$ is an equivalence and $L\eta \simeq \eta L$. Based on ideas of Adams, Bousfield [14] rigorously constructed a localization functor which forces the E-equivalences to be invertible; more precisely:

Theorem 2.1 (Bousfield, 1979) *If E is a spectrum, there is a localization functor L_E on the stable homotopy category together with a natural transformation $\eta_E\colon \mathrm{id} \to L_E$ such that, for any spectrum X, the map $\eta_E(X)\colon X \to L_E X$ exhibits $L_E X$ as the initial E-local spectrum with a map from X. The functor L_E is called Bousfield localization at E and the fiber C_E of η_E is called E-acyclization.*

It follows formally that $\eta_E(X)$ is also the terminal E-equivalence out of X. The proof of this theorem relies on verifying the existence of a set of suitable generators for the category of E-acyclics. It is an open problem [30, Conj. 9.1] whether every localization functor on the stable homotopy category arises as localization with respect to some spectrum E.

The fiber sequence $C_E \to \mathrm{id} \to L_E$ can be thought of as providing a way to decompose the stable homotopy category into two subcategories in a well-behaved way. We might therefore ask for a classification of all Bousfield localizations. The first result in this direction was proven by Ohkawa [41]. To state it, recall that two spectra E and F are said to be Bousfield equivalent if they have identical categories of acyclics, i.e., $\ker(L_E) = \ker(L_F)$. The corresponding equivalence class of E is denoted by $\langle E \rangle$, so we have $\langle E \rangle = \langle F \rangle$ if and only if $L_E \simeq L_F$. As usual, we define $\langle E \rangle \vee \langle F \rangle = \langle E \vee F \rangle$ and $\langle E \rangle \wedge \langle F \rangle = \langle E \wedge F \rangle$.

Theorem 2.2 (Ohkawa) *The collection of Bousfield classes of spectra forms a set of cardinality at least* $2^{\aleph_0}$ *and at most* $2^{2^{\aleph_0}}$.

In light of this result, a classification of all Bousfield localizations does not seem to be feasible, see [30] for some partial results. Instead, we will single out two particularly well-behaved families among all Bousfield localizations:

Definition 2.3 A localization functor L is called smashing if it commutes with set-indexed direct sums or, equivalently, if the natural transformation $X \wedge LS^0 \to LX$ is an equivalence for all spectra X. Moreover, L is finite if there exists a collection of finite spectra that generates the category $\ker(L)$ of L-acyclics.

Miller [37] proves that any finite localization is smashing and any smashing localization functor L is equivalent to Bousfield localization at LS^0 by [43], so from now on all localization functors we consider are assumed to be Bousfield localizations.

3 The Telescope Conjecture

We start with some examples of finite and smashing localizations; as before, everything is implicitly localized at a prime p. Let $K(n)$ and $E(n)$ be the nth Morava K-theory and nth Johnson–Wilson theory, respectively, with coefficients

$$K(n)_* = \mathbb{F}_p[v_n^{\pm 1}] \quad \text{and} \quad E(n)_* = \mathbb{Z}_{(p)}[v_1, \ldots, v_{n-1}, v_n][v_n^{-1}],$$

where v_i is of degree $2(p^i - 1)$. By [43], if a finite spectrum F is $K(n)$-acyclic, then it is also $K(n-1)$-acyclic[1]; since $\langle E(n) \rangle = \langle \bigvee_{i=0}^{n} K(i) \rangle$, this spectrum F is then also $E(n)$-acyclic. A finite spectrum F is of type n if n is minimal with the

[1]In fact, as long as $n > 1$, this result has been extended to all suspension spectra by Bousfield [15]. For $n = 1$, a counterexample is given by $K(\mathbb{Z}, 3)$.

property that $K(n)_*(F) \neq 0$, and such a finite number n exists for any nontrivial finite spectrum. By the periodicity theorem [27], any finite type n spectrum F admits an (essentially unique) v_n-self map, and we write $\mathrm{Tel}(F) = F[v_n^{-1}]$ for the associated telescope. It then follows from the thick subcategory theorem [27] that the Bousfield class of $\mathrm{Tel}(F)$ depends only on n, so we will also write $\mathrm{Tel}(n)$ for $\mathrm{Tel}(F)$.

Definition 3.1 Let $n \geq 0$, then we define two localization functors on the stable homotopy category by

$$L_n^f = L_{\mathrm{Tel}(0)\vee \mathrm{Tel}(1)\vee \ldots \vee \mathrm{Tel}(n)} \quad \text{and} \quad L_n = L_{E(n)} \simeq L_{K(0)\vee K(1)\vee \ldots \vee K(n)},$$

referred to as the finite L_n-localization and L_n-localization, respectively.

As the terminology suggests, the functors L_n^f are in fact finite localizations, with $\ker(L_n^f)$ generated by any finite type $(n+1)$-spectrum [36, 46]. It then follows from the thick subcategory theorem that any finite localization functor of the category of spectra which is not equal to the identity or the zero functor must be one of the L_n^f. Their key features are summarized in the next proposition, see [36, 37, 46].

Proposition 3.2 (Mahowald–Sadofsky, Miller, Ravenel) *For each n, the functor L_n^f is a finite and thus smashing localization. If F is a finite type n spectrum then $L_n^f F \simeq \mathrm{Tel}(F)$.*

Having classified all finite localizations, we now turn to the a priori larger set of smashing localizations. The smash product theorem [44] and its proof establish the first part of the next result:

Theorem 3.3 (Hopkins–Ravenel) *For any $n \geq 0$, the localization functor L_n is smashing.*

There is a natural transformation $L_n^f \to L_n$ which is an equivalence on all MU-module spectra and all L_i-local spectra for any $i \geq 0$, as shown in [28, 31]. In other words, there is a close relationship between the functors L_n and their finite counterparts L_n^f. As explained in [44], if the two localizations were in fact equivalent for all n, then two naturally arising filtrations on the stable homotopy groups of spheres would coincide, making the computation of $\pi_* S^0$ more amenable to algebraic techniques. This led Ravenel [43] to:

Conjecture 3.4 (Telescope conjecture) *For any $n \geq 0$, the natural map $L_n^f \to L_n$ is an equivalence.*

For $n = 0$, both L_0^f and L_0 identify with rationalization. Based on explicit computations of the homotopy groups of $L_1 S^0/p$ and $L_1^f S^0/p = \mathrm{Tel}(S^0/p)$ by Mahowald ($p = 2$, [34]) and Miller ($p > 2$, [38]), Bousfield [14] deduced:

Theorem 3.5 (Bousfield, Mahowald, Miller) *The telescope conjecture holds at height $n = 1$.*

One might thus hope for an inductive approach to the telescope conjecture, passing from height $n-1$ to height n. The corresponding relative version admits a number of equivalent formulations, see [35]:

Proposition 3.6 *Let $n \geq 1$ and suppose F is finite of type n, then the following are equivalent:*

1. *If $L_{n-1}^f \simeq L_{n-1}$, then $L_n^f \simeq L_n$.*
2. $\mathrm{Tel}(F) \simeq L_n F$.
3. $\langle \mathrm{Tel}(F) \rangle = \langle K(n) \rangle$.
4. *The Adams–Novikov spectral sequence for* $\mathrm{Tel}(F)$ *converges to* $\pi_* \mathrm{Tel}(F)$.

Note that, by the thick subcategory theorem, a single example or counterexample that is finite of type n is enough to settle the passage from height $n-1$ to height n.

For $n = 2$ and $p \geq 5$, Ravenel [45] began the analogue of Miller's height 1 calculation for $V(1) = S^0/(p, v_1)$, attempting to show that the telescope conjecture is false in these cases, but this computation has not yet been completed due to its considerable complexity. In [35], Mahowald, Ravenel, and Shick describe an alternative approach based on a spectrum $Y(n)$ such that $\pi_* L_n Y(n)$ is finitely generated over $R(n)_* = K(n)_*[v_{n+1}, \ldots, v_{2n}]$, but $\pi_* L_n^f Y(n)$ can only be finitely generated over $R(n)_*$ if there is a "bizarre pattern of differentials" in the corresponding localized Adams spectral sequence. Thus, if these patterns could be ruled out, we would disprove the telescope conjecture at heights $n \geq 2$. At this time, the telescope conjecture is still open for all $n \geq 2$ and all p, and generally believed to be false.

4 Classification of Smashing Bousfield Localizations

This section discusses the classification of smashing Bousfield localizations of the (p-local) stable homotopy category. In particular, we prove that the telescope conjecture for all heights n is equivalent to the so-called generalized telescope conjecture (or generalized smashing conjecture). Since this material is more technical than the rest of this survey, the reader may want to skip ahead to the conclusion at the end of this section. We start with two lemmas, the first of which is reminiscent of the type classification of finite spectra.

Lemma 4.1 *Let L be a smashing localization functor on the stable homotopy category. If $LK(n) \not\simeq 0$, then $LK(n-1) \not\simeq 0$.*

Proof Suppose $LK(n) \not\simeq 0$. Since $K(n) \wedge LS^0 \simeq LK(n)$ is a module over $K(n)$ and hence splits into a wedge of shifted copies of $K(n)$, we see that $K(n)$ is L-local and thus the canonical map $K(n) \to LK(n)$ is an equivalence. This implies that $\langle LS^0 \rangle \geq \langle K(n) \rangle$: Indeed, if $X \wedge LS^0 \simeq 0$, then $0 \simeq X \wedge LS^0 \wedge K(n) \simeq X \wedge K(n)$ as well.

The next claim is that $\langle LS^0\rangle \geq \langle\bigvee_{i=0}^n K(i)\rangle$. To this end, note that $L_{K(n)}S^0$ is $K(n)$-local, hence LS^0-local. Because L is smashing, we get an equality $\langle LS^0 \wedge L_{K(n)}S^0\rangle = \langle L_{K(n)}S^0\rangle$, which then yields

$$\langle LS^0\rangle \geq \langle LS^0 \wedge L_{K(n)}S^0\rangle = \langle L_{K(n)}S^0\rangle = \langle\bigvee_{i=0}^n K(i)\rangle, \tag{2}$$

where the last equality is [28, Cor. 2.4]. Therefore, we have $LK(n-1) \not\simeq 0$.

The proof of the next lemma requires the nilpotence theorem.

Lemma 4.2 *Suppose L is a smashing localization and $n \geq 0$, then $LK(n) \simeq 0$ if and only if any finite spectrum of type at least n is in* $\ker(L)$.

Proof Suppose $LK(n) \simeq 0$ and let F be a finite spectrum of type at least n. Replacing F with $\mathrm{End}(F) \simeq DF \wedge F$ if necessary, we may assume that F and thus LF are ring spectra. By the nilpotence theorem, it thus suffices to show that $K(i) \wedge LF \simeq 0$ for all $0 \leq i \leq \infty$. Since L is smashing, $K(i) \wedge LF \simeq LK(i) \wedge F \simeq 0$ for $n \leq i \leq \infty$ using the assumption and Theorem 4.1, while the hypothesis on F guarantees that it also vanishes for $0 \leq i < n$.

Conversely, let F be a finite type n spectrum so that $LF \simeq 0$. It follows that $F \wedge LK(n) \simeq 0$. But $K(n)$ is a retract of $LK(n)$ provided $LK(n) \not\simeq 0$, so $F \wedge K(n) \simeq 0$ as well, contradicting the assumption on F. Therefore, $LK(n) \simeq 0$.

As the next proof shows, we can use Theorem 4.1 to detect smashing localizations.

Proposition 4.3 *If L is a smashing localization which is neither* 0 *nor the identity functor, then there exists an $n \geq 0$ such that* $\ker(L_n^f) \subseteq \ker(L) \subseteq \ker(L_n)$.

Proof By Theorem 4.1, any smashing localization functor L belongs to one of the following three classes:

1. $LK(n) = 0$ for all n, or
2. there exists an n such that $LK(n) \not\simeq 0$ and $LK(m) \simeq 0$ for all $m > n$, or
3. $LK(n) \simeq K(n)$ for all n.

In Case (1), $\ker(L)$ contains the sphere spectrum S^0 by Theorem 4.2, so $L \simeq 0$. If L belongs to the second class, then Theorem 4.2 shows that $\ker(L_n^f) \subseteq \ker(L)$, so it remains to show that $\ker(L) \subseteq \ker(L_n)$. To this end, let $X \in \ker(L)$. Because L is smashing, this implies $LS^0 \wedge X \simeq 0$ and thus $\bigvee_{i=0}^n K(i) \wedge X \simeq 0$ by (2). Therefore, $X \in \ker(L_n)$ as desired.

Finally, if $LK(n) \simeq K(n)$ for all n, then (2) implies that any L_n-local spectrum is L-local, so $S^0 \simeq \lim_n L_nS^0$ is L-local by the chromatic convergence theorem. Therefore, L must be equivalent to the identity functor, again using that L is smashing.

Corollary 4.4 *The telescope conjecture holds for all n if and only if all smashing localization functors on the stable homotopy category are finite.*

This latter formulation, originally due to Bousfield [14, Conj. 3.4], generalizes well to other compactly generated triangulated categories where it has been studied extensively, see for example [32, 33].

5 The Chromatic Splitting Conjecture

The chromatic splitting conjecture describes how the localizations $L_n S^0$ for varying n assemble into S^0 via the chromatic tower (1), working p-locally as before. Informally speaking, it asserts that this gluing process is as simple as it can be without being trivial, but there are various refinements of its statement. We will focus on the weakest form here and refer the interested reader to [28] for further details.

For each $n \geq 1$ there is a map of fiber sequences, where the right square—known as the chromatic fracture square—is a homotopy pullback:

$$\begin{array}{ccccc} F(L_{n-1}S^0, L_nX) & \longrightarrow & L_nX & \longrightarrow & L_{K(n)}X \\ \downarrow \simeq & & \downarrow & \overset{\alpha_n}{\swarrow} & \downarrow \\ F(L_{n-1}S^0, L_nX) & \xrightarrow[\beta_n]{} & L_{n-1}X & \underset{\gamma_n}{\rightleftarrows} & L_{n-1}L_{K(n)}X. \end{array} \tag{3}$$

Consider the question whether there exists a map α_n as indicated making the top triangle in the chromatic fracture square commute. By chasing the diagram, such a map exists if and only if β_n is nullhomotopic, which in turn is equivalent to the existence of a map γ_n splitting the map $L_{n-1}X \to L_{n-1}L_{K(n)}X$. Based on explicit computations of the cohomology of Morava stabilizer groups as well as of $\pi_* L_{K(n)}S^0$ for small n, Hopkins (see [28]) arrived at the following:

Conjecture 5.1 (Chromatic splitting conjecture) *If X is the p-completion of a finite spectrum, then a splitting γ_n exists for all n.*

The finiteness assumption on X is essential in this conjecture: Indeed, Devinatz [18] proves that, for $X = BP_p$ the p-completion of the Brown–Peterson spectrum, the map $L_{n-1}BP_p \to L_{n-1}L_{K(n)}BP_p$ splits if and only if $n = 1$. If the chromatic splitting conjecture holds for a finite spectrum X, then we obtain the following consequences:

1. The canonical map $X_p \to \prod_n L_{K(n)}X_p$ is the inclusion of a summand, as proven in [28].
2. Taking the limit over the compositions $L_{K(n+1)}X \xrightarrow{\alpha_{n+1}} L_nX \to L_{K(n)}X$ gives an equivalence $X_p \to \lim_n L_{K(n)}X$. This follows follows from the chromatic convergence theorem by cofinality.

In other words, the chromatic splitting conjecture implies that a finite spectrum X can be recovered from its monochromatic pieces $L_{K(n)}X$.

We now review what is known about the chromatic splitting conjecture for S_p^0, the p-complete sphere spectrum. Take $n = 1$ and $p > 2$, then a classical computation with complex K-theory, originally due to Adams and Baird [1] and then revisited by Ravenel [43], shows that

$$\pi_i L_{K(1)} S_p^0 \cong \begin{cases} \mathbb{Z}_p & \text{for } i \in \{-1, 0\}, \\ \mathbb{Z}/p^{s+1} & \text{for } i = 2(p-1)p^s m - 1 \text{ with } p \nmid m, \\ 0 & \text{otherwise.} \end{cases}$$

Thus $\pi_* L_0 L_{K(1)} S_p^0 \cong \mathbb{Q}_p$ for $i = 0$ and $i = -1$ and is 0 otherwise; of course, $\pi_* L_0 S_p^0$ is isomorphic to $\mathbb{Q}_p$ in degree 0. One can then see that $L_0 L_{K(1)} S_p^0$ splits as $L_0 S_p^0 \vee L_0 S_p^{-1}$. Replacing complex K-theory by real K-theory yields the same conclusion for $n = 1$ and $p = 2$. The analogous computations at height $n = 2$ are considerably more complex and are the subject of extensive work by Shimomura–Yabe [47] ($p \geq 5$), Goerss–Henn–Mahowald–Rezk [23, 24] ($p = 3$), and Beaudry–Goerss–Henn [9] ($p = 2$). Their results can be summarized as follows:

Theorem 5.2 (Beaudry–Goerss–Henn–Mahwald–Rezk–Shimomura–Yabe) *The chromatic splitting conjecture holds for $n = 2$ and all p. If $p \geq 3$, then*

$$L_1 L_{K(2)} S_p^0 \simeq L_1(S_p^0 \vee S_p^{-1}) \vee L_0(S_p^{-3} \vee S_p^{-4}),$$

while for $p = 2$, we have

$$L_1 L_{K(2)} S_p^0 \simeq L_1(S_p^0 \vee S_p^{-1} \vee S_p^{-2}/p \vee S_p^{-3}/p) \vee L_0(S_p^{-3} \vee S_p^{-4}).$$

There is a stronger version of Theorem 5.1, also due to Hopkins, which additionally describes how the fiber term $F(L_{n-1}S^0, L_n X)$ in (3) decomposes into spectra of the form $L_i X$ with $0 \leq i \leq n-1$; we refer the interested reader to [28] for the details. A compact formulation, encoding the combinatorics of the conjectured decomposition for all heights simultaneously in terms of a single generating function, has been given by Morava in [40]. However, in light of the $p = 2$ case of the previous theorem proven by Beaudry, Goerss, and Henn, the original conjecture requires a modification accounting for additional terms, at least for $n = 2$ and $p = 2$.

If correct, the strong version of the chromatic splitting conjecture (both in its original or modified form) would imply [6] that the stable homotopy groups of $L_{K(n)} S^0$ are finitely generated over $\mathbb{Z}_p$ for $n \geq 1$, another major open problem in chromatic homotopy theory, see [20] for partial results. This conjecture is open for all heights $n \geq 3$ and primes p; there are hints [8, 42] that the problem might at least be approachable for large primes with respect to the height n.

We end this section with the following result by Minami [39], which provides some evidence for the chromatic splitting conjecture at general heights. He introduces a class of so-called robust spectra including finite spectra as well as BP and proves:

Theorem 5.3 (Minami) *Fix a height n and prime p. If X is a robust spectrum and m and k are positive integers satisfying $m - k \geq n + s_0 + 1$ where s_0 is the vanishing line intercept of the $E(n)$-based Adams–Novikov spectral sequence for S^0, then the map $L_m X \to L_n X$ factors through $L_{K(k+1)\vee \ldots \vee K(m)} X$.*

6 An Algebraic Analogue

We conclude this survey by discussing an algebraic analogue of the stable homotopy category in which algebraic versions of the generating hypothesis, the telescope conjecture, as well as the chromatic splitting conjecture have been settled. This is just one instance of the observation that the chromatic programme and consequently the above chromatic conjectures can be formulated in many other contexts, thereby providing a plethora of test cases as well as motivation for a fruitful transfer of techniques. Other examples include derived categories of quasi-coherent sheaves on schemes or stacks, stable equivariant homotopy categories, motivic categories, or categories arising in non-commutative geometry, see [2] for an overview.

Let G be a finite group, let k be a field of characteristic p, and write kG for the associated group algebra. Recall that the stable module category StMod_{kG} is the quotient of Mod_{kG} by the projectives and that it comes equipped with the structure of a symmetric monoidal triangulated category with tensor unit k. As in [12] we write $\mathrm{Proj}(H^*(G;k))$ for the projective variety of the Noetherian graded commutative ring $H^*(G;k)$; the underlying set of $\mathrm{Proj}(H^*(G;k))$ consists of the homogeneous prime ideals in $H^*(G;k)$ different from the ideal of all positive degree elements.

The finite localization functors on StMod_{kG} have been classified in the work of Benson, Carlson, and Rickard [10]. As a result of a series of papers culminating in [12], Benson, Iyengar, and Krause generalized this to a complete classification of all localization functors: They develop a theory of support and employ it to establish a bijection between the set of localizing tensor ideals of StMod_{kG} and arbitrary subsets of $\mathrm{Proj}(H^*(G;k))$. Their theory yields in particular a proof of the telescope conjecture in this context.

Theorem 6.1 (Benson–Iyengar–Krause) *The generalized telescope conjecture holds in* StMod_{kG}*, i.e., the category of acyclics of any smashing localization functor is generated by compact objects. Furthermore, the smashing localization functors on* StMod_{kG} *are in bijection with specialization closed*[2] *subsets of* $\mathrm{Proj}(H^*(G;k))$.

In fact, they establish an analogous classification for the larger category Stable_{kG} of unbounded complexes of injective kG-modules up to homotopy, which fits into a recollement between StMod_{kG} and the derived category of kG-modules [13]. In this case, the role of the parametrizing variety is played by $\mathrm{Spec}^h(H^*(G;k))$, the Zariski spectrum of all homogeneous prime ideals of $H^*(G;k)$. In particular, any specialization closed subset $\mathcal{V} \subseteq \mathrm{Spec}^h(H^*(G;k))$ gives rise to a localization functor $L_{\mathcal{V}}$ on Stable_{kG}. For example, if $\mathfrak{p}$ is a homogeneous prime ideal, then $\mathcal{V}(\mathfrak{p}) = \{\mathfrak{q} \mid \mathfrak{p} \subseteq \mathfrak{q}\} \subseteq \mathrm{Spec}^h(H^*(G;k))$ is specialization closed, and thus provides a localization functor $L_{\mathcal{V}(\mathfrak{p})}$ and a completion functor $\Lambda^{\mathfrak{p}}$. These functors should be thought of as algebraic analogues of the functor L_{n-1} and $L_{K(n)}$.

Before we can state the analogue of the chromatic splitting conjecture in this context, we need to introduce some terminology: To emphasize the analogy to stable

[2] A subset $\mathcal{V}$ is called specialization closed if $\mathfrak{p} \in \mathcal{V}$ and $\mathfrak{p} \subseteq \mathfrak{q}$ imply $\mathfrak{q} \in \mathcal{V}$.

homotopy category, we write $\pi_* M$ for the graded abelian group of homotopy classes of maps from k to M in Stable_{kG}. Call two prime ideals $\mathfrak{p}, \mathfrak{p}' \in \mathrm{Spec}^h(H^*(G;k))$ adjacent if $\mathfrak{p}' \subsetneq \mathfrak{p}$ and this chain does not refine, i.e., there does not exist $\mathfrak{q} \in \mathrm{Spec}^h(H^*(G;k))$ such that $\mathfrak{p}' \subsetneq \mathfrak{q} \subsetneq \mathfrak{p}$. Furthermore, a module $M \in \mathrm{Stable}_{kG}$ is said to be $\mathfrak{p}$-local if $\pi_* M$ is a $\mathfrak{p}$-local $H^*(G;k)$-module, and a compact M is said to be of type $\mathfrak{p}'$ if $\pi_* M$ is $\mathfrak{p}'$-torsion as a graded $H^*(G;k)$-module.

Theorem 6.2 ([7]) *Suppose G is a finite p-group. Let $\mathfrak{p}, \mathfrak{p}' \in \mathrm{Spec}^h(H^*(G;k))$ be adjacent prime ideals and let $M \in \mathrm{Stable}_{kG}$ be $\mathfrak{p}$-local. There is a homotopy pullback square*

$$\begin{array}{ccc} M & \longrightarrow & \Lambda^{\mathfrak{p}} M \\ \downarrow & & \downarrow \\ L_{\mathcal{V}(\mathfrak{p})} M & \longrightarrow & L_{\mathcal{V}(\mathfrak{p})} \Lambda^{\mathfrak{p}} M. \end{array}$$

If M is compact and of type $\mathfrak{p}'$, then the bottom map in this square is split.

Finally, we consider the analogue of the generating hypothesis in StMod_{kG}, which asserts that a map $f : M \to N$ between finitely generated modules is nullhomotopic, i.e., factors through a projective module, if and only if $\pi_* f = 0$. Based on earlier work of [11] in the p-group case, [16] gives a complete answer:

Theorem 6.3 (Benson–Carlson–Chebolu–Christensen–Mináč) *The generating hypothesis holds for StMod_{kG} if and only if the p-Sylow subgroup of G is isomorphic to either C_2 or C_3.*

The techniques used in their proof, namely Auslander–Reiten theory, carry over to the chromatic setting to establish the failure of a $K(n)$-local analogue of the generating hypothesis [4], thereby bringing us back to our starting point.

Acknowledgements I would like to thank Mike Hopkins and the participants of Talbot 2013 for several useful discussions on this topic as well as Agnès Beaudry, Malte Leip, Doug Ravenel, Gabriel Valenzuela, and the referee for comments on an earlier draft of this document. Furthermore, I am grateful to Haynes Miller and Norihiko Minami for encouraging me to revise my original talk notes.

References

1. Adams, J.F.: Operations of the nth Kind in K-Theory, and What We Don't Know About RP^∞. London Mathematical Society Lecture Note Series, vol. 11, pp. 1–9 (1974)
2. Balmer, P.: Tensor triangular geometry. In: Proceedings of the International Congress of Mathematicians, vol. II, pp. 85–112. Hindustan Book Agency, New Delhi (2010)
3. Barthel, T.: Chromatic completion. Proc. Am. Math. Soc. **144**(5), 2263–2274 (2016)
4. Barthel, T.: Auslander-Reiten sequences, Brown-Comenetz duality, and the $K(n)$-local generating hypothesis. Algebr. Represent. Theory **20**(3), 569–581 (2017)

5. Barthel, T., Beaudry, A.: Chromatic structures in stable homotopy theory (2019). arXiv e-prints arXiv:1901.09004
6. Barthel, T., Beaudry, A., Peterson, E.: The homology of inverse limits and the chromatic splitting conjecture. Forthcoming
7. Barthel, T., Heard, D., Valenzuela, G.: The algebraic chromatic splitting conjecture for Noetherian ring spectra. Math. Z. **290**(3–4), 1359–1375 (2018)
8. Barthel, T., Schlank, T., Stapleton, N.: Chromatic homotopy theory is asymptotically algebraic. arXiv e-prints (2017). arXiv:1711.00844
9. Beaudry, A., Goerss, P.G., Henn, H.W.: Chromatic splitting for the $K(2)$-local sphere at $p = 2$. arXiv e-prints (2017). arXiv:1712.08182
10. Benson, D.J., Carlson, J.F., Rickard, J.: Thick subcategories of the stable module category. Fund. Math. **153**(1), 59–80 (1997)
11. Benson, D.J., Chebolu, S.K., Christensen, J.D., Mináč, J.: The generating hypothesis for the stable module category of a p-group. J. Algebra **310**(1), 428–433 (2007)
12. Benson, D.J., Iyengar, S.B., Krause, H.: Stratifying modular representations of finite groups. Ann. Math. (2) **174**(3), 1643–1684 (2011)
13. Benson, D.J., Krause, H.: Complexes of injective kG-modules. Algebra Number Theory **2**(1), 1–30 (2008)
14. Bousfield, A.K.: The localization of spectra with respect to homology. Topology **18**(4), 257–281 (1979)
15. Bousfield, A.K.: On $K(n)$-equivalences of spaces. In: Homotopy Invariant Algebraic Structures (Baltimore, MD, 1998). Contemporary Mathematics, vol. 239, pp. 85–89. American Mathematical Society Providence, RI (1999)
16. Carlson, J.F., Chebolu, S.K., Mináč, J.: Freyd's generating hypothesis with almost split sequences. Proc. Am. Math. Soc. **137**(8), 2575–2580 (2009)
17. Devinatz, E.S.: K-theory and the generating hypothesis. Am. J. Math. **112**(5), 787–804 (1990)
18. Devinatz, E.S.: A counterexample to a BP-analogue of the chromatic splitting conjecture. Proc. Am. Math. Soc. **126**(3), 907–911 (1998)
19. Devinatz, E.S.: The generating hypothesis revisited. In: Stable and Unstable Homotopy (Toronto, ON, 1996). Fields Institute Communications, vol. 19, pp. 73–92. American Mathematical Society, Providence, RI (1998)
20. Devinatz, E.S.: Towards the finiteness of $\pi_* L_{K(n)} S^0$. Adv. Math. **219**(5), 1656–1688 (2008)
21. Devinatz, E.S., Hopkins, M.J., Smith, J.H.: Nilpotence and stable homotopy theory. I. Ann. Math. (2) **128**(2), 207–241 (1988)
22. Freyd, P.: Stable homotopy. In: Proceedings of the Conference Categorical Algebra (La Jolla, Calif., 1965), pp. 121–172. Springer, New York (1966)
23. Goerss, P., Henn, H.-W., Mahowald, M., Rezk, C.: A resolution of the $K(2)$-local sphere at the prime 3. Ann. Math. (2) **162**(2), 777–822 (2005)
24. Goerss, P.G., Henn, H.-W., Mahowald, M.: The rational homotopy of the $K(2)$-local sphere and the chromatic splitting conjecture for the prime 3 and level 2. Doc. Math. **19**, 1271–1290 (2014)
25. Hopkins, M.J., Gross, B.H.: The rigid analytic period mapping, Lubin-Tate space, and stable homotopy theory. Bull. Am. Math. Soc. (N.S.) **30**(1), 76–86 (1994)
26. Hopkins, M.J., Mahowald, M., Sadofsky, H.: Constructions of elements in Picard groups. In: Topology and representation theory (Evanston, IL, 1992). Contemporary Mathematics, vol. 158, pp. 89–126. American Mathematical Society, Providence, RI (1994)
27. Hopkins, M.J., Smith, J.H.: Nilpotence and stable homotopy theory. II. Ann. Math. (2) **148**(1), 1–49 (1998)
28. Hovey, M.: Bousfield localization functors and Hopkins' omatic splitting conjecture. In: The Čech centennial (Boston, MA, 1993). Contemporary Mathematics, vol. 181, pp. 225–250. American Mathematical Society, Providence, RI (1995)
29. Hovey, M.: On Freyd's generating hypothesis. Q. J. Math. **58**(1), 31–45 (2007)
30. Hovey, M., Palmieri, J.H.: The structure of the Bousfield lattice. In: Homotopy Invariant Algebraic Structures (Baltimore, MD, 1998). Contemporary Mathematics, vol. 239, pp. 175–196. American. Mathematical Society, Providence, RI (1999)

31. Hovey, M.A., Strickland, N.P.: Morava K-theories and localisation. Mem. Am. Math. Soc. **139**(666), viii+100–100 (1999)
32. Keller, B.: A remark on the generalized smashing conjecture. Manuscripta Math. **84**(2), 193–198 (1994)
33. Krause, H.: Smashing subcategories and the telescope conjecture–an algebraic approach. Invent. Math. **139**(1), 99–133 (2000)
34. Mahowald, M.: bo-resolutions. Pacific J. Math. **92**(2), 365–383 (1981)
35. Mahowald, M., Ravenel, D., Shick, P.: The triple loop space approach to the telescope conjecture. In: Homotopy Methods in Algebraic Topology (Boulder, CO, 1999). Contemporary Mathematics, vol. 271, pp. 217–284. American Mathematical Society, Providence, RI (2001)
36. Mahowald, M., Sadofsky, H.: v_n telescopes and the Adams spectral sequence. Duke Math. J. **78**(1), 101–129 (1995)
37. Miller, H.: Finite localizations. Bol. Soc. Mat. Mexicana (2) **37**(1-2), 383–389 (1992). Papers in honor of José Adem (Spanish)
38. Miller, H.R.: On relations between Adams spectral sequences, with an application to the stable homotopy of a Moore space. J. Pure Appl. Algebra **20**(3), 287–312 (1981)
39. Minami, N.: On the chromatic tower. Am. J. Math. **125**(3), 449–473 (2003)
40. Morava, J.: A remark on Hopkins' chromatic splitting conjecture (2014). arXiv e-prints arXiv:1406.3286
41. Ohkawa, T.: The injective hull of homotopy types with respect to generalized homology functors. Hiroshima Math. J. **19**(3), 631–639 (1989)
42. Pstrągowski, P.: Chromatic homotopy is algebraic when $p>n^2+n+1$ (2018). arXiv e-prints arXiv:1810.12250
43. Ravenel, D.C.: Localization with respect to certain periodic homology theories. Am. J. Math. **106**(2), 351–414 (1984)
44. Ravenel, D.C.: Nilpotence and Periodicity in Stable Homotopy Theory. Annals of Mathematics Studies, vol. 128. Princeton University Press, Princeton, NJ (1992). Appendix C by Jeff Smith
45. Ravenel, D.C.: Progress report on the telescope conjecture. In: Adams Memorial Symposium on Algebraic Topology, 2 (Manchester, 1990). London Mathematical Society Lecture Note Series, vol. 176, pp. 1–21. Cambridge University Press, Cambridge (1992)
46. Ravenel, D.C.: Life after the telescope conjecture. In: Algebraic K-theory and algebraic topology (Lake Louise, AB, 1991). NATO Advanced Science Institutes Series C: Mathematical and Physical Sciences, vol. 407, pp. 205–222. Kluwer Academic. Publications, Dordrecht (1993)
47. Shimomura, K., Yabe, A.: The homotopy groups $\pi_*(L_2S^0)$. Topology **34**(2), 261–289 (1995)

Spectral Algebra Models of Unstable v_n-Periodic Homotopy Theory

Mark Behrens and Charles Rezk

Abstract We give a survey of a generalization of Quillen–Sullivan rational homotopy theory which gives spectral algebra models of unstable v_n-periodic homotopy types. In addition to describing and contextualizing our original approach, we sketch two other recent approaches which are of a more conceptual nature, due to Arone-Ching and Heuts. In the process, we also survey many relevant concepts which arise in the study of spectral algebra over operads, including topological André-Quillen cohomology, Koszul duality, and Goodwillie calculus.

Keywords v_n-periodic homtopy theory · Bousfield–Kuhn functor · Topological André-Quillen cohomology

1 Introduction

In his seminal paper [80], Quillen showed that there are equivalences of homotopy categories

$$\mathrm{Ho}(\mathrm{Top}_{\mathbb{Q}}^{\geq 2}) \simeq \mathrm{Ho}(\mathrm{DGCoalg}_{\mathbb{Q}}^{\geq 2}) \simeq \mathrm{Ho}(\mathrm{DGLie}_{\mathbb{Q}}^{\geq 1})$$

between simply connected rational spaces, simply connected rational differential graded commutative coalgebras, and connected rational differential graded Lie algebras. In particular, given a simply connected space X, there are *models* of its rational homotopy type

M. Behrens
Department of Mathematics, University of Notre Dame, Notre Dame, IN 46556, USA
e-mail: mbehren1@nd.edu

C. Rezk (✉)
Department of Mathematics, University of Illinois at Urbana-Champaign, Urbana, IL 61801, USA
e-mail: rezk@illinois.edu

T. Ohsawa and N. Minami (eds.), *Bousfield Classes and Ohkawa's Theorem*,
Springer Proceedings in Mathematics & Statistics 309,
https://doi.org/10.1007/978-981-15-1588-0_10

$$C_{\mathbb{Q}}(X) \in \mathrm{DGCoalg}_{\mathbb{Q}},$$
$$L_{\mathbb{Q}}(X) \in \mathrm{DGLie}_{\mathbb{Q}}$$

such that

$$H_*(C_{\mathbb{Q}}(X)) \cong H_*(X; \mathbb{Q}) \qquad \text{(isomorphism of coalgebras)},$$
$$H_*(L_{\mathbb{Q}}(X)) \cong \pi_{*+1}(X) \otimes \mathbb{Q} \qquad \text{(isomorphism of Lie algebras)}.$$

In the case where the space X is of finite type, one can also extract its rational homotopy type from the dual $C_{\mathbb{Q}}(X)^\vee$, regarded as a differential graded commutative algebra. This was the perspective of Sullivan [89], whose notion of minimal models enhanced the computability of the theory.

The purpose of this paper is to give a survey of an emerging generalization of this theory where unstable rational homotopy is replaced by v_n-periodic homotopy.

Namely, the Bousfield–Kuhn functor $\Phi_{K(n)}$ is a functor from spaces to spectra, such that the homotopy groups of $\Phi_{K(n)}(X)$ are a version of the unstable v_n-periodic homotopy groups of X. We say that a space X is $\Phi_{K(n)}$-*good* if the Goodwillie tower of $\Phi_{K(n)}$ converges at X. A theorem of Arone-Mahowald [8] proves spheres are $\Phi_{K(n)}$-good.

The main result is the following theorem (Theorem 6.4, Corollary 8.3).

Theorem 1.1 *There is a natural transformation (the "comparison map")*

$$c_X^{K(n)} : \Phi_{K(n)}(X) \to \mathrm{TAQ}_{S_{K(n)}}(S^X_{K(n)})$$

which is an equivalence on finite $\Phi_{K(n)}$*-good spaces.*

Here the target of the comparison map is the topological André-Quillen cohomology of the $K(n)$-local Spanier-Whitehead dual of X (regarded as a non-unital commutative algebra over the $K(n)$-local sphere), where $K(n)$ is the nth Morava K-theory spectrum. We regard $S^X_{K(n)}$ as a commutative algebra model of the unstable v_n-periodic homotopy type of X, and the theorem is giving a means of extracting the unstable v_n-periodic homotopy groups of X from its commutative algebra model. A result of Ching [26] implies that the target of the comparison map is an algebra over a spectral analog of the Lie operad. As such, we regard the target as a Lie algebra model for the unstable v_n-periodic homotopy type of X.

The original results date back to 2012, and are described in a preprint of the authors [22] which has (still?) not been published. The paper is very technical, and the delay in publication is due in part to difficulties in getting these technical details correct. In the mean-time, Arone-Ching [1] and Heuts [44] have announced proofs which reproduce and expand on the authors' results using more conceptual techniques.

The idea of this survey is to provide a means to disseminate the authors' original work until the original account is published. As [22] is more of a forced march than a reflective ramble, it also seemed desirable to have a discussion which explained the main ideas without getting bogged down in the inevitable details one must contend with (which involve careful work with the Morava E-theory Dyer-Lashof algebra, amongst other things). The approach of Arone-Ching uses a localized analog of their classification theory for Taylor towers, together with Ching's Koszul duality for modules over an operad. Heuts' approach is a byproduct of his theory of polynomial approximations of ∞-categories. Both of these alternatives, as we mentioned before, are more conceptual than our computational approach, but require great care to make precise.

This survey, by contrast, is written to convey the *ideas* behind all three approaches, without delving into many details. We also attempt to connect the theory with many old and new developments in spectral algebra. We hope that the interested reader will consult cited sources for more careful treatments of the subjects herein. In particular, all constructions are implicitly derived/homotopy invariant, and we invite the reader to cast them in his/her favorite model category or ∞-category.

Organization of the Paper.

Section 2*:* We describe the general notion of stabilization of a homotopy theory, and the Hess/Lurie theory of homotopy descent as a way of encoding unstable homotopy theory as "stable homotopy theory with descent data".

Section 3*:* The equivalence between rational differential graded Lie algebras and rational differential graded commutative coalgebras is an instance of Koszul duality. We describe the theory of Koszul duality, which provides a correspondence between algebras over an operad, and coalgebras over its Koszul dual.

Section 4*:* We revisit rational homotopy theory and recast it in spectral terms. We also describe Mandell's work, which gives commutative algebra models of p-adic homotopy types.

Section 5*:* We give an overview of chromatic (v_n-periodic) homotopy theory, both stable and unstable, and review the Bousfield–Kuhn functor.

Section 6*:* We define the comparison map, and state the main theorem in the case where X is a sphere.

Section 7*:* We give an overview of the proof of the main theorem in the case where X is a sphere. The proof involves Goodwillie calculus and the Morava E-theory Dyer-Lashof algebra, both of which we review in this section.

Section 8*:* We explain how the main theorem extends to all finite $\Phi_{K(n)}$-good spaces. We also discuss computational consequences of the theorem, most notably the work of Wang and Zhu.

Section 9*:* After summarizing Ching's Koszul duality for modules over an operad, we give an exposition of the Arone-Ching theory of fake Taylor towers, and their classification of polynomial functors. We then explain how they use this theory, in the localized context, to give a different proof (and strengthening) of the main theorem.

Section 10*:* We summarize Heuts' theory of polynomial approximations of ∞-categories, and his general theory of coalgebra models of homotopy types. We discuss Heuts' application of his general theory to Koszul duality, and to unstable v_n-periodic homotopy, where his theory also reproves and strengthens the main theorem.

Conventions.

- For a commutative (E_∞) ring spectrum R, we shall let Mod_R denote the category of R-module spectra, with symmetric monoidal structure given by $\wedge_R$. For X, Y in Mod_R, we will let $F_R(X, Y)$ denote the spectrum of R-module maps from X to Y, and $X^\vee := F_R(X, R)$ denotes the R-linear dual. For a pointed space X, We shall let R^X denote the function spectrum $F(\Sigma^\infty X, R)$.
- For X a space or spectrum, we shall use $X_p^\wedge$ to denote its p-completion with respect to a prime p, X_E to denote its Bousfield localization with respect to a spectrum E, and $X^{\geq n}$ to denote its $(n-1)$-connected cover.
- For all but the last section, our homotopical framework will always implicitly take place in the context of relative categories: a category $\mathcal{C}$ with a subcategory $\mathcal{W}$ of "equivalences" [29] (in the last section we work in the context of ∞-categories). The homotopy category will be denoted $\mathrm{Ho}(\mathcal{C})$, and refers to the localization $\mathcal{C}[\mathcal{W}^{-1}]$. Functors between homotopy categories are always implicitly derived. We shall use $\mathcal{C}(X, Y)$ to refer to the maps in $\mathcal{C}$, and $[X, Y]_\mathcal{C}$ to denote the maps in $\mathrm{Ho}(\mathcal{C})$. We shall use $\underline{\mathcal{C}}(X, Y)$ to denote the derived mapping space.
- Top_* denotes the category of pointed spaces (with equivalences the weak homotopy equivalences), Sp the category of spectra (with equivalences the stable equivalences), and for a spectrum E, $(\mathrm{Top}_*)_E$ and Sp_E denote the variants where we take the equivalences to be the E-homology isomorphisms.
- All operads $\mathcal{O}$ in Mod_R are assumed to be reduced, in the sense that $\mathcal{O}_0 = *$ and $\mathcal{O}_1 = R$. We shall let $\mathrm{Alg}_\mathcal{O}$ denote the category of $\mathcal{O}$-algebras. As spelled out in greater detail in Sect. 3, $\mathrm{TAQ}^\mathcal{O}$ will denote topological André-Quillen homology, and $\mathrm{TAQ}_\mathcal{O}$ will denote topological André-Quillen cohomology (its R-linear dual). In the case where $\mathcal{O} = \mathrm{Comm}_R$, the (reduced) commutative operad in Mod_R, we shall let TAQ^R (respectively TAQ_R) denote the associated topological André-Quillen homology (respectively cohomology).[1]

2 Models of "Unstable Homotopy Theory"

The approach to unstable homotopy theory we are considering fits into a general context, which we will now describe.

Stable homotopy theories. As Quillen points out in [79], any pointed model category $\mathcal{C}$ comes equipped with a notion of suspension $\Sigma_\mathcal{C}$ and loops $\Omega_\mathcal{C}$, given by

$$\Sigma_\mathcal{C} X = \mathrm{hocolim}(* \leftarrow X \rightarrow *),$$
$$\Omega_\mathcal{C} X = \mathrm{holim}(* \rightarrow X \leftarrow *).$$

[1]This is slightly non-standard, as Comm-algebras are the same thing as *non-unital* commutative algebras in Mod_R. However, as we explain in Sect. 3, the category of such is equivalent to the category of augmented commutative R-algebras.

This gives the notion of a category $\mathrm{Sp}(\mathcal{C})$ of spectra in $\mathcal{C}$. With hypotheses on $\mathcal{C}$, and a suitable notion of stable equivalence (see, for example, [48, 86]), $\mathrm{Sp}(\mathcal{C})$ is a model for the *stabilization* of $\mathcal{C}$ (in the sense of [63]). There are adjoint functors

$$\Sigma^\infty_{\mathcal{C}} : \mathrm{Ho}(\mathcal{C}) \leftrightarrows \mathrm{Ho}(\mathrm{Sp}(\mathcal{C})) : \Omega^\infty_{\mathcal{C}}. \tag{2.1}$$

We regard $\mathrm{Ho}(\mathcal{C})$ as the unstable homotopy theory of $\mathcal{C}$, and $\mathrm{Ho}(\mathrm{Sp}(\mathcal{C}))$ as the stable homotopy theory of $\mathcal{C}$.

The fundamental question. Typically, the unstable homotopy theory is *more complicated* than the stable homotopy theory. One would therefore like to think that an unstable homotopy type is a stable homotopy type with extra structure. More specifically:

Question 2.2 *Is there an algebraic structure "?" on* $\mathrm{Sp}(\mathcal{C})$ *and functors:*

$$\mathfrak{A} : \mathrm{Ho}(\mathcal{C}) \leftrightarrows \mathrm{Ho}(\mathrm{Alg}_?(\mathrm{Sp}(\mathcal{C}))) : \mathfrak{E}$$

so that $X \simeq \mathfrak{E}\mathfrak{A}(X)$ *(natural isomorphism in the homotopy category)?*

If so, we say that ?-algebras model the unstable homotopy types of $\mathcal{C}$.

Remark 2.3

(1) Often, one must restrict attention to certain subcategories of $\mathrm{Ho}(\mathcal{C})$, $\mathrm{Ho}(\mathrm{Alg}_?)$ to get something like this (e.g. 1-connected rational unstable homotopy types).
(2) One can hope for more: is $\mathfrak{A}$ fully faithful? Can we then characterize the essential image?
(3) When $(\mathfrak{A}, \mathfrak{E})$ form an adjoint pair, we can say something sharper: in this case, there is always a *canonical* equivalence between the full subcategories

$$\mathrm{Ho}\{X \in \mathcal{C} \text{ s.t. } X \simeq \mathfrak{E}\mathfrak{A}(X)\} \simeq \mathrm{Ho}\{A \in \mathrm{Alg}_?(\mathrm{Sp}(\mathcal{C})) \text{ s.t. } A \simeq \mathfrak{A}\mathfrak{E}(A)\}.$$

This identifies both the "good" subcategory of $\mathrm{Ho}(\mathcal{C})$ and its essential image under $\mathfrak{A}$, and shows that $\mathfrak{A}$ is fully faithful on this subcategory.

Example 2.4 In the case of $\mathcal{C} = (\mathrm{Top}_*)_{\mathbb{Q}}$—rational pointed spaces—the stabilization is rational spectra $\mathrm{Sp}_{\mathbb{Q}}$. We have

$$\mathrm{Ho}(\mathrm{Sp}_{\mathbb{Q}}) \simeq \mathrm{Ho}(\mathrm{Ch}_{\mathbb{Q}}),$$

where $\mathrm{Ch}_{\mathbb{Q}}$ denotes rational $\mathbb{Z}$-graded chain complexes. In this context Quillen's work provides two answers to Question 2.2: the algebraic structure can be taken to be either commutative coalgebras or Lie algebras.

Homotopy descent. The theory of homotopy decent of Hess [43] and Lurie [63] (see also [3]) provides a canonical candidate answer to Question 2.2. Namely, the adjunction (2.1) gives rise to a comonad $\Sigma^\infty_{\mathcal{C}}\Omega^\infty_{\mathcal{C}}$ on $\mathrm{Sp}(\mathcal{C})$, and for any $X \in \mathcal{C}$, the spectrum $\Sigma^\infty_{\mathcal{C}} X$ is a coalgebra for this comonad.[2] Thus one can regard the functor $\Sigma^\infty_{\mathcal{C}}$ as refining to a functor

$$\mathfrak{A} : \mathrm{Ho}(\mathcal{C}) \to \mathrm{Ho}(\mathrm{Coalg}_{\Sigma^\infty_{\mathcal{C}}\Omega^\infty_{\mathcal{C}}}).$$

Asking for this to be an equivalence is asking for the adjunction to be "comonadic". It is typically only reasonable to expect that one gets an equivalence between suitable subcategories of these two categories. Even then, this may be of little use if there is no explicit understanding of what it means to be a $\Sigma^\infty_{\mathcal{C}}\Omega^\infty_{\mathcal{C}}$-coalgebra.

Example 2.5 Suppose that $\mathcal{C} = \mathrm{Top}_*$, the category of pointed spaces. Then there is always a map

$$X \to C(\Omega^\infty, \Sigma^\infty\Omega^\infty, \Sigma^\infty X) \tag{2.2}$$

where $C(-,-,-)$ denote the comonadic cobar construction. Explicitly,

$$C(\Omega^\infty, \Sigma^\infty\Omega^\infty, \Sigma^\infty X) = \mathrm{Tot}(QX \Rightarrow QQX \Rrightarrow \cdots),$$

the Bousfield–Kan Q-completion of X. It follows that the map (2.2) is an equivalence for X nilpotent, and for nilpotent spaces the unstable homotopy type can be recovered from the $\Sigma^\infty\Omega^\infty$-comonad structure on $\Sigma^\infty X$. But what does it mean explicitly to endow a spectrum with a $\Sigma^\infty\Omega^\infty$-coalgebra structure? This seems to be a difficult question, but Arone, Klein, Heuts, and others have partial information (see [45, 54]). Rationally, however, $\Sigma^\infty\Omega^\infty$ is equivalent (on connected spaces) to the free commutative coalgebra functor, and coalgebras for this comonad are therefore rationally equivalent to commutative coalgebras.

3 Koszul Duality

The equivalence

$$\mathrm{Ho}(\mathrm{DGCoalg}^{\geq 2}_{\mathbb{Q}}) \simeq \mathrm{Ho}(\mathrm{DGLie}^{\geq 1}_{\mathbb{Q}})$$

mentioned in the introduction is an instance of *Koszul duality* [5, 24, 31, 33, 36, 37, 63]. In this section we will attempt to summarize the current state of affairs to the best of our abilities.

Let R be a commutative ring spectrum, and let $\mathcal{O}$ be an operad in Mod_R. *All operads $\mathcal{O}$ in this paper are assumed to be* **reduced**: $\mathcal{O}_0 = *$ and $\mathcal{O}_1 = R$.

We shall let $\mathrm{Alg}_{\mathcal{O}} = \mathrm{Alg}_{\mathcal{O}}(\mathrm{Mod}_R)$ denote the category of $\mathcal{O}$-algebras. An equivalence of $\mathcal{O}$-algebras is a map of $\mathcal{O}$-algebras whose underlying map of spectra is an

[2] One should regard this coalgebra structure as "descent data".

equivalence.[3] Note that since the operad $\mathcal{O}$ is reduced, the category $\mathrm{Alg}_{\mathcal{O}}$ is pointed, with $*$ serving as both the initial and terminal object. There is a free-forgetful adjunction

$$\mathcal{F}_{\mathcal{O}} : \mathrm{Mod}_R \leftrightarrows \mathrm{Alg}_{\mathcal{O}} : \mathcal{U}$$

where

$$\mathcal{F}_{\mathcal{O}}(X) = \bigvee_i \left(\mathcal{O}_i \wedge_R X^{\wedge_R i}\right)_{\Sigma_i} \tag{3.1}$$

is the free $\mathcal{O}$-algebra generated by X. We shall abusively also use $\mathcal{F}_{\mathcal{O}}$ to denote the associated monad on Mod_R, so that $\mathcal{O}$-algebras are the same thing as $\mathcal{F}_{\mathcal{O}}$-algebras:

$$\mathrm{Alg}_{\mathcal{O}} \simeq \mathrm{Alg}_{\mathcal{F}_{\mathcal{O}}}.$$

Topological André-Quillen homology. Because $\mathcal{O}$ is reduced, there is a natural transformation of monads

$$\epsilon : \mathcal{F}_{\mathcal{O}} \to \mathrm{Id}.$$

For A an $\mathcal{O}$-algebra, its module of *indecomposables* QA is defined to be the coequalizer of ϵ and the $\mathcal{F}_{\mathcal{O}}$-algebra structure map:

$$\mathcal{F}_{\mathcal{O}}(A) \rightrightarrows A \to QA.$$

The functor Q has a right adjoint

$$Q : \mathrm{Alg}_{\mathcal{O}} \leftrightarrows \mathrm{Mod}_R : \mathrm{triv}$$

where, for an R-module X, the $\mathcal{O}$-algebra $\mathrm{triv}X$ is given by endowing X with $\mathcal{O}$-algebra structure maps:

$$\mathcal{O}_1 \wedge_R X = R \wedge_R X \xrightarrow{\approx} X,$$
$$\mathcal{O}_n \wedge_R X^n \xrightarrow{*} X, \qquad n \neq 1.$$

The *topological André-Quillen homology* of A is defined to be the left derived functor

$$\mathrm{TAQ}^{\mathcal{O}}(A) := \mathbb{L}QA.$$

It is effectively computed as the realization of the monadic bar construction:

$$\mathrm{TAQ}^{\mathcal{O}}(A) \simeq B(\mathrm{Id}, \mathcal{F}_{\mathcal{O}}, A).$$

[3] We refer the reader to [46] for a thorough treatment of the homotopy theory of $\mathcal{O}$-algebras suitable for our level of generality. We advise the reader that some of the technical details in this reference are correctly dealt with in [55, 76].

The Topological André-Quillen cohomology is defined to be the R-linear dual of $\mathrm{TAQ}^{\mathcal{O}}$:

$$\mathrm{TAQ}_{\mathcal{O}}(A) := \mathrm{TAQ}^{\mathcal{O}}(A)^{\vee}.$$

Suppose $R = Hk$ is the Eilenberg-MacLane spectrum associated to a $\mathbb{Q}$-algebra k, $\mathcal{O}$ is the commutative operad (see Example 3.2 below), and A is the Eilenberg-MacLane $\mathcal{O}$-algebra associated to an ordinary augmented commutative k-algebra. Then we can regard $\mathrm{TAQ}^{\mathcal{O}}$ as being an object of the derived category of k under the equivalence

$$\mathrm{Ho}(\mathrm{Mod}_{Hk}) \simeq \mathrm{Ho}(\mathrm{Ch}_k)$$

and we recover classical André-Quillen homology. Basterra defined TAQ for commutative R-algebras for arbitrary commutative ring spectra R, and showed that the monadic bar construction gives a formula for it [9]. The case of general topological operads was introduced in [14]; this work was extended to the setting of spectral operads in [42] (see also [35]).

The important properties of $\mathrm{TAQ}^{\mathcal{O}}$ are:

(1) $\mathrm{TAQ}^{\mathcal{O}}$ is excisive—it takes homotopy pushouts of $\mathcal{O}$-algebras to homotopy pullbacks of R-modules (which are the same as homotopy pushouts in this case),
(2) $\mathrm{TAQ}^{\mathcal{O}}(\mathcal{F}_{\mathcal{O}}(X)) \simeq X$—this is a consequence of the fact that $Q\mathcal{F}_{\mathcal{O}}X \approx X$.

(1) and (2) above imply that if A is built out of free $\mathcal{O}$-algebra cells, $\mathrm{TAQ}^{\mathcal{O}}(A)$ is built out of R-module cells in the same dimensions. In this way, TAQ provides information on the "cell structure" of an $\mathcal{O}$-algebra.

Example 3.2 The (reduced) commutative operad $\mathrm{Comm} = \mathrm{Comm}_R$ is given by

$$\mathrm{Comm}_i = \begin{cases} *, & i = 0, \\ R, & i \geq 1. \end{cases}$$

A Comm_R-algebra is a *non-unital* commutative R-algebra. The category of non-unital commutative R-algebras is equivalent to the category of augmented commutative R-algebras:

$$\mathrm{Alg}_{\mathrm{Comm}_R} \simeq (\mathrm{Alg}_R)_{/R}.$$

Given an augmented commutative R-algebra A, the augmentation ideal IA given by the fiber

$$IA \to A \xrightarrow{\epsilon} R$$

is the associated non-unital commutative algebra. In this setting, we have

$$\mathrm{TAQ}^{\mathrm{Comm}_R}(IA) \simeq \mathrm{TAQ}^R(A)$$

where $\mathrm{TAQ}^R(-)$ is the TAQ of [9].

The stable homotopy theory of $\mathcal{O}$-algebras. The following theorem was first proven in the context of simplicial commutative rings in [86], in the context of R arbitrary and $\mathcal{O} = \mathrm{Comm}$ in [13, 14], and R and $\mathcal{O}$ arbitrary in [74] (see also [31], [63, Thm. 7.3.4.13]).

Theorem 3.3 *There is an equivalence of categories*

$$\mathrm{Ho}(\mathrm{Sp}(\mathrm{Alg}_{\mathcal{O}})) \simeq \mathrm{Ho}(\mathrm{Mod}_R).$$

Under this equivalence, the functors

$$\Sigma^{\infty}_{\mathrm{Alg}_{\mathcal{O}}} : \mathrm{Ho}(\mathrm{Alg}_{\mathcal{O}}) \leftrightarrows \mathrm{Ho}(\mathrm{Mod}_R) : \Omega^{\infty}_{\mathrm{Alg}_{\mathcal{O}}}$$

are given by

$$\Sigma^{\infty}_{\mathrm{Alg}_{\mathcal{O}}} A \simeq \mathrm{TAQ}^{\mathcal{O}}(A),$$
$$\Omega^{\infty}_{\mathrm{Alg}_{\mathcal{O}}} X \simeq \mathrm{triv} X.$$

The adjunction above extends to derived mapping spaces, and gives the following (compare with [9]).

Corollary 3.4 *The spaces of the* $\mathrm{TAQ}_{\mathcal{O}}$*-spectrum are given by*

$$\Omega^{\infty}\Sigma^n \mathrm{TAQ}_{\mathcal{O}}(A) \simeq \underline{\mathrm{Alg}}_{\mathcal{O}}(A, \mathrm{triv}\Sigma^n R).$$

Proof We have

$$\begin{aligned}\underline{\mathrm{Alg}}_{\mathcal{O}}(A, \mathrm{triv}\Sigma^n R) &\simeq \underline{\mathrm{Mod}}_R(\mathrm{TAQ}^{\mathcal{O}}(A), \Sigma^n R)\\ &\simeq \underline{\mathrm{Mod}}_R(R, \Sigma^n \mathrm{TAQ}_{\mathcal{O}}(A))\\ &\simeq \Omega^{\infty}\Sigma^n \mathrm{TAQ}_{\mathcal{O}}(A).\end{aligned}$$

□

Divided power coalgebras. Question 2.2 clearly has a tautological answer when $\mathcal{C} = \mathrm{Alg}_{\mathcal{O}}$: it consists of $\mathcal{O}$-algebras in $\mathrm{Sp}(\mathcal{C}) \simeq \mathrm{Mod}_R$. However, this is *not* the canonical spectral algebra model given by the theory of homotopy descent of Sect. 2—we should be considering the $\Sigma^{\infty}_{\mathrm{Alg}_{\mathcal{O}}}\Omega^{\infty}_{\mathrm{Alg}_{\mathcal{O}}}$-coalgebra $\mathrm{TAQ}^{\mathcal{O}}(A)$ as a candidate spectral algebra model for A.

But what does it mean to be a $\Sigma^{\infty}_{\mathrm{Alg}_{\mathcal{O}}}\Omega^{\infty}_{\mathrm{Alg}_{\mathcal{O}}}$-coalgebra? The answer, according to [24, 31], is a *divided power coalgebra over the Koszul dual* $B\mathcal{O}$. Let us unpack what this means.

For any symmetric sequence $\mathcal{Y} = \{\mathcal{Y}_i\}$ of R-modules, one can use (3.1) to define a functor

$$\mathcal{F}_{\mathcal{Y}} : \mathrm{Mod}_R \to \mathrm{Mod}_R.$$

The category of symmetric sequences of R-modules possesses a monoidal structure $\circ$ called the composition product, such that

$$\mathcal{F}_{\mathcal{Y}} \circ \mathcal{F}_{\mathcal{Z}} = \mathcal{F}_{\mathcal{Y} \circ \mathcal{Z}}.$$

The monoids associated to the composition product are precisely the operads in Mod_R. The unit for this monoidal structure is the symmetric sequence 1_R with

$$(1_R)_i = \begin{cases} R, & i = 1, \\ *, & i \neq 1. \end{cases}$$

Every reduced operad $\mathcal{O}$ in Mod_R is augmented over 1_R. The *Koszul dual* of $\mathcal{O}$ is the symmetric sequence obtained by forming the bar construction with respect to the composition product

$$B\mathcal{O} := B(1_R, \mathcal{O}, 1_R) = |1 \Leftarrow \mathcal{O} \Lleftarrow \mathcal{O} \circ \mathcal{O} \cdots|.$$

Ching showed that $B\mathcal{O}$ admits a cooperad structure [26].

Example 3.5 Suppose $R = H\mathbb{Q}$, so we can replace Mod_R with $\mathrm{Ch}_{\mathbb{Q}}$. Take $\mathcal{O} = \mathrm{Lie}_{\mathbb{Q}}$, the Lie operad. Then we have $B\mathrm{Lie}_{\mathbb{Q}} = s\mathrm{Comm}_{\mathbb{Q}}^{\vee}$ the suspension of the commutative cooperad [26, 37].[4]

Example 3.6 In the case of $R = S$, the sphere spectrum, and $\mathcal{O}$ the commutative operad, Ching showed that

$$B\mathrm{Comm}_S \simeq (\partial_* \mathrm{Id}_{\mathrm{Top}_*})^{\vee}$$

the duals of the Goodwillie derivatives of the identity functor on Top_*[5] [26]. He also showed that with respect to the resulting operad structure on $\partial_* \mathrm{Id}_{\mathrm{Top}_*}$, we have

$$sH_* \partial_* \mathrm{Id}_{\mathrm{Top}_*} \cong \mathrm{Lie}_{\mathbb{Z}}.$$

As such, we will *define* the shifted spectral Lie operad as

$$s^{-1}\mathrm{Lie}_S := \partial_* \mathrm{Id}_{\mathrm{Top}_*}.$$

Following [24], we have for an R-module X:

---

[4]In general, for a (co)operad $\mathcal{O}$, the *suspension* of the (co)operad $s\mathcal{O}$ is a new (co)operad for which $(s\mathcal{O})_i \simeq \Sigma^{i-1}\mathcal{O}_i$ (nonequivaraintly), with the property that an $s\mathcal{O}$-(co)algebra structure on X is the same thing as an $\mathcal{O}$-(co)algebra structure on ΣX [7, 73].

[5]This identification used the computation of $\partial_* \mathrm{Id}_{\mathrm{Top}_*}$ of [8, 53] as input.

$$\begin{aligned}\Sigma^\infty_{\mathrm{Alg}_{\mathcal{O}}}\Omega^\infty_{\mathrm{Alg}_{\mathcal{O}}} X &\simeq \mathrm{TAQ}^{\mathcal{O}}(\mathrm{triv} X) \\ &\simeq B(\mathrm{Id}, \mathcal{F}_{\mathcal{O}}, \mathrm{triv} X) \\ &\simeq \mathcal{F}_{B\mathcal{O}} X.\end{aligned}$$

If R and $\mathcal{O}$ are connective, and X is connected, we have

$$\mathcal{F}_{B\mathcal{O}} X \simeq \prod_i \left(B\mathcal{O}_i \wedge_R X^{\wedge_R i}\right)_{\Sigma_i}.$$

Thus, at least on the level of the homotopy category, the data of a $\Sigma^\infty_{\mathrm{Alg}_{\mathcal{O}}}\Omega^\infty_{\mathrm{Alg}_{\mathcal{O}}}$-coalgebra C corresponds to the existence of a collection of coaction maps:

$$\psi_i : C \to \left(B\mathcal{O}_i \wedge_R C^{\wedge_R i}\right)_{\Sigma_i}.$$

The term *divided power* comes from the fact that a standard coalgebra over a cooperad consists of coaction maps into the Σ_i-fixed points rather than the Σ_i-orbits.

The general notion of a divided power (co)algebra over a (co)operad goes back to Fresse (see [32, 33]). For a precise definition of divided power coalgebras in the present homotopy-coherent context, we refer the reader to [31, 45]. In this language, we have functors

$$\mathrm{TAQ}^{\mathcal{O}} : \mathrm{Ho}(\mathrm{Alg}_{\mathcal{O}}) \leftrightarrows \mathrm{Ho}(\mathrm{d.p.Coalg}_{B\mathcal{O}}) : \mathfrak{C}. \tag{3.2}$$

Instances of Koszul duality. The following "Koszul Duality" theorem (a special case of a general conjecture of Francis-Gaitsgory [31]) generalizes Quillen's original theorem, as well as subsequent work in the algebraic context [33, 36, 37, 87].

Theorem 3.8 (Ching-Harper [24]) *In the case where R and $\mathcal{O}$ are connective, the functors (3.2) restrict to give an equivalence of categories*

$$\mathrm{Ho}(\mathrm{Alg}^{\geq 1}_{\mathcal{O}}) \simeq \mathrm{Ho}(\mathrm{d.p.Coalg}^{\geq 1}_{B\mathcal{O}}).$$

Example 3.9 Returning to the context of $R = H\mathbb{Q}$, and $\mathcal{O} = \mathrm{Lie}_{\mathbb{Q}}$ of Example 3.5, Theorem 3.8 recovers Quillen's original theorem:

$$\mathrm{Ho}(\mathrm{Alg}^{\geq 1}_{\mathrm{Lie}_{\mathbb{Q}}}) \simeq \mathrm{Ho}(\mathrm{Coalg}^{\geq 1}_{s\mathrm{Comm}^\vee_{\mathbb{Q}}}) \simeq \mathrm{Ho}(\mathrm{Coalg}^{\geq 2}_{\mathrm{Comm}^\vee_{\mathbb{Q}}}).$$

Note that we have not mentioned divided powers. This is because, rationally, coinvariants and invariants with respect to finite groups are isomorphic via the norm map, so *every* rational coalgebra is a divided power coalgebra.

4 Models of Rational and p-Adic Homotopy Theory

In this section we will return to Quillen–Sullivan theory, and a p-adic analog studied by Kriz, Goerss, Mandell, and Dwyer-Hopkins.

Rational homotopy theory, again. We begin by recasting Quillen–Sullivan theory into the language of spectral algebra. This in some sense defeats the original purpose of the theory—which was to encode rational homotopy theory in an *algebraic* category where you can literally write down the models in terms of generators, relations and differentials, but our recasting of the theory will motivate what follows.

Consider the functors

$$H\mathbb{Q} \wedge - : \mathrm{Ho}((\mathrm{Top}_*)_{\mathbb{Q}}) \to \mathrm{Ho}(\mathrm{Coalg}_{\mathrm{Comm}^\vee_{H\mathbb{Q}}}),$$
$$H\mathbb{Q}^- : \mathrm{Ho}((\mathrm{Top}_*)_{\mathbb{Q}})^{\mathrm{op}} \to \mathrm{Ho}(\mathrm{Alg}_{\mathrm{Comm}_{H\mathbb{Q}}}).$$

Essentially, for $X \in \mathrm{Top}_*$, $H\mathbb{Q} \wedge X$ is a spectral model for the reduced chains on X, and $H\mathbb{Q}^X$ is a spectral model for the reduced cochains of X. The commutative coalgebra/algebra structures come from the diagonal

$$\Delta : X \to X \wedge X.$$

The two functors are related by $H\mathbb{Q}^X = (H\mathbb{Q} \wedge X)^\vee$. If X is of finite type, there is no loss of information in using the cochains $H\mathbb{Q}^X$. There is a definite advantage to working with algebras rather than coalgebras if you like model categories.[6]

Quillen's theorem implies these functors restrict to give equivalences of categories:

$$H\mathbb{Q} \wedge (-) : \mathrm{Ho}(\mathrm{Top}_{\mathbb{Q}}^{\ge 2}) \xrightarrow{\simeq} \mathrm{Ho}(\mathrm{Coalg}^{\ge 2}_{\mathrm{Comm}^\vee_{H\mathbb{Q}}}),$$
$$H\mathbb{Q}^{(-)} : \mathrm{Ho}(\mathrm{Top}_{\mathbb{Q}}^{\ge 2,\mathrm{f.t.}}) \xrightarrow{\simeq} \mathrm{Ho}(\mathrm{Alg}^{\le -2,\mathrm{f.t.}}_{\mathrm{Comm}_{H\mathbb{Q}}}).$$

His Lie algebra models then come from applying Koszul duality (see Example 3.9).

p-adic homotopy theory. Fix a prime p. Analogous approaches to p-adic homotopy theory using cosimplicial commutative algebras, simplicial commutative coalgebras, E_∞-algebras in chain complexes, and commutative algebras in spectra were developed respectively by Kříž [56], Goerss [39], Mandell [66], and Dwyer-Hopkins (see [66]). We will focus on the spectral algebra setting, which is closely tied to Mandell's algebraic setting.

[6]For suitable monads $\mathbb{M}$ on cofibrantly generated model categories C it is typically straightforward to place induced model structures on $\mathrm{Alg}_{\mathbb{M}}$ [47]—coalgebras over comonads are more difficult to handle. This may be an instance where there is a definite advantage in working with ∞-categories. However, we also point out that Hess-Shipley [52] give a useful framework which in practice can often give model category structures on categories of coalgebras over comonads.

The basic idea in these approaches is to replace the role of $H\mathbb{Q}$ with the role of $H\bar{\mathbb{F}}_p$. Consider the cochain functor with $\bar{\mathbb{F}}_p$-coefficients on p-complete spaces:

$$H\bar{\mathbb{F}}_p^{(-)} : \mathrm{Ho}((\mathrm{Top}_*)_{\mathbb{Z}_p})^{\mathrm{op}} \to \mathrm{Ho}(\mathrm{Alg}_{\mathrm{Comm}_{H\bar{\mathbb{F}}_p}}).$$

Theorem 4.1 (Mandell [66]) *The $\bar{\mathbb{F}}_p$-cochains functor gives a fully faithful embedding*

$$H\bar{\mathbb{F}}_p^{(-)} : \mathrm{Ho}((\mathrm{Top}_*)_{\mathbb{Z}_p}^{\mathrm{nilp,f.t.}})^{\mathrm{op}} \hookrightarrow \mathrm{Ho}(\mathrm{Alg}_{\mathrm{Comm}_{H\bar{\mathbb{F}}_p}}). \tag{4.1}$$

of the homotopy category of nilpotent p-complete spaces of finite type into the homotopy category of commutative $H\bar{\mathbb{F}}_p$-algebras.

Remark 4.3 Actually, the functor (4.1) induces an equivalence on derived mapping spaces. Mandell also computes the effective image of this functor.

Remark 4.4 The approach of [39] suggests that the finite type hypothesis could be removed if one worked with $H\bar{\mathbb{F}}_p$-coalgebras.

What goes wrong when using $H\mathbb{F}_p$ instead of $H\bar{\mathbb{F}}_p$? Because the $\bar{\mathbb{F}}_p$-cochains are actually defined over $\mathbb{F}_p$, there is a continuous action of

$$\mathrm{Gal} := \mathrm{Gal}(\bar{\mathbb{F}}_p/\mathbb{F}_p) \cong \widehat{\mathbb{Z}}$$

on $H\bar{\mathbb{F}}_p^X$, with homotopy fixed points:

$$(H\bar{\mathbb{F}}_p^X)^{h\mathrm{Gal}(\bar{\mathbb{F}}_p/\mathbb{F}_p)} \simeq H\mathbb{F}_p^X.$$

It follows that for X, Y nilpotent and of finite type, we have

$$\begin{aligned}\underline{\mathrm{Alg}}_{\mathrm{Comm}_{H\mathbb{F}_p}}(H\mathbb{F}_p^Y, H\mathbb{F}_p^X) &\simeq \underline{\mathrm{Alg}}_{\mathrm{Comm}_{H\bar{\mathbb{F}}_p}}(H\bar{\mathbb{F}}_p^Y, H\bar{\mathbb{F}}_p^X)^{h\mathrm{Gal}} \\ &\simeq \underline{\mathrm{Top}}_*(X_p^\wedge, Y_p^\wedge)^{h\mathrm{Gal}}.\end{aligned}$$

However, the action of Gal on $\underline{\mathrm{Top}}_*(X_p^\wedge, Y_p^\wedge)$ is trivial, so we have

$$\begin{aligned}\underline{\mathrm{Top}}_*(X_p^\wedge, Y_p^\wedge)^{h\mathrm{Gal}} &\simeq \underline{\mathrm{Top}}_*(X_p^\wedge, Y_p^\wedge)^{B\mathbb{Z}} \\ &\simeq L\underline{\mathrm{Top}}_*(X_p^\wedge, Y_p^\wedge) \quad \text{(the free loop space).}\end{aligned}$$

In unpublished work (closely related to [67]), Mandell has shown the same holds for $H\mathbb{F}_p$ replaced by S_p, the p-adic sphere spectrum when X and Y are additionally assumed to be finite:

$$\underline{\mathrm{Alg}}_{\mathrm{Comm}}(S_p^Y, S_p^X) \simeq L\underline{\mathrm{Top}}_*(X_p^\wedge, Y_p^\wedge).$$

In fact, Mandell has shown the integral cochains functors gives a faithful embedding of the integral homotopy category into the category of integral E_∞-algebras [67]

$$\mathrm{Ho}(\mathrm{Top}_*^{\mathrm{nilp,f.t.}})^{op} \hookrightarrow \mathrm{Ho}(\mathrm{Alg}_{E_\infty}(\mathrm{Ch}_{\mathbb{Z}}))$$

Medina has recently proven a related statement using E_∞-coalgebras [69], and Blomquist-Harper have recently announced another setup using coalgebra structures on integral chains [10]. In unpublished work, Mandell has a similar result for commutative S-algebras: the Spanier-Whitehead dual functor gives a faithful embedding:

$$S^{(-)} : \mathrm{Ho}(\mathrm{Top}_*^{\mathrm{nilp,finite}})^{op} \hookrightarrow \mathrm{Ho}(\mathrm{Alg}_{\mathrm{Comm}}(\mathrm{Sp})).$$

Where are the p-adic Lie algebras? There is no known "Lie algebra model" for unstable p-adic homotopy theory. One of the problems is that, unlike the rational case, commutative $H\bar{\mathbb{F}}_p$-coalgebras do not automatically come equipped with divided power structures, so Koszul duality does not seem to apply (cf. the rational analogue of Example 3.5). Applying Koszul duality in the other direction, to get a "divided power Lie coalgebra model" (via a Koszul duality equivalence with commutative algebras) is fruitless as well, since $\mathrm{TAQ}^{H\bar{\mathbb{F}}_p}(H\bar{\mathbb{F}}_p^X) \simeq *$ for any finite-type nilpotent X (Thm. 3.4 of [67]).

One indication that one should not expect a Lie algebra model for p-adic homotopy types is that rationally, the composite

$$\mathrm{Ho}(\mathrm{Sp}_{\mathbb{Q}}^{\geq 2}) \xrightarrow{\Omega^\infty(-)} \mathrm{Ho}(\mathrm{Top}_{\mathbb{Q}}^{\geq 2}) \underset{\simeq}{\xrightarrow{L_{\mathbb{Q}}}} \mathrm{Ho}(\mathrm{Alg}_{\mathrm{Lie}_{\mathbb{Q}}}^{\geq 1})$$

is given by

$$L_{\mathbb{Q}}(\Omega^\infty Z) \simeq \mathrm{triv}\Sigma^{-1}Z.$$

where we give the spectrum $\Sigma^{-1}Z$ the *trivial* Lie bracket. This, strangely, means that a simply connected rational homotopy type is an infinite loop space if and only if its associated Lie algebra is equivalent to one with a trivial bracket. There is thus a functor

$$\Phi_0 : \mathrm{Ho}(\mathrm{Top}_{\mathbb{Q}}^{\geq 2}) \to \mathrm{Ho}(\mathrm{Sp}_{\mathbb{Q}}) \tag{4.2}$$

given by forgetting the Lie algebra structure on $L_{\mathbb{Q}}$. For a 1-connected spectrum Z, we have

$$\Phi_0\Omega^\infty Z \simeq Z_{\mathbb{Q}},$$

i.e., we can recover the rationalization of the spectrum from its 0th space. It follows that rationally, simply connected infinite loop spaces have *unique* deloopings. An analogous fact does not hold for p-adic infinite loop spaces.

5 v_n-Periodic Homotopy Theory

In both the case of rational homotopy theory, and p-adic homotopy theory, there are notions of "homotopy groups" and "homology groups". In the rational case, we have

$$\text{rational homotopy} = \pi_*(X) \otimes \mathbb{Q},$$
$$\text{rational homology} = H_*(X; \mathbb{Q}).$$

The appropriate analogs in the p-adic case are

$$\text{mod } p \text{ homotopy} = \pi_*(X; M(p)) := [\Sigma^* M(p), X]_{\mathrm{Top}_*},$$
$$\text{mod } p \text{ homology} = H_*(X; \mathbb{F}_p).$$

For 1-connected spaces, a map is a rational homotopy isomorphism if and only if it is a rational homology isomorphism, and similarly, a map is a mod p homotopy isomorphism if and only if it is a mod p homology isomorphism.

The idea of chromatic homotopy theory is that a p-local homotopy type is built out of monochromatic (or v_n-periodic) layers, and that elements of p-local homotopy groups fit into periodic families of different frequencies. The v_n-periodic homotopy groups isolate the elements in a particular frequency. The associated homology theory is the nth Morava K-theory.

Stable v_n-periodic homotopy theory. We begin with the stable picture. v_n-periodic stable homotopy theory has its own notion of homotopy and homology groups. The appropriate homology theory is the *nth Morava K-theory spectrum* $K(n)$, with

$$K(n)_* = \mathbb{F}_p[v_n^{\pm}], \quad |v_n| = 2(p^n - 1)$$

(for $n = 0$ we have $K(0) = H\mathbb{Q}$ and $v_0 = p$). The appropriate notion of homotopy groups are the *v_n-periodic homotopy groups*, defined as follows. A finite p-local spectrum V is called *type n* if it is $K(n-1)$-acyclic, and not $K(n)$-acyclic. The periodicity theorem of Hopkins-Smith [51] states that V has an asymptotically unique *v_n self-map*: a $K(n)$-equivalence

$$v : \Sigma^k V \to V$$

(with $k > 0$ if $n > 0$). The v_n-periodic homotopy groups (with coefficients in V) of a spectrum Z are defined to be

$$v_n^{-1}\pi_*(Z; V) := v^{-1}[\Sigma^* V, Z]_{\mathrm{Sp}}.$$

For $n > 0$ these groups are periodic, of period dividing k, the degree of the chosen self-map v. Note these groups do not depend on the choice of v_n self-map (by asymptotic uniqueness) but they do depend on the choice of finite type n spectrum V. However, for any two such spectra V, V', it turns out that a map is a $v_n^{-1}\pi_*(-; V)$

isomorphism if and only if it is a $v_n^{-1}\pi_*(-; V')$ isomorphism. It is straightforward to check that if we take $T(n)$ to be the "telescope"

$$T(n) = v_n^{-1}V := \mathrm{hocolim}(V \xrightarrow{v} \Sigma^{-k}V \xrightarrow{v} \Sigma^{-2k}V \xrightarrow{v} \cdots)$$

then a $v_n^{-1}\pi_*$-isomorphism is the same thing as a $T(n)_*$-isomorphism.

For maps of spectra it can be shown that

$$v_n^{-1}\pi_*\text{-isomorphism} \Rightarrow K(n)_*\text{-isomorphism}.$$

Ravenel's *telescope conjecture* [81] predicts the converse is true. This is easily verified in the case of $n = 0$, and deep computational work of Mahowald [64] and Miller [70] implies the conjecture is valid for $n = 1$. It is believed to be false for $n \geq 2$, but the problem remains open despite the valiant efforts of many researchers [71].

As such, there are potentially two *different* stable v_n-periodic categories, $\mathrm{Sp}_{T(n)}$ and $\mathrm{Sp}_{K(n)}$, corresponding to the localizations with respect to the two potentially different notions of equivalence. $K(n)$-localization gives a functor

$$(-)_{K(n)} : \mathrm{Ho}(\mathrm{Sp}_{T(n)}) \to \mathrm{Ho}(\mathrm{Sp}_{K(n)}).$$

Remark 5.1 Arguably localization with respect to $T(n)$ is more fundamental, but there are no known computations of $\pi_* Z_{T(n)}$ for a finite spectrum Z and $n \geq 2$ (if we had such a computation, we probably would have resolved the telescope conjecture for that prime p and chromatic level n). By contrast, the whole motivation of the chromatic program is that the homotopy groups $\pi_* Z_{K(n)}$ are essentially computable (though in practice these computations get quite involved, and little has been done for $n \geq 3$).

The stable chromatic tower. p-local stable homotopy types are assembled from the stable v_n-periodic categories in the following manner. Let $L_n^f\mathrm{Sp}$ denote the category of spectra which are $\bigoplus_{i=0}^n v_i^{-1}\pi_*$-local, and let $L_n\mathrm{Sp}$ denote the category of spectra which are $\bigoplus_{i=0}^n K(i)_*$-local, with associated (and potentially different) localization functors L_n^f, L_n. A spectrum Z has two potentially different *chromatic towers*

$$\cdots \to L_2^f Z \to L_1^f Z \to L_0^f Z,$$
$$\cdots \to L_2 Z \to L_1 Z \to L_0 Z.$$

Under favorable circumstances (for example, when Z is finite [49]) we have chromatic convergence: the map

$$Z_{(p)} \to \underset{n}{\mathrm{holim}}\, L_n Z$$

is an equivalence. Presumably one can expect similar results for L_n^f, though the authors are not aware of any work on this.

The *monochromatic layers* are the fibers

$$M_n^f Z \to L_n^f Z \to L_{n-1}^f Z,$$
$$M_n Z \to L_n Z \to L_{n-1} Z.$$

Let $M_n^f \mathrm{Sp}$ (respectively $M_n \mathrm{Sp}$) denote the subcategory of $L_n^f \mathrm{Sp}$ (respectively $L_n \mathrm{Sp}$) consisting of the image of the functor M_n^f (respectively M_n). Then the pairs of functors

$$(-)_{T(n)} : \mathrm{Ho}(M_n^f \mathrm{Sp}) \leftrightarrows \mathrm{Ho}(\mathrm{Sp}_{T(n)}) : M_n^f,$$
$$(-)_{K(n)} : \mathrm{Ho}(M_n \mathrm{Sp}) \leftrightarrows \mathrm{Ho}(\mathrm{Sp}_{K(n)}) : M_n$$

give equivalences between the respective homotopy categories (see, for example, [18]). We have

$$v_n^{-1} V \simeq M_n^f V \simeq V_{T(n)}$$

and

$$v_n^{-1} \pi_*(Z; V) \cong [\Sigma^* M_n^f V, M_n^f Z]_{\mathrm{Sp}} \cong [\Sigma^* V_{T(n)}, Z_{T(n)}]_{\mathrm{Sp}}.$$

$T(n)$-local Tate spectra. For G a finite group, and Z a spectrum with a G-action, there is a natural transformation

$$N : Z_{hG} \to Z^{hG}$$

called the *norm map* [38]. The cofiber is called the *Tate spectrum*:

$$Z^{tG} := \mathrm{cof}(Z_{hG} \to Z^{hG}).$$

The following theorem is due to Hovey-Sadofsky [50] in the $K(n)$-local case, and was strengthened by Kuhn [58] to the $T(n)$-local case (see also [27, 41, 72]).

Theorem 5.2 (Greenlees-Sadofsky, Kuhn) *If Z is $T(n)$-local, then the spectrum Z^{tG} is $T(n)$-acyclic, and the norm map is a $T(n)$-equivalence.*

In the case of $n = 0$, this reduces to the familiar statement that rationally, invariants and coinvariants with respect to a finite group are isomorphic via the norm. In general, this theorem implies that $T(n)$-local coalgebras, $T(n)$-locally, admit unique divided power structures. In some sense, Theorem 5.2 will be the primary mechanism which will allow unstable v_n-periodic homotopy types to admit Lie algebra models.

Unstable v_n-periodic homotopy theory. Perhaps the most illuminating approach to *unstable* v_n-periodic homotopy theory is that of [18], which we follow here. This approach builds on previous work of Davis, Mahowald, Dror Farjoun, and many others. Like the stable case, there will be two potentially different notions of unstable v_n-periodic equivalence: one based on unstable v_n-periodic homotopy groups, and one based on $K(n)$-homology.

The appropriate unstable analogs of v_n-periodic homotopy groups are defined as follows. The periodicity theorem implies that unstably, a finite type n complex admits a v_n-self map

$$v : \Sigma^{k(N_0+1)} V \to \Sigma^{kN_0} V$$

for some $N_0 \gg 0$. For any $X \in \mathrm{Top}_*$, its v_n-periodic homotopy groups (with coefficients in V) are defined by

$$v_n^{-1}\pi_*(X; V) := v^{-1}[\Sigma^* V, X]_{\mathrm{Top}_*}.$$

for $n > 0$ (v_0-periodic homotopy is taken to be rational homotopy). For $n > 0$ this definition only makes sense for $* \gg 0$, but because the result is k-periodic, one can define these groups for all $* \in \mathbb{Z}$. These give the notion of a *$v_n^{-1}\pi_*$-equivalence of spaces*. Bousfield argues in [18] that the appropriate notion of unstable v_n-periodic homology equivalence is that of a *virtual $K(n)$-equivalence*—a map of spaces $X \to Y$ for which the induced map

$$(\Omega X)^{\geq n+3} \to (\Omega Y)^{\geq n+3}$$

is a $K(n)_*$-isomorphism.[7] Rather than try to explain why this is the appropriate notion we will simply point out that Bousfield proves that if the telescope conjecture is true, then virtual $K(n)$-equivalences are $v_n^{-1}\pi_*$-isomorphisms.

We will focus on the version of unstable v_n-periodic homotopy theory based on $v_n^{-1}\pi_*$-equivalences. The authors do not know if any attempt has been made to systematically study the unstable theory based on virtual $K(n)$-equivalences (in case the telescope conjecture is false).

Bousfield defines $L_n^f \mathrm{Top}_*$ to be the nullification of Top_* with respect to

$$\Sigma V_{n+1} \vee \bigvee_{\ell \neq p} M(\mathbb{Z}/\ell, 2),$$

where V_{n+1} is a type $n+1$ complex of minimal connectivity (say it is $(d_n - 3)$-connected). Let L_n^f denote the associated localization functor. When restricted to $\mathrm{Top}_*^{\geq d_n}$, L_n^f is localization with respect to $\bigoplus_{i=0}^n v_i^{-1}\pi_*$-equivalences. For a space X there is an unstable chromatic tower

$$\cdots \to L_2^f X \to L_1^f X \to L_0^f X.$$

The unstable chromatic tower actually always converges to $X_{(p)}$ for a trivial reason: the sequence d_n is non-decreasing and unbounded [16].

The nth monochromatic layer is defined to be the homotopy fiber

$$M_n^f X \to L_n^f X \to L_{n-1}^f X.$$

---

[7] A variant of this definition is explored by Kuhn in [59].

Bousfield defines the nth unstable monochromatic category $M_n^f \mathrm{Top}_*$ to be the full subcategory of Top_* consisting of the spaces of the form $(M_n^f X)^{\geq d_n}$. Bousfield's work in [18] implies the equivalences in $M_n^f \mathrm{Top}_*$ are precisely the $v_n^{-1}\pi_*$-equivalences. Furthermore, for any type n complex V with an unstable v_n-self map

$$v : \Sigma^k V \to V$$

the v_n-periodic homotopy groups are in fact the V-based homotopy groups as computed in $\mathrm{Ho}(M_n^f \mathrm{Top}_*)$:

$$v_n^{-1}\pi_*(X; V) \cong [\Sigma^* V, M_n^f X]_{\mathrm{Top}_*}.$$

The Bousfield–Kuhn functor. Bousfield and Kuhn [18, 61] observe v_n-periodic homotopy groups are the homotopy groups of a spectrum $\Phi_V(X)$. The kNth space of this spectrum is given by

$$\Phi_V(X)_{kN} = \underline{\mathrm{Top}}_*(V, X)$$

with spectrum structure maps generated by the maps

$$\Phi_V(X)_{kN} = \underline{\mathrm{Top}}_*(V, X) \xrightarrow{v^*} \underline{\mathrm{Top}}_*(\Sigma^k V, X) \simeq \Omega^k \Phi_V(X)_{k(N+1)}.$$

It follows that

$$\pi_* \Phi_V(X) \cong v_n^{-1}\pi_*(X; V).$$

The above definition only depended on $\Sigma^{kN} V$ for N large. As a result, it only depends on the stable homotopy type $\Sigma^\infty V$. One can therefore take a suitable inverse system V_i of finite type n spectra so that

$$\operatorname*{holim}_i v_n^{-1} V_i \simeq S_{T(n)}.$$

The *Bousfield–Kuhn* functor

$$\Phi_n : \mathrm{Ho}(\mathrm{Top}_*) \to \mathrm{Ho}(\mathrm{Sp}_{T(n)})$$

is given by

$$\Phi_n(X) = \operatorname*{holim}_i \Phi_{V_i^\vee}(X).$$

We *define* the completed unstable v_n-periodic homotopy groups (without coefficients in a type n complex) by[8]

---

[8]These should not be confused with the "uncompleted" unstable v_n-periodic homotopy groups studied by Bousfield, Davis, Mahowald, and others. These are given as the homotopy groups of $M_n^f \Phi_n(X)$ (see [60]).

$$v_n^{-1}\pi_*(X)^\wedge := \pi_*\Phi_n(X).$$

The Bousfield–Kuhn functor enjoys many remarkable properties:

(1) For $X \in \mathrm{Top}_*$ and a type n spectrum V we have

$$[\Sigma^* V, \Phi_n(X)]_{\mathrm{Sp}} \simeq v_n^{-1}\pi_*(X; V).$$

(2) Φ_n preserves fiber sequences.
(3) For $Z \in \mathrm{Sp}$ there is a natural equivalence

$$\Phi_n \Omega^\infty Z \simeq Z_{T(n)}.$$

Property (3) above is the strangest property of all: it implies (since by (2) Φ_n commutes with Ω) that a $T(n)$-local spectrum is *determined* by any one of the spaces in its Ω-spectrum, *independent* of the infinite loop space structure.

Relation between stable and unstable v_n-periodic homotopy. The category $\mathrm{Ho}(\mathrm{Sp}_{T(n)})$ serves as the "stable homotopy category" of the unstable v_n-periodic homotopy category $\mathrm{Ho}(M_n^f \mathrm{Top}_*)$, with adjoint functors [18]

$$(\Sigma^\infty -)_{T(n)} : \mathrm{Ho}(M_n^f \mathrm{Top}_*) \leftrightarrows \mathrm{Ho}(\mathrm{Sp}_{T(n)}) : (\Omega^\infty M_n^f -)^{\ge d_n}.$$

Analogously to the rational situation, it is shown in [18] that the composite

$$\mathrm{Ho}(\mathrm{Sp}_{T(n)}) \xrightarrow{(\Omega^\infty M_n^f -)^{\ge d_n}} \mathrm{Ho}(M_n^f \mathrm{Top}_*) \xrightarrow{\Phi_n} \mathrm{Ho}(\mathrm{Sp}_{T(n)})$$

is naturally isomorphic to the identity functor. Thus the stable v_n-periodic homotopy category admits a fully faithful embedding into the unstable v_n-periodic homotopy category. This leads one to expect that there is a "Lie algebra" model of unstable v_n-periodic homotopy, where the infinite loop spaces correspond to the Lie algebras with trivial Lie structure.

The $K(n)$-local variant. There is a variant of the Bousfield–Kuhn functor

$$\Phi_{K(n)} : \mathrm{Ho}(\mathrm{Top}_*) \to \mathrm{Ho}(\mathrm{Sp}_{K(n)})$$

defined by

$$\Phi_{K(n)}(X) \simeq \Phi_n(X)_{K(n)}.$$

We then have

$$\Phi_{K(n)} \Omega^\infty Z \simeq Z_{K(n)}.$$

There is a corresponding variant of completed unstable v_n-periodic homotopy groups which (probably to the chagrin of many) we will denote:

$$v_{K(n)}^{-1}\pi_*(X)^\wedge := \pi_*\Phi_{K(n)}(X).$$

Of course if the telescope conjecture is true, $\Phi_n(X) \simeq \Phi_{K(n)}(X)$, and the two versions of unstable v_n-periodic homotopy agree. If the telescope conjecture is not true, the groups $v_{K(n)}^{-1}\pi_*$ will likely be far more computable than $v_n^{-1}\pi_*$.

6 The Comparison Map

Motivated by rational and p-adic homotopy theory, one could ask: to what degree is an unstable homotopy type $X \in M_n^f \mathrm{Top}_*$ modeled by the $T(n)$-local Comm-algebra $S_{T(n)}^X$ (the "$S_{T(n)}$-valued cochains")? I.e., what can be said of the functor:

$$S_{T(n)}^{(-)} : \mathrm{Ho}(M_n^f \mathrm{Top}_*)^{op} \to \mathrm{Ho}(\mathrm{Alg}_{\mathrm{Comm}}(\mathrm{Sp}_{T(n)}))?$$

The first thing to check is to what degree the unstable v_n-periodic homotopy groups of X can be recovered from the algebra $S_{T(n)}^X$: i.e. for an unstable type n complex V with v_n-self map

$$v : \Sigma^k V \to V$$

what can be said of the following composite?

$$v_n^{-1}\pi_*(X; V) \cong [\Sigma^* V, M_n^f(X)]_{\mathrm{Top}_*} \to \left[S_{T(n)}^X, S_{T(n)}^{\Sigma^* V}\right]_{\mathrm{Alg}_{\mathrm{Comm}}} \tag{6.1}$$

We begin with the observation, which we learned from Mike Hopkins, that the Comm-algebra $S_{T(n)}^V$ is actually an "infinite loop object" in the category $\mathrm{Alg}_{\mathrm{Comm}}$:

Proposition 6.2 *There is an equivalence of* Comm-*algebras*

$$S_{T(n)}^V \simeq \mathrm{triv}(V^\vee).$$

Proof The existence of the v_n-self map v shows that $S_{T(n)}^V$ is an infinite loop object of $\mathrm{Alg}_{\mathrm{Comm}}$:

$$S_{T(n)}^V \xrightarrow[\simeq]{(v^N)^*} S_{T(n)}^{\Sigma^{Nk} V} \simeq \Omega^{Nk} S_{T(n)}^V.$$

The result follows from the fact that the infinite loop objects in $\mathrm{Alg}_{\mathrm{Comm}}$ are the trivial algebras on the underlying spectra. □

Using Corollary 3.4, we now deduce:

Corollary 6.3 *We have*

$$\underline{\mathrm{Alg}}_{\mathrm{Comm}}(S_{T(n)}^X, S_{T(n)}^{\Sigma^* V}) \simeq \Omega^\infty \Sigma^* \mathrm{TAQ}_{S_{T(n)}}(S_{T(n)}^X) \wedge V^\vee.$$

We deduce that (6.1) refines to a natural transformation

$$c_X^V : \Phi_V(X) \to \mathrm{TAQ}_{S_{T(n)}}(S_{T(n)}^X) \wedge V^\vee.$$

Taking a suitable homotopy inverse limit of these natural transformations gives a natural transformation

$$c_X : \Phi_n(X) \to \mathrm{TAQ}_{S_{T(n)}}(S_{T(n)}^X)$$

which we will call *the comparison map*. A variant, which involves replacing $S_{T(n)}$ with $S_{K(n)}$, everywhere, is defined in [22]:

$$c_X^{K(n)} : \Phi_{K(n)}(X) \to \mathrm{TAQ}_{S_{K(n)}}(S_{K(n)}^X).$$

The main theorem of [22] is

Theorem 6.4 *The comparison map $c_X^{K(n)}$ is an equivalence for X a sphere.*

It follows formally from this theorem that the comparison map is an equivalence for a larger class of spaces: the class of finite $\Phi_{K(n)}$-*good* spaces. This will be discussed in Sect. 8. In the case of $n = 1$, Theorem 6.4 was originally proven by French [34].

It is shown in [26] that cobar constructions for $\mathcal{O}$-coalgebras get a $C\mathcal{O}$-algebra structure (where C denotes the cooperadic cobar construction). The spectrum

$$\mathrm{TAQ}_{S_{T(n)}}(S_{T(n)}^X)$$

is therefore an algebra over $s^{-1}\mathrm{Lie}_S$ (see Example 3.6). We might regard this as a candidate for a "Lie algebra model" for the unstable v_n-periodic homotopy type of X, though this is probably only reasonable for X finite, as will be explained in Sect. 10.

7 Outline of the Proof of the Main Theorem

Our approach to Theorem 6.4 is essentially computational in nature, and uses the Morava E-theory Dyer-Lashof algebra in an essential way. Unfortunately, the proof given in [22] is necessarily technical, and consequently is not optimized for leisurely reading. In this section we give an overview of the main ideas of our proof. As we will explain in Sects. 9 and 10, Arone-Ching [1] and Heuts [44] have announced more abstract approaches to prove Theorem 6.4, with stronger consequences. Perhaps the situation is comparable to the early work on p-adic homotopy theory of Kříž and Goerss [39, 56]: Kriz's approach (like that of [66]) is computational, based on the Steenrod algebra, whereas Goerss' is abstract, based on Galois descent and model category theory. Both approaches offer insight into the theory of using commutative algebras/coalgebras to model p-adic homotopy types. We hope the same is true of the two approaches to model unstable v_n-periodic homotopy.

Goodwillie towers. The proof of 6.4 involves induction up the Goodwillie towers of both the source and target of the comparison map. The key fact that the argument hinges on is an observation of Kuhn [57]: the layers of both of these towers are abstractly equivalent.

For our application of Goodwillie calculus to the situation, we point out that, in the context of model categories, Pereira [74] has shown that Goodwillie's calculus of functors (as developed in [40]) applies to homotopy functors

$$F : \mathcal{C} \to \mathcal{D}$$

between arbitrary model categories with fairly minimal hypotheses (see also [21, 63]).[9] For simplicity we shall assume that $\mathcal{C}$ and $\mathcal{D}$ are pointed, and restrict attention to reduced F (i.e. $F(*) \simeq *$).

Associated to F is its *Goodwillie tower*, a series of *k-excisive approximations*

$$P_k F : \mathcal{C} \to \mathcal{D}$$

which form a tower under F:

$$F \to \cdots \to P_k F \to P_{k-1} F \to \cdots \to P_1 F.$$

We say the Goodwillie tower *converges* at X if the map

$$F(X) \to \operatorname*{holim}_k P_k F(X)$$

is an equivalence. The *layers* of the Goodwillie tower are the fibers

$$D_k F \to P_k F \to P_{k-1} F.$$

If F is finitary (i.e. preserves filtered homotopy colimits), the layers take the form

$$D_k F(X) \simeq \Omega^\infty_{\mathcal{D}} \mathrm{cr}^{lin}_k(F)(\Sigma^\infty_{\mathcal{C}} X, \cdots, \Sigma^\infty_{\mathcal{C}} X)_{h\Sigma_k}$$

where

$$\mathrm{cr}^{lin}_k(F) : \mathrm{Sp}(\mathcal{C})^{\times k} \to \mathrm{Sp}(\mathcal{D})$$

is a certain symmetric multilinear functor called the *multilinearized cross-effect*. In the case where $\mathrm{Sp}(\mathcal{C})$, $\mathrm{Sp}(\mathcal{D})$ are Quillen equivalent to $\mathrm{Sp} = \mathrm{Sp}(\mathrm{Top}_*)$, the multilinearized cross effect is given by

$$\mathrm{cr}^{lin}_k(F)(Z_1, \cdots, Z_k) \simeq \partial_k F \wedge Z_1 \wedge \cdots \wedge Z_k$$

---

[9] Yet another general treatment of homotopy calculus can be found in [12], but at present this approach only applies to functors which take values in spectra.

where $\partial_k F$ is a spectrum with Σ_k-action (the *kth derivative of* F), and we have

$$D_k F(X) \simeq \Omega^\infty_{\mathcal{D}} \left(\partial_k F \wedge_{h\Sigma_k} (\Sigma^\infty_{\mathcal{C}} X)^{\wedge k}\right).$$

The Goodwillie tower is an analog for functors of the Taylor series of a function, with $D_k(F)$ playing the role of the kth term of the Taylor series.

We consider the Goodwillie towers of the functors

$$\Phi_{K(n)} : \mathrm{Top}_* \to \mathrm{Sp}_{K(n)}$$
$$\mathrm{TAQ}_{S_{K(n)}}(S^{(-)}_{K(n)}) : \mathrm{Top}_* \to \mathrm{Sp}_{K(n)}.$$

Note that the second of these functors is not finitary ($\Phi_{K(n)}$ is actually finitary, as long as the corresponding homotopy colimit is taken in the category $\mathrm{Sp}_{K(n)}$). In the case of $\Phi_{K(n)}$, it is fairly easy to see that its Goodwillie tower is closely related to the Goodwillie tower of the identity functor

$$\mathrm{Id} : \mathrm{Top}_* \to \mathrm{Top}_*.$$

Lemma 7.1 *There are equivalences*

$$P_k \Phi_{K(n)} \simeq \Phi_{K(n)} P_k \mathrm{Id}.$$

Proof This follows easily from observing that the fibers of the RHS are given by

$$\Phi_{K(n)} D_k \mathrm{Id}(X) \simeq (s^{-1}\mathrm{Lie}_k \wedge_{h\Sigma_k} X^{\wedge k})_{K(n)}$$

and are therefore homogeneous of degree k. □

More subtly, Kuhn constructed a filtration on TAQ^R [57] which results in a tower

$$\mathrm{TAQ}_R(A) \to \cdots \to F_k\, \mathrm{TAQ}_R(A) \to F_{k-1}\, \mathrm{TAQ}_R(A) \to \cdots. \tag{7.1}$$

For all A we have an equivalence

$$\mathrm{TAQ}_R(A) \xrightarrow{\simeq} \mathrm{holim}\, F_k\, \mathrm{TAQ}_R(A) \tag{7.2}$$

for the simple reason that Kuhn's filtration of TAQ^R is exhaustive.

Theorem 7.4 (Kuhn [57]) *The fibers of the tower (7.1) are given by*

$$s^{-1}\mathrm{Lie}_k \wedge^{h\Sigma_k} (A^{\wedge_R k})^\vee \to F_k\, \mathrm{TAQ}_R(A) \to F_{k-1}\, \mathrm{TAQ}_R(A).$$

Corollary 7.5 *For finite* X *the Goodwillie tower of the functor* $\mathrm{TAQ}_{S_{K(n)}}(S^{(-)}_{K(n)})$ *is given by*

$$P_k(\mathrm{TAQ}_{S_{K(n)}}(S^{(-)}_{K(n)}))(X) \simeq F_k\, \mathrm{TAQ}_{S_{K(n)}}(S^X_{K(n)}).$$

Proof Combining Theorem 5.2 with Theorem 7.4 shows the layers of the RHS are equivalent to

$$(s^{-1}\mathrm{Lie}_k \wedge_{h\Sigma_k} X^{\wedge k})_{K(n)}.$$

In particular, they are homogeneous of degree k. □

It follows that the comparison map actually induces a natural transformation of towers

$$P_n(c_X^{K(n)}) : \Phi_{K(n)} P_k \mathrm{Id}(X) \to F_k \, \mathrm{TAQ}_{S_{K(n)}}(S^X_{K(n)})$$

when restricted to finite X. In fact, the proofs of Lemma 7.1 and Corollary 7.5 actually imply that for X finite, the layers of these towers are abstractly equivalent. Thus, to show that the maps $P_n(c_X^{K(n)})$ are equivalences, we just need to show that they *induce* equivalences on the layers (which we already know are equivalent)! This will be accomplished computationally using

The Morava E-theory Dyer-Lashof algebra. Let E_n denote the nth Morava E-theory spectrum, with

$$(E_n)_* \cong W(\mathbb{F}_{p^n})[[u_1, \ldots, u_{n-1}]][u^{\pm}].$$

The ring $(E_n)_0$ has a unique maximal ideal $\mathfrak{m}$. We shall let

$$(E_n^{\wedge})_* Z := \pi_*(E_n \wedge Z)_{K(n)}$$

denote the completed E-homology of a spectrum Z. If the uncompleted Morava E_n-homology is flat over $(E_n)_*$, the completed E-homology is the $\mathfrak{m}$-completion of the uncompleted homology. Let K_n denote the 2-periodic version of $K(n)$, with

$$(K_n)_* \cong (E_n)_*/\mathfrak{m} \cong \mathbb{F}_{p^n}[u^{\pm}].$$

In [83], the second author defined a monad[10]

$$\mathbb{T} : \mathrm{Mod}_{(E_n)_*} \to \mathrm{Mod}_{(E_n)_*}$$

such that the completed E-homology of a Comm-algebra has the structure of a $\mathbb{T}$-algebra. A $\mathbb{T}$-algebra is basically an algebra over the Morava E-theory Dyer-Lashof algebra Γ_n. For an $(E_n)_*$-module M, the value of the functor $\mathbb{T}M$ is the free Γ_n-algebra on M (for a precise description of what is meant by this, consult [83]).

The work of Strickland [88] basically determines the structure of the dual of Γ_n in terms of rings of functions on the formal schemes of subgroups of the Lubin-Tate formal groups. In the case of $n = 1$, the corresponding Morava E-theory is p-adic K-theory, and Γ_1 is generated by the Adams operation ψ^p with no relations. In the case of $n = 2$, the explicit structure of Γ_2 was determined by the second author in

[10]The monad denoted $\mathbb{T}$ here is actually a non-unital variant of the monad $\mathbb{T}$ of [83].

[82] for $p = 2$, and mod p for all primes in [84]. An integral presentation of Γ_2 has recently been determined by Zhu [95]. Very little is known about the explicit structure of Γ_n for $n \geq 3$ except that it is Koszul [85] in the sense of Priddy [78].

For the purpose of our discussion of Theorem 6.4, the only thing we really need to know about $\mathbb{T}$ is the following theorem of the second author (see [83]):

Theorem 7.6 *If $(E_n^\wedge)_*Z$ is flat over $(E_n)_*$, then the natural transformation*

$$\mathbb{T}(E_n^\wedge)_*Z \to (E_n^\wedge)_*\mathcal{F}_{\mathrm{Comm}}Z$$

induces an isomorphism

$$(\mathbb{T}(E_n^\wedge)_*Z)^\wedge_{\mathfrak{m}} \xrightarrow{\cong} (E_n^\wedge)_*\mathcal{F}_{\mathrm{Comm}}Z.$$

There is a "completed" variant of the functor $\mathcal{F}_{\mathrm{Comm}}$:

$$\widehat{\mathcal{F}}_{\mathrm{Comm}}(Z) := \prod_i Z^i_{h\Sigma_i}.$$

The following lemma of [22] is highly non-trivial, as completed Morava E-theory in general behaves badly with respect to products.

Lemma 7.7 *There is a completed variant of the free $\mathbb{T}$-algebra functor:*

$$\widehat{\mathbb{T}} : \mathrm{Mod}_{E_*} \to \mathrm{Alg}_{\mathbb{T}}$$

and for spectra Z a natural transformation

$$\widehat{\mathbb{T}}(E_n^\wedge)_*Z \to (E_n^\wedge)_*\widehat{\mathcal{F}}_{\mathrm{Comm}}Z$$

which is an isomorphism if $(E_n^\wedge)_*Z$ is flat and finitely generated.

In [22] we construct a version of the Basterra spectral sequence for E-theory: for a $K(n)$-local Comm-algebra A whose E_n-homology satisfies a flatness hypothesis, the spectral sequence takes the form

$$AQ^{*,*}_{\mathbb{T}}((E_n^\wedge)_*A; (K_n)_*) \Rightarrow (K_n)_* \mathrm{TAQ}_{S_{K(n)}}(A). \quad (7.3)$$

Here $AQ^{*,*}_{\mathbb{T}}(-; M)$ denotes Andre-Quillen cohomology of $\mathbb{T}$-algebras with coefficients in an E_*-module M (see [22] for a precise definition—these cohomology groups are closely related to those defined in [35]).

The comparison map on QX. The next step in the proof of Theorem 6.4 is to prove the following key proposition.

Proposition 7.9 *There is a non-negative integer N so that for all N-fold suspension spaces X with $(E_n^\wedge)_*X$ free and finitely generated over $(E_n)_*$, the comparison map*

$$(\Sigma^\infty X)_{K(n)} \simeq \Phi_{K(n)}(QX) \xrightarrow{c^{K(n)}_{QX}} \mathrm{TAQ}_{S_{K(n)}}(S^{QX}_{K(n)})$$

is an equivalence.

We will prove this proposition by showing that the comparison map induces an isomorphism in Morava $K(n)$-homology. The first step is to compute the $K(n)_*$-homology of the the RHS. This is accomplished in [22] with the following technical lemma:

Lemma 7.10 *For X satisfying the hypotheses of Proposition 7.9, there is a map of $(E_n)_*$-modules*

$$(E_n^\wedge)_* S^{QX}_{K(n)} \to \widehat{\mathbb{T}}\tilde{E}_n^* X$$

which is an isomorphism of $\mathbb{T}$-algebras mod $\mathfrak{m}$*, in the sense that it is an isomorphism mod* $\mathfrak{m}$*, and commutes with the* $\mathbb{T}$*-action mod* $\mathfrak{m}$.

Heuristically, this lemma might seem to follow from Theorem 5.2 and the Snaith splitting:

$$\begin{aligned} S^{QX}_{K(n)} &\simeq S^{\bigvee_i X^i_{h\Sigma_i}}_{K(n)} \\ &\simeq \prod_i \left(S^{X^i}_{K(n)}\right)^{h\Sigma_i} \\ &\simeq \left(\prod_i \left(S^{X^i}_{K(n)}\right)_{h\Sigma_i}\right)_{K(n)} \\ &\simeq \left(\widehat{\mathcal{F}}_{\mathrm{Comm}} S^X_{K(n)}\right)_{K(n)}. \end{aligned}$$

However, as was pointed out to us by Nick Kuhn, this is *not* an equivalence of Comm-algebras (or even non-unital H_∞-ring spectra)! Nevertheless, Lemma 7.10 establishes that on Morava E-theory, this sequence of equivalences induces an isomorphism of $\mathbb{T}$-algebras mod $\mathfrak{m}$.

Proof of Proposition 7.9. The natural transformation

$$\Sigma^\infty QX = \Sigma^\infty \Omega^\infty \Sigma^\infty X \to \Sigma^\infty X$$

induces a natural transformation

$$S^X_{K(n)} \to S^{QX}_{K(n)}$$

of spectra, hence a natural transformation

$$\mathcal{F}_{\mathrm{Comm}_{S_{K(n)}}} S^X_{K(n)} \to S^{QX}_{K(n)}$$

of Comm-algebras. We thus get a natural transformation

$$\begin{aligned}\mathrm{TAQ}_{S_{K(n)}}(S_{K(n)}^{QX}) \xrightarrow{\eta_X} \mathrm{TAQ}_{S_{K(n)}}(\mathcal{F}_{\mathrm{Comm}_{S_{K(n)}}} S_{K(n)}^X) \\ \simeq (\Sigma^\infty X)_{K(n)} \\ \simeq \Phi_{K(n)}(QX).\end{aligned}$$

It can be shown that $\eta_X \circ c_{QX}^{K(n)} \simeq \mathrm{Id}$. Since $(\widetilde{K}_n)_* X$ is finite, it suffices to show that $(K_n)_* \mathrm{TAQ}_{S_{K(n)}}(S^X_{K(n)})$ is abstractly isomorphic to $(\widetilde{K}_n)_* X$. This is proven using the Basterra spectral sequence (7.3). The spectral sequence collapses to the desired result as we have (using Lemma 7.10)

$$\begin{aligned}AQ^{s,*}_{\mathbb{T}}((E_n^\wedge)_* S^{QX}_{K(n)}; (K_n)_*) &\cong AQ^{s,*}_{\mathbb{T}}(\widehat{\mathbb{T}} \widetilde{E}_n^* X; (K_n)_*) \\ &\cong AQ^{s,*}_{\mathbb{T}}(\mathbb{T} \widetilde{E}_n^* X; (K_n)_*) \\ &\cong \begin{cases} (\widetilde{K}_n)_* X, & s = 0, \\ 0, & s > 0. \end{cases}\end{aligned}$$

□

The comparison map on spheres. We now outline the proof of Theorem 6.4. Let $X = S^q$. The following strong convergence theorem of Arone-Mahowald [8] is crucial.

Theorem 7.11 (Arone-Mahowald) *The natural transformation*

$$\Phi_{K(n)}(X) \to \Phi_{K(n)} P_k \mathrm{Id}(X)$$

is an equivalence for q odd and $k = p^n$, or q even and $k = 2p^n$.

The basic strategy is to attempt to apply Proposition 7.9 to the Bousfield–Kan cosimplicial resolution

$$X \to Q^{\bullet+1} X = (QX \Rightarrow QQX \Rrightarrow \cdots).$$

We first assume that the dimension q of the sphere $X = S^q$ is large and odd. Unfortunately, for $s \geq 1$, $Q^s X$ does not satisfy the finiteness hypotheses of Proposition 7.9 required to deduce that the comparison map is an equivalence. We instead consider the diagram

$$\begin{array}{ccc} \Phi_{K(n)}(X) & \longrightarrow & \mathrm{Tot}\, \Phi_{K(n)} P_{p^n}(Q^{\bullet+1})(X) \\ \Big\downarrow{\scriptstyle c_X^{K(n)}} & & \Big\downarrow{\scriptstyle \simeq} \\ \mathrm{TAQ}_{S_{K(n)}}(S^X_{K(n)}) & \longrightarrow & \mathrm{Tot}\, \mathrm{TAQ}_{S_{K(n)}}\left(S_{K(n)}^{P_{p^n}(Q^{\bullet+1})(X)}\right) \end{array} \tag{7.4}$$

In the above diagram, the right vertical map is an equivalence using Proposition 7.9: the Snaith splitting may be iterated to give an equivalence [6]

$$P_{p^n}(Q^{s+1})(X) \simeq QY^s$$

where the space Y^s does satisfy the hypotheses of Proposition 7.9. Using finiteness properties of the cosimplicial space $Y^\bullet$, we show in [22] that the top horizontal map

$$\Phi_{K(n)}(X) \simeq \Phi_{K(n)} P_{p^n}\mathrm{Id}(X) \to \mathrm{Tot}\, \Phi_{K(n)} P_{p^n}(Q^{\bullet+1})(X)$$

of (7.4) is an equivalence. It follows that the comparison map has a weak retraction when restricted to large dimensional odd spheres X:

$$\begin{array}{ccc} \Phi_{K(n)}X & \xrightarrow{\quad\simeq\quad} & \Phi_{K(n)}X \\ & {\scriptstyle c_X^{K(n)}} \searrow \quad \nearrow & \\ & \mathrm{TAQ}_{S_{K(n)}}(S_{K(n)}^X) & \end{array}$$

Using standard methods of Goodwillie calculus (or more specifically, Weiss calculus [93] in this case) it follows that for X a large dimensional odd sphere, the induced map on Goodwillie towers

$$\{P_k \Phi_{K(n)}(X)\}_k \xrightarrow{c^{K(n)}} \{F_k\, \mathrm{TAQ}_{S_{K(n)}}(S_{K(n)}^X)\}_k \tag{7.5}$$

has a weak retraction. The theorem (for X a large dimensional odd sphere) follows from the fact that (1) the layers of the towers are abstractly equivalent, and (2) the layers of the towers have finite $K(n)$-homology. Since Goodwillie derivatives are determined by the values of the functors on large dimensional spheres, it follows that the induced map of symmetric sequences

$$\partial_* \Phi_{K(n)} \xrightarrow{c^{K(n)}} \partial_*(\mathrm{TAQ}_{S_{K(n)}}(S_{K(n)}^{(-)})) \tag{7.6}$$

is an equivalence. It follows that the map (7.5) is actually an equivalence of towers for *all* spheres X. The theorem now follows from Theorem 7.11 and (7.2).

8 Consequences

We begin this section by explaining how our result for spheres actually implies that the comparison map is an equivalence on the larger class of finite $\Phi_{K(n)}$-*good spaces*. We also survey some computational applications of our theory, and end the section with some questions.

$\Phi_{K(n)}$-good spaces. We observe that our method of proving Theorem 6.4 actually yields a stronger result.

Theorem 8.1 *For X any finite complex, the comparison map gives an equivalence of towers*

$$\{P_k \Phi_{K(n)}(X)\}_k \xrightarrow[\simeq]{c^{K(n)}} \{F_k \operatorname{TAQ}_{S_{K(n)}}(S^X_{K(n)})\}_k$$

and therefore an equivalence

$$c_X^{K(n)} : P_\infty \Phi_{K(n)}(X) \xrightarrow{\simeq} \operatorname{TAQ}_{S_{K(n)}}(S^X_{K(n)}).$$

Proof This follows from the equivalence (7.6). Note the restriction to finite complexes is necessary as the target functor is not finitary. □

We will say that a space X is $\Phi_{K(n)}$-*good* if the map

$$\Phi_{K(n)}(X) \to \operatorname{holim}_k P_k(\Phi_{K(n)})(X) \tag{8.1}$$

is an equivalence.

Corollary 8.3 *A finite space X is $\Phi_{K(n)}$-good if and only if the comparison map*

$$c_X^{K(n)} : \Phi_{K(n)}(X) \to \operatorname{TAQ}_{S_{K(n)}}(S^X_{K(n)})$$

is an equivalence.

Theorem 7.11 clearly implies spheres are $\Phi_{K(n)}$-good. The functor $\Phi_{K(n)}$ preserves all fiber sequences, but it seems the target of the comparison map is not as robust.

Lemma 8.4 *The functor* $\operatorname{TAQ}_{S_{K(n)}}(S^{(-)}_{K(n)})$ *preserves products of finite spaces.*

Proof This follows from the fact that TAQ is excisive, together with the fact that there is an equivalence of augmented commutative S-algebras

$$S^{X \times Y_+} \simeq S^{X_+} \wedge S^{Y_+}.$$ □

Corollary 8.5 *The product of finite $\Phi_{K(n)}$-good spaces is $\Phi_{K(n)}$-good.*

We shall say that a fiber sequence of finite spaces

$$F \to E \to B$$

is $K(n)$-*cohomologically Eilenberg-Moore* if the map of augmented commutative S-algebras

$$S^{E_+} \wedge_{S^{B_+}} S \to S^{F_+}$$

is a $K(n)$-equivalence. The motivation behind this terminology is that with this condition the associated cohomological Eilenberg-Moore spectral sequence converges [30, Sect. IV.6]

$$\mathrm{Tor}^{*,*}_{K(n)^*(B)}(K(n)^*(F), K(n)^*) \Rightarrow K(n)^*(E).$$

The following lemma follows immediately from the excisivity of TAQ.

Lemma 8.6 *Suppose that*

$$F \to E \to B$$

is a fiber sequence of finite spaces which is $K(n)$-cohomologically Eilenberg-Moore. Then the induced sequence

$$\mathrm{TAQ}_{S_{K(n)}}(S^F_{K(n)}) \to \mathrm{TAQ}_{S_{K(n)}}(S^E_{K(n)}) \to \mathrm{TAQ}_{S_{K(n)}}(S^B_{K(n)})$$

is a fiber sequence.

Since $\Phi_{K(n)}$ preserves fiber sequences, we deduce the following.

Corollary 8.7 *Suppose that*

$$F \to E \to B$$

is a fiber sequence of finite spaces which is $K(n)$-cohomologically Eilenberg-Moore. Then if any two of the spaces in the sequence are $\Phi_{K(n)}$-good, so is the third.

Using this we can give examples of $\Phi_{K(n)}$-good spaces which are not spheres (or finite products of spheres).

Proposition 8.8 *The special unitary groups $SU(k)$ and symplectic groups $Sp(k)$ are $\Phi_{K(n)}$-good.*

Proof For simplicity we treat the special unitary groups; the symplectic case is essentially identical. Petrie [77] showed that additively there is an isomorphism

$$MU_*SU(k) \cong \Lambda_{MU_*}[y_3, y_5, \dots, y_{2k-1}].$$

It follows from the collapsing universal coefficient spectral sequence that there is an additive isomorphism

$$K(n)^*SU(k) \cong \Lambda_{K(n)_*}[x_3, x_5, \dots, x_{2k-1}]. \tag{8.2}$$

The Atiyah-Hirzebruch spectral sequence for $K(n)^*SU(k)$ must therefore collapse (any differentials would otherwise make the rank of $K(n)^*SU(k)$ too small). There are no possible extensions, as the exterior algebra is free as a graded-commutative algebra. Therefore (8.2) is an isomorphism of $K(n)_*$-algebras. This can than be used to show that the fiber sequences

$$SU(k-1) \to SU(k) \to S^{2k-1}$$

are $K(n)$-cohomologically Eilenberg-Moore. The result follows by induction (using Corollary 8.7). □

Not all spaces are $\Phi_{K(n)}$-good. Brantner and Heuts have recently shown that wedges of spheres of dimension greater than 1, and mod p Moore spaces, are examples of non-$\Phi_{K(n)}$-good spaces [11].

Some computations. The target of the comparison map should be regarded as computable, and the source should be regarded as mysterious. Because of this, our theorem has important computational consequences. We take a moment to mention some things that have already been done.

In [22], we show that the Morava E-theory of the layers of the Goodwillie tower for $\Phi_{K(n)}$ evaluated on S^1 are given by the cohomology of the second author's *modular isogeny complex* [84]. Theorem 8.1 was applied by the authors in [22] to compute the Morava E-theory of the attaching maps between the consecutive non-trivial layers of this Goodwillie tower. Iterating the double suspension, these computations then restrict to give an approach to computing the Morava E-theory of the Goodwillie tower of $\Phi_{K(n)}$ evaluated on all odd dimensional spheres.

We envision this as a step in the program of Arone-Mahowald [8, 60] to compute the unstable $v_{K(n)}$-periodic homotopy groups of spheres (and other $\Phi_{K(n)}$-good spaces) using stable $v_{K(n)}$-periodic homotopy groups and Goodwillie calculus. This would generalize a number of known calculations in the case of $n = 1$. These computations include those of Mahowald [65] and Thompson [90] for spheres, and would generalize Bousfield's technology [17, 19, 20], for computations for spherically resolved spaces. Bousfield's theory was applied successfully by Don Davis and his collaborators to compute v_1-periodic homotopy groups of various compact Lie groups (see [28], where the previous work on this subject, by Bendersky, Davis, Mahowald, and Mimura is summarized[11]).

To this end, Zhu has used his explicit computation of the Morava E-theory Dyer-Lashof algebra at $n = 2$ [95] to compute the Morava E-theory of $\Phi_{K(2)}(S^q)$ for q odd [94].

Using our technology, but employing BP-theory instead of Morava E-theory, Wang has computed the groups $v_{K(2)}^{-1}\pi_*(S^3)^\wedge$ for $p \geq 5$ [92]. Wang has also computed the monochromatic Hopf invariants of the β-family at these primes. These are the analogs of the classical Hopf invariants, but computed in the category $M_2^f \mathrm{Top}_*$.

Theorem 8.10 (Wang [91]) *The monochromatic Hopf invariant of $\beta_{i/j,k}$ is $\beta_{i-j/k}$.*

[11]Technically, the previous computations used the unstable Adams-Novikov spectral sequence, but were simplified using Bousfield's results.

Finally, Brantner has recently computed the algebra of power operations which naturally act on the completed E-theory of any spectral Lie algebra (such as those arising as spectral Lie algebra models of unstable v_n-periodic homotopy types) [23].

Some questions. We end this section with some questions.

Question 8.11 *Does the bracket from the s^{-1}Lie-structure on TAQ-coincide with the Whitehead product in unstable v_n-periodic homotopy?*

Question 8.12 *In [20], Bousfield introduces the notion of a $\widehat{K}\Phi$-good space. What is the relationship between this notion and the notion of being $\Phi_{K(1)}$-good?*

Question 8.13 *Is there a relationship to X being $\Phi_{K(n)}$-good and the convergence of X's unstable v_n-periodic $E_n^\wedge$-based Adams spectral sequence to $v_{K(n)}^{-1}\pi_*(X)^\wedge$?*

9 The Arone-Ching Approach

The central component of Goodwillie's theory of homotopy calculus, from which the theory derives much of its computational power, is the idea that the layers of the Goodwillie tower of a functor F are classified by its symmetric sequence of derivatives $\partial_* F$. Arone and Ching have pursued a research program which seeks to endow $\partial_* F$ with enough extra structure to recover the entire Goodwillie tower of F [2–4]. In this section we will focus on the setup of [2], and will describe their approach to give a conceptual alternative proof of Theorem 8.1. *In this section we will only consider homotopy functors*

$$F : \mathcal{C} \to \mathcal{D}$$

where $\mathcal{C}$ and $\mathcal{D}$ are either the categories of pointed spaces or spectra.[12]

Modules over operads. Let $\mathcal{O}$ be a reduced operad in Mod_R, and let $\mathcal{A} = \{\mathcal{A}_i\}$ be a symmetric sequence of R-module spectra. A left (respectively right) module structure on $\mathcal{A}$ is the structure of an associative action

$$\mathcal{O} \circ \mathcal{A} \to \mathcal{A} \qquad (\text{resp. } \mathcal{A} \circ \mathcal{O} \to \mathcal{A}).$$

One similarly has the notion of a left/right comodule structure. Explicitly, a left $\mathcal{O}$-module structure on $\mathcal{A}$ is encoded in structure maps

$$\mathcal{O}_k \wedge_R \mathcal{A}_{n_1} \wedge_R \cdots \wedge_R \mathcal{A}_{n_k} \to \mathcal{A}_{n_1+\cdots+n_k},$$

and a right $\mathcal{O}$-module structure is encoded in structure maps

$$\mathcal{A}_k \wedge_R \mathcal{O}_{n_1} \wedge_R \cdots \wedge_R \mathcal{O}_{n_k} \to \mathcal{A}_{n_1+\cdots+n_k}.$$

[12] Later in this section we will also allow $\mathcal{D}$ to be $\mathrm{Sp}_{T(n)}$.

The structure maps for left/right comodules are obtained simply by reversing the direction of the above arrows.

Suppose that A is an $\mathcal{O}$-algebra. Regarding A as the symmetric sequence

$$(A, *, *, \cdots)$$

with A in the 0th spot, the $\mathcal{O}$-algebra structure on A can also be regarded as a left $\mathcal{O}$-module structure on A. Less obviously, the $\mathcal{O}$-algebra structure can also be encoded in a right *comodule* structure on the symmetric sequence[13]

$$A^{\wedge_R *} := (*, A, A^2, A^3, \cdots).$$

For simplicity, assume that each of the R-module spectra $\mathcal{O}_i$ are strongly dualizible. Then $\mathcal{O}^\vee$ is a cooperad, and the $\mathcal{O}$-algebra structure on $\mathcal{A}$ is encoded in a right $\mathcal{O}^\vee$-comodule structure on $A^{\wedge_R *}$

$$A^{n_1+\cdots+n_k} \to A^k \wedge_R \mathcal{O}^\vee_{n_1} \wedge_R \cdots \wedge_R \mathcal{O}^\vee_{n_k}.$$

These comodule structure maps are adjoint to the maps

$$\mathcal{O}_{n_1} \wedge_R \cdots \wedge_R \mathcal{O}_{n_k} \wedge_R A^{n_1+\cdots+n_k} \to A^k$$

obtained by smashing together k algebra structure maps.

Koszul duality, again. *In this subsection, all symmetric sequences $\mathcal{A}$ are assumed to satisfy $\mathcal{A}_0 = *$.* With this hypothesis, Ching's construction of the cooperad structure on the operadic bar construction

$$B\mathcal{O} = B(1_R, \mathcal{O}, 1_R)$$

extends to give $B\mathcal{O}$-comodule structures [26]. Specifically, suppose that $\mathcal{M}$ is a right $\mathcal{O}$-module. Then

$$B\mathcal{M} := B(\mathcal{M}, \mathcal{O}, 1_R)$$

gets the structure of a right $B\mathcal{O}$-comodule. Similarly, for a left $\mathcal{O}$-module $\mathcal{N}$,

$$B\mathcal{N} := B(1_R, \mathcal{O}, \mathcal{N})$$

gets the structure of a left $B\mathcal{O}$-comodule. There are dual statements which endow cobar constructions of comodules with module structures.

---

[13]It is more natural to define the 0th space of the symmetric sequence $A^{\wedge_R *}$ to be R, but it makes no difference as we are assuming $\mathcal{O}$ is reduced. For the purposes of the rest of the section this convention will be more useful.

In this manner the operadic bar and cobar constructions give functors

$$B : \text{lt.Mod}_{\mathcal{O}} \leftrightarrows \text{lt.Comod}_{B\mathcal{O}} : C,$$
$$B : \text{rt.Mod}_{\mathcal{O}} \leftrightarrows \text{rt.Comod}_{B\mathcal{O}} : C.$$

Some of the key ideas in the following Koszul duality theorem can be found in [2], but a proof of the full statement should appear in [25].

Theorem 9.1 (Ching) *The bar/cobar constructions give an equivalence of homotopy categories of right (co)modules*

$$B : \text{Ho}(\text{rt.Mod}_{\mathcal{O}}) \leftrightarrows \text{Ho}(\text{rt.Comod}_{B\mathcal{O}}) : C.$$

In the case of left modules, the bar construction gives a fully faithful embedding

$$B : \text{Ho}(\text{lt.Mod}_{\mathcal{O}}) \hookrightarrow \text{Ho}(\text{lt.Comod}_{B\mathcal{O}}).$$

Remark 9.2 Ching expects that one should also get an equivalence of homotopy categories for left modules, but presently do not know how to prove this.

Remark 9.3 In both the case of left and right modules, the bar construction induces equivalences of derived mapping spaces

$$\underline{\text{lt./rt.Mod}}_{\mathcal{O}}(\mathcal{M}, \mathcal{N}) \xrightarrow{\simeq} \underline{\text{lt./rt.Comod}}_{B\mathcal{O}}(B\mathcal{M}, B\mathcal{N}).$$

The reader may be startled that the Koszul duality in Theorem 9.1 applies to the full categories of modules, and not some suitable subcategory, and makes no mention of "divided power structures" (as was the case of the instances of Koszul duality of Sect. 3). It seems that one should rather think of Theorem 9.1 as an extension of Koszul duality for (co)operads, rather than Koszul duality for (co)algebras over (co)operads. Indeed, regarding an $\mathcal{O}$-algebra structure on A as a left $\mathcal{O}$-module structure on A, Theorem 9.1 does not apply, as the symmetric sequence $(A, *, *, \cdots)$ does not have trivial 0th spectrum. Theorem 9.1 (with dualizability hypotheses on $\mathcal{O}$) does encode an $\mathcal{O}$-algebra structure on A in a $(B\mathcal{O})^\vee$-comodule structure on $CA^{\wedge_R *}$, but the latter does not translate into anything like a $B\mathcal{O}$-coalgebra structure.

Remark 9.4 Ching does have a *different* Koszul duality Quillen adjunction

$$Q : \text{lt./rt.Comod}_{B\mathcal{O}} \leftrightarrows \text{lt./rt.Mod}_{\mathcal{O}} : \text{Prim} \tag{9.1}$$

which *does not* in general give an equivalence of homotopy categories, but which *does* restrict (in the case of right modules) to give the usual Koszul duality between $(B\mathcal{O})^\vee$-algebras and $\mathcal{O}^\vee$-coalgebras. The monad and comonad of this adjunction encode divided power module and comodule structures, which extend the previously established notions of divided power structures for algebras and coalgebras.

The fake Taylor tower. In [2], Arone and Ching establish that the derivatives of a functor

$$F : \mathcal{C} \to \mathcal{D}$$

have the structure of a $\partial_* \mathrm{Id}_{\mathcal{D}}$-$\partial_* \mathrm{Id}_{\mathcal{C}}$-bimodule. Note that in the case where either $\mathcal{C}$ or $\mathcal{D}$ is the category Sp of spectra, $\partial_* \mathrm{Id}_{\mathrm{Sp}} = 1$, and a left or right $\partial_* \mathrm{Id}_{\mathrm{Sp}}$-module structure amounts to no additional structure.

A key tool, introduced in [2] is the notion of the *fake Taylor tower* of the functor F. The fake Taylor tower is the closest approximation to the Goodwillie tower which can be formed using only the bimodule structure of $\partial_* F$, and is defined as follows.

For $X \in \mathcal{C}$, let R_X denote the corepresentable functor

$$R_X : \mathcal{C} \to \mathcal{D}$$

given by

$$R_X(Z) = [\Sigma^\infty]\underline{\mathcal{C}}(X, Z)$$

(where the Σ^∞ in the above formula is only used if $\mathcal{D} = \mathrm{Sp}$). Then the fake Taylor tower $\{P_n^{\mathit{fake}} F\}$ is the tower of functors under F given by (in the case where X is finite[14])

$$P_n^{\mathit{fake}} F(X) := {}_{\partial_* \mathrm{Id}_{\mathcal{D}}}\underline{\mathrm{Bimod}}_{\partial_* \mathrm{Id}_{\mathcal{C}}}(\partial_* R_X, \tau_n \partial_* F).$$

Here, for a symmetric sequence $\mathcal{A}$, we are letting $\tau_n \mathcal{A}$ denote its nth truncation

$$\tau_n \mathcal{A}_k := \begin{cases} \mathcal{A}_k, & k \le n, \\ *, & k > n. \end{cases} \tag{9.2}$$

With the hypothesis that all symmetric sequences have trivial 0th term, it is easy to see that operad and module structures on $\mathcal{A}$ induce corresponding structures on $\tau_n \mathcal{A}$.

The layers of the fake Taylor tower given by the fibers

$$D_n^{\mathit{fake}} F \to P_n^{\mathit{fake}} F \to P_{n-1}^{\mathit{fake}} F$$

take the form

$$D_n^{\mathit{fake}} F(X) \simeq \Omega_{\mathcal{D}}^\infty \left(\partial_n F \wedge \Sigma_{\mathcal{C}}^\infty X^n\right)^{h\Sigma_n}.$$

The following theorem is essentially proven in [2].

Theorem 9.7 (Arone-Ching) *There is a natural transformation of towers*

$$\{P_n F\} \to \{P_n^{\mathit{fake}} F\}$$

[14]For X infinite, one must regard R_X as a pro-functor.

such that the induced map on fibers is given by the norm map

$$N : \Omega^\infty_{\mathcal{D}} \left(\partial_n F \wedge \Sigma^\infty_{\mathcal{C}} X^n\right)_{h\Sigma_n} \to \Omega^\infty_{\mathcal{D}} \left(\partial_n F \wedge \Sigma^\infty_{\mathcal{C}} X^n\right)^{h\Sigma_n}.$$

Thus, in general, the map from the Goodwillie tower to the fake Taylor tower is not an equivalence, and the difference is measured by the Tate spectra

$$\Omega^\infty_{\mathcal{D}} \left(\partial_n F \wedge \Sigma^\infty_{\mathcal{C}} X^n\right)^{t\Sigma_n}.$$

Although we do not need it for what follows, we pause to mention that Arone and Ching have a refinement of this theory which recovers the Goodwillie tower from descent data on the derivatives. Observe that the fake Taylor tower only depends on the bimodule $\partial_* F$. The following is proven in [3].

Theorem 9.8 (Arone-Ching) *The limit of the fake Taylor tower is right adjoint to the derivatives functor:*

$$\partial_* : \mathrm{Funct}(\mathcal{C}, \mathcal{D}) \leftrightarrows {}_{\partial_*\mathrm{Id}_{\mathcal{D}}}\mathrm{Bimod}_{\partial_*\mathrm{Id}_{\mathcal{C}}} : P^{fake}_\infty.$$

In particular, one can now employ the comonadic descent theory of Sect. 2 to regard the derivatives as taking values in $\partial_* \circ P^{fake}_\infty$-comodules.

Theorem 9.9 (Arone-Ching [3]) *The Goodwillie tower of a functor F can be recovered using the comonadic cobar construction*

$$P_n F \simeq C(P^{fake}_\infty, \partial_* \circ P^{fake}_\infty, \tau_n \partial_* F).$$

In the case of functors from spectra to spectra, this theorem reduces to McCarthy's classification of polynomial functors [68].

Application to the Bousfield–Kuhn functor. We now summarize Arone and Ching's approach to Theorem 8.1. Actually, their method proves something stronger, as it applies to the functor Φ_n instead of $\Phi_{K(n)}$. Call a space Φ_n-*good* if the map

$$\Phi_n X \to \operatorname*{holim}_k \Phi_n P_k \mathrm{Id}_{\mathrm{Top}_*}(X)$$

is an equivalence.

Theorem 9.10 (Arone-Ching) *For all finite X, the comparison map*

$$c_X : P_\infty \Phi_n(X) \to \mathrm{TAQ}_{S_{T(n)}}(S^X_{T(n)})$$

is an equivalence. Thus for all finite Φ_n-good spaces, the comparison map gives an equivalence

$$c_X : \Phi_n(X) \xrightarrow{\simeq} \mathrm{TAQ}_{S_{T(n)}}(S^X_{T(n)}).$$

Proof The basic strategy is to analyze the fake taylor tower of the functor

$$\Phi_n : \mathrm{Top}_* \to \mathrm{Sp}_{T(n)}.$$

The argument used in Lemma 7.1 applies equally well to Φ_n, and it follows that we have

$$\partial_* \Phi_n \simeq s^{-1}\mathrm{Lie}_{T(n)}$$

with right $\partial_* \mathrm{Id} = s^{-1}\mathrm{Lie}$ structure given by localization of the right action of this operad on itself. By Theorem 5.2, the map

$$\begin{aligned} D_k \Phi_n(X) &= \left[\left((s^{-1}\mathrm{Lie}_k)_{T(n)} \wedge X^k \right)_{h\Sigma_k} \right]_{T(n)} \\ &\xrightarrow{N} \left((s^{-1}\mathrm{Lie}_k)_{T(n)} \wedge X^k \right)^{h\Sigma_k} \\ &= D_k^{fake} \Phi_n(X) \end{aligned}$$

of Theorem 9.7 is an equivalence. Thus in the $T(n)$-local context, the fake Taylor tower agrees with the Goodwillie tower. Using [3, Lemma 6.14] and Theorem 9.1, we have

$$\begin{aligned} P_\infty \Phi_n(X) &\simeq \underline{\mathrm{rt.Mod}}_{s^{-1}\mathrm{Lie}}(\partial_* R_X, s^{-1}\mathrm{Lie}_{T(n)}) \\ &\simeq \underline{\mathrm{rt.Mod}}_{s^{-1}\mathrm{Lie}}(B(\Sigma^\infty X^{\wedge *}, \mathrm{Comm}, 1)^\vee, s^{-1}\mathrm{Lie}_{T(n)}) \\ &\simeq \underline{\mathrm{rt.Mod}}_{s^{-1}\mathrm{Lie}}\left(C(S_{T(n)}^{X^{\wedge *}}, \mathrm{Comm}_{T(n)}^\vee, 1), C(1, \mathrm{Comm}_{T(n)}^\vee, 1) \right) \\ &\simeq \underline{\mathrm{rt.Comod}}_{\mathrm{Comm}^\vee}(S_{T(n)}^{X^{\wedge *}}, 1_{S_{T(n)}}) \\ &\simeq \mathrm{Alg}_{\mathrm{Comm}}(S_{T(n)}^X, \mathrm{triv} S_{T(n)}) \\ &\simeq \mathrm{TAQ}_{S_{T(n)}}(S_{T(n)}^X). \end{aligned}$$

□

10 The Heuts Approach

The approach of Arone and Ching described in the last section arose from a classification theory of Goodwillie towers. In this section we describe Heuts' general theoretical framework, which arises from classifying unstable homotopy theories with a fixed stabilization [45]. Our goal is simply to give enough of the idea of the theory to sketch Heuts' proof of Theorem 8.1. We refer the reader to the source material for a proper and more rigorous treatment.

Like the approach of Arone-Ching, Heuts' proof is more conceptual than ours, and his results have the potential to be slightly more general that Theorem 9.10, in that they seem to indicate that by modifying the comparison map to have target derived primitives of a coalgebra, the comparison map c_X may be an equivalence for *all* Φ_n-good spaces (not just finite spaces—see Question 10.19 and Remark 10.20).

Unlike the previous sections, where we worked in a setting of actual categories with weak equivalences, in this section we work in the setting of ∞-categories. For the purposes of this section, $\mathcal{C}$ will always denote an arbitrary pointed compactly generated ∞-category.

∞-operads and cross-effects. The adjunction

$$\Sigma^\infty_{\mathcal{C}} : \mathcal{C} \leftrightarrows \mathrm{Sp}(\mathcal{C}) : \Omega^\infty_{\mathcal{C}}$$

gives rise to a comonad $\Sigma^\infty_{\mathcal{C}}\Omega^\infty_{\mathcal{C}}$ on $\mathrm{Sp}(\mathcal{C})$. Lurie [63] observes that the multilinearized cross effects

$$\otimes^n_{\mathcal{C}} := \mathrm{cr}^{lin}_n(\Sigma^\infty_{\mathcal{C}}\Omega^\infty_{\mathcal{C}}) : \mathrm{Sp}(\mathcal{C})^n \to \mathrm{Sp}(\mathcal{C})$$

get an additional piece of algebraic structure: they corepresent a symmetric multicategory structure on $\mathrm{Sp}(\mathcal{C})$ in the sense that the mapping spaces

$$\underline{\mathrm{Sp}(\mathcal{C})}\left(\otimes^n_{\mathcal{C}}(Y_1, \dots, Y_n), Y\right)$$

endow $\mathrm{Sp}(\mathcal{C})$ with the structure of a symmetric multicategory enriched in spaces.

If $\mathrm{Sp}(\mathcal{C}) \simeq \mathrm{Sp}$, then (as discussed in the beginning of Sect. 7) we have

$$\otimes^n_{\mathcal{C}}(Y_1, \dots, Y_n) \simeq \partial_n(\Sigma^\infty_{\mathcal{C}}\Omega^\infty_{\mathcal{C}}) \wedge Y_1 \wedge \cdots \wedge Y_n.$$

Saying that the cross-effects $\otimes^n_{\mathcal{C}}$ corepresent a symmetric multicategory is equivalent to saying that the derivatives $\partial_*(\Sigma^\infty_{\mathcal{C}}\Omega^\infty_{\mathcal{C}})$ form a cooperad. In this context, this fact was first observed by Arone and Ching [2], who proved that the derivatives of any comonad on Sp form a cooperad.

Remark 10.1 In the language of Lurie, $(\mathrm{Sp}(\mathcal{C}), \otimes^*_{\mathcal{C}})$ forms a *stable ∞-operad*. This terminology comes from the fact that a symmetric multicategory is the same thing as a (colored) operad. We will deliberately avoid this terminology in our treatment, as it may seem somewhat confusing that a stable ∞-operad on Sp is encoded by a *co*operad in Sp.

The linearizations of the diagonals in $\mathcal{C}$

$$\Delta^n : X \to X^{\times n}$$

gives rise to Σ_n-equivariant maps

$$\Delta^n : \Sigma^\infty_{\mathcal{C}} X \to \otimes^n_{\mathcal{C}}(\Sigma^\infty_{\mathcal{C}} X, \cdots, \Sigma^\infty_{\mathcal{C}} X) =: (\Sigma^\infty_{\mathcal{C}} X)^{\otimes_{\mathcal{C}} n}$$

which yield maps

$$\Delta^n : \Sigma^\infty_{\mathcal{C}} X \to \left((\Sigma^\infty_{\mathcal{C}} X)^{\otimes_{\mathcal{C}} n}\right)^{h\Sigma_n}.$$

Composing out to the Tate spectrum gives maps

$$\delta^n_{\mathcal{C}} : \Sigma^\infty_{\mathcal{C}} X \to \left((\Sigma^\infty_{\mathcal{C}} X)^{\otimes_{\mathcal{C}} n}\right)^{t\Sigma_n}. \tag{10.1}$$

Heuts [45] refers to these maps as *Tate diagonals*. In the context of $\mathcal{C} = \mathrm{Top}_*$, these natural transformations are well studied: their target is closely related to Jones-Wegmann homology (see [15, II.3]) and the topological Singer construction of Lunøe-Nielsen-Rognes [62].

Polynomial approximations of ∞-categories. Heuts constructs *polynomial approximations $P_n\mathcal{C}$*: these are ∞-categories equipped with adjunctions

$$\Sigma^\infty_{\mathcal{C},n} : \mathcal{C} \leftrightarrows P_n\mathcal{C} : \Omega^\infty_{\mathcal{C},n}$$

so that

$$P_n\mathrm{Id}_{\mathcal{C}}(X) \simeq \Omega^\infty_{\mathcal{C},n}\Sigma^\infty_{\mathcal{C},n} X.$$

The ∞-categories $P_n\mathcal{C}$ are determined by universal properties which we will not specify here. We do point out that the identity functor $\mathrm{Id}_{P_n\mathcal{C}}$ is n-excisive. We have $P_1\mathcal{C} \simeq \mathrm{Sp}(\mathcal{C})$. For $n \le m$ we have

$$P_nP_m\mathcal{C} \simeq P_n\mathcal{C}$$

and therefore we get a tower

$$\begin{array}{ccccc} \mathcal{C} & & & & \\ {\scriptstyle \Sigma^\infty_{\mathcal{C},1}}\downarrow & \searrow^{\Sigma^\infty_{\mathcal{C},2}} & & & \\ P_1\mathcal{C} & \xleftarrow[\Sigma^\infty_{P_2\mathcal{C},1}]{} & P_2\mathcal{C} & \xleftarrow[\Sigma^\infty_{P_3\mathcal{C},2}]{} & \cdots \end{array}$$

We shall say that an object X of $\mathcal{C}$ is *convergent* if the Goodwillie tower of $\mathrm{Id}_{\mathcal{C}}$ converges at X. Heuts proves that the induced functor

$$\mathcal{C} \to P_\infty\mathcal{C} := \operatorname*{holim}_n P_n\mathcal{C}$$

restricts to a full and faithful embedding on the full ∞-subcategory $\mathcal{C}^{\mathrm{conv}}$ of convergent objects.

Let $\mathcal{C}^{\text{n-conv}}$ denote the full ∞-subcategory of $\mathcal{C}$ consisting of objects for which the map

$$X \to P_n\mathrm{Id}_{\mathcal{C}}(X)$$

is an equivalence. Then we have

Lemma 10.3 *The functor*

$$\Sigma^\infty_{\mathcal{C},n} : \mathcal{C}^{\text{n-conv}} \to P_n\mathcal{C}$$

is fully faithful.

Proof We have for X and Y in $\mathcal{C}^{\text{n-conv}}$:

$$\begin{aligned}\underline{\mathcal{C}}(X, Y) &\simeq \underline{\mathcal{C}}(X, \Omega^\infty_{\mathcal{C},n}\Sigma^\infty_{\mathcal{C},n}Y)\\ &\simeq \underline{P_n\mathcal{C}}(\Sigma^\infty_{\mathcal{C},n}X, \Sigma^\infty_{\mathcal{C},n}Y).\end{aligned}$$

□

The natural transformations

$$\Sigma^\infty_{\mathcal{C}}\Omega^\infty_{\mathcal{C}} \simeq \Sigma^\infty_{P_n\mathcal{C}}\Sigma^\infty_{\mathcal{C},n}\Omega^\infty_{\mathcal{C},n}\Omega^\infty_{P_n\mathcal{C}} \to \Sigma^\infty_{P_n\mathcal{C}}\Omega^\infty_{P_n\mathcal{C}}$$

induce natural transformations of cross-effects

$$\otimes^k_{\mathcal{C}} \to \otimes^k_{P_n\mathcal{C}}.$$

For $k \leq n$ these natural transformations are equivalences.

As the source and target of the Tate diagonals (10.1) are $(n-1)$-excisive functors of X (see [58]), the Tate diagonals extend to give natural transformations of functors $P_{n-1}\mathcal{C} \to \mathrm{Sp}(\mathcal{C})$:

$$\delta^n_{\mathcal{C}} : \Sigma^\infty_{P_{n-1}\mathcal{C}}X \to \left((\Sigma^\infty_{P_{n-1}\mathcal{C}}X)^{\otimes_{\mathcal{C}} n}\right)^{t\Sigma_n}.$$

We emphasize that, as the notation suggests, the Tate diagonals $\{\delta^n_{\mathcal{C}}\}_n$ depend not only on the functors $\otimes^*_{\mathcal{C}}$ on $\mathrm{Sp}(\mathcal{C})$, but also on the unstable category $\mathcal{C}$ itself.

A spectral algebra model for $P_n\mathcal{C}$. Heuts gives a model for $P_n\mathcal{C}$ as a certain category of coalgebras in $\mathrm{Sp}(\mathcal{C})$. As the theory of homotopy descent of Sect. 2 would have us believe, a good candidate spectral algebra model would be to consider $\Sigma^\infty_{P_n\mathcal{C}}\Omega^\infty_{P_n\mathcal{C}}$-coalgebras. We must analyze what it means for $Y \in \mathrm{Sp}(\mathcal{C})$ to have a coalgebra structure map

$$Y \to \Sigma^\infty_{P_n\mathcal{C}}\Omega^\infty_{P_n\mathcal{C}}Y.$$

This is closely related to having a structure map

$$Y \to P_n(\Sigma^\infty_{\mathcal{C}}\Omega^\infty_{\mathcal{C}})Y.$$

A general theorem of McCarthy [68], as formulated by [58],[15] applies to the functor $\Sigma^\infty_{\mathcal{C}}\Omega^\infty_{\mathcal{C}}$ to give a homotopy pullback

[15]To be precise, this is established by McCarthy and Kuhn in the case where $\mathcal{C} = \mathrm{Top}_*$.

$$\begin{array}{ccc} P_n(\Sigma^\infty_C \Omega^\infty_C)(Y) & \longrightarrow & (Y^{\otimes_C n})^{h\Sigma_n} \\ \downarrow & & \downarrow \\ P_{n-1}(\Sigma^\infty_C \Omega^\infty_C)(Y) & \longrightarrow & (Y^{\otimes_C n})^{t\Sigma_n} \end{array}$$

Thus inductively a $\Sigma^\infty_{P_nC}\Omega^\infty_{P_nC}$-coalgebra is determined by the data of a map

$$Y \to P_{n-1}(\Sigma^\infty_C \Omega^\infty_C)(Y)$$

and a lifting[16]

$$\begin{array}{ccccc} & & & \dashrightarrow & (Y^{\otimes_C n})^{h\Sigma_n} \\ & & & & \downarrow \\ Y & \longrightarrow & P_{n-1}(\Sigma^\infty_C \Omega^\infty_C)(Y) & \longrightarrow & (Y^{\otimes_C n})^{t\Sigma_n} \end{array}$$

The bottom composite agrees with the Tate diagonal δ^n_C for $Y = \Sigma^\infty_{C,n-1}X$.

We will refer to these coalgebras as *Tate-compatible* $\otimes^{\leq n}_C$*-coalgebras*, and denote the ∞-category of such

$$\mathrm{TateCoalg}_{\otimes^{\leq n}_C}.$$

Roughly speaking, a Tate-compatible $\otimes^{\leq n}_C$-coalgebra is an object $Y \in \mathrm{Sp}(C)$ equipped with inductively defined structure consisting of coaction maps

$$\Delta^k : Y \to (Y^{\otimes_C k})^{h\Sigma_k}$$

for $k \leq n$, and homotopies H_k making the following diagrams homotopy commute

$$\begin{array}{ccc} & & (Y^{\otimes_C n})^{h\Sigma_k} \\ & \overset{\Delta^k}{\nearrow} & \downarrow \\ Y & \underset{\delta^k_C}{\longrightarrow} & (Y^{\otimes_C k})^{t\Sigma_k} \end{array}$$

The coaction maps Δ^k and the homotopies H_k are required to satisfy compatibility conditions which we will not (and likely cannot!) explicitly specify.[17] The maps Δ^k and homotopies H_k for $k \leq n$ then induce the $(n+1)$st Tate diagonal

$$\delta^{n+1}_C : Y \to (Y^{\otimes_C n+1})^{t\Sigma_{n+1}}$$

---

[16]This is something the first author learned from Arone.

[17]Heuts is able to circumvent the need to explicitly spell out these compatibility conditions by defining the ∞-categories $\mathrm{TateCoalg}_{\otimes^{\leq n}_C}$ via an inductive sequence of fibrations of ∞-categories.

and the process continues. Note that the Tate diagonal $\delta_{\mathcal{C}}^{n+1}$ depends not only on the structure maps Δ^k and H_k for $k \leq n$, but also the unstable category $\mathcal{C}$ itself (more precisely, it depends only on the polynomial approximation $P_n\mathcal{C}$).

Theorem 10.4 (Heuts) *There is an equivalence of ∞-categories*

$$P_n\mathcal{C} \simeq \mathrm{TateCoalg}_{\otimes_{\mathcal{C}}^{\leq n}}.$$

Question 10.5 *In the case where $F = \mathrm{Id}$, how is Arone-Ching's reconstruction theorem (Theorem 9.9) related to the framework of Heuts?*

Remark 10.6 In [45], Heuts also considers the question: what data on the stable ∞-category $\mathrm{Sp}(\mathcal{C})$ determines the tower of unstable categories $\{P_n\mathcal{C}\}$? As should be heuristically clear from Theorem 10.4, Heuts proves the tower is determined by the cross-effects $\{\otimes_{\mathcal{C}}^n\}$ and the Tate diagonals $\{\delta_{\mathcal{C}}^n\}$. In particular, given a stable ∞-category $\mathcal{D}$, a tower of polynomial approximations of an unstable theory is determined by specifying a sequence of symmetric multilinear functors

$$\otimes^n : \mathcal{D}^n \to \mathcal{D}$$

which corepresent a symmetric multicategory structure on $\mathcal{D}$, as well as a sequence of inductively defined (and suitably compatible) Tate diagonals

$$\delta^n : \Sigma^\infty_{P_{n-1}\mathcal{C}} X \to (\Sigma^\infty_{P_{n-1}\mathcal{C}} X^{\otimes n})^{t\Sigma_n}.$$

Koszul duality, yet again. Let R be a commutative ring spectrum, and let $\mathcal{O}$ be a reduced operad in Mod_R. Following [45], we run the general theory in the case $\mathcal{C} = \mathrm{Alg}_{\mathcal{O}}$. The cooperads representing the symmetric multilinear functors $\otimes^*_{\mathrm{Alg}_{\mathcal{O}}}$ on $\mathrm{Sp}(\mathrm{Alg}_{\mathcal{O}}) \simeq \mathrm{Mod}_R$ are determined by the following

Theorem 10.7 (Francis-Gaitsgory [31, Lem. 3.3.4]) *There is an equivalence of cooperads*[18]

$$\partial_*(\Sigma^\infty_{\mathrm{Alg}_{\mathcal{O}}} \Omega^\infty_{\mathrm{Alg}_{\mathcal{O}}}) \simeq B\mathcal{O}.$$

Therefore a $\otimes^*_{\mathrm{Alg}_{\mathcal{O}}}$-coalgebra A is simply a $B\mathcal{O}$-coalgebra. The Tate diagonals on Mod_R turn out to be null in this case, so a Tate compatible structure on a $B\mathcal{O}$-coalgebra A is a compatible choice of liftings of the coaction maps

$$\begin{array}{ccc} & & (B\mathcal{O}_i \wedge_R A^i)_{h\Sigma_i} \\ & \nearrow & \downarrow \\ A & \longrightarrow & (B\mathcal{O}_i \wedge_R A^i)^{h\Sigma_i} \end{array}$$

[18] This relies on the treatment of Koszul duality of monoids in [63]. In Lurie's ∞-categorical treatment, the coalgebra structure on $B\mathcal{O}$ making this theorem true is only coherently homotopy associative. Presumably it can be strictified to an actual point-set level operad structure on a model of $B\mathcal{O}$, but the authors are not knowledgeable enough to know the feasibility of this, nor do they know if this cooperad structure is equivalent to that of Ching [26].

Thus a Tate compatible structure is the same thing as a divided power structure (or perhaps one can take this as a definition of a divided power structure). We shall denote the ∞-category of such (with structure maps as above for $i \le n$) by $\text{d.p.Coalg}_{B\mathcal{O}^{\le n}}$.

Theorem 10.8 (Heuts) *There are equivalences of ∞-categories*

$$P_n \text{Alg}_{\mathcal{O}} \simeq \text{d.p.Coalg}_{B\mathcal{O}^{\le n}}.$$

Heuts recovers the following weak Koszul duality result.

Corollary 10.9 (Heuts) *There is a fully faithful embedding*

$$\text{TAQ}^{\mathcal{O}} : \text{Alg}_{\mathcal{O}}^{\text{conv}} \hookrightarrow \operatorname*{holim}_n \text{d.p.Coalg}_{B\mathcal{O}^{\le n}}.$$

To determine the convergent objects of $\text{Id}_{\text{Alg}_{\mathcal{O}}}$, it is helpful to know the structure of this Goodwillie tower. The following result was suggested by Harper and Hess [46], was proven in the case of the commutative operad by Kuhn [59], and was proven by Pereira [75].

Theorem 10.10 (Pereira) *The Goodwillie tower of* $\text{Id}_{\text{Alg}_{\mathcal{O}}}$ *is given by*

$$P_n \text{Id}_{\text{Alg}_{\mathcal{O}}}(A) = B(\mathcal{F}_{\tau_n \mathcal{O}}, \mathcal{F}_{\mathcal{O}}, A).$$

Here $\tau_n \mathcal{O}$ denotes the truncation (9.2).

In particular, connectivity estimates of Harper and Hess [46] imply that if R and $\mathcal{O}$ are connective, and A is connected, then A is convergent. Thus Corollary 10.9 recovers half of Theorem 3.8. Another important case are operads for which $\mathcal{O} = \tau_n \mathcal{O}$. Then every $\mathcal{O}$-algebra is convergent, and Corollary 10.9 recovers a theorem of Cohn.

Application to unstable v_n-periodic homotopy. To recover and generalize Theorem 8.1, Heuts applies his general framework to the unstable v_n-periodic homotopy category. Unfortunately, the ∞-category modeling $M_n^f \text{Top}_*$ of Sect. 5 seems to fail to be compactly generated. To rectify this, Heuts works with a slightly different ∞-category, which we will denote $v_n^{-1}\text{Top}_*$. This is the full ∞-subcategory of $L_n^f \text{Top}_*$ consisting of colimits of finite $(d_n - 1)$-connected type n complexes. The categories $v_n^{-1}\text{Top}_*$ and $M_n^f \text{Top}_*$ are very closely related. The Bousfield–Kuhn functor factors as

$$\begin{array}{ccc} \text{Top}_* & \xrightarrow{\quad \Phi_n \quad} & \text{Sp}_{T(n)} \\ & \searrow \qquad \nearrow_{\Phi_n'} & \\ & v_n^{-1}\text{Top}_* & \end{array} \tag{10.2}$$

and detects the equivalences in $v_n^{-1}\text{Top}_*$.

We have $\mathrm{Sp}(v_n^{-1}\mathrm{Top}_*) \simeq \mathrm{Sp}_{T(n)}$. The multilinear cross-effects are given by the commutative cooperad:

$$\partial_*(\Sigma^\infty\Omega^\infty) \simeq \mathrm{Comm}^\vee.$$

In this context Theorem 5.2 implies that the Tate diagonals are trivial, and Tate compatible commutative coalgebras are the same thing as commutative coalgebras. Heuts deduces (using Theorems 10.4 and 10.8):

Theorem 10.12 (Heuts) *There are equivalences of ∞-categories*

$$P_k(v_n^{-1}\mathrm{Top}_*) \simeq \mathrm{Coalg}_{\mathrm{Comm}^{\leq k}}(\mathrm{Sp}_{T(n)}) \simeq P_k(\mathrm{Alg}_{s^{-1}\mathrm{Lie}}(\mathrm{Sp}_{T(n)})).$$

In a sense made precise in the corollary below, this gives two spectral algebra models of $\mathcal{C}$.

Corollary 10.13 *There are fully faithful embeddings of ∞-categories*

$$(v_n^{-1}\mathrm{Top}_*)^{\mathrm{conv}} \hookrightarrow \operatorname*{holim}_k \mathrm{Coalg}_{(\mathrm{Comm}^\vee)^{\leq k}}(\mathrm{Sp}_{T(n)}),$$
$$(v_n^{-1}\mathrm{Top}_*)^{\mathrm{conv}} \hookrightarrow P_\infty \mathrm{Alg}_{s^{-1}\mathrm{Lie}}(\mathrm{Sp}_{T(n)}).$$

We can be explicit about the functors giving these spectral algebra models. In general there is an adjunction

$$\mathrm{triv} : \mathrm{Mod}_R \leftrightarrows \mathrm{Coalg}_{\mathrm{Comm}_R^\vee} : \mathrm{Prim}$$

where triv Y is the coalgebra with trivial coproduct, and $\mathrm{Prim}(A)$ is the *derived primitives* of a coalgebra A, given by the comonadic cobar construction:

$$\mathrm{Prim}(A) := C(\mathrm{Id}, \mathcal{F}_{\mathrm{Comm}_R^\vee}, A).$$

For A a Comm_R-algebra finite as an R-module, we have

$$\mathrm{TAQ}_R(A) \simeq \mathrm{Prim}(A^\vee). \tag{10.3}$$

Ching's work endows $\mathrm{Prim}(A)$ with the structure of an s^{-1}Lie-algebra.

The functors of Theorem 10.12 are induced from the functors

$$v_n^{-1}\mathrm{Top}_* \xrightarrow{(\Sigma^\infty -)_{T(n)}} \mathrm{Coalg}_{\mathrm{Comm}^\vee}(\mathrm{Sp}_{T(n)}) \xrightarrow{\mathrm{Prim}} \mathrm{Alg}_{s^{-1}\mathrm{Lie}}(\mathrm{Sp}_{T(n)}).$$

An argument following the same lines as Sect. 6 gives a refined comparison map

$$\widetilde{c}_X : \Phi_n(X) \to \mathrm{Prim}(\Sigma^\infty X)_{T(n)}.$$

Under the equivalence (10.3), this agrees with the comparison map c_X for X finite, and for such X gives $c_X^{K(n)}$ after $K(n)$-localization. From Theorem 10.12, Heuts deduces that for a space X, the comparison map refines to an equivalence of towers

$$\widetilde{c}_X : \Phi_n P_k \mathrm{Id}_{\mathrm{Top}_*} X \xrightarrow{\simeq} \mathrm{Prim}\, \Omega^\infty_{\mathrm{Coalg},k} \Sigma^\infty_{\mathrm{Coalg},k} (\Sigma^\infty X)_{T(n)}. \tag{10.4}$$

Using Theorem 7.11, Heuts obtains the following refinement of Theorem 6.4.

Corollary 10.16 (Heuts) *The comparison map $\widetilde{c}_X$ is an equivalence for X a sphere.*

Question 10.17 *What is the relationship between the ∞-subcategory $(v_n^{-1}\mathrm{Top}_*)^{\mathrm{conv}} \subseteq v_n^{-1}\mathrm{Top}_*$ and the ∞-subcategory consisting of the images of Φ_n-good spaces?*

Remark 10.18 If we knew that the functor Φ'_n of (10.2) preserved homotopy limits, then it is fairly easy to check (using the fact that Φ'_n detects equivalences) that the two ∞-subcategories of Question 10.17 would in fact coincide. As already remarked in Sect. 5, Φ_n also factors through a related functor

$$\Phi''_n : M_n^f \mathrm{Top}_* \to \mathrm{Sp}_{T(n)}.$$

Bousfield produces a left adjoint for Φ''_n in [18], and it therefore follows that Φ''_n commutes with homotopy limits.

It would seem that for X an infinite CW complex, the coalgebra $(\Sigma^\infty X)_{T(n)}$ is a more appropriate model for the unstable v_n-periodic homotopy type X than the algebra $S^X_{T(n)}$. To this end we ask the following

Question 10.19 *Is $\widetilde{c}_X$ an equivalence for* all *Φ_n-good spaces X?*

Remark 10.20 We expect the answer to Question 10.19 should be "yes", as the tower which is the target of (10.4) should be an analog for primitives of the Kuhn filtration, and hence should converge without hypotheses.

Acknowledgements The authors benefited greatly from conversations with Greg Arone, Michael Ching, Bill Dwyer, Rosona Eldred, Sam Evans, John Francis, John Harper, Gijs Heuts, Mike Hopkins, Nick Kuhn, Jacob Lurie, Mike Mandell, Akhil Mathew, Anibal Medina, Lennart Meier, Luis Alexandre Pereira, and Yifei Zhu. The authors are grateful to Norihiko Minami for encouraging this submission to these conference proceedings, honoring the memory of Tetsusuke Ohkawa. The authors would also like to thank the referee for his/her many useful comments and corrections. Both authors were supported by grants from the NSF.

References

1. Arone, G., Ching, M.: Localized Taylor Towers (in Preparation)
2. Arone, G., Ching, M.: Operads and chain rules for the calculus of functors. Astérisque, no. 338, vi+158 (2011)
3. Arone, G., Ching, M.: A classification of Taylor towers of functors of spaces and spectra. Adv. Math. **272**, 471–552 (2015)

4. Arone, G., Ching, M.: Cross-effects and the classification of Taylor towers. Geom. Topol. **20**(3), 1445–1537 (2016)
5. Ayala, D., Francis, J.: Zero-pointed manifolds. arXiv:1409.2857 (2015)
6. Arone, G., Kankaanrinta, M.: A functorial model for iterated Snaith splitting with applications to calculus of functors. Stable and unstable homotopy (Toronto, ON, 1996), Fields Inst. Commun. Am. Math. Soc. Providence, RI, **19**, 1–30 (1998)
7. Arone, G., Kankaanrinta, M.: The sphere operad. Bull. Lond. Math. Soc. **46**(1), 126–132 (2014)
8. Arone, G., Mahowald, M.: The Goodwillie tower of the identity functor and the unstable periodic homotopy of spheres. Invent. Math. **135**(3), 743–788 (1999)
9. Basterra, M.: André-Quillen cohomology of commutative S-algebras. J. Pure Appl. Algebr. **144**(2), 111–143 (1999)
10. Blomquist, J., Harper, J.: An integral chains analog of Quillen's rational homotopy theory equivalence. arXiv:1611.04157 (2016)
11. Brantner, L., Heuts, G.: The v_n-periodic Goodwillie tower on wedges and cofibers. arXiv:1612.02694 (2016)
12. Bauer, K., Johnson, B., McCarthy, R.: Cross effects and calculus in an unbased setting. Trans. Am. Math. Soc. **367**(9), 6671–6718 (With an appendix by Rosona Eldred) (2015)
13. Basterra, M., McCarthy, R.: Γ-homology, topological André-Quillen homology and stabilization. Topol. Appl. **121**(3), 551–566 (2002)
14. Basterra, M., Mandell, M.A.: Homology and cohomology of E_∞ ring spectra. Math. Z. **249**(4), 903–944 (2005)
15. Bruner, R.R., May, J.P., McClure, J.E., Steinberger, M.: H_∞ Ring Spectra and Their Applications. Lecture Notes in Mathematics, vol. 1176. Springer, Berlin (1986)
16. Bousfield, A.K.: Localization and periodicity in unstable homotopy theory. J. Am. Math. Soc. **7**(4), 831–873 (1994)
17. Bousfield, A.K.: The K-theory localizations and v_1-periodic homotopy groups of H-spaces. Topology **38**(6), 1239–1264 (1999)
18. Bousfield, A.K.: On the telescopic homotopy theory of spaces. Trans. Am. Math. Soc. **353**(6), 2391–2426 (2001) (electronic)
19. Bousfield, A.K.: On the 2-primary v_1-periodic homotopy groups of spaces. Topology **44**(2), 381–413 (2005)
20. Bousfield, A.K.: On the 2-adic K-localizations of H-spaces. Homol. Homotopy Appl. **9**(1), 331–366 (2007)
21. Biedermann, G., Röndigs, O.: Calculus of functors and model categories, II. Algebr. Geom. Topol. **14**(5), 2853–2913 (2014)
22. Behrens, M., Rezk, C.: The Bousfield-Kuhn functor and topological André-Quillen cohomology. Available at www.nd.edu/~mbehren1/papers (2015)
23. Brantner, L.: The Lubin-Tate Theory of Spectral Lie Algebras (in Preparation)
24. Ching, M., Harper, J.E.: Derived koszul duality and TQ-homology completion of structured ring spectra. arXiv:1502.06944 (2015)
25. Ching, M.: Koszul duality for modules and comodules over operads of spectra (in Preparation)
26. Ching, M.: Bar constructions for topological operads and the Goodwillie derivatives of the identity. Geom. Topol. **9**, 833–933 (2005) (electronic)
27. Clausen, D., Mathew, A.: A short proof of telescopic Tate vanishing. Proc. Am. Math. Soc. (2017) (To appear)
28. Davis, D.M.: From representation theory to homotopy groups. Mem. Am. Math. Soc. **160**(759), viii+50 (2002)
29. Dwyer, W.G., Hirschhorn, P.S., Kan, D.M., Smith, J.H.: Homotopy Limit Functors on Model Categories and Homotopical Categories. Mathematical Surveys and Monographs, vol. 113. American Mathematical Society, Providence, RI (2004)
30. Elmendorf, A.D., Kriz, I., Mandell, M.A., May, J.P.: Rings, Modules, and Algebras in Stable Homotopy Theory. Mathematical Surveys and Monographs, vol. 47. American Mathematical Society, Providence, RI, With an appendix by M. Cole (1997)
31. Francis, J., Gaitsgory, D.: Chiral Koszul duality. Selecta Math. (N.S.) **18**(1), 27–87 (2012)

32. Fresse, B.: On the homotopy of simplicial algebras over an operad. Trans. Am. Math. Soc. **352**(9), 4113–4141 (2000)
33. Fresse, B.: Koszul duality of operads and homology of partition posets. Homotopy theory: relations with algebraic geometry, group cohomology, and algebraic K-theory. Contemp. Math. Am. Math. Soc., vol. 346, pp. 115–215, Providence, RI (2004)
34. French, J.: Derived mapping spaces as models for localizations. Ph.D. thesis, M.I.T. (2010)
35. Goerss, P.G., Hopkins, M.J.: André-Quillen (co)-homology for simplicial algebras over simplicial operads, Une dégustation topologique [Topological morsels]: homotopy theory in the Swiss Alps (Arolla, 1999). Contemp. Math. Am. Math. Soc. Providence, RI **265**, 41–85 (2000)
36. Getzler, E., Jones, J.D.S.: Operads, homotopy algebra and iterated integrals for double loop spaces. arXiv:hep-th/9403055
37. Ginzburg, V., Kapranov, M.: Koszul duality for operads. Duke Math. J. **76**(1), 203–272 (1994)
38. Greenlees, J.P.C., May, J.P.: Generalized Tate cohomology. Mem. Am. Math. Soc. **113**(543), viii+178 (1995)
39. Goerss, P.G.: Simplicial chains over a field and p-local homotopy theory. Math. Z. **220**(4), 523–544 (1995)
40. Goodwillie, T.G.: Calculus. III. Taylor series. Geom. Topol. **7**, 645–711 (2003) (electronic)
41. Greenlees, J.P.C., Sadofsky, H.: The Tate spectrum of v_n-periodic complex oriented theories. Math. Z. **222**(3), 391–405 (1996)
42. Harper, J.E.: Bar constructions and Quillen homology of modules over operads. Algebr. Geom. Topol. **10**(1), 87–136 (2010)
43. Hess, K.: A general framework for homotopic descent and codescent. arXiv:1001.1556 (2010)
44. Heuts, G.: Periodicity in unstable homotopy (in Preparation)
45. Heuts, G.: Goodwillie approximations to higher categories. arXiv:1510.03304 (2016)
46. Harper, J.E., Hess, K.: Homotopy completion and topological Quillen homology of structured ring spectra. Geom. Topol. **17**(3), 1325–1416 (2013)
47. Hirschhorn, P.S.: Model Categories and Their Localizations. Mathematical Surveys and Monographs, vol. 99. American Mathematical Society, Providence, RI (2003)
48. Hovey, M.: Spectra and symmetric spectra in general model categories. J. Pure Appl. Algebr. **165**(1), 63–127 (2001)
49. Hopkins, M.J., Ravenel, D.C.: Suspension spectra are harmonic. Bol. Soc. Mat. Mexicana (2) **37**(1–2), 271–279 (1992) (Papers in honor of José Adem (Spanish))
50. Hovey, M., Sadofsky, H.: Tate cohomology lowers chromatic Bousfield classes. Proc. Am. Math. Soc. **124**(11), 3579–3585 (1996)
51. Hopkins, M.J., Smith, J.H.: Nilpotence and stable homotopy theory. II Ann. Math. (2) **148**(1), 1–49 (1998)
52. Hess, K., Shipley, B.: The homotopy theory of coalgebras over a comonad. Proc. Lond. Math. Soc. (3) **108**(2), 484–516 (2014)
53. Johnson, B.: The derivatives of homotopy theory. Trans. Am. Math. Soc. **347**(4), 1295–1321 (1995)
54. Klein, J.R.: Moduli of suspension spectra. Trans. Am. Math. Soc. **357**(2), 489–507 (2005)
55. Kuhn, N.J., Pereira, L.A.: Operad bimodules and composition products on André-Quillen filtrations of algebras. Algebr. Geom. Topol. **17**(2), 1105–1130 (2017)
56. Kříž, I.: p-adic homotopy theory. Topol. Appl. **52**(3), 279–308 (1993)
57. Kuhn, N.J.: The McCord model for the tensor product of a space and a commutative ring spectrum. Categorical decomposition techniques in algebraic topology (Isle of Skye, 2001). Progr. Math. Birkhäuser, Basel **215**, 213–236 (2004)
58. Kuhn, N.J.: Tate cohomology and periodic localization of polynomial functors. Invent. Math. **157**(2), 345–370 (2004)
59. Kuhn, N.J.: Localization of André-Quillen-Goodwillie towers, and the periodic homology of infinite loopspaces. Adv. Math. **201**(2), 318–378 (2006)
60. Kuhn, N.J.: Goodwillie towers and chromatic homotopy: an overview. In: Proceedings of the Nishida Fest (Kinosaki 2003). Geom. Topol. Monogr., vol. 10, Geom. Topol. Publ., Coventry, pp. 245–279 (2007)

61. Kuhn, N.J.: A guide to telescopic functors. Homol. Homotopy Appl. **10**(3), 291–319 (2008)
62. Lunøe-Nielsen, S., Rognes, J.: The topological Singer construction. Doc. Math. **17**, 861–909 (2012)
63. Lurie, J.: Higher algebra. Available at www.math.harvard.edu/~lurie/ (2016)
64. Mahowald, M.: bo-resolutions. Pac. J. Math. **92**(2), 365–383 (1981)
65. Mahowald, M.: The image of J in the EHP sequence. Ann. Math. (2) **116**(1), 65–112 (1982)
66. Mandell, M.A.: E_∞ algebras and p-adic homotopy theory. Topology **40**(1), 43–94 (2001)
67. Mandell, M.A.: Cochains and homotopy type. Publ. Math. Inst. Hautes Études Sci. no. 103, 213–246 (2006)
68. McCarthy, R.: Dual calculus for functors to spectra. Homotopy methods in algebraic topology (Boulder, CO, 1999). Contemp. Math. Am. Math. Soc. Providence, RI **271**, 183–215 (2001)
69. Medina, A.: E_∞-comodules and topological manifolds
70. Miller, H.R.: On relations between Adams spectral sequences, with an application to the stable homotopy of a Moore space. J. Pure Appl. Algebr. **20**(3), 287–312 (1981)
71. Mahowald, M., Ravenel, D., Shick, P.: The triple loop space approach to the telescope conjecture. Homotopy methods in algebraic topology (Boulder, CO, 1999). Contemp. Math. Am. Math. Soc. Providence, RI, **271**, 217–284 (2001)
72. Mahowald, M., Shick, P.: Root invariants and periodicity in stable homotopy theory. Bull. Lond. Math. Soc. **20**(3), 262–266 (1988)
73. Markl, M., Shnider, S., Stasheff, J.: Operads in Algebra, Topology and Physics. Mathematical Surveys and Monographs, vol. 96. American Mathematical Society, Providence, RI (2002)
74. Pereira, L.A.: A general context for goodwillie calculus. arXiv:1301.2832 (2013)
75. Pereira, L.A.: Goodwillie calculus in algebras over a spectral operad. Available at http://www.faculty.virginia.edu/luisalex/research (2015)
76. Pereira, L.A.: Cofibrancy of operadic constructions in positive symmetric spectra. Homol. Homotopy Appl. **18**(2), 133–168 (2016)
77. Petrie, T.: The weakly complex bordism of Lie groups. Ann. Math. (2) **88**, 371–402 (1968)
78. Priddy, S.B.: Koszul resolutions. Trans. Am. Math. Soc. **152**, 39–60 (1970)
79. Quillen, D.G.: Homotopical Algebra. Lecture Notes in Mathematics, No. 43. Springer, Berlin (1967)
80. Quillen, D.: Rational homotopy theory. Ann. Math. (2) **90**, 205–295 (1969)
81. Ravenel, D.C.: Localization with respect to certain periodic homology theories. Am. J. Math. **106**(2), 351–414 (1984)
82. Rezk, C.: Power operations for Morava E-theory of height 2 at the prime 2. arXiv:0812.1320 (2008)
83. Rezk, C.: The congruence criterion for power operations in Morava E-theory. Homol. Homotopy Appl. **11**(2), 327–379 (2009)
84. Rezk, C.: Modular isogeny complexes. Algebr. Geom. Topol. **12**(3), 1373–1403 (2012)
85. Rezk, C.: Rings of power operations for Morava E-theories are Koszul. arXiv:1204.4831 (2012)
86. Schwede, S.: Spectra in model categories and applications to the algebraic cotangent complex. J. Pure Appl. Algebr. **120**(1), 77–104 (1997)
87. Schlessinger, M., Stasheff, J.: The Lie algebra structure of tangent cohomology and deformation theory. J. Pure Appl. Algebr. **38**(2–3), 313–322 (1985)
88. Strickland, N.P.: Morava E-theory of symmetric groups. Topology **37**(4), 757–779 (1998)
89. Sullivan, D.: Infinitesimal computations in topology. Inst. Hautes Études Sci. Publ. Math. (1977), no. 47, 269–331 (1978)
90. Thompson, R.D.: The v_1-periodic homotopy groups of an unstable sphere at odd primes. Trans. Am. Math. Soc. **319**(2), 535–559 (1990)
91. Wang, G.: The monochromatic HOPF invariant. arXiv:1410.7292 (2014)
92. Wang, G.: Unstable chromatic homotopy theory. Ph.D. thesis, M.I.T. (2015)
93. Weiss, M.: Orthogonal calculus. Trans. Am. Math. Soc. **347**(10), 3743–3796 (1995)
94. Zhu, Y.: Morava E-homology of Bousfield-Kuhn functors on odd-dimensional spheres (preprint)
95. Zhu, Y.: Modular equations for Lubin-Tate formal groups at chromatic level 2. arXiv:1508.03358 (2015)

On Quasi-Categories of Comodules and Landweber Exactness

Takeshi Torii

Abstract In this paper we study quasi-categories of comodules over coalgebras in a stable homotopy theory. We show that the quasi-category of comodules over the coalgebra associated to a Landweber exact $\mathbb{S}$-algebra depends only on the height of the associated formal group. We also show that the quasi-category of $E(n)$-local spectra is equivalent to the quasi-category of comodules over the coalgebra $A \otimes A$ for any Landweber exact $\mathbb{S}_{(p)}$-algebra A of height n at a prime p. Furthermore, we show that the category of module objects over a discrete model of the Morava E-theory spectrum in $K(n)$-local discrete symmetric $\mathbb{G}_n$-spectra is a model of the $K(n)$-local category, where $\mathbb{G}_n$ is the extended Morava stabilizer group.

Keywords Landweber exactness · Quasi-Category · Comodule · Stable homotopy theroy · Complex oriented spectrum · $K(n)$-local category

1 Introduction

It is known that the stable homotopy category of spectra is intimately related to the theory of formal groups through complex cobordism and the Adams–Novikov spectral sequence by the works of Morava [29], Miller–Ravenel–Wilson [28], Devinatz–Hopkins–Smith [9], Hopkins–Smith [12], Hovey–Strickland [17] and many others. The E_2-page of the Adams–Novikov spectral sequence is described as the derived functor of taking primitives in the abelian category of graded comodules over the co-operation Hopf algebroid associated to the complex cobordism spectrum.

We also have a localized version of the Adams–Novikov spectral sequence. For example, for a Landweber exact spectrum E of height n at a prime p, we have an E-based Adams–Novikov spectral sequence abutting to the homotopy groups of E-local spectra. In this case the E-localization and the E_2-page of the E-based Adams–Novikov spectral sequence depend only on the height n of the associated

T. Torii (✉)
Department of Mathematics, Okayama University, Okayama 700–8530, Japan
e-mail: torii@math.okayama-u.ac.jp

T. Ohsawa and N. Minami (eds.), *Bousfield Classes and Ohkawa's Theorem*, Springer Proceedings in Mathematics & Statistics 309,
https://doi.org/10.1007/978-981-15-1588-0_11

formal group at p. There are many results that the derived functor describing the E_2-page of the E-based Adams–Novikov spectral sequence depends only on the substack of the moduli stack of formal groups [14, 15, 18, 30].

These results suggest that there may be an intimate relationship between localized quasi-categories of spectra and quasi-categories of comodules over co-operation coalgebras. In this paper we investigate this relationship. We show that the quasi-category of comodules over a coalgebra associated to a Landweber exact $\mathbb{S}$-algebra depends only on the height of the associated formal group and that the quasi-category of comodules over a coalgebra associated to a Landweber exact $\mathbb{S}_{(p)}$-algebra of height n at a prime p is equivalent to the quasi-category of $E(n)$-local spectra, where $E(n)$ is the nth Johnson-Wilson spectrum at p.

First, we introduce a quasi-category of comodules over a coalgebra associated to an algebra object of a stable homotopy theory $\mathscr{C}$. In this paper we regard coalgebra objects as algebra objects of the opposite monoidal quasi-category of A-A-bimodule objects for an algebra object A of $\mathscr{C}$. We regard comodule objects over a coalgebra Γ as module objects over Γ in the opposite quasi-category of A-module objects in $\mathscr{C}$. In particular, we show that $A \otimes A$ is a coalgebra object for an algebra object A of $\mathscr{C}$ and we can consider the quasi-category

$$\mathrm{LComod}_{\Gamma(A)}(\mathscr{C})$$

of left comodules over $A \otimes A$ in $\mathscr{C}$, where $\Gamma(A)$ represents the pair $(A, A \otimes A)$. For a map $A \to B$ of algebra objects of $\mathscr{C}$, we have the extension of scalars functor $B \otimes_A (-) : \mathrm{LMod}_A(\mathscr{C}) \to \mathrm{LMod}_B(\mathscr{C})$, where $\mathrm{LMod}_A(\mathscr{C})$ and $\mathrm{LMod}_B(\mathscr{C})$ are the quasi-categories of left A-modules and B-modules, respectively. We show that the extension of scalars functor extends to a functor

$$B \otimes_A (-) : \mathrm{LComod}_{\Gamma(A)}(\mathscr{C}) \longrightarrow \mathrm{LComod}_{\Gamma(B)}(\mathscr{C}).$$

of quasi-categories of comodules.

Next, we consider Landweber exact $\mathbb{S}$-algebras in the quasi-category of spectra Sp, where $\mathbb{S}$ is the sphere spectrum. We show that, if A is a Landweber exact $\mathbb{S}$-algebra, then the quasi-category of comodules over the coalgebra $A \otimes A$ depends only on the height of the associated formal group.

Theorem 1 (cf. Theorem 8) *If A and B are Landweber exact $\mathbb{S}$-algebras with the same height at all primes p, then there is an equivalence of quasi-categories*

$$\mathrm{LComod}_{\Gamma(A)}(\mathrm{Sp}) \simeq \mathrm{LComod}_{\Gamma(B)}(\mathrm{Sp}).$$

We also show that the quasi-category of comodules over $A \otimes A$ is equivalent to the quasi-category $L_n\mathrm{Sp}$ of $E(n)$-local spectra if A is a Landweber exact $\mathbb{S}_{(p)}$-algebra of height n at a prime p.

Theorem 2 (cf. Theorem 9) *If A is a Landweber exact $\mathbb{S}_{(p)}$-algebra of height n at a prime p, then there is an equivalence of quasi-categories*

$$L_n \mathrm{Sp} \simeq \mathrm{LComod}_{\Gamma(A)}(\mathrm{Sp}).$$

As an application of the results in this paper we show that the model category constructed in [33] is a model of the $K(n)$-local category, where $K(n)$ is the nth Morava K-theory spectrum at a prime p. We denote by $\Sigma\mathrm{Sp}$ the model category of symmetric spectra and by $\Sigma\mathrm{Sp}_{K(n)}$ the left Bousfield localization of $\Sigma\mathrm{Sp}$ with respect to $K(n)$. The nth extended Morava stabilizer group $\mathbb{G}_n$ is a profinite group and we can consider the model category $\Sigma\mathrm{Sp}(\mathbb{G}_n)$ of discrete symmetric $\mathbb{G}_n$-spectra and its Bousfield localization $\Sigma\mathrm{Sp}(\mathbb{G}_n)_{K(n)}$ with respect to $K(n)$. We have a commutative monoid object F_n in $\Sigma\mathrm{Sp}(\mathbb{G}_n)_{K(n)}$ constructed by Davis [7] and Behrens–Davis [3], which is a discrete model of the nth Morava E-theory spectrum E_n. In [33] we showed that the extension of scalars functor

$$L_{K(n)}(F_n \otimes (-)) : \Sigma\mathrm{Sp}(\mathbb{G}_n)_{K(n)} \longrightarrow \mathrm{LMod}_{F_n}(\Sigma\mathrm{Sp}(\mathbb{G}_n)_{K(n)}),$$

which is a left Quillen functor, is homotopically fully faithful, that is, it induces a weak homotopy equivalence between mapping spaces for any two objects in $\Sigma\mathrm{Sp}(\mathbb{G}_n)_{K(n)}$. In this paper we show that this functor is actually a left Quillen equivalence and hence we can consider the category $\mathrm{LMod}_{F_n}(\Sigma\mathrm{Sp}(\mathbb{G}_n)_{K(n)})$ to be a model of the $K(n)$-local category.

Theorem 3 (cf. Theorem 11) *The extension of scalars functor*

$$L_{K(n)}(F_n \otimes (-)) : \Sigma\mathrm{Sp}(\mathbb{G}_n)_{K(n)} \longrightarrow \mathrm{LMod}_{F_n}(\Sigma\mathrm{Sp}(\mathbb{G}_n)_{K(n)})$$

is a left Quillen equivalence.

The organization of this paper is as follows: In Sect. 2 we fix some notation we use throughout this paper. In Sect. 3 we review the theory of quasi-categories. We try to explain that quasi-categories are very flexible models for $(\infty, 1)$-categories and that the theory of quasi-categories is an appropriate setting to study coalgebras and comodules of spectra. In Sect. 4 we study opposite coCartesian fibrations, opposite monoidal quasi-categories, and opposite tensored quasi-categories. In particular, we show that a lax monoidal right adjoint functor between monoidal quasi-categories induces a lax monoidal right adjoint functor between the opposite monoidal quasi-categories. In Sect. 5 we introduce a quasi-category of comodules over a coalgebra in a stable homotopy theory. We define a cotensor product of a right comodule and a left comodule over a coalgebra as a limit of the cobar construction. We study the relationship between localizations of a stable homotopy theory and quasi-categories of comodules. In Sect. 6 we study comodules in spectra over a coalgebra associated to a Landweber exact $\mathbb{S}$-algebra. First, we study the Bousfield–Kan spectral sequence associated to the two-sided cobar construction. Next, we show that the quasi-category of comodules over the coalgebra associated to a Landweber exact $\mathbb{S}$-algebra depends only on the height of the associated formal group. Finally, we show that the model category of modules over F_n in $K(n)$-local discrete symmetric $\mathbb{G}_n$-spectra is a model of the $K(n)$-local category. In Sect. 7 we give a proof of Proposition 1 (stated in

Sect. 4), which is technical but important for constructing a canonical map between opposite coCartesian fibrations.

2 Notation

For a category $\mathscr{C}$, we denote by $\mathrm{Hom}_{\mathscr{C}}(x, y)$ the set of all morphisms from x to y in $\mathscr{C}$ for $x, y \in \mathscr{C}$.

We denote by sSet the category of simplicial sets. For a simplicial set K, we denote by K^{op} the opposite simplicial set (see [22, Sect. 1.2.1]). If K is a quasi-category, then K^{op} is also a quasi-category. For simplicial sets X, Y, we denote by $\mathrm{Fun}(X, Y)$ the simplicial mapping space from X to Y. For a simplicial set X equipped with a map $\pi : X \to S$ of simplicial sets, we denote by X_s the fiber of π over $s \in S$. If X and Y are simplicial sets over a simplicial set S, then we denote by $\mathrm{Fun}_S(X, Y)$ the simplicial set of maps from X to Y over S.

For a small (simplicial) category $\mathscr{C}$, we denote by $N(\mathscr{C})$ the simplicial set obtained by applying the (simplicial) nerve functor $N(-)$ to $\mathscr{C}$ (see [22, Sect. 1.1.5]). We denote by Cat_∞ the quasi-category of (small) quasi-categories (see [22, Sect. 3]).

We denote by $\Sigma\mathrm{Sp}$ the category of symmetric spectra equipped with the stable model structure (see [16]). We denote by Sp the quasi-category of spectra, which is the underlying quasi-category of the simplicial model category $\Sigma\mathrm{Sp}$. We denote by Ho(Sp) the stable homotopy category of spectra. We denote by $\mathbb{S}$ the sphere spectrum. For a spectrum $X \in \mathrm{Sp}$, we write X_* for the homotopy groups $\pi_* X$. For spectra $X, Y \in \mathrm{Sp}$, we write $X \otimes Y$ for the smash product of X and Y.

3 Review of Quasi-Categories

In this section we review the theory of quasi-categories which are models for $(\infty, 1)$-categories. Quasi-categories were introduced by Boardman–Vogt [4] as weak Kan complexes, and developed by Joyal [19] and Lurie [22, 23]. An (∞, n)-category is an ∞-category with invertible k-morphisms for all $k > n$. There are many models for $(\infty, 1)$-categories, including relative categories, topological categories, simplicial categories, Segal categories, complete Segal spaces, and so on. Topological categories are intuitively easy to understand but they are actually difficult to work with. Quasi-categories are yet another model for $(\infty, 1)$-categories. Quasi-categories are simplicial sets which satisfy some extension property. The meaning of the definition of quasi-categories is difficult to understand at first glance but quasi-categories are sufficiently flexible because they are closed under various categorical operations.

One of the motivations for the definition of quasi-categories comes from the fact that we can embed the category of small categories into the category of simplicial sets. Another is the fact that Kan complexes are models for $(\infty, 0)$-categories.

Roughly speaking, a category consists of a collection of objects, and sets of morphisms between objects equipped with associative and unital composition laws. We can regard a small category as a simplicial set. In fact, there is a fully faithful functor from the category of small categories to the category of simplicial sets, which is called the nerve functor.

Let $\mathscr{C}$ be a small category. The nerve $N(\mathscr{C})$ of $\mathscr{C}$ is a simplicial set, in which the set of n-simplices is the set $\mathrm{Hom}_{\mathrm{Cat}}([n], \mathscr{C})$ of functors from $[n]$ to $\mathscr{C}$, where Cat is the category of small categories and $[n]$ is the category associated to the ordered set $\{0 < 1 < \cdots < n\}$. The nerve functor is fully faithful and the essential image is characterized as follows. A simplicial set K is in the essential image of the nerve functor if and only if there is a unique extension $\Delta^n \to K$ for any map $\Lambda^n_i \to K$ of simplicial sets with $0 < i < n$:

$$\begin{array}{ccc} \Lambda^n_i & \longrightarrow & K \\ \big\downarrow & \nearrow & \\ \Delta^n, & & \end{array}$$

where Δ^n is the standard simplicial n-simplex and Λ^n_i is the ith horn of Δ^n.

A Kan complex is defined to be a simplicial set which has the following extension property. A simplicial set K is a Kan complex if and only if there is an extension $\Delta^n \to K$ of any map $\Lambda^n_i \to K$ of simplicial sets with $0 \le i \le n$. A Kan complex is a fibrant object in the category of simplicial sets equipped with the Kan model structure. It is known that the homotopy theory of Kan complexes is equivalent to the homotopy theory of topological spaces. We can regard topological spaces as models for $(\infty, 0)$-categories, which are also called ∞-groupoids. To a topological space X, we can associate the fundamental groupoid $\pi_{\le 1}(X)$. Objects of $\pi_{\le 1}(X)$ are points of X, and morphisms are homotopy classes of paths in X. Although the fundamental groupoid $\pi_{\le 1}(X)$ contains only the information of the 1-type of X, we can generalize this construction and obtain the fundamental ∞-groupoid $\pi_{\le\infty}(X)$ of a topological space X, which contains the information of the homotopy type of X. Roughly speaking, objects of $\pi_{\le\infty}(X)$ are points of X, 1-morphisms are paths in X, 2-morphisms are homotopies between paths, and higher morphisms are higher homotopies. A generally accepted principal of higher category theory says that the homotopy theory of ∞-groupoids is equivalent to the homotopy theory of topological spaces via the construction of fundamental ∞-groupoids. Since the homotopy theory of Kan complexes is equivalent to the homotopy theory of topological spaces, we can regard Kan complexes as models for $(\infty, 0)$-categories.

Ordinary categories and $(\infty, 0)$-categories are examples of $(\infty, 1)$-categories. Thus, models for $(\infty, 1)$-categories should be generalizations of ordinary categories and topological spaces in some sense. Quasi-categories are models for $(\infty, 1)$-categories. A quasi-category is a simplicial set satisfying the following extension property which is a generalization of the extension properties of both small categories and Kan complexes.

Definition 1 Let X be a simplicial set. We say that X is a quasi-category if for any $i, n \in \mathbb{Z}$, such that $0 < i < n$, any map $\Lambda_i^n \to X$ of simplicial sets can be extended to a map $\Delta^n \to X$.

By definition, Kan complexes and nerves of small categories are quasi-categories. Furthermore, we have a generalization of the nerve functor, which is called the simplicial nerve functor. The simplicial nerve functor assigns a simplicial set to a simplicial category. In particular, when we regard a small category $\mathscr{C}$ as a simplicial category with discrete mapping simplicial sets, the simplicial nerve of $\mathscr{C}$ is isomorphic to the nerve of $\mathscr{C}$. If a simplicial category $\mathscr{C}$ is fibrant—that is, the mapping simplicial sets $\mathrm{Map}_{\mathscr{C}}(x, y)$ are Kan complexes for all objects x, y of $\mathscr{C}$, then the simplicial nerve $N(\mathscr{C})$ is a quasi-category. Furthermore, it is known that the homotopy theory of simplicial categories is equivalent to the homotopy theory of quasi-categories via the simplicial nerve functor.

There is a quasi-category $\mathscr{S}$ of spaces, which is obtained by taking the simplicial nerve of the simplicial category of Kan complexes. The quasi-category $\mathscr{S}$ of spaces is of great importance in higher category theory because Kan complexes are models for $(\infty, 0)$-categories. Roughly speaking, $(\infty, 1)$-categories are regarded as categories enriched in $\mathscr{S}$, and the theory of $(\infty, 1)$-categories is obtained from the theory of categories by replacing the category of sets with the quasi-category $\mathscr{S}$ of spaces.

For a quasi-category X, we can define objects and morphisms of X. The objects of X are the 0-simplices, and the morphisms are the 1-simplices of X. Furthermore, we can define the mapping space $\mathrm{Map}_X(x, y)$ for any objects $x, y \in X$. The composition law of the mapping spaces is unital and associative up to higher homotopy. Taking π_0 of the mapping spaces, we obtain the homotopy category $\mathrm{Ho}(X)$ of X, which is an ordinary category. A morphism of a quasi-category X is said to be an equivalence if it represents an isomorphism in the homotopy category $\mathrm{Ho}(X)$.

A functor between quasi-categories is defined to be a map of simplicial sets. This is good news because the definition of functors between quasi-categories is very simple, whereas the correct ∞-categorical definition of functors between topological or simplicial categories is quite difficult. A functor $F : X \to Y$ between quasi-categories is fully faithful if it induces an equivalence $\mathrm{Map}_X(x, x') \to \mathrm{Map}_Y(F(x), F(x'))$ of mapping spaces in $\mathscr{S}$ for all objects $x, x' \in X$. The functor F is said to be essentially surjective if any object $y \in Y$ is equivalent to an object of the form $F(x)$ for some $x \in X$. A functor F is an equivalence of quasi-categories if it is fully faithful and essentially surjective.

There is a model structure on the category of simplicial sets which is called the Joyal model structure. Any simplicial set is a cofibrant object in the Joyal model structure as in the case of the Kan model structure. But a simplicial set is a fibrant object in the Joyal model structure if and only if it is a quasi-category, and weak equivalences between fibrant objects are equivalences of quasi-categories. In general, if we have a simplicial model category, then by taking the simplicial nerve of the full subcategory consisting of cofibrant-fibrant objects, we obtain the underlying quasi-category of the simplicial model category. Unfortunately, the Joyal model structure is not compatible with simplicial enrichment, that is, it is not a simplicial model category. But we have a simplicial model category whose underlying quasi-category is the quasi-category of (small) quasi-categories. A marked simplicial set is a pair (X, M), where X is a simplicial set and M is a set of 1-simplices of X that contains all degenerate 1-simplices. There is a simplicial model structure on the category of marked simplicial sets. Any marked simplicial set is cofibrant. A marked simplicial set (X, M) is fibrant if and only if X is a quasi-category and M is the set of all equivalences. Taking the simplicial nerve of the full subcategory of fibrant-cofibrant marked simplicial sets, we obtain a quasi-category, which is called the quasi-category Cat_∞ of (small) quasi-categories.

The classical stable homotopy category of spectra is an analogue of the derived category of abelian groups. We have an enhancement of the classical stable homotopy category, that is, a quasi-category Sp of spectra, whose homotopy category is the classical stable homotopy category. The quasi-category Sp is defined to be the limit of the tower

$$\cdots \xrightarrow{\Omega} \mathscr{S}_* \xrightarrow{\Omega} \mathscr{S}_* \xrightarrow{\Omega} \mathscr{S}_*$$

in the quasi-category $\widehat{\mathrm{Cat}}_\infty$ of (not necessarily small) quasi-categories, where $\mathscr{S}_*$ is the quasi-category of pointed spaces and $\Omega : \mathscr{S}_* \to \mathscr{S}_*$ is the loop functor.

The quasi-category Sp of spectra is an example of stable quasi-categories. The framework of stable quasi-categories gives a way to do homological algebra. Actually, the homotopy category of a stable quasi-category is a triangulated category in a canonical way. Thus, stable quasi-categories are regarded as an enhancement of triangulated categories. The quasi-category Sp of spectra has a universal property in the realm of stable quasi-categories in the following sense. We can consider the quasi-category $\mathscr{P}r^{\mathrm{L}}_{\mathrm{st}}$ of presentable stable quasi-categories. The quasi-category $\mathscr{P}r^{\mathrm{L}}_{\mathrm{st}}$ has a symmetric monoidal structure, and a commutative monoid object of $\mathscr{P}r^{\mathrm{L}}_{\mathrm{st}}$ is a symmetric monoidal quasi-category which is stable and presentable and whose tensor product is colimit-preserving in each variable. We can consider the quasi-category $\mathrm{CAlg}(\mathscr{P}r^{\mathrm{L}}_{\mathrm{st}})$ of commutative monoid objects of $\mathscr{P}r^{\mathrm{L}}_{\mathrm{st}}$, and the quasi-category Sp of spectra is an initial object of $\mathrm{CAlg}(\mathscr{P}r^{\mathrm{L}}_{\mathrm{st}})$. In particular, Sp is a stable presentable symmetric monoidal quasi-category whose tensor product is colimit-preserving in each variable.

Since the quasi-category Sp of spectra is symmetric monoidal, we can consider various algebraic objects in Sp. Many algebraic structures can be controlled by operads, and Lurie developed the theory of ∞-operads in [23], which control many types of algebraic structures in quasi-categories. In particular, in Sp we can consider associative algebras, commutative algebras, and modules over an associative algebra. Associative algebras in Sp correspond to A_∞-ring spectra, and commutative algebras in Sp correspond to E_∞-ring spectra. We can regard modules over an associative algebra in Sp as modules over the corresponding A_∞-ring spectrum. Furthermore, we can use the theory of quasi-categories and ∞-operads to define quasi-categories of coalgebras and comodules over a coalgebra in Sp. It seems to the author that the theory of quasi-categories is an appropriate setting to define and study coalgebras and comodules over a coalgebra of spectra. In this paper we study comodules of spectra related to complex oriented cohomology theories.

4 Opposite Monoidal Quasi-Categories and Opposite Tensored Quasi-Categories over Monoidal Quasi-Categories

In this section we study the opposite quasi-categories of monoidal quasi-categories and the opposite quasi-categories of tensored quasi-categories over monoidal quasi-categories. The author thinks that the results in this section are well-known to experts but he decided to include this section because he is not aware of appropriate references.

In Sect. 4.1 we recall a model of opposite coCartesian fibrations by Barwick–Glasman–Nardin [2] and study maps between opposite coCartesian fibrations. In Sect. 4.2 we study the opposite quasi-category of a monoidal quasi-category and show that a lax monoidal right adjoint functor between monoidal quasi-categories induces a lax monoidal right adjoint functor between the opposite monoidal quasi-categories. In Sect. 4.3 we study the opposite of a tensored quasi-category over a monoidal quasi-category. We show that a lax tensored right adjoint functor between tensored quasi-categories induces a lax tensored right adjoint functor between the opposites of the tensored quasi-categories.

4.1 Opposite CoCartesian Fibrations

For a coCartesian fibration we have the opposite coCartesian fibration whose fibers are the opposite quasi-categories of the fibers of the original coCartesian fibration. In this subsection we recall the explicit model of opposite coCartesian fibrations due to Barwick–Glasman–Nardin [2]. We show that a map between coCartesian fibrations whose restriction to every fiber admits a left adjoint induces a map between the opposite coCartesian fibrations.

First, we recall the explicit model of opposite coCartesian fibrations by Barwick–Glasman–Nardin [2].

Let S be a simplicial set and let $p : X \to S$ be a coCartesian fibration with small fibers. We denote by X_s the quasi-category that is the fiber of p over $s \in S$. Let Cat_∞ be the quasi-category of small quasi-categories. By [22, Sect. 3.3.2], the coCartesian fibration p is classified by a functor $\mathbf{X} : S \to \mathrm{Cat}_\infty$. There is an involution

$$R : \mathrm{Cat}_\infty \longrightarrow \mathrm{Cat}_\infty$$

carrying a quasi-category to its opposite. The composite functor $R\mathbf{X}$ classifies a coCartesian fibration $Rp : RX \to S$ in which the fiber $(RX)_s$ of Rp over $s \in S$ is equivalent to the opposite quasi-category $(X_s)^{\mathrm{op}}$ for all $s \in S$. We call $Rp : RX \to S$ the opposite coCartesian fibration of $p : X \to S$. In the following of this subsection we assume that the base simplicial set S is a quasi-category.

To describe the model of opposite coCartesian fibrations, we recall the twisted arrow quasi-category. The twisted arrow quasi-category $\widetilde{\mathscr{O}}(K)$ for a quasi-category K is the simplicial set in which the set of n-simplices is given by

$$\widetilde{\mathscr{O}}(K)_n = \mathrm{Hom}_{\mathrm{s}}\mathrm{Set}((\Delta^n)^{\mathrm{op}} \star \Delta^n, K)$$

with obvious structure maps. The simplicial set $\widetilde{\mathscr{O}}(K)$ is actually a quasi-category (see [24, Prop. 4.2.3]). Note that the inclusions $\Delta^n \hookrightarrow (\Delta^n)^{\mathrm{op}} \star \Delta^n$ and $(\Delta^n)^{\mathrm{op}} \hookrightarrow (\Delta^n)^{\mathrm{op}} \star \Delta^n$ induce maps of simplicial sets $\widetilde{\mathscr{O}}(K) \to K$ and $\widetilde{\mathscr{O}}(K) \to K^{\mathrm{op}}$, respectively.

We use the twisted arrow category $\widetilde{\mathscr{O}}(\Delta^n)$ for the n-simplex Δ^n for $n \geq 0$ to describe the model of opposite coCartesian fibrations. The twisted arrow quasi-category $\widetilde{\mathscr{O}}(\Delta^n)$ for Δ^n is the nerve of $\widetilde{[n]}$, where $\widetilde{[n]}$ is the ordered set of all pairs (i, j) of integers with $0 \leq i \leq j \leq n$ equipped with order relation $(i, j) \leq (i', j')$ if and only if $i \geq i'$ and $j \leq j'$. The ordered set $\widetilde{[n]}$ is depicted as follows

$$\begin{array}{ccccccccccc}
00 & \to & 01 & \to & 02 & \to & \cdots & \to & 0\bar{1} & \to & 0\bar{0} \\
 & & \uparrow & & \uparrow & & & & \uparrow & & \uparrow \\
 & & 11 & \to & 12 & \to & \cdots & \to & 1\bar{1} & \to & 1\bar{0} \\
 & & & & \uparrow & & & & \uparrow & & \uparrow \\
 & & & & 22 & \to & \cdots & \to & 2\bar{1} & \to & 2\bar{0} \\
 & & & & & & & & \uparrow & & \uparrow \\
 & & & & & & \ddots & & \vdots & & \vdots \\
 & & & & & & & & \uparrow & & \uparrow \\
 & & & & & & & & \bar{1}\bar{1} & \to & \bar{1}\bar{0} \\
 & & & & & & & & & & \uparrow \\
 & & & & & & & & & & \bar{0}\bar{0},
\end{array} \tag{1}$$

where $\bar{k} = n - k$.

By functoriality of $\widetilde{\mathscr{O}}(-)$, we have a cosimplicial simplicial set $\widetilde{\mathscr{O}}(\Delta^\bullet)$. For a simplicial set K over S, we define a simplicial set $H(K)$ over S as follows. The simplicial set $H(K)$ is a simplicial subset of $\mathrm{Hom}_{\mathrm{sSet}}(\widetilde{\mathscr{O}}(\Delta^\bullet), K)$. A map $\varphi : \widetilde{\mathscr{O}}(\Delta^n) \to K$ is an n-simplex of $H(K)$ for $n \geq 0$ if the j-simplex $\varphi(jj) \to \cdots \to \varphi(1j) \to \varphi(0j)$ covers a totally degenerate j-simplex of S, that is, a j-simplex in the image of the map $S_0 \to S_j$, for all $0 \leq j \leq n$. Assigning to an n-simplex φ of $H(K)$ the n-simplex $p\varphi(00) \to p\varphi(01) \to \cdots \to p\varphi(0n)$ of S, we obtain a map $H(K) \to S$.

Let $p : X \to S$ be a coCartesian fibration, where S is a quasi-category. We define a simplicial set RX as follows. The simplicial set RX is a simplicial subset of $H(X)$. A map $\varphi : \widetilde{\mathscr{O}}(\Delta^n) \to X$ is an n-simplex of RX for $n \geq 0$ if the following two conditions are satisfied:

1. The j-simplex $\varphi(jj) \to \cdots \to \varphi(1j) \to \varphi(0j)$ covers a totally degenerate j-simplex of S for all $0 \leq j \leq n$.
2. The 1-simplex $\varphi(ij) \to \varphi(ik)$ is a p-coCartesian edge for all $0 \leq i \leq j \leq k \leq n$.

As in $H(X)$, we have a map

$$Rp : RX \to S,$$

which is a coCartesian fibration. The fiber $(RX)_s$ over $s \in S$ is equivalent to the opposite quasi-category $(X_s)^{\mathrm{op}}$ of the fiber X_s for all $s \in S$. An edge $\varphi \in \mathrm{Hom}_{\mathrm{sSet}}(\Delta^1, RX)$ is Rp-coCartesian if and only if the edge $\varphi(11) \to \varphi(01)$ is an equivalence in the fiber X_s, where $s = p\varphi(11)$. The coCartesian fibration $Rp : RX \to S$ is a model of the opposite coCartesian fibration corresponding to the composite

$$R\mathbf{X} : S \stackrel{\mathbf{X}}{\longrightarrow} \mathrm{Cat}_\infty \stackrel{R}{\longrightarrow} \mathrm{Cat}_\infty,$$

where $\mathbf{X} : S \to \mathrm{Cat}_\infty$ is the map corresponding to the coCartesian fibration $p : X \to S$, and $R : \mathrm{Cat}_\infty \to \mathrm{Cat}_\infty$ is the functor which assigns to a quasi-category its opposite quasi-category.

Next, we consider a map between coCartesian fibrations which admits a left adjoint for each fibers. We show that the map induces a canonical map in the opposite direction between the opposite coCartesian fibrations.

Let $p : X \to S$ and $q : Y \to S$ be coCartesian fibrations over a quasi-category S. Suppose we have a map $G : Y \to X$ over S. Note that we do not assume that G preserves coCartesian edges. The map $G : Y \to X$ over S induces a functor $G_s : Y_s \to X_s$ between the quasi-categories of fibers for each $s \in S$.

We shall define a simplicial set $\mathscr{R}$ over S equipped with maps $\pi_X : \mathscr{R} \to RX$ and $\pi_Y : \mathscr{R} \to RY$ over S. For a simplicial set K and X, we denote by $\mathrm{Fun}(K, X)$ the mapping simplicial set from K to X. The map $p : X \to S$ induces a map $p_* : \mathrm{Fun}(\Delta^1, X) \to \mathrm{Fun}(\Delta^1, S)$. We regard S as a simplicial subset of $\mathrm{Fun}(\Delta^1, S)$ via constant maps. We denote by $\mathrm{Fun}^S(\Delta^1, X)$ the pullback of p_* along the inclusion $S \hookrightarrow \mathrm{Fun}(\Delta^1, S)$. The inclusion $\Delta^{\{i\}} \hookrightarrow \Delta^1$ induces a map $\mathrm{Fun}^S(\Delta^1, X) \to X$ over S for $i = 0, 1$.

We have the inclusion $RX \hookrightarrow H(X) \cong H(\mathrm{Fun}(\Delta^{\{0\}}, X))$. The map $G : Y \to X$ over S induces a map $RY \hookrightarrow H(Y) \stackrel{G_*}{\longrightarrow} H(X) \cong H(\mathrm{Fun}(\Delta^{\{1\}}, X))$. The inclusion $\Delta^{\{i\}} \hookrightarrow \Delta^1$ induces a map $H(\mathrm{Fun}^S(\Delta^1, X)) \to H(\mathrm{Fun}(\Delta^{\{i\}}, X))$ for $i = 0, 1$. Using these maps, we define a simplicial set $\mathscr{R}$ by

$$\mathscr{R} = RX \times_{H(\mathrm{Fun}(\Delta^{\{0\}}, X))} H(\mathrm{Fun}^S(\Delta^1, X)) \times_{H(\mathrm{Fun}(\Delta^{\{1\}}, X))} RY.$$

We have a map $\mathscr{R} \to S$ and projections $\pi_X : \mathscr{R} \to RX$ and $\pi_Y : \mathscr{R} \to RY$ over S.

Now we assume that the functor $G_s : Y_s \to X_s$ admits a left adjoint F_s for all $s \in S$. Then an object x of X with $s = p(x)$ determines an object $(x, u_x, F_s(x))$ of $\mathscr{R}$, where $u_x : x \to G_s F_s(x)$ is the unit map of the adjunction (F_s, G_s) at x. We define $\mathscr{R}^0$ to be the full subcategory of $\mathscr{R}$ spanned by $\{(x, u_x, F_s(x))\}$ for all $x \in X$, where $s = p(x)$. Let

$$\pi_X^0 : \mathscr{R}^0 \longrightarrow RX.$$

be the restriction of π_X to $\mathscr{R}^0$.

Proposition 1 *The map $\pi_X^0 : \mathscr{R}^0 \to RX$ is a trivial Kan fibration.*

We defer the proof of Proposition 1 to Sect. 7.

We take a section T_0 of π_X^0, which is unique up to contractible space of choices. Let $\pi_Y^0 : \mathscr{R}^0 \to RY$ be the restriction of π_Y to $\mathscr{R}^0$. We define a functor

$$RF : RX \longrightarrow RY$$

to be $\pi_Y^0 T_0$.

We would like to describe some properties of the section T_0. Let $s \in S$. We consider the restriction of T_0 to $(RX)_s$. The fiber $\mathscr{R}_s$ is described as

$$\mathscr{R}_s = (RX)_s \times_{H(\mathrm{Fun}(\Delta^{\{0\}}, X_s))} H(\mathrm{Fun}(\Delta^1, X_s)) \times_{H(\mathrm{Fun}(\Delta^{\{1\}}, X_s))} (RY)_s,$$

and the fiber $\mathscr{R}_s^0$ is a full subcategory of $\mathscr{R}_s$. The composition $(RX)_s \hookrightarrow H(X_s) \stackrel{(F_s)_*}{\longrightarrow} H(Y_s)$ factors through $(RY)_s$. We denote by $R(F_s)$ the induced functor $(RX)_s \to (RY)_s$. The unit map $u_s : 1_{X_s} \to G_s F_s$ in $\mathrm{Fun}(X_s, X_s)$ can be identified with a map $u_s : X_s \to \mathrm{Fun}(\Delta^1, X_s)$. We obtain a map $Hu_s : (RX)_s \to H(\mathrm{Fun}(\Delta^1, X_s))$ by the composition $(RX)_s \hookrightarrow H(X_s) \stackrel{(u_s)_*}{\longrightarrow} H(\mathrm{Fun}(\Delta^1, X_s))$. Note that Hu_s followed by $H(\mathrm{Fun}(\Delta^1, X_s)) \to H(\mathrm{Fun}(\Delta^{\{0\}}, X_s))$ is the inclusion $(RX)_s \hookrightarrow H(X_s)$, and the map Hu_s followed by $H(\mathrm{Fun}(\Delta^1, X_s)) \to H(\mathrm{Fun}(\Delta^{\{1\}}, X_s))$ is the composition of $R(F_s) : (RX)_x \to (RY)_s$ followed by the inclusion $(RY)_s \hookrightarrow H(X_s)$. Hence we obtain a section of $\mathscr{R}_s^0$ over $(RX)_s$:

$$(1_{(RX)_s}, Hu_s, R(F_s)) : (RX)_s \longrightarrow \mathscr{R}_s^0.$$

Proposition 2 *We have*

$$T_0|_{(RX)_s} \simeq (1_{(RX)_s}, Hu_s, R(F_s))$$

for any $s \in S$

Proof Restricting π_X^0 to the fibers over $s \in S$, we obtain a trivial Kan fibration $(\pi_X^0)_s : \mathscr{R}_s^0 \to (RX)_s$. The restriction of the section T_0 to $(RX)_s$ is a section of $(\pi_X^0)_s$. The map $(1_{(RX)_s}, Hu_s, R(F_s))$ is also a section of $(\pi_X^0)_s$. Hence we have $T_0|_{(RX)_s} \simeq (1_{(RX)_s}, Hu_s, R(F_s))$. □

Next, we consider the image of edges of RX under the section T_0. Let φ be an edge of RX over $e : s \to s'$ in S represented by a p-coCartesian edge $\varphi(00) \to \varphi(01)$ in X and an edge $\varphi(11) \to \varphi(01)$ in the fiber $X_{s'}$. We take a q-coCartesian edge $\psi : F_s\varphi(00) \to y'$ in Y over e. Since $\varphi(00) \to \varphi(01)$ is p-coCartesian, we obtain an edge $\varphi(01) \to G_{s'}y'$ in $X_{s'}$, which makes the following diagram commute

$$\begin{array}{ccc} \varphi(00) & \longrightarrow & \varphi(01) \\ {\scriptstyle u}\downarrow & & \vdots \\ G_sF_s\varphi(00) & \xrightarrow{G\psi} & G_{s'}y', \end{array} \qquad (2)$$

where u is the unit map of the adjunction (F_s, G_s) at $\varphi(00)$. Let $w : F_{s'}\varphi(11) \to y'$ be the map in $Y_{s'}$ obtained from $\varphi(11) \to \varphi(01)$ by applying $F_{s'}$ followed by the adjoint map of $\varphi(01) \to G_{s'}y'$. We denote by $R\varphi$ the edge of RY over e represented by

$$F_s\varphi(00) \xrightarrow{\psi} y' \xleftarrow{w} F_{s'}\varphi(11).$$

Since the composite $\varphi(11) \to \varphi(01) \to G_{s'}y'$ is adjoint to $w : F_{s'}\varphi(11) \to y'$, we have an edge $H\varphi : \varphi \to G(R\varphi)$ of $H(\mathrm{Fun}^S(\Delta^1, X))$ represented by the following commutative diagram

$$\begin{array}{ccccc} \varphi(00) & \longrightarrow & \varphi(01) & \longleftarrow & \varphi(11) \\ \downarrow{\scriptstyle u} & & \downarrow & & \downarrow{\scriptstyle u} \\ G_sF_s\varphi(00) & \xrightarrow{G\psi} & G_{s'}y' & \xleftarrow{G_{s'}w} & G_{s'}F_{s'}\varphi(11). \end{array}$$

Proposition 3 *For any edge φ of RX, we have*

$$T_0(\varphi) \simeq (\varphi, H\varphi, R\varphi).$$

Proof Let $\pi^0_\varphi : \mathscr{R}^0_\varphi \to \Delta^1$ be the trivial Kan fibration obtained by the pullback of π^0_X along the map $\varphi : \Delta^1 \to RX$. The triple $(\varphi, H\varphi, R\varphi)$ determines a section of π^0_φ. Hence $T_0(\varphi) \simeq (\varphi, H\varphi, R\varphi)$. □

The main result in this subsection is the following theorem.

Theorem 4 *Let $p : X \to S$ and $q : Y \to S$ be coCartesian fibrations over a quasi-category S. Suppose we have a map $G : Y \to X$ over S. If G_s admits a left adjoint F_s for all $s \in S$, then there exists a canonical map $RF : RX \to RY$ over S up to contractible space of choices. We have $(RF)_s \simeq F_s^{\mathrm{op}}$ for all $s \in S$ and $RF(\varphi) \simeq R\varphi$ for any edge φ of RX.*

Proof The theorem follows from Propositions 1, 2, and 3. □

Now we consider which coCartesian edge of RX is preserved by the functor RF. Let $e : s \to s'$ be a 1-simplex of S. Since $q : Y \to S$ and $p : X \to S$ are coCartesian fibrations, we have functors $e_!^Y : Y_s \to Y_{s'}$ and $e_!^X : X_s \to X_{s'}$ associated to e. The map $G : Y \to X$ over S induces a diagram $\partial(\Delta^1 \times \Delta^1) \to \mathrm{Cat}_\infty$ depicted as

$$\begin{array}{ccc} Y_s & \xrightarrow{G_s} & X_s \\ \downarrow{\scriptstyle e_!^Y} & & \downarrow{\scriptstyle e_!^X} \\ Y_{s'} & \xrightarrow{G_{s'}} & X_{s'} \end{array} \tag{3}$$

and a natural transformation

$$e_!^X G_s \longrightarrow G_{s'} e_!^Y. \tag{4}$$

If natural transformation (4) is an equivalence, then $G : Y \to X$ preserves coCartesian edges over e.

We recall the definition of left adjointable diagram (see [23, Def. 4.7.5.13]). Suppose we are given a diagram of quasi-categories

$$\begin{array}{ccc} \mathscr{C} & \xrightarrow{G} & \mathscr{D} \\ {\scriptstyle U}\downarrow & & \downarrow{\scriptstyle V} \\ \mathscr{C}' & \xrightarrow{G'} & \mathscr{D}' \end{array}$$

which commutes up to a specified equivalence $\alpha : VG \simeq G'U$. We say that this diagram is left adjointable if the functors G and G' admit left adjoints F and F', respectively, and if the composite transformation

$$F'V \to F'VGF \stackrel{\alpha}{\simeq} F'G'UF \to UF$$

is an equivalence, where the first map is induced by the unit map of the adjunction (F, G), and the third map is induced by the counit map of the adjunction (F', G').

Proposition 4 *Let e be a 1-simplex of S. If natural transformation* (4) *is an equivalence and diagram* (3) *equipped with this equivalence is left adjointable, then $RF : RX \to RY$ preserves coCartesian edges over e.*

Proof Let φ be an Rp-coCartesian edge of RX over e represented by $\varphi(00) \to \varphi(01) \leftarrow \varphi(11)$, where $\varphi(00) \to \varphi(01)$ is a p-coCartesian edge of X over e and $\varphi(11) \to \varphi(01)$ is an equivalence in $X_{s'}$. We can regard $\varphi(11)$ as $e_!^X\varphi(00)$.

Suppose that the edge $R\varphi$ of RY over e is represented by

$$F_s\varphi(00) \xrightarrow{\psi} y' \xleftarrow{w} F_{s'}\varphi(11),$$

where ψ is a q-coCartesian edge of Y over e and w is an edge of $Y_{s'}$. We have to show that w is an equivalence of $Y_{s'}$.

We can regard y' as $e_!^Y F_s\varphi(00)$ and $F_{s'}\varphi(11)$ as $F_{s'}e_!^X\varphi(00)$. The morphism w is the adjoint of the morphism $\varphi(11) \to \varphi(01) \to G_{s'}y'$, which can be identified with $e_!^X\varphi(00) \to G_{s'}e_!^Y F_s\varphi(00)$. By the assumption that diagram (3) is left adjointable, we see that w is an equivalence. □

4.2 Opposite Monoidal Quasi-Categories

In this subsection we study the opposite monoidal quasi-category of a monoidal quasi-category. We show that a lax right adjoint functor between monoidal quasi-categories induces a lax right adjoint functor between the opposite monoidal quasi-categories.

First, we recall the definition of monoidal quasi-categories. Let $p : M \to N(\Delta)^{\mathrm{op}}$ be a coCartesian fibration of simplicial sets. For any $n \geq 0$, the inclusion $[1] \cong \{i -$

$1, i\} \hookrightarrow [n]$ induces a functor $p_i : X_{[n]} \to X_{[1]}$ of quasi-categories for $i = 1, \ldots, n$. We say that p is a monoidal quasi-category if the functor

$$p_1 \times \cdots \times p_n : X_{[n]} \longrightarrow \overbrace{X_{[1]} \times \cdots \times X_{[1]}}^{n}$$

is a categorical equivalence for all $n \geq 0$. The fiber $M_{[1]}$ of p over $[1] \in \Delta$ is said to be the underlying quasi-category of the monoidal category p.

Let $p : M \to N(\Delta)^{\mathrm{op}}$ be a monoidal quasi-category. Since p is a coCartesian fibration by definition, we have a functor $\mathbf{X} : N(\Delta)^{\mathrm{op}} \to \mathrm{Cat}_\infty$ classifying p. We have the opposite coCartesian fibration $Rp : RM \to N(\Delta)^{\mathrm{op}}$ that is classified by the functor $R\mathbf{X}$. We easily see that $Rp : RM \to N(\Delta)^{\mathrm{op}}$ is a monoidal quasi-category. Note that the fiber $(RM)_{[n]}$ is equivalent to $(M_{[n]})^{\mathrm{op}} \simeq (M_{[1]}^{\mathrm{op}})^n$ for any $n \geq 0$. We say that RM is the opposite monoidal quasi-category of M.

A map $[m] \to [n]$ in Δ is said to be convex if it is injective and the image is $\{i, i+1, \ldots, i+m\}$ for some i. Let $p : M \to N(\Delta)^{\mathrm{op}}$ and $q : N \to N(\Delta)^{\mathrm{op}}$ be monoidal quasi-categories. A lax monoidal functor $G : N \to M$ between the monoidal quasi-categories is a map of simplicial sets over $N(\Delta)^{\mathrm{op}}$ which carries p-coCartesian edges over convex morphisms in $N(\Delta)^{\mathrm{op}}$ to q-coCartesian edges.

Lemma 1 *If $G_{[1]} : N_{[1]} \to M_{[1]}$ admits a left adjoint $F_{[1]}$, then there is a canonical functor $RF : RM \to RN$ over $N(\Delta)^{\mathrm{op}}$ up to contractible space of choices. We have $(RF)_{[n]} \simeq (F_{[1]}^{\mathrm{op}})^n$ for all $n \geq 0$.*

Proof For any $n \geq 0$, we have equivalences $M_{[n]} \simeq (M_{[1]})^n$ and $N_{[n]} \simeq (N_{[1]})^n$. Since G is a lax monoidal functor, we see that $G_{[n]}$ is equivalent to $(G_{[1]})^n$ under the above equivalences. Hence $G_{[n]}$ admits a left adjoint for all $n \geq 0$. The lemma follows from Theorem 4. □

Proposition 5 *If $G : N \to M$ is a lax monoidal functor between monoidal quasi-categories such that $G_{[1]} : N_{[1]} \to M_{[1]}$ admits a left adjoint, then the functor $RF : RM \to RN$ is also a lax monoidal functor between the opposite monoidal quasi-categories.*

Proof We have to show that RF preserves coCartesian edges over convex morphisms. Let $\alpha : [m] \to [n]$ be a convex morphism in Δ. Since $G : N \to M$ is a lax monoidal functor, we have a commutative diagram

$$\begin{array}{ccc} N_{[n]} & \xrightarrow{G_{[n]}} & M_{[n]} \\ {\scriptstyle \alpha_!^N}\downarrow & & \downarrow{\scriptstyle \alpha_!^M} \\ N_{[m]} & \xrightarrow{G_{[m]}} & M_{[m]} \end{array}$$

in Cat_∞. Since $\alpha_!^M : M_{[n]} \to M_{[m]}$ and $\alpha_!^N : N_{[n]} \to N_{[m]}$ are equivalent to projections $M_{[1]}^n \to M_{[1]}^m$ and $N_{[1]}^n \to N_{[1]}^m$, respectively, we see that the diagram is left adjointable. The proposition follows from Proposition 4. □

4.3 Opposites of Tensored Quasi-Categories Over Monoidal Quasi-Categories

In this subsection we study the opposite of a tensored quasi-category over a monoidal quasi-category. We show that the opposite of a lax tensored right adjoint functor between tensored quasi-categories induces a lax tensored right adjoint functor between the opposites of the tensored quasi-categories.

First, we recall the definition of left tensored quasi-category over a monoidal quasi-category. Let $p : X \to N(\Delta)^{\mathrm{op}} \times \Delta^1$ be a coCartesian fibration of simplicial sets. For any $n \geq 0$, the identity $\mathrm{id}_{[n]} : [n] \to [n]$ in Δ and the edge $\{0\} \to \{1\}$ in Δ^1 induces a morphism $([n], 0) \to ([n], 1)$ in $N(\Delta)^{\mathrm{op}} \times \Delta^1$, and hence we obtain a functor of quasi-categories $\alpha_n : X_{([n],0)} \to X_{([n],1)}$. For any $n \geq 0$, the inclusion $[0] \cong \{n\} \hookrightarrow [n]$ in Δ and the identity $\mathrm{id}_{\{0\}} : \{0\} \to \{0\}$ in Δ^1 induces a morphism $([0], 0) \to ([n], 0)$ in $N(\Delta)^{\mathrm{op}} \times \Delta^1$, and hence we obtain a functor of quasi-categories $\beta_n : X_{([n],0)} \to X_{([0],0)}$. If the base change of p along the inclusion $N(\Delta)^{\mathrm{op}} \times \{1\} \hookrightarrow N(\Delta)^{\mathrm{op}} \times \Delta^1$ is a monoidal quasi-category, and the functor

$$\alpha_n \times \beta_n : X_{([n],0)} \longrightarrow X_{([n],1)} \times X_{([0],0)}$$

is a categorical equivalence for all $n \geq 0$, then we say that p is a left tensored quasi-category.

Now suppose $p : X \to N(\Delta)^{\mathrm{op}} \times \Delta^1$ is a left tensored quasi-category. We set $\mathscr{M} = X_{([1],1)}$ and $\mathscr{C} = X_{([0],0)}$. Note that $\mathscr{M}$ is the underlying quasi-category of a monoidal quasi-category $p|_{N(\Delta)^{\mathrm{op}} \times \{1\}}$. We say that $\mathscr{C}$ is left tensored over the monoidal quasi-category $\mathscr{M}$.

Let Σ be a set of edges of $N(\Delta)^{\mathrm{op}} \times \Delta^1$ consisting of edges of the forms

$$([n], 0) \longrightarrow ([m], 0),$$

where $[m] \to [n]$ is a convex morphism in Δ that carries m to n, and

$$([n], i) \longrightarrow ([m], 1)$$

for $i = 0, 1$, where $[m] \to [n]$ is convex.

Suppose that $p : X \to N(\Delta)^{\mathrm{op}} \times \Delta^1$ and $q : Y \to N(\Delta)^{\mathrm{op}} \times \Delta^1$ are left tensored quasi-categories. We say that a functor $G : Y \to X$ over $N(\Delta)^{\mathrm{op}} \times \Delta^1$ is a lax left tensored functor if G carries p-coCartesian edges over Σ to q-coCartesian edges.

Lemma 2 *Let $p : X \to N(\Delta)^{\mathrm{op}} \times \Delta^1$ and $q : Y \to N(\Delta)^{\mathrm{op}} \times \Delta^1$ be left tensored quasi-categories. If $G : Y \to X$ is a lax left tensored functor such that $G_{([0],0)}$ and $G_{([1],1)}$ admit left adjoints $F_{([0],0)}$ and $F_{([1],1)}$, respectively, then there is a canonical functor $RF : RX \to RY$ over $N(\Delta)^{\mathrm{op}} \times \Delta^1$ up to contractible space of choices. We have $(RF)_{([n],0)} \simeq (F^{\mathrm{op}}_{([1],1)})^n \times F^{\mathrm{op}}_{([0],0)}$ and $(RF)_{([n],1)} \simeq (F^{\mathrm{op}}_{[1],1})^n$ for all $n \geq 0$.*

Proof For any $n \geq 0$, we have equivalences $X_{([n],0)} \simeq (X_{([1],1)})^n \times X_{([0],0)}$ and $Y_{([n],0)} \simeq (Y_{([1],1)})^n \times Y_{([0],0)}$. Since G is a lax left tensored functor, we see that

$G_{([n],0)}$ is equivalent to $(G_{([1],1)})^n \times G_{([0],0)}$. In the same way, we see that $G_{([n],1)}$ is equivalent to $(G_{([1],1)})^n$ for any $n \geq 0$. Hence G_s admits a left adjoint for all $s \in N(\Delta)^{\mathrm{op}} \times \Delta^1$. The lemma follows from Theorem 4. □

Proposition 6 *Let $p : X \to N(\Delta)^{\mathrm{op}} \times \Delta^1$ and $q : Y \to N(\Delta)^{\mathrm{op}} \times \Delta^1$ be left tensored quasi-categories. If $G : Y \to X$ is a lax left tensored functor such that $G_{([0],0)}$ and $G_{([1],1)}$ admit left adjoints, then the functor $RF : RX \to RY$ is also a lax left tensored functor.*

Proof We can prove the proposition in the same way as Proposition 5. We have to show that RF preserves coCartesian edges over Σ. Let $\alpha : s \to s'$ be an edge in Σ. Since $G : Y \to X$ is a lax left tensored functor, we have a commutative diagram

$$\begin{array}{ccc} Y_s & \xrightarrow{G_s} & X_s \\ {\scriptstyle \alpha_!^Y}\downarrow & & \downarrow{\scriptstyle \alpha_!^X} \\ Y_{s'} & \xrightarrow{G_{s'}} & X_{s'} \end{array}$$

in Cat_∞. Since $\alpha_!^Y : Y_s \to Y_{s'}$ and $\alpha_!^X : X_s \to X_{s'}$ are equivalent to the projections, we see that the diagram is left adjointable. The proposition follows from Proposition 4. □

5 Quasi-Categories of Comodules

In this section we introduce a quasi-category of comodules over a coalgebra in a stable homotopy theory $\mathscr{C}$. We regard a coalgebra as an algebra object of the opposite monoidal quasi-category of A-A-bimodule objects, where A is an algebra object of $\mathscr{C}$. We regard a comodule object over a coalgebra Γ as a module object over Γ in the opposite quasi-category of A-module objects. We define a cotensor product of a right comodule and a left comodule over a coalgebra as a limit of the cobar construction. Using these formulations, we study the functor from the localization of $\mathscr{C}$ with respect to A to the quasi-category of comodules over the coalgebra $A \otimes A$.

5.1 Monoidal Structure on ${}_A\mathbf{BMod}_A(\mathscr{C})^{\mathbf{op}}$

In this subsection we introduce a quasi-category of coalgebras and a quasi-category of comodules over a coalgebra in a stable homotopy theory.

Let $\mathscr{M}^{\otimes}$ be a monoidal quasi-category. We denote by $\mathscr{M}$ the underlying quasi-category of the monoidal quasi-category $\mathscr{M}^{\otimes}$. For algebra objects A and B of $\mathscr{M}$, we denote by ${}_A\mathrm{BMod}_B(\mathscr{M})$ the quasi-category of A-B-bimodule objects in $\mathscr{M}$. If B is the monoidal unit $\mathbf{1}$ in $\mathscr{M}$, we abbreviate the quasi-category ${}_A\mathrm{BMod}_{\mathbf{1}}(\mathscr{M})$ of A-$\mathbf{1}$-bimodule objects in $\mathscr{M}$ as ${}_A\mathrm{BMod}(\mathscr{M})$. Let $\mathscr{N}$ be a quasi-category left tensored over $\mathscr{M}^{\otimes}$. For an algebra object A of $\mathscr{M}$, we denote by $\mathrm{LMod}_A(\mathscr{N})$ the quasi-category of left A-module objects in $\mathscr{N}$. Note that there is a natural equivalence $\mathrm{LMod}_A(\mathscr{M}) \simeq {}_A\mathrm{BMod}(\mathscr{M})$ of quasi-categories.

Let $(\mathscr{C}, \otimes, \mathbf{1})$ be a stable homotopy theory in the sense of [25, Def. 2.1], that is, $\mathscr{C}$ is a presentable stable quasi-category which is the underlying quasi-category of a symmetric monoidal quasi-category $\mathscr{C}^{\otimes}$, where the tensor product commutes with all colimits separately in each variable. For an algebra object A of $\mathscr{C}$, we denote by ${}_A\mathrm{BMod}_A(\mathscr{C})$ the quasi-category of A-A-bimodules in $\mathscr{C}$, which is the underlying quasi-category of the monoidal quasi-category ${}_A\mathrm{BMod}_A(\mathscr{C})^{\otimes}$, where the tensor product is given by the relative tensor product $\otimes_A$ and the unit is the A-A-bimodule A (see [23, 4.3 and 4.4]). Note that the relative tensor product $\otimes_A$ commutes with all colimits separately in each variable by [23, Cor. 4.4.2.15]. For algebra objects A and B of $\mathscr{C}$, we denote by ${}_A\mathrm{BMod}_A(\mathscr{C})$ the quasi-category of A-B-bimodules, which is presentable by [23, Cor. 4.3.3.10].

If $\mathscr{M}^{\otimes}$ is a monoidal quasi-category, then the opposite quasi-category $(\mathscr{M}^{\otimes})^{\mathrm{op}}$ also carries a monoidal structure. Since ${}_A\mathrm{BMod}_A(\mathscr{C})$ is the underlying quasi-category of the monoidal quasi-category ${}_A\mathrm{BMod}_A(\mathscr{C})^{\otimes}$ for an algebra object A of $\mathscr{C}$, the opposite quasi-category ${}_A\mathrm{BMod}_A(\mathscr{C})^{\mathrm{op}}$ is the underlying quasi-category of the opposite monoidal quasi-category $({}_A\mathrm{BMod}_A(\mathscr{C})^{\otimes})^{\mathrm{op}}$. We regard an algebra object Γ of ${}_A\mathrm{BMod}_A(\mathscr{C})^{\mathrm{op}}$ as a coalgebra object of ${}_A\mathrm{BMod}_A(\mathscr{C})$. We define the quasi-category ${}_A\mathrm{CoAlg}_A(\mathscr{C})$ of coalgebra objects of ${}_A\mathrm{BMod}_A(\mathscr{C})$ to be the opposite of the quasi-category of algebra objects of ${}_A\mathrm{BMod}_A(\mathscr{C})^{\mathrm{op}}$:

$$ {}_A\mathrm{CoAlg}_A(\mathscr{C}) = \mathrm{Alg}({}_A\mathrm{BMod}_A(\mathscr{C})^{\mathrm{op}})^{\mathrm{op}}. $$

For a quasi-category $\mathscr{Y}$ left tensored over a monoidal category $\mathscr{M}^{\otimes}$, the opposite quasi-category $\mathscr{Y}^{\mathrm{op}}$ carries the structure of left tensored quasi-category over the opposite monoidal quasi-category $(\mathscr{M}^{\otimes})^{\mathrm{op}}$.

The quasi-category ${}_A\mathrm{BMod}(\mathscr{C}) \simeq \mathrm{LMod}_A(\mathscr{C})$ is left tensored over ${}_A\mathrm{BMod}_A(\mathscr{C})^{\otimes}$ by the relative tensor product $\otimes_A$ for an algebra object A of $\mathscr{C}$. Hence the opposite quasi-category ${}_A\mathrm{BMod}(\mathscr{C})^{\mathrm{op}}$ is left tensored over the opposite monoidal quasi-category $({}_A\mathrm{BMod}_A(\mathscr{C})^{\otimes})^{\mathrm{op}}$. Let Γ be a coalgebra object of ${}_A\mathrm{BMod}_A(\mathscr{C})$, that is, an algebra object of ${}_A\mathrm{BMod}_A(\mathscr{C})^{\mathrm{op}}$. We regard a left Γ-module in ${}_A\mathrm{BMod}(\mathscr{C})^{\mathrm{op}}$ as a left Γ-comodule in ${}_A\mathrm{BMod}(\mathscr{C})$. We define the quasi-category of left Γ-comodules $\mathrm{LComod}_{(A,\Gamma)}(\mathscr{C})$ to be the opposite of the quasi-category of left Γ-module objects in ${}_A\mathrm{BMod}_A(\mathscr{C})^{\mathrm{op}}$:

$$ \mathrm{LComod}_{(A,\Gamma)}(\mathscr{C}) = (\mathrm{LMod}_{\Gamma}({}_A\mathrm{BMod}(\mathscr{C})^{\mathrm{op}}))^{\mathrm{op}}. $$

Note that $\mathrm{LComod}_{(A,\Gamma)}(\mathscr{C})$ is right tensored over $\mathscr{C}$ and there is a forgetful functor

$$ \mathrm{LComod}_{(A,\Gamma)}(\mathscr{C}) \longrightarrow {}_A\mathrm{BMod}(\mathscr{C}) \simeq \mathrm{LMod}_A(\mathscr{C}), $$

which is a map of quasi-categories right tensored over $\mathscr{C}$.

In the same way as $\mathrm{LComod}_{(A,\Gamma)}(\mathscr{C})$, we can define the quasi-category of right Γ-comodules $\mathrm{RComod}_{(A,\Gamma)}(\mathscr{C})$ in $\mathscr{C}$ for a coalgebra Γ of ${}_A\mathrm{BMod}_A(\mathscr{C})$.

5.2 Comparison Maps

In this subsection we construct a functor between quasi-categories of comodules for a map of algebra objects. We also show that the definition of a quasi-category of comodules in this paper is consistent with the definition in [33].

Suppose $(\mathscr{C}, \otimes, \mathbf{1})$ be a stable homotopy theory. Let $f : A \to B$ be a map of algebra objects of $\mathscr{C}$. We have the functor

$$(f, f)^* : {}_B\mathrm{BMod}_B(\mathscr{C}) \longrightarrow {}_A\mathrm{BMod}_A(\mathscr{C})$$

which is obtained by restriction of scalars through f. The functor $(f, f)^*$ extends to a lax monoidal functor

$$((f, f)^*)^{\otimes} : {}_B\mathrm{BMod}_B(\mathscr{C})^{\otimes} \longrightarrow {}_A\mathrm{BMod}_A(\mathscr{C})^{\otimes}.$$

Furthermore, the functor $(f, f)^*$ admits a left adjoint

$$(f, f)_! : {}_A\mathrm{BMod}_A(\mathscr{C}) \longrightarrow {}_B\mathrm{BMod}_B(\mathscr{C}),$$

which assigns to an A-A-bimodule X the B-B-bimodule $B \otimes_A X \otimes_A B$. By Proposition 5, we obtain the following lemma.

Lemma 3 *If $f : A \to B$ is a map of algebra objects of $\mathscr{C}$, then the induced functor*

$$(f, f)_!^{\mathrm{op}} : {}_A\mathrm{BMod}_A(\mathscr{C})^{\mathrm{op}} \longrightarrow {}_B\mathrm{BMod}_B(\mathscr{C})^{\mathrm{op}}$$

can be extended to a lax monoidal functor.

In the remainder of this paper, for simplicity, we say that the underlying quasi-category $\mathscr{M}$ of a monoidal category $\mathscr{M}^{\otimes}$ is a monoidal category and that the underlying functor $F : \mathscr{M} \to \mathscr{N}$ of a (lax) monoidal functor $F^{\otimes} : \mathscr{M}^{\otimes} \to \mathscr{N}^{\otimes}$ is a (lax) monoidal functor.

For a map of algebra objects $f : A \to B$ in $\mathscr{C}$, the lax monoidal functor $(f, f)_!^{\mathrm{op}}$ induces a map of quasi-categories of algebra objects

$$(f, f)_!^{\mathrm{op}} : \mathrm{Alg}({}_A\mathrm{BMod}_A(\mathscr{C}))^{\mathrm{op}}) \longrightarrow \mathrm{Alg}({}_B\mathrm{BMod}_B(\mathscr{C})^{\mathrm{op}})$$

and hence we obtain a map of quasi-categories of coalgebra objects

$$(f, f)_! : {}_A\mathrm{CoAlg}_A(\mathscr{C}) \longrightarrow {}_B\mathrm{CoAlg}_B(\mathscr{C}).$$

Therefore, for a coalgebra object Γ of ${}_A\mathrm{BMod}_A(\mathscr{C})$, we obtain a coalgebra object

$$B \otimes_A \Gamma \otimes_A B = (f, f)_!(\Gamma)$$

of ${}_B\mathrm{BMod}_B(\mathscr{C})$. In particular, since the monoidal unit A in ${}_A\mathrm{BMod}_A(\mathscr{C})$ is a coalgebra object, we see that

$$B \otimes_A B = (f, f)_!(A)$$

is a coalgebra object of ${}_B\mathrm{BMod}_B(\mathscr{C})$.

In particular, since the monoidal unit $\mathbf{1}$ is a coalgebra object of ${}_\mathbf{1}\mathrm{BMod}_\mathbf{1}(\mathscr{C}) \simeq \mathscr{C}$, we have a coalgebra object

$$A \otimes A = (f, f)_!(\mathbf{1})$$

of ${}_A\mathrm{BMod}_A(\mathscr{C})$, where $f : \mathbf{1} \to A$ is the unit map of A. We write

$$\Gamma(A) = (A, A \otimes A)$$

for simplicity and we call $A \otimes A$-comodules $\Gamma(A)$-comodules interchangeably.

Let $f : A \to B$ be a map of algebra objects of $\mathscr{C}$. We denote by $f^* : \mathrm{LMod}_B(\mathscr{C}) \to \mathrm{LMod}_A(\mathscr{C})$ the restriction of scalars functor. Recall that f^* is a right adjoint to the extension of scalars functor

$$f_! : \mathrm{LMod}_A(\mathscr{C}) \to \mathrm{LMod}_B(\mathscr{C}),$$

which is given by $f_!(M) \simeq B \otimes_A M$.

Theorem 5 *Let Γ be a coalgebra object in ${}_A\mathrm{BMod}_A(\mathscr{C})$ and let $f : A \to B$ be a map of algebra objects of $\mathscr{C}$. The map f induces a functor of quasi-categories*

$$f_! : \mathrm{LComod}_{(A,\Gamma)}(\mathscr{C}) \longrightarrow \mathrm{LComod}_{(B,\Sigma)}(\mathscr{C})$$

which covers the functor $f_! : \mathrm{LMod}_A(\mathscr{C}) \to \mathrm{LMod}_B(\mathscr{C})$ through the forgetful functors, where $\Sigma = (f, f)_!\Gamma$.

Proof This follows from Proposition 6. □

Suppose we have a map $f : A \to B$ of algebra objects of $\mathscr{C}$. This induces an adjunction of functors

$$f_! : \mathrm{LMod}_A(\mathscr{C}) \rightleftarrows \mathrm{LMod}_B(\mathscr{C}) : f^*.$$

Taking the opposite quasi-categories, we obtain an adjunction of functors

$$f^{*\mathrm{op}} : \mathrm{LMod}_B(\mathscr{C})^{\mathrm{op}} \rightleftarrows \mathrm{LMod}_A(\mathscr{C})^{\mathrm{op}} : f_!^{\mathrm{op}}.$$

By this adjunction, we obtain an endomorphism monad

$$T \in \mathrm{Alg}(\mathrm{End}(\mathrm{LMod}_B(\mathscr{C})^{\mathrm{op}})),$$

and a quasi-category

$$\mathrm{LMod}_T(\mathrm{LMod}_B(\mathscr{C})^{\mathrm{op}}).$$

of left T-modules in $\mathrm{LMod}_B(\mathscr{C})^{\mathrm{op}}$ (see [23, Sect. 4.7.4]).

The following theorem shows that the definition of a quasi-category of comodules is consistent with the definition in [33].

Theorem 6 *There is an equivalence of quasi-categories*

$$\mathrm{LComod}_{(B,B\otimes_A B)}(\mathscr{C}) \simeq \mathrm{LMod}_T(\mathrm{LMod}_B(\mathscr{C})^{\mathrm{op}})^{\mathrm{op}}.$$

Proof Put $\mathscr{A} = \mathrm{LMod}_A(\mathscr{C})^{\mathrm{op}}$ and $\mathscr{B} = \mathrm{LMod}_B(\mathscr{C})^{\mathrm{op}}$. We have an adjunction of functors $F : \mathscr{B} \rightleftarrows \mathscr{A} : G$, where $F = f^{*\,\mathrm{op}}$ and $G = f_!^{\,\mathrm{op}}$. Since $\mathrm{LMod}_B(\mathscr{C}) \simeq {}_B\mathrm{BMod}(\mathscr{C})$, we can regard $B \otimes_A B$ as an algebra object of $\mathrm{End}(\mathscr{B})$. By [23, Prop. 4.7.4.3], we can lift G to $\overline{G} \in \mathrm{LMod}_{B\otimes_A B}(\mathrm{Fun}(\mathscr{B}, \mathscr{A}))$. We can verify that the composition

$$B \otimes_A B \longrightarrow (B \otimes_A B) \circ G \circ F \longrightarrow G \circ F$$

is an equivalence in $\mathrm{End}(\mathscr{B})$, where the first map is induced by the unit map of the adjunction (F, G), and the second map is induced by the left $B \otimes_A B$-action on G in $\mathrm{Fun}(\mathscr{B}, \mathscr{A})$. By [23, Prop. 4.7.4.3], we see that $B \otimes_A B$ is equivalent to the endomorphism monad T. Hence we obtain an equivalence between $\mathrm{LMod}_T(\mathrm{LMod}_B(\mathscr{C})^{\mathrm{op}})$ and $\mathrm{LMod}_{B\otimes_A B}({}_B\mathrm{BMod}(\mathscr{C})^{\mathrm{op}})$. □

5.3 Cotensor Products for Comodules in Quasi-Categories

Let $(\mathscr{C}, \otimes, \mathbf{1})$ be a stable homotopy theory. In this subsection we define a (derived) cotensor product of comodules in $\mathscr{C}$. In particular, we define a (derived) functor of taking primitives of comodules. We also study a comodule structure on cotensor products.

Let A be an algebra object of $\mathscr{C}$. Suppose Γ is a coalgebra object of the quasi-category ${}_A\mathrm{BMod}_A(\mathscr{C})$ of A-A-bimodules in $\mathscr{C}$, that is, Γ is an algebra object of the opposite monoidal quasi-category ${}_A\mathrm{BMod}_A(\mathscr{C})^{\mathrm{op}}$.

For a right Γ-comodule M and a left Γ-comodule N, we shall define a cotensor product

$$M \Box_\Gamma N.$$

We regard M as an object in $\mathrm{RMod}_\Gamma(\mathrm{BMod}_A(\mathscr{C})^{\mathrm{op}})$ and N as an object in $\mathrm{LMod}_\Gamma({}_A\mathrm{BMod}(\mathscr{C})^{\mathrm{op}})$. We can construct a two-sided bar construction

$$B_\bullet(M, \Gamma, N),$$

which is a simplicial object in $\mathscr{C}^{\mathrm{op}}$. The simplicial object $B_\bullet(M, \Gamma, N)$ has the nth term given by

$$B_n(M, \Gamma, N) \simeq M \otimes_A \overbrace{\Gamma \otimes_A \cdots \otimes_A \Gamma}^{n} \otimes_A N$$

with the usual structure maps. We regard $B_\bullet(M, \Gamma, N)$ as a cosimplicial object

$$C^\bullet(M, \Gamma, N)$$

in $\mathscr{C}$ and define the cotensor product $M\square_\Gamma N$ to be the limit of the cosimplicial object $C^\bullet(M, \Gamma, N)$:

$$M\square_\Gamma N = \lim_{N(\Delta)} C^\bullet(M, \Gamma, N).$$

Now we regard A as a right A-module and suppose that A is a right Γ-comodule via $\eta_R : A \to A \otimes_A \Gamma \simeq \Gamma$. We define a functor

$$P : \mathrm{LComod}_{(A,\Gamma)} \longrightarrow \mathscr{C}$$

by

$$P(N) = A\square_\Gamma N.$$

We consider the functor P is a derived functor of taking primitives in N.

Suppose we have algebra objects A, B, C of $\mathscr{C}$. The quasi-category of B-A-bimodules ${}_B\mathrm{BMod}_A(\mathscr{C})$ in $\mathscr{C}$ is right tensored over the monoidal quasi-category ${}_A\mathrm{BMod}_A(\mathscr{C})$ and the quasi-category of A-C-bimodules ${}_A\mathrm{BMod}_C(\mathscr{C})$ in $\mathscr{C}$ is left tensored over the monoidal quasi-category ${}_A\mathrm{BMod}_A(\mathscr{C})$. Let Γ be a coalgebra object of ${}_A\mathrm{BMod}_A(\mathscr{C})$. We can define right Γ-comodule objects of ${}_B\mathrm{BMod}_A(\mathscr{C})$ and left Γ-comodule objects of ${}_A\mathrm{BMod}_C(\mathscr{C})$ in the same way as Γ-comodule objects of ${}_A\mathrm{BMod}_A(\mathscr{C})$. Suppose we have a right Γ-comodule M of ${}_B\mathrm{BMod}_A(\mathscr{C})$ and a left Γ-comodule N of ${}_A\mathrm{BMod}_C(\mathscr{C})$. We can form the cobar construction $C^\bullet(M, \Gamma, N)$ in ${}_B\mathrm{BMod}_C(\mathscr{C})$. Hence the cotensor product $M\square_\Gamma N$ is a B-C-bimodule

$$M\square_\Gamma N \in {}_B\mathrm{BMod}_C(\mathscr{C}).$$

Let Σ be a coalgebra object of ${}_B\mathrm{BMod}_B(\mathscr{C})$. Now suppose M is a (Σ, Γ)-bicomodule object of ${}_B\mathrm{BMod}_A(\mathscr{C})$, that is, M is a (Σ, Γ)-bimodule object of ${}_B\mathrm{BMod}_A(\mathscr{C})^{\mathrm{op}}$. In general, the cotensor product $M\square_\Gamma N$ does not support a left Σ-comodule structure. The following proposition gives us a sufficient condition for $M\square_\Gamma N$ to be a left Σ-comodule object of ${}_B\mathrm{BMod}_C(\mathscr{C})$ induced by the left Σ-comodule structure on M.

Proposition 7 *Let M be a (Σ, Γ)-bicomodule object of ${}_B\mathrm{BMod}_A(\mathscr{C})$ and let N be a left Γ-comodule object of ${}_A\mathrm{BMod}_C(\mathscr{C})$. If the canonical map*

$$\overbrace{\Sigma \otimes_B \cdots \otimes_B \Sigma}^{r} \otimes_B (M\square_\Gamma N) \longrightarrow (\overbrace{\Sigma \otimes_B \cdots \otimes_B \Sigma}^{r} \otimes_B M)\square_\Gamma N$$

is an equivalence in ${}_B\mathrm{BMod}_C(\mathscr{C})$ for all $r > 0$, then the left Σ-comodule structure on M induces a left Σ-comodule structure on $M\square_\Gamma N$.

In order to prove Proposition 7, we need the following lemma.

Lemma 4 *Let $\mathcal{M}$ be a monoidal quasi-category, A an algebra object of $\mathcal{M}$, and $\mathcal{D}$ a quasi-category left-tensored over $\mathcal{M}$. Suppose we have a diagram $X : K \to \mathrm{LMod}_A(\mathcal{D})$, where K is a simplicial set. We set $Y = \pi \circ X : K \to \mathcal{D}$, where $\pi : \mathrm{LMod}_A(\mathcal{D}) \to \mathcal{D}$ is the forgetful functor. We assume that there exists a colimit $\mathrm{colim}^{\mathcal{D}}_K(A^{\otimes r} \otimes Y)$ in $\mathcal{D}$ for all $r \geq 0$. If the canonical map $\mathrm{colim}^{\mathcal{D}}_K(A^{\otimes r} \otimes Y) \to A^{\otimes r} \otimes \mathrm{colim}^{\mathcal{D}}_K Y$ is an equivalence for all $r > 0$, then there exists a colimit of X in $\mathrm{LMod}_A(\mathcal{D})$ and the forgetful functor $\pi : \mathrm{LMod}_A(\mathcal{D}) \to \mathcal{D}$ preserves the colimit.*

Proof We use the notation in [23, Sect. 4.2.2]. Let $\mathcal{D}^{\circledast}$ and $\mathcal{M}^{\circledast}$ be quasi-categories defined in [23, Notation 4.2.2.16]. We have maps $\mathcal{D}^{\circledast} \xrightarrow{q} \mathcal{M}^{\circledast} \xrightarrow{p} N(\Delta)^{\mathrm{op}}$, where p and $p \circ q$ are coCartesian fibrations by [23, Remark. 4.2.2.24]. Furthermore, q is a categorical fibration by [23, Remark. 4.2.2.18] and a locally coCartesian fibration by [23, Lem. 4.2.2.19]. Note that there is an equivalence of quasi-categories $\mathcal{D}^{\circledast}_{[s]} \simeq \mathcal{M}^{\circledast}_{[s]} \times \mathcal{D}$ and the restriction $q_{[s]} : \mathcal{D}^{\circledast}_{[s]} \to \mathcal{M}^{\circledast}_{[s]}$ is the projection for any $[s] \in N(\Delta)^{\mathrm{op}}$.

We have simplicial models of algebra and module objects in quasi-categories (see [23, Sects. 4.1.2 and 4.2.2]). We have a full subcategory ${}^{\Delta}\mathrm{Alg}(\mathcal{M})$ of the quasi-category $\mathrm{Fun}_{N(\Delta)^{\mathrm{op}}}(N(\Delta)^{\mathrm{op}}, \mathcal{M}^{\circledast})$ which is equivalent to the quasi-category $\mathrm{Alg}(\mathcal{M})$ of algebra objects of $\mathcal{M}$ (see [23, Def. 4.1.2.14 and Prop. 4.1.2.15]). We denote by $A' : N(\Delta)^{\mathrm{op}} \to \mathcal{M}^{\circledast}$ the corresponding simplicial object of $\mathcal{M}^{\circledast}$ to $A \in \mathrm{Alg}(\mathcal{M})$.

We form a pullback diagram

$$\begin{array}{ccc} \mathcal{N} & \xrightarrow{j} & \mathcal{D}^{\circledast} \\ {\scriptstyle q'}\downarrow & & \downarrow{\scriptstyle q} \\ N(\Delta)^{\mathrm{op}} & \xrightarrow{A'} & \mathcal{M}^{\circledast}, \end{array}$$

where q' is a locally coCartesian fibration and a categorical fibration. Note that the fiber $\mathcal{N}_{[n]}$ of q' over $[n]$ is equivalent to $\mathcal{D}$ for all $[n] \in N(\Delta)^{\mathrm{op}}$. We have a full subcategory ${}^{\Delta}\mathrm{LMod}_{A'}(\mathcal{D})$ of $\mathrm{Fun}_{N(\Delta)^{\mathrm{op}}}(N(\Delta)^{\mathrm{op}}, \mathcal{N})$, which is equivalent to $\mathrm{LMod}_A(\mathcal{D})$ (see [23, Cor. 4.2.2.15]). An object G of $\mathrm{Fun}_{N(\Delta)^{\mathrm{op}}}(N(\Delta)^{\mathrm{op}}, \mathcal{N})$ belongs to $\mathrm{LMod}_{A'}(\mathcal{D})$ if and only if the edge $(j \circ G)(\alpha^{\mathrm{op}}) : (j \circ G)([n]) \to (j \circ G)([m])$ is $p \circ q$-coCartesian for any convex map $\alpha : [m] \to [n]$ in Δ such that $\alpha(m) = n$.

We denote by $f : K \times N(\Delta)^{\mathrm{op}} \to \mathcal{N}$ the map corresponding to the diagram $X \in \mathrm{Fun}(K, \mathrm{LMod}_A(\mathcal{D}))$. We let $g : K^{\triangleright} \times N(\Delta)^{\mathrm{op}} \to N(\Delta)^{\mathrm{op}}$ be the projection. We have a commutative diagram

$$\begin{array}{ccc} K \times N(\Delta)^{\mathrm{op}} & \xrightarrow{f} & \mathcal{N} \\ \downarrow & & \downarrow{\scriptstyle q'} \\ K^{\triangleright} \times N(\Delta)^{\mathrm{op}} & \xrightarrow{g} & N(\Delta)^{\mathrm{op}}, \end{array}$$

where the left vertical arrow is the inclusion. We shall show that there is a q'-left Kan extension $\overline{f} : K^{\triangleright} \times N(\Delta)^{\mathrm{op}} \to \mathscr{N}$ which makes the whole diagram commutative, and that the adjoint map gives rise to a colimit diagram $K^{\triangleright} \to \mathrm{LMod}_{A'}(\mathscr{D}) \simeq \mathrm{LMod}_A(\mathscr{D})$.

Let $f_{[n]} : K \to \mathscr{N}_{[n]} \simeq \mathscr{D}$ be the restriction of the map f over $[n] \in N(\Delta)^{\mathrm{op}}$, which is equivalent to Y. Since Y has a colimit in $\mathscr{D}$ by the assumption, we obtain a colimit diagram $\overline{f}_{[n]} : K^{\triangleright} \to \mathscr{N}_{[n]} \simeq \mathscr{D}$ that is an extension of f.

Let $\alpha : [n] \to [m]$ be an edge in $N(\Delta)^{\mathrm{op}}$. Since q' is a locally coCartesian fibration, α induces a functor $\alpha_! : \mathscr{N}_{[n]} \to \mathscr{N}_{[m]}$. The composition $\alpha_! \circ f_{[n]} : K \to \mathscr{N}_{[m]} \simeq \mathscr{D}$ is equivalent to $A^{\otimes r} \otimes Y$ for some $r \geq 0$. This implies that $\alpha_! \circ \overline{f}_{[n]}$ is a colimit diagram in $\mathscr{N}_{[m]} \simeq \mathscr{D}$ by the assumption that the canonical map $\mathrm{colim}^{\mathscr{D}}_K(A^{\otimes r} \otimes Y) \to A^{\otimes r} \otimes \mathrm{colim}^{\mathscr{D}}_K Y$ is an equivalence. Hence we see that $i_{[n]} \circ \overline{f}_{[n]}$ is a q'-colimit diagram by [22, Prop. 4.3.1.10], where $i_{[n]} : \mathscr{N}_{[n]} \hookrightarrow \mathscr{N}$ is the inclusion.

By the dual of [23, Lem. 3.2.2.9(1)], there exists a q'-left Kan extension $\overline{f} : K^{\triangleright} \times N(\Delta)^{\mathrm{op}} \to \mathscr{N}$ of f such that $q' \circ \overline{f} = g$. The restriction of $\overline{f}$ to $K^{\triangleright} \times \{[n]\}$ is equivalent to $i_{[n]} \circ \overline{f}_{[n]}$ for all $[n] \in N(\Delta)^{\mathrm{op}}$.

We consider the adjoint map $K^{\triangleright} \to \mathrm{Fun}(N(\Delta)^{\mathrm{op}}, \mathscr{N})$ of f. By the dual of [23, Lem. 3.2.2.9(2)], this map is a $(q')^{N(\Delta)^{\mathrm{op}}}$-colimit diagram, where $(q')^{N(\Delta^{\mathrm{op}})} : \mathrm{Fun}(N(\Delta)^{\mathrm{op}}, \mathscr{N}) \to \mathrm{Fun}(N(\Delta)^{\mathrm{op}}, N(\Delta)^{\mathrm{op}})$ is induced by q'. Since g is the projection, we see that it factors through $\mathrm{Fun}_{N(\Delta)^{\mathrm{op}}}(N(\Delta)^{\mathrm{op}}, \mathscr{N})$ and we obtain a map $\widehat{f} : K^{\triangleright} \to \mathrm{Fun}_{N(\Delta)^{\mathrm{op}}}(N(\Delta)^{\mathrm{op}}, \mathscr{N})$. By [22, Prop. 4.3.1.5(4)], we see that $\widehat{f}$ is a colimit diagram.

We shall show that $\widehat{f}$ factors through $\mathrm{LMod}_{A'}(\mathscr{D})$. Note that the restriction of $\widehat{f}$ to K factors through $\mathrm{LMod}_{A'}(\mathscr{D})$. Let $F = \widehat{f}(\infty) \in \mathrm{Fun}_{N(\Delta)^{\mathrm{op}}}(N(\Delta)^{\mathrm{op}}, \mathscr{N})$, where ∞ is the cone point of $K^{\triangleright}$. Since $\overline{f}_{[n]}$ is a colimit diagram extending Y, we have $F([n]) \simeq \mathrm{colim}^{\mathscr{D}}_K Y$ in $\mathscr{N}_{[n]} \simeq \mathscr{D}$ for any $[n] \in N(\Delta)^{\mathrm{op}}$. Let $\alpha : [m] \to [n]$ be a convex map in Δ such that $\alpha(m) = n$. The induced functor $\alpha_! : \mathscr{N}_{[n]} \to \mathscr{N}_{[m]}$ is identified with the identity functor of $\mathscr{D}$. This implies that $(j \circ F)(\alpha^{\mathrm{op}}) : (j \circ F)([n]) \to (j \circ F)([m])$ is a $p \circ q$-coCartesian edge. Hence $\widehat{f}$ factors through the full subcategory ${}^{\Delta}\mathrm{LMod}_{A'}(\mathscr{D})$ and the map $\widehat{f} : K^{\triangleright} \to {}^{\Delta}\mathrm{LMod}_{A'}(\mathscr{D})$ is a colimit diagram.

By the construction of $\widehat{f}$, the composition $\pi \circ \widehat{f} : K^{\triangleright} \to \mathscr{D}$ is also a colimit diagram, where $\pi : {}^{\Delta}\mathrm{LMod}_{A'}(\mathscr{D}) \to \mathscr{D}$ is the forgetful functor. This completes the proof. □

Proof (*Proof of Proposition* 7) We shall apply Lemma 4. We have the monoidal quasi-category ${}_B\mathrm{BMod}_B(\mathscr{C})^{\mathrm{op}}$, the algebra object Σ of ${}_B\mathrm{BMod}_B(\mathscr{C})^{\mathrm{op}}$, and the quasi-category ${}_B\mathrm{BMod}_C(\mathscr{C})^{\mathrm{op}}$ left tensored over ${}_B\mathrm{BMod}_B(\mathscr{C})^{\mathrm{op}}$. By the bar construction, we have a simplicial object $B_\bullet(M, \Gamma, N)$ of $\mathrm{LMod}_\Sigma({}_B\mathrm{BMod}_C(\mathscr{C})^{\mathrm{op}})$. By the assumption, the canonical map

$$\mathrm{colim}_{N(\Delta)^{\mathrm{op}}} B_\bullet(\Sigma^{\otimes_B r} \otimes_B M, \Gamma, N) \to \Sigma^{\otimes_B r} \otimes_B \mathrm{colim}_{N(\Delta)^{\mathrm{op}}} B_\bullet(M, \Gamma, N)$$

is an equivalence in ${}_B\mathrm{BMod}_C(\mathscr{C})^{\mathrm{op}}$ for all $r > 0$. By Lemma 4, there exists a colimit of $B_\bullet(M, \Gamma, N)$ in $\mathrm{LMod}_\Sigma({}_B\mathrm{BMod}_C(\mathscr{C})^{\mathrm{op}})$ and the colimit is created in ${}_B\mathrm{BMod}_C(\mathscr{C})^{\mathrm{op}}$. Hence we see that the cosimplicial object $C^\bullet(M, \Gamma, N)$ has a limit in $\mathrm{LComod}_{\Sigma}({}_{B}\mathrm{BMod}_{C}(\mathscr{C}))$ and the underlying object of the limit is $M\square_{\Gamma}N$ in ${}_{B}\mathrm{BMod}_{C}(\mathscr{C})$. □

5.4 *Equivalence of Quasi-Categories of Comodules*

Let $(\mathscr{C}, \otimes, \mathbf{1})$ be a stable homotopy theory and let A be an algebra object of $\mathscr{C}$. In this subsection we study the relationship between the localization of $\mathscr{C}$ with respect to A and the quasi-category of $\Gamma(A)$-comodules in $\mathscr{C}$.

We regard A as a right A-module and the map of right A-modules $\eta_R : A \simeq S \otimes A \to A \otimes A$ induces a right $\Gamma(A)$-comodule structure on A. Since $\mathrm{RComod}_{\Gamma(A)}(\mathscr{C})$ is left tensored over $\mathscr{C}$, we have $X \otimes A \in \mathrm{RComod}_{\Gamma(A)}(\mathscr{C})$ for any $X \in \mathscr{C}$.

Recall that we have a cosimplicial object

$$C^{\bullet}(A, A \otimes A, M)$$

in $\mathscr{C}$ by cobar construction for $M \in \mathrm{LComod}_{\Gamma(A)}(\mathscr{C})$. The totalization of $C^{\bullet}(A, A \otimes A, M)$ is $P(M)$. In the same way as in [33, Prop. 5.1], we obtain an adjunction of functors

$$A \otimes (-) : \mathscr{C} \rightleftarrows \mathrm{LComod}_{\Gamma(A)}(\mathscr{C}) : P.$$

Let $C^{\bullet} : N(\Delta) \to \mathscr{C}$ be a cosimplicial object in $\mathscr{C}$. We recall the Tot tower associated to $C^{\bullet}$. For $r \geq 0$, we denote by $\Delta^{\leq r}$ the full subcategory of Δ spanned by $\{[0], [1], \ldots, [r]\}$. We denote by $C^{\bullet}|_{N(\Delta^{\leq r})}$ the restriction of $C^{\bullet}$ to $N(\Delta^{\leq r})$. We recall that $\mathrm{Tot}_r(C^{\bullet})$ is defined to be the limit of $C^{\bullet}|_{N(\Delta^{\leq r})}$ in $\mathscr{C}$. The inclusion $\Delta^{\leq r} \hookrightarrow \Delta^{\leq r+1}$ induces a map $\mathrm{Tot}_{r+1}(C^{\bullet}) \to \mathrm{Tot}_r(C^{\bullet})$ for $r \geq 0$ and we obtain a tower $\{\mathrm{Tot}_r C^{\bullet}\}_{r\geq 0}$. Note that the limit of the tower is equivalent to $\mathrm{Tot}(C^{\bullet})$:

$$\mathrm{Tot}(C^{\bullet}) \simeq \lim_{r} \mathrm{Tot}_r(C^{\bullet}).$$

If there is a coaugmentation $D \to C^{\bullet}$, then we obtain a map of towers $c(D) \to \{\mathrm{Tot}_r(C^{\bullet})\}_{r\geq 0}$, where $c(D)$ is the constant tower on D.

We denote by $\mathrm{Pro}(\mathscr{C})$ the quasi-category of pro-objects in $\mathscr{C}$ (see [25, Sect. 3]). We have a fully faithful embedding $\mathscr{C} \hookrightarrow \mathrm{Pro}(\mathscr{C})$. We say that an object of $\mathrm{Pro}(\mathscr{C})$ is constant if it is equivalent to an object in the image of the embedding $\mathscr{C} \hookrightarrow \mathrm{Pro}(\mathscr{C})$.

Lemma 5 *For any $M \in \mathrm{LComod}_{\Gamma(A)}(\mathscr{C})$, the cosimplicial object $A \otimes C^{\bullet}(A, A \otimes A, M)$ is split, and hence the tower $\{\mathrm{Tot}_r(A \otimes C^{\bullet}(A, A \otimes A, M))\}$ associated to the cosimplicial object $A \otimes C^{\bullet}(A, A \otimes A, M)$ is equivalent to the constant object M in $\mathrm{Pro}(\mathscr{C})$.*

Proof We have an isomorphism of cosimplicial objects

$$A \otimes C^{\bullet}(A, A \otimes A, M) \cong C^{\bullet}(A \otimes A, A \otimes A, M).$$

The lemma follows from the fact that $C^\bullet(A \otimes A, A \otimes A, M)$ is a split cosimplicial object. □

A full subcategory $\mathscr{I} \subset \mathscr{C}$ is said to be an ideal if $X \otimes Y \in \mathscr{I}$ whenever $X \in \mathscr{C}$ and $Y \in \mathscr{I}$ (cf. [25, Definition 2.16]). A full subcategory $\mathscr{D} \subset \mathscr{C}$ is said to be thick if $\mathscr{D}$ is closed under finite limits and colimits and under retracts. If, furthermore, $\mathscr{D}$ is an ideal, we say that $\mathscr{D}$ is a thick tensor ideal (cf. [25, Definition 3.16]). Given a collection of objects in $\mathscr{C}$, the thick tensor ideal generated by them is the smallest thick tensor ideal containing the collection. Let A be an algebra object of $\mathscr{C}$. An object $X \in \mathscr{C}$ is said to be A-nilpotent if X belongs to the thick tensor ideal generated by A.

Lemma 6 *Let A be an algebra object of $\mathscr{C}$. If the unit* $\mathbf{1}$ *is A-nilpotent, then the tower associated to the cosimplicial object $C^\bullet(A, A \otimes A, M)$ is equivalent to the constant object $P(M)$ in* $\mathrm{Pro}(\mathscr{C})$ *for any $M \in \mathrm{LComod}_{\Gamma(A)}(\mathscr{C})$.*

Proof Let $\mathscr{I}$ be the class of objects X in $\mathscr{C}$ such that the tower associated to the cosimplicial object $X \otimes C^\bullet(A, A \otimes A, M)$ is equivalent to a constant object in $\mathrm{Pro}(\mathscr{C})$. We see that $\mathscr{I}$ is a thick tensor ideal of $\mathscr{C}$ and contains A by Lemma 5. Hence $\mathscr{I}$ contains the unit $\mathbf{1}$ by the assumption. This implies that the tower associated to $C^\bullet(A, A \otimes A, M)$ is equivalent to the constant object $\lim_{N(\Delta)} C^\bullet(A, A \otimes A, M) \simeq P(M)$ in $\mathrm{Pro}(\mathscr{C})$. □

For any $X \in \mathscr{C}$ and $M \in \mathrm{LComod}_{\Gamma(A)}(\mathscr{C})$, we have an equivalence of cosimplicial objects $X \otimes C^\bullet(A, A \otimes A, M) \simeq C^\bullet(X \otimes A, A \otimes A, M)$. This induces a natural map

$$X \otimes P(M) \longrightarrow (X \otimes A)\square_{A\otimes A} M.$$

Proposition 8 *Let A be an algebra object of $\mathscr{C}$. If the unit* $\mathbf{1}$ *is A-nilpotent, then the natural map $X \otimes P(M) \to (X \otimes A)\square_{A\otimes A} M$ is an equivalence for any $X \in \mathscr{C}$ and $M \in \mathrm{Comod}_{\Gamma(A)}(\mathscr{C})$.*

Proof By Lemma 6, the tower associated to the cosimplicial object $C^\bullet(A, A \otimes A, M)$ is equivalent to the constant object $P(M)$ in $\mathrm{Pro}(\mathscr{C})$. This implies that the tower

$$\{\mathrm{Tot}_r(X \otimes C^\bullet(A, A \otimes A, M))\}$$

associated to the cosimplicial object $X \otimes C^\bullet(A, A \otimes A, M)$ is also equivalent to the constant object $X \otimes P(M)$ in $\mathrm{Pro}(\mathscr{C})$. By the equivalence of cosimplicial objects $X \otimes C^\bullet(A, A \otimes A, M) \simeq C^\bullet(X \otimes A, A \otimes A, M)$, we obtain the equivalence $X \otimes P(M) \simeq (X \otimes A)\square_{A\otimes A} M$. □

Corollary 1 *Let A be an algebra object of $\mathscr{C}$. If the unit* $\mathbf{1}$ *is A-nilpotent, then the counit map*

$$A \otimes P(M) \to M$$

is an equivalence for any $M \in \mathrm{LComod}_{\Gamma(A)}(\mathscr{C})$.

Proof By Proposition 8, we have a natural equivalence $A \otimes P(M) \simeq (A \otimes A) \Box_{A \otimes A} M$. The corollary follows from the fact that $(A \otimes A) \Box_{A \otimes A} M \simeq M$. □

We consider the localization of $\mathscr{C}$ with respect to A. A morphism $f : X \to Y$ in $\mathscr{C}$ is said to be an A-equivalence if $A \otimes f$ is an equivalence. We denote by $L_A \mathscr{C}$ the localization of $\mathscr{C}$ with respect to the class of A-equivalences.

The following theorem is a slight generalization of [25, Prop. 3.21].

Theorem 7 *Let A be an algebra object of $\mathscr{C}$. If the unit $\mathbf{1}$ is A-nilpotent, then $L_A \mathscr{C}$ is equivalent to $\mathrm{LComod}_{\Gamma(A)}(\mathscr{C})$. We have an adjoint equivalence*

$$A \otimes (-) : L_A \mathscr{C} \rightleftarrows \mathrm{LComod}_{\Gamma(A)}(\mathscr{C}) : P,$$

and we can identify the functor $A \otimes (-) : \mathscr{C} \to \mathrm{LComod}_{\Gamma(A)}(\mathscr{C})$ with the localization $\mathscr{C} \to L_A \mathscr{C}$.

Proof We have an adjoint pair of functors $A \otimes (-) : \mathscr{C} \rightleftarrows \mathrm{LComod}_{\Gamma(A)}(\mathscr{C}) : P$. Clearly, $A \otimes f$ is an equivalence in $\mathrm{LComod}_{\Gamma(A)}(\mathscr{C})$ if and only if f is an A-equivalence for any morphism f in $\mathscr{C}$. Hence it suffices to show that the right adjoint P is fully faithful. The counit map $\varepsilon : A \otimes P(M) \to M$ is an equivalence for any $M \in \mathrm{LComod}_{\Gamma(A)}(\mathscr{C})$ by Corollary 1. Hence we see that P is fully faithful. □

6 Comodules in the Quasi-Category of Spectra

In this section we study quasi-category of comodules in spectra. Using the Bousfield–Kan spectral sequence and the results in [18], we show that the quasi-category of comodules associated to a Landweber exact $\mathbb{S}$-algebra depends only on its height. We also show that the $E(n)$-local category is equivalent to the quasi-category of comodules over the coalgebra $E(n) \otimes E(n)$. In [33] we considered the model category of F_n-modules in the category of discrete symmetric $\mathbb{G}_n$-spectra, where $\mathbb{G}_n$ is the extended Morava stabilizer group and F_n is a discrete model of the Morava E-theory spectrum E_n. We show that the category of F_n-modules in the discrete symmetric $\mathbb{G}_n$-spectra models the $K(n)$-local category.

6.1 Cotensor Product and Its Derived Functor in Algebraic Setting

In this subsection we recall some properties of the category of comodules over a coalgebra in algebraic setting. We study the derived functor of cotensor product of comodules and show that the derived functor can be described by the cobar complex in some situations. The content in this section is not new. Our main reference is [31, Appendix A1.2].

Let Ab_* be the category of graded abelian groups. Let A be a monoid object in Ab_*. We denote by ${}_A\mathrm{BMod}_A(\mathrm{Ab}_*)$ the category of A-A-bimodules in Ab_*. The category ${}_A\mathrm{Bmod}_A(\mathrm{Ab}_*)$ is a monoidal category with the tensor product $\otimes_A$ and the unit A. We denote by ${}_A\mathrm{BMod}_A(\mathrm{Ab}_*)^{\mathrm{op}}$ the opposite monoidal category. A coalgebra in ${}_A\mathrm{BMod}_A(\mathrm{Ab}_*)$ is defined to be a monoid object in ${}_A\mathrm{BMod}_A(\mathrm{Ab}_*)^{\mathrm{op}}$. In other word, a coalgebra Γ is an A-A-bimodule equipped with maps

$$\begin{aligned} \psi &: \Gamma \longrightarrow \Gamma \otimes_A \Gamma, \\ \varepsilon &: \Gamma \longrightarrow A \end{aligned}$$

in ${}_A\mathrm{BMod}_A(\mathrm{Ab}_*)$ satisfying the coassociativity and counit conditions.

We denote by ${}_A\mathrm{CoAlg}_A(\mathrm{Ab}_*)$ the category of coalgebras in ${}_A\mathrm{BMod}_A(\mathrm{Ab}_*)$. By definition, we have an equivalence

$${}_A\mathrm{CoAlg}_A(\mathrm{Ab}_*) \simeq \mathrm{Alg}({}_A\mathrm{BMod}_A(\mathrm{Ab}_*)^{\mathrm{op}})^{\mathrm{op}}.$$

We denote by $\mathrm{LMod}_A(\mathrm{Ab}_*)$ the category of left A-modules and by $\mathrm{RMod}_A(\mathrm{Ab}_*)$ the category of right A-modules, respectively. Let $\Gamma \in {}_A\mathrm{CoAlg}_A(\mathrm{Ab}_*)$. A left Γ-comodule is defined to be a left A-module M equipped with a map

$$\psi : M \longrightarrow \Gamma \otimes_A M$$

in $\mathrm{LMod}_A(\mathrm{Ab}_*)$ satisfying the coassociativity and counit conditions. A right Γ-comodule is defined in the similar fashion. We denote by $\mathrm{LComod}_{(A,\Gamma)}(\mathrm{Ab}_*)$ the category of left Γ-comodules and by $\mathrm{RComod}_{(A,\Gamma)}(\mathrm{Ab}_*)$ the category of right Γ-comodules, respectively.

The following lemma is obtained in the same way as in [31, Thm. A1.1.3 and Lem. A1.2.2].

Lemma 7 *If Γ is flat as a right A-module, then* $\mathrm{LComod}_{(A,\Gamma)}(\mathrm{Ab}_*)$ *is an abelian category with enough injectives.*

In the following of this subsection we assume that a coalgebra $\Gamma \in {}_A\mathrm{CoAlg}_A(\mathrm{Ab}_*)$ is flat as a right A-module and a left A-module. Hence we can do homological algebra in $\mathrm{LComod}_{(A,\Gamma)}(\mathrm{Ab}_*)$. We abbreviate $\mathrm{Hom}_{\mathrm{LComod}_{(A,\Gamma)}(\mathrm{Ab}_*)}(-,-)$ as $\mathrm{Hom}_\Gamma(-,-)$. For a left Γ-comodule M, we define

$$\mathrm{Ext}^i_\Gamma(M,-)$$

to be the ith right derived functor of

$$\mathrm{Hom}_\Gamma(M,-) : \mathrm{LComod}_{(A,\Gamma)}(\mathrm{Ab}_*) \longrightarrow \mathrm{Ab}_*.$$

For $M \in \mathrm{RComod}_{(A,\Gamma)}(\mathrm{Ab}_*)$ and $N \in \mathrm{LComod}_{(A,\Gamma)}(\mathrm{Ab}_*)$, we denote by $M \square_\Gamma N$ the cotensor product of M and N over Γ (see, for example, [31, Definition A1.1.4]). We consider the functor

$$M\square_{\Gamma}(-) : \mathrm{LComod}_{(A,\Gamma)}(\mathrm{Ab}_*) \longrightarrow \mathrm{Ab}_*.$$

We define

$$\mathrm{Cotor}^i_{\Gamma}(M, -)$$

to be the ith right derived functor of $M\square_{\Gamma}(-)$. Note that if M is flat as a right A-module, then $\mathrm{Cotor}^0_{\Gamma}(M, N) \cong M\square_{\Gamma}N$ since $M\square_{\Gamma}(-)$ is left exact in this case.

Let M be a left Γ-comodule that is finitely generated and projective as a left A-module. There is a right Γ-comodule structure on $\mathrm{Hom}_A(M, A)$ and we have a natural isomorphism

$$\mathrm{Hom}_{\Gamma}(M, N) \cong \mathrm{Hom}_A(M, A)\square_{\Gamma}N$$

for any $N \in \mathrm{LComod}_{(A,\Gamma)}(\mathrm{Ab}_*)$ (cf. [31, Lem. A1.1.6]). This implies that there is a natural isomorphism

$$\mathrm{Ext}^i_{\Gamma}(M, N) \cong \mathrm{Cotor}^i_{\Gamma}(\mathrm{Hom}_A(M, A), N)$$

for any $i \geq 0$. In particular, we have a natural isomorphism

$$\mathrm{Ext}^i_{\Gamma}(A, N) \cong \mathrm{Cotor}^i_{\Gamma}(A, N)$$

for any $N \in \mathrm{LComod}_{(A,\Gamma)}(\mathrm{Ab}_*)$ and $i \geq 0$.

For a right Γ-comodule M and a left Γ-comodule N, we have a cosimplicial object

$$C^{\bullet}(M, \Gamma, N)$$

in Ab_* obtained by the cobar construction. In particular, we have

$$C^r(M, \Gamma, N) = M \otimes_A \Gamma^{\otimes_A r} \otimes_A N$$

for $r \geq 0$. The cobar complex $C^*(M, \Gamma, N)$ is the associated cochain complex. The normalized cobar complex $\overline{C}^*(M, \Gamma, N)$ is a subcomplex of $C^*(M, \Gamma, N)$ that is given by

$$\overline{C}^r(M, \Gamma, N) = M \otimes_A \overline{\Gamma}^{\otimes_A r} \otimes_A N,$$

for $r \geq 0$, where $\overline{\Gamma} = \ker \varepsilon$.

We say that a left Γ-comodule N is relatively injective if N is a direct summand of $\Gamma \otimes_A N'$ as a left Γ-comodule for some left A-module N'. For a left Γ-comodule N, the map $\psi : N \to \Gamma \otimes_A N$ induces an augmentation $N \to C^*(\Gamma, \Gamma, N)$. This gives a resolution of N in $\mathrm{LComod}_{(A,\Gamma)}(\mathrm{Ab}_*)$ by relative injectives. Note that the resolution is split in $\mathrm{LMod}_A(\mathrm{Ab}_*)$. The splitting is given by

$$\varepsilon \otimes 1^{\otimes r} \otimes 1 : \Gamma \otimes_A \Gamma^{\otimes_A r} \otimes_A N \longrightarrow A \otimes_A \Gamma^{\otimes_A r} \otimes_A N \cong \Gamma \otimes_A \Gamma^{\otimes_A (r-1)} \otimes_A N.$$

Similarly, we have a resolution $N \to \overline{C}^*(\Gamma, \Gamma, N)$ of N in $\mathrm{LComod}_{(A,\Gamma)}(\mathrm{Ab}_*)$ by relative injectives that is split in $\mathrm{LMod}_A(\mathrm{Ab}_*)$. By the proof of [31, Lemma A1.2.9], the cobar complex $C^*(\Gamma, \Gamma, N)$ is cochain homotopy equivalent to the normalized cobar complex $\overline{C}^*(\Gamma, \Gamma, N)$. Since $C^*(M, \Gamma, N) \cong M\Box_\Gamma C^*(\Gamma, \Gamma, N)$ and $\overline{C}^*(M, \Gamma, N) \cong M\Box_\Gamma \overline{C}^*(\Gamma, \Gamma, N)$, this implies that

$$H^*(C^*(M, \Gamma, N)) \cong H^*(\overline{C}^*(M, \Gamma, N))$$

for any $M \in \mathrm{RComod}_{(A,\Gamma)}(\mathrm{Ab}_*)$ and $N \in \mathrm{LComod}_{(A,\Gamma)}(\mathrm{Ab}_*)$.

The following proposition is obtained in the same way as in [31, Cor. A1.2.12].

Proposition 9 *If M is flat as a right A-module, then*

$$H^*(C^*(M, \Gamma, N)) \cong \mathrm{Cotor}^*_\Gamma(M, N).$$

In particular, we have

$$H^*(C^*(A, \Gamma, N)) \cong \mathrm{Ext}^*_\Gamma(A, N).$$

6.2 Bousfield–Kan Spectral Sequences

In this subsection we work in the quasi-category of spectra Sp and study the Bousfield–Kan spectral sequence abutting to the homotopy groups of cotensor products of comodules.

Note that Sp is a presentable stable symmetric monoidal category in which the tensor product commutes with all colimits separately in each variable. We use $\otimes$ for the tensor product in Sp instead of $\wedge$. We denote by $\mathbb{S}$ the sphere spectrum that is the monoidal unit.

We would like to compute the homotopy groups of cotensor products of comodules. Since a cotensor product of comodules is a limit of a cosimplicial object, we have the Bousfield–Kan spectral sequence abutting to the homotopy groups of the cotensor product.

First, we recall the Bousfield–Kan spectral sequence associated to a cosimplicial object in Sp. Let $X^\bullet : N(\Delta) \to \mathrm{Sp}$ be a cosimplicial object in Sp. Since the quasi-category Sp of spectra is the underlying quasi-category of the combinatorial simplicial model category $\Sigma\mathrm{Sp}$ of symmetric spectra, we can take a cosimplicial object $Y^\bullet : \Delta \to \Sigma\mathrm{Sp}^\circ$ such that $N(Y^\bullet) \simeq X^\bullet$ by [22, Prop. 4.2.4.4.], where $\Sigma\mathrm{Sp}^\circ$ is the simplicial full subcategory of $\Sigma\mathrm{Sp}$ consisting of objects that are both fibrant and cofibrant. Then the limit $\lim_{N(\Delta)} X^\bullet$ in Sp is represented by the homotopy limit $\mathrm{holim}_\Delta Y^\bullet$.

We recall that $\mathrm{Tot}_r(X^\bullet)$ is defined to be the limit of $X^\bullet|_{N(\Delta^{\le r})}$ in Sp for $r \ge 0$, where $\Delta^{\le r}$ is the full subcategory of Δ spanned by $\{[0], [1], \dots, [r]\}$. The inclusion $\Delta^{\le r} \hookrightarrow \Delta^{\le r+1}$ induces a map $\mathrm{Tot}_{r+1}(X^\bullet) \to \mathrm{Tot}_r(X^\bullet)$ for $r \ge 0$. We have a tower

$\{\mathrm{Tot}_r X^\bullet\}_{r\geq 0}$ and the limit of the tower is equivalent to $\mathrm{Tot}(X^\bullet)$:

$$\mathrm{Tot}(X^\bullet) \simeq \lim_r \mathrm{Tot}_r(X^\bullet).$$

Let $F_r(X^\bullet)$ be the fiber of the map $\mathrm{Tot}_r(X^\bullet) \to \mathrm{Tot}_{r-1}(X^\bullet)$ for $r \geq 0$, where $\mathrm{Tot}_{-1}(X^\bullet) = 0$. Associated to the tower $\{\mathrm{Tot}_r(X^\bullet)\}_{r\geq 0}$, by applying the homotopy groups, we obtain the Bousfield–Kan spectral sequence

$$E_1^{s,t} \cong \pi_{t-s} F_s(X^\bullet) \Longrightarrow \pi_{t-s}\mathrm{Tot}(X^\bullet)$$

(see [6, Chap. IX, Sect. 4]). We can identify the E_2-page of the spectral sequence with the cohomotopy groups of the cosimplicial graded abelian group $\pi_*(X^\bullet)$:

$$E_2^{s,t} \cong \pi^s \pi_t(X^\bullet)$$

(see [6, Chap. X, Sect. 7]).

Next, we construct a spectral sequence that computes the homotopy groups of cotensor products of comodules. Let A be an algebra object of Sp and Γ a coalgebra object of ${}_A\mathrm{BMod}_A(\mathrm{Sp})$. Recall that the cotensor product $M\square_\Gamma N$ is defined to be the limit of the cosimplicial object $C^\bullet(M, \Gamma, N)$ for a right Γ-comodule M and a left Γ-comodule N. Hence we obtain the Bousfield–Kan spectral sequence abutting to the homotopy groups of the cotensor product $M\square_\Gamma N$:

$$E_2^{s,t} \Longrightarrow \pi_{t-s}(M\square_\Gamma N),$$

where the E_2-page is given by

$$E_2^{s,t} \cong \pi^s \pi_t\, C^\bullet(M, \Gamma, N).$$

For a spectrum $X \in \mathrm{Sp}$, we write X_* for the homotopy groups $\pi_* X$ for simplicity. Now we suppose that Γ_* is flat as a left A_*-module and a right A_*-module. Since $C^r(M, \Gamma, N) \simeq M \otimes_A \overbrace{\Gamma \otimes_A \cdots \otimes_A \Gamma}^{r} \otimes_A N$ for all $r \geq 0$, we see that

$$\pi_* C^\bullet(M, \Gamma, N) \cong C^\bullet(M_*, \Gamma_*, N_*)$$

if M_* is a flat right A_*-module or N_* is a flat left A_*-module.

Proposition 10 *If M_* is a flat right A_*-module or N_* is a flat left A_*-module, then we have the Bousfield–Kan spectral sequence abutting to the homotopy groups of the cotensor product $M\square_\Gamma N$:*

$$E_2^{s,t} \Longrightarrow \pi_{t-s}(M\square_\Gamma N).$$

The E_2-page of the spectral sequence is given by

$$E_2^{s,t} \cong \mathrm{Cotor}^s_{\Gamma_*}(M_*, N_*)_t.$$

Now we regard A as a right A-module and suppose A is a right Γ-comodule via η_R. In this case A_* is a right Γ_*-comodule via η_{R*}. Then we can regard A_* as a left Γ_*-comodule by using the isomorphism $A_* \cong \mathrm{Hom}_{A_*}(A_*, A_*)$ of left A_*-modules, where $\mathrm{Hom}_{A_*}(A_*, A_*)$ is the graded abelian group of graded homomorphisms of right A_*-modules. Hence we can form $\mathrm{Ext}^*_{\Gamma_*}(A_*, N_*)$ for any left Γ-comodule N.

We consider the Bousfield–Kan spectral sequence associated to $C^\bullet(A, \Gamma, N)$. Note that the limit of $C^\bullet(A, \Gamma, N)$ is $P(N) = A\square_\Gamma N$.

Corollary 2 *We assume that the right A-module A is a right Γ-comodule via η_R. Then we have the Bousfield–Kan spectral sequence abutting to the homotopy groups of $P(N)$:*

$$E_2^{s,t} \Longrightarrow \pi_{t-s}P(N),$$

where the E_2-page is given by

$$E_2^{s,t} \cong \mathrm{Ext}^s_{\Gamma_*}(A_*, N_*)_t.$$

In the following of this subsection we study the relationship between Bousfield–Kan spectral sequences and Adams spectral sequences.

For an $\mathbb{S}$-algebra A, we have the coalgebra $A \otimes A$ in ${}_A\mathrm{BMod}_A(\mathrm{Sp})$. We write $\Gamma(A) = (A, A \otimes A)$ for simplicity and we call $A \otimes A$-comodules $\Gamma(A)$-comodules interchangeably. We can regard A as a left $\Gamma(A)$-comodule via $\eta_L : A \simeq A \otimes \mathbb{S} \stackrel{\mathrm{id}_A \otimes u}{\longrightarrow} A \otimes A$ and as a right $\Gamma(A)$-comodule via $\eta_R : A \simeq \mathbb{S} \otimes A \stackrel{u \otimes \mathrm{id}_A}{\longrightarrow} A \otimes A$, where $u : \mathbb{S} \to A$ is the unit map.

By Theorem 5, we have a left $\Gamma(A)$-comodule $A \otimes X$ for any $X \in \mathrm{Sp}$. We consider the cobar construction

$$C^\bullet(A, A \otimes A, A \otimes X),$$

where $X \in \mathrm{Sp}$. Note that we have a coaugmentation $X \to C^\bullet(A, A \otimes A, A \otimes X)$, which is given by $X \simeq \mathbb{S} \otimes X \stackrel{u \otimes \mathrm{id}_X}{\longrightarrow} A \otimes X \simeq C^0(A, A \otimes A, A \otimes X)$. This induces a map

$$X \to P(A \otimes X) = \lim_{N(\Delta)} C^\bullet(A, A \otimes A, A \otimes X).$$

We have an equivalence $C^\bullet(A, A \otimes A, A \otimes X) \simeq C^\bullet(A, A \otimes A, A) \otimes X$. We see that the cobar construction $C^\bullet(A, A \otimes A, A)$ is the Amitsur complex in Sp given by

$$C^r(A, A \otimes A, A) \simeq \overbrace{A \otimes \cdots \otimes A}^{r+1}$$

for any $r \geq 0$ with the usual structure maps. The Bousfield–Kan spectral sequence of the cobar construction $C^\bullet(A, A \otimes A, A \otimes X)$ is related to the A-Adams spectral sequence of X. Although this may be well-known to experts, we briefly review this relation for the reader's convenience (see, for example, [26, Sect. 2.1]).

The coaugmented cosimplicial object $X \to C^\bullet(A, A \otimes A, A \otimes X)$ induces a tower

$$\{\mathrm{Tot}_r C^\bullet(A, A \otimes A, A \otimes X)\}_{r \geq 0},$$

and a map of towers $c(X) \to \{\mathrm{Tot}_r C^\bullet(A, A \otimes A, A \otimes X)\}_{r \geq 0}$ for any $X \in \mathrm{Sp}$. This tower is related to the A-Adams tower of X.

Let $\overline{A}$ be the fiber of the unit map $u : \mathbb{S} \to A$. We have a canonical map $\overline{A} \to \mathbb{S}$. For $r \geq 0$, we set

$$T_r(A, X) = \overbrace{\overline{A} \otimes \cdots \otimes \overline{A}}^{r} \otimes X,$$

where we understand $\overline{A}^{\otimes 0} = \mathbb{S}$. Using the canonical map $\overline{A} \to \mathbb{S}$, we define a map $T_{r+1}(A, X) \to T_r(A, X)$ for $r \geq 0$ by

$$T_{r+1}(A, X) \simeq \overline{A} \otimes \overline{A}^{\otimes r} \otimes X \to \mathbb{S} \otimes \overline{A}^{\otimes r} \otimes X \simeq T_r(A, X).$$

With these maps, we obtain a tower $\{T_r(A, X)\}_{r \geq 0}$ and a map $\{T_r(A, X)\}_{r \geq 0} \to c(X)$ of towers.

Let $G_r(A, X)$ be the cofiber of the map $T_{r+1}(A, X) \to T_r(A, X)$ for $r \geq 0$. Associate to the tower $\{T_r(A, X)\}_{r \geq 0}$, by applying the homotopy groups, we obtain the A-Adams spectral sequence of X. The E_1-page of the spectral sequence is given by

$$E_1^{s,t} \cong \pi_{t-s} G_s(A, X)$$

(see, for example, [31, Chap. 2.2]).

We set $C^\bullet = C^\bullet(A, A \otimes A, A \otimes X)$. By [26, Prop. 2.14], the cofiber of the map $T_{r+1}(A, X) \to X$ is equivalent to $\mathrm{Tot}_r(C^\bullet)$ for all $r \geq 0$. Hence we obtain a natural cofiber sequence of towers

$$\{T_{r+1}(A, X)\}_{r \geq 0} \to c(X) \to \{\mathrm{Tot}_r(C^\bullet)\}_{r \geq 0}.$$

In particular, we see that $G_r(A, X)$ is equivalent to the fiber $F_r(C^\bullet)$ of the map $\mathrm{Tot}_r(C^\bullet) \to \mathrm{Tot}_{r-1}(C^\bullet)$. Comparing the spectral sequences, we see that the A-Adams spectral sequence of X coincides with the Bousfield–Kan spectral sequence associated to the cobar construction $C^\bullet(A, A \otimes A, A \otimes X)$.

We recall that the map $X \to P(A \otimes X)$ is an A-nilpotent completion in Ho(Sp) in the sense of Bousfield [5], where Ho(Sp) is the stable homotopy category of spectra.

Let R be a ring spectrum in Ho(Sp). A spectrum W is said to be R-nilpotent if W lies in the thick ideal of Ho(Sp) generated by R. An R-nilpotent resolution of a spectrum Z is a tower $\{W_r\}_{r \geq 0}$ equipped with a map of towers $c(Z) \to \{W_r\}_{r \geq 0}$ in

Ho(Sp) such that W_r is R-nilpotent for all $r \geq 0$ and the map

$$\mathrm{colim}_r\, \mathrm{Hom}_{\mathrm{Ho(Sp)}}(W_r, N) \to \mathrm{Hom}_{\mathrm{Ho(Sp)}}(Z, N)$$

is an isomorphism for any R-nilpotent spectrum N. An R-nilpotent completion of Z is defined to be the map $Z \to \mathrm{holim}_r W_r$ for an R-nilpotent resolution $\{W_r\}_{r\geq 0}$ of Z.

We shall show that the tower $\{\mathrm{Tot}_r(C^\bullet)\}_{r\geq 0}$ is an A-nilpotent resolution of X, where $C^\bullet = C^\bullet(A, A\otimes A, A\otimes X)$. For any $r \geq 0$, the fiber $F_r(C^\bullet)$ of the map $\mathrm{Tot}_r(C^\bullet) \to \mathrm{Tot}_{r-1}(C^\bullet)$ is equivalent to $G_r(A, X)$. Since $G_r(A, X) \simeq A \otimes \overline{A}^{\otimes r} \otimes X$ is a left A-module, $G_r(A, X)$ is A-nilpotent for all $r \geq 0$. By induction on r and the fact that $\mathrm{Tot}_0(C^\bullet) = A \otimes X$, we see that $\mathrm{Tot}_r(C^\bullet)$ is A-nilpotent for all $r \geq 0$.

Recall that the fiber of the map $X \to \mathrm{Tot}_r(C^\bullet)$ is equivalent to $T_{r+1}(A, X)$ for all $r \geq 0$. The map $\overline{A} \to \mathbb{S}$ is null in Ho(Sp) after tensoring with A since the unit map $\mathbb{S} \to A$ has a left inverse after tensoring with A. Hence we see that the map $T_{r+1}(A, X) \to T_r(A, X)$ is null in Ho(Sp) after tensoring with A. This implies that the map $\mathrm{colim}_r\, \mathrm{Hom}_{\mathrm{Ho(Sp)}}(\mathrm{Tot}_r(C^\bullet), A \otimes Y) \to \mathrm{Hom}_{\mathrm{Ho(Sp)}}(X, A \otimes Y)$ is an isomorphism for any spectrum Y. Since the class of A-nilpotent spectra coincides with the thick subcategory generated by the class $\{A \otimes Z |\ Z \in \mathrm{Sp}\}$, we see that the map $\mathrm{colim}_r\, \mathrm{Hom}_{\mathrm{Ho(Sp)}}(\mathrm{Tot}_r(C^\bullet), N) \to \mathrm{Hom}_{\mathrm{Ho(Sp)}}(X, N)$ is an isomorphism for any A-nilpotent spectrum N.

Therefore, the tower $\{\mathrm{Tot}_r(C^\bullet)\}_{r\geq 0}$ is an A-nilpotent resolution of X. Since $P(A \otimes X) \simeq \lim_r \mathrm{Tot}_r(C^\bullet)$, we see that the map $X \to P(A \otimes X)$ is an A-nilpotent completion in Ho(Sp).

6.3 Complex Oriented Spectra

In this subsection we study quasi-categories of comodules over coalgebras associated to Landweber exact $\mathbb{S}$-algebras. We show that the quasi-category of comodules over the coalgebra associated to a Landweber exact $\mathbb{S}$-algebra depends only on the height of the underlying MU_*-algebra.

Let MU be the complex cobordism spectrum. The coefficient ring of MU is a polynomial ring over the ring $\mathbb{Z}$ of integers with infinitely many variables

$$\pi_* MU = \mathbb{Z}[x_1, x_2, \ldots]$$

with degree $|x_i| = 2i$ for $i \geq 1$. We assume that the Chern numbers of x_{p^n-1} are all divisible by p for all positive integers n and all prime numbers p. In this case the ideals $I_{p,n} = (p, x_{p-1}, \ldots, x_{p^{n-1}-1})$ are invariant and independent of the choice of generators. We set $I_{p,0} = (0)$ and $I_{p,\infty} = \bigcup_{n\geq 0} I_{p,n}$. The ideals $I_{p,n}$ for $0 \leq n \leq \infty$ and all primes p are the only invariant prime ideals in MU_* (see [20]).

For a graded commutative MU_*-algebra R_*, we say that R_* is Landweber exact if $p, x_{p-1}, \ldots, x_{p^n-1}, \ldots$ is a regular sequence in R_* for all prime numbers p.

If $E^*(-)$ is a complex oriented cohomology theory represented by a spectrum E, then there is a ring spectrum map $f : MU \to E$ in the stable homotopy category Ho(Sp) of spectra. We say E is Landweber exact if E_* is a graded commutative ring and Landweber exact via the graded ring homomorphism $f_* : MU_* \to E_*$.

We consider an $\mathbb{S}$-algebra that is Landweber exact.

Definition 2 We say that A is a Landweber exact $\mathbb{S}$-algebra if A is an $\mathbb{S}$-algebra spectrum equipped with a map $f : MU \to A$ of ring spectra in Ho(Sp) such that A_* is a graded commutative ring and Landweber exact via the graded ring homomorphism $f_* : MU_* \to A_*$.

Let p be a prime number and let $\mathbb{S}_{(p)}$ be the localization of the sphere spectrum $\mathbb{S}$ at p. We can consider a Landweber exact $\mathbb{S}_{(p)}$-algebra in the same way. If A is a Landweber exact $\mathbb{S}$-algebra, then the localization $A_{(p)}$ is a Landweber exact $\mathbb{S}_{(p)}$-algebra at any prime number p.

Example 1 For any prime number p and any positive integer n, the Johnson-Wilson spectrum $E(n)$ at p is a complex oriented Landweber exact spectrum. By [1, Proposition 4.1], $E(n)$ admits an $MU_{(p)}$-algebra spectrum structure. Hence, in particular, $E(n)$ is a Landweber exact $\mathbb{S}_{(p)}$-algebra.

Definition 3 For a Landweber exact graded commutative ring A_*, we denote by $\mathrm{ht}_p\, A_*$ the height of A_* at a prime p in the sense of [18, Definition 7.2], that is, the largest number n such that $A_*/I_{p,n}$ is nonzero, or ∞ if $A_*/I_{p,n}$ is nonzero for all n. For a Landweber exact $\mathbb{S}$-algebra A, we denote by $\mathrm{ht}_p\, A$ the height of A_* at p.

For Landweber exact $\mathbb{S}$-algebras E and F, we have an isomorphism

$$F_*(E) \cong F_* \otimes_{MU_*} MU_*(MU) \otimes_{MU_*} E_*.$$

By abuse of notation, for graded commutative Landweber exact MU_*-algebras A_* and B_*, we set $B_*(A) = B_* \otimes_{MU_*} MU_*(MU) \otimes_{MU_*} A_*$. We denote by $\Gamma(A_*)$ the pair $(A_*, A_*(A))$, which forms a graded Hopf algebroid (see [31, Appendix A.1]). We can consider the categories of graded $\Gamma(A_*)$-comodules $\mathrm{LComod}_{\Gamma(A_*)}(\mathrm{Ab}_*)$ which is an abelian category since $\Gamma(A_*)$ is a flat Hopf algebroid.

The canonical map $MU_*(MU) \to A_*(A)$ induces a map of graded Hopf algebroids $\Phi(A) : \Gamma(MU_*) \to \Gamma(A_*)$. We consider the functor

$$\Phi(A)_* : \mathrm{LComod}_{\Gamma(MU_*)}(\mathrm{Ab}_*) \longrightarrow \mathrm{LComod}_{\Gamma(A_*)}(\mathrm{Ab}_*)$$

given by $\Phi(A)_*(M_*) = A_* \otimes_{MU_*} M_*$ for $M_* \in \mathrm{LComod}_{\Gamma(MU_*)}(\mathrm{Ab}_*)$. The functor $\Phi(A)_*$ has the right adjoint $\Phi(A)^* : \mathrm{LComod}_{\Gamma(A_*)}(\mathrm{Ab}_*) \to \mathrm{LComod}_{\Gamma(MU_*)}(\mathrm{Ab}_*)$ given by

$$\Phi(A)^*(N_*) = MU_*(A) \Box_{A_*(A)} N_*$$

for $N_* \in \mathrm{LComod}_{\Gamma(A_*)}(\mathrm{Ab}_*)$ (see [18, Lem. 2.4 and Remark after its proof]). Let $\mathscr{T}_{A_*}$ be the class of all graded $\Gamma(MU_*)$-comodules M_* such that $A_* \otimes_{MU_*} M_*$ is trivial.

By [18, Thm. 2.5], the adjoint pair $(\Phi(A)_*, \Phi(A)^*)$ induces an adjoint equivalence of categories between $\mathrm{LComod}_{\Gamma(A_*)}(\mathrm{Ab}_*)$ and the localization of $\mathrm{LComod}_{\Gamma(MU_*)}$ with respect to $\mathscr{T}_{A_*}$.

Let A_* and B_* be graded commutative Landweber exact MU_*-algebras. We recall that $B_*(A) = B_* \otimes_{MU_*} MU_*(MU) \otimes_{MU_*} A_*$. Note that $B_*(A)$ is a graded left $\Gamma(B_*)$-comodule and a graded right $\Gamma(A_*)$-comodule. We can define a functor

$$G_{B_*,A_*} : \mathrm{LComod}_{\Gamma(A_*)}(\mathrm{Ab}_*) \longrightarrow \mathrm{LComod}_{\Gamma(B_*)}(\mathrm{Ab}_*)$$

which assigns to a graded left $\Gamma(A_*)$-comodule M_* the graded left $\Gamma(B_*)$-comodule $G_{B_*,A_*}(M_*)$ given by

$$G_{B_*,A_*}(M_*) = B_*(A)\Box_{A_*(A)} M_*.$$

Lemma 8 *If* $\mathrm{ht}_p A_* = \mathrm{ht}_p B_*$ *for all primes* p*, then the functor* G_{B_*,A_*} *gives an equivalence of categories between* $\mathrm{LComod}_{\Gamma(A_*)}(\mathrm{Ab}_*)$ *and* $\mathrm{LComod}_{\Gamma(B_*)}(\mathrm{Ab}_*)$.

Proof Note that the functor G_{B_*,A_*} is the composition $\Phi(B)_*\Phi(A)^*$. The functor $\Phi(A)^*$ induces an equivalence of categories from $\mathrm{LComod}_{\Gamma(A_*)}(\mathrm{Ab}_*)$ to the localization of $\mathrm{LComod}_{\Gamma(MU_*)}(\mathrm{Ab}_*)$ with respect to $\mathscr{T}_{A_*}$ and $\Phi(B)_*$ induces an equivalence of categories from the localization of $\mathrm{LComod}_{\Gamma(MU_*)}(\mathrm{Ab}_*)$ with respect to $\mathscr{T}_{B_*}$ to $\mathrm{LComod}_{\Gamma(B_*)}(\mathrm{Ab}_*)$. By [18, Thm. 7.3], the assumption that A_* and B_* have the same heights for all p implies that $\mathscr{T}_{A_*} = \mathscr{T}_{B_*}$. Hence we see that G_{B_*,A_*} gives an equivalence of categories. $\Box$

Lemma 9 *Let* A, B, C *be Landweber exact* $\mathbb{S}$*-algebras. We assume that* $\mathrm{ht}_p A = \mathrm{ht}_p B = \mathrm{ht}_p C$ *for all primes* p*. Then, for any* $\Gamma(A)$*-comodule* M*, the canonical map*

$$C \otimes ((B \otimes A)\Box_{A\otimes A} M) \longrightarrow (C \otimes B \otimes A) \underset{A\otimes A}{\Box} M$$

is an equivalence.

Proof Let R be a Landweber exact $\mathbb{S}$-algebra which has the same height at all p as A. First, we consider the homotopy groups of the cotensor product $(R \otimes A)\Box_{A\otimes A} M$. The Bousfield–Kan spectral sequence abutting to the homotopy groups of $(R \otimes A)\Box_{A\otimes A} M$ has the E_2-page given by

$$E_2^{s,t} \cong \mathrm{Cotor}^s_{A_*(A)}(R_*(A), M_*)_t.$$

We have

$$H^s(C(A_*(A), A_*(A), M_*)) = 0$$

for $s > 0$. Note that the cobar complex $C^*(A_*(A), A_*(A), M_*)$ is a cochain complex in the abelian category $\mathrm{LComod}_{\Gamma(A_*)}(\mathrm{Ab}_*)$. Applying the functor G_{R_*,A_*} to $C^*(C(A_*(A), A_*(A), M_*))$, we obtain

$$H^s(C(R_*(A), A_*(A), M_*)) = 0$$

for $s > 0$ by Lemma 8. Hence the Bousfield–Kan spectral sequence abutting to the homotopy groups of $(R \otimes A)\Box_{A\otimes A}M$ collapses from the E_2-page and we obtain an isomorphism

$$\pi_*((R \otimes A)\Box_{A\otimes A}M) \cong R_*(A)\Box_{A_*(A)}M_*.$$

In particular, since $C \otimes B$ is Landweber exact through the map $MU \to B \to C \otimes B$ in Ho(Sp), we obtain an isomorphism

$$\pi_*((C \otimes B \otimes A)\Box_{A\otimes A}M) \cong (C \otimes B)_*(A)\Box_{A_*(A)}M_*.$$

Since we have an isomorphism

$$(C \otimes B)_*(A) \cong C_*(B) \otimes_{B_*} B_*(A),$$

we obtain an isomorphism

$$\pi_*((C \otimes B \otimes A)\Box_{A\otimes A}M)) \cong C_*(B) \otimes_{B_*} (B_*(A)\Box_{A_*(A)}M_*).$$

On the other hand, we have isomorphisms

$$\pi_*(C \otimes ((B \otimes A)\Box_{A\otimes A}M)) \cong C_*(B) \otimes_{B_*} \pi_*((B \otimes A)\Box_{A\otimes A}M)$$

and

$$\pi_*((B \otimes A)\Box_{A\otimes A}M) \cong B_*(A)\Box_{A_*(A)}M_*.$$

Hence we see that the canonical map $C \otimes (B \otimes A)\Box_{A\otimes A}M \to (C \otimes B \otimes A)\Box_{A\otimes A}M$ induces an isomorphism of homotopy groups. This completes the proof. □

Corollary 3 *Let A and B be Landweber exact $\mathbb{S}$-algebras. We assume that* $\mathrm{ht}_p A = \mathrm{ht}_p B$ *for all primes p. For any left $\Gamma(A)$-comodule M, the left $\Gamma(B)$-comodule structure on B induces a left $\Gamma(B)$-comodule structure on $(B \otimes A)\Box_{A\otimes A}M$.*

Proof For $r > 0$, $B^{\otimes r}$ is a Landweber exact $\mathbb{S}$-algebra and has the same height at all p as B. By Lemma 9, the canonical map $B^{\otimes r} \otimes ((B \otimes A)\Box_{A\otimes A}M) \to (B^{\otimes r} \otimes B \otimes A)\Box_{A\otimes A}M$ is an equivalence for all $r > 0$. Applying Lemma 4 for the simplicial object $B_\bullet(B \otimes A, A \otimes A, M)$ in $\mathrm{LMod}_{B\otimes B}(\mathrm{LMod}_B(\mathrm{Sp})^{\mathrm{op}})$, we see that the cosimplicial object $C^\bullet(B \otimes A, A \otimes A, M)$ has a limit in $\mathrm{LComod}_{\Gamma(B)}(\mathrm{Sp})$ and the forgetful functor $\mathrm{LComod}_{\Gamma(B)}(\mathrm{Sp}) \to \mathrm{LMod}_B(\mathrm{Sp})$ preserves the limit. □

Using Corollary 3, we can define a functor

$$F_{B,A} : \mathrm{Comod}_{\Gamma(A)}(\mathrm{Sp}) \longrightarrow \mathrm{Comod}_{\Gamma(B)}(\mathrm{Sp})$$

by assigning to $M \in \mathrm{LComod}_{\Gamma(A)}(\mathrm{Sp})$ the $\Gamma(B)$-comodule $F_{B,A}(M)$ given by

$$F_{B,A}(M) = (B \otimes A) \underset{A\otimes A}{\Box} M.$$

Theorem 8 *Let A and B be Landweber exact $\mathbb{S}$-algebras. We assume that A and B have the same height at all p. Then the functor $F_{B,A}$ gives an equivalence of quasi-categories*

$$\mathrm{LComod}_{\Gamma(A)}(\mathrm{Sp}) \simeq \mathrm{LComod}_{\Gamma(B)}(\mathrm{Sp}).$$

Proof For any left $\Gamma(A)$-comodule M, we have

$$\pi_* F_{B,A}(M) \cong B_*(A)\Box_{A_*(A)} M_*.$$

Since the functor $G_{B_*,A_*} = B_*(A)\Box_{A_*(A)}(-)$ gives an equivalence of categories by Lemma 8, we see that the functor $F_{B,A}$ gives an equivalence of quasi-categories between $\mathrm{LComod}_{\Gamma(A)}(\mathrm{Sp})$ and $\mathrm{LComod}_{\Gamma(B)}(\mathrm{Sp})$. □

Proposition 11 *Let A and B be Landweber exact $\mathbb{S}$-algebras. We assume that $\mathrm{ht}_p A = \mathrm{ht}_p B$ for all primes p. Then the following diagram is commutative*

$$\begin{array}{ccc} & \mathrm{Sp} & \\ {}^{A\otimes(-)}\swarrow & & \searrow^{B\otimes(-)} \\ \mathrm{LComod}_{\Gamma(A)} & \xrightarrow{F_{B,A}} & \mathrm{LComod}_{\Gamma(B)}. \end{array}$$

Proof Since $B \otimes \mathbb{S} \simeq B \otimes_{\mathbb{S}} \mathbb{S} \otimes \mathbb{S}$ is an extended right $\Gamma(\mathbb{S})$-comodule, we have a natural equivalence

$$B \otimes X \simeq (B \otimes \mathbb{S})\Box_{\mathbb{S}\otimes\mathbb{S}}(\mathbb{S} \otimes X)$$

for any $X \in \mathrm{Sp}$. The unit map $\mathbb{S} \to A$ induces a natural map

$$(B \otimes \mathbb{S})\Box_{\mathbb{S}\otimes\mathbb{S}}(\mathbb{S} \otimes X) \longrightarrow (B \otimes A)\Box_{A\otimes A}(A \otimes X)$$

and hence we obtain a natural map

$$f : B \otimes X \longrightarrow (B \otimes A)\Box_{A\otimes A}(A \otimes X).$$

Note that f is a map of left $\Gamma(B)$-comodule. Since

$$\pi_*((B \otimes A)\Box_{A\otimes A}(A \otimes X)) \cong B_*(A)\Box_{A_*(A)} A_*(X),$$

we see that f induces an isomorphism of homotopy groups. This completes the proof. □

6.4 *The $E(n)$-Local Category*

In this subsection we study the quasi-category of $E(n)$-local spectra, where $E(n)$ is the nth Johnson-Wilson spectrum at a prime p. We show that the quasi-category of $E(n)$-local spectra is equivalent to the quasi-category of comodules over the coalgebra $A \otimes A$ for any Landweber exact $\mathbb{S}_{(p)}$-algebra of height n at p.

In this subsection we fix a non-negative integer n and a prime number p. Let $L_n\mathrm{Sp}$ be the Bousfield localization of the quasi-category of spectra with respect to the nth Johnson-Wilson spectrum $E(n)$ at p. We denote by $L_n : \mathrm{Sp} \to \mathrm{Sp}$ the associated localization functor. The quasi-category $L_n\mathrm{Sp}$ is a stable homotopy theory with the tensor product in Sp and the unit $L_n\mathbb{S}$, where $L_n\mathbb{S}$ is the $E(n)$-localization of the sphere spectrum $\mathbb{S}$.

We recall that $E(n)$ is a Landweber exact $\mathbb{S}_{(p)}$-algebra with height $\mathrm{ht}_p\, E(n) = n$. For simplicity, we set $\Gamma(n) = (E(n), E(n) \otimes E(n))$. The functor

$$E(n) \otimes (-) : \mathrm{Sp} \to \mathrm{LComod}_{\Gamma(n)}(\mathrm{Sp})$$

factors through $L_n\mathrm{Sp}$ and we obtain a functor

$$E(n) \otimes (-) : L_n\mathrm{Sp} \longrightarrow \mathrm{LComod}_{\Gamma(n)}(\mathrm{Sp}).$$

Since any $N \in \mathrm{LMod}_{E(n)}(\mathrm{Sp})$ is $E(n)$-local, we see that $P(M)$ lies in $L_n\mathrm{Sp}$ for any $M \in \mathrm{LComod}_{\Gamma(n)}(\mathrm{Sp})$, and we obtain an adjunction of functors

$$E(n) \otimes (-) : L_n\mathrm{Sp} \rightleftarrows \mathrm{LComod}_{\Gamma(n)}(\mathrm{Sp}) : P.$$

Proposition 12 *The pair of functors*

$$E(n) \otimes (-) : L_n\mathrm{Sp} \rightleftarrows \mathrm{LComod}_{\Gamma(n)}(\mathrm{Sp}) : P$$

is an adjoint equivalence.

Proof By [15, Theorem 5.3], the unit $L_n\mathbb{S}$ is $E(n)$-nilpotent (see, also, [32, Chap. 8]). By Theorem 7, we obtain the proposition. □

Let A be a Landweber exact $\mathbb{S}_{(p)}$-algebra with $\mathrm{ht}_p\, A = n$. Since A is Bousfield equivalent to $E(n)$ by [13, Corollary 1.12], we have a canonical equivalence of stable homotopy theories

$$L_A\mathrm{Sp} \simeq L_n\mathrm{Sp},$$

where $L_A\mathrm{Sp}$ is the Bousfield localization of Sp with respect to A. In the same way as $E(n)$, we have an adjunction of functors

$$A \otimes (-) : L_A\mathrm{Sp} \rightleftarrows \mathrm{LComod}_{\Gamma(A)}(\mathrm{Sp}) : P.$$

Recall that we have the functor

$$F_{A,E(n)} : \mathrm{LComod}_{\Gamma(n)}(\mathrm{Sp}) \to \mathrm{LComod}_{\Gamma(A)}(\mathrm{Sp})$$

given by

$$F_{A,E(n)}(M) = (A \otimes E(n))\Box_{E(n)\otimes E(n)} M$$

for $M \in \mathrm{LComod}_{\Gamma(n)}(\mathrm{Sp})$. By Proposition 11, we see that there is a commutative diagram of quasi-categories

$$\begin{array}{ccc} L_n\mathrm{Sp} & \xrightarrow{E(n)\otimes(-)} & \mathrm{LComod}_{\Gamma(n)}(\mathrm{Sp}) \\ \downarrow & & \downarrow F_{A,E(n)} \\ L_A\mathrm{Sp} & \xrightarrow{A\otimes(-)} & \mathrm{LComod}_{\Gamma(A)}(\mathrm{Sp}), \end{array}$$

where the left vertical arrow is the canonical equivalence.

Theorem 9 *If A is a Landweber exact $\mathbb{S}_{(p)}$-algebra of height n at p, then the pair of functors*

$$A \otimes (-) : L_A\mathrm{Sp} \rightleftarrows \mathrm{Comod}_{A\otimes A}(\mathrm{Sp}) : P$$

is an adjoint equivalence.

Proof The theorem follows from Theorem 8 and Proposition 12. □

6.5 Connective Cases

In this subsection we consider the quasi-category of comodules over $A \otimes A$ for a connective $\mathbb{S}$-algebra A. We show that the quasi-category of connective A-local spectra is equivalent to the quasi-category of connective comodules over $A \otimes A$ under some conditions.

We say that a spectrum X is connective if $\pi_i X = 0$ for all $i < 0$. We denote by $\mathrm{Sp}^{\geq 0}$ the full subcategory of Sp consisting of connective objects. In this subsection we let A be a connective $\mathbb{S}$-algebra and assume that $A_*(A)$ is flat as a left and right A_*-module.

We consider the condition that the multiplication map induces an isomorphism $\pi_0(A \otimes A) \xrightarrow{\cong} \pi_0 A$. Note that there is an isomorphism $\pi_0(A \otimes A) \cong \pi_0 A \otimes \pi_0 A$.

Let R be a (possibly non-commutative) ring with identity 1. The core cR of R is defined to be the subring

$$cR = \{r \in R | \, r \otimes 1 = 1 \otimes r \text{ in } R \otimes R\}$$

(see [5, 6.4]). The core cR is a commutative ring and the multiplication map $cR \otimes cR \to cR$ is an isomorphism. We see that, if the multiplication map induces an isomorphism $\pi_0(A \otimes A) \xrightarrow{\cong} \pi_0 A$, then $c\pi_0 A = \pi_0 A$ and, in particular, $\pi_0 A$ is a commutative ring.

Lemma 10 *Let M be a connective left $\Gamma(A)$-comodule in* Sp. *If the multiplication map $A \otimes A \to A$ induces an isomorphism $\pi_0(A \otimes A) \xrightarrow{\cong} \pi_0 A$, then $P(M)$ is connective.*

Proof We have the Bousfield–Kan spectral sequence

$$E_2^{s,t} \cong \pi^s \pi_t C^\bullet(A, A \otimes A, M) \Longrightarrow \pi_{t-s} P(M).$$

This is an upper half plane spectral sequence. Since $A_*(A)$ is flat as a left and right A_*-module, there is an isomorphism

$$\pi_* C^r(A, A \otimes A, M) \cong C^r(A_*, A_*(A), M_*)$$

for any $r \geq 0$. Hence we obtain an isomorphism

$$E_2^{s,*} \cong H^s(\overline{C}^*(A_*, A_*(A), M_*)).$$

Let $\overline{\Gamma}$ be the kernel of the map $A_*(A) \to A_*$ induced by the multiplication. By the assumption that $\pi_0(A \otimes A) \xrightarrow{\cong} \pi_0(A)$, $\overline{\Gamma}_t = 0$ for $t \leq 0$. This implies that

$$\overline{C}^s(A_*, A_*(A), M_*)_t = 0$$

for $t < s$. Hence $E_2^{s,t} = 0$ for $t < s$. The lemma follows from [6, Lemma X.7.3]. □

We recall that L_ASp is the Bousfield localization of Sp with respect to A and L_A : Sp → Sp is the associated localization functor and that we have the adjoint pair of functors

$$A \otimes (-) : L_A\mathrm{Sp} \rightleftarrows \mathrm{LComod}_{\Gamma(A)}(\mathrm{Sp}) : P.$$

We denote by $L_A\mathrm{Sp}^{\geq 0}$ the full subcategory of L_ASp consisting of connective objects. We also denote by $\mathrm{LComod}_{\Gamma(A)}(\mathrm{Sp})^{\geq 0}$ the full subcategories of $\mathrm{LComod}_{\Gamma(A)}(\mathrm{Sp})$ consisting of connective objects.

By Lemma 10, we see that the functor $P : \mathrm{LComod}_{\Gamma(A)}(\mathrm{Sp}) \to L_A\mathrm{Sp}$ restricted to the full subcategory $\mathrm{LComod}_{\Gamma(A)}(\mathrm{Sp})^{\geq 0}$ factors through $L_A\mathrm{Sp}^{\geq 0}$ when the multiplication map induces an isomorphism $\pi_0(A \otimes A) \xrightarrow{\cong} \pi_0 A$. Hence we obtain the following corollary.

Corollary 4 *If the multiplication map induces an isomorphism $\pi_0(A \otimes A) \xrightarrow{\cong} \pi_0 A$, then there is an adjunction of functors*

$$A \otimes (-) : L_A\mathrm{Sp}^{\geq 0} \rightleftarrows \mathrm{LComod}_{\Gamma(A)}(\mathrm{Sp})^{\geq 0} : P.$$

We would like to show that the pair of functors $(A \otimes (-), P)$ is an adjoint equivalence under some conditions. In order to show that $A \otimes (-)$ is fully faithful, we have to show that the unit map $X \to P(A \otimes X)$ is an equivalence for any $X \in L_A\mathrm{Sp}^{\geq 0}$.

We recall that the map $X \to P(A \otimes X)$ is an A-nilpotent completion in Ho(Sp) (see Sect. 6.2). The relation between the A-localization and the A-nilpotent completion was studied by Bousfield [5].

Lemma 11 *We assume that $\pi_0 A$ is isomorphic to $\mathbb{Z}/n\mathbb{Z}$ for some $n \geq 2$ or $\mathbb{Z}[J^{-1}]$ for some set J of primes. Then the functor $A \otimes (-) : L_A \mathrm{Sp}^{\geq 0} \to \mathrm{LComod}_{\Gamma(A)}(\mathrm{Sp})^{\geq 0}$ is fully faithful.*

Proof Note that the multiplication map induces an isomorphism $\pi_0(A \otimes A) \xrightarrow{\cong} \pi_0(A)$ under the assumption and hence we have the adjunction of functors $A \otimes (-) : L_A \mathrm{Sp}^{\geq 0} \rightleftarrows \mathrm{LComod}_{\Gamma(A)}(\mathrm{Sp})^{\geq 0} : P$ by Corollary 4.

We have to show that the unit map $X \to P(A \otimes X)$ of the adjunction $(A \otimes (-), P)$ is an equivalence for any $X \in L_A \mathrm{Sp}^{\geq 0}$. By [5, Thm. 6.5 and 6.6], the A-localization $Y \to L_A Y$ is equivalent to the A-nilpotent completion $Y \to P(A \otimes Y)$ for any connective spectrum Y under the assumption. This implies that the map $X \to P(A \otimes X)$ is an equivalence for any $X \in L_A \mathrm{Sp}^{\geq 0}$. □

In order to show that the left adjoint $A \otimes (-)$ is essentially surjective, we need the following lemma.

Lemma 12 *We assume that the multiplication map induces an isomorphism $\pi_0(A \otimes A) \xrightarrow{\cong} \pi_0 A$. Let M be a connective left $\Gamma(A)$-comodule. If $P(M) \simeq 0$, then $M \simeq 0$.*

Proof We shall show that $P(M) \not\simeq 0$ if $M \not\simeq 0$. Suppose that $\pi_i M = 0$ for $i < n$ and $\pi_n M \neq 0$. Let $E_r^{s,t}$ be the Bousfield–Kan spectral sequence abutting to $\pi_* P(M)$. We have $E_2^{s,*} \cong H^s(\overline{C}^*(A_*, A_*(A), M_*))$. By the assumption, $E_2^{s,t} = 0$ for $t - s < n$ and $E_2^{0,n} \cong \pi_n M$. In particular, we have $E_\infty^{0,n} \neq 0$ and $\lim_r^1 E_r^{s,s+n} = 0$ for all $s \geq 0$. By [6, Lemma IX.5.4], $E_\infty^{0,n} \cong \mathrm{Im}(\pi_n P(M) \to \pi_n M)$. Hence we obtain $\pi_n P(M) \neq 0$. □

Theorem 10 *Let A be a connective $\mathbb{S}$-algebra such that $A_*(A)$ is flat as a left and right A_*-module. We assume that $\pi_0 A$ is isomorphic to $\mathbb{Z}/n$ for some $n \geq 2$ or $\mathbb{Z}[J^{-1}]$ for some set J of primes. Then there is an adjoint equivalence of quasi-categories*

$$A \otimes (-) : L_A \mathrm{Sp}^{\geq 0} \rightleftarrows \mathrm{LComod}_{\Gamma(A)}(\mathrm{Sp})^{\geq 0} : P.$$

Proof By Corollary 4, we have the adjunction of functors $A \otimes (-) : L_A \mathrm{Sp}^{\geq 0} \rightleftarrows \mathrm{LComod}_{\Gamma(A)}(\mathrm{Sp})^{\geq 0} : P$. By Lemma 11, the left adjoint $A \otimes (-)$ is fully faithful. Hence it suffices to show that $A \otimes (-)$ is essentially surjective.

Let $M \in \mathrm{Comod}_{\Gamma(A)}(\mathrm{Sp})^{\geq 0}$. By the counit of the adjunction, we have a map $\varepsilon : A \otimes P(M) \to M$. Let N be the cofiber of ε. Since the unit map $P(M) \to P(A \otimes P(M))$ is an equivalence, we see that $P(N) \simeq 0$. By Lemma 12, we obtain $N \simeq 0$ and hence $A \otimes P(M) \simeq M$. This shows that the left adjoint $A \otimes (-)$ is essentially surjective. □

In the following of this subsection we shall consider some examples.

First, we consider the complex cobordism spectrum MU. The spectrum MU admits a commutative $\mathbb{S}$-algebra structure by [27]. Since $\pi_* MU = \mathbb{Z}[x_1, x_2, \ldots]$ with $|x_i| = 2i$ for $i > 0$, MU is a connective commutative $\mathbb{S}$-algebra, the multiplication map induces an isomorphism $\pi_0(MU \otimes MU) \xrightarrow{\cong} \pi_0 MU$, and $c\pi_0 MU \cong \mathbb{Z}$. Note that the localization $L_{MU}X$ coincides with X for a connective spectrum X by [5, Thm. 3.1] and hence $L_{MU}\mathrm{Sp}^{\geq 0}$ is equivalent to $\mathrm{Sp}^{\geq 0}$. By Theorem 10, we obtain the following corollary.

Corollary 5 (cf. [11, 6.1.2]) *There is an adjoint equivalence*

$$MU \otimes (-) : \mathrm{Sp}^{\geq 0} \rightleftarrows \mathrm{LComod}_{\Gamma(MU)}(\mathrm{Sp})^{\geq 0} : P.$$

Next, we consider the Brown-Peterson spectrum BP at a prime p. The spectrum BP admits an $\mathbb{S}$-algebra structure by [21, Sect. 2]. The coefficient ring of BP is a polynomial ring over the ring $\mathbb{Z}_{(p)}$ of integers localized at p with infinitely many variables

$$\pi_* BP = \mathbb{Z}_{(p)}[v_1, v_2, \ldots],$$

with degree $|v_i| = 2(p^i - 1)$ for $i \geq 1$. Hence BP is a connective $\mathbb{S}$-algebra, the multiplication map induces an isomorphism $\pi_0(BP \otimes BP) \xrightarrow{\cong} \pi_0 BP$, and $c\pi_0 BP \cong \mathbb{Z}_{(p)}$. Note that the localization $L_{BP}X$ coincides with the p-localization $X_{(p)}$ for a connective spectrum X by [5, Thm. 3.1] and hence $L_{BP}\mathrm{Sp}^{\geq 0}$ is equivalent to the full subcategory $\mathrm{Sp}^{\geq 0}_{(p)}$ of p-local spectra in $\mathrm{Sp}^{\geq 0}$. By Theorem 10, we obtain the following corollary.

Corollary 6 *There is an adjoint equivalence*

$$BP \otimes (-) : \mathrm{Sp}^{\geq 0}_{(p)} \rightleftarrows \mathrm{LComod}_{\Gamma(BP)}(\mathrm{Sp})^{\geq 0} : P.$$

Finally, we consider the mod p Eilenberg–Mac Lane spectrum $H\mathbb{F}_p$ for a prime p. We know that $H\mathbb{F}_p$ is a connective commutative $\mathbb{S}$-algebra. Since $\pi_0 H\mathbb{F}_p \cong \mathbb{F}_p$, we see that the multiplication induces an isomorphism $\pi_0(H\mathbb{F}_p \otimes H\mathbb{F}_p) \xrightarrow{\cong} \pi_0 H\mathbb{F}_p$ and $c\pi_0 H\mathbb{F}_p \cong \mathbb{F}_p$. If X is a connective spectrum, then $L_{H\mathbb{F}_p}X$ is equivalent to the p-completion of X by [5, Thm. 3.1], and hence $L_{H\mathbb{F}_p}\mathrm{Sp}^{\geq 0}$ is equivalent to the full subcategory $(\mathrm{Sp}^{\wedge}_p)^{\geq 0}$ of p-complete spectra in $\mathrm{Sp}^{\geq 0}$. By Theorem 10, we obtain the following corollary.

Corollary 7 ([cf. [11, 6.1.1]) *There is an adjoint equivalence*

$$H\mathbb{F}_p \otimes (-) : (\mathrm{Sp}^{\wedge}_p)^{\geq 0} \rightleftarrows \mathrm{LComod}_{\Gamma(H\mathbb{F}_p)}(\mathrm{Sp})^{\geq 0} : P.$$

6.6 A Model of the $K(n)$-Local Category

Let $K(n)$ be the nth Morava K-theory spectrum at a prime p and $\mathbb{G}_n$ the nth Morava stabilizer group. In this subsection we show that the category of module objects over F_n in the $K(n)$-local discrete symmetric $\mathbb{G}_n$-spectra models the $K(n)$-local category, where F_n is a discrete model of the nth Morava E-theory spectrum.

The $K(n)$-local category is the Bousfield localization of the stable homotopy category of spectra with respect to $K(n)$. It is known that the $K(n)$-local categories for various n and p are fundamental building blocks of the stable homotopy category of spectra. Thus it is important to understand the $K(n)$-local category.

Let E_n be the nth Morava E-theory spectrum at p. The Morava E-theory spectrum E_n is a commutative ring spectrum in the stable homotopy category of spectra and $\mathbb{G}_n$ is identified with the group of multiplicative automorphisms of E_n. By Goerss–Hopkins [10], it was shown that the commutative ring spectrum structure on E_n can be lifted to a unique E_∞-ring spectrum structure up to homotopy. Furthermore, it was shown that $\mathbb{G}_n$ acts on E_n in the category of E_∞-ring spectra. There is a $K(n)$-local E_n-based Adams spectral sequence abutting to the homotopy groups of the $K(n)$-local sphere whose E_2-page is the continuous cohomology groups of $\mathbb{G}_n$ with coefficients in the homotopy groups of E_n. This suggests that the $K(n)$-local sphere may be the $\mathbb{G}_n$-homotopy fixed points of E_n. Motivated by this observation, Devinatz–Hopkins [8] constructed a $K(n)$-local E_∞-ring spectrum E_n^{dhU} for any open subgroup U of $\mathbb{G}_n$, which has expected properties of the homotopy fixed points spectrum.

Davis [7] constructed a discrete $\mathbb{G}_n$-spectrum F_n which is defined by

$$F_n = \operatorname{colim}_U E_n^{dhU},$$

where U ranges over the open subgroups of $\mathbb{G}_n$. The spectrum F_n is a discrete model of E_n and actually we can recover E_n from F_n by the $K(n)$-localization as

$$L_{K(n)} F_n \simeq E_n.$$

Furthermore, Behrens–Davis [3] upgraded the discrete $\mathbb{G}_n$-spectrum F_n to a commutative monoid object in the category $\Sigma\mathrm{Sp}(\mathbb{G}_n)$ of discrete symmetric $\mathbb{G}_n$-spectra, and showed that the unit map $L_{K(n)}\mathbb{S} \to F_n$ is a consistent $K(n)$-local $\mathbb{G}_n$-Galois extension.

We can give a model structure on the category $\Sigma\mathrm{Sp}(\mathbb{G}_n)$ of discrete symmetric $\mathbb{G}_n$-spectra and consider the left Bousfield localization $\Sigma\mathrm{Sp}(\mathbb{G}_n)_{K(n)}$ with respect to $K(n)$ (see [3]). The category $\Sigma\mathrm{Sp}(\mathbb{G}_n)_{K(n)}$ is a left proper, combinatorial, $\Sigma\mathrm{Sp}$-model category.

The unit map $\mathbb{S} \to F_n$ induces a symmetric monoidal $\Sigma\mathrm{Sp}$-Quillen adjunction

$$\mathrm{Ex} : \Sigma\mathrm{Sp}_{K(n)} \rightleftarrows \mathrm{LMod}_{F_n}(\Sigma\mathrm{Sp}(\mathbb{G}_n)_{K(n)}) : \mathrm{Re},$$

where $\Sigma\mathrm{Sp}_{K(n)}$ is the left Bousfield localization of the category $\Sigma\mathrm{Sp}$ of symmetric spectra with respect to $K(n)$. In [33] we showed that the total left derived functor $\mathbb{L}\mathrm{Ex}$ of Ex is fully faithful as an Ho(Sp)-enriched functor. In this subsection we shall show that the adjunction is actually a Quillen equivalence and hence we can regard $\mathrm{LMod}_{F_n}(\Sigma\mathrm{Sp}(\mathbb{G}_n)_{K(n)})$ as a model of the $K(n)$-local category.

We denote by $\mathrm{Sp}_{K(n)}$ the underlying quasi-category of the simplicial model category $\Sigma\mathrm{Sp}_{K(n)}$. The quasi-category $\mathrm{Sp}_{K(n)}$ is a stable homotopy theory with the tensor product $L_{K(n)}(-\otimes -)$ and the unit $L_{K(n)}\mathbb{S}$. Since we can regard E_n as an algebra object of $\mathrm{Sp}_{K(n)}$, we can consider the coalgebra $L_{K(n)}(E_n \otimes E_n)$ in ${}_{E_n}\mathrm{BMod}_{E_n}(\mathrm{Sp}_{K(n)})$ and the quasi-category of left $\Gamma(E_n)$-comodules

$$\mathrm{LComod}_{\Gamma(E_n)}(\mathrm{Sp}_{K(n)}),$$

where $\Gamma(E_n) = (E_n, L_{K(n)}(E_n \otimes E_n))$.

Proposition 13 *We have an equivalence of quasi-categories*

$$L_{K(n)}(E_n \otimes (-)) : \mathrm{Sp}_{K(n)} \xrightarrow{\simeq} \mathrm{LComod}_{\Gamma(E_n)}(\mathrm{Sp}_{K(n)}).$$

Proof We shall apply Theorem 7 for the stable homotopy theory $\mathrm{Sp}_{K(n)}$ and the algebra object E_n. The unit object $L_{K(n)}\mathbb{S}$ is E_n-nilpotent in $\mathrm{Sp}_{K(n)}$ by [8, Prop. A.3]. Note that a map $f : X \to Y$ in $\mathrm{Sp}_{K(n)}$ is an equivalence if and only if $L_{K(n)}(E_n \otimes f)$ is an equivalence since $K(n) \otimes E_n$ is a wedge of copies of $K(n)$. Hence $L_{E_n}(\mathrm{Sp}_{K(n)}) \simeq \mathrm{Sp}_{K(n)}$ and the proposition follows from Theorem 7. □

We have an adjunction of quasi-categories

$$\mathrm{Sp}_{K(n)} \rightleftarrows \mathrm{LMod}_{E_n}(\mathrm{Sp}_{K(n)}), \tag{5}$$

where the left adjoint is given by smashing with E_n in $\mathrm{Sp}_{K(n)}$ and the right adjoint is the forgetful functor. By Theorem 6, we have an equivalence of quasi-categories

$$\mathrm{LComod}_{\Gamma(E_n)}(\mathrm{Sp}_{K(n)}) \simeq \mathrm{LComod}_{\Theta}(\mathrm{LMod}_{E_n}(\mathrm{Sp}_{K(n)})^{\mathrm{op}})^{\mathrm{op}},$$

where Θ is the comonad associated to adjunction (5). Hence we see that the forgetful functor $\mathrm{LMod}_{E_n}(\mathrm{Sp}_{K(n)}) \to \mathrm{Sp}_{K(n)}$ exhibits the quasi-category $\mathrm{Sp}_{K(n)}$ as comonadic over $\mathrm{LMod}_{E_n}(\mathrm{Sp}_{K(n)})$ (see also [25, Prop. 10.10]).

Let $\mathrm{Sp}(\mathbb{G}_n)_{K(n)}$ be the underlying quasi-category of $\Sigma\mathrm{Sp}(\mathbb{G}_n)_{K(n)}$. We can regard F_n as an algebra object of $\mathrm{Sp}(\mathbb{G}_n)_{K(n)}$, and form a quasi-category

$$\mathrm{LMod}_{F_n}(\mathrm{Sp}(\mathbb{G}_n)_{K(n)})$$

of left module objects over F_n in $\mathrm{Sp}(\mathbb{G}_n)_{K(n)}$. Note that $\mathrm{LMod}_{F_n}(\mathrm{Sp}(\mathbb{G}_n)_{K(n)})$ is equivalent to the underlying quasi-category of the symmetric monoidal $\Sigma\mathrm{Sp}$-model category $\mathrm{LMod}_{F_n}(\Sigma\mathrm{Sp}(\mathbb{G}_n)_{K(n)})$. The adjunction (Ex, Re) of the $\Sigma\mathrm{Sp}$-model categories induces an adjunction of the underlying quasi-categories

$$\mathscr{E}x : \mathrm{Sp}_{K(n)} \rightleftarrows \mathrm{LMod}_{F_n}(\mathrm{Sp}(\mathbb{G}_n)_{K(n)}) : \mathscr{R}e.$$

Let $U : \Sigma\mathrm{Sp}(\mathbb{G}_n)_{K(n)} \to \Sigma\mathrm{Sp}_{K(n)}$ be the forgetful functor. We can regard UF_n as a commutative monoid object in $\Sigma\mathrm{Sp}_{K(n)}$, and the unit map $\mathbb{S} \to UF_n$ induces a symmetric monoidal $\Sigma\mathrm{Sp}$-Quillen adjunction

$$\Sigma\mathrm{Sp}_{K(n)} \rightleftarrows \mathrm{LMod}_{UF_n}(\Sigma\mathrm{Sp}_{K(n)}),$$

where the left adjoint is given by smashing with UF_n and the right adjoint is given by the forgetful functor. This induces an adjunction of the underlying quasi-categories

$$\mathrm{Sp}_{K(n)} \rightleftarrows \mathrm{LMod}_{UF_n}(\mathrm{Sp}_{K(n)}). \tag{6}$$

Hence we can consider the quasi-category of comodules

$$\mathrm{Comod}_{(UF_n,\Theta)}(\mathrm{Sp}_{K(n)})$$

associated to the adjunction and a map of quasi-categories

$$\mathrm{Coex} : \mathrm{Sp}_{K(n)} \longrightarrow \mathrm{Comod}_{(UF_n,\Theta)}(\mathrm{Sp}_{K(n)}).$$

In [33] we showed that there is an equivalence of quasi-categories

$$\mathrm{LMod}_{F_n}(\mathrm{Sp}(\mathbb{G}_n)_{K(n)}) \simeq \mathrm{Comod}_{(UF_n,\Theta)}(\mathrm{Sp}_{K(n)}),$$

and there is an equivalence of functors

$$\mathscr{E}x \simeq \mathrm{Coex}$$

under this equivalence.

Since the canonical map $UF_n \to E_n$ of commutative algebras is a weak equivalence in $\Sigma\mathrm{Sp}_{K(n)}$, we have a Quillen equivalence

$$\mathrm{LMod}_{UF_n}(\Sigma\mathrm{Sp}_{K(n)}) \xrightarrow{\simeq} \mathrm{LMod}_{E_n}(\Sigma\mathrm{Sp}_{K(n)}),$$

and hence we obtain an equivalence of the underlying quasi-categories

$$\mathrm{LMod}_{UF_n}(\mathrm{Sp}_{K(n)}) \xrightarrow{\simeq} \mathrm{LMod}_{E_n}(\mathrm{Sp}_{K(n)}).$$

Under this equivalence, we can identify two adjunctions (5) and (6), and hence the forgetful functor $\mathrm{LMod}_{UF_n}(\mathrm{Sp}_{K(n)}) \to \mathrm{Sp}_{K(n)}$ exhibits $\mathrm{Sp}_{K(n)}$ as comonadic over $\mathrm{LMod}_{UF_n}(\mathrm{Sp}_{K(n)})$, that is, the functor Coex is an equivalence of quasi-categories.

The adjunction $(\mathscr{E}x, \mathscr{R}e)$ of quasi-categories induces an adjunction of the homotopy categories

$$\mathrm{Ho}(\mathrm{Sp}_{K(n)}) \rightleftarrows \mathrm{Ho}(\mathrm{LMod}_{F_n}(\mathrm{Sp}(\mathbb{G}_n)_{K(n)})),$$

which is identified with the derived adjunction of the Quillen adjunction (Ex, Re). Since the functor Coex is equivalent to $\mathscr{E}x$ and is an equivalence of quasi-categories, the total left derived functor $\mathbb{L}$Ex is an equivalence of categories. Hence we obtain the following theorem.

Theorem 11 *The adjunction*

$$\mathrm{Ex} : \Sigma\mathrm{Sp}_{K(n)} \rightleftarrows \mathrm{Mod}_{F_n}(\Sigma\mathrm{Sp}(\mathbb{G}_n)_{K(n)}) : \mathrm{Re}$$

is a Quillen equivalence and hence the category $\mathrm{Mod}_{F_n}(\Sigma\mathrm{Sp}(\mathbb{G}_n)_{K(n)})$ *models the* $K(n)$*-local category.*

7 Proof of Proposition 1

In this section we prove Proposition 1 stated in Sect. 4.1, which is technical but important for constructing a canonical map between opposite coCartesian fibrations. First, we give some basic examples of inner anodyne maps and study opposite marked anodyne maps. In Sect. 7.3 we introduce a marked simplicial set $\widetilde{\mathscr{O}}(\Delta^n)^+$ in which the underlying simplicial set is $\widetilde{\mathscr{O}}(\Delta^n)$ and study inclusions of subcomplexes of the marked simplicial sets $\widetilde{\mathscr{O}}(\Delta^n)^+$ and $(\widetilde{\mathscr{O}}(\Delta^n)^+ \times (\Delta^{\{0\}})^\flat) \cup (\widetilde{\mathscr{O}}(\Delta^n)^\flat \times (\Delta^1)^\flat)$. In Sect. 7.4 we give a proof of Proposition 1.

7.1 Examples of Inner Anodyne Maps

In this subsection we give some basic examples of inner anodyne maps.

A map of simplicial sets is said to be inner anodyne if it has the left lifting property with respect to all inner fibrations. The class of inner anodyne maps is the smallest weakly saturated class of morphisms generated by all horn inclusions $\Lambda^n_i \hookrightarrow \Delta^n$ for $0 < i < n$.

For a sequence $i_1, \ldots, i_k$ of integers such that $0 \le i_1 < \ldots < i_k \le n$, we denote by $\Lambda^n(i_1, \ldots, i_k)$ the subcomplex $\bigcup_{i \neq i_1, \ldots, i_k} d_i \Delta^{n-1}$ of Δ^n.

Lemma 13 *The inclusion* $\Lambda^n(i_1, \ldots, i_k) \hookrightarrow \Delta^n$ *is an inner anodyne map for* $k > 0$ *and* $0 < i_1 < \ldots < i_k < n$.

Proof We shall prove the lemma by induction on k. When $k = 1$, the inclusion is the map $\Lambda^n_i \hookrightarrow \Delta^n$ for $0 < i = i_1 < n$ and hence it is an inner anodyne map. Suppose the lemma holds for $k-1$ and we shall prove the lemma for k. The subcomplex $\Lambda^n(i_1, \ldots, i_k) \cap d_{i_k}\Delta^{n-1}$ of $d_{i_k}\Delta^{n-1}$ is isomorphic to the subcomplex $\Lambda^{n-1}(i_1, \ldots, i_{k-1})$ of Δ^{n-1}. By the hypothesis of induction, the inclusion $\Lambda^{n-1}(i_1, \ldots, i_{k-1}) \hookrightarrow \Delta^{n-1}$ is an inner anodyne map. By the cobase change of the inclusion $\Lambda^{n-1}(i_1,\ldots,i_{k-1}) \hookrightarrow \Delta^{n-1}$ along the map $\Lambda^{n-1}(i_1,\ldots,i_{k-1}) \cong \Lambda^n(i_1,\ldots,i_k) \cap d_{i_k}\Delta^{n-1} \hookrightarrow \Lambda^n(i_1,\ldots,i_k)$, we see that the inclusion $\Lambda^n(i_1,\ldots,i_k) \hookrightarrow \Lambda^n(i_1,\ldots,i_{k-1})$ is an inner anodyne map. By the hypothesis of induction, the inclusion $\Lambda^n(i_1,\ldots,i_{k-1}) \hookrightarrow \Delta^n$ is an inner anodyne map. Hence the composition $\Lambda^n(i_1,\ldots,i_k) \hookrightarrow \Lambda^n(i_1,\ldots,i_{k-1}) \hookrightarrow \Delta^n$ is also an inner anodyne map. □

Lemma 14 *The inclusion $\Lambda^n(0,i_1,\ldots,i_k) \hookrightarrow \Delta^n$ is an inner anodyne map for $k > 0$ and $1 < i_1 < \ldots < i_k < n$.*

Proof The subcomplex $\Lambda^n(0,i_1,\ldots,i_k) \cap d_0\Delta^{n-1}$ of $d_0\Delta^{n-1}$ is isomorphic to the subcomplex $\Lambda^{n-1}(i_1-1,\ldots,i_k-1)$ of Δ^{n-1}. Since $0 < i_1 - 1 < \cdots < i_k - 1 < n-1$, the inclusion $\Lambda^{n-1}(i_1-1,\ldots,i_k-1) \hookrightarrow \Delta^{n-1}$ is an inner anodyne map by Lemma 13. By the cobase change of $\Lambda^{n-1}(i_1-1,\ldots,i_k-1) \hookrightarrow \Delta^{n-1}$ along the map $\Lambda^{n-1}(i_1-1,\ldots,i_k-1) \cong \Lambda^n(0,i_1,\ldots,i_k) \cap d_0\Delta^{n-1} \hookrightarrow \Lambda^n(0,i_1,\ldots,i_k)$, we see that the inclusion $\Lambda^n(0,i_1,\ldots,i_k) \hookrightarrow \Lambda^n(i_1,\ldots,i_k)$ is an inner anodyne map. Since the inclusion $\Lambda^n(i_1,\ldots,i_k) \hookrightarrow \Delta^n$ is an inner anodyne map by Lemma 13, the composition $\Lambda^n(0,i_1,\ldots,i_k) \hookrightarrow \Lambda^n(i_1,\ldots,i_k) \hookrightarrow \Delta^n$ is also an inner anodyne map. □

7.2 Opposite Marked Anodyne Maps

In this subsection we study opposite marked anodyne maps.

A marked simplicial set is a pair $(K, \mathscr{E})$, where K is a simplicial set and $\mathscr{E}$ is a set of edges of K that contains all degenerate edges. A map of marked simplicial sets $(K, \mathscr{E}) \to (L, \mathscr{E}')$ is a map of simplicial set $f : K \to L$ such that $f(\mathscr{E}) \subset \mathscr{E}'$. We denote by sSet^+ the category of marked simplicial sets.

For a simplicial set K, we denote by $K^\flat$ the marked simplicial set $(K, s_0(K_0))$, where $s_0(K_0)$ is the set of all degenerate edges of K, and by $K^\sharp$ the marked simplicial set (K, K_1), where K_1 is the set of all edges of K.

For a marked simplicial set $(K, \mathscr{E})$, we have the opposite marked simplicial set $(K, \mathscr{E})^{\mathrm{op}} = (K^{\mathrm{op}}, \mathscr{E}^{\mathrm{op}})$, where K^{op} is the opposite simplicial set of K and $\mathscr{E}^{\mathrm{op}}$ is the corresponding set of edges of K^{op}.

We say that a map of marked simplicial sets $K \to L$ is an opposite marked anodyne map if the opposite $K^{\mathrm{op}} \to L^{\mathrm{op}}$ is a marked anodyne map defined in [22, Def. 3.1.1.1]. The class of opposite marked anodyne maps in sSet^+ is the smallest weakly saturated class of morphisms with the following properties:

1. For each $0 < i < n$, the inclusion $(\Lambda^n_i)^\flat \hookrightarrow (\Delta^n)^\flat$ is opposite marked anodyne.
2. For every $n > 0$, the inclusion

$$(\Lambda^n_0, (\Lambda^n_0)_1 \cap \mathscr{E}) \hookrightarrow (\Delta^n, \mathscr{E})$$

   is opposite marked anodyne, where $\mathscr{E}$ denotes the set of all degenerate edges of Δ^n together with the initial edge $\Delta^{\{0,1\}}$.

3. The inclusion
$$(\Lambda^2_1)^\sharp \coprod_{(\Lambda^2_1)^\flat} (\Delta^2)^\flat \hookrightarrow (\Delta^2)^\sharp$$
is opposite marked anodyne.
4. For every Kan complex K, the map $K^\flat \to K^\sharp$ is opposite marked anodyne.

Lemma 15 *The inclusion $(\Lambda^n_0)^\sharp \hookrightarrow (\Delta^n)^\sharp$ is opposite marked anodyne for $n > 0$.*

Proof When $n = 1$, the lemma holds by property 2 of the class of opposite marked anodyne maps. We consider the case $n = 2$. By [22, Cor. 3.1.1.7], the inclusion $(\Lambda^2_0)^\sharp \coprod_{(\Lambda^2_0)^\flat} (\Delta^2)^\flat \hookrightarrow (\Delta^2)^\sharp$ is opposite marked anodyne. The inclusion $(\Lambda^2_0, (\Lambda^2_0)_1 \cap \mathscr{E}) \hookrightarrow (\Delta^2, \mathscr{E})$ is opposite marked anodyne by property 2 of the class of opposite marked anodyne maps, where $\mathscr{E}$ is the set of edges of Δ^2 consisting of all degenerate edges together with $\Delta^{\{0,1\}}$. Taking the pushout of $(\Lambda^2_0, (\Lambda^2_0)_1 \cap \mathscr{E}) \hookrightarrow (\Delta^2, \mathscr{E})$ along the map $(\Lambda^2_0, (\Lambda^2_0)_1 \cap \mathscr{E}) \to (\Lambda^2_0)^\sharp$, we see that the inclusion $(\Lambda^2_0)^\sharp \hookrightarrow (\Lambda^2_0)^\sharp \coprod_{(\Lambda^2_0)^\flat} (\Delta^2)^\flat$ is opposite marked anodyne. Hence the composition $(\Lambda^2_0)^\sharp \hookrightarrow (\Lambda^2_0)^\sharp \coprod_{(\Lambda^2_0)^\flat} (\Delta^2)^\flat \hookrightarrow (\Delta^2)^\sharp$ is also opposite marked anodyne.

Now we consider the case $n \geq 3$. The inclusion $(\Lambda^n_0, (\Lambda^n_0)_1 \cap \mathscr{E}) \hookrightarrow (\Delta^n, \mathscr{E})$ is opposite marked anodyne by property 2 of the class of opposite marked anodyne maps, where $\mathscr{E}$ is the set of edges of Δ^n consisting of all degenerate edges together with $\Delta^{\{0,1\}}$. Taking the pushout of $(\Lambda^n_0, (\Lambda^n_0)_1 \cap \mathscr{E}) \hookrightarrow (\Delta^n, \mathscr{E})$ along the map $(\Lambda^n_0, (\Lambda^n_0)_1 \cap \mathscr{E}) \to (\Lambda^n_0)^\sharp$, we see that the inclusion $(\Lambda^n_0)^\sharp \hookrightarrow (\Delta^n)^\sharp$ is opposite marked anodyne. □

Lemma 16 *Let $K = (\Delta^n \times \partial\Delta^1) \cup (\Lambda^n_0 \times \Delta^1)$ be the subcomplex of $\Delta^n \times \Delta^1$ for $n \geq 1$. Let $\mathscr{E}$ be the set of edges of $\Delta^n \times \Delta^1$ consisting of all degenerate edges together with $\Delta^{\{0,1\}} \times \Delta^{\{0\}}$. The inclusion $(K, K_1 \cap \mathscr{E}) \hookrightarrow (\Delta^n \times \Delta^1, \mathscr{E})$ is an opposite marked anodyne map.*

Proof Put $L(i) = (\Delta^{\{0,\dots,i\}} \times \Delta^{\{0\}}) \star (\Delta^{\{i,\dots,n\}} \times \Delta^{\{1\}})$ for $0 \leq i \leq n$. We set $\overline{L}(i) = K \cup (\cup^i_{j=0} L(j))$ for $0 \leq i \leq n$. Note that $\overline{L}(n) = \Delta^n \times \Delta^1$.

First, we show that the inclusion $K \hookrightarrow \overline{L}(n-1)$ is inner anodyne. Since $L(0) \cap K$ is isomorphic to Λ^{n+1}_1 in $L(0) \cong \Delta^{n+1}$, we see that the inclusion $K \hookrightarrow \overline{L}(0)$ is inner anodyne. For $0 < i < n$, since $L(i) \cap \overline{L}(i-1)$ is isomorphic to $\Lambda^{n+1}(0, i+1)$ in $L(i) \cong \Delta^{n+1}$, we see that the inclusion $\overline{L}(i-1) \hookrightarrow \overline{L}(i)$ is inner anodyne by Lemma 14. Hence the composition $K \hookrightarrow \overline{L}(0) \hookrightarrow \cdots \hookrightarrow \overline{L}(n-1)$ is also inner anodyne.

Since the class of inner anodyne maps is stable under the opposite, the inclusion $K^{\mathrm{op}} \hookrightarrow \overline{L}(n-1)^{\mathrm{op}}$ is also inner anodyne. By [22, Remark 3.1.1.4], we see that $K^\flat \hookrightarrow \overline{L}(n-1)^\flat$ is an opposite marked anodyne map. This implies that the inclusion $(K, K_1 \cap \mathscr{E}) \hookrightarrow (\overline{L}(n-1), \overline{L}(n-1)_1 \cap \mathscr{E})$ is opposite marked anodyne.

Now we consider the inclusion $\overline{L}(n-1) \hookrightarrow \overline{L}(n)$. We see that $L(n) \cap \overline{L}(n-1)$ is isomorphic to Λ^{n+1}_0 in $L(n) \cong \Delta^{n+1}$. We can identify $(L(n), L(n)_1 \cap \mathscr{E})$ with

$(\Delta^{n+1}, \mathscr{E}')$, where $\mathscr{E}'$ is the set of edges of Δ^{n+1} consisting of all degenerate edges together with $\Delta^{\{0,1\}}$. Since the map $(\Lambda_0^{n+1}, (\Lambda_0^{n+1})_1 \cap \mathscr{E}') \hookrightarrow (\Delta^{n+1}, \mathscr{E}')$ is opposite marked anodyne, we see that $(\overline{L}(n-1), \overline{L}(n-1)_1 \cap \mathscr{E}) \to (\overline{L}(n), \mathscr{E})$ is opposite marked anodyne.

Therefore, the composition $(K, K_1 \cap \mathscr{E}) \hookrightarrow (\overline{L}(n-1), \overline{L}(n-1)_1 \cap \mathscr{E}) \hookrightarrow (\overline{L}(n), \mathscr{E})$ is also an opposite marked anodyne map. This completes the proof. □

7.3 The Marked Simplicial Set $\widetilde{\mathscr{O}}(\Delta^n)^+$

In this subsection we introduce a marked simplicial set $\widetilde{\mathscr{O}}(\Delta^n)^+$ in which the underlying simplicial set is $\widetilde{\mathscr{O}}(\Delta^n)$. We study inclusions of subcomplexes of the marked simplicial sets $\widetilde{\mathscr{O}}(\Delta^n)^+$ and $(\widetilde{\mathscr{O}}(\Delta^n)^+ \times (\Delta^{\{0\}})^\flat) \cup (\widetilde{\mathscr{O}}(\Delta^n)^\flat \times (\Delta^1)^\flat)$.

Let $\widetilde{\mathscr{E}}$ be the set of edges of $\widetilde{\mathscr{O}}(\Delta^n)$ consisting of all non-degenerate edges together with edges $ij \to ik$ for $0 \le i \le j \le k \le n$. We regard the pair $(\widetilde{\mathscr{O}}(\Delta^n), \widetilde{\mathscr{E}})$ as a marked simplicial set. For a subcomplex K of $\widetilde{\mathscr{O}}(\Delta^n)$, we set $\widetilde{\mathscr{E}}_K = \widetilde{\mathscr{E}} \cap K_1$ and denote by K^+ the marked simplicial set $(K, \widetilde{\mathscr{E}}_K)$.

For $n > 0$, we let M_n be the subcomplex of $\widetilde{\mathscr{O}}(\Delta^n)$ that contains all non-degenerate k-simplices for $0 \le k \le n$ except for the n-simplex corresponding to $nn \to \cdots \to 0n$.

Lemma 17 *The inclusion $\widetilde{\mathscr{O}}(\partial\Delta^n)^+ \hookrightarrow M_n^+$ is an opposite marked anodyne map for all $n > 0$.*

Proof First, we consider the case $n = 1$. We let B_0 be the 1-simplex corresponding to $00 \to 01$. The subcomplex $B_0 \cap \widetilde{\mathscr{O}}(\partial\Delta^1)$ is isomorphic to Λ_0^1 in $B_0 \cong \Delta^1$. The inclusion $(\Lambda_0^1)^\sharp \hookrightarrow (\Delta^1)^\sharp$ is opposite marked anodyne by Lemma 15. Taking the pushout of $(\Lambda_0^1)^\sharp \hookrightarrow (\Delta^1)^\sharp$ along the map $(\Lambda_0^1)^\sharp \cong B_0^+ \cap \widetilde{\mathscr{O}}(\partial\Delta^1)^+ \to \widetilde{\mathscr{O}}(\partial\Delta^1)^+$, we see that the inclusion $\widetilde{\mathscr{O}}(\partial\Delta^1)^+ \hookrightarrow M_1^+$ is opposite marked anodyne.

Next, we consider the case $n = 2$. Let B_0 be the 2-simplex in $\widetilde{\mathscr{O}}(\Delta^2)$ corresponding to $00 \to 01 \to 02$. The subcomplex $B_0^+ \cap \widetilde{\mathscr{O}}(\partial\Delta^2)^+$ is isomorphic to $(\Lambda_0^2)^\sharp$ in $B_0^+ \cong (\Delta^2)^\sharp$. By Lemma 15, the inclusion $(\Lambda_0^2)^\sharp \hookrightarrow (\Delta^2)^\sharp$ is opposite marked anodyne. Taking the pushout of $(\Lambda_0^2)^\sharp \hookrightarrow (\Delta^2)^\sharp$ along the map $(\Lambda_0^2)^\sharp \cong B_0^+ \cap \widetilde{\mathscr{O}}(\partial\Delta^2)^+ \to \widetilde{\mathscr{O}}(\partial\Delta^2)^+$, we obtain an opposite marked anodyne map $\widetilde{\mathscr{O}}(\partial\Delta^2)^+ \hookrightarrow \widetilde{\mathscr{O}}(\partial\Delta^2)^+ \cup B_0^+$. Let $B_1(0)$ be the 2-simplex in $\widetilde{\mathscr{O}}(\Delta^2)$ corresponding to $11 \to 01 \to 02$. The subcomplex $B_1(0) \cap (\widetilde{\mathscr{O}}(\partial\Delta^2) \cup B_0)$ is isomorphic to Λ_1^2 in Δ^2. The inclusion $(\Lambda_1^2)^\flat \hookrightarrow (\Delta^2)^\flat$ is opposite marked anodyne. Taking the pushout of $(\Lambda_1^2)^\flat \hookrightarrow (\Delta^2)^\flat$ along the map $(\Lambda_1^2)^\flat \to \widetilde{\mathscr{O}}(\partial\Delta^2)^+ \cup B_0^+$, we obtain an opposite marked anodyne map $\widetilde{\mathscr{O}}(\partial\Delta^2)^+ \cup B_0^+ \to \widetilde{\mathscr{O}}(\partial\Delta^2)^+ \cup B_0^+ \cup B_1(0)^+$. Let $B_1(1)$ be the 2-simplex in $\widetilde{\mathscr{O}}(\Delta^2)$ corresponding to $11 \to 12 \to 02$. The subcomplex $B_1(1) \cap (\widetilde{\mathscr{O}}(\partial\Delta^2) \cup B_0 \cup B_1(0))$ is isomorphic to Λ_0^2 in Δ^2. The inclusion $(\Lambda_0^2, (\Lambda_0^2)_1 \cap \mathscr{E}') \hookrightarrow (\Delta^2, \mathscr{E}')$ is opposite marked anodyne, where $\mathscr{E}'$ is the set of edges of Δ^2 consisting of all degenerate edges together with $\Delta^{\{0,1\}}$. Taking the pushout of $(\Lambda_0^2, (\Lambda_0^2)_1 \cap \mathscr{E}') \hookrightarrow (\Delta^2, \mathscr{E}')$ along the map $(\Lambda_0^2, (\Lambda_0^2)_1 \cap \mathscr{E}') \cong B_1(1)^+ \cap (\widetilde{\mathscr{O}}(\partial\Delta^2)^+ \cup B_0^+ \cup B_1(0)^+) \to \widetilde{\mathscr{O}}(\partial\Delta^2)^+ \cup B_0^+ \cup B_1(0)^+$, we see that the inclusion $\widetilde{\mathscr{O}}(\partial\Delta^2)^+ \cup B_0^+ \cup$

$B_1(0)^+ \to \widetilde{\mathscr{O}}(\partial\Delta^2)^+ \cup B_0^+ \cup B_1(0)^+ \cup B_1(1)^+$ is opposite marked anodyne. Hence the composition $\widetilde{\mathscr{O}}(\partial\Delta^2)^+ \to \widetilde{\mathscr{O}}(\partial\Delta^2)^+ \cup B_0^+ \to \cdots \to \widetilde{\mathscr{O}}(\partial\Delta^2)^+ \cup B_0^+ \cup B_1(0)^+ \cup B_1(1)^+ = M_2^+$ is also opposite marked anodyne.

Now we assume $n \geq 3$. In this case we note that all edges of $\widetilde{\mathscr{O}}(\Delta^n)$ are included in $\widetilde{\mathscr{O}}(\partial\Delta^n)$. Let B_0^+ be the n-simplex in $\widetilde{\mathscr{O}}(\Delta^n)^+$ corresponding to $00 \to 01 \to \cdots \to 0n$. The subcomplex $B_0^+ \cap \widetilde{\mathscr{O}}(\partial\Delta^n)^+$ of B_0^+ is isomorphic to $(\Lambda_0^n)^\sharp$ in $B_0^+ \cong (\Delta^n)^\sharp$. By Lemma 15, the inclusion $(\Lambda_0^n)^\sharp \hookrightarrow (\Delta^n)^\sharp$ is opposite marked anodyne. Taking the pushout of $(\Lambda_0^n)^\sharp \hookrightarrow (\Delta^n)^\sharp$ along the map $(\Lambda_0^n)^\sharp \cong B_0^+ \cap \widetilde{\mathscr{O}}(\partial\Delta^n)^+ \to \widetilde{\mathscr{O}}(\partial\Delta^n)^+$, we see that the inclusion $\widetilde{\mathscr{O}}(\partial\Delta^n)^+ \hookrightarrow \widetilde{\mathscr{O}}(\partial\Delta^n)^+ \cup B_0^+$ is opposite marked anodyne.

For $0 \leq i < n$, we let $\mathscr{L}(i)$ be the set of all paths from ii to $0n$ in diagram (1). To a path $l \in \mathscr{L}(i)$, we assign a sequence of integers $J(l) = (j_i, j_{i-1}, \ldots, j_1)$ with $i \leq j_i \leq j_{i-1} \leq \cdots \leq j_1 \leq n$ such that l is depicted as

$$\begin{array}{ccccccc}
 & & & & & 0j_1 \to \cdots \to 0n \\
 & & & & & \uparrow \\
 & & & & \cdots \to & 1j_1 \\
 & & & \cdots & & \\
 & & (i-1)j_i \to \cdots & & & \\
 & & \uparrow & & & \\
ii \to \cdots \to & & ij_i & & &
\end{array}$$

We give $\{J(l) | l \in \mathscr{L}(i)\}$ the lexicographic order, and write $l < l'$ if $J(l) < J(l')$. This gives rise to a total order on $\mathscr{L}(i)$. For example, the path $ii \to \cdots \to 0i \to \cdots \to 0n$ is the smallest and the path $ii \to \cdots \to in \to \cdots \to 0n$ is the largest. For $l \in \mathscr{L}(i)$, we denote by $B(l)$ the n-simplex in $\widetilde{\mathscr{O}}(\Delta^n)$ corresponding to l. Note that $\mathscr{L}(0)$ consists of a unique element l_0 and that $B(l_0) = B_0$. We set $B_i = \cup_{l \in \mathscr{L}(i)} B(l)$ and $\overline{B}_i = \widetilde{\mathscr{O}}(\partial\Delta^n) \cup \bigcup_{j=0}^i B_j$. We shall show that the inclusion $\overline{B}_{i-1}^+ \hookrightarrow \overline{B}_i^+$ is opposite marked anodyne for $0 < i < n$.

For $0 < i < n$ and $l \in \mathscr{L}(i)$, we set $\overline{B}(l) = \overline{B}_{i-1} \cup \bigcup_{l' \leq l} B(l')$ and $\overline{B}(l)^\circ = \overline{B}_{i-1} \cup \bigcup_{l' < l} B(l')$. It suffices to show that the inclusion $\overline{B}(l)^{\circ +} \hookrightarrow \overline{B}(l)^+$ is opposite marked anodyne for all $l \in \mathscr{L}(i)$.

Let l_i be the path $ii \to \cdots \to 0i \to \cdots \to 0n$ for $0 < i < n$. The subcomplex $B(l_i) \cap \overline{B}_{i-1}$ of $B(l_i)$ is isomorphic to Λ_i^n of Δ^n. The inclusion $(\Lambda_i^n)^\flat \hookrightarrow (\Delta^n)^\flat$ is opposite marked anodyne. Taking the pushout of $(\Lambda_i^n)^\flat \hookrightarrow (\Delta^n)^\flat$ along the map $(\Lambda_i^n)^\flat \cong B(l_i)^\flat \cap \overline{B}_{i-1}^+ \to \overline{B}_{i-1}^+$, we see that the inclusion $\overline{B}_{i-1}^+ \hookrightarrow \overline{B}_{i-1}^+ \cup B(l_i)^+$ is opposite marked anodyne.

Let l_i' be the path $ii \to \cdots \to in \to \cdots \to 0n$. We take $l \in \mathscr{L}(i)$ such that $l_i < l < l_i'$. Let $\{\alpha_1, \ldots, \alpha_k\}$ $(0 < \alpha_1 < \ldots < \alpha_k < n)$ be the set of integers such that the sub-path $l(\alpha_t - 1) \to l(\alpha_t) \to l(\alpha_t + 1)$ of l is depicted as

$$\begin{array}{cc}
a, b & \to a, b+1 \\
\uparrow & \\
a+1, b &
\end{array} \tag{7}$$

for $t = 1, \dots, k$. We consider the subcomplex $B(l) \cap \overline{B}(l)^{\circ}$ of $B(l)$. There are two cases. (1) If the first edge of l is $ii \to (i-1)i$, then the subcomplex $B(l) \cap \overline{B}(l)^{\circ}$ of $B(l)$ is isomorphic to the subcomplex $\Lambda^n(\alpha_1, \dots, \alpha_k)$ of Δ^n. Since the inclusion $\Lambda^n(\alpha_1, \dots, \alpha_k) \hookrightarrow \Delta^n$ is inner anodyne by Lemma 13, the inclusion $(\Lambda^n(\alpha_1, \dots, \alpha_k))^{\flat} \hookrightarrow (\Delta^n)^{\flat}$ is opposite marked anodyne. Taking the pushout of $(\Lambda^n(\alpha_1, \dots, \alpha_k))^{\flat} \hookrightarrow (\Delta^n)^{\flat}$ along the map $(\Lambda^n(\alpha_1, \dots, \alpha_k))^{\flat} \cong B(l)^{\flat} \cap \overline{B}(l)^{\circ +} \to \overline{B}(l)^{\circ +}$, we see that the inclusion $\overline{B}(l)^{\circ +} \hookrightarrow \overline{B}(l)^{+}$ is opposite marked anodyne. (2) If the first edge of l is $ii \to i(i+1)$, then the subcomplex $B(l) \cap \overline{B}(l)^{\circ}$ of $B(l)$ is isomorphic to the subcomplex $\Lambda^n(0, \alpha_1, \dots, \alpha_k)$ of Δ^n. Note that $\alpha_1 > 1$ in this case. Since the inclusion $\Lambda^n(0, \alpha_1, \dots, \alpha_k) \hookrightarrow \Delta^n$ is inner anodyne by Lemma 14, the inclusion $(\Lambda^n(0, \alpha_1, \dots, \alpha_k))^{\flat} \hookrightarrow (\Delta^n)^{\flat}$ is opposite marked anodyne. Taking the pushout of $(\Lambda^n(0, \alpha_1, \dots, \alpha_k))^{\flat} \hookrightarrow (\Delta^n)^{\flat}$ along the map $(\Lambda^n(0, \alpha_1, \dots, \alpha_k))^{\flat} \cong B(l)^{\flat} \cap \overline{B}(l)^{\circ +} \to \overline{B}(l)^{\circ +}$, we see that the inclusion $\overline{B}(l)^{\circ +} \hookrightarrow \overline{B}(l)^{+}$ is opposite marked anodyne.

Finally, we shall show that $\overline{B}(l'_i)^{\circ +} \hookrightarrow \overline{B}(l'_i)^{+}$ is opposite marked anodyne for $0 < i < n$. The subcomplex $B(l'_i) \cap \overline{B}(l'_i)^{\circ}$ is isomorphic to the subcomplex Λ^n_0 of Δ^n. Note that $ii \to i(i+1)$ is a marked edge, which corresponds to $\Delta^{\{0,1\}}$ under the isomorphism $\Lambda^n_0 \cong B(l'_i) \cap \overline{B}(l'_i)^{\circ}$. The inclusion $(\Lambda^n_0, (\Lambda^n_0) \cap \mathscr{E}') \hookrightarrow (\Delta^n, \mathscr{E}')$ is opposite marked anodyne, where $\mathscr{E}'$ is the set of edges of Δ^n consisting of all degenerate edges together with $\Delta^{\{0,1\}}$. Taking the pushout of $(\Lambda^n_0, (\Lambda^n_0) \cap \mathscr{E}') \hookrightarrow (\Delta^n, \mathscr{E}')$ along the map $(\Lambda^n_0, (\Lambda^n_0) \cap \mathscr{E}') \to B(l'_i)^{+} \cap \overline{B}(l'_i)^{\circ +} \to \overline{B}(l'_i)^{\circ +}$, we see that the inclusion $\overline{B}(l'_i)^{\circ +} \hookrightarrow \overline{B}(l'_i)^{+}$ is opposite marked anodyne. This completes the proof. □

For $n > 0$, we let

$$\begin{aligned} \widetilde{A} &= (\widetilde{\mathscr{O}}(\Delta^n) \times \Delta^{\{0\}}) \cup (\widetilde{\mathscr{O}}(\partial \Delta^n) \times \Delta^1), \\ \widetilde{B} &= \widetilde{A} \cup (M_n \times \Delta^{\{1\}}), \\ \widetilde{C} &= \widetilde{A} \cup (M_n \times \Delta^1) \end{aligned}$$

be the subcomplexes of $\widetilde{\mathscr{O}}(\Delta^n) \times \Delta^1$. We denote by $(\widetilde{\mathscr{O}}(\Delta^n) \times \Delta^1)^{+}$ the marked simplicial set $(\widetilde{\mathscr{O}}(\Delta^n)^{+} \times (\Delta^{\{0\}})^{\flat}) \cup (\widetilde{\mathscr{O}}(\Delta^n)^{\flat} \times (\Delta^1)^{\flat})$. For a subcomplex K of $\widetilde{\mathscr{O}}(\Delta^n) \times \Delta^1$, we denote by K^{+} the subcomplex of the marked simplicial set $(\widetilde{\mathscr{O}}(\Delta^n) \times \Delta^1)^{+}$ in which the underlying simplicial set is K.

Lemma 18 *The inclusion $\widetilde{B}^{+} \hookrightarrow \widetilde{C}^{+}$ is an opposite marked anodyne map.*

Proof We use the notation in the proof of Lemma 17. Recall that B_0 is the n-simplex in $\widetilde{\mathscr{O}}(\Delta^n)$ corresponding to $00 \to 01 \to \cdots \to 0n$. Since the subcomplex $B_0 \cap \widetilde{\mathscr{O}}(\partial \Delta^n)$ of B_0 is isomorphic to Λ^n_0 in Δ^n, the subcomplex $\widetilde{B} \cap (B_0 \times \Delta^1)$ of $B_0 \times \Delta^1$ is isomorphic to the subcomplex $(\Delta^n \times \partial \Delta^1) \cup (\Lambda^n_0 \times \Delta^1)$ of $\Delta^n \times \Delta^1$. Since $00 \to 01$ is a marked edge of $\widetilde{\mathscr{O}}(\Delta^n)^{+}$, we see that the inclusion $\widetilde{B}^{+} \hookrightarrow \widetilde{B}^{+} \cup (B_0 \times \Delta^1)^{+}$ is opposite marked anodyne by using Lemma 16.

We set $C_i = \widetilde{B} \cup (\overline{B}_i \times \Delta^1)$. We shall show that the inclusion $C^{+}_{i-1} \hookrightarrow C^{+}_i$ is opposite marked anodyne for $0 < i < n$. For this purpose, it suffices to show that the inclusion $C_{i-1}^{+} \cup (\overline{B}(l)^{\circ} \times \Delta^1)^{+} \hookrightarrow C_{i-1}^{+} \cup (\overline{B}(l) \times \Delta^1)^{+}$ is opposite marked anodyne for all $l \in \mathscr{L}(i)$.

Recall that l_i is the path $ii \to \cdots \to 0i \to \cdots \to 0n$ and that the subcomplex $B(l_i) \cap \overline{B}_{i-1}$ of $B(l_i)$ is isomorphic to Λ_i^n of Δ^n. This implies that $(B(l_i) \times \Delta^1) \cap C_{i-1}$ is isomorphic to $(\Lambda_i^n \times \Delta^1) \cup (\Delta^n \times \partial\Delta^1)$. The inclusion $(\Lambda_i^n \times \Delta^1) \cup (\Delta^n \times \partial\Delta^1) \hookrightarrow (\Delta^n \times \Delta^1)$ is inner anodyne for $0 < i < n$ by [22, Cor. 2.3.2.4]. This implies that $(\Lambda_i^n \times \Delta^1)^{\flat} \cup (\Delta^n \times \partial\Delta^1)^{\flat} \hookrightarrow (\Delta^n \times \Delta^1)^{\flat}$ is opposite marked anodyne. Hence we see that the inclusion $C_{i-1}^{+} \hookrightarrow C_{i-1}^{+} \cup (B(l_i) \times \Delta^1)^{+}$ is opposite marked anodyne.

We take $l \in \mathscr{L}(i)$ such that $l_i < l < l_i'$, where l_i' is the path $ii \to \cdots \to in \to \cdots \to 0n$. Let $\{\alpha_1, \ldots, \alpha_k\}$ $(0 < \alpha_1 < \ldots < \alpha_k < n)$ be the set of integers such that the sub-path $l(\alpha_t - 1) \to l(\alpha_t) \to l(\alpha_t + 1)$ is $a + 1, b \to a, b \to a, b + 1$ for some a, b.

Recall that the subcomplex $B(l) \cap \overline{B}(l)^{\circ}$ of $B(l)$ is isomorphic to the subcomplex $\Lambda^n(\alpha_1, \ldots, \alpha_k)$ of Δ^n, if the first edge of l is $ii \to (i-1)i$. This implies that $(B(l) \times \Delta^1) \cap (C_{i-1} \cup (\overline{B}(l)^{\circ} \times \Delta^1))$ is isomorphic to $(\Lambda^n(\alpha_1, \ldots, \alpha_k) \times \Delta^1) \cup (\Delta^n \times \partial\Delta^1)$. In this case the inclusion $(\Lambda^n(\alpha_1, \ldots, \alpha_k) \times \Delta^1) \cup (\Delta^n \times \partial\Delta^1) \hookrightarrow \Delta^n \times \Delta^1$ is inner anodyne by Lemma 13 and [22, Cor. 2.3.2.4]. This implies that $(\Lambda^n(\alpha_1, \ldots, \alpha_k) \times \Delta^1)^{\flat} \cup (\Delta^n \times \partial\Delta^1)^{\flat} \hookrightarrow (\Delta^n \times \Delta^1)^{\flat}$ is opposite marked anodyne. Hence we see that $C_{i-1}^{+} \cup (\overline{B}(l)^{\circ} \times \Delta^1)^{+} \hookrightarrow C_{i-1}^{+} \cup (\overline{B}(l) \times \Delta^1)^{+}$ is opposite marked anodyne in this case.

If the first edge of l is $ii \to i(i+1)$, then the subcomplex $B(l) \cap \overline{B}(l)^{\circ}$ of $B(l)$ is isomorphic to the subcomplex $\Lambda^n(0, \alpha_1, \ldots, \alpha_k)$ of Δ^n, where $\alpha_1 > 1$. This implies that $(B(l) \times \Delta^1) \cap (C_{i-1} \cup (\overline{B}(l)^{\circ} \times (\Delta^1)))$ is isomorphic to $(\Lambda^n(0, \alpha_1, \ldots, \alpha_k) \times \Delta^1) \cup (\Delta^n \times \partial\Delta^1)$. In this case the inclusion $(\Lambda^n(0, \alpha_1, \ldots, \alpha_k) \times \Delta^1) \cup (\Delta^n \times \partial\Delta^1) \hookrightarrow \Delta^n \times \Delta^1$ is inner anodyne by Lemma 14 and [22, Cor. 2.3.2.4]. This implies that $(\Lambda^n(0, \alpha_1, \ldots, \alpha_k) \times \Delta^1)^{\flat} \cup (\Delta^n \times \partial\Delta^1)^{\flat} \hookrightarrow (\Delta^n \times \Delta^1)^{\flat}$ is opposite marked anodyne. Hence we see that $C_{i-1}^{+} \cup (\overline{B}(l)^{\circ} \times \Delta^1)^{+} \hookrightarrow C_{i-1}^{+} \cup (\overline{B}(l) \times \Delta^1)^{+}$ is also opposite marked anodyne in this case.

Finally, we shall show that $C_{i-1}^{+} \cup (\overline{B}(l_i')^{\circ} \times \Delta^1)^{+} \hookrightarrow C_{i-1}^{+} \cup (\overline{B}(l_i') \times \Delta^1)^{+}$ is opposite marked anodyne. Since the subcomplex $B(l_i') \cap \overline{B}(l_i')^{\circ}$ is isomorphic to the subcomplex Λ_0^n of Δ^n, the subcomplex $(B(l_i') \times \Delta^1) \cap (C_{i-1} \cup (\overline{B}(l_i')^{\circ} \times \Delta^1))$ is isomorphic to $(\Lambda_0^n \times \Delta^1) \cup (\Delta^n \times \partial\Delta^1)$. Since $ii \to i(i+1)$ is a marked edge of $\widetilde{\mathscr{O}}(\Delta^n)$, we see that $C_{i-1}^{+} \cup (\overline{B}(l_i')^{\circ} \times \Delta^1)^{+} \hookrightarrow C_{i-1}^{+} \cup (\overline{B}(l_i') \times \Delta^1)^{+}$ is opposite marked anodyne by using Lemma 16. This completes the proof. □

7.4 *Proof of Proposition 1*

In this subsection we give a proof of Proposition 1. For this purpose, we show that the map $\pi_X : \mathscr{R} \to RX$ has right lifting property with respect to the maps $\partial\Delta^n \hookrightarrow \Delta^n$ for $n > 0$ if the final vertex $\Delta^{\{n\}}$ goes to an object of $\mathscr{R}^0$.

Let $p : X \to S$ and $q : Y \to S$ be coCartesian fibrations over a quasi-category S. Suppose we have a map $G : Y \to X$ over S such that G_s admits a left adjoint F_s for all $s \in S$.

We recall that

$$\mathscr{R} = RX \times_{H(\mathrm{Fun}(\Delta^{\{0\}},X))} H(\mathrm{Fun}^S(\Delta^1, X)) \times_{H(\mathrm{Fun}(\Delta^{\{1\}},X))} RY.$$

We have the projection map $\pi_X : \mathscr{R} \to RX$.

We identify objects of RX with objects of X. For $x \in X$ with $s = p(x)$, we have an object $(x, u_x, F_s(x))$ of $\mathscr{R}$ over s, where $u_x : x \to G_s F_s(x)$ is the unit map of the adjunction (F_s, G_s) at x.

The following is a key lemma.

Lemma 19 *Suppose we have a commutative diagram*

$$\begin{array}{ccc} \partial\Delta^n & \xrightarrow{f} & \mathscr{R} \\ \big\downarrow & \nearrow & \big\downarrow \pi_X \\ \Delta^n & \xrightarrow{g} & RX \end{array}$$

for $n > 0$, where the left vertical arrow is the inclusion. We put $x = g(\Delta^{\{n\}})$ and $s = p(x)$. If $f(\Delta^{\{n\}}) = (x, u_x, F_s(x))$, then there exists a dotted arrow $\Delta^n \to \mathscr{R}$ making the whole diagram commutative.

Proof (*Proof of Proposition* 1) By Lemma 19, the map $\pi_X^0 : \mathscr{R}^0 \to RX$ has the right lifting property with respect to the maps $\partial\Delta^n \hookrightarrow \Delta^n$ for all $n \geq 0$. Hence π_X^0 is a trivial Kan fibration. □

In order to prove Lemma 19, we consider the following situation.

Let h be a map $\widetilde{\mathscr{O}}(\Delta^n) \to S$ for $n > 0$ such that $h(ii) \to \cdots \to h(0i)$ is a totally degenerate simplex in S for all $0 \leq i \leq n$. We set $\overline{h} = h\pi$, where $\pi : \widetilde{\mathscr{O}}(\Delta^n) \times \Delta^1 \to \widetilde{\mathscr{O}}(\Delta^n)$ is the projection.

Let $X^\natural$ be the marked simplicial set in which the simplicial set is X and the set of marked edges consists of all p-coCartesian edges. Suppose that we have an n-simplex in RX that is represented by $g : \widetilde{\mathscr{O}}(\Delta^n) \to X$ covering h. Note that we can regard g as a map of marked simplicial sets $\widetilde{\mathscr{O}}(\Delta^n)^+ \to X^\natural$.

Furthermore, we suppose that we have a map $\partial\Delta^n \to \mathscr{R}$ that is represented by a triple of maps (g', k, f), where $g' : \widetilde{\mathscr{O}}(\partial\Delta^n) \to X$, $k : \widetilde{\mathscr{O}}(\partial\Delta^n) \to \mathrm{Fun}^S(\Delta^1, X)$, and $f : \widetilde{\mathscr{O}}(\partial\Delta^n) \to Y$. We assume that g' is the restriction of g. Then the maps g', k, and f cover h, respectively.

Let $Y^\natural$ be the marked simplicial set defined in the same way as $X^\natural$. We can regard f as a map of marked simplicial sets $\widetilde{\mathscr{O}}(\partial\Delta^n)^+ \to Y^\natural$. There is an extension $\widetilde{f}$ of f to M_n^+ covering h by Lemma 17.

We recall that $\widetilde{A}$, $\widetilde{B}$, and $\widetilde{C}$ are subcomplexes of $\widetilde{\mathscr{O}}(\Delta^n) \times \Delta^1$ given by $\widetilde{A} = (\widetilde{\mathscr{O}}(\Delta^n) \times \Delta^{\{0\}}) \cup (\widetilde{\mathscr{O}}(\partial\Delta^n) \times \Delta^1)$, $\widetilde{B} = \widetilde{A} \cup (M_n \times \Delta^{\{1\}})$, $\widetilde{C} = \widetilde{A} \cup (M_n \times \Delta^1)$.

Using the maps g, k, and $G(\widetilde{f})$, we obtain a map of marked simplicial sets $\widetilde{B}^+ \to X^\natural$ over $\overline{h}$. Furthermore, by Lemma 18, we can extend this map to a map of marked simplicial sets $w : \widetilde{C}^+ \to X^\natural$ covering $\overline{h}$.

Let D be the n-simplex of $\widetilde{\mathscr{O}}(\Delta^n)$ corresponding to $nn \to \cdots \to 0n$. By restricting w to $(D \times \Delta^{\{0\}}) \cup (\partial D \times \Delta^1)$, we obtain a map $v : (D \times \Delta^{\{0\}}) \cup (\partial D \times \Delta^1) \to X_s$, where $s = h(nn)$. We denote by g_D the restriction of g to D and by $\widetilde{f}_{\partial D}$ the restriction of $\widetilde{f}$ to ∂D. Note that the restriction of v to $D \times \Delta^{\{0\}}$ is identified with g_D and that the restriction of v to $\partial D \times \Delta^{\{1\}}$ is $G_s(\widetilde{f}_{\partial D})$.

We would like to have maps $\widetilde{f}_D : D \to Y_s$ and $\overline{v} : D \times \Delta^1 \to X_s$ such that $\widetilde{f}_D$ is an extension of $\widetilde{f}_{\partial D}$, $\overline{v}$ is an extension of v, and the restriction of $\overline{v}$ to $D \times \Delta^{\{1\}}$ is $G(\widetilde{f}_D)$. Hence, in order to prove Lemma 19, it suffices to prove the following lemma.

Lemma 20 *Let $L : \mathscr{C} \rightleftarrows \mathscr{D} : R$ be an adjunction of quasi-categories. Suppose we have maps $f : (\Delta^n \times \Delta^{\{0\}}) \cup (\partial\Delta^n \times \Delta^1) \to \mathscr{C}$ and $g : \partial\Delta^n \to \mathscr{D}$ for $n > 0$ such that $Rg = f|_{\partial\Delta^n \times \Delta^{\{1\}}}$. We put $c = f(\Delta^{\{0\}} \times \Delta^{\{0\}})$ and $d = g(\Delta^{\{0\}})$. If $g(d) = L(c)$ and $f(\Delta^{\{0\}} \times \Delta^1)$ is the unit map $c \to RL(c)$ of the adjunction (L, R) at c, then there exist maps $F : \Delta^n \times \Delta^1 \to \mathscr{C}$ and $G : \Delta^n \to \mathscr{D}$ such that F is an extension of f, G is an extension of g, and $RG = F|_{\Delta^n \times \Delta^{\{1\}}}$.*

Proof Let $\pi : \mathscr{M} \to \Delta^1$ be a map associated to the adjunction (L, R), which is a coCartesian fibration and a Cartesian fibration. We may assume that the fibers $\mathscr{M}_{\{0\}}$ and $\mathscr{M}_{\{1\}}$ over $\{0\}$ and $\{1\}$ are isomorphic to $\mathscr{C}$ and $\mathscr{D}$, respectively. We regard f as a map $(\Delta^n \times \Delta^{\{0\}}) \cup (\partial\Delta^n \times \Delta^1) \to \mathscr{M}_{\{0\}}$ and g as a map $\partial\Delta^n \to \mathscr{M}_{\{1\}}$.

Since $\mathscr{M} \to \Delta^1$ is a Cartesian fibration, we can extend the map g to a map $h : \partial\Delta^n \times \Delta^1 \to \mathscr{M}$ such that $h|_{\partial\Delta^n \times \Delta^{\{0\}}} = Rg$, $h|_{\partial\Delta^n \times \Delta^{\{1\}}} = g$, and $h(\Delta^{\{i\}} \times \Delta^1)$ is a π-Cartesian edge over Δ^1 for all $i = 0, 1, \ldots, n$. By the assumption that $Rg = f|_{\partial\Delta^n \times \Delta^{\{1\}}}$, we obtain a map $k : \partial\Delta^n \times \Lambda^2_1 \to \mathscr{M}$ such that $k|_{\partial\Delta^n \times \Delta^{\{0,1\}}} = f|_{\partial\Delta^n \times \Delta^{\{0,1\}}}$ and $k|_{\partial\Delta^n \times \Delta^{\{1,2\}}} = h$.

By the assumptions that $g(d) = L(c)$ and $f(\Delta^{\{0\}} \times \Delta^1)$ is the unit map $c \to RL(c)$, we have a map $l : \Delta^{\{0\}} \times \Delta^2 \to \mathscr{M}$ such that $l|_{\Delta^{\{0\}} \times \Delta^{\{0,1\}}} = f|_{\Delta^{\{0\}} \times \Delta^1}$, $l|_{\Delta^{\{0\}} \times \Delta^{\{1,2\}}} = k|_{\Delta^{\{0\}} \times \Delta^{\{1,2\}}}$, and $l(\Delta^{\{0\}} \times \Delta^{\{0,2\}})$ is π-coCartesian.

Hence we obtain a map $k \cup l : (\partial\Delta^n \times \Lambda^2_1) \cup (\Delta^{\{0\}} \times \Delta^2) \to \mathscr{M}$. Let $\sigma : \Delta^n \times \Delta^2 \to \Delta^1$ be the projection $\Delta^n \times \Delta^2 \to \Delta^2$ followed by $s^0 : \Delta^2 \to \Delta^1$, where $s^0(\{0\}) = s^0(\{1\}) = \{0\}$ and $s^0(\{2\}) = \{1\}$. We shall show that $k \cup l$ extends to a map on $\Delta^n \times \Delta^2$ covering σ.

Since $\Lambda^2_1 \hookrightarrow \Delta^2$ is inner anodyne, $(\partial\Delta^n \times \Lambda^2_1) \cup (\Delta^{\{0\}} \times \Delta^2) \to \partial\Delta^n \times \Delta^2$ is also inner anodyne by [22, Cor. 2.3.2.4]. Hence there is an extension $m : \partial\Delta^n \times \Delta^2 \to \mathscr{M}$ of $k \cup l : (\partial\Delta^n \times \Lambda^2_1) \cup (\Delta^{\{0\}} \times \Delta^2) \to \mathscr{M}$ covering σ.

We have the map $f'|_{\Delta^n \times \Delta^{\{0\}}} \cup m|_{\partial\Delta^n \times \Delta^{\{0,2\}}} : (\Delta^n \times \Delta^{\{0\}}) \cup (\partial\Delta^n \times \Delta^{\{0,2\}}) \to \mathscr{M}$. Since $m(\Delta^{\{0\}} \times \Delta^{\{0,2\}})$ is a π-coCartesian edge over Δ^1, there is an extension $p(0,2) : \Delta^n \times \Delta^{\{0,2\}} \to \mathscr{M}$ of $f'|_{\Delta^n \times \Delta^{\{0\}}} \cup m|_{\partial\Delta^n \times \Delta^{\{0,2\}}}$ covering σ by [22, Prop. 2.4.1.8].

We have the map $p(0,2)|_{\Delta^n \times \Delta^{\{2\}}} \cup m|_{\partial\Delta^n \times \Delta^{\{1,2\}}} : (\Delta^n \times \Delta^{\{2\}}) \cup (\partial\Delta^n \times \Delta^{\{1,2\}}) \to \mathscr{M}$. Since $m(\Delta^{\{n\}} \times \Delta^{\{1,2\}})$ is a π-Cartesian edge over Δ^1, there is an extension $p(1,2) : \Delta^n \times \Delta^{\{1,2\}} \to \mathscr{M}$ of $P(0,2)|_{\Delta^n \times \Delta^{\{2\}}} \cup m|_{\partial\Delta^n \times \Delta^{\{1,2\}}}$ covering σ by the dual of [22, Prop. 2.4.1.8].

Hence we obtain a map $q = m \cup p(1,2) \cup p(0,2) : (\partial\Delta^n \times \Delta^2) \cup (\Delta^n \times \Lambda^2_2) \to \mathscr{M}$ covering σ. We note that $q(\Delta^{\{i\}} \times \Delta^{\{1,2\}})$ is π-Cartesian for all $i = 0, 1, \ldots, n$.

Let $\mathscr{E}$ be the set of edges of Δ^2 consisting of all degenerate edges together with $\Delta^{\{1,2\}}$. We denote by $(\Delta^2)^+$ the marked simplicial set $(\Delta^2, \mathscr{E})$ and by $(\Lambda^2_2)^+$ the marked simplicial set $(\Lambda^2_2, \mathscr{E} \cap (\Lambda^2_2)_1)$. The map of marked simplicial sets $(\Lambda^2_2)^+ \to (\Delta^2)^+$ is marked anodyne by [22, Def. 3.1.1.1]. This implies that $(\Delta^n)^\flat \times (\Lambda^2_2)^+ \cup (\partial\Delta^n)^\flat \times (\Delta^2)^+ \to (\Delta^n)^\flat \times (\Delta^2)^+$ is also marked anodyne by [22, Prop. 3.1.2.3].

Let $(\Delta^1)^\sharp$ be the marked simplicial set Δ^1 equipped with the set of all edges, and let $\mathscr{M}^\natural$ be the marked simplicial set $\mathscr{M}$ equipped with the set of all π-Cartesian edges. Since $q(\Delta^{\{i\}} \times \Delta^{\{1,2\}})$ is a π-Cartesian edge for all $i = 0, 1, \ldots, n$, we have a map of marked simplicial sets $q : (\Delta^n)^\flat \times (\Lambda^2_2)^+ \cup (\partial\Delta^n)^\flat \times (\Delta^2)^+ \to \mathscr{M}^\natural$. We consider the following commutative diagram of marked simplicial sets

$$\begin{array}{ccc} (\Delta^n)^\flat \times (\Lambda^2_2)^+ \cup (\partial\Delta^n)^\flat \times (\Delta^2)^+ & \longrightarrow & \mathscr{M}^\natural \\ \Big\downarrow & \overset{r}{\nearrow} & \Big\downarrow \pi \\ (\Delta^n)^\flat \times (\Delta^2)^+ & \xrightarrow{\ \sigma\ } & (\Delta^1)^\sharp, \end{array}$$

where the upper horizontal arrow is q. Since the left vertical arrow is marked anodyne, there is a dotted arrow r which makes the whole diagram commutative by [22, Prop. 3.1.1.6]. The proof is completed by setting $F = r|_{\Delta^n \times \Delta^{\{0,1\}}}$ and $G = r|_{\Delta^n \times \Delta^{\{2\}}}$. □

Acknowledgements The author would like to thank the anonymous referee for careful reading of the manuscript and helpful comments.

References

1. Baker, A., Jeanneret, A.: Brave new Hopf algebroids and extensions of MU-algebras. Homol. Homotopy Appl. **4**(1), 163–173 (2002)
2. Barwick, C., Glasman, S., Nardin, D.: Dualizing cartesian and cocartesian fibrations. Theory Appl. Categ. **33**(4), 67–94 (2018)
3. Behrens, M., Davis, D.G.: The homotopy fixed point spectra of profinite Galois extensions. Trans. Amer. Math. Soc. **362**(9), 4983–5042 (2010)
4. Boardman, J.M., Vogt, R.M.: Homotopy Invariant Algebraic Structures on Topological Spaces. Lecture Notes in Mathematics, vol. 347. Springer, Berlin (1973)
5. Bousfield, A.K.: The localization of spectra with respect to homology. Topology **18**(4), 257–281 (1979)
6. Bousfield, A.K., Kan, D.M.: Homotopy Limits, Completions and Localizations. Lecture Notes in Mathematics, vol. 304. Springer, Berlin (1972)
7. Davis, D.G.: Homotopy fixed points for $L_K(n)(E_n \wedge X)$ using the continuous action. J. Pure Appl. Algebra **206**(3), 322–354 (2006)
8. Devinatz, E.S., Hopkins, M.J.: Homotopy fixed point spectra for closed subgroups of the Morava stabilizer groups. Topology **43**(1), 1–47 (2004)
9. Devinatz, E.S., Hopkins, M.J., Smith, J.H.: Nilpotence and stable homotopy theory. I. Ann. Math. **128** (2), 207–241 (1988)

10. Goerss, P.G., Hopkins, M.J.: Moduli Spaces of Commutative Ring Spectra, Structured Ring Spectra. London Mathematical Society Lecture Note Series, vol. 315, pp. 151–200. Cambridge University Press, Cambridge (2004)
11. Hess, K.: A general framework for homotopic descent and codescent (2010). arXiv:1001.1556
12. Hopkins, M.J., Smith, J.H.: Nilpotence and stable homotopy theory. II. Ann. Math. **148**(2), no. 1, 1–49 (1998)
13. Hovey, M.: Bousfield localization functors and Hopkins' chromatic splitting conjecture. In: The Čech Centennial (Boston, MA, 1993). Contemporary Mathematics, vol. 181, pp. 225–250. American Mathematical Society, Providence (1995)
14. Hovey, M.: Morita theory for Hopf algebroids and presheaves of groupoids. Amer. J. Math. **124**(6), 1289–1318 (2002)
15. Hovey, M., Sadofsky, H.: Invertible spectra in the $E(n)$-local stable homotopy category. J. London Math. Soc. **60**(2), no. 1, 284–302 (1999)
16. Hovey, M., Shipley, B., Smith, J.: Symmetric spectra. J. Amer. Math. Soc. **13**(1), 149–208 (2000)
17. Hovey, M., Strickland, N.P.: Morava K-theories and localisation. Mem. Amer. Math. Soc. 139(666) (1999)
18. Hovey, M., Strickland, N.: Comodules and Landweber exact homology theories. Adv. Math. **192**(2), 427–456 (2005)
19. Joyal, A.: Quasi-categories and Kan complexes. J. Pure Appl. Algebra **175**(1–3), 207–222 (2002)
20. Landweber, P.S.: Homological properties of comodules over $MU_*(MU)$ and $BP_*(BP)$. Amer. J. Math. **98**(3), 591–610 (1976)
21. Lazarev, A.: Towers of MU-algebras and the generalized Hopkins-Miller theorem. Proc. London Math. Soc. **87**(3), no. 2, 498–522 (2003)
22. Lurie, J.: Higher Topos Theory. Annals of Mathematics Studies, vol. 170. Princeton University Press, Princeton (2009)
23. Lurie, J.: Higher algebra (version 9/14/2014). http://www.math.harvard.edu/~lurie/
24. Lurie, J.: Derived algebraic geometry X: formal moduli problems (2011). http://www.math.harvard.edu/~lurie/
25. Mathew, A.: The Galois group of a stable homotopy theory. Adv. Math. **291**, 403–541 (2016)
26. Mathew, A., Naumann, N., Noel, J.: Nilpotence and descent in equivariant stable homotopy theory. Adv. Math. **305**, 994–1084 (2017)
27. May, J.P.: E_∞ Ring Spaces and E_∞ Ring Spectra. Lecture Notes in Mathematics, vol. 577. Springer, Berlin (1977)
28. Miller, H.R., Ravenel, D.C., Wilson, W.S.: Periodic phenomena in the Adams-Novikov spectral sequence. Ann. Math. **106**(2), no. 3, 469–516 (1977)
29. Morava, J.: Noetherian localisations of categories of cobordism comodules. Ann. Math. **121**(2), no. 1, 1–39 (1985)
30. Naumann, N.: The stack of formal groups in stable homotopy theory. Adv. Math. **215**(2), 569–600 (2007)
31. Ravenel, D.C.: Complex Cobordism and Stable Homotopy Groups of Spheres. Pure and Applied Mathematics, vol. 121. Academic Press Inc, Orlando (1986)
32. Ravenel, D.C.: Nilpotence and Periodicity in Stable Homotopy Theory. Annals of Mathematics Studies, vol. 128. Princeton University Press, Princeton (1992)
33. Torii, T.: Discrete G-Spectra and embeddings of module spectra. J. Homotopy Relat. Struct. **12**(4), 853–899 (2017)

Koszul Duality for E_n-Algebras in a Filtered Category

Takuo Matsuoka

Abstract We describe the use of filtration for algebra, in particular, for the Koszul duality, in a stable $(\infty, 1)$-category, while illustrating how simple arguments with filtrations lead to finding nice behaviour of very basic constructions in homotopical algebra.

Keywords Koszul duality · Higher morita category · E_n-algebra · Filtration · Completeness · Homotopical algebra · Stable category

1 Introduction

1.1 Overview

A classical instance of the Koszul duality was described by Quillen. Namely, he essentially established an equivalence of the $(\infty, 1)$-categories of reduced differential graded Lie algebras and 2-reduced differential graded commutative coalgebras over $\mathbb{Q}$ (both localized, of course, with respect to weak equivalences) [19]. This connects, by algebraic means, algebraic models of simply connected rational homotopy types, one given by Sullivan [22] and another given by Quillen [19].

In terms of the Koszul duality for operads, developed later by Ginzburg and Kapranov [9], Quillen's equivalence reflects the Koszul duality between the Lie and the commutative operads. On the other hand, the associative operad is self Koszul dual in a way, so there is a similar equivalence between some (augmented, or equivalently up to categorical equivalence, non-unital) associative algebras and corresponding (augmented/non-unital) associative coalgebras. For an associative algebra A, an associative colagebra C corresponding to A under this equivalence is said to be Koszul dual to A. In good cases, certain parts (of the $(\infty, 1)$-categories) of A-modules and

T. Matsuoka (✉)
Intage Technosphere Inc., Intage Akihabara Bldg., 3 Kanda-Neribeicho, Chiyoda-Ku, Tokyo, Japan
e-mail: motogeomtop@gmail.com

T. Ohsawa and N. Minami (eds.), *Bousfield Classes and Ohkawa's Theorem*,
Springer Proceedings in Mathematics & Statistics 309,
https://doi.org/10.1007/978-981-15-1588-0_12

of C-comodules will be equivalent. Instead of C-comodules, one could think of the modules over the linear dual $C^\vee$. The augmented/non-unital associative algebra $C^\vee$ is also called the Koszul dual of A, and A will in turn be Koszul dual to $C^\vee$. A suitable equivalence between the derived categories of modules over Koszul dual algebras were established by Beilinson, Ginsburg and Schechtman [1]. An instance of this is a canonical equivalence between derived categories of graded modules over symmetric and exterior algebra, established by Bernstein, I. M. Gelfand and S. I. Gelfand [4], which had inspired Beilinson, Ginsburg and Schechtman.

What we have called the "self" Koszul duality of the associative operad, was generalized for E_n-operad by Fresse [8], from the case $n = 1$. (The commutative–Lie correspondence is also recovered if we let $n \to \infty$.) A consequence would be an equivalence of certain augmented/non-unital E_n-algebras and E_n-coalgebras. However, in view of Dunn's result that the structure of an E_n-algebra is essentially the n-fold iteration of the structure of an E_1-algebra [7], it should be possible to establish a similar equivalence in a simple manner by establishing and iterating the Koszul duality for E_1-algebras. The purpose of this survey is to describe a set of simple techniques which are useful for examining how far such a result holds, and also seem useful for application of homotopical algebra in general if the version of the Koszul duality here does not serve the reader's purposes well.

We shall follow [13], which was influenced by Costello [5]. There are also related works by Positselski [18] in addition to the already mentioned work by Beilinson, Ginsburg and Schechtman [1].

Following Quillen and some others, we consider a correspondence between algebraic and *co*algebraic structures. While this simply happened to be as much as we needed for the applications we had, this will separate from our work some complexities arising from the process of taking the linear duals.

1.2 Basic Constructions

A traditional version of the Koszul dual construction assumes a differential graded structure. We shall start with the description of a perhaps less traditional version, which can be made for an associative (co-)algebra object in a quite general symmetric monoidal $(\infty, 1)$-category. We shall use this construction throughout our work since it will be easy to iterate as well as to analyse.

Let C be an augmented associative coalgebra in a symmetric monoidal $(\infty, 1)$-category $\mathscr{A}$, or a "coaugmented coassociative" one, to emphasize the variance. In our terminology, we shall more often not emphasize the variance when a confusion seems unlikely to arise. Associativity in a symmetric monoidal *$(\infty, 1)$-category* means data for homotopy coherent associativity, which in particular is a structure rather than a property.

Given such C, its *Koszul dual* is an augmented associative algebra $C^!$ described as follows.

First of all, its underlying object is $\mathbf{1}\square_C\mathbf{1}$, where the unit object $\mathbf{1}$ is given the structure of a C-comodule coming through the augmentation map $\varepsilon\colon \mathbf{1} \to C$ from the comodule structure of $\mathbf{1}$ over the unit coalgebra, and $\square_C$ denotes the cotensor product operation over C. The contensor product (as well as all other constuctions) is relative to the structures of $(\infty, 1)$-categories, so it should be understood to be isomorphic (in the homotopy category) to the suitably *derived* and homotopy invariant version of the construction if the $(\infty, 1)$-categories come e.g., from model categories.

Equivalently, the underlying object of $C^!$ is an object representing the presheaf $\mathscr{A}^{\mathrm{op}} \to \mathrm{Space}$, $X \mapsto \mathrm{Map}_{\mathrm{Comod}_C}(X \otimes \mathbf{1}, \mathbf{1})$, where $X \otimes \mathbf{1}(= X)$ is made into a C-comodule by the action of C on the factor $\mathbf{1}$. Here, Space denotes the standard $(\infty, 1)$-category of "spaces" or infinity groupoids. The structure of an associative algebra of $C^!$ results from this, and we take as the augmentation the map $\eta^!\colon C^! \to \mathbf{1}^! = \mathbf{1}$ for the unit $\eta\colon C \to \mathbf{1}$.

From this description, $C^!$ represents the presheaf on the $(\infty, 1)$-category of augmented associative algebras which maps an object A to the space of A-module structures on the C-comodule $\mathbf{1}$, lifting the A-module structure on the underlying object $\mathbf{1}$ given by the augmentation map of A. In particular, $\mathrm{Map}_{\mathrm{Alg}_*}(A, C^!) = \mathrm{Map}_{\mathrm{Coalg}_*}(A^!, C)$, where $A^! = \mathbf{1} \otimes_A \mathbf{1}$ is the augmented associative coalgebra Koszul dual to A. The subscripts $*$ here indicates that the categories are those of *augmented* algebras and coalgebras. (For example, the map $A^! \xrightarrow{\eta} \mathbf{1} \xrightarrow{\varepsilon} C$ corresponds to the map $A \xrightarrow{\varepsilon} \mathbf{1} \xrightarrow{\eta} C^!$.)

Another term for an associative algebra is "E_1-algebra". We adopt the iterative definition of an *E_n-algebra* for $n \geq 2$, that the symmetric monoidal $(\infty, 1)$-category $\mathrm{Alg}_{E_n}(\mathscr{A})$ of E_n-algebras in $\mathscr{A}$ is inductively determined as $\mathrm{Alg}_{E_1}\big(\mathrm{Alg}_{E_{n-1}}(\mathscr{A})\big)$, where the symmetric monoidal structure on associative algebras is given on the underlying objects. This conforms with Dunn's theorem [7]. The $(\infty, 1)$-category of coalgebras is simply $\mathrm{Coalg}_{E_n}(\mathscr{A}) = \mathrm{Alg}_{E_n}(\mathscr{A}^{\mathrm{op}})^{\mathrm{op}}$.

We consider the iterated Koszul duality functor $(\ )^!\colon \mathrm{Alg}_{E_n *}(\mathscr{A}) \to \mathrm{Coalg}_{E_n *}(\mathscr{A})$ defined inductively as the composite

$$\mathrm{Alg}_{E_1 *}(\mathrm{Alg}_{E_{n-1} *}) \longrightarrow \mathrm{Coalg}_{E_1 *}(\mathrm{Alg}_{E_{n-1} *}) \longrightarrow \mathrm{Coalg}_{E_1 *}(\mathrm{Coalg}_{E_{n-1} *}),$$

where the first map is the associative Koszul duality construction, and the next map is induced from the inductively defined E_{n-1}-Koszul duality functor, which is canonically op-lax symmetric monoidal by induction. By applying this for $\mathscr{A}^{\mathrm{op}}$, one also obtains a functor $(\ )^!\colon \mathrm{Coalg}_{E_n *}(\mathscr{A}) \to \mathrm{Alg}_{E_n *}(\mathscr{A})$.

1.3 Specific Results

A *stable* $(\infty, 1)$-category ([12], see Toën–Vezzosi [23] for a discussion of the origin of the notion) is an important and very reasonable place in which to do algebra. For example, homological algebra takes place in the stable $(\infty, 1)$-category of chain complexes (with quasi-isomorphisms inverted). Algebra in the stable $(\infty, 1)$-category of spectra is one formalization of the idea of brave new algebra proposed by Waldhausen. The purpose of this survey is to give an overview of a setting where the Koszul duality and some other basic algebraic constructions in a symmetric monoidal stable $(\infty, 1)$-category $\mathscr{A}$ behave nicely to lead to such results as an equivalence of large classes of augmented algebras and coalgebras by the Koszul duality, to be described shortly (Theorem 2). Specifically, we consider a symmetric monoidal stable $(\infty, 1)$-category $\mathscr{A}$, equipped with a *filtration* with respect to which $\mathscr{A}$ becomes *complete*, or at least can be completed.

The primary example will be given by the $(\infty, 1)$-category of complete filtered objects (Sect. 2). In fact, the influence to the author's work on the present subject came from the use of complete filtered objects in a related context in Costello's [5] (see also the appendix of Costello–Gwilliam [6]). Filtration and completeness are also used in the work of Positselski on the Koszul duality [18]. The approach to be described here, despite its slight abstractness, has the advantage of including a few more examples such as the filtration given by a t-structure, and hopefully of clarifying some logic.

Remark 1 Beilinson, Ginsburg and Schechtman considers filtration, or "f-structure" in their terminology, of a triangulated category [1]. A triangulated category is conceptually close to a stable $(\infty, 1)$-category, so our approach is very close to theirs. The most notable difference of their approach to ours is that an f-structure is similar to but is *different* from and excludes a t-structure, while filtration in our sense includes an f-structure as well as a t-structure. See Remark 17 for details. Their focus is in fact on a mixture of an f- and a t- structure which are compatible with each other.

To illustrate the usefulness of the setting, in the presence of complete filtration satisfying some mild conditions to be explained in the later sections, we have, as follows from the associativity result Lemma 5 below, that the functor $C \mapsto C^!$ is symmetric monoidal when restricted for E_n-coalgebras C satisfying some positivity condition with respect to the filtration, which we call "copositivity" (Definition 64). In practice, one often has similar associativity as Lemma 5 for free for the tensor product over algebras, but usually not simultaneously also over coalgebras. In the presence of a complete filtration, associativity holds (Lemma 65) also over "positive" algebras (Definition 64).

In particular, if A is a positive augmented E_{n+1}-algebra, then the Koszul dual $\mathbf{1} \otimes_A \mathbf{1}$ of its underlying associative algebra becomes an E_n-algebra in the $(\infty, 1)$-category of augmented associative coalgebras. Moreover, by another consequence, Proposition 52, of the filtration, this E_n-algebra is equivalent to the tensor product $\mathbf{1} \otimes_A \mathbf{1}$ *taken in the* $(\infty, 1)$*-category of* E_n*-algebras*. Similar consequences can be observed also for copositive E_n-coalgebras.

Let $\mathrm{Alg}_{E_n}(\mathscr{A})_+$ denote the $(\infty, 1)$-category of positive augmented E_n-algebras in $\mathscr{A}$, and similarly, Coalg_+ for copositive coalgebras. One obtains the following under additional mild conditions to be also explained in the later sections.

Theorem 2 ([13]) *Let $\mathscr{A}$ be a monoidal complete filtered stable $(\infty, 1)$-category with uniformly bounded sequential limits and translational looping (Definitions 36, 68). Then the constructions of Koszul duals give inverse equivalences*

$$\mathrm{Alg}_{E_n}(\mathscr{A})_+ \overset{\sim}{\longleftrightarrow} \mathrm{Coalg}_{E_n}(\mathscr{A})_+.$$

In the statement above, we have used a more descriptive term than in [13].

Remark 3 This theorem can be considered as a special case of a similar theorem for locally constant factorization algebras on a manifold M [15], obtained by combining the methods here with the subject of the article [16] in these proceedings. Namely, the theorem for factorization algebras specializes to Theorem 2 in the case $M = \mathbb{R}^n$. The theorem for factorization algebras is in analogy with a common generalization by Lurie [12, Remark 5.5.6.11] of the iterated loop space theory and the Verdier duality.

1.4 Further Consequences

One also obtains a *Morita theoretic* functoriality of the Koszul duality.

To explain what this is, in [11], Lurie has outlined a generalization for E_n-algebras of the "Morita" category due to Bénabou [2]. By collecting suitable versions of bimodules, one obtains an $(\infty, n+1)$-category $\mathrm{Alg}_n(\mathscr{A})$, in which

- an object is an E_n-algebra in $\mathscr{A}$,
- a 1-morphism is an E_{n-1}-algebra in $\mathscr{A}$ equipped with the structure of a suitable kind of bimodule,
- a 2-morphisms is an E_{n-2}-algebra in $\mathscr{A}$ equipped with the structure of a suitable kind of bimodule,

and so on, generalizing the 2-category of associative algebras and bimodules.

In order to make the construction of this work, one usually assumes that the monoidal multiplication functors preserve geometric realizations variablewise. However, unless the monoidal multiplication also preserve totalizations, one cannot have both algebraic and coalgebraic versions of this in the same way. Despite these difficulties, it turns out that, in a complete filtered symmetric monoidal $(\infty, 1)$-category satisfying some mild conditions, the construction works for both positive augmented algebras and copositive augmented coalgebras at the same time.

Remark 4 In a quite different context, the construction of the both higher Morita categories work in a Cartesian closed symmetric monoidal $(\infty, 1)$-category which is closed under the geometric realization and the finite limits. The coalgebraic higher Morita category in this context was identified by Ben-Zvi and Nadler with the $(\infty, n+1)$-category of iterated correspondences [3, Remark 1.17]. The Koszul duality in the $(\infty, 1)$-category of spaces is given by the iterated looping and delooping constructions, and is understood very well through the iterated loop space theory.

For example, in order for the construction of the coalgebraic Morita category, even in the case $n = 1$, to work, we would like the following.

Let C_i, $i = 0, 1, 2$, be coalgebras in $\mathscr{A}$, and let $K_{i,i+1}$ for $i = 0, 1$ be a left C_i- right C_{i+1}-bicomodule. Then we would first like $K_{01}\square_{C_1}K_{12}$ to be a C_0–C_2-bicomodule in a natural way.

We have this in the following case. Assume $\mathscr{A}$ to be a monoidal complete filtered stable $(\infty, 1)$-category. We assume the mild condition to be explained in the later sections, that "loops and sequential limits are uniformly bounded" with respect to the filtration in $\mathscr{A}$ (Definitions 36, 41). We also assume that C_1 is a copositive augmented coalgebra, and $K_{i,i+1}$ are "bounded below" in the filtration (Definition 19). Then for any bounded below object L, the canonical map

$$(K_{01}\square_{C_1}K_{12}) \otimes L \longrightarrow K_{01}\square_{C_1}(K_{12} \otimes L)$$

can be shown to be an equivalence using Proposition 49 below.

It follows that if C_0 and C_2 are copositive and in particular, bounded below, then the bicomodule structures of $K_{i,i+1}$, $i = 0, 1$ induce a structure of a C_0–C_2-bicomodule on the cotensor product. In fact, the resulting bicomodule has the universal property to be expected of the cotensor product.

For the construction of the Morita category, we would further like the following to hold.

Lemma 5 *Let $\mathscr{A}$ be a monoidal complete filtered stable $(\infty, 1)$-category with uniformly bounded loops and sequential limits. Let C_i, $i = 0, 1, 2, 3$, be copositive augmented associative coalgebras in $\mathscr{A}$, and let $K_{i,i+1}$ be a left C_i- right C_{i+1}-bicomodule for $i = 0, 1, 2$, whose underlying object is bounded below.*

Then the resulting map

$$(K_{01}\square_{C_1}K_{12})\square_{C_2}K_{23} \longrightarrow K_{01}\square_{C_1}K_{12}\square_{C_2}K_{23}$$

is an equivalence of C_0–C_3-comodules, where the target denotes the totalization of the obvious bicosimplicial (co)bar construction (each cosimplicial index coming from the actions of each of the coalgebras C_1, C_2).

Proof will be discussed in Sect. 7.

Under similar mild conditions and positivity, one can further check that all other basic constructions also behave nicely in a complete filtered symmetric monoidal stable $(\infty, 1)$-category. Let us denote the $(\infty, n+1)$-categories we obtain by $\mathrm{Alg}_n^+(\mathscr{A})$ and $\mathrm{Coalg}_n^+(\mathscr{A})$ respectively.

Theorem 6 ([13]) *Let $\mathscr{A}$ be a symmetric monoidal complete filtered stable $(\infty, 1)$-category with uniformly bounded sequential limits and translational looping (Definitions 36, 41). Then for every n, the construction of the Koszul dual define a symmetric monoidal functor*

$$(\,)^!\colon \mathrm{Alg}_n^+(\mathscr{A}) \longrightarrow \mathrm{Coalg}_n^+(\mathscr{A}).$$

It is an equivalence with inverse given by the Koszul duality construction.

Corollary 7 *Let $\mathscr{A}$ be as in Theorem 6. Then any object of the symmetric monoidal $(\infty, 1)$-category* $\mathrm{Coalg}_n^+(\mathscr{A})$ *is n-dualizable.*

Indeed, for a reasonable symmetric monoidal $(\infty, 1)$-category $\mathscr{A}$, Lurie has given a description of the n-dimensional fully extended topological field theory in the Morita $(n+1)$-category $\mathrm{Alg}_n(\mathscr{A})$ associated to any object $A \in \mathrm{Alg}_n(\mathscr{A})$ [11], using the topological chiral homology (to be reviewed in the article [16] in these proceedings). His description also works in $\mathrm{Alg}_n^+(\mathscr{A})$ here, so any object of $\mathrm{Alg}_n^+(\mathscr{A})$ and hence any object of $\mathrm{Coalg}_n^+(\mathscr{A})$ is n-dualizable. Given $A \in \mathrm{Alg}_n^+(\mathscr{A})$, the associated topological field theory in $\mathrm{Coalg}_n^+(\mathscr{A})$ can in fact be described using the *compactly supported* topological chiral homology of A, as a consequence of the Poincaré duality for the topological chiral homology [15], which is analogous to the "nonabelian" Poincaré duality theorem of Lurie [12] and closely related earlier results of Segal [21], McDuff [17] and Salvatore [20].

1.5 Outline

Our plan for the rest of this article is to first describe basic notions and facts on symmetric monoidal filtered stable $(\infty, 1)$-categories, and then to show how these can be used for the study of the Koszul duality.

2 Filtration of a Stable Category

2.1 Complementary Localizations of a Stable Category

A filtration of a stable $(\infty, 1)$-category will be defined by pairs of *localizations* which are *complementary* to each other.

Definition 8 Let $\mathscr{C}$ be a $(\infty, 1)$-category.

A functor $\mathscr{C} \to \mathscr{D}$ is a **left localization** if it has a fully faithful functor as a right adjoint.

A full subcategory $\mathscr{D}$ of $\mathscr{C}$ is a **left localization** of $\mathscr{C}$ if the inclusion functor $\mathscr{D} \hookrightarrow \mathscr{C}$ has a left adjoint.

Right localization is defined similarly, so it is just left localization in the opposite variance.

We consider the following situation. Let $\mathscr{A}$ be a stable $(\infty, 1)$-category, and let $\mathscr{A}_\ell \subset \mathscr{A}$ be a full subcategory which is a left localization of $\mathscr{A}$. Denote by $(\ )_\ell$ the localization functor $\mathscr{A} \to \mathscr{A}_\ell$. By abuse of notation, we also denote by $(\ )_\ell$ the composite

$$\mathscr{A} \xrightarrow{(\ )_\ell} \mathscr{A}_\ell \hookrightarrow \mathscr{A}.$$

Definition 9 A right localization $\mathscr{A}_r$ of $\mathscr{A}$ is **complementary** to the left localization $\mathscr{A}_\ell$ of $\mathscr{A}$ as above if for every $X \in \mathscr{A}_r$ and $Y \in \mathscr{A}_\ell$, the space $\mathrm{Map}(X, Y)$ is contractible, and the sequence

$$(\ )_r \xrightarrow{\varepsilon} \mathrm{id} \xrightarrow{\eta} (\ )_\ell : \mathscr{A} \longrightarrow \mathscr{A},$$

where $(\ )_r$ is the right localization functor considered as $\mathscr{A} \to \mathscr{A}$, and the maps are the counit and the unit maps for the respective adjunctions, is a fibre sequence (by the unique null homotopy of the composite $\eta\varepsilon$).

As a full subcategory of $\mathscr{A}$, $\mathscr{A}_r$ consists of objects $X \in \mathscr{A}$ for which the counit $\varepsilon \colon X_r \to X$ is an equivalence, or equivalently, $X_\ell \simeq \mathbf{0}$. It follows that given any left localization $\mathscr{A}_\ell$ of $\mathscr{A}$, if it has a complementary right localization, then the right localization is characterized as the right localization to the full subcategory of $\mathscr{A}$ consisting of objects $X \in \mathscr{A}$ for which $X_\ell \simeq \mathbf{0}$.

Given any right localization, its **complementary** *left* localization is defined in the opposite way. It is immediate that if a left localization has a complementary right localization, then this left localization is left complementary to its right complement.

Remark 10 Given a pair of complementary localizations $\mathscr{A}_\ell$, $\mathscr{A}_r$, their homotopy categories $\mathrm{ho}(\mathscr{A}_\ell)$, $\mathrm{ho}(\mathscr{A}_r)$ form a pair of localizations of the triangulated category $\mathrm{ho}\,\mathscr{A}$ which are complementary to each other in the similar sense. Note the full subcategories $\mathrm{ho}(\mathscr{A}_\ell)$, $\mathrm{ho}(\mathscr{A}_r)$ of $\mathrm{ho}\,\mathscr{A}$ determine the full subcategories $\mathscr{A}_\ell$, $\mathscr{A}_r$ of $\mathscr{A}$. Conversely, a pair of complementary localizations of $\mathscr{A}$ is always obtained from a complementary pair of localization of $\mathrm{ho}\,\mathscr{A}$. Cf. Lurie [12, Proposition 1.2.1.5].

The pair $\mathrm{ho}\,\mathscr{A}_\ell$, $\mathrm{ho}\,\mathscr{A}_r$ of localizations of $\mathrm{ho}\,\mathscr{A}$ defines a *step* in the sense of Beilinson, Ginsburg and Schechtman [1, Definition 1.2.1] *only* if $\mathrm{ho}\,\mathscr{A}_\ell$ and $\mathrm{ho}\,\mathscr{A}_r$ are triangulated themselves. This is when $\mathscr{A}_\ell$ and $\mathscr{A}_r$ are stable.

Even though a localization of a stable $(\infty, 1)$-category will not necessarily be stable itself (a sufficient condition will be given in Proposition 26 below), complementary localizations will be additive at least. Let us first see that they are pointed, namely, have zero objects.

Lemma 11 *Let $\mathscr{A}$ be a stable $(\infty, 1)$-category, and let $\mathscr{A}_\ell$, $\mathscr{A}_r$ be left and right localizations of $\mathscr{A}$ respectively which are complementary to each other. Then $\mathscr{A}_\ell$ is pointed, and dually for $\mathscr{A}_r$.*

Proof $\mathbf{0}_{r\ell} \simeq \mathbf{0}$ implies $\mathbf{0} \in \mathscr{A}_\ell$, which is then a zero object of $\mathscr{A}_\ell$. □

All inclusion and localization functors will preserve the zero objects.

The following lemma gives a useful way to detect local equivalences.

Lemma 12 *Let $\mathscr{A}$ be a stable $(\infty, 1)$-category with complementary left and right localizations $(\ )_\ell\colon \mathscr{A} \to \mathscr{A}_\ell$ and $(\ )_r\colon \mathscr{A} \to \mathscr{A}_r$ respectively. Then, for a cofibre sequence*

$$W \longrightarrow X \longrightarrow Y$$

in $\mathscr{A}$, if W belongs to the full subcategory $\mathscr{A}_r$ of $\mathscr{A}$, then the localized map $X_\ell \to Y_\ell$ is an equivalence.

Proof W belongs to $\mathscr{A}_r$ if and only if $W_\ell \simeq \mathbf{0}$.

By applying the localization functor $(\ )_\ell\colon \mathscr{A} \to \mathscr{A}_\ell$ to the given cofibre sequence, we obtain a cofibre sequence in $\mathscr{A}_\ell$. If $W_\ell \simeq \mathbf{0}$, then the map $X_\ell \to Y_\ell$ in the sequence is an equivalence. □

Corollary 13 *In the situation of Lemma 12, X belongs to $\mathscr{A}_r$ if and only if Y belongs to $\mathscr{A}_r$.*

The additivity will follow from the following closure property with respect to the formation of limits (and colimits for a right localization).

Lemma 14 *Let $\mathscr{A}$ be a stable $(\infty, 1)$-category. Then, if a left localization $\mathscr{A}_\ell$ of $\mathscr{A}$ has a complementary right localization, then $\mathscr{A}_\ell$ is closed in $\mathscr{A}$ under any limit which exists in $\mathscr{A}$.*

Proof This follows since $\mathscr{A}_\ell$ is the full subcategory of $\mathscr{A}$ consisting of $X \in \mathscr{A}$ for which $X_r \simeq \mathbf{0}$, and since the functor $(\ )_r\colon \mathscr{A} \to \mathscr{A}_r$ is a right adjoint, and hence preserves any limit. □

Note also that the limit taken in $\mathscr{A}$ of a diagram lying in the full subcategory $\mathscr{A}_\ell$ (which in fact belongs to $\mathscr{A}_\ell$, according to the above) will be a limit in $\mathscr{A}_\ell$ of the diagram. On the other hand, since the inclusion $\mathscr{A}_\ell \hookrightarrow \mathscr{A}$ preserves limits, if a limit of a diagram $\mathscr{A}_\ell$ exists in the $(\infty, 1)$-category $\mathscr{A}_\ell$, then it also will be a limit in $\mathscr{A}$.

One obtains the following 'additivity'.

Corollary 15 *If a left localization $\mathscr{A}_\ell$ has a complementary right localization, then $\mathscr{A}_\ell$ is closed in $\mathscr{A}$ under the finite coproduct in $\mathscr{A}$.*

Proof Let X, Y be object of $\mathscr{A}$ which belong to $\mathscr{A}_\ell$. Then the coproduct $X \amalg Y$ in $\mathscr{A}$ is equivalent to the product $X \times Y$ in $\mathscr{A}$, which belongs to $\mathscr{A}_\ell$ by Lemma 14. □

All inclusion and localization functors will preserve the finite products and coproduct, which coincide in each of the three $(\infty, 1)$-categories.

2.2 *Filtration*

Definition 16 A **filtration** of a stable $(\infty, 1)$-category $\mathscr{A}$ is a sequence of full subcategories

$$\mathscr{A} \supset \cdots \supset \mathscr{A}_{\geq r} \supset \mathscr{A}_{\geq r+1} \supset \cdots$$

indexed by integers, each of which is the inclusion of a right localization which has a complementary left localization, denoted by $(\ )^{<r} : \mathscr{A} \to \mathscr{A}^{<r}$.

A **filtered** stable $(\infty, 1)$-category is a stable $(\infty, 1)$-category which is equipped with a filtration.

Remark 17 Our definition of a filtration as a sequence of pairs of complementary localizations, is similar to Beilinson, Ginsburg and Schechtman's definition of an *f-structure* on a triangulated category as a sequence of steps [1, Definition 1.3.1]. The difference is that pair of complementary localizations is a less restrictive notion than the notion of step, as explained in Remark 10. As a consequence, filtration in our sense includes a t-structure for instance (Example 20), which f-structure is meant to *not* overlap.

Remark 18 The notion of a filtration on a stable $(\infty, 1)$-category is self-dual in the following sense. Namely, if a stable $(\infty, 1)$-category $\mathscr{A}$ is given a filtration, then $\mathscr{B} := \mathscr{A}^{\mathrm{op}}$ has a filtration given by $\mathscr{B}_{\geq r} := (\mathscr{A}^{\leq -r})^{\mathrm{op}}$, where $\mathscr{A}^{\leq s} := \mathscr{A}^{<s+1}$.

Therefore, all notions and statements we formulate will have dual versions, which we shall speak about freely without further notices.

Definition 19 Let $\mathscr{A}$ be a filtered stable $(\infty, 1)$-category. Then an object X of $\mathscr{A}$ is said to be **bounded below** in the filtration if there exists an integer r such that X belongs to the full subcategory $\mathscr{A}_{\geq r}$ of $\mathscr{A}$.

Let us see a few examples of filtrations.

Example 20 Let $\mathscr{A}$ be a stable $(\infty, 1)$-category equipped with a *t-structure* [12]. For example, $\mathscr{A}$ may be the stable $(\infty, 1)$-category of chain complexes over a ring, or more generally, of suitablly structured spectra, such as modules over a connective ring spectrum, where the t-structure is defined by the connectivity (and coconnectivity) of the underlying spectra.

Then the usual sequences of full subcategories defines a filtration on $\mathscr{A}$. In the case of a t-structure by connectivity, $\mathscr{A}_{\geq r}$ will consist of objects whose underlying spectrum has connectivity at least r, and $\mathscr{A}^{<r}$ will consist of objects with coconectivity less than r. In fact, a t-structure can be characterized as a filtration satisfying a simple condition. See Example 69.

Let us see another typical example.

Let $\mathbb{Z}$ be the category

$$\cdots \longleftarrow n \longleftarrow n+1 \longleftarrow \cdots$$

defined by the poset of integers.

Definition 21 A **filtered** object of an $(\infty, 1)$-category $\mathscr{C}$ is a functor $\mathbb{Z} \to \mathscr{C}$.

We shall typically express a filtered object $X \in \mathrm{Fun}(\mathbb{Z}, \mathscr{C})$ as a sequence

$$\cdots \longleftarrow F_n X \longleftarrow F_{n+1} X \longleftarrow \cdots$$

in $\mathscr{C}$.

Let $\mathscr{B}$ be a stable $(\infty, 1)$-category. Then the stable $(\infty, 1)$-category $\mathscr{A} = \mathrm{Fun}(\mathbb{Z}, \mathscr{B})$ of filtered objects in $\mathscr{B}$, is filtered as follows.

We let $\mathscr{A}_{\geq r}$ be the $(\infty, 1)$-category of sequences

$$F_r X \longleftarrow F_{r+1} X \longleftarrow \cdots,$$

and $(\ )_{\geq r} : \mathscr{A} \to \mathscr{A}_{\geq r}$ to be the functor which forgets objects $F_n X$ for $n < r$. $(\ )_{\geq r}$ is a right localization which has a complementary left localization, which we denote by $(\ )^{<r} : \mathscr{A} \to \mathscr{A}^{<r}$.

Remark 22 Remark 18 allows us to consider this filtration also as a filtration on $\mathscr{A}^{\mathrm{op}}$. We obtain another filtration on $\mathscr{A}^{\mathrm{op}}$ by replacing $\mathscr{B}$ by $\mathscr{B}^{\mathrm{op}}$ in the construction above. Note $\mathscr{A}^{\mathrm{op}} = \mathrm{Fun}(\mathbb{Z}, \mathscr{B}^{\mathrm{op}})$. These are different filtrations. Namely, the $(\infty, 1)$-category of filtered objects have two distinct filtrations (unless it is the trivial category).

Without loss of generality, we normally discuss only the former filtration, but sometimes on $\mathscr{A}$ *and* sometimes on $\mathscr{A}^{\mathrm{op}}$.

Before giving the next example of a filtration, it will be convenient to be able to tell when a localization has a complement.

Proposition 23 ([13]) *A left localization $(\ )_\ell : \mathscr{A} \to \mathscr{A}_\ell$ of a stable $(\infty, 1)$-category $\mathscr{A}$ has a complementary right localization if and only if*

$$\left(\mathrm{Fibre}[\eta : \mathrm{id} \to (\ )_\ell]\right)_\ell \simeq \mathbf{0}.$$

Example 24 This condition is satisfied if the left localization is *exact* in the following sense.

Definition 25 A left localization of a stable $(\infty, 1)$-category $\mathscr{A}$ is **exact** if the localization functor $(\ )_\ell : \mathscr{A} \to \mathscr{A}_\ell$ (and equivalently, $(\ )_\ell : \mathscr{A} \to \mathscr{A}$) preserves finite limits.

A right localization is **exact** if the localization functor is exact.

One further obtains the following for an exact localization.

Proposition 26 *Let $\mathscr{A}$ be a stable $(\infty, 1)$-category, and let $(\ )_\ell : \mathscr{A} \to \mathscr{A}_\ell$ be an exact left localization of $\mathscr{A}$. Then the $(\infty, 1)$-category $\mathscr{A}_\ell$ is stable, the inclusion functor $\mathscr{A}_\ell \hookrightarrow \mathscr{A}$ is also exact, and the complementary right localization (see Example 24) is also exact.*

Example 27 Let $\mathscr{A}$ be the functor category into a stable $(\infty, 1)$-category, and assume it admits some version of the Goodwillie calculus [10]. Then it has a filtration in which $\mathscr{A}^{<r}$ is the full subcategory consisting of $(r-1)$-excisive functors. The left localization $\mathscr{A} \to \mathscr{A}^{<r}$ is given by the universal $(r-1)$-excisive approximation of functors, which is exact as follows e.g., from the construction.

3 Completion

Let us define what it means for a category to be *complete* with respect to a filtration. The property will turn out very useful.

Let $\mathscr{A}$ be a filtered stable $(\infty, 1)$-category. Then define

$$\mathscr{A}_{\geq\infty} := \lim_r \mathscr{A}_{\geq r} = \bigcap_r \mathscr{A}_{\geq r},$$

the intersection taken in $\mathscr{A}$. We obtain the sequence

$$\mathscr{A}_{\geq\infty} \longrightarrow \mathscr{A} \longrightarrow \lim_r \mathscr{A}^{<r}$$

as the limit of the sequence

$$\mathscr{A}_{\geq r} \longrightarrow \mathscr{A} \longrightarrow \mathscr{A}^{<r}.$$

Let us denote by τ the functor $\mathscr{A} \to \lim_r \mathscr{A}^{<r}$ here. If $\mathscr{A}$ is closed under the sequential limit, then this has a right adjoint which we shall denote by lim. For an object $X = (X_r)_r$ of $\lim_r \mathscr{A}^{<r}$, it is given by

$$\lim X = \lim_r X_r,$$

where the limit on the right hand side is taken in $\mathscr{A}$.

Definition 28 Let $\mathscr{A}$ be a filtered stable $(\infty, 1)$-category which is closed under the sequential limit. Then we denote $\lim \tau X$ by $\hat{X}$. We say that X is **complete** if the unit map $\eta \colon X \to \lim \tau X = \hat{X}$ for the adjunction is an equivalence.

We denote by $\hat{\mathscr{A}}$ the full subcategory of $\mathscr{A}$ consisting of complete objects.

Example 29 For every r, $\mathscr{A}^{<r} \subset \hat{\mathscr{A}}$ in $\mathscr{A}$.

Definition 30 We say that a filtered stable $(\infty, 1)$-category $\mathscr{A}$ is **complete** if it is closed under the sequential limit, and $\hat{\mathscr{A}}$ is the whole of $\mathscr{A}$, namely, if every object of $\mathscr{A}$ is complete.

It will be useful to be able to *complete* a stable $(\infty, 1)$-category with respect to a filtration. We unfortunately do not know a general definition of the completion of a filtered stable $(\infty, 1)$-category $\mathscr{A}$. However, we can write down what seems to be a sufficient condition for the obvious candidate $\hat{\mathscr{A}}$ to correctly be the completion of $\mathscr{A}$.

Definition 31 Let $\mathscr{A}$ be a filtered stable $(\infty, 1)$-category. Then we say that $\hat{\mathscr{A}}$ is the **completion** of $\mathscr{A}$ if the following conditions are satisfied.

0 $\mathscr{A}$ is closed under the sequential limit, so we have $\hat{\mathscr{A}}$ defined.
1 The functor $\hat{(\,)}\colon \mathscr{A} \to \mathscr{A}$ preserves sequential limits.
2 $\hat{(\,)}$ lands in $\hat{\mathscr{A}}$.
3 The map $\eta\colon \mathrm{id} \to \hat{(\,)}$ makes $\hat{(\,)}$ a left localization for the full subcategory $\hat{\mathscr{A}}$.

If $\mathscr{A}$ has $\hat{\mathscr{A}}$ as its completion in this sense, then we call the localization functor the **completion** functor. In this case, we call η the **completion** map.

The following easy lemma is a part of the motivation for Definition 31.

Lemma 32 *If $\hat{\mathscr{A}}$ is the completion of $\mathscr{A}$, then the sequential limits exists in $\hat{\mathscr{A}}$, and the completion functor preserves sequential limits.*

Proposition 33 ([13]) *Let $\mathscr{A}$ be a filtered stable $(\infty, 1)$-category with $\hat{\mathscr{A}}$ its completion. Then the full subcategory $\mathscr{A}_{\geq\infty}$ of $\mathscr{A}$ is a right localization complementary to the left localization $\hat{\mathscr{A}}$.*

Outline of proof It can be proved separately that the completion of an object X of $\mathscr{A}$ vanishes if and only if X belongs to $\mathscr{A}_{\geq\infty}$. Therefore, it suffices to show that completion has a complementary right localization. Existence of the complement follows from Proposition 23 and Lemma 34 below since the fibre of the completion map is $\lim_r X_{\geq r}$. □

The following will be useful.

Lemma 34 *Let $\mathscr{A}$ be a filtered stable $(\infty, 1)$-category with $\hat{\mathscr{A}}$ its completion. Suppose given an inverse system*

$$\cdots \longleftarrow X_i \longleftarrow X_{i+1} \longleftarrow \cdots$$

in $\mathscr{A}$, and suppose there is a sequence $(r_i)_i$ of integers, tending to ∞ as $i \to \infty$, such that X_i belongs to $\mathscr{A}_{\geq r_i}$ for every i.

Then $\lim_i X_i$ belongs to $\mathscr{A}_{\geq\infty}$.

Proof As stated during the proof of Proposition 33, it suffices to prove that its completion vanishes. However,

$$\widehat{\lim_i} X_i = \lim_i \hat{X}_i = \lim_r \lim_i X_i^{<r} \simeq \lim_r \mathbf{0} = \mathbf{0}.$$

□

Let us next give a sufficient condition for $\hat{\mathscr{A}}$ to be the completion of $\mathscr{A}$ which is easy to check in practice.

Proposition 35 ([13]) *Let $\mathscr{A}$ be a filtered stable $(\infty, 1)$-category which is closed under the sequential limit. If sequential limits in $\mathscr{A}$ are* ***uniformly bounded*** *in the sense of Definition 36 below, then $\hat{\mathscr{A}}$ is the completion of $\mathscr{A}$.*

The following condition will turn out useful also for other purposes.

Definition 36 Let $\mathscr{A}$ be a filtered stable $(\infty, 1)$-category which is closed under the sequential limit. Then we say that **sequential limits are uniformly bounded** in $\mathscr{A}$ if there exists an integer d such that for every integer r, and for every inverse sequence in the full subcategory $\mathscr{A}_{\geq r}$ of $\mathscr{A}$, the limit of the sequence taken in $\mathscr{A}$, belongs to $\mathscr{A}_{\geq r+d}$. We refer to such d as a **uniform lower bound** for sequential limits in $\mathscr{A}$.

Remark 37 $\mathscr{A}$ is assumed to have finite limits and sequential limits, so it has countable products at least, and if sequential limits are uniformly bounded, then so are countable products in the similar sense. In the case where the filtration is given by a t-structure, if countable products in $\mathscr{A}$ are uniformly bounded below by b, then the familiar computation of a sequential limit in terms of countable products by Milnor shows that sequential limits will be bounded below by $b - 1$.

In the case of a filtered $(\infty, 1)$-category of filtered objects, as well as Goodwillie's filtration (Example 27), sequential limits are bounded below by 0 assuming that the object-wise sequential limits exist.

One also obtains the following.

Proposition 38 ([13]) *Let $\mathscr{A}$ be a filtered stable $(\infty, 1)$-category with uniformly bounded sequential limits. Then the functor* $\lim\colon \lim_r \mathscr{A}^{<r} \to \mathscr{A}$ *induces an equivalence* $\lim_r \mathscr{A}^{<r} \xrightarrow{\sim} \hat{\mathscr{A}}$.

4 The Completion as a Complete Category

When $\hat{\mathscr{A}}$ is the completion of a filtered stable $(\infty, 1)$-category $\mathscr{A}$, then it will be useful if the completion is itself a complete filtered stable $(\infty, 1)$-category. We would like to first give a sufficient condition for the completion to be a *stable* $(\infty, 1)$-category. A sufficient condition for a general localization was given in Proposition 26.

Definition 39 Let $\mathscr{A}$ be a filtered stable $(\infty, 1)$-category with $\hat{\mathscr{A}}$ its completion. Then we say that the completion is **exact** if $\hat{\mathscr{A}}$ is an exact left localization of $\mathscr{A}$.

Proposition 40 ([13]) *Let $\mathscr{A}$ be a filtered stable $(\infty, 1)$-category with $\hat{\mathscr{A}}$ its completion. If loops are* ***uniformly bounded*** *in $\mathscr{A}$ in the sense of Definition 41 below, then the completion is exact.*

The following condition will turn out useful also for other purposes.

Definition 41 Let $\mathscr{A}$ be a filtered stable $(\infty, 1)$-category. An integer ω is said to be a **uniform lower bound for loops** in $\mathscr{A}$ if for every integer r, and for every object of the full subcategory $\mathscr{A}_{\geq r}$ of $\mathscr{A}$, its loop in $\mathscr{A}$ belongs to $\mathscr{A}_{\geq r+\omega}$. We say that **loops are uniformly bounded** in $\mathscr{A}$ if loops in $\mathscr{A}$ have a uniform lower bound.

Example 42 If the filtration is a t-structure on $\mathscr{A}$, then ω can be taken as -1.

ω can be taken as 0 for the filtered $(\infty, 1)$-category of filtered objects, as well as for Goodwillie's filtration. In fact, all localizations are exact in these filtrations.

Remark 43 An integer $\omega \gneq 0$ cannot be a uniform lower bound for loops unless $\mathscr{A}_{\geq r}$ for all r are the same subcategory of $\mathscr{A}$. Indeed, $\Omega^{-1} = \Sigma$ maps $\mathscr{A}_{\geq r}$ into $\mathscr{A}_{\geq r}$ by Lemma 14.

Remark 44 By Corollary 13, ω is a uniform lower bound for loops if and only if it is a uniform lower bound for *fibres* in the similar sense. Indeed, if $W \to X \to Y$ is a fibre sequence in $\mathscr{A}$, then there is a fibre sequence $\Omega Y \to W \to X$.

It follows again from Corollary 13, that the uniform lower bound of fibres more generally bounds fibre products.

Definition 45 Let $\mathscr{A}$, $\mathscr{B}$ be filtered stable $(\infty, 1)$-categories, and let $F\colon \mathscr{A} \to \mathscr{B}$ be an exact functor. Then we say that an integer b is a **lower bound** of F if for every r, F takes the full subcategory $\mathscr{A}_{\geq r}$ of the source to the full subcategory $\mathscr{B}_{\geq r+b}$ of the target.

We say that F is **bounded below** if it has a lower bound.

Upper bound/boundedness of F is defined as the lower bound/boundedness of $F\colon \mathscr{A}^{\mathrm{op}} \to \mathscr{A}^{\mathrm{op}}$ with respect to the dual filtration on $\mathscr{A}^{\mathrm{op}}$ (Remark 18).

Thus, uniformly boundedness of loops in $\mathscr{A}$ means boundedness below of the functor $\Omega\colon \mathscr{A} \to \mathscr{A}$. Ω also has 0 as an upper bound by Lemma 14.

We obtain from the following, that a uniform lower bound for loops also gives an upper bound of the suspension functor.

Lemma 46 *Let $\mathscr{A}$, $\mathscr{B}$ be filtered stable $(\infty, 1)$-categories, and let $F\colon \mathscr{A} \to \mathscr{B}$ be a functor which has a right adjoint G. Then an integer b is a lower bound of F if and only if $-b$ is an upper bound of G.*

Proof For an integer b, the composite

$$\mathscr{A}_{\geq r} \hookrightarrow \mathscr{A} \xrightarrow{F} \mathscr{B} \xrightarrow{(\ )^{<r+b}} \mathscr{B}^{<r+b}$$

is null if and only if the composite of the right adjoints

$$\mathscr{A}_{\geq r} \xleftarrow{(\ )_{\geq r}} \mathscr{A} \xleftarrow{G} \mathscr{B} \hookleftarrow \mathscr{B}^{<r+b}$$

is null, since either adjoint of a null functor is null. □

Proposition 47 *Let $\mathscr{A}$ be a filtered stable $(\infty, 1)$-category with $\hat{\mathscr{A}}$ its completion. If the completion is exact, then the canonical tower*

$$\hat{\mathscr{A}} \longrightarrow \cdots \longrightarrow \mathscr{A}^{<r} \longrightarrow \mathscr{A}^{<r-1} \longrightarrow \cdots$$

makes $\hat{\mathscr{A}}$ into a complete filtered stable $(\infty, 1)$-category.

Proof As we have remarked in Example 29, for every r, $\mathscr{A}^{<r} \subset \hat{\mathscr{A}}$ as full subcategories of $\mathscr{A}$. It follows that the restriction to $\hat{\mathscr{A}}$ of the localization functor $\mathscr{A} \to \mathscr{A}^{<r}$ is a left localization. A complementary right localization to this is given by $\mathscr{A}_{\geq r} \cap \hat{\mathscr{A}}$.

It is easy to verify that every object of $\hat{\mathscr{A}}$ is complete with respect to this filtration of $\hat{\mathscr{A}}$. □

Lemma 48 *Let $\mathscr{A}$ be a filtered stable $(\infty, 1)$-category with $\hat{\mathscr{A}}$ its exact completion. Then any class of limits which exist in $\mathscr{A}$ (and therefore also in $\hat{\mathscr{A}}$ by Lemma 14) and are uniformly bounded, have the same uniform lower bound in $\hat{\mathscr{A}}$.*

Proof Lemma 14 in fact states that $\hat{\mathscr{A}}$ is closed under the limits which exists in $\mathscr{A}$. The result follows since the full subcategory $\hat{\mathscr{A}}_{\geq r}$ in the filtration of $\hat{\mathscr{A}}$ is just $\mathscr{A}_{\geq r} \cap \hat{\mathscr{A}}$ as a full subcategory of $\mathscr{A}$. □

5 Totalization in a Filtered Category

Let $\boldsymbol{\Delta}_f$ denote the subcategory of the category $\boldsymbol{\Delta}$ of combinatorial simplices (or finite non-empty totally ordered sets), where only face maps (maps *strictly* preserving the order of vertices) are included. A covariant functor $X^\bullet \colon \boldsymbol{\Delta}_f \to \mathscr{A}$ is a cosimplicial object 'without degeneracies' of $\mathscr{A}$. Its *totalization* Tot $X^\bullet$ is by definition, the limit over $\boldsymbol{\Delta}_f$ of the diagram $X^\bullet$.

The following important result gives a useful sufficient condition for the preservation of the totalization of a cosimplicial object without degeneracies.

Proposition 49 *Let $\mathscr{A}$, $\mathscr{B}$ be filtered stable $(\infty, 1)$-categories which have sequential limits, and let $F \colon \mathscr{A} \to \mathscr{B}$ be an exact functor which is bounded below. Assume that loops and sequential limits are uniformly bounded in $\mathscr{A}$, and $\hat{\mathscr{B}}$ is the completion of $\mathscr{B}$.*

Let $X^\bullet \colon \boldsymbol{\Delta}_f \to \mathscr{A}$ be such that there exists a sequence $r = (r_n)_n$ of integers, tending to ∞ as $n \to \infty$, such that for a uniform lower bound ω for loops, and for every n, X^n belongs to $\mathscr{A}_{\geq -\omega n + r_n}$. Then the canonical map

$$F(\mathrm{Tot}\, X^\bullet) \longrightarrow \mathrm{Tot}\, F X^\bullet$$

is an equivalence after completion.

Proof According to the sequence of full subcategories

$$\boldsymbol{\Delta}_f \supset \cdots \supset \boldsymbol{\Delta}_f^{\leq n} \supset \boldsymbol{\Delta}_f^{\leq n-1} \supset \cdots,$$

where objects of $\boldsymbol{\Delta}_f^{\leq n}$ are simplices of dimension at most n, we have the sequence

$$\mathrm{Tot}\, X^\bullet \longrightarrow \cdots \longrightarrow \mathrm{sk}_n\, \mathrm{Tot}\, X^\bullet \longrightarrow \mathrm{sk}_{n-1}\, \mathrm{Tot}\, X^\bullet \longrightarrow \cdots$$

such that $\mathrm{Tot}\, X^\bullet = \lim_n \mathrm{sk}_n\, \mathrm{Tot}\, X^\bullet$, where "$\mathrm{sk}_n$ Tot" is a single symbol representing the operation of taking the limit over $\boldsymbol{\Delta}_f^{\leq n}$.

It is standard that the fibre of the map $\mathrm{sk}_n\, \mathrm{Tot}\, X^\bullet \to \mathrm{sk}_{n-1}\, \mathrm{Tot}\, X^\bullet$ is equivalent to $\Omega^n X^n$. It follows from our assumption that this belongs to $\mathscr{A}_{\geq r_n}$. It follows that the fibre of the map $\mathrm{Tot}\, X^\bullet \to \mathrm{sk}_n\, \mathrm{Tot}\, X^\bullet$ belongs to $\mathscr{A}_{\geq r_n+d}$ for d a uniform bound for sequential limits.

It follows that the fibre of the map $F(\mathrm{Tot}\, X^\bullet) \to \mathrm{sk}_n\, \mathrm{Tot}\, FX^\bullet$ belongs to $\mathscr{B}_{\geq r_n+d+b}$ for a bound b of F. By taking the limit over n, we obtain the result from Lemma 51. □

Remark 50 The fibre of the map $\mathrm{sk}_n\, \mathrm{Tot}\, X^\bullet \to \mathrm{sk}_{n-1}\, \mathrm{Tot}\, X^\bullet$ for a usual cosimplicial object $X^\bullet$ with degeneracies is slightly more complicated to describe.

Lemma 51 *Let $\mathscr{A}$ be a filtered stable $(\infty, 1)$-category with $\hat{\mathscr{A}}$ its completion.*

Suppose given a map of inverse systems

$$\begin{array}{ccccccc} \cdots \longleftarrow & X_i & \longleftarrow & X_{i+1} & \longleftarrow \cdots \\ & {\scriptstyle f_i}\downarrow & & \downarrow{\scriptstyle f_{i+1}} & \\ \cdots \longleftarrow & Y_i & \longleftarrow & Y_{i+1} & \longleftarrow \cdots \end{array}$$

in $\mathscr{A}$, and suppose there is a sequence $(r_i)_i$ of integers, tending to ∞ as $i \to \infty$, such that the fibre of f_i belongs to $\mathscr{A}_{\geq r_i}$ for every i.

Then the map $\lim_i f_i \colon \lim_i X_i \to \lim_i Y_i$ is an equivalence after completion.

Proof This follows from Lemma 34, Proposition 33 and Lemma 12. □

Proposition 52 *Let $\mathscr{A}$, $\mathscr{B}$ be filtered stable $(\infty, 1)$-categories, and let $F \colon \mathscr{A} \to \mathscr{B}$ be an exact functor which is bounded below. Assume that $\hat{\mathscr{B}}$ is the completion of $\mathscr{B}$. Let $X_\bullet \colon \boldsymbol{\Delta}_f^{\mathrm{op}} \to \mathscr{A}$ be such that there exists a sequence $r = (r_n)_n$ of integers, tending to ∞ as $n \to \infty$, such that for every n, X^n belongs to $\mathscr{A}_{\geq r_n}$. Then the canonical map*

$$|FX_\bullet| \longrightarrow F|X_\bullet|$$

is an equivalence after completion.

Proof The proof of this is simpler. One simply notes that the full subcategory $\mathscr{A}_{\geq r}$ of $\mathscr{A}$ is closed under any colimit by Lemma 14, and similarly in $\mathscr{B}$. It follows that the fibre of the map in question belongs to $\mathscr{B}_{\geq \infty}$, and we conclude by applying Proposition 33 and Lemma 12. □

6 Monoidal Structure on a Filtered Category

6.1 Monoidal Filtered Category

By a **monoidal structure** on a stable $(\infty, 1)$-category $\mathscr{A}$, we mean a monoidal structure on the underlying $(\infty, 1)$-category of $\mathscr{A}$ *whose multiplication operations are exact* in each variable.

Definition 53 Let $\mathscr{A}$ be a filtered stable $(\infty, 1)$-category, and let $\otimes$ be a monoidal structure on the stable $(\infty, 1)$-category (underlying) $\mathscr{A}$. We say that the monoidal structure is **compatible** with the filtration on $\mathscr{A}$ if for every finite totally ordered set I, and every sequence $r = (r_i)_{i\in I}$ of integers, the functor $\bigotimes_I \colon \mathscr{A}^I \to \mathscr{A}$ takes the full subcategory $\prod_{i\in I} \mathscr{A}_{\geq r_i}$ of the source, to the full subcategory $\mathscr{A}_{\geq \sum_I r}$ of the target.

We call a filtered stable $(\infty, 1)$-category $\mathscr{A}$ equipped with a compatible monoidal structure a **monoidal** filtered stable $(\infty, 1)$-category. If the monoidal structure is symmetric, then it will just be a **symmetric** monoidal filtered $(\infty, 1)$-category.

Remark 54 Even though both filtration and monoidal structure are self-dual notion on a stable $(\infty, 1)$-category, compatibility of these two kinds of structures is not self-dual. Indeed, boundedness below in $\mathscr{A}^{\mathrm{op}}$ means boundedness *above* in $\mathscr{A}$. Instead, Lemma 46 implies that the internal hom functor would have suitable boundedness above on a symmetric monoidal stable $(\infty, 1)$-category.

Example 55 Let k be a connective E_∞-ring in either chain complexes or spectra, and let $\mathscr{A}$ be the stable $(\infty, 1)$-category of k-modules. Then the monoidal structure on $\mathscr{A}$ by the (derived) tensor product over k, is compatible with the filtration defined by the connectivity of the underlying spectra (Example 20).

Example 56 In the case where $\mathscr{A}$ is a functor category with Goodwillie's filtration, if the target category of the functors is a symmetric monoidal stable $(\infty, 1)$-category, then the pointwise symmetric monoidal structure *on* $\mathscr{A}^{\mathrm{op}}$ is compatible with the filtration. Note Remark 18.

Let $\mathscr{B}$ be a symmetric monoidal stable $(\infty, 1)$-category. Assume the following.

Assumption 57

- $\mathscr{B}$ has all small colimits.
- The monoidal multiplication functors on $\mathscr{B}$ preserve colimits variable-wise.

Then, the $(\infty, 1)$-category $\mathscr{A} = \mathrm{Fun}(\mathbb{Z}, \mathscr{B})$ of filtered objects can be equipped with a symmetric monoidal structure by the Day convolution, using the symmetric monoidal structure of the poset $\mathbb{Z}$ (see Sect. 2) given by the operations of addition.

Explicitly, if $X = (F_n X)_n$ and $Y = (F_n Y)_n$ are objects of $\mathscr{A}$, then we have $X \otimes Y \in \mathscr{A}$ defined by

$$F_n(X \otimes Y) = \mathrm{colim}_{i+j \geq n}(F_i X \otimes F_j Y).$$

This monoidal multiplication preserves colimits variablewise.

The filtration on $\mathscr{A}$ is obviously compatible with the monoidal structure *on* $\mathscr{A}^{\mathrm{op}}$. Note Remark 22.

The following can also be verified.

Proposition 58 ([13]) *The symmetric monoidal structure on $\mathscr{A}$ is compatible with the filtration on $\mathscr{A}$.*

6.2 Completion of a Monoidal Structure

Definition 59 Let $\mathscr{A}$ be a monoidal filtered stable $(\infty, 1)$-category with $\hat{\mathscr{A}}$ completing the filtration. Then we say that the monoidal structure is **completable** if there is a monoidal structure on $\hat{\mathscr{A}}$ such that the completion functor $\mathscr{A} \to \hat{\mathscr{A}}$ is monoidal.

Remark 60 Together with a monoidal structure of the completion functor, the monoidal structure on $\hat{\mathscr{A}}$ will be uniquely determined. For example, the monoidal operations on $\hat{\mathscr{A}}$ in the completable case will be describable as the composites

$$\hat{\mathscr{A}}^n \hookrightarrow \mathscr{A}^n \xrightarrow{\otimes} \mathscr{A} \xrightarrow{\hat{()}} \hat{\mathscr{A}}.$$

Proposition 61 ([13]) *Let $\mathscr{A}$ be a filtered stable $(\infty, 1)$-category with $\hat{\mathscr{A}}$ being its exact localization. Then a monoidal structure of $\mathscr{A}$ is completable if and only if for every integer $n \geq 0$, the monoidal product $\bigotimes_{i=0}^{n} X_i$ for a sequence X_i, $0 \leq i \leq n$, of objects of $\mathscr{A}$ necessarily belongs to the full subcategory $\mathscr{A}_{\geq \infty}$ of $\mathscr{A}$ whenever $X_i \in \mathscr{A}_{\geq \infty}$ for some i.*

Proposition 62 ([13]) *Let $\mathscr{A}$ be a monoidal filtered stable $(\infty, 1)$-category with $\hat{\mathscr{A}}$ its exact completion. If the monoidal structure on $\mathscr{A}$ is completable, then $\hat{\mathscr{A}}$ with the induced structures is a monoidal (complete) filtered stable $(\infty, 1)$-category.*

The Day symmetric monoidal structure on the $(\infty, 1)$-category of filtered objects $\mathscr{A} = \mathrm{Fun}(\mathbb{Z}, \mathscr{B})$ is obviously completable *in* $\mathscr{A}^{\mathrm{op}}$. Note Remark 22.

The following can also be verified.

Proposition 63 ([13]) *The monoidal structure on the filtered $(\infty, 1)$-category of filtered objects in $\mathscr{B}$ as in Proposition 58, is completable in the case where $\mathscr{B}$ is closed under the sequential limits.*

7 Applications to the Koszul Duality

7.1 Notation

Results in this section will depend on analysis of cobar constructions. We denote the cobar construction by B, instead of Ω, which is traditional but can easily be confused with the looping functor in our exposition.

7.2 Fundamental Results

We would like to discuss applications of the notions and results which we have described in the previous sections.

Given an augmented E_n-algebra A, its **augmentation ideal** is by definition, the fibre of the augmentation map $\varepsilon\colon A \to \mathbf{1}$. In particular, this definition applied in the opposite category specifies what the augmentation ideal of an augmented coalgebra is.

Definition 64 Let $\mathscr{A}$ be a symmetric monoidal filtered stable $(\infty, 1)$-category. An augmented E_n-algebra A is said to be **positive** if its augmentation ideal belongs to $\mathscr{A}_{\geq 1}$.

An augmented E_n-coalgebra C in $\mathscr{A}$ is said to be **copositive** if there is a uniform bound ω for loops in $\mathscr{A}$ such that the augmentation ideal of C belongs to $\mathscr{A}_{\geq 1-n\omega}$.

Note also Definition 19.

Proof of Lemma 5 Denote the augmentation ideal of C_i by I_i. We express the cotensor product $K_{01}\square_{C_1}K_{12}$ etc. as the totalization of the cobar construction $B^\bullet(K_{01}, I_1, K_{12})$ etc. *without* degeneracies (in the sense that it is a diagram over $\boldsymbol{\Delta}_f$), associated to the actions of the non-unital coalgebra I_1 etc. See Sect. 5. It is easy to check that the usual bar construction, with degeneracies, associated to the unital coalgebra C_1 etc., is the right Kan extension of the version here, so the totalizations are equivalent.

The source then can be written as $\operatorname{Tot} B^\bullet(K_{01}\square_{C_1}K_{12}, I_2, K_{23})$.

For every n, the functor $-\otimes I_2^{\otimes n}\otimes K_{23}$ is bounded below, so the assumptions and Proposition 49 implies that

$$B^\bullet(K_{01}\square_{C_1}K_{12}, I_2, K_{23}) = \operatorname{Tot} B^\bullet(B^*(K_{01}, I_1, K_{12}), I_2, K_{23}),$$

where the totalization is in the variable $*$.

However, the totalization of this is nothing but the target. □

The proof of the following is similar.

Lemma 65 ([13]) *Let $\mathscr{A}$ be a monoidal complete filtered stable $(\infty, 1)$-category. Let A_i, $i = 0, 1, 2, 3$, be positive augmented algebras in $\mathscr{A}$, and let $K_{i,i+1}$ be a left A_i- right A_{i+1}-bimodule for $i = 0, 1, 2$, whose underlying object is bounded below.*

Then the resulting map

$$K_{01} \otimes_{A_1} K_{12} \otimes_{A_2} K_{23} \longrightarrow (K_{01} \otimes_{A_1} K_{12}) \otimes_{A_2} K_{23}$$

is an equivalence, where the source denotes the realization of the obvious bisimplicial bar construction (dual to the corresponding construction in Lemma 5).

One uses Proposition 52 instead of Proposition 49.

7.3 Positivity of the Koszul Dual

Let us turn to the Koszul duality.

Let $\mathscr{A}$ be a monoidal filtered stable $(\infty, 1)$-category with uniformly bounded loops and sequential limits. Then iterated application of the following lemma implies for a copositive augmented E_n-coalgebra C in $\mathscr{A}$, that its Koszul dual algebra is positive.

Lemma 66 *Let $\mathscr{A}$ be a monoidal filtered stable $(\infty, 1)$-category with uniformly bounded loops and sequential limits. Let C be an augmented associative coalgebra in $\mathscr{A}$. Assume that, for an integer $r \geq 1$ and a uniform bound ω for loops in $\mathscr{A}$, the augmentation ideal of C belongs to the full subcategory $\mathscr{A}_{\geq r-\omega}$ of $\mathscr{A}$.*

Let $\eta\colon C \to \mathbf{1}$ be the unit map of C. Then the fibre of the map $\eta^!\colon C^! \to \mathbf{1}^! = \mathbf{1}$ in $\mathscr{A}$ belongs to the full subcategory $\mathscr{A}_{\geq r}$ of $\mathscr{A}$.

Proof Let J be the augmentation ideal of C, so $J \in \mathscr{A}_{\geq r-\omega}$. Since we have Corollary 13, it suffices to prove that the fibre of each of the following obvious maps belongs to $\mathscr{A}_{\geq r}$:

$$\begin{aligned} C^! = \operatorname{Tot} B^\bullet(\mathbf{1}, J, \mathbf{1}) &\longrightarrow \operatorname{sk}_{-d} \operatorname{Tot} B^\bullet(\mathbf{1}, J, \mathbf{1}) \\ &\longrightarrow \operatorname{sk}_0 \operatorname{Tot} B^\bullet(\mathbf{1}, J, \mathbf{1}) = \mathbf{1}, \end{aligned}$$

where $d \leq 0$ is a uniform lower bound for sequential limits in $\mathscr{A}$. We shall prove

$$\operatorname{Fibre}[\operatorname{Tot} B^\bullet(\mathbf{1}, J, \mathbf{1}) \to \operatorname{sk}_{-d} \operatorname{Tot} B^\bullet(\mathbf{1}, J, \mathbf{1})] \in \mathscr{A}_{\geq r}. \tag{66}$$

The other part is simpler.

In order to prove (66), by the Definition 36 of a uniform lower bound for sequential limits, it suffices to prove that the fibre of the map

$$\operatorname{sk}_n \operatorname{Tot} B^\bullet(\mathbf{1}, J, \mathbf{1}) \longrightarrow \operatorname{sk}_{-d} \operatorname{Tot} B^\bullet(\mathbf{1}, J, \mathbf{1})$$

belongs to $\mathscr{A}_{\geq r-d}$ for all $n \geq -d+1$. However, this follows from Corollary 13, since for every $k \geq -d+1$, the fibre $\Omega^k B^k(\mathbf{1}, J, \mathbf{1}) = \Omega^k J^{\otimes k}$ of the map $\operatorname{sk}_k \operatorname{Tot} B^\bullet(\mathbf{1}, J, \mathbf{1}) \to \operatorname{sk}_{k-1} \operatorname{Tot} B^\bullet(\mathbf{1}, J, \mathbf{1})$ belongs to $\mathscr{A}_{\geq kr} \subset \mathscr{A}_{\geq r-d}$. □

With similar but simpler arguments using the dual of Lemma 14, one obtains the following.

Lemma 67 *Let $\mathscr{A}$ be a monoidal filtered stable $(\infty, 1)$-category. If looping translates the filtration of $\mathscr{A}$ in the sense of Definition 68 below, then the Koszul dual of a positive augmented E_n-algebra in $\mathscr{A}$ is a copositive augmented E_n-coalgebra.*

Definition 68 Let $\mathscr{A}$ be a filtered stable $(\infty, 1)$-category. Then we say that looping **translates the filtration** of $\mathscr{A}$, or looping is **translational** in $\mathscr{A}$, if there is a uniform lower bound ω for loops in $\mathscr{A}$ for which $-\omega$ is a lower bound of the functor $\Sigma = \Omega^{-1} \colon \mathscr{A} \to \mathscr{A}$. Equivalently (by Lemma 46), ω which is also an upper bound of the functor Ω.

Example 69 A t-structure on a stable $(\infty, 1)$-category is equivalent to a filtration with respect to which the loop functor is bounded above and below by -1. See Example 20.

In the $(\infty, 1)$-category of filtered objects and in a functor category with Goodwillie's filtration (Sect. 2), the loop functor is bounded above and below by 0.

Remark 70 In general, if looping translates the filtration of $\mathscr{A}$, and if there exists an integer r for which $\mathscr{A}_{\geq r+1}$ is a *proper* subcategory of $\mathscr{A}_{\geq r}$, and equivalently, $\mathscr{A}^{<r}$ is a proper subcategory of $\mathscr{A}^{<r+1}$, then a lower bound ω of $\Omega \colon \mathscr{A} \to \mathscr{A}$ for which $-\omega$ is an upper bound of $\Sigma \colon \mathscr{A} \to \mathscr{A}$, must be the greatest lower bound of Ω. It follow by duality, that ω must also be the least upper bound of Ω. Note Lemma 46.

The following proposition might be clarifying.

Proposition 71 *Let $\mathscr{A}$ be a filtered stable $(\infty, 1)$-category. Then an integer ω is a lower and an upper bound of the functor $\Omega \colon \mathscr{A} \to \mathscr{A}$ if and only if for every integer r and every object $X \in \mathscr{A}$, we have an equivalence $(\Omega X)^{<r+\omega} \simeq \Omega(X^{<r})$ in $\mathscr{A}$.*

Proof If ω is a lower and upper bound of Ω, then, since in the cofibre sequence

$$\Omega(X_{\geq r}) \longrightarrow \Omega X \longrightarrow \Omega(X^{<r}),$$

the fibre and the cofibre will respectively be in $\mathscr{A}_{\geq r+\omega}$ and be in $\mathscr{A}^{<r+\omega}$, we have that the map $\Omega X \to \Omega(X^{<r})$ induces an equivalence $(\Omega X)^{<r+\omega} \xrightarrow{\sim} \Omega(X^{<r})$ by Lemma 12.

Conversely, suppose we have equivalences $(\Omega X)^{<r+\omega} \simeq \Omega(X^{<r})$. Then for $X \in \mathscr{A}$ belonging to $\mathscr{A}^{<r}$, this implies that $\Omega X = \Omega(X^{<r})$ belongs to $\mathscr{A}^{<r+\omega}$, so Ω takes the full subcategory $\mathscr{A}^{<r}$ of $\mathscr{A}$ to the full subcategory $\mathscr{A}^{<r+\omega}$. For X belonging to $\mathscr{A}_{\geq r}$, we obtain $(\Omega X)^{<r+\omega} \simeq \Omega(X^{<r}) \simeq \mathbf{0}$, so Ω takes the full subcategory $\mathscr{A}_{\geq r}$ to $\mathscr{A}_{\geq r+\omega}$. □

7.4 Koszul Duality

The case $n = 1$ of Theorem 2 is implied by Theorems 72 and 73 below. The case for an arbitrary n is also obtained by iterating the arguments.

Let A be an augmented associative algebra. Then, for a right A-module K, we define a right $A^!$-comodule $\mathbb{D}_A K$ as $K \otimes_A \mathbf{1}$. Dually, if C is an augmented associative coalgebra, then for a right C-comodule L, we have a right $C^!$-module $\mathbb{D}_C L = L\Box_C \mathbf{1}$.

Theorem 72 ([13]) *Let $\mathscr{A}$ be a monoidal complete filtered stable $(\infty, 1)$-category with uniformly bounded sequential limits and translational looping.*

Let A be a positive augmented associative algebra in $\mathscr{A}$, and K be a right A-module which is bounded below. Then the canonical map $K \to \mathbb{D}_{A^!}\mathbb{D}_A K$ is an equivalence (of A-modules). In particular, the canonical map $A \to A^{!!}$ (of augmented associative algebras) is an equivalence.

Theorem 73 ([13]) *Let $\mathscr{A}$ be a monoidal complete filtered stable $(\infty, 1)$-category with uniformly bounded loops and sequential limits. Let C be a copositive augmented associative coalgebra in $\mathscr{A}$, and let K be a right C-comodule which is bounded below. Then the canonical map $\mathbb{D}_{C^!}\mathbb{D}_C K \to K$ is an equivalence (of C-comodules). In particular, the canonical map $C^{!!} \to C$ (of augmented associative coalgebras) is an equivalence.*

We recognize that the following result is obtained.

Theorem 74 (Cf. [13]) *Let $\mathscr{A}$ be a monoidal complete filtered stable $(\infty, 1)$-category with uniformly bounded sequential limits and translational looping. Let C be a copositive augmented associative coalgebra in $\mathscr{A}$. Then the functor*

$$\mathbb{D}_C\colon \mathrm{Comod}_{C,>-\infty}(\mathscr{A}) \longrightarrow \mathrm{Mod}_{C^!,>-\infty}(\mathscr{A})$$

is an equivalence with inverse $\mathbb{D}_{C^!}$, where $\mathrm{Comod}_{C,>-\infty}(\mathscr{A})$ denotes the $(\infty, 1)$-category of bounded below right C-comodules in $\mathscr{A}$.

These theorems follow from Lemmata 75 and 76 below either by using the Barr–Beck arguments [12] or by more concrete arguments as in [13].

Lemma 75 *Let $\mathscr{A}$ be a monoidal complete filtered stable $(\infty, 1)$-category with uniformly bounded loops and sequential limits. Let C be a copositive augmented associative coalgebra in $\mathscr{A}$. Then the functor*

$$-\Box_C\mathbf{1}\colon \mathrm{Comod}_{C,>-\infty}(\mathscr{A}) \to \mathscr{A}$$

reflects equivalences.

For an associative algebra A in $\mathscr{A}_{\geq 0}$, let $\mathrm{Mod}_{A,\geq r}(\mathscr{A})$ denote the $(\infty, 1)$-category of right A-modules in $\mathscr{A}_{\geq r}$.

Lemma 76 *Let A be a positive augmented associative algebra in a monoidal complete filtered stable $(\infty, 1)$-category $\mathscr{A}$. Then, the functor* $- \otimes_A \mathbf{1}\colon \mathrm{Mod}_{A,\geq r}(\mathscr{A}) \to \mathscr{A}_{\geq r}$ *reflects equivalences.*

The proofs of these lemmata are similar to each other. We shall first look at the proof of Lemma 76, which requires fewer assumptions.

We shall use the next lemma. Let $\mathscr{A}$ be a monoidal complete filtered stable $(\infty, 1)$-category, and let A be a positive augmented associative algebra in $\mathscr{A}$. Then we define the *powers* of the augmentation ideal I of A by $I^r := I^{\otimes_A r}$. This is unambiguous in view of Lemma 65. Note that multiplication of A gives an A-bimodule map $I^r \to I^s$ whenever $r \geq s$. Denote the cofibre of this map by I^s/I^r. When $s = 0$, this, A/I^r, is an A-algebra.

Lemma 77 *Let $\mathscr{A}$ be a monoidal filtered stable category with $\hat{\mathscr{A}}$ completing it, and let A be a positive augmented associative algebra in $\mathscr{A}$.*

Let K be a right A-module which is bounded below. Then the map $K \to \lim_r K \otimes_A A/I^r$ *is an equivalence after completion.*

Proof Since the fibre of the map $A \to A/I^r$ (namely I^r) belongs to $\mathscr{A}_{\geq r}$, the result follows from Lemma 51. (Write K as $K \otimes_A A$.) □

Proof of Lemma 76 Suppose an A-module K in $\mathscr{A}_{\geq r}$ satisfies $K \otimes_A \mathbf{1} \simeq \mathbf{0}$. We want to show that $K \simeq \mathbf{0}$.

In order to do this, it suffices, from the previous lemma, to prove $K \otimes_A (I^s/I^{s+1}) \simeq \mathbf{0}$ for all $s \geq 0$. However, $I^s/I^{s+1} \simeq \mathbf{1} \otimes_A I^s$ as a left A-module. □

Proof of Lemma 75 We would like to apply the arguments of the proof of Lemma 76. We simply need to establish the counterpart of Lemma 77. This will be simpler except that we need to use Lemma 78 below. □

The proof of the following lemma is similar to the proof of Lemma 66, but is simpler.

Lemma 78 ([13]) *Let $\mathscr{A}$ be a monoidal filtered stable $(\infty, 1)$-category with uniformly bounded loops and sequential limits. Let C be a copositive augmented coalgebra, K a right C-comodule, and L a left C-comodule, all in $\mathscr{A}$. If for integers r and s, (the underlying object of) K belongs to $\mathscr{A}_{\geq r}$, and L belongs to $\mathscr{A}_{\geq s}$, then $K \square_C L$ belongs to $\mathscr{A}_{\geq r+s}$.*

Techniques which establishes Theorem 6 are not very different.

7.5 Constructions of Positive Algebras

Let us see examples of positive algebras. In the case of a filtration by connectivity (Example 20), the conditions are simply on the connectivity. We shall give examples in filtered objects.

Example 79 Let $\mathscr{B}$ be a symmetric monoidal stable $(\infty, 1)$-category satisfying Assumption 57. Given an augmented associative algebra A in $\mathscr{B}$, the construction of the powers I^r of the augmentation ideal leads to the following associative algebra in the filtered symmetric monoidal $(\infty, 1)$-category $\mathscr{A} := \mathrm{Fun}(\mathbb{Z}, \mathscr{B})$:

$$\cdots \xleftarrow{=} A \xleftarrow{=} \cdots \xleftarrow{=} A = F_0A \longleftarrow F_1A \longleftarrow \cdots \longleftarrow F_rA \longleftarrow \cdots,$$

where $F_rA = I^r$ for $r \geq 0$. Since $F_\bullet \mathbf{1} = \mathbf{1}$, the augmentation of A induces an augmentation on $F_\bullet A$, and $F_\bullet A$ becomes a positive augmented associative algebra in $\mathscr{A}$.

Since the monoidal structure on $\mathscr{A}$ was completable by Proposition 63, we obtain a positive augmented associative algebra $\hat{F}_\bullet A$ in the symmetric monoidal complete filtered stable $(\infty, 1)$-category $\hat{\mathscr{A}}$.

This construction can be described in a more systematic manner, and generalizes for augmented locally constant factorization algebras. We refer the reader to [15].

Let us discuss another example.

Let $\mathscr{B}$ be the standard symmetric monoidal stable $(\infty, 1)$-category of chain complexes over a field k of characteristic 0. Let $\mathfrak{g}$ be a dg Lie-algebra over k. Then the Chevalley–Eilenberg complex $C_\bullet\mathfrak{g} = (\mathrm{Sym}^*(\Sigma\mathfrak{g}), d)$ (where $\Sigma = (\ )[1]$ is the suspension functor, and the differential d is the sum of the internal differential from $\mathfrak{g}$ and the Chevalley–Eilenberg differential) can be refined to give a filtered object of $\mathscr{B}$:

$$\cdots \longrightarrow \mathbf{0} \longrightarrow \cdots \longrightarrow \mathbf{0} = F_1C_\bullet\mathfrak{g} \longrightarrow F_0C_\bullet\mathfrak{g} \longrightarrow \cdots \longrightarrow F_{-r}C_\bullet\mathfrak{g} \longrightarrow \cdots,$$

where $F_{-r}C_\bullet\mathfrak{g} := (\mathrm{Sym}^{\leq r}(\Sigma\mathfrak{g}), d)$, so $C_\bullet\mathfrak{g} = \mathrm{colim}_{r\to\infty} F_{-r}C_\bullet\mathfrak{g}$.

It turns out that the construction $F_*C_\bullet$ is a symmetric monoidal functor between symmetric monoidal $(\infty, 1)$-categories

$$\mathrm{dgLie} \longrightarrow \mathrm{Fun}(\mathbb{Z}, \mathscr{B}) =: \mathscr{A},$$

where the symmetric monoidal structure on dg Lie algebras is given by the direct sum operations (and quasi-isomorphisms are inverted) [15]. In particular, $F_*C_\bullet\mathfrak{g}$ is an augmented commutative coalgebra in $\mathscr{A}$. In $\mathscr{A}^{\mathrm{op}}$, $F_*C_\bullet\mathfrak{g}$ is a positive augmented algebra. Note Remark 22.

The E_n-Koszul dual of this in $\mathscr{A}^{\mathrm{op}}$ is $F_*C_\bullet(\Omega^n\mathfrak{g})$, where the structure of an E_n-algebra of $F_*C_\bullet(\Omega^n\mathfrak{g})$ in $\mathscr{A}$ (the structure of a coalgebra in $\mathscr{A}^{\mathrm{op}}$) comes from the standard structure of an E_n-coalgebra on the sphere S^n as a pointed space.

In fact, there is a more general version of this [15] in the context of the Koszul duality *for locally constant factorization algebras*. The E_n-Koszul duality can be seen as the 'local' case of the Koszul duality for factorization algebra through the correspondence of Theorem 8 of [16] in these proceedings, established by Lurie [12]. We refer the reader to Lurie [12, Remark 5.5.6.11] for the basic ideas.

Acknowledgements The author is grateful to the anonymous referee for the fair criticism and helpful suggestions.

References

1. Beilinson, A.A., Ginsburg, V.A., Schechtman, V.V., Koszul duality. J. Geom. Phys. **5.3**, 317–350 (1988). https://doi.org/10.1016/0393-0440(88)90028-9
2. Bénabou, J.: Introduction to bicategories, Reports of the Midwest Category Seminar, pp. 1–77. Springer Berlin, (1967)
3. Ben-Zvi, D., Nadler, D.: Nonlinear traces. arXiv:1305.7175
4. Bernstein, I.N., Gelfand, I.M., Gelfand, S.I.: Algebraic bundles over $\mathbb{P}^n$ and problems of linear algebra. (Russian) Funkts. Anal. Prilozh. **12**(3), 66–67 (1978). (English) Funct. Anal. Appl. **12**, 212–214 (1979)
5. Costello, K.: Supersymmetric gauge theory and the Yangian. arXiv:1303.2632
6. Costello, K., Gwilliam, O.: Factorization algebras in quantum field theory. draft available http://www.math.northwestern.edu/~costello/
7. Dunn, G.: Tensor product of operads and iterated loop spaces. J. Pure Appl. Algebra **50**(3), 237–258 (1988)
8. Fresse, B.: Koszul duality of E_n -operads. Sel. Math. New Ser. **17**(2), 363–434 (2011)
9. Ginzburg, V., Kapranov, M.: Koszul duality for operads. Duke Math. J. **76**(1), 203–272 (1994)
10. Goodwillie, T.G.: Calculus. III. Taylor series. Geom. Topol. **7**, 645–711 (2003) (electronic)
11. Lurie, J.: On the classification of topological field theories. Current Developments in Mathematics, vol. 2008, pp. 129–280 International Press, Somerville (2008)
12. Lurie, J.: Higher Algebra (2017) available http://www.math.harvard.edu/~lurie/
13. Matsuoka, T.: Koszul duality between E_n-algebras and coalgebras in a filtered category. arXiv:1409.6943
14. Matsuoka, T.: Descent properties of the topological chiral homology. Münster J. Math. **10**, 83–118. Mathematical Reviews MR3624103 (2017). Available via http://www.math.unimuenster.de/mjm/vol10.html
15. Matsuoka, T.: Koszul duality for locally constant factorization algebras. Serdica Math. J. **41**(4), 369–414. (2015). Special issue on the International Conference "Mathematics Days in Sofia". Open access via http://www.math.bas.bg/serdica/
16. Matsuoka, T.: Some technical aspects of factorization algebras on manifolds, in these proceedings
17. McDuff, D.: Configuration spaces of positive and negative particles. Topology **14**, 91–107 (1975)
18. Positselski, L.: Two kinds of derived categories, Koszul duality, and comodule-contramodule correspondence. Mem. Am. Math. Soc. **212**(996), vi+133 (2011). ISBN: 978-0-8218-5296-5
19. Quillen, D.: Rational homotopy theory. Ann. Math. **2**(90), 205–295 (1969)
20. Salvatore, P.: Configuration spaces with summable labels. In: Cohomological Methods in Homotopy Theory. Progress in Mathematics, vol. 196, pp. 375–396 (2001)
21. Segal, G.: Configuration-spaces and iterated loop-spaces. Inventiones Math. **21**(3), 213–221 (1973)
22. Sullivan, D.: Infinitesimal computations in topology. Publ. Math. Inst. Hautes Étud. Sci. **47**, 269–331 (1977)
23. Toën, B., Vezzosi, G.: A remark on K-theory and S-categories. Topology **43**(4), 765–791 (2004)

Some Technical Aspects of Factorization Algebras on Manifolds

Takuo Matsuoka

Abstract We describe the basic ideas of factorization algebras on manifolds and topological chiral homology, with emphasis on their gluing properties.

Keywords Factorization algebra · Topological chiral homology · E_n-algebra · Homotopical algebra

1 Introduction

The notion of *chiral algebra* and equivalently, of *factorization algebra* was introduced by Beilinson and Drinfeld on algebraic curves [3]. There is an interesting counterpart of this on manifolds. Following some of the pioneers of the research of these objects on manifolds, we call them *factorization algebras*.

One motivation for studying factorization algebras on manifolds comes from the central role which they play in quantum field theory, generalizing the role of chiral algebras for conformal field theory. Namely, observables of a quantum (or a classical) field theory having locality form a factorization algebra, and this is the structure in terms of which one can rigorously understand quantization of a physical theory (in perturbative sense) [6], analogously to the deformation quantization of the classical mechanics [10].

For an approach to *locally constant* factorization algebras, namely, factorization algebras with topological invariance, Lurie has introduced and studied the *topological chiral homology* [11, Chap. 5]. Similar functor has several other names; in particular, the "factorization homology" (without necessarily requiring "topological" invariance of the "coefficients") [2, 6], or the "higher order Hochschild homology" (at least for coefficients in a commutative algebra) in [13], in which the work of Anderson [1] is mentioned for an earlier appearance of the notion.

T. Matsuoka (✉)
Intage Technosphere Inc., Intage Akihabara Bldg., 3 Kanda-Neribeicho, Chiyoda-Ku, Tokyo 101-0022, Japan
e-mail: motogeomtop@gmail.com

T. Ohsawa and N. Minami (eds.), *Bousfield Classes and Ohkawa's Theorem*, Springer Proceedings in Mathematics & Statistics 309,
https://doi.org/10.1007/978-981-15-1588-0_13

Factorization algebras are defined by a certain gluing property, and topological chiral homology is another process of gluing. In the locally constant setting, more general forms of gluing property are implied, and this leads to useful reformulations of the basic notions. Even though this subject is inevitably technical, the results turn out to be useful. We plan to describe some fundamental results with their implications.

We rely mostly on Lurie [11], Costello and Gwilliam [6], and the present author [12]. Other references, Ayala and Francis [2], Calaque [5], Ginot, Tradler and Zeinalian [9], Ginot [8], on factorization algebras have also been useful.

Since the subject of factorization algebras is relatively young, we expect that the overview which we give of some technical issues and solutions might be useful for the reader who foresees use of factorization algebras in their research.

2 Prefactorization Algebras

A factorization algebra will be analogous to a sheaf. We first define what will correspond to a presheaf in the definition of a factorization algebra.

Given a manifold M, let us denote by $\mathrm{Open}(M)$ the collection of all open subsets of M. It is a category, in fact a poset, under the inclusion of open sets in M. We moreover consider it as a multicategory, where, for a finite family of open sets $U = (U_s)_s$ of M, and an open set V, we let there be exactly one multimap $U \to V$ if U_s's are pairwise disjointly included in V in M, and otherwise, we let there be no multimap. The previous poset underlies this multicategory as the category formed with unary multimaps as morphisms.

Following Lurie [11] we shall refer to the homotopically enriched version of a multicategory (i.e., "coloured operad") as an *infinity operad*. Recall that a symmetric monoidal $(\infty, 1)$-category $\mathscr{A}$ has an underlying infinity operad, whose objects are objects of (the underlying $(\infty, 1)$-category of) $\mathscr{A}$, and in which, for a finite set S, an S-ary multimap $X \to Y$, where $X = (X_s)_{s \in S}$ and $X_s, Y \in \mathscr{A}$, is a map $\bigotimes_S X \to Y$ in $\mathscr{A}$.

Definition 0 Let M be a manifold, and let $\mathscr{A}$ be a symmetric monoidal $(\infty, 1)$-category. Then a **prefactorization algebra** on M in $\mathscr{A}$ is a functor from $\mathrm{Open}(M)$ to the underlying infinity operad of $\mathscr{A}$.

3 Assumption on the Target Category

For the rest of this survey, we assume that the target $(\infty, 1)$-category $\mathscr{A}$ of prefactorization algebras has colimits, and the monoidal multiplication functor on $\mathscr{A}$ preserves colimits variable-wise.

Note that a colimit in an $(\infty, 1)$-category is isomorphic in the homotopy category to the homotopy colimit if the $(\infty, 1)$-category comes e.g., from a model category, and the latter notion is defined appropriately.

4 Factorization Algebras

A factorization algebra will be required to satisfy the gluing property for the following class of covers.

Definition 1 (Costello and Gwilliam [6]) Let M be a manifold. A collection of open sets $\mathscr{U} = \{U_s\}_{s \in S}$ of M is a **Weiss cover** of M if for any finite subset x of M, there is an element $s \in S$ such that $x \subset U_s$.

In an earlier version of the draft [6], the authors were using what they called *factorizing* covers instead of Weiss covers. The definition will be recalled shortly. There is no difference between the two definitions in many practical situations where the cover is closed under taking finite disjoint union. Namely, in this case, the cover is factorizing if and only if it is Weiss.

In the case where the assumption is *not* satisfied, one can easily force the condition by replacing S by the set of finite subsets T of S for which U_t are pairwise disjoint for $t \in T$ (and by letting $U_T = \bigcup_{t \in T} U_t \subset M$). Let us denote this new cover $\{U_T\}_T$ obtained from $\mathscr{U}$ by $\mathscr{U}_\sqcup$.

Definition 2 A collection of open sets $\mathscr{U} = (U_s)_{s \in S}$ of a manifold M is a **factorizing cover** of M if the cover $\mathscr{U}_\sqcup$ of M by the unions of finite numbers of pair-wise disjoint open sets from $\mathscr{U}$, is Weiss in the sense of Definition 1.

Given a cover, the gluing in a prefactorization algebra can be formulated in terms of the usual Čech complex which is either a chain complex under presence of the Dold–Kan equivalence, or (the geometric realization of) a simplicial object in general. (For an explicit description in the case at hand, we refer the reader to [6] for the chain complex, and to e.g., Example 16 below (see also Remark 17) for the simplicial object.) Following Costello and Gwilliam we shall denote it by $\check{C}(\mathscr{U}, A)$. For a factorizing (rather than Weiss) cover, one should instead consider $\check{C}(\mathscr{U}_\sqcup, A)$.

Definition 3 (Cf. Costello and Gwillam [6]) A **factorization algebra** on a manifold M in a symmetric monoidal $(\infty, 1)$-category $\mathscr{A}$, is a prefactorization algebra A on M in $\mathscr{A}$, with the following properties.

- For every open set $U \subset M$ and Weiss cover $\mathscr{U}$ of U, the canonical map
$$\check{C}(\mathscr{U}, A) \longrightarrow A(U)$$
is an equivalence.
- A is *monoidal* in the sense that, for every disjoint open sets $U, V \subset M$, the induced map
$$A(U \sqcup V) \longleftarrow A(U) \otimes A(V) \tag{4}$$

is an equivalence, as well as the canonical map

$$A(\varnothing) \longleftarrow \mathbf{1}. \tag{5}$$

We denote the $(\infty, 1)$-category of factorization algebras on M in $\mathscr{A}$ by $\mathrm{Alg}_M(\mathscr{A})$.

Remark 6 Costello and Gwilliam refer to a factorization algebra in the above sense as a *strict* factorization algebra [6], as oppose to a *lax* factorization algebra, for which the monoidality is not required. In this survey, we only consider the strict version of the notion.

As one consequence of the definition of a factorization algebra, Costello and Gwilliam discuss how the $(\infty, 1)$-category of factorization algebras satisfy the sheaf axiom as the base manifold varies. (In that discussion [6], they mean strict factorization algebra by "factorization algebra".)

In Sect. 6, we shall look at the corresponding result, Theorem 25, for "locally constant" factorization algebras (Definition 22). The question is more involved in this case since one needs to understand in what sense the local constancy may be a local property of a factorization algebra, which is less trivial than in the case of a sheaf.

5 Topological Chiral Homology

Within factorization algebras, we would like to consider an analogue of locally constant sheaves. A locally constant sheaf taking values in an $(\infty, 1)$-category is determined by the local system formed by its stalks. Indeed, its sections (of which the ones which we consider are always the 'derived' ones) can be computed as the local coefficient cohomology.

Topological chiral homology, introduced by Lurie [11], is an analogue of the local coefficient cohomology, which takes "coefficients" in a "locally constant algebra" over disks in a manifold M, and produces a factorization algebra on M. Let us describe it.

Let M be a manifold. Then, following Lurie, we denote by $\mathrm{Disk}(M)$, the full submulticategory of $\mathrm{Open}(M)$ consisting of open submanifolds $U \subset M$ homeomorphic to an open disk (by an unspecified homeomorphism). In particular, every prefactorization algebra restricts to a $\mathrm{Disk}(M)$-algebra.

Definition 7 (Lurie [11] Definition 2.3.3.20) Let $\mathscr{E}$ be a multicategory, and let $\mathscr{A}$ be a symmetric monoidal $(\infty, 1)$-category or more generally an infinity operad. Then an $\mathscr{E}$-algebra $A\colon \mathscr{E} \to \mathscr{A}$ is said to be **locally constant** if it takes every unary multimap of $\mathscr{E}$ to an equivalence in the underlying $(\infty, 1)$-category of $\mathscr{A}$. We shall denote the $(\infty, 1)$-category of locally constant $\mathscr{E}$-algebras in $\mathscr{A}$ as $\mathrm{Alg}^{\mathrm{loc}}_{\mathscr{E}}(\mathscr{A})$.

Given a factorization algebra A on M, if the restriction of A to $\mathrm{Disk}(M)$ is locally constant over $\mathrm{Disk}(M)$ in this sense, then we consider $A|_{\mathrm{Disk}(M)}$ as an analogue in the analogy with sheaves, of the local system formed by the stalks of a locally constant sheaf. In this analogue of a local system, the stalk has more structure than just being an object of $\mathscr{A}$. Indeed, the analogue of the stalk would be the restriction of the $\mathrm{Disk}(M)$-algebra through an open embedding $\mathbb{R}^n \hookrightarrow M$, and there is the following theorem.

Theorem 8 (Lurie [11]) *The $(\infty, 1)$-category of locally constant* $\mathrm{Disk}(\mathbb{R}^n)$*-algebra is equivalent to the $(\infty, 1)$-category of E_n-algebras.*

Following Lurie, let us denote by $\mathrm{Disj}(M)$ the full subposet of $\mathrm{Open}(M)$ generated under the disjoint union by the objects of $\mathrm{Disk}(M) \subset \mathrm{Open}(M)$. Namely, an open submanifold $U \subset M$ belongs to $\mathrm{Disj}(M)$ if and only if it is homeomorphic (by an unspecified homeomorphism) to the disjoint union of a finite number of disks. $\mathrm{Disj}(M)$ has a partially defined monoidal structure given by the disjoint union in M, and every $\mathrm{Disk}(M)$-algebra in a symmetric monoidal category $\mathscr{A}$ extends uniquely to a symmetric monoidal functor $\mathrm{Disj}(M) \to \mathscr{A}$, so we may identify these two notions.

Definition 9 (Lurie [11]) Let M be a manifold, and let $\mathscr{A}$ be a symmetric monoidal $(\infty, 1)$-category. Given a locally constant $\mathrm{Disk}(M)$-algebra A (extending to a symmetric monoidal functor $A\colon \mathrm{Disj}(M) \to \mathscr{A}$), the **topological chiral homology** of M with coefficients in A is the object

$$\int_M A := \mathrm{colim}_{\mathrm{Disj}(M)} A$$

of $\mathscr{A}$, where the colimit is over the underlying category (poset) of $\mathrm{Disj}(M)$.

Remark 10 By the results of Sects. 5.4.5 and 5.5.2 of [11], the colimit in the definition of the topological chiral homology can in fact be written as a sifted colimit. In fact, all constructions and results which we describe on topological chiral homology and *locally constant* factorization algebras will be valid under only the assumption of Sect. 3 on the target category of prefactorization algebras *for sifted colimits* rather than for all small colimits as stated there.

Since Disj as a functor is symmetric monoidal with respect to the disjoint union of manifolds, the association

$$\mathrm{Open}(M) \ni U \longmapsto \mathrm{HF}_\bullet(U, A) := \int_U A \in \mathscr{A}, \tag{11}$$

namely, the left Kan extension to $\mathrm{Open}(M)$ of the functor $A\colon \mathrm{Disj}(M) \to \mathscr{A}$, has a unique symmetric monoidal structure which extends that of A on $\mathrm{Disj}(M)$.

Theorem 12 (Ginot–Tradler–Zeinalian [9]) *The prefactorization algebra* (11) *satisfies the descent for factorizing covers.*

In fact, a theorem below of Lurie leads to a stronger form of gluing property for the topological chiral homology, which simultaneously generalizes the descent for a Weiss cover and the definition of the topological chiral homology itself.

Consider a diagram of open sets of a topological space X, given by a functor $\chi : \mathscr{C} \to \mathrm{Open}(X)$, where $\mathscr{C}$ is a small category. In this situation, one defines for every point $x \in X$, the full subcategory

$$\mathscr{C}_x := \{i \in \mathscr{C} \mid \chi(i) \ni x\}$$

(with an abuse of notation) of $\mathscr{C}$. For example, the collection $\{\chi(i)\}_{i\in\mathscr{C}}$ of open sets of X is a cover of X if $\mathscr{C}_x \neq \varnothing$ for every point x.

The following theorem is a generalization of the Seifert–van Kampen theorem.

Theorem 13 (Lurie [11], Theorem A.3.1) *Let X be a topological space. Let $\mathscr{C}$ be a small category and let $\chi : \mathscr{C} \to \mathrm{Open}(X)$ be a functor. Assume that the following condition* (14) *is satisfied.*

$$\textit{For every point } x \in X, \mathscr{C}_x \textit{ has contractible classifying space.} \tag{14}$$

Denote the forgetful functor $\mathrm{Open}(X) \to \mathrm{Space}$ *by U, where* Space *denotes the standard $(\infty, 1)$-category of topological spaces (with weak homotopy equivalences inverted). Then the canonical map*

$$\operatorname*{colim}_{\mathscr{C}}(U\chi) \longrightarrow X$$

is an equivalence in Space.

Example 15 For a manifold M, the inclusion $\mathrm{Disk}(M) \hookrightarrow \mathrm{Open}(M)$ satisfies the condition (14) for Theorem 13 since for every $x \in M$, the poset $\mathrm{Disk}(M)_x$ is directed. Therefore, the natural map

$$\operatorname*{hocolim}_{D\in\mathrm{Disk}(M)} D \longrightarrow M$$

is a weak homotopy equivalence. Note that the source is equivalent to $B\mathrm{Disk}(M)$ since the homotopy colimit is over a diagram which is weakly homotopy equivalent to the terminal diagram.

This manner of considering a cover in terms of a diagram of open sets is actually useful.

Example 16 Suppose given an open cover $\mathscr{U} = \{U_s\}_{s\in S}$ of M indexed by a set S.

Denote by $\boldsymbol{\Delta}_{/S}$ the category of combinatorial simplices whose vertices are labeled by elements of S. Namely, its objects are finite non-empty ordinal I equipped with a set map $s : I \to S$. Then the cover determines a functor $\chi : (\boldsymbol{\Delta}_{/S})^{\mathrm{op}} \to \mathrm{Open}(M)$ by

$$(I, s : I \to S) \longmapsto U_s := \bigcap_{i\in I} U_{s(i)}.$$

Given a prefactorization algebra A on M, the Čech object $\check{\mathrm{C}}(\mathscr{U}, A)$ for $\mathscr{U}$ is equivalent to $\mathrm{colim}_{(\boldsymbol{\Delta}_{/S})^{\mathrm{op}}} A$.

Remark 17 In Example 16, the left Kan extension of χ (with target category replaced e.g., by the $(\infty, 1)$-category of presheaves on M) along the forgetful functor $(\boldsymbol{\Delta}_{/S})^{\mathrm{op}} \to \boldsymbol{\Delta}^{\mathrm{op}}$ is the usual hypercover associated to the original cover, and the Čech complex is the usual geometric realization for this hypercover.

The mentioned gluing property is given by the following, 'factorizing' analogue of Theorem 13. Given a functor $\chi : \mathscr{C} \to \mathrm{Open}(M)$, we consider for a subset $x \subset M$, the full subcategory

$$\mathscr{C}_x := \{i \in \mathscr{C} \mid x \subset \chi(i)\}$$

of $\mathscr{C}$.

Theorem 18 ([12]) *Let M be a manifold. Let $\mathscr{C}$ be a small category and let $\chi : \mathscr{C} \to \mathrm{Open}(M)$ be a functor. Assume that the following condition is satisfied.*

For any non-empty finite subset $x \subset M$, $\mathscr{C}_x$ has contractible classifying space. (19)

Let A be a locally constant $\mathrm{Disk}(M)$*-algebra in a symmetric monoidal $(\infty, 1)$-category $\mathscr{A}$ satisfying our conditions stated in Sect. 3. Then the canonical map*

$$\int_M A \longleftarrow \mathrm{colim}_{i \in \mathscr{C}} \int_{\chi(i)} A$$

is an equivalence.

Example 20 The open cover of M given by the inclusion $\mathrm{Disj}(M) \hookrightarrow \mathrm{Open}(M)$ satisfies the condition (19). Theorem 18 agrees in this case with the definition of the topological chiral homology.

Example 21 Let $\mathscr{U}$ be an open cover of a manifold M, and consider the functor χ of Example 16. Then it is immediate to see that χ satisfies the condition (19) if (and only if) the cover is Weiss.

Theorem 18 in this case (applied on every open $U \subset M$) gives an alternative proof that the prefactorization algebra $\mathrm{HF}_\bullet(-, A)$ constructed from a locally constant $\mathrm{Disk}(M)$-algebra A is a factorization algebra.

We shall denote the $(\infty, 1)$-category of locally constant $\mathrm{Disk}(M)$-algebras in a symmetric monoidal category or an infinity operad $\mathscr{A}$ as $\mathrm{Alg}_M^{\mathrm{loc}}(\mathscr{A})$. In the case where $\mathscr{A}$ satisfies the assumptions of Sect. 3, we consider $\mathrm{Alg}_M^{\mathrm{loc}}(\mathscr{A})$ as a full subcategory of prefactorization algebras on M by the embedding $A \mapsto \mathrm{HF}_\bullet(-, A)$, and then $\mathrm{Alg}_M^{\mathrm{loc}}(\mathscr{A})$ is contained in $\mathrm{Alg}_M(\mathscr{A})$.

Definition 22 A factorization algebra on a manifold M in a symmetric monoidal category $\mathscr{A}$ is said to be **locally constant** if it belongs to the full subcategory $\mathrm{Alg}_M^{\mathrm{loc}}(\mathscr{A})$ of $\mathrm{Alg}_M(\mathscr{A})$.

Here is another important example. Let I denotes the interval $[-\infty, \infty]$, and let $p\colon M \to I$ be a smooth map such that the restriction $p\colon p^{-1}(-\infty, \infty) \to$

$(-\infty, \infty)$ is the projection of a locally trivial smooth fibre bundle. Then $M = p^{-1}[-\infty, 0] \cup_{p^{-1}\{0\}} p^{-1}[0, \infty]$. Thus, the codimension 1 submanifold $N := p^{-1}\{0\} \subset M$ divides M into two pieces, and the normal bundle of N, identified with the tubular neighbourhood $T := p^{-1}(-\infty, \infty)$ of N in M, can be trivialized so that $p\colon T \to \mathbb{R}^1$ will be the projection. Conversely, p may be constructed essentially from such data (of division and trivialization).

Let $\mathrm{Disj}(I)$ denote the poset of open subsets of I which is the disjoint union of a finite number of intervals (each open *in* I), but is not equal to the whole I. Then Theorem 18 apply to the cover of M given by the functor

$$p^{-1}\colon \mathrm{Disj}(I) \longrightarrow \mathrm{Open}(M),$$

so, given a locally constant factorization algebra A on M, the canonical map

$$\operatorname*{colim}_{D\in \mathrm{Disj}(I)} A(p^{-1}D) \longrightarrow A(M) \tag{23}$$

will be an equivalence. This colimit has a concrete description as follows.

In this situation, $A\big(p^{-1}(-\infty, \infty)\big)$ acquires the structure of an associative algebra in $\mathscr{A}$, which acts on $A\big(p^{-1}[-\infty, \infty)\big)$ from the right (say, depending on the conventions), and on $A\big(p^{-1}(-\infty, \infty]\big)$ from the other side. (We in fact have a factorization algebra $p_*A := A \circ p^{-1}$ on I which is locally constant along each of the strata $\{\pm\infty\}$, $\mathbb{R}^1$ of I. The mentioned structure results from this. For example, the associative algebra is $p_*A|_{\mathbb{R}^1}$. Note Theorem 8.)

Theorem 24 (Cf. Lurie [11], Theorem 5.5.3.11) *The colimit in* (23) *is naturally equivalent to* $p_*A[-\infty, \infty) \otimes_{p_*A(-\infty,\infty)} p_*A(-\infty, \infty]$.

We shall see a simple example of this in Sect. 6

6 Descent Properties of Factorization Algebras

In this section, we would like to explain how the following theorem can be proved. Let Man denote the category of manifolds and open embeddings.

Theorem 25 ([12], cf. Costello and Gwilliam [6]) *The presheaf* $M \mapsto \mathrm{Alg}^{\mathrm{loc}}_M(\mathscr{A})$ *on* Man *of* $(\infty, 1)$*-categories is a sheaf.*

As mentioned in Sect. 4, the similar theorem in the non-locally constant setting is obtained by Costello–Gwilliam [6]. In locally constant setting, we need to solve an additional problem of whether local constancy of a factorization algebra is a 'local' property in some useful manner. We shall describe useful theorems obtained in answering this question.

Theorem 26 *Let M be a manifold, and let $\mathscr{E} \subset \mathrm{Disj}(M)$ $(\subset \mathrm{Open}(M))$ be a full subposet which is* ***solid*** *in the sense that if V, $W \in \mathrm{Disj}(M)$ are disjoint and $V \sqcup W \in$*

$\mathscr{E}$, then $V \in \mathscr{E}$ (and $W \in \mathscr{E}$). Assume the Hypotheses 29 below. Then a symmetric monoidal functor $A\colon \mathrm{Open}(M) \to \mathscr{A}$ is a locally constant factorization algebra on M if and only if it satisfies the following.

Let $\mathscr{E}_1 := \mathscr{E} \cap \mathrm{Disk}(M)$.

1 A sends every morphism in $\mathscr{E}_1$ to an equivalence.
2 The underlying functor of A is a left Kan extension of its restriction to $\mathscr{E}$.

In other words, any pair $\mathscr{E}_1 \subset \mathscr{E}$ satisfying the hypotheses can replace the pair $\mathrm{Disk}(M) \subset \mathrm{Disj}(M)$ in the definition of a locally constant factorization algebra/topological chiral homology.

Remark 27 $\mathscr{E}$ is *not* assumed to be closed under the disjoint union in $\mathrm{Disj}(M)$.

In order to formulate the necessary hypotheses, for a solid full subposet $\mathscr{E} \subset \mathrm{Disj}(M)$ as in Theorem 26, consider the following (not full) submulticategory of $\mathrm{Disk}(M)$, which we shall denote by $\overline{\mathscr{E}}_1$. The objects of $\overline{\mathscr{E}}_1$ are the objects of $\mathscr{E}_1$, and $\overline{\mathscr{E}}_1$ is generated by the following form of multimaps in $\mathrm{Disk}(M)$ between objects belonging to $\mathscr{E}_1$. Namely, for a finite set S, the multimap $D \to E$ in $\mathrm{Disk}(M)$ for $D = (D_s)_{s \in S}$, D_s, $E \in \mathscr{E}_1$ (thus D is pair-wise disjoint, and $\bigsqcup_S D \subset E$ in M) should belong to $\overline{\mathscr{E}}_1$ if $\bigsqcup_S D \in \mathscr{E}$, and we let $\overline{\mathscr{E}}_1$ be the smallest submulticategory of $\mathrm{Disk}(M)$ containing all these multimaps. Note that the poset $\mathscr{E}_1$ underlies this multicategory. Solidness of $\mathscr{E}$ implies that any map in $\mathscr{E}$ is the disjoint union of a finite number of generating multimaps $\bigsqcup_S D \to E$ of $\overline{\mathscr{E}}_1$.

For an object $V \in \mathscr{E}_1$ and a finite set S, denote by $\mathscr{E}_S(V)$, the full subposet of the direct product $(\mathscr{E}_1)^S$ consisting of S-labeled families $D \in (\mathscr{E}_1)^S$ for which there is an S-ary multimap $D \to V$ in $\overline{\mathscr{E}}_1$.

Example 28 If $\mathscr{E}$ is the whole $\mathrm{Disj}(M)$, then $\overline{\mathscr{E}}_1 = \mathrm{Disk}(M)$, and $\mathscr{E}_S(V) = \mathrm{Disj}_S(V)$, where, for any manifold U, we denote by $\mathrm{Disj}_S(U)$, the poset of pair-wise disjoint S-labeled families of open disks in U. In the proof of [11, Lemma 5.4.5.11] of Lurie, it is proved (using the generalized Seifert–van Kampen theorem 13) that the classifying space of $\mathrm{Disj}_S(U)$ is equivalent to the labeled (by S) configuration space of U.

Hypothesis 29 a. For every $U \in \mathrm{Open}(M)$, the inclusion functor $\mathscr{E}_{/U} \to \mathrm{Open}(M)_{/U} = \mathrm{Open}(U)$ satisfies the instance on U of the condition (19).

b. The inclusion $\mathscr{E}_1 \hookrightarrow \mathrm{Disk}(M)$ induces an equivalence $B\mathscr{E}_1 \xrightarrow{\sim} B\mathrm{Disk}(M) \simeq M$; see Example 15.

c. For every object $V \in \mathscr{E}_1 \subset \mathscr{E}$ and a finite set S, the inclusion $\mathscr{E}_S(V) \hookrightarrow \mathrm{Disj}_S(M)_{/V} = \mathrm{Disj}_S(V)$ induces an equivalence on the classifying space.

Remark 30 • It follows from Example 28, that the condition (c) is automatically satisfied for S of cardinality up to 1.

• It follows from Theorem 13 that a sufficient condition for the condition (b) is that the inclusion $\chi : \mathscr{C} := \mathscr{E}_1 \hookrightarrow \mathrm{Open}(M)$ satisfies the condition (14).

There is a sufficient condition for the condition (c) which can be checked easily in practice. For every object $V \in \mathscr{E}_1$ and an injection $x\colon S \hookrightarrow V$, $S \ni s \mapsto x_s \in V$, define the full subposet $\mathscr{E}_S(V, x) \subset \mathscr{E}_S(V)$ as consisting of S-labeled family $D = (D_s)_{s\in S} \in \mathscr{E}_S(V)$, $D_s \in \mathscr{E}_1$, such that $x_s \in D_s$ in V for every $s \in S$.

Proposition 31 *The condition (c) is satisfied if, for every injection $x\colon S \hookrightarrow V$, the poset $\mathscr{E}_S(V, x)$ has contractible classifying space.*

This follows from the generalized Seifert–van Kampen theorem 13. See the proof of [11, Lemma 5.4.5.11] of Lurie.

The following is a situation where the hypotheses are satisfied.

Example 32 We can take $\mathscr{E}$ as follows. Suppose M is the interior of a compact manifold $\overline{M}$ with boundary. We let $\mathscr{E}$ be such that $U \in \mathrm{Disj}(M)$ belongs to $\mathscr{E}$ if and only if there exists an smooth immersion $i\colon D \to \overline{M}$, where D is the coproduct of a finite number of closed disks, such that i restricts to a diffeomorphism from the interior of D to U.

The key to the proof of Theorem 26 is the following theorem, which follows immediately from Theorem 2.3.3.23 and (the proof of) Theorem 5.4.5.9, of Lurie's book [11].

Theorem 33 *Let M be a manifold, and let $\mathscr{E}$ be a solid full subposet of* $\mathrm{Disj}(M)$ *(see the formulation of Theorem 26). Assume the conditions (b), (c) of Hypothesis 29. Then, for every multicategory $\mathscr{M}$, the following restriction functor is an equivalence:*

$$\mathrm{Alg}_M^{\mathrm{loc}}(\mathscr{M}) \longrightarrow \mathrm{Alg}_{\mathscr{E}_1}^{\mathrm{loc}}(\mathscr{M}); \tag{34}$$

see Definition 7.

Proof *(Proof of Theorem 26)* Necessity of the assumptions follows from Theorem 18.

For sufficiency, the most non-trivial point is that the assumptions implies that A is locally constant (on the whole $\mathrm{Disk}(M)$). However, Theorem 33 gives a *locally constant* factorization algebra which coincides with A on $\mathscr{E}$, and then this will coincide with A as a prefactorization algebra by the necessity part of the proof. □

(Outline of proof of Theorem 25) Let a cover of a manifold M be given by $\mathscr{U} = (U_s)_{s\in S}$ where S is an indexing set. Let $\mathscr{C} := (\boldsymbol{\Delta}_{/S})^{\mathrm{op}}$ be as in Example 21, and define $\chi\colon \mathscr{C} \to \mathrm{Open}(M)$ in the way described there. We would like to prove that the restriction functor

$$\mathrm{Alg}_M^{\mathrm{loc}}(\mathscr{A}) \longrightarrow \lim_{i\in\mathscr{C}} \mathrm{Alg}_{\chi(i)}^{\mathrm{loc}}(\mathscr{A}) \tag{35}$$

is an equivalence.

We can write the limit in the target as an algebra over the following multicategory $\mathscr{E}_1$, and then apply Theorem 33.

For an open disk $D \in \mathrm{Disk}(M)$, define

$$\mathscr{C}_D := \{i \in \mathscr{C} \mid D \subset \chi(i)\}.$$

Then this is either empty or has contractible classifying space. Indeed, $\mathscr{C}_D = (\boldsymbol{\Delta}_{/S_D})^{\mathrm{op}}$, where $S_D := \{s \in S \mid D \in U_s\}$.

Define $\mathscr{E}_1$ to be the full submulticategory of $\mathrm{Disk}(M)$ consisting of disks D such that $\mathscr{C}_D$ is non-empty. Then Hypothesis 29 is satisfied where $\mathscr{E} \subset \mathrm{Disj}(M)$ consists of disjoint unions of the disks belonging to $\mathscr{E}_1$. □

Example 36 Take distinct two points x_i, $i = 0, 1$ on a circle S^1, and cover S^1 by the open sets $U_i := S^1 - \{x_i\}$. Using a framing of S^1, U_i or each component of $U_0 \cap U_1$, which we shall denote by V and W, can be identified with $\mathbb{R}^1$ up to a contractible space of choices of framed diffeomorphisms. Therefore, locally constant factorization algebras on each of these manifolds can canonically be identified with associative algebras. Therefore, a locally constant factorization algebra on S^1 is obtained by giving associative algebras A_i on U_i and identifications of them on V and on W.

Let us consider $A_0 = A_1 =: A$ by using the identification on V. Then the identification on W gives us an automorphism $\tau\colon A \to A$. For this factorization algebra on S^1, which we shall denote by B, we obtain from Theorem 24 (by cutting S^1 into two pieces, on the side of x_0, and on the side of x_1) an equivalence

$$B(S^1) \simeq A_0 \otimes_{A \otimes A^{\mathrm{op}}} A_1 \simeq \mathrm{HH}_\bullet(A, A^\tau),$$

where the right hand side is the Hochschild homology object in $\mathscr{A}$ with coefficients in A^τ, the A–A-bimodule A *twisted* by τ. In other words, we have

$$\int_{S^1} B = \mathrm{HH}_\bullet(A, A^\tau)$$

where B on the left hand side denotes the corresponding locally constant $\mathrm{Disk}(S^1)$-algebra.

7 Product Formulae on Factorization Algebras

We would like to give further illustration of use of Theorem 33.

Let $\mathscr{A}$ be a symmetric monoidal $(\infty, 1)$-category satisfying the assumption stated in Sect. 3.

Theorem 37 (Cf. Ginot [8], Calaque [5]) *Let B, F be manifolds. Then, the restriction functor*

$$\mathrm{Alg}^{\mathrm{loc}}_{F\times B}(\mathscr{A}) \longrightarrow \mathrm{Alg}^{\mathrm{loc}}_B(\mathrm{Alg}^{\mathrm{loc}}_F(\mathscr{A}))$$

is an equivalence of symmetric monoidal $(\infty, 1)$-categories.

Remark 31 If one swaps the factors of $B \times F$, then on the side of algebras, one recovers the canonical equivalence $\mathrm{Alg}^{\mathrm{loc}}_B(\mathrm{Alg}^{\mathrm{loc}}_F) \simeq \mathrm{Alg}^{\mathrm{loc}}_F(\mathrm{Alg}^{\mathrm{loc}}_B)$.

(Outline of proof of Theorem 37) To first give a technical remark, the $(\infty, 1)$-category $\mathrm{Alg}_F^{\mathrm{loc}}$ has sifted colimits, and they are preserved by the tensor product (since these are the same colimits and tensor product on the underlying objects). Therefore, in view of Remark 10, locally constant factorization algebras in $\mathrm{Alg}_F^{\mathrm{loc}}(\mathscr{A})$ are within our framework.

The restriction functor is symmetric monoidal since the symmetric monoidal structures on the $(\infty, 1)$-categories of algebras are value-wise, so it suffices to prove that it is an equivalence of $(\infty, 1)$-categories. We obtain this by applying Theorem 33.

Let $M := F \times B$. We let $\mathscr{E}$ be the solid full subposet of $\mathrm{Disj}(M)$ consisting of those object D for which there exists objects D' of $\mathrm{Disj}(B)$ and D'' of $\mathrm{Disj}(F)$, such that any component of D is a component of $D' \times D'' \subset M$.

Theorem 33 applies to this $\mathscr{E}$, and we obtain that the restriction functor $\mathrm{Alg}_M^{\mathrm{loc}} \to \mathrm{Alg}_{\overline{\mathscr{E}}_1}^{\mathrm{loc}}$ is an equivalence.

However, the restriction functor $\mathrm{Alg}_{\overline{\mathscr{E}}_1}^{\mathrm{loc}} \to \mathrm{Alg}_{\mathrm{Disk}(B)}^{\mathrm{loc}}(\mathrm{Alg}_{\mathrm{Disk}(F)}^{\mathrm{loc}})$ can directly be seen to be an equivalence.

For example, a locally constant factorization algebra on $\mathbb{R}^2$ is the same as an associative algebra in the $(\infty, 1)$-category of associative algebras since a locally constant factorization algebra on $\mathbb{R}^1$ can be directly seen to be the same as an associative algebra.

Inductively, a locally constant factorization algebra on $\mathbb{R}^n$ is an iterated associative algebra object.

Remark 39 The proof by Ginot [8] of Theorem 37 is by relying on a theorem of Dunn [7] on Boardman–Vogt's "little cubes" [4]. The proof outlined above (where we have essentially followed [12]) is independent of Dunn's theorem.

Remark 40 On a product manifold $M = B \times F$, Theorem 26 applies also to the solid full subposet of $\mathrm{Disj}(M)$ consisting of the disjoint unions of disks in M of the form $D' \times D''$ for disks D' in B and D'' in F. The result we obtain is another description of the $(\infty, 1)$-category $\mathrm{Alg}_M^{\mathrm{loc}}$, namely as the $(\infty, 1)$-category of 'locally constant' algebras on this basis of topology for M.

Iterating this, one finds a description of the $(\infty, 1)$-category of locally constant algebras on $\mathbb{R}^n$ which identifies it essentially with the $(\infty, 1)$-category of algebras over the little cubes. Therefore, Theorem 37 also proves Dunn's theorem.

Remark 41 A version of Theorem for general (i.e., not assumed locally Formulae) factorization algebras is described by Calaque in [5] with a (sketch of) proof by a strategy similar to ours. Namely, he introduces the notion of *factorizing basis* as Definition 42 below, using the similar condition as the condition (a) of Hypothesis 29. Then he applies a theorem [5, Theorem 2.1.9] which corresponds in a way, to Theorem 26.

We remark that the theorem for locally constant algebras may not be a corollary of this since comparison of the "locally constant" objects through Calaque's equivalence would perhaps not be straightforward.

Definition 42 (Claque [5]) A **factorizing basis** of M is a collection $\mathscr{B}$ of open sets of M such that for any $U \in \mathrm{Open}(M)$, the collection $\mathscr{B}_{/U} := \{V \in \mathscr{B} \mid V \subset U\}$ is a *factorizing* cover of U (Definition 2).

Remark 43 This is *different* from the notion of factorizing basis used by Costello and Gwilliam [6].

Finally, we shall state a natural generalization of Theorem 37, in which the source and the target of the algebras are twisted.

Namely, we consider algebras on the total space E of a *fibre bundle* taking values in a locally constant factorization *algebra* $\mathscr{A}$ of $(\infty, 1)$-categories on E. Then, as a twisted version of Alg_F in Theorem 37, one can construct a locally constant factorization algebra $\mathrm{Alg}_{E/B}(\mathscr{A})$ of $(\infty, 1)$-categories on the base manifold B of the fibre bundle.

Theorem 44 ([12]) *Let B be a manifold, and let $E \to B$ be a smooth fibre bundle over B. For a locally constant factorization algebra $\mathscr{A}$ on E of $(\infty, 1)$-categories, there is a natural equivalence*

$$\mathrm{Alg}_E(\mathscr{A}) \xrightarrow{\sim} \mathrm{Alg}_B(\mathrm{Alg}_{E/B}(\mathscr{A}))$$

of $(\infty, 1)$-categories, given by a suitable 'restriction' functor.

Remark 45 For this theorem, no assumption on sifted colimits are needed for $\mathscr{A}$. If $\mathscr{A}$ is instead a single fixed symmetric monoidal $(\infty, 1)$-category, there is actually a slight difference between an algebra in $\mathscr{A}$ (for which Theorem 37 may fail without assumption on sifted colimits), and an algebra taking values in the 'constant' algebra at $\mathscr{A}$ (to which Theorem 44 *always* applies). The assumption on sifted colimits simply ensures equivalence of these two notions of an algebra.

Acknowledgements The author is grateful to the anonymous referee for the fair criticism and helpful suggestions.

Appendix

A Factorization Algebra on an Orbifold

1 Let $\mathscr{A}$ be as in Sect. 3. Then Theorem 25 implies that the contravariant functor $M \mapsto \mathrm{Alg}^{\mathrm{loc}}_M(\mathscr{A})$ on the category Man of manifolds and open embeddings, extends uniquely to a sheaf on the category (enriched in groupoids) of orbifolds and local diffeomorphisms (or "étale" maps) between them. Indeed the $(\infty, 1)$- (or 2-) categories of sheaves (of e.g., $(\infty, 1)$-categories) on these categories are equivalent.

For an orbifold X, it would perhaps make sense to refer to the value associated to X by this extended sheaf, as the $(\infty, 1)$-category "of locally constant factorization algebras on X". This defines a notion of locally constant factorization algebra on X.

The purpose of this appendix is to give a concrete description of this notion of locally constant factorization algebra on an orbifold, which also leads to a very simple description of the functoriality of the $(\infty, 1)$-category of locally constant factorization algebras with respect to local diffeomorphisms.

2 Let us denote by Orb, the category enriched in groupoids of orbifolds with *local diffeomorphisms* as morphisms. We denote by $\overline{\text{Man}}$, the category of manifolds with local diffeomorphisms as morphisms. We have a non-full and full inclusions

$$\text{Man} \hookrightarrow \overline{\text{Man}} \hookrightarrow \text{Orb}.$$

(As any other category, we treat all these categories as $(\infty, 1)$-categories.)

Let X be an orbifold. Then by $\text{LocDiff}(X)$, we mean $\text{Man}_{/X}$. The coCartesian symmetric monoidal structure on $\overline{\text{Man}}_{/X}$ (i.e., the symmetric monoidal structure given by the finite coproduct operations) restricts to a symmetric monoidal structure on $\text{LocDiff}(X)$.

Given an object $(M, f) \in \text{LocDiff}(X)$, where M is a manifold, and $f : M \to X$ is a local diffeomorphism, we obtain an induced symmetric monoidal functor $f_! : \text{Open}(M) \to \text{LocDiff}(X)$. Thus, we obtain from a symmetric monoidal functor $A : \text{LocDiff}\, X \to \mathscr{A}$, a prefactorization algebra $f^*A := A \circ f_!$ on M.

One sees from the definitions, that a locally constant factorization algebra on X is equivalent as a datum to a symmetric monoidal functor $\text{LocDiff}(X) \to \mathscr{A}$ for which f^*A is a locally constant factorization algebra on M for every object (M, f) of $\text{LocDiff}(X)$. Equivalently, A should be locally constant in disks over X (i.e., on $\overline{\text{Disk}}(X)$ defined below), and such that the canonical map $A(M, f) \leftarrow \text{colim}_{\text{Disj}(M)} f^*A$ is an equivalence for every (M, f).

3 We can also express a locally constant factorization algebra on X as a locally constant algebra over suitable disks over X.

Let $\overline{\text{Disk}}(X)$ denote the full submulticategory of (the underlying multicategory of) $\text{LocDiff}(X)$, where the object $(U, i) \in \text{LocDiff}(X)$ for a manifold U and a local diffeomorphism $i : U \to X$, belongs to $\overline{\text{Disk}}(X)$ if there exists a diffeomorphism of U with a finite dimensional Euclidean space.

Now, given a general object (M, f) of $\text{LocDiff}(X)$, locally constant factorization algebras on M were equivalent to locally constant $\text{Disk}(M)$-algebras, but the functor $f_!$ identifies multimaps in $\text{Disk}(M)$ with multimaps in $\overline{\text{Disk}}(X)$. It follows that a locally constant factorization on X is equivalent as a datum to a locally constant algebra on $\overline{\text{Disk}}(X)$.

4 Let $f : X \to Y$ be local diffeomorphism of orbifolds. For a locally constant factorization algebra A on Y, we obtain a simple description of the pull-back f^*A of A by f. Indeed, we obtain the induced symmetric monoidal functor $f_! : \text{LocDiff}(X) \to \text{LocDiff}(Y)$, and, it follows from the above description of locally constant factorization algebras on orbifolds, that there is a natural equivalence

$$f^*A = A \circ f_! \tag{46}$$

of symmetric monoidal functors on $\mathrm{LocDiff}(X)$.

Proposition 47 *Let $p\colon U \to X$ be a surjective local diffeomorphism of orbifolds. Then a symmetric monoidal functor $A\colon \mathrm{LocDiff}(X) \to \mathscr{A}$ is a locally constant factorization algebra on X if and only if $A \circ p_!\colon \mathrm{LocDiff}(U) \to \mathscr{A}$ is a locally constant factorization algebra on U.*

Proof The necessity is clear from the equivalence (46).

For the converse, assume that $A \circ p_!$ is a locally constant factorization algebra on U. Then, for a manifold M and a local diffeomorphism $f\colon M \to X$, we need to prove that the prefactorization algebra f^*A on M is a locally constant factorization algebra.

The assumption implies that $U_M := U \times_X M$ is a manifold, and the projection $p_M\colon U_M \to M$ is a surjective local diffeomorphism. Therefore, it suffices by Theorem 25, to prove for every $V \in \mathrm{Open}(U_M) \times_{\mathrm{LocDiff}(M)} \mathrm{Open}(M)$, that $(f^*A)|_V$ is a locally constant factorization algebra on V.

However, the composite $\mathrm{Open}(V) \hookrightarrow \mathrm{Open}(M) \xrightarrow{f_!} \mathrm{LocDiff}(X)$ is isomorphic to the composite

$$\mathrm{Open}(V) \xrightarrow{g_!} \mathrm{LocDiff}(U) \xrightarrow{p_!} \mathrm{LocDiff}(X),$$

where $g\colon V \to U$ denotes the inclusion $V \hookrightarrow U_M$ followed by the projection $U_M \to U$, which is a local diffeomorphism.

References

1. Anderson, D.W.: Chain functors and homology theories. In: Symposium Algebraic Topology. Lecture Notes in Mathematics, vol. 249, pp. 1–12 (1971)
2. Ayala, D., Francis, J.: Factorization homology of topological manifolds. J. Topol. **8**(4), 1045–1084 (2015)
3. Beilinson, A., Drinfeld, V.: Chiral Algebras. American Mathematical Society Colloquium Publications. vol. 51, vi+375 pp. American Mathematical Society, Providence, RI (2004). ISBN: 0-8218-3528-9
4. Boardman, J.M., Vogt, R.M.: Homotopy invariant algebraic structures on topological spaces. Lecture Notes in Mathematics, vol. 347. Springer, Berlin-New York (1973)
5. Calaque, D.: Around Hochschild (co)homology. Habilitation thesis, Université Claude Bernard Lyon 1 (2013)
6. Costello, K., Gwilliam, O.: Factorization algebras in quantum field theory. Draft available at http://www.math.northwestern.edu/~costello/
7. Dunn, G.: Tensor product of operads and iterated loop spaces. J. Pure Appl. Algebra **50**(3), 237–258 (1988)
8. Ginot, G.: Notes on factorization algebras, factorization homology and applications. Calaque, D. et al. (ed.) Mathematical aspects of quantum field theories. Springer. Mathematical Physics Studies, pp. 429–552 (2015)
9. Ginot, G., Tradler, T., Zeinalian, M.: Higher Hochschild homology, topological chiral homology and factorization algebras. Commun. Math. Phys. **326**(3), 635–686 (2014)

10. Kontsevich, M.: Operads and motives in deformation quantization. Mosh Flato (1937–1998). Lett. Math. Phys. **48**(1), 35–72 (1999)
11. Lurie, J.: Higher Algebra (2017). http://www.math.harvard.edu/~lurie/
12. Matsuoka, T.: Descent properties of the topological chiral homology. Münster J. Math. **10**, 83–118 (2017). Mathematical Reviews MR3624103. Available via https://www.uni-muenster.de/FB10/mjm/vol10.html
13. Pirashvili, T.: Hodge decomposition for higher order Hochschild homology. Ann. Sci. Éc. Norm. Supér. (4) **33**(2), 151–179 (2000)

A Role of the L^2 Method in the Study of Analytic Families

Takeo Ohsawa

Abstract An expository account is given on the L^2 method for the $\bar{\partial}$ equation, L^2 extension theorems and the Bergman kernel focusing on the recent applications to analytic families.

Keywords Analytic family

2010 Mathematics Subject Classification Primary 32E40 · Secondary 32T05

Introduction

Geometric invariants of complex manifolds are encoded in the space of L^2 space of holomorphic sections of vector bundles. They are compressed in the Bergman kernel as the works of Kodaira [28], Hörmander [24] and Fefferman [17] have shown, so that relations between analysis and geometry on complex manifolds are accumulated in the results on the Bergman kernels. There are such instances in the study of analytic families.

Given a smooth analytic family of complex manifolds, say $p : M \to T$, the parameter dependence of the diagonalized Bergman kernel $K_t = K_{M_t}$ of $M_t = p^{-1}(t)$ reflects how the complex structure of M_t deforms. It was proved by Berndtsson [3] that $\log K_t$ depends plurisubharmonically in t if M is Stein. This property was strengthened in [4] and was later applied in [5] to give an alternate proof of

The author would like to express his sincere thanks to Norihiko Minami and Stefan Nemirovski for giving him strong motivations to write this paper. He also thanks to the referee for the fast, elaborate and useful comments for improving the presentation.

T. Ohsawa (✉)
Graduate School of Mathematics, Nagoya University, Furocho, Chikusaku, Nagoya 464-8602, Japan
e-mail: ohsawa@math.nagoya-u.ac.jp

T. Ohsawa and N. Minami (eds.), *Bousfield Classes and Ohkawa's Theorem*, Springer Proceedings in Mathematics & Statistics 309,
https://doi.org/10.1007/978-981-15-1588-0_14

the optimalized version of an L^2 extension theorem which had been obtained in [6, 20]. Another consequence of Berndtsson's theorem is that such a family is locally analytically trivial if $\log K_t \in C^\infty$ and $\partial\bar\partial \log K_t$ annihilates a horizontal distribution (a subbundle of the holomorphic tangent bundle $T_M^{1,0}$ of M which bijects to $T_T^{1,0}$ by p). This generalizes a result of Maitani and Yamaguchi [32] for pseudoconvex families of Jordan domains in $\mathbb{C}$. Roughly speaking, the Bergman kernel detects the rigidity of analytic families. On the other hand, it was proved by Nishino [37] that a Stein submersion over the unit disc is trivial if the fibers are $\mathbb{C}$. Although this rigidity does not follow directly from $K_{\mathbb{C}} \equiv 0$, it turned out that an L^2 extension theorem in [50] is available to give its alternate proof (cf. [48]). For the family of $\mathbb{C}^n$ with $n \geq 2$, a rigidity criterion can be proved by a similar method (cf. [49]). The purpose of the present article is to give an expository account on [48, 49] providing with some backgrounds and supplementary remarks. In particular, it will be shown that an analytic family $M \to T$ is locally trivial if M admits a complete Kähler metric and $M_t \cong \mathbb{C}$ for all $t \in T$.

1 L^2 Method of Solving the $\bar\partial$ Equation

First we recall a basic existence theorem for the $\bar\partial$ equation with L^2 norm estimates on complete Kähler manifolds.

Let M be an n-dimensional connected complex manifold equipped with a complete Kähler metric g. We note that M admits a complete Kähler metric of the form $\partial\bar\partial\lambda(\varphi)$ for some $\lambda : \mathbb{R} \to \mathbb{R}$ if there exists a proper C^∞ map $\varphi : M \to (-\infty, \infty)$ satisfying $\partial\bar\partial\varphi > 0$. Here, by an abuse of notation, $\partial\bar\partial\varphi$ stands also for the complex Hessian of φ as well as the complex exterior derivatives applied to φ. By virtue of Grauert [21, 22] we know accordingly that the complement of an analytic set in a Stein manifold admits a complete Kähler metric. Recall that the L^2 norm $\|h\| (= \|h\|_g)$ (resp. the weighted L^2 norm $\|h\|_\Phi (= \|h\|_{\Phi,g})$ of a measurable (p, q)-form h on M) is defined as

$$\left(\int_M |h|^2 dV\right)^{\frac{1}{2}} \quad \left(\text{resp. } \left(\int_M e^{-\Phi}|h|^2 dV\right)^{\frac{1}{2}}\right)$$

where $|h|$ and dV respectively stand for the length of h and the volume form with respect to the metric g. The space of L^2 forms with respect to $\|\,\|_\Phi$ will be denoted by $L_\Phi^{p,q}(M)$. Recall also that

$$\|h\|^2 = \int_M h \wedge \overline{*h} \quad (\text{resp. } \|h\|_\Phi^2 = \int_M e^{-\Phi} h \wedge \overline{*h}),$$

where $*$ denotes Hodge's star operator, and that $L_\Phi^{n,0}(M)$ does not depend on the choice of the metric g. Based on the method originated in [27] and developed in [1, 2, 24], the following was proved in [44].

Theorem 1.1 (cf. [44], Theorem 1.5 and Corollary 1.6) *Let M be as above and let Φ be a strictly plurisubharmonic function of class C^4 on M. Then, for any $\bar\partial$ closed $(n,1)$-form f on M satisfying*

$$\int_M e^{-\Phi} f \wedge \overline{*_{\partial\bar\partial\Phi} f} < \infty,$$

there exists an $(n,0)$-form h satisfying $\bar\partial h = f$ and

$$i^{n^2} \int_M e^{-\Phi} h \wedge \overline{h} \leq \int_M e^{-\Phi} f \wedge \overline{*_{\partial\bar\partial\Phi} f}.$$

Here $*_{\partial\bar\partial\Phi}$ denotes Hodge's star operator with respect to $\partial\bar\partial\Phi$.

Recall that the proof of Theorem 1.1 in [44] is an application of the Riesz representation theorem based on the estimate

$$\left| \int_M e^{-\Phi} f \wedge \overline{*_{\partial\bar\partial\Phi+g} u} \right|^2 \leq \|\bar\partial^* u\|^2_{\Phi,\partial\bar\partial\Phi+g} \int_M e^{-\Phi} f \wedge \overline{*_{\partial\bar\partial\Phi} f} \tag{1.1}$$

which holds for any u in the domain of the adjoint $\bar\partial^*$ of $\bar\partial$ with respect to the weighted norm $\| \, \|_{\Phi,\partial\bar\partial\Phi+g}$. That g is a complete Kähler metric is used substantially in the proof of (1.1).

By a standard limiting argument for a sequence $\partial\bar\partial\Phi + \epsilon g$ $(\searrow 0)$, Theorem 1.1 can be generalized as a vanishing theorem for the L^2 $\bar\partial$ cohomology groups for higher degrees. The result further generalizes for the cohomology with coefficients in semipositive vector bundles (cf. [12, 44]). This approach turned out to be effective to strengthen the Kodaira vanishing theorem (cf. [10, 14, 31, 33, 45]). In many situations which will be discussed also in Sect. 3, Theorem 1.1 is simply applied in the following way.

Let M and Φ be as above, let $x \in M$ be any point and let $z = (z_1, \ldots, z_n)$ be a local coordinate around x which maps a neighborhood U of x onto $\mathbb{D}^n$, where $\mathbb{D} = \{\zeta \in \mathbb{C}; |\zeta| < 1\}$. Let $\chi : M \to [0,1]$ be a C^∞ function satisfying $\operatorname{supp}\chi \subset U$ and $\chi \equiv 1$ on a neighborhood of x, let α be a C^∞ $(n,0)$-form on M satisfying $\alpha = \chi dz_1 \wedge \cdots \wedge dz_n$ on U, and let Ψ be a C^∞ function on $M \setminus \{x\}$ satisfying $\operatorname{supp}\Psi \subset U$ and $\Psi = 2n\chi \log \|z\|$ on $U \setminus \{x\}$, where $\|z\|^2 = \sum_{j=1}^n |z_j|^2$. Clearly $\Psi + m\Phi$ is strictly plurisubharmonic on $M \setminus \{x\}$ for sufficiently large m. Then, by Theorem 1.1 one can find for such m an $(n,0)$-form u on $M \setminus \{x\}$ such that $\bar\partial u = \bar\partial\alpha$ and

$$i^{n^2} \int_M e^{-\Psi - m\Phi} u \wedge \overline{u} < \infty. \tag{1.2}$$

Then, by (1.2), $\alpha - u$ extends to a holomorphic n-form on M say $\tilde{\alpha}$ such that $\tilde{\alpha}(x) \neq 0$. Similarly, for any two distinct points $x, y \in M$, one can find a holomorphic n-form β on M satisfying $\beta(x) = 0$ and $\beta(y) \neq 0$.

2 L^2 Extension Theorems and Suita Conjecture

From now on, the set of holomorphic functions (resp. plurisubharmonic functions) on M will be denoted by $\mathcal{O}(M)$ (resp. $\mathrm{PSH}(M)$). Refining the argument in Sect. 1 together with the computation that yields (1.1), one has an extension theorem for L^2 holomorphic top forms. To state it we assume that M is a Stein manifold of dimension n and take any $s \in \mathcal{O}(M)$ such that ds is not identically 0 on every irreducible component of $s^{-1}(0)$. We put $X = s^{-1}(0)$ and $X_0 := \{x \in X;\ ds(x) \neq 0\}$. Let $\varphi \in \mathrm{PSH}(M)$.

Theorem 2.1 (cf. [50]) *In the above situation, let f be a holomorphic $(n-1)$-form on X_0 satisfying $|\int_{X_0} e^{-\varphi} f \wedge \overline{f}| < \infty$. Then there exists a holomorphic n-form F on M such that*

$$F = f \wedge ds \quad \textit{holds at every point of } X_0 \tag{2.1}$$

and

$$\left| \int_M e^{-\varphi}(1 + |s|^2)^{-2} F \wedge \overline{F} \right| \leq C_0 \left| \int_{X_0} e^{-\varphi} f \wedge \overline{f} \right|. \tag{2.2}$$

Here $C_0 = 1620\pi$.

Corollary *Let D be a bounded pseudoconvex domain in $\mathbb{C}^n$ with Lipschitz continuous boundary. Then the Bergman kernel function $k_D(z, w)$ of D satisfies $\inf_{z \in D} \delta_D(z)^2 k_D(z, z) > 0$. Here $\delta_D(z) := \inf_{w \notin D} \|z - w\|$.*

Theorem 2.1 was refined in [46, 47] in such a way that it entails an application to a question posed by Suita [53]. For any Riemann surface R, Suita conjectured that the diagonalized Bergman kernel $K_R = k_R(z, z)|dz|^2$ of R with respect to the L^2 holomorphic 1-forms satisfies $\pi K_R > c_\beta^2 |dz|^2$ unless $R = \mathbb{D} \setminus E$ for some E of logarithmic capacity 0 where the equality holds. Here $c_\beta = c_{\beta,R}$ is defined coordinatewise by

$$c_\beta(z) = \exp\left(\lim_{w \to z} (g_R(w, z) - \log|z - w|)\right),$$

where $g_R : R \times R \to [-\infty, 0)$ denotes the Green function of R. The quantity $\lim_{w \to z}(g_R(w, z) - \log|z - w|)$ is known as the Robin function. By refining the argument of producing the top forms as in Sect. 1, it was shown that $750\pi K_R \geq c_\beta^2 |dz|^2$ in [46] and $512\pi K_R \geq c_\beta^2 |dz|^2$ in [47].

Błocki [6] has proved the following.

Theorem 2.2 *Let Ω be a bounded domain in $\mathbb{C}$ containing 0, let D be a pseudoconvex domain contained in $\mathbb{C}^{n-1} \times \Omega$ and let $D' = D \cap \{z_n = 0\}$. Then, for any $\varphi \in \mathrm{PSH}(D)$ and for any $f \in \mathcal{O}(D')$, there exists a holomorphic extension F of f to D satisfying*

$$\int_D e^{-\varphi}|F|^2 d\lambda_n \leq \frac{\pi}{(c_{\beta,\Omega}(0))^2} \int_{D'} e^{-\varphi}|f|^2 d\lambda_{n-1}.$$

Here $d\lambda_m$ denotes the Lebesgue measure on $\mathbb{C}^m$.

Corollary $\pi k_\Omega(z, z) \geq c_\beta(z)^2$ *holds for any bounded domain Ω in $\mathbb{C}$.*

Suita conjecture has been completely settled by Guan and Zhou in [20]. We note that Guan–Zhou's variant of Theorem 2.2 is the following.

Theorem 2.3 *Let M be an n dimensional Stein manifold, let $\varphi, \psi \in \mathrm{PSH}(M)$ and take $w \in \mathcal{O}(M)$ such that $\sup_M (\psi + 2\log|w|) \leq 0$ and dw does not vanish identically on each irreducible component of $w^{-1}(0)$. Let $H = w^{-1}(0)$ and $H_0 = \{x \in H; dw(x) \neq 0\}$. Then, for any holomorphic $(n-1)$ form f on H_0 satisfying*

$$\left| \int_{H_0} e^{-\varphi-\psi} f \wedge \overline{f} \right| < \infty,$$

there exists a holomorphic n form F on M such that $F = f \wedge dw$ holds at every point of H_0 and satisfies

$$\left| \int_M e^{-\varphi} F \wedge \overline{F} \right| \leq 2\pi \left| \int_{H_0} e^{-\varphi-\psi} f \wedge \overline{f} \right|.$$

See also [16] for an alternate proof for the equality case.

3 Bergman Kernel in Analytic Families

After the decisive works [6, 20] very busy years went rushing by us in the L^2 extension theory. Besides the existence theorems with precise bounds, new connections between several basic results have been found. The most remarkable example is a theorem of Berndtsson and Lempert in [B-L] which generalizes Theorem 2.1 with $C_0 = 2\pi$ and also Theorem 2.2 by a completely new method based on the plurisubharmonicity of certain functions associated to a Stein family of domains in M (resp. in $\mathbb{C}^n$) connecting M (resp. D) and X (resp. D') by a decreasing sequence of subdomains. Their proof is based on Berndtsson's work [3, 4] which is a generalization of a theorem of Maitani and Yamaguchi [32]. Substantially, it is a slight generalization of Lempert's alternate proof of the inequality part of Suita conjecture, which is short enough to be sketched below as well as in the introduction of Błocki's another paper [7].

Proof $\pi K \geq c_\beta^2|dz|^2$: Given a Riemann surface R and a point $z_0 \in R$ such that $g_R \not\equiv -\infty$, we consider a domain $\mathcal{R}$ in $\mathbb{D} \times R$ defined by

$$\mathcal{R} = \{(t, z); g_R(z, z_0) < \log|t| \text{ or } t = 0\}.$$

Since $\log|t| - g_R(z, z_0)$ is plurisubharmonic in (t, z) on $\mathbb{D} \times (R \setminus \{z_0\})$, $\mathcal{R}$ is a Stein manifold which connects R $(|t| = 1)$ and $\{z_0\}$ $(t = 0)$. Let $p : \mathcal{R} \to \mathbb{D}$ be the restriction of the projection $\mathbb{D} \times R \to \mathbb{D}$ and let $R_t = p^{-1}(t)$. Then, by virtue of [32] one has

$$\phi(z, t) := \log k_{R_t}(z) - 2(g_R(z, z_0) - \log|z - z_0| - \log|t|) \in \mathrm{PSH}(\mathcal{R}),$$

so that $\phi(z_0, t)$ is a convex function of $\log|t|$. Clearly

$$\lim_{t \to 0} \phi(z_0, t) = -\log\pi$$

and

$$\lim_{t \to 1} \phi(z_0, t) = \log k_R(z_0) - 2\log c_\beta(z_0).$$

Hence $\phi(z_0, t)$ is an increasing function in $\log|t|$ so that

$$-\log\pi \leq \log k_R(z_0) - 2\log c_\beta(z_0).$$

Thus we obtain $\pi k_R(z_0) \geq c_\beta(z_0)^2$. □

In the above proof the log-subharmonicity of the Bergman kernel with respect to the parameter t plays an essential role. This property was generalized in [3, 4] in full generality as a curvature property of the direct image sheaf of the relative canonical sheaf twisted by a semipositive vector bundle under a Stein submersion or a proper Kähler submersion. For a Stein submersion $p : M \to \mathbb{D}$, its concise variant for the Bergman kernel $K_{M_t} = k_t(z)|dz_1 \wedge \cdots dz_n|^2$ is as follows.

Theorem 3.1 (cf. [3]) $\log k_t(z) \in \mathrm{PSH}(M)$.

We note that Theorem 3.1 had been observed in [25] when $M_t \cong \mathbb{D}$ by analyzing the corresponding family of the Riemann mappings. For the holomorphic motions of Riemann surfaces, a variational formula for $\log k_t(z)$ was obtained in [31, 33], which was generalized in [32] for any smooth Stein families of Riemann surfaces. As was mentioned in the introduction, the plurisubharmonicity of $\log k_t$ was recognized in the context of the deformation theory in these early works. In fact, [31–33] were preceded by [57], where a variational formula for the Robin function was established in order to simplify Nishino's proof of the rigidity theorem in [37] which was mentioned in the introduction (see also [58]). A new viewpoint was brought by Guan and Zhou in [20], where they found a relation between Theorem 3.1 and the optimal L^2 extension theorem by giving an alternate proof of Theorem 3.1 as a corollary of Theorem 2.3. Thus [B-L] was a revenge in some sense.

Quite recently a counter-revenge appeared in [15] asserting in particular that Theorem 2.1 naturally implies Theorem 3.1 and its subsequent generalizations in [4]. Namely, everything was already there in Theorem 2.1 with C_0 independent of φ and f. The point is to read an approximation theorem of Demailly for plurisubharmonic functions, which was the first unexpected application of Theorem 2.1, as a characterization of plurisubharmonicity. To state Demailly's result we put

$$\mathcal{O}_\varphi(D) = \{f \in \mathcal{O}(D); \int_D e^{-\varphi}|f|^2 d\lambda_n < \infty\}$$

for a domain $D \subset \mathbb{C}^n$ and $\varphi \in \mathrm{PSH}(D)$.

Theorem 3.2 (cf. [13]) *Let D be a pseudoconvex domain in $\mathbb{C}^n$ and let $\varphi \in \mathrm{PSH}(\Omega)$ such that $\mathcal{O}_\varphi(D) \neq \{0\}$. For any complete orthogonal system $\{f_\mu\}$ of $\mathcal{O}_{m\varphi}(D)$ ($m \in \mathbb{N}$), put $\varphi_m = \frac{1}{m}\log\sum_\mu |f_\mu|^2$. Then there exist constants C_1, C_2 which are independent of m such that*

$$\varphi(z) - \frac{C_1}{m} \leq \varphi_m(z) \leq \sup_{\|\zeta - z\| < r} \varphi(\zeta) + \frac{1}{m}\log\frac{C_2}{r^n}, \quad z \in D,\ r < \delta_D(z) \tag{3.1}$$

holds.

The idea of [15] is that (3.1) implies the plurisubharmonicity of φ and that Theorem 3.1 follows from Theorem 2.1 similarly. It must be noted here that Theorem 3.2 is preceded by a result of Bremermann [9] asserting that plurisubharmonic functions on pseudoconvex domains can be approximated by convex combinations of $\log|f|$ and their upper envelopes for holomorphic f. This was an immediate consequence of the solution of the Levi problem in [51] (see also [42] and [8]). The approximation in terms of the weighted Bergman kernels originates substantially in [35, 55] in the context of polarized Kähler metrics.

Anyway, one can symbolize the relations between [3, 4, 6, 20, 50], [B-L], [15, 32] as follows:

References [6, 20] ⇒ [50].
Reference[20] ⇒ [3, 4, 32].
References [3, 4] ⇒ [6, 20] (by [B-L]).
Reference [50] ⇒ [3, 4] (by [15]).

On the other hand, being also awared of the interface between the deformation theory and the L^2 extension theory, the author tried to apply Theorem 2.1 to Nishino's theory. The subsequent results in [48, 49] will be reported below.

4 Rigidity Theorems by the L^2 Technique

In 1969, T. Nishino established the following as a basic result in the classification of entire functions of two variables.

Theorem 4.1 ("Lemme fondamental" in [37]) *Let M be a two dimensional Stein manifold and let π be a holomorphic submersion from M onto the unit disc $\mathbb{D} = \{t \in \mathbb{C}; |t| < 1\}$. Assume that every fiber of π is holomorphically equivalent to $\mathbb{C}$. Then π is holomorphically equivalent to the projection $\mathbb{C} \times \mathbb{D} \to \mathbb{D}$.*

What Nishino aimed at was to discribe the properties of an entire function f by the conformal structures of $f^{-1}(t)$.[1]

We recall that the rigidity of this kind was shown by Fischer and Grauert [18] for the families of compact complex manifolds in the context of the deformation theory of Kodaira and Spencer [30].

Given $f \in \mathcal{O}(\mathbb{C}^2)$ such that the irreducible components of $f^{-1}(t)$ are compactifiable, Theorem 4.1 implies the existence of a compactification of $\mathbb{C}^2$ associated to f. About the compactifications of $\mathbb{C}^2$, Kodaira [29] showed that they are rational and Morrow [34] completed the classification. Those took place in the days of [36]$\sim$[41].[2] Such a relationship between Nishino's theory and affine algebraic geometry is reflected in the works of Suzuki [54] and Ueda [56].[3]

Since every analytic $\mathbb{C}$ bundle over $\mathbb{D}$ is trivial, it suffices to prove the equivalence locally. Thus we may assume in advance that there exists a holomorphic section $s : \mathbb{D} \to M$. The original proof of Theorem 4.1 consists of an elaborate study of a canonically defined map $\psi : M \to \mathbb{C}^2$ univalent on each fiber. The map ψ is associated to s and any nowhere zero holomorphic vector field say ξ on a neighborhood of $s(\mathbb{D})$ which is tangent to the fibers of π, in such a way that $\psi|_{s(\mathbb{D})} = 0$ and $\xi\psi|_{s(\mathbb{D})} = 1$. Even the continuity of such a function ψ is not evident but naturally follows from the classical Koebe distortion theorem. The subtlety is analyticity of ψ. The argument in [37] for that is quite technical, which was later simplified by Yamaguchi [57].[4] Yamaguchi's method is to deduce the analyticity of ψ from the plurisubharmonicity of $\log |\psi|$. This approach was later extended in [32] and eventually gave rise to a new perspective to the L^2 extension problem in [B-L].

In [48] it turned out that Theorem 4.1 is a direct consequence of Theorem 2.1. The idea is to identify the exterior derivative of the reciprocal of ψ as a relative canonical form with a pole of order 2 along $s(\mathbb{D})$. It is clear that Theorem 2.1 is applicable because

---

[1] [37, Theorem II]: $f \in \mathcal{O}(\mathbb{C}^2)$ and $\mathrm{cap}\{t; \mathbb{C} \subset f^{-1}(t)\} > 0 \Rightarrow f \in \mathcal{O}(\mathbb{C}) \circ \mathrm{Aut}\mathbb{C}^2$ ($\circ$ denotes the composite). Reference [40, Théorème principal]: {irreducible components of $f^{-1}(t)\} \subset \{\Sigma \setminus \Gamma$; Σ is compact and Γ is a finite set} $\Rightarrow f \in \mathcal{O}(\mathbb{C}) \circ \mathbb{C}[z, w] \circ \mathrm{Aut}\mathbb{C}^2$.

[2] See also [19].

[3] Reference [54]: Every polynomial embedding of $\mathbb{C}$ into $\mathbb{C}^2$ can be linearized by $\mathrm{Aut}\mathbb{C}^2$. Reference [56]: Every compactification of $\mathbb{C} \times (\mathbb{C} \setminus \{0\})$ is rational.

[4] Chirka [11] also gave an alternate proof of Theorem 4.1 by applying holomorphic motions and Teichmüller theory.

$$f \in \mathbb{C} \cdot \frac{dz}{z^2} \iff f \in L^{1,0}_{\alpha \log^+ \frac{1}{|z|}}(\mathbb{C}\setminus\{0\}) \cap \mathrm{Ker}\bar{\partial} \quad \text{for some } 2 \leq \alpha < 4.$$

A map from[5] M to $\mathbb{D} \times (\mathbb{C} \cup \{\infty\})$ is given by a primitive of an L^2 extension of dz/z^2 from $\pi^{-1}(0)$. The primitive can be defined as an integral along the paths in the fibers of π starting from $s'(\mathbb{D})$ for some holomorphic section $s' : \mathbb{D} \to M$ such that $s(\mathbb{D}) \cap s'(\mathbb{D}) = \varnothing$. We note that the Steinness of M is also necessary to conclude the analyticity of the complement of the image by virtue of Hartogs [23]. The method of [48] was extended to Stein families of $\mathbb{C}^n$ in [49] as a generalization of the following assertion which is essentially equivalent to Theorem 4.1.

Theorem 4.2 *Let $\pi : M \to \mathbb{D}$ be a Stein submersion with fibers equivalent to $\mathbb{C}$. Assume that it admits a holomorphic section $s : \mathbb{D} \to M$. Then, for any family of biholomorphic maps $\beta_t : \mathbb{C} \to M_t$ with $\beta_t(0) = \sigma(t)$, there exists a function $\gamma : \mathbb{D} \to \mathbb{C}$ such that the map $(z, t) \mapsto \beta_t(\gamma(t)z)$ is a biholomorphism from $\mathbb{C} \times \mathbb{D}$ to M.*

The generalization is as follows.

Theorem 4.3 *Let π be a holomorphic submersion from an $(n+1)$-dimensional Stein manifold M onto $\mathbb{D}$. Suppose that there exists a proper holomorphic embedding*

$$\sigma : \{z = (z_1, \ldots, z_n) \in \mathbb{C}^n; z_1 \cdots z_n = 0\} \times \mathbb{D} \hookrightarrow M$$

satisfying the following.

(1) $\pi \circ \sigma = pr_{\mathbb{D}}$.

(2) *There exist plurisubharmonic functions φ_j $(1 \leq j \leq n)$ on $M \setminus \sigma(\{z : z_j = 0\} \times \mathbb{D})$ such that for each j and $t \in \mathbb{D}$ there exist a biholomorphic map $\beta_t : \mathbb{C}^n \to M_t$ and a constant $C_t > 0$ satisfying*

$$\varphi_j(\beta_t(z)) - C_t \leq \log^+ \frac{1}{|z_j|} \leq \varphi_j(\beta_t(z)) + C_t \tag{4.1}$$

on $\mathbb{C}^n \setminus \{z_j = 0\}$.

Then there exist biholomorphic maps $\gamma_t : \mathbb{C}^n \to \mathbb{C}^n$ $(t \in \mathbb{D})$ such that the map $(z, t) \mapsto \beta_t(\gamma_t(z))$ is a biholomorphism from $\mathbb{C}^n \times \mathbb{D}$ to M.

Now look at the argument of section one that produces L^2 holomorphic top forms. Then it just tells us that Theorem 1.1 suffices for ensuring the existence of a non-zero holomorphic 2-form F on $M \setminus s(\mathbb{D})$ in Theorem 4.2 which has a pole of order 2 along

[5] For the notation $L^{1,0}_{\log^+ \frac{1}{|z|}}$, see section one.

$s(\mathbb{D})$ and square integrable on the complement of some subset of M lying properly over $\mathbb{D}$.[6] Accordingly, it is not hard to generalize Theorem 3.2 as follows.

Theorem 4.4 *Let M be a complete Kähler manifold of dimension 2 with a holomorphic submersion onto $\mathbb{D}$ whose fibers are $\mathbb{C}$. Then $M \cong \mathbb{D} \times \mathbb{C}$.*

For the proof, what remains to show is the following counterpart of [23].

Lemma *Let Γ be the graph of a continuous function $z = f(t)$ from $\mathbb{D}$ to $\mathbb{C}$. Then f is holomorphic if $\mathbb{D} \times \mathbb{C} \setminus \Gamma$ has a complete Kähler metric.*[7]

Proof Since the problem is local on Γ, we may assume in advance that $\Gamma \subset \mathbb{D} \times \mathbb{D}$. Let $\gamma : M \to \mathbb{D} \times \mathbb{D}$ be the topological double covering branched along Γ. Then $M \setminus \gamma^{-1}(\Gamma)$ is an unbranched double covering of $\mathbb{D} \times \mathbb{D} \setminus \Gamma$ so that it is a complex manifold admitting a complete Kähler metric. Let $\sigma \in \mathrm{Aut}(M \setminus \gamma^{-1}(\Gamma))$ be the covering transformation without fixed point. Then, similarly as in section one, one has a square integrable holomorphic 2-form $h \neq 0$ on $M \setminus \gamma^{-1}(\Gamma)$ satisfying $h(x) = 0$ and $h(y) \neq 0$ for a pair of points x and y satisfying $\sigma(x) = y$. Then, by putting $\hat{h} = h - \sigma^* h$ one has a nonzero holomorphic 2-form $\hat{h}$ satisfying $\sigma^* \hat{h} = -\hat{h}$. Then we put

$$\rho = \frac{\hat{h}}{\gamma^*(dt \wedge dz)}.$$

Clearly, for every $t \in \mathbb{D}$, ρ extends to a meromorphic function on $\gamma^{-1}(\{t\} \times \mathbb{D})$ $(\cong \mathbb{D})$ whose order of zero or pole at the point $\gamma^{-1}(f(t))$ is an odd integer. Hence Γ is contained in the divisor of $\rho \cdot \sigma^* \rho$, so that it must be analytic. □

5 A Splitting Theorem

Finally, we would like to add a remark that the L^2 extension theorem is still useful to study a question similar to Nishino's rigidity theorem.

We shall say that a closed connected analytic subset N of pure codimension m in a complex manifold M is *plumbed* if there exists a neighborhood $V \supset N$ and a holomorphic map $\pi : V \to \mathbb{C}^m$ with connected fibers such that $\pi^{-1}(\{x; |x| \leq \epsilon\})$ is nonempty and closed in M for all sufficiently small positive number ϵ. We shall call $\{p; |\pi(p)| \leq \epsilon\}$ *a plumbing neighborhood* of N for such ϵ and $\pi^{-1}(t)$ with $|t| < \epsilon$ *a fiber of the plumbing*.

---

[6] Actually the argument is slightly more delicate because we need to have $m\Phi \leq (4 - \epsilon) \log^+ \frac{1}{|z|}$ $(0 < \epsilon \ll 1)$ near $z = 0$. For that one may use $\log(|z|^3 + |t|^6)$ instead of $2\log(|z|^2 + |t|^2)$ to define Ψ (cf. [49]).

[7] Shcherbina [52] proved that f is holomorphic if and only if $\Gamma \subset \varphi^{-1}(-\infty)$ holds for some $\varphi(\not\equiv -\infty) \in \mathrm{PSH}(\mathbb{D} \times \mathbb{C})$. The lemma was proved in [43] under a restrictive assumption that f is continuously differentiable.

Theorem 5.1 (cf. [49]) *Let S be a connected Stein surface containing a plumbed curve C. Assume that there exists a plumbed curve C' intersecting with C at one point transversally such that the fibers of the plumbing are biholomorphic to $\mathbb{C}$. Then S is biholomorphically equivalent to $\mathbb{C} \times C$.*

For the proof, the following variant of Theorem 2.1 is applied.

Theorem 5.2 (cf. [47, Theorem 4]) *Let M be a Stein manifold of dimension n, let φ be a nonnegative plurisubharmonic function on M and let N be a closed nonsingular complex hypersurface in M such that one can find a neighborhood $W \supset N$ and a holomorphic function s on W satisfying $s^{-1}(0) = N$ and $ds \neq 0$. Assume that there exists a continuous function $G : M \to [-\infty, 0)$ such that $N = G^{-1}(-\infty)$, $\varphi + G$ is plurisubharmonic on M and $G - \log |s|^2$ is bounded on $W \setminus N$. Then, for any holomorphic $(n-1)$-form ω on N satisfying*

$$\left| \int_N e^{-\varphi} \omega \wedge \overline{\omega} \right| < \infty,$$

and for any $\delta > 0$, there exists a holomorphic n-form $\tilde{\omega}$ on M satisfying

$$\left| \int_M e^{-(1+\delta)\varphi} \tilde{\omega} \wedge \overline{\tilde{\omega}} \right| < \infty$$

such that $\tilde{\omega} = \omega \wedge ds$ holds at every point of N.

References

1. Andreotti, A., Vesentini, E.: Sopra un teorema di Kodaira. Ann. Scuola Norm. Sup. Pisa (3) **15**, 283–309 (1961)
2. Andreotti, A., Vesentini, E.: Carleman estimates for the Laplace–Beltrami equation on complex manifolds. Inst. Hautes Études Sci. Publ. Math. **25**, 81–130 (1965)
3. Berndtsson, B.: Subharmonicity properties of the Bergman kernel and some other functions associated to pseudoconvex domains. Ann. Inst. Fourier (Grenoble) **56**, 1633–1662 (2006)
4. Berndtsson, B.: Curvature of vector bundles associated to holomorphic fibrations. Ann. Math. **169**, 531–560 (2009)
5. Berndtsson, B., Lempert, L.: A proof of the Ohsawa–kegoshi theorem with sharp estimates. J. Math. Soc. Japan **68**(4), 1461–1472 (2016)
6. Błocki, Z.: Suita conjecture and the Ohsawa–Takegoshi extension theorem. Invent. Math. **193**, 149–158 (2013)
7. Błocki, Z.: Bergman kernel and pluripotential theory, In: Analysis, Complex Geometry, and Mathematical Physics: In Honor of Duong H. Phong, 1–10, Contemporary Mathematics, vol. 644. American Mathematical Society, Providence, RI (2015)
8. Bremermann, H.J.: Die Charakterisierung von Regülaritätsgebieten durch pseudokonvexe Funktionen. Schr. Math. Inst. Univ. Münster, **5**, i+92 (1951)
9. Bremermann, H.J.: On the conjecture of the equivalence of the plurisubharmonic functions and the Hartogs functions. Math. Ann. **131**, 76–86 (1956)

10. Cao, J.-Y.: Numerical dimension and a Kawamata-Viehweg-Nadel-type vanishing theorems on compact Kähler manifolds. Compos. Math. **150**, 1869–1902 (2014)
11. Chirka, E.M.: Holomorphic motions and the uniformization of holomorphic families of Riemann surfaces, (Russian) Uspekhi Mat. Nauk **67** (2012), 6(408), 125–202 (Translation in Russian Math. Surveys **67**(6), 1091–1165 (2012))
12. Demailly, J.-P.: Estimations L^2 pour l'opérateur $\bar{\partial}$ d'un fibré vectoriel holomorphe semi-positif au-dessus d'une variété kählérienne complète. Ann. Sci. École Norm. Sup. **15**, 457–511 (1982)
13. Chirka, E.M.: Regularization of closed positive currents and intersection theory. J. Algebraic Geom. **1**, 361–409 (1992)
14. Chirka, E.M.: Analytic Methods in Algebraic Geometry. Higher Education Press, Beijing (2010)
15. Deng, F.-S., Wang, Z.-W., Zhang, L.-Y., Zhou, X.-Y.: New characterizations of plurisubharmonic functions and positivity of direct image sheaves. arXiv:1809.10371
16. Dong, R.X.: Equality in Suita's conjecture. arXiv:1807.05537v1
17. Fefferman, C.: Monge–Ampère equations, the Bergman kernel, and geometry of pseudoconvex domains. Ann. Math. **103**, 395–416 (1976)
18. Fischer, W., Grauert, H.: Lokal-triviale Familien kompakter komplexer Mannigfaltigkeiten, Nachr. Akad. Wiss. Gttingen Math.-Phys. Kl. II, 89-94 (1965)
19. Furushima, M., Nobe, M., Ohshima, Y.: A note on minimal normal compactifications of $\mathbb{C}^2$. Kumamoto J. Math. **27**, 5–21 (2014)
20. Guan, Q.-A., Zhou, X.-Y.: A solution of an L^2 extension problem with optimal estimate and applications. Ann. Math. **181**, 1139–1208 (2015)
21. Guan, Q.-A., Zhou, X.-Y.: Charakterisierung der Holomorphiegebiete durch die vollständige Kählersche Metrik. Math. Ann. **131**, 38–75 (1956)
22. Guan, Q.-A., Zhou, X.-Y.: On Levi's problem and the imbedding of real-analytic manifolds. Ann. Math. **68**, 460–472 (1958)
23. Hartogs, F.: Über die aus den singulären Stellen einer analytishcen Funktion mehrerer Veränderlichen bestehenden Gebilde. Acta Math. **32**, 57–79 (1909)
24. Hörmander, L.: L^2 estimates and existence theorems for the $\bar{\partial}$ operator. Acta Math. **113**, 89–152 (1965)
25. Hössjer, G.: Über die konforme Abbildungeines Veränderlichen Bereiches (Transactions of Chalmers University of Technology Gothenburg, Sweden) vol. 10, pp. 2–15 (1942)
26. Hosono, G.: The optimal jet L^2 extension of Ohsawa-Takegoshi type, arXiv:1706.08725 [math.CV]
27. Kodaira, K.: On a differential-geometric method in the theory of analytic stacks. Proc. Nat. Acad. Sci. U.S.A. **39**, 1268–1273 (1953)
28. Kodaira, K.: On Kähler varieties of restricted type. Ann. Math. **60**, 28–48 (1954)
29. Kodaira, K.: Holomorphic mappings of polydiscs into compact complex manifolds. J. Differ. Geom. **6**, 33–46 (1971/1972)
30. Kodaira, K., Spencer, D.C.: On deformations of complex analytic structures, I-II. Ann. Math. **67**, 328–466 (1958)
31. Maitani, F.: Variations of meromorphic differentials under quasiconformal deformations. J. Math. Kyoto Univ. **24**, 49–66 (1984)
32. Maitani, F., Yamaguchi, H.: Variation of Bergman metrics on Riemann surfaces. Math. Ann. **330**, 477–489 (2004)
33. Matsumura, S.: A vanishing theorem of Kollár–Ohsawa type. Math. Ann. **366**, 1451–1465 (2016)
34. Morrow, J.: Minimal normal compactifications of $\mathbb{C}^*)^2$, Complex analysis, 1972. In: Proceedings of the Conference Rice University, Houston, Texas, 1972, Vol. I. Geometry of singularities. Rice University Studies, vol. 59, Issue 1, pp. 97–112 (1973)
35. Nadel, A.: Multiplier ideal sheaves and Kähler-Einstein metrics of positive scalar curvature. Ann. Math. **132**, 549–596 (1990)
36. Nishino, T.: Nouvelles recherches sur les fonctions entières de plusieurs variables complexes. I. J. Math. Kyoto Univ. **8**, 49–100 (1968)

37. Nishino, T.: Nouvelles recherches sur les fonctions entières de plusieurs variables complexes. II. Fonctions entières qui se réduisent à celles d'une variable. J. Math. Kyoto Univ. **9**, 221–274 (1969)
38. Nishino, T.: Nouvelles recherches sur les fonctions entières de plusiers variables complexes. III. Sur quelques propriétés topologiques des surfaces premières. J. Math. Kyoto Univ. **10**, 245–271 (1970)
39. Nishino, T.: Nouvelles recherches sur les fonctions entières de plusieurs variables complexes. IV. Types de surfaces premières. J. Math. Kyoto Univ. **13**, 217–272 (1973)
40. Nishino, T.: Nouvelles recherches sur les fonctions entières de plusieurs variables complexes. V. Fonctions qui se réduisent aux polynômes. J. Math. Kyoto Univ. **15**(3), 527–553 (1975)
41. Nishino, T.: Value distribution of analytic functions of two variables (Japanese). Sūgaku **32**(3), 230–246 (1980)
42. Norguet, F.: Sur les domaines d'holomorphie des fonctions uniformes de plusieurs variables complexes (Passage du local au global). Bull. Soc. Math. France **82**, 137–159 (1954)
43. Ohsawa, T.: Analyticity of complements of complete Kähler domains. Proc. Japan Acad. Ser. A Math. Sci. **56**, 484–487 (1980)
44. Ohsawa, T.: On complete Kähler domains with C^1-boundary. Publ. Res. Inst. Math. Sci. **16**(3), 929–940 (1980)
45. Ohsawa, T.: Vanishing theorems on complete Kähler manifolds. Publ. Res. Inst. Math. Sci. **20**(1), 21–38 (1984)
46. Ohsawa, T.: On the Bergman kernel of hyperconvex domains. Nagoya Math. J. **129**, 43–52 (1993). (Addendum, Nagoya Math. J. **137**, 145–148 (1995))
47. Ohsawa, T.: On the extension of L^2 holomorphic functions V. Effect of generalization. Nagoya Math. J. **161**, 1–21 (2001)
48. Ohsawa, T.: L^2 proof of Nishino's rigidity theorem, to appear in Kyoto J. Math
49. Ohsawa, T.: Generalizations of theorems of Nishino and Hartogs by the L^2 method, in preparation
50. Ohsawa, T., Takegoshi, K.: On the extension of L^2 holomorphic functions. Math. Z. **195**, 197–204 (1987)
51. Oka, K.: Sur les fonctions de plusieurs variables. IX. Domaines finis sans point critique intérieur, Jap. J. Math. **23**, 97–155 (1953)
52. Shcherbina, N.: Pluripolar graphs are holomorphic. Acta Math. **194**, 203–216 (2005)
53. Suita, N.: Capacities and kernels on Riemann surfaces. Arch. Ration. Mech. Anal. **46**, 212–217 (1972)
54. Suzuki, M.: Propriétés topologiques des polynômes de deux variables complexes, et automorphismes algébriques de l'espace $\mathbb{C}^2$. J. Math. Soc. Jpn. **26**, 241–257 (1974)
55. Tian, G.: On a set of polarized Kähler metrics on algebraic manifolds. J. Diff. Geom. **32**, 99–130 (1990)
56. Ueda, T.: Compactifications of $\mathbb{C} \times \mathbb{C}^*$ and $(\mathbb{C}^*)^2$. Tôhoku Math. J. 2 **31**(1), 81–90 (1979)
57. Yamaguchi, H.: Parabolicité d'une fonction entière. J. Math. Kyoto Univ. **16**, 71–92 (1976)
58. Yamaguchi, H.: Complex and vector potential theory, (Japanese). Sūgaku **50**(3), 225–247

FSC
www.fsc.org
MIX
Papier aus verantwortungsvollen Quellen
Paper from responsible sources
FSC® C105338

If you have any concerns about our products,
you can contact us on
ProductSafety@springernature.com

In case Publisher is established outside the EU,
the EU authorized representative is:
Springer Nature Customer Service Center GmbH
Europaplatz 3, 69115 Heidelberg, Germany

Printed by Libri Plureos GmbH
in Hamburg, Germany